# THE PLATES OF EARTH'S LITHOSPHERE

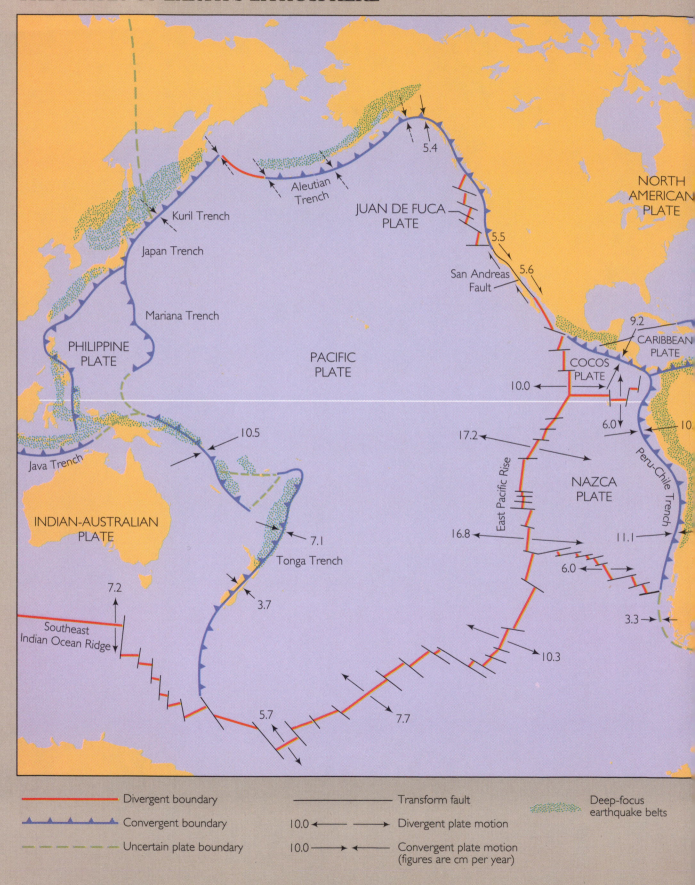

NORTH AMERICAN PLATE

JUAN DE FUCA PLATE

5.4

Aleutian Trench

Kuril Trench

Japan Trench

Mariana Trench

PHILIPPINE PLATE

PACIFIC PLATE

San Andreas Fault

5.5

5.6

9.2

CARIBBEAN PLATE

COCOS PLATE

10.0 ←

6.0

10.

17.2 ←

Peru-Chile Trench

Java Trench

10.5

East Pacific Rise

NAZCA PLATE

16.8 ←

11.1 →

INDIAN-AUSTRALIAN PLATE

7.1

Tonga Trench

6.0 ←

3.7

3.3 →

7.2

10.3

Southeast Indian Ocean Ridge

5.7

7.7

**Divergent boundary**

**Convergent boundary**

**Uncertain plate boundary**

**Transform fault**

10.0 ← → **Divergent plate motion**

10.0 → ← **Convergent plate motion** (figures are cm per year)

**Deep-focus earthquake belts**

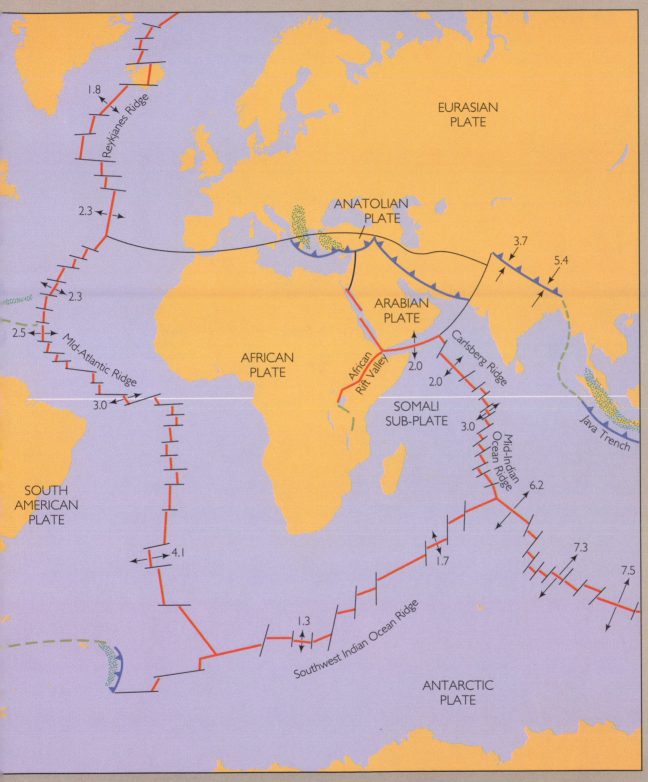

EURASIAN
PLATE

ANATOLIAN
PLATE

1.8

Reykjanes Ridge

2.3

2.3

3.7

5.4

2.5

Mid-Atlantic Ridge

ARABIAN
PLATE

AFRICAN
PLATE

African Rift Valley

Carlsberg Ridge

2.0

2.0

SOMALI
SUB-PLATE

3.0

3.0

Mid-Indian Ocean Ridge

Java Trench

SOUTH
AMERICAN
PLATE

6.2

7.3

4.1

7.5

1.7

1.3

Southwest Indian Ocean Ridge

ANTARCTIC
PLATE

# UNDERSTANDING EARTH

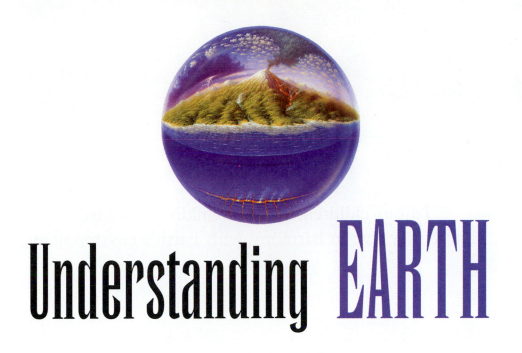

# Understanding EARTH

**FRANK PRESS**
Carnegie Institution of Washington

**RAYMOND SIEVER**
Harvard University

W. H. FREEMAN AND COMPANY

NEW YORK

To our children, and our children's children;
may they live in harmony with Earth's environment

Cover illustration © 1994 by Matt Zumbo

Interior illustrations by Ian Worpole, Network Graphics, and Tomo Narashima

LIBRARY OF CONGRESS CATALOGING-IN-PUBLICATION DATA

Press, Frank.
    Understanding earth / Frank Press and Raymond Siever.
        p.    cm.
    Includes index.
    ISBN 0-7167-2239-9
    1. Earth sciences.   I. Siever, Raymond.   II. Title.
QE28.P9      1993
550—dc20                                                                93-4967
                                                                        CIP

PRINTED   IN   THE   UNITED   STATES   OF   AMERICA

2 3 4 5 6 7 8 9 0   RRD   9 9 8 7 6 5 4

# BRIEF CONTENTS

1 BUILDING A PLANET 2

2 MINERALS: BUILDING BLOCKS OF ROCKS 22

3 ROCKS: RECORDS OF GEOLOGIC PROCESSES 50

4 IGNEOUS ROCKS: SOLIDS FROM MELTS 64

5 VOLCANISM 90

6 WEATHERING AND EROSION 118

7 SEDIMENTS AND SEDIMENTARY ROCKS 142

8 METAMORPHIC ROCKS 168

9 THE ROCK RECORD AND THE GEOLOGIC TIME SCALE 186

10 FOLDS, FAULTS, AND OTHER RECORDS OF ROCK DEFORMATION 210

11 MASS WASTING 230

12 THE HYDROLOGIC CYCLE AND GROUNDWATER 250

13 RIVERS: TRANSPORT TO THE OCEANS 276

14 WINDS AND DESERTS 304

15 GLACIERS: THE WORK OF ICE 326

16 LANDSCAPE EVOLUTION 352

17 THE OCEANS 368

18 EARTHQUAKES 402

19 EXPLORING EARTH'S INTERIOR 428

20 PLATE TECTONICS: THE UNIFYING THEORY 446

21 DEFORMATION OF THE CONTINENTAL CRUST 478

22 ENERGY RESOURCES FROM THE EARTH 500

23 MINERAL RESOURCES FROM THE EARTH 522

# CONTENTS

Preface                                                                  xiii

**PART 1**
**UNDERSTANDING THE EARTH SYSTEM**                                        1

**1 BUILDING A PLANET**                                                   2
Aspects of Geology                                                       4
The Origin of the System of Planets                                      6
Earth as an Evolving Planet                                              9
Plate Tectonics: A Modern Paradigm
    for Geological Science                                               13
Geologic Time                                                           18
The Scientific Method                                                   18
*Summary / Key Terms and Concepts / Exercises*
*Thought Questions / Suggested Readings*

**2 MINERALS:**
**BUILDING BLOCKS OF ROCKS**                                            22
Minerals Make Up Rocks                                                  24
What Are Minerals?                                                      24
The Atomic Structure of Matter                                          25
Chemical Reactions                                                      27
Chemical Bonds                                                          31
The Atomic Structure of Minerals                                        32
The Rock-Forming Minerals                                               35
Physical Properties of Minerals                                         40
Chemical Properties of Minerals                                         46
*Summary / Key Terms and Concepts*
*Mineral Names to Remember / Exercises*
*Thought Questions / Suggested Readings*

**3 ROCKS: RECORDS OF
GEOLOGIC PROCESSES**                                                     50
Igneous Rocks                                                           53
Sedimentary Rocks                                                       54
Metamorphic Rocks                                                       55
The Chemical Composition of Rocks                                       56
Where We See Rocks                                                      57
The Rock Cycle                                                          59
*Summary / Key Terms and Concepts / Exercises*
*Thought Questions / Suggested Readings*

**4 IGNEOUS ROCKS:
SOLIDS FROM MELTS**                                                      64
Major Types of Igneous Rocks                                            66
The Origins of Magmas                                                   72
Magmatic Differentiation                                                76
Forms of Magmatic Intrusions                                            82
Plutonism and Plate Tectonics                                           87
*Summary / Key Terms and Concepts / Exercises*
*Thought Questions / Suggested Readings*

**5 VOLCANISM**                                                          90
Volcanic Deposits                                                       93
Eruptive Styles                                                         97
The Global Pattern of Volcanism                                        108
Volcanism and Human Affairs                                            114
*Summary / Key Terms and Concepts / Exercises*
*Thought Questions / Suggested Readings*

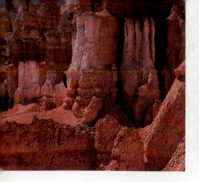

## 6 WEATHERING AND EROSION 118

Geologic Factors That Control
  Weathering 120
Chemical Weathering 122
Physical Weathering 132
Soil: The Residue of Weathering 134
Weathering Makes the Raw Material
  of Sediment 139

*Summary / Key Terms and Concepts / Exercises
Thought Questions / Suggested Readings*

## 7 SEDIMENTS AND SEDIMENTARY ROCKS 142

The Raw Material of Sediment:
  Particles and Dissolved Substances 145
Transportation of Sediment 146
Sedimentation: The End of the Line 148
Diagenesis and Lithification 152
Classification of Sediments and
  Sedimentary Rocks 154
Bedding and Sedimentary Structures 156
Clastic Sediments and Sedimentary
  Rocks 158
Chemical and Biochemical Sediments
  and Sedimentary Rocks 161

*Summary / Key Terms and Concepts / Exercises
Thought Questions / Suggested Readings*

## 8 METAMORPHIC ROCKS 168

Physical and Chemical Factors
  Controlling Metamorphism 171
Kinds of Metamorphism 172
Metamorphic Textures 174

Regional Metamorphism and
  Metamorphic Grade 179
Contact Metamorphic Zones 181
Plate Tectonics and Metamorphism 183

*Summary / Key Terms and Concepts / Exercises
Thought Questions / Suggested Readings*

## 9 THE ROCK RECORD AND THE GEOLOGIC TIME SCALE 186

Timing the Earth 188
Absolute Time and the Geologic
  Time Scale 197
Estimating the Rates of Very Slow
  Earth Processes 204
An Overview of Geologic Time 207

*Summary / Key Terms and Concepts / Exercises
Thought Questions / Suggested Readings*

## 10 FOLDS, FAULTS, AND OTHER RECORDS OF ROCK DEFORMATION 210

Interpreting Field Data 212
How Rocks Become Deformed 215
Folds 217
How a Rock Fractures:
  Joints and Faults 220
Unraveling Geologic History 225
Expression of Deformation in
  the Landform 226

*Summary / Key Terms and Concepts / Exercises
Thought Questions / Suggested Readings*

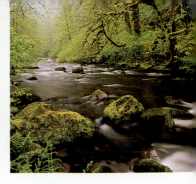

PART 2
SURFACE PROCESSES    229

11 MASS WASTING    230
What Makes Masses Move?    232
Classification of Mass Movements    238
Catastrophic Mass Movements    244
*Summary / Key Terms and Concepts / Exercises
Thought Questions / Suggested Readings*

12 THE HYDROLOGIC CYCLE
AND GROUNDWATER    250
Flows and Reservoirs    252
Hydrology and Climate    254
The Hydrology of Runoff    257
Groundwater    259
Water Resources from Major
   Aquifers    266
Erosion by Groundwater    267
Water Quality    270
Water Deep in the Crust    272
*Summary / Key Terms and Concepts / Exercises
Thought Questions / Suggested Readings*

13 RIVERS:
TRANSPORT TO THE OCEANS    276
How Stream Waters Flow    278
Stream Loads and Sediment
   Movement    279
How Running Water Erodes
   Solid Rock    282
Stream Valleys, Channels,
   and Floodplains    283
How Streams Change with Time
   and Distance    287

Drainage Networks    294
Deltas: The Mouths of Rivers    298
*Summary / Key Terms and Concepts / Exercises
Thought Questions / Suggested Readings*

14 WINDS AND DESERTS    304
Wind as a Flow of Air    306
Wind as a Transport Agent    307
Wind as an Agent of Erosion    310
Wind as a Depositional Agent    312
The Desert Environment    317
*Summary / Key Terms and Concepts / Exercises
Thought Questions / Suggested Readings*

15 GLACIERS: THE WORK OF ICE    326
Ice as a Material    328
What Is a Glacier?    329
Glacial Budgets: How Glaciers Form,
   Grow, and Shrink    330
How Glaciers Move    333
Glacial Landscapes    336
Ice Ages: The Pleistocene Glaciation    344
*Summary / Key Terms and Concepts / Exercises
Thought Questions / Suggested Readings*

16 LANDSCAPE EVOLUTION    352
Topography, Elevation, and Relief    354
Landforms: The Components
   of Landscape    356
Factors That Control Landscape    360
The Face of North America    362
The Evolution of Landscape    364
*Summary / Key Terms and Concepts / Exercises
Thought Questions// Suggested Readings*

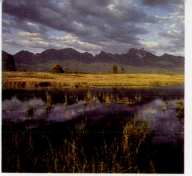
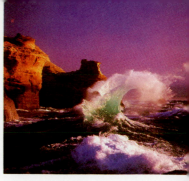

## 17 THE OCEANS   368

The Edge of the Sea: Waves and
Tides   370
Shorelines   377
Sensing the Floor of the Ocean   383
Profiles of Two Oceans   384
Continental Margins   390
The Floor of the Deep Ocean   392
Sedimentation in the Sea   397
Differences in the Geology of
Oceans and Continents   398
*Summary / Key Terms and Concepts / Exercises*
*Thought Questions / Suggested Readings*

## PART 3
# INTERNAL PROCESSES, EXTERNAL EFFECTS   401

## 18 EARTHQUAKES   402

What Is an Earthquake?   405
Studying Earthquakes   407
The Big Picture: Earthquakes and
Plate Tectonics   414
Earthquake Destructiveness   416
*Summary / Key Terms and Concepts / Exercises*
*Thought Questions / Suggested Readings*

## 19 EXPLORING EARTH'S INTERIOR   428

Exploring the Interior with
Seismic Waves   430
Earth's Internal Heat   437

The Interior Revealed by Earth's
Magnetic Field   440
*Summary / Key Terms and Concepts / Exercises*
*Thought Questions / Suggested Readings*

## 20 PLATE TECTONICS: THE UNIFYING THEORY   446

From Controversial Hypothesis to
Respectable Theory   448
Overview   450
The Mosaic of Plates   451
Rates of Plate Motion   454
The Geometry of Plate Motion   460
Rock Assemblages and Plate
Tectonics   462
Microplate Terranes and
Plate Tectonics   470
The Grand Reconstruction   470
The Driving Mechanism of
Plate Tectonics   473
*Summary / Key Terms and Concepts / Exercises*
*Thought Questions / Suggested Readings*

## 21 DEFORMATION OF THE CONTINENTAL CRUST   478

Some Regional Tectonic Structures   480
The Stable Interior   482
Orogenic Belts   483
Coastal Plain and Continental Shelf   493
Regional Vertical Movement   494
*Summary / Key Terms and Concepts / Exercises*
*Thought Questions / Suggested Readings*

PART 4
BOUNTIFUL EARTH                          499

22 ENERGY RESOURCES FROM
THE EARTH                                500
Resources and Reserves                   502
Energy                                   503
Oil and Natural Gas                      505
Coal                                     509
Oil Shale and Tar Sands                  512
The Future of Fossil Fuels              512
Nuclear Energy Fueled by Uranium         513
Solar Energy                             515
Geothermal Energy                        517
Conservation                             519
Energy Policy                            519
*Summary / Key Terms and Concepts / Exercises*
*Thought Questions / Suggested Readings*

23 MINERAL RESOURCES
FROM THE EARTH                           522
Minerals as Economic Resources           524
The Geology of Mineral Deposits          529
Ore Deposits and Plate Tectonics         534
Finding New Mineral Deposits             540
*Summary / Key Terms and Concepts / Exercises*
*Thought Questions / Suggested Readings*

APPENDIX 1 Conversion Factors            546
APPENDIX 2 Numerical Data
    Pertaining to Earth                  549
APPENDIX 3 Properties of the
    Most Common Minerals of
    Earth's Crust                        551
APPENDIX 4 Topographic and
    Geologic Maps                        555
Glossary                                 561
Index                                    575

## BOXES

Box 2.1  Asbestos and Health         44–45
Box 4.1  Japan: A Growing Island Arc  86
Box 5.1  Kilauea: An Instrumented
         Volcano                         99
Box 5.2  The Explosion of Krakatoa
                                    104–105
Box 5.3  Mount St. Helens: Dangerous
         But Predictable            112–113
Box 5.4  Reducing the Risks of
         Hazardous Volcanoes            115
Box 6.1  Acid Rain                  126–127
Box 6.2  Soil Erosion               136–137
Box 9.1  Interpreting the Grand Canyon
         Sequence                   198–199

Box 9.2  Radon: An Environmental
         Threat                     202–203
Box 9.3  Photosynthesis and Organic
         Carbon                     204–205
Box 11.1  Preventing Landslides         237
Box 11.2  Reducing Loss from Landslides
                                    244–245
Box 12.1  Water, A Precious Resource:
          Who Gets It?                  256
Box 12.2  Depleted Groundwaters and
          Water Resources               268
Box 13.1  The Development of Cities on
          Floodplains                   288
Box 13.2  Historic Floods and Flood
          Control                   290–291
Box 14.1  Droughts and Dust Bowls       309

**Box 14.2** Desertification in the Sahel   319

**Box 15.1** Future Changes in Sea Level
and the Next Glaciation   347

**Box 15.2** Carbon Dioxide and the
Greenhouse Effect   348–349

**Box 17.1** Preserving Our Beaches
380–381

**Box 17.2** Hot Springs on the Seafloor
394

**Box 17.3** The Oceans as a Deep Waste
Repository   395

**Box 18.1** Earthquake Magnitudes,
Ground Motion, and Energy
413

**Box 18.2** Tsunamis   419

**Box 18.3** Protection in an Earthquake
422–423

**Box 19.1** Finding Oil with Seismic Waves
432–433

**Box 19.2** The Uplift of Scandinavia:
Nature's Experiment with
Isostasy   435

**Box 20.1** Drilling in the Deep Sea   458

**Box 20.2** Charting the Seafloor by
Satellite   463

**Box 21.1** The Collision between India
and Eurasia   486–487

**Box 22.1** Climatic Catastrophe   514

**Box 22.2** Radioactive Waste Disposal   516

**Box 23.1** Use of Federal Lands in the
United States   528

**Box 23.2** Can We Have a Sustainable
World with Unlimited
Population Growth?   543

# PREFACE

When we wrote our first textbook, *Earth,* geology was a field flush with the excitement of new discovery. Recognition of continental drift only a decade earlier had triggered a revolution in our understanding of our planet. For the first time in the history of the discipline, an all-encompassing synthesis of much geological knowledge was being advanced. We were learning of the immense forces—turning cyclically in the core, in the surrounding mantle, in the crust, and in the air, oceans, and biosphere— keeping our planet in a constant state of change. This new picture of a dynamic Earth, and the new techniques helping to describe it, was central to our writing of that book. We wanted to share, with as many students as possible, something of the exhilaration and intellectual excitement the profession was feeling. We were enormously gratified by the warmth with which *Earth* was received and the loyalty of its users over several editions. Now, thirty years in the wake of the plate-tectonics revolution and twenty since the publication of *Earth,* geology has reached a new critical juncture. The world in which we live has changed. Understanding our planet has now become essential to humankind's survival. Our growing world population requires more resources, faces increasing losses from natural hazards, and contributes to growing pollution of air, water, and land. Human disturbances of the environment now equal or exceed many natural processes in magnitude and rate; many of these disturbances are a threat to life on the planet.

Students now entering college belong to the generation that will lead our world through the first decades of the next century. We believe the social, political, and economic issues they face will prove many times more challenging than those we have already encountered. They will have a tremendous need for scientific literacy in general, and will benefit from an understanding of geology and related earth sciences in particular, to help them make wise decisions about such issues as resource development, waste disposal, environmental protection, and land use.

We have brought this conviction to our writing of *Understanding Earth.*

Our new textbook is for today's students, especially beginning students whose one course in geology may be their sole college exposure to the physical sciences. Naturally, as geologists we still want to share many of the fascinating aspects of our discipline with students. As teachers we have tried to do this in a way that is compelling, accurate, and above all, up-to-date. We do so with the hope that learning how scientists think and work, and understanding something of the past, present, and future of the physical life of our planet, will help our readers to think more deeply and responsibly about the issues they will confront as citizens.

## Goals

*Understanding Earth* is designed for a one-term introductory course in physical geology for nonscience majors. We took it as a challenge to present the essential material, both traditional and modern, that a good geology course should cover, but in terms accessible to a student who has had no previous college science. We have deliberately emphasized a broad view, one stressing concepts and trying to show by many examples what is meant by the "scientific method." As much as possible we have tried to impart something of what motivates contemporary geologists and of the methods they use. We have tried to integrate in a natural way the newest discoveries and concerns with the traditional discussions of such basic topics as geomorphology, sedimentation, petrology, volcanism, and structural geology.

## Pedagogy

Explanatory drawings, diagrams, and photos will be of considerable value in simplifying otherwise difficult concepts. Illustrations also serve as alternative restatements of concepts presented in the text and as summaries of material covered earlier. The opportunity to use full color for the first time in drawings and photos has been exciting. We believe that the high-quality visual program newly created for *Understanding Earth* is one of its greatest learning tools and will serve to motivate the student.

The brief introductions at the beginnings of parts and chapters are designed to forecast in a general way the nature of the subject matter, how it fits together, and how it relates to other parts or chapters. Summaries in question-and-answer form appear at the end of each chapter to serve as a systematic review of major concepts. Exercises were devised to help students test their comprehension of key chapter material. Thought questions ask students to apply ideas and principles to situations not specifically covered in the text. Suggested readings point curious students toward books and articles, both popular and technical, on subjects they wish to explore more deeply. Boxed essays provide expanded perspectives on selected topics. Many of these essays focus on compelling issues from current affairs, such as the destruction of the ozone layer, acid rain, the use of federal lands by commercial interests, the storage of radioactive waste, and the controversial problems of access to water resources.

## Organization

To accommodate the many ways that different instructors may want to structure the course, we have made each chapter as self-sufficient as possible. Nevertheless, few geologic processes can be taught as wholly independent subjects; they must be seen in the larger geologic context. We have thus used the recurrence of many important topics as an opportunity for review and alternative restatement. This approach enhances learning and increases flexibility in the way the book can be used.

Relying on a consensus of views from colleagues and reviewers, we arrived at the following four-part organization for this text:

PART 1 **UNDERSTANDING THE EARTH SYSTEM.** An introductory chapter proceeds from a discussion of the origin of Earth to a first treatment of the geological cycle and elementary plate tectonics. This enables the instructor to refer in a general way to plate tectonics when discussing the many subjects cov-

ered before the chapter devoted solely to plate tectonics appears later in the text.

Part 1 provides a grounding in what we call Earth's "basic system." Rocks and minerals are followed by geologic time, stratigraphy, and structure. Part 1 could almost be considered a basic text in the foundations of geology that evolved in the first half of the century, to which are added the insights provided by the later development of plate tectonics. The section on igneous rocks is followed by the closely related chapter on volcanism, so that knowledge of the origin of igneous rocks precedes the particulars of eruptive processes. In contrast, we cover weathering and erosion in a separate chapter as a necessary prelude to the origin of sedimentary rocks in the following chapter. The instructor may choose to reverse the order of presentation of igneous and sedimentary rocks and will find that the text is adaptable and flexible in this regard. After the major rock types are covered, the student is introduced to geologic time, stratigraphy, and structure.

PART 2 **SURFACE PROCESSES.** Part 2 treats the major aspects of Earth's surface processes, essentially the major topics of geomorphology (water, rivers, wind, and ice). These subjects are the most easily comprehended parts of geology because most students can draw on personal experience with Earth's landscapes. Completing this part is a chapter on the oceans, which continue to grow in importance in geology and in global environmental issues. Here we emphasize shorelines and shallow water processes, as those most readily observed, and those aspects of the deep sea that relate most closely to plate tectonics and to the formation of marine sediments.

PART 3 **INTERNAL PROCESSES, EXTERNAL EFFECTS.** Part 3 is an exploration of Earth's interior and its dynamic interaction with the crust, which is manifest by tectonism. There is a full treatment of plate tectonics as it is understood today. Following is a chapter on the role of plate tectonics in the formation and deformation of the continents, a culmination that brings together all the elements of geologic history, which defied interpretation in the earlier decades of this century.

PART 4 BOUNTIFUL EARTH. Part 4 closes with applications of geology to mineral and energy resources. Here we discuss ore deposits, coal, oil and gas, uranium, and alternative energy resources. The discussion places economic necessity and environmental degradation in apposition and highlights the advantages of policies of sustainable development. No essential geologic subjects are included in these chapters, so they can be made optional if the instructor chooses.

## Trends and Themes

Although plate tectonics is no longer new, the concept still inspires a fast pace of activity in today's geology, geochemistry, and geophysics that should lend currency and excitement to an introductory course. And so we integrate, at an appropriate level, the flavor of today's research with the basic principles of physical geology and call attention to newer approaches to some subjects; for example, correlating sea-level changes with climate and tectonics, and unraveling continental geology in terms of microplates and ancient plate interactions.

We have introduced a number of topics that are the focus of current research and important parts of geology. Among these are the relationship of sedimentary sequences to sea level, the evolution of orogenic belts in relation to plate tectonics, and current ideas on glaciation and climate, crust-mantle interactions, and the influence of comet and asteroid impacts on Earth's surface environments.

We use illustrations to introduce students to newer technologies, such as side-scan acoustic radar used in oceanography and seismic wave tomography used in the study of the interior. We want to impart the idea that geology, like other sciences, is constantly changing, renewing itself as new ideas and instruments come to the fore.

Geology is important as a framework for understanding the physical background of the myriad forms of life that evolved over billions of years on this planet. Thus geology has become the setting for many different kinds of environmental studies, from geological hazards to the compositions of soils to water supplies that affect public health.

Global change is fast becoming one of the major areas of concentration in earth science research. We discuss climate in relation to glaciation, the greenhouse effect, and the rise of carbon dioxide in the atmosphere. We also cover the connections among climate and weathering and erosion, landscape, and sedimentation.

We discuss geologic hazards and hazard prevention in chapters on volcanism (dangerous eruptive behavior), mass wasting (landslides), rivers (floods), water (groundwater contamination), and the interior (earthquakes).

Geology has been and remains an eminently practical science. All the common metals used in industry originate as ore deposits discovered by geologists. We use sand, gravel, and limestone for construction, for cement, and for many other industrial activities. Our major sources of energy—coal, oil, and gas—are geological deposits. Exploration for oil and gas, in fact, remains one of the major occupations of geologists.

## Supplements

It is a pleasure to be able to announce the availability of a wide and innovative selection of print and visual materials to supplement *Understanding Earth*:

*Study Guide* by David M. Best, Northern Arizona University

Contains chapter-by-chapter summaries and a wide variety of practice exercises designed to enhance text coverage and lectures. In addition to the 23 chapter treatments, there are reviews and practice exercises for each of the text's four parts.

*Instructor's Manual* by Philip M. Astwood, University of South Carolina, Columbia

Chapter summaries and objectives, as well as answers to Exercises and Thought Questions in the text, facilitate lecture preparation.

Printed and computerized *Test Banks* by Simon M. Peacock, Arizona State University

Over 1100 multiple-choice questions are keyed by page number to the text. Computerized *Test Banks* are available in both Macintosh and IBM formats.

*Annotated 35mm Slide Set* by Peter L. Kresan, University of Arizona

One hundred color photographs selected from the text and from additional outside sources are only one part of this package. Informative commentaries written by a geologist-photographer have been provided for each image. Each slide's note is identified by a scanned black-and-white representation.

*Overhead Transparencies*

One hundred color diagrams are reproduced from the text. Written suggestions for incorporating transparencies into lectures and testing are included in the package.

*Geology Videodisc,* produced by Videodiscovery in cooperation with W. H. Freeman and Company

Completely correlated to *Understanding Earth,* the disc will include 300 illustrations adapted from the text, 1200 additional still photographs, 3-D animations, laboratory demonstrations, original mini-documentaries, archival motion images of surface processes, and an aerial photography survey of the geological provinces and features of North America. A complete bar code directory, quick reference card, and instructor lesson plans are included with the disc.

## Acknowledgments

It is a challenge both to teachers and to authors of geology texts to encompass the many important aspects of geology in a single course and to inspire interest and enthusiasm in the student. To meet this challenge, we have called on the advice of many colleagues teaching in all kinds of college and university settings. From the earliest planning stages of this book we relied on a consensus of views in deciding on an organization for the text and in choosing which topics to include. As we wrote and rewrote the chapters, we again relied on our colleagues to guide us in making the presentation pedagogically sound, accurate, and accessible and stimulating to students. To each and every one we are grateful:

Gary Allen
*University of New Orleans*

N. L. Archbold
*Western Illinois University*

Richard J. Arculus
*University of Michigan, Ann Arbor*

Philip M. Astwood
*University of South Carolina*

R. Scott Babcock
*Western Washington University*

Evelyn J. Baldwin
*El Camino Community College*

David M. Best
*Northern Arizona University*

Stuart Birnbaum
*University of Texas, San Antonio*

F. W. Cambray
*Michigan State University*

Ernest H. Carlson
*Kent State University*

Max F. Carman
*University of Houston*

George R. Clark II
*Kansas State University*

G. S. Clark
*University of Manitoba*

Peter Dahl
*Kent State University*

Larry E. Davis
*Washington State University*

Robert T. Dodd
*SUNY at Stony Brook*

Bruce J. Douglas
*Indiana University*

Grenville Drapper
*Florida International University*

C. Patrick Ervin
*Northern Illinois University*

Stanley Fagerlin
*Southwest Missouri State University*

Jack D. Farmer
*University of California, Los Angeles*

Stanley C. Finney
*California State University, Long Beach*

Tim Flood
*Saint Norbert College*

Richard M. Fluegeman, Jr.
*Ball State University*

Charles Frank
*Southern Illinois University*

Robert B. Furlong
*Wayne State University*

Gary H. Girty
*San Diego State University*

William D. Gosnold
*University of North Dakota*

Bryan Gregor
*Wright State University*

G. C. Grender
*Virginia Polytechnic Inst. and State Univ.*

Mickey E. Gunter
*University of Idaho*

David A. Gust
*University of New Hampshire*

Kermit M. Gustafson
*Fresno City College*

Eric Hetherington
*University of Minnesota*

J. Hatten Howard III
*University of Georgia*

Herbert J. Hudgens
*Tarrant County Junior College*

Ruth Kalamarides
*Northern Illinois University*

Phillip Kehler
*University of Arkansas, Little Rock*

Peter L. Kresan
*University of Arizona*

Albert M. Kudo
*University of New Mexico*

Don Layton
*Cerritos College*

Peter Leavens
*University of Delaware*

Barbara J. Leitner
*University of Montevallo*

John D. Longshore
*Humboldt State University*

Peter Martini
*University of Guelph*

G. David Mattison
*California State University, Chico*

Robert D. Merrill
*California State University, Fresno*

Kula C. Misra
*University of Tennessee, Knoxville*

Roger D. Morton
*University of Alberta*

Simon M. Peacock
*Arizona State University*

Donald R. Prothero
*Occidental College*

C. Nicholas Raphael
*Eastern Michigan University*

James Roche
*Louisiana State University*

Fred Schwab
*Washington and Lee University*

D. W. Shakel
*Pima Community College*

Charles R. Singler
*Youngstown State University*

David B. Slavsky
*Loyola University of Chicago*

Douglas L. Smith
*University of Florida*

Richard Smosma
*West Virginia University*

Donald K. Sprowl
*University of Kansas*

Randolph P. Steinen
*University of Connecticut*

John F. Taylor
*Indiana University of Pennsylvania*

Thomas M. Tharp
*Purdue University*

Nicholas H. Tibbs
*Southeast Missouri State University*

Jan Tullis
*Brown University*

Kenneth J. Van Dellen
*Macomb Community College*

We have also had the benefit of informal advice on content and checking for accuracy from many colleagues, including especially C. W. Burnham, S. B. Jacobsen, Jane Selverstone, and J. B. Thompson, Jr. Frank Press wishes to acknowledge the support of the Andrew W. Mellon Foundation.

Others have worked with us more directly in writing and preparing the manuscript for publication. W. Naylor Stone capably prepared the glossary and index. At our side always were the editors at W. H. Freeman and Company, especially Jerry Lyons, who guided this project from its inception and was constant in his encouragement; Sonia DiVittorio, whose concern for both the overall pattern and the hundreds of details was central to her careful and intelligent development of the text and illustrations; and Diana Siemens, who ably shepherded the manuscript through final editing and composition. The superb photographs that illuminate the text were obtained by Travis Amos, who tirelessly combed many sources looking for the best possible choices.

The quality of the finished book would not have been possible without the final, and most skillful, efforts of Alison Lew, designer; Bill Page, illustration coordinator; John Hatzakis, layout artist; and Julia DeRosa, production coordinator. We are especially indebted to our illustrators—Ian Warpole, Tomo Narashima, and Network Graphics—for transforming our often *very* rough sketches into many outstanding drawings.

September 1993

Frank Press
Raymond Siever

# UNDERSTANDING EARTH

# UNDER-STANDING THE EARTH SYSTEM

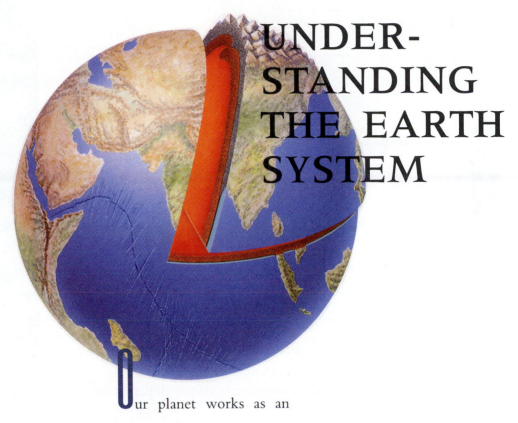

Our planet works as an interacting system of matter and energy that generates volcanoes, glaciers, mountains, lowlands, continents, and oceans. The energy that drives the system comes from Earth's internal heat, which is responsible for plate tectonics, and solar radiation, which circulates the atmosphere and oceans and powers erosion. The matter of Earth—its rocks and minerals—and its structure are the relics of Earth system dynamics evolving over 4.6 billion years of geologic time. The three great clans of rocks and their geologic structures reflect geologic processes. Igneous rocks are linked to volcanism, sedimentary rocks to weathering and erosion, and metamorphic rocks to mountain building.

# BUILDING
# A PLANET

Geology is an outdoor science with Earth as its laboratory. Many geologists are inspired solely by intellectual curiosity and a desire to understand, explain, and appreciate the world around them. In that sense geology is a pure science. But geology makes practical contributions to human life and therefore is an applied science as well. Our economic system depends on the materials and fuels extracted from the Earth. Our survival depends on learning how to live in a way that protects our environment, sustains our resources, and lessens the dangers of nature's hazards. This chapter presents the big picture: the origin of Earth and the forces that have shaped it since its birth. A few underlying principles that encompass much of geology are also introduced.

View of the Himalayas in Nepal. This mountain range, which contains some of the world's highest peaks, was created by the collision of India and Asia due to plate movements. *William Thompson.*

Some 15 or 20 billion years ago our universe began with a cosmic explosion that has come to be called the Big Bang. Before that moment, all matter was compacted into a single, inconceivably dense point. Scientists know little of what happened in the first second of the beginning of time, when the explosion hurled matter outward into an empty universe. But astronomers have acquired a general understanding of the following billions of years, when stars and galaxies formed. Geologists have concentrated their attention on the last 4.6 billion years, the time since the beginning of our solar system, with its Sun and planets—the time during which our own planet formed and evolved. It is the job of the geologist to unravel this history and write the biography of Earth.

"Civilization exists by geological consent, subject to change without notice," said the philosopher-historian Will Durant, reminding us of the remarkable circumstances that make this planet congenial to life. Earth, after all, is a very special place, and not just because we humans inhabit it. More than a million life forms have developed on this unique planet in the solar system. What led to the special conditions in which life as we know it is possible—oceans, continents, a hospitable atmosphere, and temperatures that are neither too high nor too low (Figure 1.1)? Was it a series of lucky coincidences, or did our planet evolve in a way that is predictable by the laws of science? In the study of geology, we not only explore Earth as it exists today; we also seek to know how it was formed, what it was like when it was newborn, how it evolved to become the planet of today, and, perhaps most exciting of all, what made it capable of supporting life.

## ASPECTS OF GEOLOGY

Like many other sciences, geology has both purely intellectual and practical aspects. The study of geology provides the knowledge needed to answer fundamental questions about how our planet works.

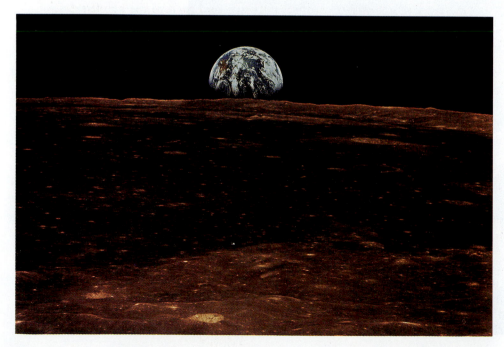

**FIGURE 1.1** Earth, with its oceans, atmosphere, and clouds, able to sustain life; geologically alive with active volcanoes, new mountains, and continents continuously renewed. Photographed by *Apollo* astronauts from the barren Moon—a planet without atmosphere or water, without life, one that "died" geologically more than 3 billion years ago. *NASA.*

**FIGURE 1.2** A landslide resulting from heavy rains destroys buildings in Hong Kong. The need for geologic advice and city planning is clear. *Director of Information Services, Hong Kong.*

Threatened by natural disasters, humans need to understand nature in order to protect themselves against its occasionally violent ways. Geologic knowledge enables us to find essential resources derived from Earth: coal, oil, natural gas, metals, chemicals, and other materials without which modern civilization would not be possible.

## Natural Disasters

As human populations multiply and become increasingly concentrated, our vulnerability to natural disasters grows. With greater urgency than ever before we seek safeguards against nature's threats: landslides, floods, droughts, volcanic eruptions, earthquakes with their destructive sea waves (Figure 1.2).

In this quest geologists have had some successes. We have learned, for example, that certain earthquakes below the seafloor can trigger giant waves that can crush homes and destroy communities on distant coasts. A warning system now provides people along the Pacific coast with a few hours' notice of the arrival of such waves. Volcanic eruptions are being predicted in some countries, so that authorities can evacuate endangered populations. Similarly, methods to predict impending earthquakes are being tested in Japan, China, the former Soviet Union, and the United States. Areas threatened by landslides and floods are being identified and rezoned to prohibit construction and thus preclude later devastation.

## Global Environment

Geologists can do much to mitigate nature's threats, but humans may be the biggest threat of all. Our species has gained the power to foul the atmosphere and oceans.

Underground aquifers, which carry much of our drinking water, are beginning to show worrisome levels of toxic chemicals. These substances originate from agricultural and industrial wastes that seep through the soil and rock to the aquifers below.

The atmosphere functions like a greenhouse, raising Earth's temperatures to levels that make life possible. The combustion of coal and oil, along with other industrial activities and the destruction of forests, releases gases that may be intensifying the greenhouse effect. It now seems quite possible that our climate may change in 50 to 100 years as a result. Unchecked, these changes could convert some fertile agricultural regions to semiarid lands and raise the level of the sea until it floods low-lying coastal cities. By studying the causes of past natural climatic changes, geologists are providing important knowledge about the possible effects of human activities on the environment.

Acid rain, a product of coal combustion and automobile exhaust, is threatening our lakes and forests. Urbanization, mining, agricultural operations, and warfare have now become more important factors than nature in modifying the surface of our planet (Figure 1.3).

**FIGURE 1.3** Oil wells in Kuwait set afire by Iraqi forces in the war of 1991. The smoke and soot and the oil that found its way into the Persian Gulf resulted in environmental damage and the loss of hundreds of millions of dollars' worth of oil. *Jim Lukoski/Black Star.*

## Economic Geology

Were it not for our ability to find and use the minerals in Earth's crust, we would have neither the energy nor the materials our modern society requires. Our cultural level would never have progressed beyond that of the Stone Age. Finding coal, oil, natural gas, and the ores from which our metals and chemicals are derived is the job of geologists. If we are to see life improve for people in poor countries, let alone maintain our own present standard of living, new mineral deposits must be found. Gone are the days when prospectors could easily find oil, iron, copper, tin, uranium, and other deposits important to the world's economies. The challenge for geologists is to reexplore the world, using new tools and techniques to search out the remaining undiscovered deposits.

The exploitation of Earth's mineral wealth has raised new and pressing concerns for geologists, principally those of protecting the environment and conserving resources. How can we most efficiently exploit nature's wealth without waste and without devastating the landscape? Somehow we must find answers to this question.

## Scholarly Geology

Geology also is a fascinating intellectual exploration in its own right. How a planet is born, how it evolves, and how it works today are only partially understood. Geologists are motivated to find the answers because, like all scientists, they have unbounded curiosity and perhaps even a sense of uneasiness when important natural phenomena remain unexplained. With the impulse of explorers, geologists will be hammering at rock outcrops, making geologic maps, probing the seafloor, and scrutinizing Moon rocks as long as our planet's features remain incompletely explained. Henry David Thoreau expressed their motivation well when he wrote: "Talk of mysteries! Think of our life in nature—daily to be shown matter, to come in contact with it—rocks, trees, wind on our cheeks! the solid earth! the actual world! . . . Who are we? where are we?"

## THE ORIGIN OF THE SYSTEM OF PLANETS

To begin at the beginning: How did our Sun and its system of planets originate? This question has attracted the attention of great philosophers and scientists for two centuries. Yet new experimental data and theoretical advances are brought to almost every meeting of geologists and astronomers, where they trigger fresh debates about the **origin of the solar system.**

An eighteenth-century philosopher suggested that the solar system formed from a rotating cloud of gas and dust. This old idea has been revived and

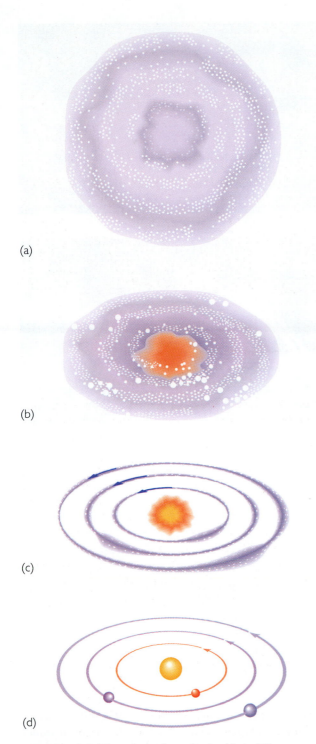

**FIGURE 1.4** The nebular hypothesis. (a) A diffuse, roughly spherical, slowly rotating nebula begins to contract. (b) As a result of contraction and rotation, a flat, rapidly rotating disk forms with matter concentrated at the center. (c) Contraction continues, the proto-Sun is formed, and rings of material are left behind. (d) The material in the rings condenses into planets revolving in orbit around the Sun. (After J. C. Brandt and S. P. Maran, *New Horizons in Astronomy*, 2d ed., San Francisco, W. H. Freeman, 1979.)

modified during the past few decades as astronomers have discovered that outer space beyond our solar system is not so empty as they once thought. Modern telescopes have found many such clouds in the universe, so we now know that these objects actually exist, and the materials that form them have been identified. The gases are mostly hydrogen and helium, the two elements that make up all but a small fraction of our Sun. The dust-sized particles are composed of materials similar to Earth's. The clouds themselves are called *nebulae* (Figure 1.4).

The nebula that formed our solar system was subjected to the force of gravity, which causes pieces of matter to be attracted to one another, smaller masses being pulled toward larger ones. As the particles were drawn together, the cloud contracted. The contraction in turn accelerated the rotation of the particles (just as ice skaters spin more rapidly when they pull in their arms), and the faster rotation flattened the cloud into a disk. Under the pull of gravity, matter began to drift toward the center, accumulating into the precursor of our Sun, or proto-Sun. The material in the proto-Sun, compressed under its own weight, became dense and hot. The internal temperature of the proto-Sun rose to about 1,000,000°C, at which point nuclear fusion began. The Sun's nuclear fusion was (and is to this day) the same nuclear reaction that occurs in a hydrogen bomb. In both cases hydrogen atoms, under intense pressure and at high temperature, combine (fuse) to form helium. Mass disappears as it is converted into energy. We experience the enormous amount of energy released as sunshine in the case of the Sun, as a great explosion in the case of the H-bomb. This conversion of mass to energy is represented by Albert Einstein's famous equation $E = mc^2$, where $E$ is the energy released by conversion of mass ($m$) and $c$ is the speed of light, about 298,000 km per second.

What about the disk of gas and dust enveloping the primitive Sun? How did it form planets, and why do the planets have different chemical compositions? Many scientists lean toward the following explanation. Initially the disk was extremely hot, so that its materials were largely in gaseous form. As the disk cooled, many of the gases condensed. That is, they changed to their liquid or solid form, just as water vapor condenses into droplets on the outside of a cold glass and as water solidifies into ice when it cools below the freezing point. The condensing material formed grains that gradually clumped together into small chunks, or planetesimals. The larger planetesimals, which had stronger gravitational attraction, pulled the smaller ones toward them to form nine planets (Figure 1.5).

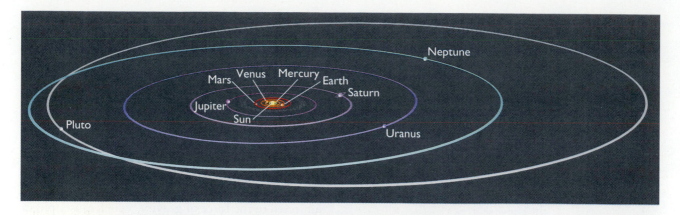

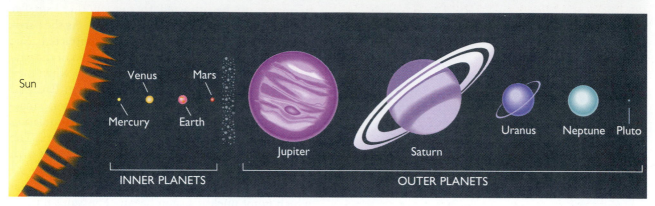

**FIGURE 1.5** The solar system. The four inner terrestrial planets are all small and rocky. The outer giant planets are gaseous. Outermost and smallest is Pluto, a snowball of methane, water, and rock.

The four *inner planets* (Mercury, Venus, Earth, and Mars) are closest to the Sun and for that reason share similar early histories. The inner planets grew where it was too hot for hydrogen, helium, water, and other light gases and liquids to be retained. These materials were mostly blown away by radiation and matter streaming from the Sun. And so the inner planets reached their full size composed of materials that remained behind, heavy metals such as iron and other heavy compounds that make up rock. Thus the inner planets became dense, rocky masses; they are called the *terrestrial* ("Earthlike") planets.

According to this scenario, most of the volatile materials—those that become gases at relatively low temperatures, such as hydrogen, helium, water, methane, and ammonia—were removed from the terrestrial planets. They were carried to the cold outer reaches of the solar system, some to accumulate on the giant *outer planets*—Jupiter, Saturn, Uranus, and Neptune—and their satellites, the rest to be swept into outer space beyond. Jupiter and Saturn

were big enough and their gravitational attraction was strong enough to enable them to hold on to all of their nebular constituents. Thus they, like the Sun, are composed mostly of hydrogen and helium and the other constituents of the original nebula. Outermost is tiny Pluto, a strange frozen mixture of gas, water, and rock.

This scenario should be taken for what it is—an explanation that many scientists think best fits the known facts. Perhaps some of these notions come close to what actually happened. We will be more certain only after much more work has been done, some of it now under way. Nebulae at various stages of development are being studied with powerful telescopes, and these observations are providing information about what goes on in the remote sections of the universe. Planetary probes carried by American and Russian spacecraft have returned data on the nature and composition of the atmospheres and surfaces of Mercury, Venus, Mars, Jupiter, Saturn, Uranus, Neptune, and the Moon. A telescope placed in orbit around Earth recently found the first direct evidence of an envelope of gas and dust, apparently the beginnings of a planet in formation, around two nearby stars. This major discovery supports the idea that our system of planets in orbit around the Sun is not a unique phenomenon. All of this scientific investigation should in time give us a clearer picture of how our own solar system started.

We have dwelt on the question of the origin of the solar system because the evolutionary course followed by a planet is set by its initial state. The current state of Earth, some 4.6 billion years later, is reasonably well known to us. Its states at these two times in the course of **planetary evolution**—the beginning and the present—have to be fitted into the explanations we develop for the changes Earth has undergone throughout its history.

## EARTH AS AN EVOLVING PLANET

How did Earth evolve from a conglomeration of chunks of matter to a planet with continents, oceans, and an atmosphere? The answer lies in the transformation of Earth from a homogeneous body to a differentiated planet—that is, one in which the interior is divided into layers that differ physically and chemically.

## The Earth Heats Up

Think of Earth as it was 4.6 billion years ago. The stage was set for the accumulation of the planet by the accretion, or gathering up, of planetesimals. Our current understanding is that three processes began to heat up the growing planet: collision of the planetesimals, compression or squeezing together of the growing planet, and radioactivity.

First, as Figure 1.6a indicates, each in-falling planetesimal hit the planet at high velocity. A moving body carries much energy of motion (think of how the energy of motion crushes a car in a collision). A planetesimal colliding with Earth at a velocity of about 11 km per second delivered as much energy as the same weight of TNT. When the planetesimals crashed into the primitive Earth, most of this energy of motion was converted to heat, which is another form of energy.[1]

Second, as the planet grew, it compressed under its own weight (Figure 1.6b). Compression also leads to a rise in temperature. A common instance of this phenomenon is the heating of the barrel of an air pump, the kind used to inflate bicycle tires. As we push down, the air is compressed quickly, too

---

[1]Every few million years a large chunk of matter still collides with Earth, sometimes with devastating effect. Some scientists believe that the impact of a large meteorite about 65 million years ago caused the extinction of many species, including the dinosaurs.

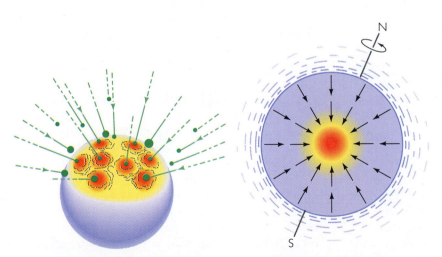
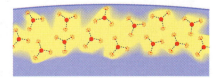

(a) Collision of planetesimals with primitive Earth

(b) Compression of planet under its own growing weight

(c) Disintegration of radioactive elements

**FIGURE 1.6** Three mechanisms that would cause the early Earth to heat up. (a) In accretion, colliding bodies bombard Earth and their energy of motion is converted to heat. (b) Gravitational compression of Earth into a smaller volume causes its interior to heat up. (c) Disintegration of radioactive elements releases atomic particles; as they are absorbed by the surrounding rock, their energy of motion is transformed to heat.

quickly for the heat to flow away, and the barrel heats up. It is likely that accretion and compression raised the internal temperature of the newly organized planet to an average of about 1000°C.

Third, several elements (uranium, for example) are radioactive. Although these elements occur in small amounts, their radioactivity has had a profound effect on Earth's evolution. Atoms of radioactive elements spontaneously disintegrate by emitting atomic particles. As these emitted particles are absorbed by the surrounding matter, their energy of motion is transformed into heat. The disintegration of the radioactive elements, depicted in Figure 1.6c, is a heat source that has persisted for billions of years. Heat generated by radioactivity also warmed the newly formed Earth, raising the interior temperatures to about 2000°C. This is the temperature at which iron melts. Iron makes up about one-third of the matter of Earth. The melting of this much iron initiated the process by which Earth became the planet we know today.

## Planetary Differentiation

Early in its history the primitive Earth was a homogeneous planet, with roughly the same kinds of material at all depths. Then at some time, possibly in the first few hundred million years, it underwent a profound reorganization after it warmed to about the melting point of iron. Because iron is heavier—that is, denser—than the other common elements, when it melted about a third of the primitive planet's mate-

rial sank to form an iron core at the center. In this catastrophic process other large portions of Earth became partially molten. There may even have been an early ocean of molten rock—a **magma** ocean—more than 100 km deep. Unlike iron, other molten materials were lighter than the parent substances from which they separated, so they floated upward to cool and form a primitive crust. The formation of a core was the first stage of the differentiation of Earth, the process in which it was converted from a homogeneous body to a zoned, or layered, body. The layers are a dense iron **core** at the center, a **crust** of lighter materials at the surface, and between them the remaining **mantle,** consisting of rocks with an intermediate density (Figure 1.7).

There are more than 100 elements, but 99 percent of Earth's mass is made up of only eight. These eight most abundant elements are listed in Figure 1.8. About 90 percent of Earth consists of only four elements: iron, oxygen, silicon, and magnesium. It is useful to compare the relative abundance of elements in the crust with their abundance in Earth as a whole. Iron makes up a full 35 percent of Earth's mass. But because most of the iron sank to the core, there is little of it in the crust, where it is only fourth in abundance. Conversely, silicon, aluminum, calcium, potassium, and sodium are among the lightest of the solid elements and therefore floated upward during differentiation; they are far more abundant in the crust than in Earth's interior. This uneven distribution of elements is a result of differentiation.

Differentiation is perhaps the most significant event in Earth's history. It led to the formation of a

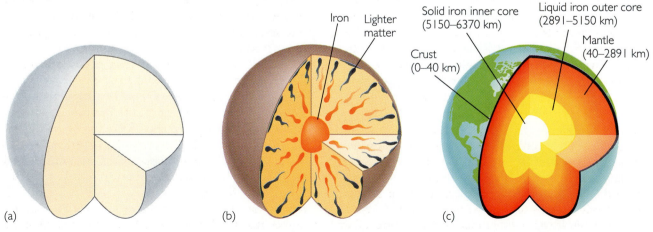

**FIGURE 1.7** Early Earth (a) was probably a homogeneous mixture with no continents or oceans. In the process of differentiation, iron sank to the center and light material floated upward to form a crust (b). As a result, Earth is a zoned planet (c) with a dense iron core, a crust of light rock, and a residual mantle between them.

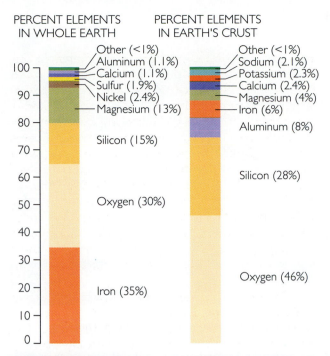

PERCENT ELEMENTS IN WHOLE EARTH

PERCENT ELEMENTS IN EARTH'S CRUST

Whole Earth:
Other (<1%)
Aluminum (1.1%)
Calcium (1.1%)
Sulfur (1.9%)
Nickel (2.4%)
Magnesium (13%)
Silicon (15%)
Oxygen (30%)
Iron (35%)

Earth's Crust:
Other (<1%)
Sodium (2.1%)
Potassium (2.3%)
Calcium (2.4%)
Magnesium (4%)
Iron (6%)
Aluminum (8%)
Silicon (28%)
Oxygen (46%)

**FIGURE 1.8** The relative abundance by weight of elements in the whole Earth and in its crust. Differentiation has created a light crust depleted of iron and rich in oxygen, silicon, aluminum, calcium, potassium, and sodium.

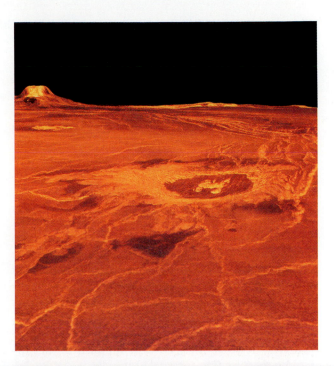

**FIGURE 1.9** Venusian crater and volcano revealed by radar on the *Magellan* spacecraft, 1992. Crater in foreground, formed by meteorite impact, is 48 km in diameter. Volcano near horizon is 3 km high. *NASA.*

crust and eventually the continents. Differentiation brought the lighter elements to the outer layers of Earth; and it initiated the escape of even lighter gases from the interior, which eventually led to the formation of the atmosphere and oceans. Gases continue to escape from Earth's interior to this day in the emissions that accompany volcanic eruptions.

But what of the other planets? Did they go through the same early history? Information transmitted from our planetary spacecraft indicates that all the terrestrial planets have undergone differentiation, but they have followed different evolutionary paths. Mercury, for example, has an iron core even larger than Earth's and a lightweight crust that has been heavily cratered. Mercury has no blanket of air to protect it from being riddled by the impacts of large meteorites over the past 4.6 billion years. It is now inactive and geologically dead—that is, without the ongoing processes of mountain making, volcanism, and earthquakes that we observe on our own planet. It has a trace of an atmosphere but no wind or water to erode and smooth its ancient, cratered surface.

Venus is Earth's twin in mass and size, but it differs from Earth profoundly in the nature of its atmosphere. Somehow—geologists do not yet understand the mechanism—Venus developed a heavy, poisonous, and incredibly hot atmosphere composed mostly of carbon dioxide. It was the terrestrial planet

we knew least about because it is shrouded by dense clouds that hide its surface from view. However, radar images from a spacecraft that reached Venus in 1990 suggest that it is geologically active, with volcanoes and mountains (Figure 1.9).

Mars, with crust and core, has a composition similar to Earth's and has experienced many of the same processes that Earth has undergone. The surface shows that volcanism and mountain making are still going on and that there was an ancient episode of flooding and erosion by water. No water is present on the surface today, and Mars's thin atmosphere is composed almost entirely of carbon dioxide.

The Moon is the best known planet other than Earth because of its proximity and the programs of manned and unmanned exploration. Early in its history it underwent differentiation to form a small core and a thick crust. Geological activity ceased some 3 billion years ago, however, and the surface we see today is very old, with craters and mountains shaped almost entirely by the ancient impacts of large meteorites. The Moon has no atmosphere. A hypothesis that has been revived and is receiving much attention argues that the Moon was blasted out of Earth early in its history when Earth collided with a Mars-sized body.

The giant gaseous outer planets will remain a puzzle for a long time. They are chemically so distinct from the terrestrial planets and so much larger that they must have followed an entirely different evolutionary course. (It has even been proposed that Jupiter and its 15 moons are akin to a small solar system whose sun—Jupiter—never got quite hot enough to shine.)

## The Formation of Continents, Oceans, and Atmosphere

From its very beginning Earth's history has been dominated by two giant engines, one internal, the other external. A heat engine—the gasoline engine of an automobile, for example—transforms heat energy released from the fuel into mechanical motion or work. Earth's internal heat engine is powered by the internal heat of its interior. The external heat engine is driven by solar energy—heat supplied to the surface by the Sun. The internal heat melts rocks, makes volcanoes, and supplies the energy to build and move continents and to thrust mountains upward. The external heat is responsible for our climate and weather, and it drives the rain and wind that erode mountains and shape our landscape.

**CONTINENTS**   We have only the most general notion of what caused the **formation of continents.** We think molten rock (magma) floated upward from the partially molten interior of Earth, cooled, and solidified to form a crust of rock. This primeval crust melted and solidified repeatedly, and the lighter materials gradually separated from the heavier ones and floated to the top, forming the primitive nucleus of a continent. The disintegration and decomposition of rock by rainwater and other components of the atmosphere broke up and altered the rocks; water, wind, and ice loosened and moved the residue of broken-down rock particles to low-lying places, such as beaches, deltas, and the bottoms of adjacent seas. Here they could pile up as thick layers. The end products were the continents, which grew as this process was repeated through countless cycles. The continents began to grow soon after differentiation and have grown throughout geologic time.

**OCEANS AND ATMOSPHERE**   Most geologists believe that the **origin of the oceans and atmosphere** can be traced to Earth itself: they came from the interior as products of Earth's heating and differentiation. Others believe that the ocean-atmosphere envelope originated outside of Earth. Comets are composed largely of ices of such substances as water plus carbon dioxide and other gases. Countless com-

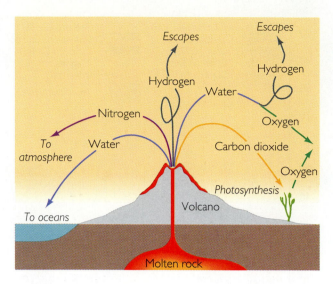

**FIGURE 1.10** Volcanism has contributed enormous amounts of water, carbon dioxide, and other gases to the atmosphere and solid materials to the continents. Photosynthesis by plants removed carbon dioxide and added oxygen to the primitive atmosphere.

ets may have bombarded Earth early in its history, carrying in water and gases that formed the early oceans and atmosphere. Geologists who believe in an internal origin argue this way: Originally the water was locked up; that is, chemically bound as oxygen and hydrogen in certain minerals, as carbon and nitrogen were. As Earth warmed and its materials partially melted, water vapor, nitrogen, carbon dioxide, and other gases were freed and carried to the surface by magmas and released through volcanoes. The amount of water vapor released by currently active volcanoes indicates that over several billion years, volcanic activity could easily have filled the oceans.

Such an early atmosphere, released from the hot interior by volcanic outgassing during the differentiation of Earth some 4 billion years ago, probably consisted of the same gases that are expelled from present-day volcanoes: water vapor, hydrogen, carbon dioxide, nitrogen, and a few other gases (Figure 1.10). The earliest atmosphere thus was entirely different from the one we live in now, which consists primarily of nitrogen and oxygen. How did the atmosphere change? The production of significant amounts of free oxygen and its persistence in the atmosphere probably came only after life had evolved at least to the complexity of photosynthetic algae. Algae are simple, one-celled forms of life. Like other organisms that employ photosynthesis, algae use carbon dioxide and water as raw materials and the energy of sunlight to manufacture organic matter, and they release oxygen as a waste product. Oxygen began to accumulate in the atmosphere and gradually built up to its present value. (Photosynthesis is explained in detail in Chapter 9, Box 9.3.)

By about 4 billion years ago, Earth had become a differentiated planet. The core was still hot and mostly molten, but the mantle was fairly well solidified, and a primitive crust and continents had developed. Oceans and atmosphere had been produced, probably by gases from the interior, and the processes we observe today were set in motion.

## Uniformitarianism

Much of what we have come to understand about the geologic past is based on observation of the present-day workings of our planet. We can observe today the growth of continents, the erosion of mountains, the eruptions of volcanoes. It is a cardinal tenet of geology, advanced in the eighteenth century by the Scottish physician and geologist James Hutton, that "the present is the key to the past." This tenet is now known as the **principle of uniformitarianism.** According to this principle, the geologic processes we see in operation as they modify Earth's crust today have worked in much the same way over geologic time. The rates of change may have varied in the past, and rare catastrophes such as the collision of a large meteorite or comet with Earth may have perturbed the processes, but no matter. Uniformitarianism, together with the laws of physics and chemistry, provides the basis for the theory and practice of geology. We will call upon uniformitarianism frequently as we attempt to decipher Earth's geologic history.

# PLATE TECTONICS: A MODERN PARADIGM FOR GEOLOGICAL SCIENCE

In the 1960s a great revolution in thinking shook the world of geologists. Physics had a comparable revolution at the beginning of the twentieth century, when the theory of relativity unified the physical laws that govern space, time, mass, and motion. Biology had a comparable revolution in the middle of this century when the discovery of DNA allowed biologists to explain how organisms transmit the information that controls their growth, development, and functioning from generation to generation. For almost 200 years geologists have supported various theories of mountain building, volcanism, and other major phenomena of Earth. No theory was general enough to explain well the whole range of geologic processes. We now have a single, all-encompassing concept that explains many of Earth's major geologic features. Furthermore, such topics as the classification and distribution of rocks and the positions and characteristics of volcanoes, earthquake belts, mountain systems, and ocean basins were formerly described more or less in isolation. Today we can treat these and other topics in the context of a unifying theory, **plate tectonics.** In the history of science simple theories that explain many observations, as this one does, are the most enduring.

Plate tectonics involves the outermost shell of our concentrically zoned planet. Earlier we described Earth's crust and mantle as chemically distinct zones that took their forms during differentiation. We will see in Chapter 20 that Earth's zones can also be characterized as strong and weak, as in Figure 1.11. We

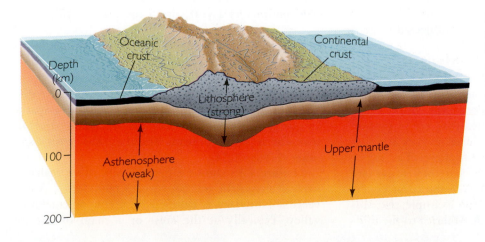

**FIGURE 1.11** Earth's outermost shell is the strong, solid lithosphere, composed of the crust and top of the mantle. It rides on a weak, partially molten region of the mantle called the asthenosphere.

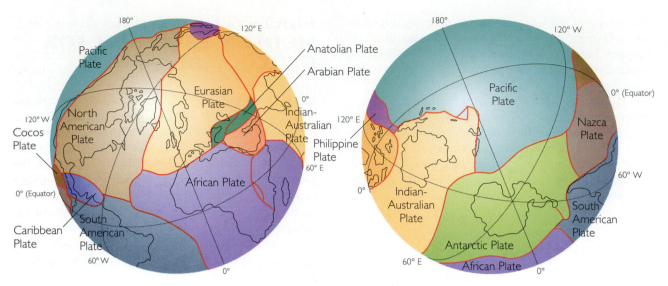

**FIGURE 1.12** Earth's plates today. On these global views, plate boundaries—where plates separate, collide, or slide past each other—are shown by red lines. Compare with the flat map of Earth's plates shown inside the front cover.

speak of "strong" and "weak" here in the sense that a ceramic is strong and a ball of wax is weak. The one is not easily deformed but can crack; the other is easily molded, like a tube of toothpaste. The **lithosphere,** which includes the crust and the top part of the mantle, is depicted in Figure 1.11 as the strong, solid, outermost shell, 50–100 km thick. The continents are raftlike inclusions embedded in the lithosphere. The lithosphere rides on the weak, partially molten **asthenosphere** (from the Greek *asthenēs,* meaning "weak"). The lithosphere is strong because it is relatively cool, being so close to the surface. The asthenosphere is weak because it is hot, almost at the melting point, and under great pressure because it is deep below the surface. The fact that the zones have different strengths determines how the lithosphere breaks apart and moves when it is subjected to geologic forces.

**THE THEORY OF PLATE TECTONICS**   According to the theory of plate tectonics, the lithosphere is not a continuous shell; it is segmented, broken into about a dozen large rigid plates that are in motion over the Earth's surface. Each plate moves as a distinct unit, riding on the asthenosphere. The major plates and the directions in which they move are sketched in Figure 1.12 and inside the front cover. The North American Plate, for example, extends from the Pacific coast of North America to the mid-

dle of the Atlantic, where it meets the Eurasian and African plates.

But why should the plates move? Because the mantle beneath the lithosphere is hot and moldable. That combination allows the materials of the mantle to move by **convection** (Figure 1.13). Convective motion occurs in a flowing material, either a liquid or a moldable solid, when hot matter rises from the bottom (because it is less dense than the matter above it) and cool matter sinks from the surface (because it is denser than the matter beneath it). We are all familiar with the circulating currents of boiling water in a pot, smoke rising from a chimney, heated air floating up to the ceiling, and cooled air sinking to the floor. The slow movement of the mantle by convection, a few centimeters a year, drags the plates along.

**PLATE BOUNDARIES**   Many large-scale geologic features occur at the boundaries of plates, where they interact with each other. There are three types of boundaries between adjacent moving plates. At **divergent boundaries** plates separate and move apart; at **convergent boundaries** plates collide; and at **transform fault boundaries** they slip past each other (Figure 1.14).

Figure 1.15 shows a geologically more realistic view of divergent and convergent boundaries. A divergent boundary is typified by a rift, or cracklike valley, typically at the crest of a chain of subsea

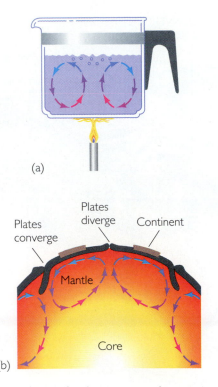

FIGURE 1.13 (a) A familiar instance of convection is seen when water is heated in a pot. (b) A simple model showing how convection currents in the deep interior may be the driving force of plate movements. Hot matter rises under the plate boundaries and flows in opposite directions, dragging the plates along and forcing them to separate. At other plate boundaries, cooled matter sinks, tending to drag the plate down.

mountains, called a **mid-ocean ridge.** These ridges wind along the bottom of the world's oceans. The Mid-Atlantic Ridge, for example, runs up the middle of the Atlantic Ocean, surfacing in several places, most extensively at Iceland (Figure 1.16). Another divergent boundary is the East Pacific Rise. Divergent boundaries are characterized by earthquake activity and by volcanism as the void between the receding plates is filled by magma that rises from below the lithosphere. The molten material solidifies as rock in the crack, and the plates of the lithosphere grow with the accretion (accumulation) of this fresh rock as they separate. As the plates separate, an ocean basin, such as the one that contains the Atlantic Ocean, can form and grow. Since new seafloor is created, this part of the process is called **seafloor spreading.**

If plates separate in one place, they must converge somewhere else, and they do. Plates grind together head-on along convergent boundaries, as shown in Figures 1.14 and 1.15. Note the profusion of geologic activities associated with a plate collision. The oceanic lithosphere descends into the mantle, beneath the continental lithosphere, a process called **subduction.** This downbuckling produces a deep-sea trench, a long, narrow feature (about 1000 by 100 km) where the ocean reaches its greatest depths (about 10 km). The edge of the overriding continental plate is crumpled and uplifted to form a mountain chain parallel to the trench. The enormous forces of such collisions produce great earthquakes. Materials

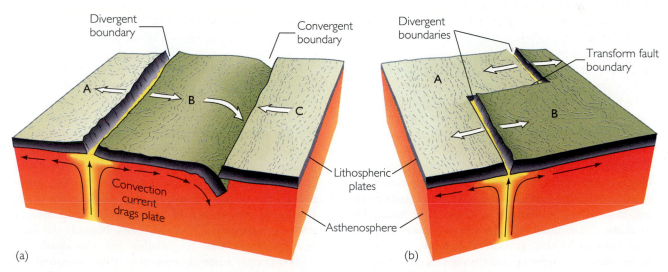

FIGURE 1.14 The three types of plate boundaries: (a) A divergent boundary, where plates A and B separate, and a convergent boundary, where plates B and C collide. (b) A transform fault boundary, where plates A and B slip past each other.

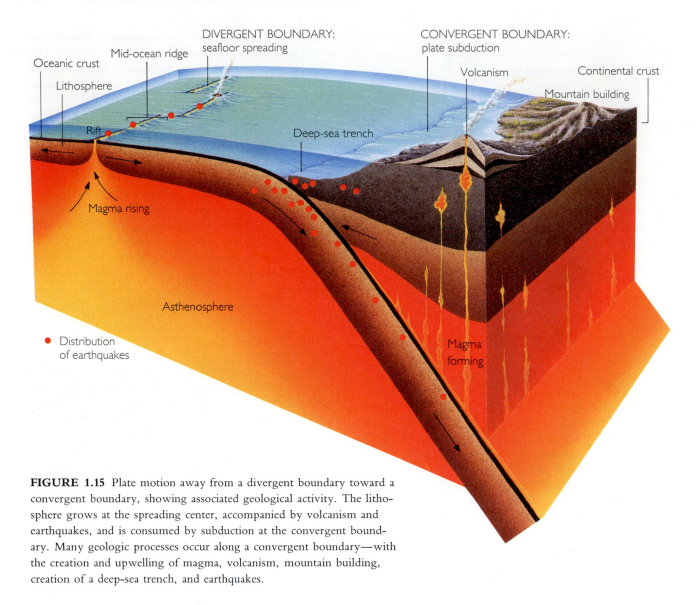

**FIGURE 1.15** Plate motion away from a divergent boundary toward a convergent boundary, showing associated geological activity. The lithosphere grows at the spreading center, accompanied by volcanism and earthquakes, and is consumed by subduction at the convergent boundary. Many geologic processes occur along a convergent boundary—with the creation and upwelling of magma, volcanism, mountain building, creation of a deep-sea trench, and earthquakes.

may be scraped off the descending slab and incorporated in the adjacent mountains. What a complicated mess for geologists to disentangle! What an exciting challenge to the science of geology!

As the oceanic plate descends into the hot mantle, parts of it may begin to melt. The rock melt, or magma, thus formed floats upward, some of it reaching the surface and erupting from volcanoes. The formation of magma in the subduction zone may be a key element in the creation of granite, one of the most common rocks found on continents.

Such boundaries of convergence, where lithosphere is consumed, have been named **subduction zones.** Recall that divergent zones are sources of new

lithosphere; subduction zones are sinks in which materials are consumed in corresponding amounts.

The west coast of South America, where the South American Plate collides with the oceanic Nazca Plate, is a subduction zone at a convergent boundary. The Andes Mountains rise on the continental side of this boundary, and the Chilean deep-sea trench lies just off the coast. In this locale deadly volcanoes are active. One of them, Nevado del Ruiz in Colombia, was responsible for the deaths of 25,000 people when it erupted in 1985. Some of the world's greatest earthquakes have been recorded along this boundary. Another subduction zone is the boundary between the small Juan de Fuca Plate and the North American Plate, just off the coasts of British Columbia, Washington, and Oregon. This convergent boundary gives rise to the volcanoes of the Cascade Range, including the dangerously active Mount St. Helens.

Some plates do not collide; they slip past each other along a transform fault. The famed San Andreas fault of California is such a boundary. There the Pacific Plate slides past the North American Plate in a northwesterly direction (Figure 1.17). Because the plates have been sliding past each other for millions of years, the rocks on the two sides are of different types and ages. The sliding is not smooth, but more of a stick-slip process; a sudden slip produces an earthquake. One such earthquake destroyed San Francisco in 1906. There is much concern that a sudden slip between the Pacific Plate and the North American Plate along the San Andreas fault near Los Angeles may be extremely destructive within the next 25 years or so.

If plates can move and continents are embedded in plates, the geography of the world may have been different in the geologic past. Some 200 million years ago all of the continents were assembled together into the supercontinent of Pangaea, as shown in Figure 1.18. (We will be discussing Pangaea in Chapter 20.)

The plate motions we have just described represent the general pattern of the work output of Earth's internal heat engine as we can see it today. We have to understand how a planet generates and gets rid of its heat if we are to understand how that planet works. Because of Earth's internal heat, plates have been created, moved, and destroyed since the large-scale differentiation of Earth ended some 4 billion years ago.

**FIGURE 1.16** The Mid-Atlantic Ridge, a plate divergence boundary, surfaces above sea level in Iceland. The cracklike valley indicates that plates are being pulled apart. *Gudmundur E. Sigvaldason, Nordic Volcanological Institute.*

**FIGURE 1.17** The view southeast along the San Andreas fault in the Carrizo Plain of California. The San Andreas is a transform fault, forming a portion of the sliding boundary between the Pacific and North American plates. *R. E. Wallace, USGS.*

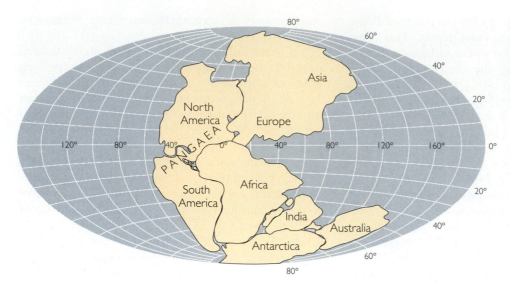

**FIGURE 1.18** Some 200 million years ago all of Earth's present continents were joined together in a single supercontinent, Pangaea ("all lands").

# GEOLOGIC TIME

When you first delve into geology, you may have to broaden your notion of time, which is based on your knowledge of history (hundreds to thousands of years) and on cycles in your own life (tens of years). We have already mentioned the age of the universe (20 billion or so years), the age of Earth and the other planets (about 4.6 billion years), and the time since the continents were formed (4 billion years) (see inside back cover). Plates move at the rate of a few centimeters a year—an amount sufficient to open ocean basins by seafloor spreading or to close them by plate collisions in hundreds of millions of years. We will see that mountains can be uplifted or eroded in millions of years and ice ages can come and go in tens of thousands of years. Volcanoes can blow their tops and earthquake faults can rupture in seconds, but they are single events in a sequence that can last hundreds of thousands or millions of years, reflecting the slow but powerful workings of the internal heat engine. The first living cells developed some 3.5 billion years ago, but human origins date back only a few million years, a few hundredths of 1 percent of Earth's existence (Figure 1.19).

# THE SCIENTIFIC METHOD

We have described many of the basic concepts of geology; but a description is not an explanation. How do we develop these ideas of Earth's evolution, of or plate tectonics, or of any scientific concept? Scientists operate within a system of open communication. They share their data and ideas at scientific meetings, in journals, in conversations almost every time they meet. Most of the great concepts of science, whether they occur as a flash of insight or after meticulous analysis, result from hundreds or thousands of such interactions. It is the essence of science that scientists build on one another's work.

When a scientist proposes a tentative explanation of a body of data derived from experiments, called a **hypothesis,** he or she offers it to the community of scientists for criticism and repeated tests against new data. A hypothesis is strengthened if it is confirmed by other workers, and especially if it successfully predicts the outcomes of new experiments. A hypothesis that has survived repeated challenges and has accumulated a substantial body of support from observations and experiments is elevated to the status of a **theory.** In its early life a theory is not cast in concrete. If new evidence indicates a theory is wrong, scientists may modify it or discard it. In this sense a scientific theory can never be proved, only falsified. The longer a theory survives all the new observations and tests and challenges, however, the greater the confidence with which it is held. This is the **scientific method,** with its goal of explaining with increasing precision the way the universe works.

*In this context plate tectonics is not a dogma but a confirmed theory whose strength lies in its simplicity and generality. Theories can be overturned, but the theories of the age of Earth, the evolution of life, and plate tectonics explain so much so well and have survived so many efforts to prove them false that most scientists treat them as facts.*

(a)

(b)

**FIGURE 1.19** Geologic phenomena can stretch over thousands of centuries or can occur with dazzling speed. (a) The peaks of the Canadian Rockies were thrust up over a span of about 10 million years. *Art Wolfe.* (b) Meteor Crater, Arizona, was formed in a few seconds by the impact of a meteorite. *F. Gohier/Explorer.*

Yet several competing hypotheses have been advanced to explain the forces that cause plates to move. Our story of the early evolution of Earth is a hypothesis, and we can expect it to change many times because of the difficulty of recovering the information contained in the oldest rocks—information that may have been largely destroyed in the violent process of differentiation.

Because the information available is incomplete and scientists' ability to observe nature is limited, some questions can be answered no better by scientists than by philosophers or poets. What, for instance, existed before the Big Bang? Or, as Walt Whitman asked:

> Great is the Earth, and the way it became
>   what it is,
> Do you imagine it is stopped at this?

In this chapter we have made many statements without providing supporting observations or an underlying rationale. We have done so in order to preview the big picture, reserving for later chapters all of the substantiation called for by the scientific method. We began with the origin of the universe, but most of our discussion has dealt with the events that led to the formation and evolution of our planet to the point where continents, oceans, and an atmosphere developed. Now that we have outlined plate tectonics, the paradigm that has shaped geologic thought in the second half of the twentieth century, we can consider continents and ocean basins and interpret information contained in the various kinds of rocks within that conceptual framework. Although plate tectonics does not explain everything, it gives us by all odds the best foundation on which to structure Earth's story.

## SUMMARY

**How did our solar system originate?** The Sun and its family of planets probably formed when a primeval cloud of gas and dust condensed about 4.6 billion years ago. The planets vary in chemical composition in accordance with their distance from the Sun and with their size.

**How did Earth form and evolve over time?**
Earth probably grew to its present size by accretion of small chunks of matter. At its birth it was probably a homogeneous body, warmed by the processes of accretion and compaction. Because of radioactivity, Earth began to heat up, and within a few hundred million years the temperature probably reached the melting point of iron. Drops of molten iron sank to Earth's center, and lighter matter floated up to form the outer layers that became the continents. Outgassing gave rise to the oceans and a primitive atmosphere. In this way Earth was transformed to a differentiated planet with chemically distinct zones: an iron core; a magnesium–iron-silicate mantle; and a crust rich in oxygen, silicon, aluminum, calcium, sodium, and radioactive elements.

**What are the basic elements of plate tectonics?** The lithosphere, or Earth's outermost shell, is broken into about a dozen large, rigid plates. These plates move over the weak asthenosphere below, driven by convection currents in the mantle. The plates jostle each other as they move in their individual courses. The boundaries of these plates are zones of intense activity. These are the sites of mountain building, volcanoes, creation and destruction of seafloor, and earthquakes.

**How do geologists use the principle of uniformitarianism?** Geologists believe that the processes that have shaped Earth have not changed over geologic time, and therefore that the key to understanding the past lies in observing how those processes work today.

**Explain the scientific method in terms of the way scientists work and develop hypotheses and theories.** Scientists share the data they develop and check one another's work. A hypothesis is a tentative explanation of a body of data. If it is confirmed repeatedly by other scientists' experiments, it may be elevated to a theory. Many theories are abandoned when subsequent experimental work shows them to be false. Confidence grows in those theories that withstand repeated tests and are able to predict the results of new experiments.

# KEY TERMS AND CONCEPTS

origin of the solar system (p. 6)

planetary evolution (p. 9)

magma (p. 10)

core (p. 10)

crust (p. 10)

mantle (p. 10)

formation of continents (p. 12)

origin of the oceans and atmosphere (p. 12)

principle of uniformitarianism (p. 13)

plate tectonics (p. 13)

lithosphere (p. 14)

asthenosphere (p. 14)

convection (p. 14)

divergent boundaries (p. 14)

convergent boundaries (p. 14)

transform fault boundaries (p. 14)

mid-ocean ridge (p. 15)

seafloor spreading (p. 15)

subduction (p. 15)

subduction zones (p. 16)

hypothesis (p. 18)

theory (p. 18)

scientific method (p. 18)

# EXERCISES

1. What factors have made Earth a particularly congenial place for life to develop?

2. How and why do the terrestrial planets differ from the giant outer planets?

3. How does the chemical composition of Earth's crust differ from that of its deeper interior?

4. What caused Earth to differentiate, and what was the result?

5. Describe the central idea of plate tectonics.

6. Describe and explain the large-scale geologic activities associated with plate boundaries.

7. How would you recognize a plate boundary?

8. What is the difference between an experiment, a hypothesis, a theory, and a fact?

# THOUGHT QUESTIONS

1. How does the discovery of solid matter around other stars contribute to the debate about the possibility of life elsewhere in the cosmos? What are the implications of the existence of life on the planets of other stars?

2. If you were an astronaut exploring another planet, what evidence would you look for to decide whether the planet was differentiated and whether it was still geologically active?

3. What are the advantages and disadvantages of living on a differentiated planet? On a geologically active planet?

4. Speculate on what life would be like today if the ancient continent of Pangaea had remained a single landmass instead of breaking up into Eurasia, Africa, and the Americas.

5. According to the Bible, Earth is some 5000 years old. Is that figure a hypothesis, a theory, a fact, or an article of faith? What is the difference between a statement based on faith and a scientific theory?

# SUGGESTED READINGS

Allegre, Claude. 1992. *From Stone to Star*. Cambridge, Mass.: Harvard University Press.

Brandt, J. C., and S. P. Maran. 1979. *New Horizons in Astronomy*, 2d ed. San Francisco: W. H. Freeman.

*Exploring Space*. 1990. Special issue of *Scientific American*.

*Managing Planet Earth*. 1989. Special issue of *Scientific American* (September).

May, Robert H. 1992. How many species inhabit Earth? *Scientific American* (April):42–48.

National Academy of Sciences. 1992. *Science and Creationism*. Washington, D.C.: National Academy Press.

National Research Council. 1990. *The Search for Life's Origins*. Washington, D.C.: National Academy Press.

National Research Council. 1993. *Solid-Earth Sciences and Society*. Washington, D.C.: National Academy Press.

Press, Frank, and Raymond Siever. 1986. The planets: A summary of current knowledge. Chap. 22 in *Earth*, 4th ed. New York: W. H. Freeman.

Stanley, Steven M. 1993. *Exploring Earth and Life Through Time*. New York: W. H. Freeman.

Walter, William J. 1992. *Space Age*. New York: Random House.

Weiner, Jonathan. 1986. *Planet Earth*. New York: Bantam Books.

Westbroek, Peter. 1991. *Life as a Geologic Force*. New York: W. W. Norton.

# MINERALS: BUILDING BLOCKS OF ROCKS

T he key to understanding the materials and composition of the Earth lies in understanding how the chemical elements are organized into minerals. Minerals can be looked at in two complementary ways: as crystals that we can see with the naked eye, and as assemblages of submicroscopic atoms organized in a definite array. Understanding the structure of atoms allows us to predict how chemical elements will react with one another and form new crystal structures. The silicate minerals that are so abundant in the Earth are built on different arrangements of the silica tetrahedra that are their fundamental building blocks. The crystal structures of all minerals are reflected in their physical properties: hardness, the ability to break or fracture along regular or irregular surfaces, color, density, and luster.

Crystals of amethyst, a variety of quartz. The planar surfaces are crystal faces, which reflect the underlying arrangement of the atoms making up the crystals. *Chip Clark.*

We have seen how plate tectonics describes Earth's large-scale structure and dynamics but have touched only briefly on the wide variety of rocks and minerals—the materials of the Earth—that are found in various plate-tectonic settings. The nature of these materials, which are formed and altered by geological processes, is an important guide to the ways those processes work. The kinds of minerals found in a volcanic rock, for example, give us a good idea of the temperature of the molten rock, perhaps as much as 1000°C, from which it was formed and the explosive violence of the eruption that brought the molten rock to Earth's surface. Such a volcano is the hallmark of the convergence of two plates along a subduction zone. From the mineral composition of another rock we can deduce that it was formed deep in the Earth's crust under conditions associated with mountain building. These conditions, temperatures as high as 700°C and pressures more than 10,000 times higher than at Earth's surface, arise where two continental plates collide and form high mountain ranges, such as the Himalayas.

The materials of the Earth are important to the economy of the world. We have all become aware of the economic and political effects of trade in the resources drawn from Earth for conversion to energy: oil, gas, coal, and uranium. Civilization has depended on mineral resources from the days before the Bronze Age, when metals were first smelted from ore minerals (minerals from which we can recover metals), to the computer age, which depends on ultrapure silicon refined from quartz for computer chips. Quartz, a hard mineral, is used for sandpaper. Table salt is refined from naturally occurring rock salt, the mineral halite.

To understand the nature of the materials that make up the solid Earth, we turn in this chapter and the next to branches of geology on a smaller scale.

**Mineralogy,** the focus of this chapter, is the science of minerals. To a geologist, minerals are the building blocks of rocks. If we are to understand rocks—which will be discussed in Chapter 3—we must first learn about the minerals that make them up.

## MINERALS MAKE UP ROCKS

A few rocks, such as limestone, are composed of a single kind of mineral. The others, such as the common rock granite, consist of several kinds of minerals. The essential difference between a mineral and a rock is homogeneity: a mineral cannot be divided into smaller components by mechanical means; rocks, with the proper tools and effort, can be divided into their constituent minerals.

To know how rocks are formed, we must understand how the minerals that make them up were produced. We use minerals to identify and classify the many kinds of rocks found in the Earth, and thus we are able to map rock formations in the field, which is essential in understanding the geology of a region and in exploring for mineral resources.

## WHAT ARE MINERALS?

Geologists define a mineral as *a naturally occurring crystalline solid substance, generally inorganic, with a specific chemical composition* (Figure 2.1). To qualify as a mineral, a substance must be found in nature. Synthetic versions of natural materials, such as artificial

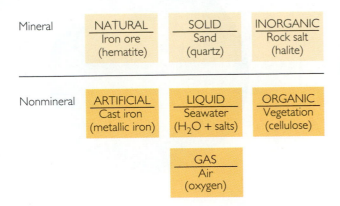

| Mineral | NATURAL<br>Iron ore<br>(hematite) | SOLID<br>Sand<br>(quartz) | INORGANIC<br>Rock salt<br>(halite) |
|---|---|---|---|
| Nonmineral | ARTIFICIAL<br>Cast iron<br>(metallic iron) | LIQUID<br>Seawater<br>($H_2O$ + salts) | ORGANIC<br>Vegetation<br>(cellulose) |
| | | GAS<br>Air<br>(oxygen) | |

**FIGURE 2.1** Minerals are distinguished from other materials in being naturally occurring, solid, inorganic substances with specific chemical compositions.

diamonds or the thousands of laboratory products invented by chemists, are not considered true minerals.

When we say that a mineral is crystalline we mean that the tiny particles of matter—atoms—that compose it are arranged in an orderly, repeating, three-dimensional array. Solid materials that have no such orderly arrangement are referred to as **glassy** or *amorphous* (without form). Window glass is amorphous, as are some natural glasses formed during volcanic eruptions.

The stipulation that minerals are inorganic substances follows historical usage and excludes the organic substances that make up plant and animal bodies. Decaying vegetation in a swamp may be geologically transformed into coal, for example, but though it is found as a natural deposit, coal is not considered a mineral. Many minerals, however, are made biologically, such as the calcite that forms the shells of many organisms. The calcite of these shells, which constitute the bulk of many limestones, fits the definition of a mineral because it is inorganic and crystalline.

What makes each mineral unique is the combination of its chemical composition and the arrangement of its atoms in an internal structure. The chemical composition either is fixed or varies within a defined range. The mineral quartz, for example, no matter what kind of rock it is found in, always consists of a fixed ratio of two atoms of oxygen to one of silicon. A slightly more complex mineral, olivine, always has a fixed ratio of the sum of iron and magnesium atoms to the number of silicon atoms, although the ratio of iron to magnesium atoms may vary.

# THE ATOMIC STRUCTURE OF MATTER

A modern dictionary lists many meanings for the word *atom* and its derivatives. One of the first is "anything considered as the smallest possible unit of any material." To the ancient Greeks, *atomos* meant "indivisible." To John Dalton (1766–1844), an English chemist and the father of the atomic theory, **atoms** were small particles of matter of several kinds that composed all substances. The particles were so small that they could not be seen with any microscope. In 1805 Dalton hypothesized that the various chemical elements consist of different kinds of atoms, that all atoms of any given element are identical, and that chemical compounds are formed by various combinations of atoms in definite proportions. By the early twentieth century, physicists, chemists, and mineralogists came to understand the structure of matter much as we know it today. We now know that atoms are the small units of matter that combine in chemical reactions, but atoms are divisible into even smaller units. At the center of every atom is a dense **nucleus** that contains practically all the mass of the atom (Figure 2.2b). The nucleus contains two kinds of particles, protons and neutrons. By convention, each of these particles is taken to have a mass of 1 atomic weight unit. A **proton** has a positive electrical charge, taken to be +1. A **neutron,** as the name implies, is electrically neutral; that is, uncharged.

Surrounding the nucleus is a cloud of moving **electrons,** each with a mass so small that it is conventionally taken to have no mass; each carries an electrical charge of −1. The number of protons in the nucleus of any atom is balanced by the same number of electrons in the cloud outside, so an atom is electrically neutral.

Modern models of atomic structure give the locations of electrons around the nucleus as *orbitals,* paths of travel around the nucleus that are not fixed orbits but can better be described as blurred regions where an electron is likely to be found, such as the electron cloud in Figure 2.2a. For convenience, we usually depict these orbitals as concentric spherical shells around the nucleus (see Figure 2.2c).

Each chemical element has a unique number of protons. That number is called the **atomic number.** All atoms of the same element have the same number of protons and therefore the same atomic number. All atoms with eight protons, for example, are oxygen atoms (atomic number 8). Since all the protons are balanced by an equal number of electrons, each element has a distinctive number of electrons, too.

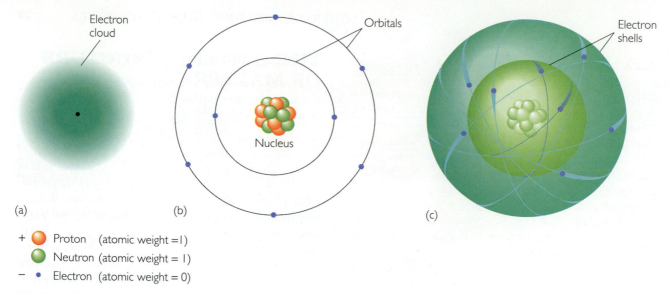

| + | ⬤ | Proton (atomic weight =1) |
|---|---|---|
| | ⬤ | Neutron (atomic weight = 1) |
| − | • | Electron (atomic weight = 0) |

**FIGURE 2.2** Electron structure of hydrogen and oxygen atoms. (a) The position of the single electron of the simplest element, hydrogen, shown as an electron cloud surrounding the nucleus. (b) Electrons of an oxygen atom are represented as lying in definite orbits around the nucleus, which contains eight protons, each with a charge of +1, and eight neutrons, each with zero charge. Eight electrons are found in two concentric shells, an inner one with two electrons and an outer one with six. (c) A more realistic representation of the electron shells of an oxygen atom. The size of the nucleus is greatly exaggerated; it is much too small to show on a true scale.

The atomic number of an element determines how it will react chemically with other elements.

The **atomic weight** of an element is the sum of the masses of its protons and neutrons. (Electrons, because they have essentially no mass, are not included in this sum.) Although the number of protons is constant, atoms of the same chemical element may have different numbers of neutrons and, therefore, different atomic weights (Figure 2.3). These various kinds of atoms are called **isotopes.** Isotopes of the element carbon, for example, all with six protons, exist with six, seven, and eight neutrons, giving atomic weights of 12, 13, and 14. In nature, the chemical elements occur as mixtures of isotopes, so their atomic weights are never whole numbers. Carbon's atomic weight, for example, is 12.011. It is close to 12 because the isotope carbon-12 is overwhelmingly abundant.

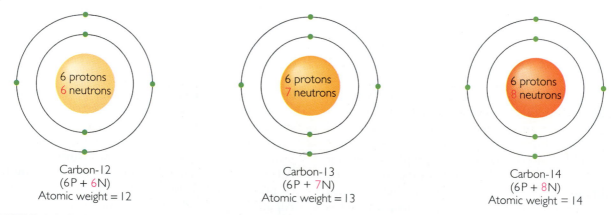

**FIGURE 2.3** Carbon isotopes. All have the same number of protons and thus the same atomic number, 6, but each isotope has a different number of neutrons and thus a different atomic weight.

# CHEMICAL REACTIONS

The structure of any particular kind of atom determines its chemical reactions with other atoms. **Chemical reactions** are interactions of the atoms of two or more chemical elements in certain fixed proportions that produce new chemical substances—chemical compounds. When two hydrogen atoms combine with one oxygen atom, they form a new chemical compound that we call water ($H_2O$). The properties of a chemical compound formed in the course of a reaction may be entirely different from those of its constituent elements. For example, when an atom of sodium, a metal, combines with an atom of chlorine, a noxious gas, it forms the chemical compound sodium chloride, better known as table salt. We represent this compound by the chemical formula NaCl, the symbol Na standing for the element sodium and the symbol Cl for the element chlorine. (Every chemical element has been assigned its own symbol, which we use as a kind of shorthand for writing chemical formulas and equations.)

Chemical reactions take place primarily through the interactions of electrons. To understand those interactions we need to know the number of electrons in an atom and how they are arranged in electron shells.

## Gaining and Losing Electrons

Electrons surround the nucleus of an atom in a unique set of concentric spheres called electron shells. In the chemical reactions of most elements, only the electrons in the outermost shells interact. In the reaction between sodium (Na) and chlorine (Cl) that forms NaCl, the sodium atom loses an electron from its outer shell of electrons and the chlorine atom gains an electron in its outer shell (Figure 2.4).

**IONS**   After **gain or loss of an electron,** the atoms in the new chemical compound are no longer electrically neutral. When the sodium atom loses an electron, it becomes a sodium **ion,** with an electrical charge of $+1$, because it still has the same number of protons but one less electron. It is now represented by the symbol $Na^+$. The chlorine atom, in gaining an electron, has become a chlorine ion with an electrical charge of $-1$, written as $Cl^-$. Positive ions, such as sodium, are called **cations;** negative ions, such as chlorine, are called **anions.** The compound NaCl itself remains electrically neutral because the positive charge on $Na^+$ is exactly balanced by the negative charge on $Cl^-$.

Groups of ions may join to form *complex ions*, such as the common sulfate ion ($SO_4^{2-}$), a component of the mineral anhydrite ($CaSO_4$) and an abun-

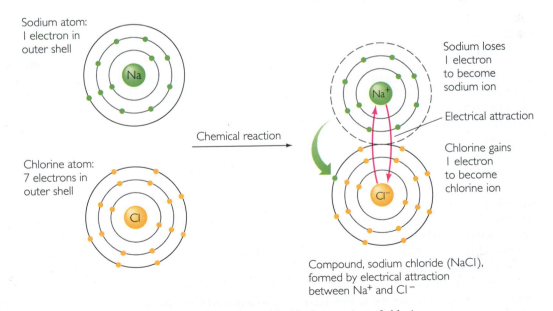

Sodium atom:
I electron in
outer shell

Na

Chlorine atom:
7 electrons in
outer shell

Cl

Chemical reaction

Sodium loses
I electron
to become
sodium ion

$Na^+$

Electrical attraction

Chlorine gains
I electron
to become
chlorine ion

$Cl^-$

Compound, sodium chloride (NaCl),
formed by electrical attraction
between $Na^+$ and $Cl^-$

**FIGURE 2.4** Formation of a chemical compound, sodium chloride, by reaction of chlorine and sodium atoms.

dant constituent of seawater. The sulfate ion is a unit made up of one sulfur ion with a $+6$ charge and four oxygen ions, each with a $-2$ charge, the net charge adding to $-2$.

**ELECTRON SHELLS AND ION STABILITY**  Before reacting with chlorine, the sodium atom has one electron in its outer shell. When it loses that electron, its outer shell is eliminated and the next shell inward, which has eight electrons (the maximum this shell can hold), becomes the outer shell. The original chlorine atom had seven electrons in its outer shell, with room for a total of eight. By gaining an electron, its outer shell is filled. There is a strong tendency for many elements to acquire a full outer electron shell, some by gaining electrons and some by losing them in the course of a chemical reaction. The stability of ions with fully occupied outer shells is related to the interactions of electrons in various orbitals around the nucleus.

Many chemical reactions involve gains and losses of several electrons as two or more elements combine. The element calcium (Ca), for example, becomes a doubly charged cation, $Ca^{2+}$, as it reacts with two chlorine atoms to form calcium chloride. (In the chemical formula for calcium chloride, $CaCl_2$, the presence of two chlorine ions is symbolized by the subscript 2. Chemical formulas thus show the relative proportion of atoms or ions in a compound. The subscript 1 for the single Ca in $CaCl_2$ and other formulas is omitted.)

## Electron Sharing

Not all chemical elements react by gaining or losing electrons. Some have a strong tendency to combine chemically by engaging in **electron sharing** with atoms of the same or a different element to achieve a stable configuration of electrons. Carbon and silicon, two of the abundant elements of Earth's crust, are elements of this kind.

An electron that is shared cannot be considered to have been gained or lost. In a sense both nuclei have "gained" the electron for whatever part of the time it can be visualized as belonging to the outer shell of one or the other atom. Nevertheless, because the atoms still have their original number of electrons, we do not refer to them as ions.

One compound made by electron sharing is methane ($CH_4$), the main component of natural gas.

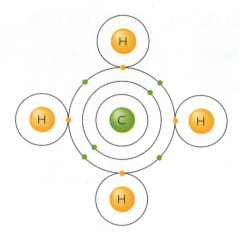

**FIGURE 2.5** Electron sharing in methane ($CH_4$) between one carbon atom, which has four electrons (green) in its outer shell, and four hydrogen atoms, each of which has one electron (yellow) in its single shell.

To make methane, four hydrogen (H) atoms react with one carbon (C) atom (Figure 2.5). The carbon has four electrons in its outer shell, while each hydrogen has one. When these electrons are shared, all five atoms act as if each had a full complement of eight electrons in its outer shell.

## The Periodic Table of the Elements

Chemists have long known that some groups of chemical elements have similar chemical properties, such as boiling and melting points and tendencies to react chemically with other elements. Other groups of elements vary drastically. As the atomic structure of atoms became known, these chemical properties proved to correspond to the electron shell patterns of the elements.

The periodic table (Figure 2.6) organizes the elements (from left to right along a row) in order of atomic number and increasing numbers of electrons in the outer shell. The third row from the top, for example, starts at the left with sodium (atomic number 11), which has one electron in its outer shell. The next is magnesium (atomic number 12), which has two electrons in its outer shell, followed by aluminum (atomic number 13), with three, and silicon (atomic number 14), with four; then come phosphorus (atomic number 15), with five, sulfur (atomic number 16), with six, and chlorine (atomic number 17), with seven. The last element in this row is argon

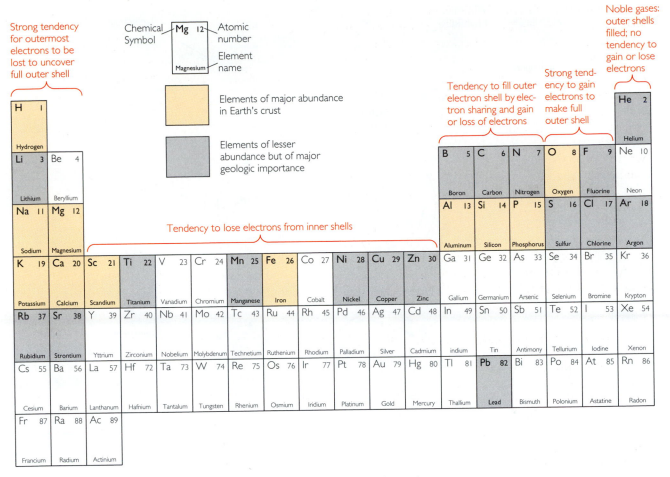

**FIGURE 2.6** The periodic table of the elements. (Two special groups of rare elements, those with atomic numbers 58–71 and those with atomic numbers 90–103, have been omitted.) Characteristics associated with different electron shell patterns are indicated at the top of the table.

(atomic number 18), with eight electrons, the maximum possible, in its outer shell.

Each column in the table groups elements that have similar electron shell patterns. All of the elements in the leftmost column have a single electron in their outer shells and have a strong tendency to lose that electron in chemical reactions (Figure 2.7). Of this group, hydrogen (H), sodium (Na), and potassium (K) are found in major abundance at Earth's surface and in its crust.

The second column from the left includes two more elements of major abundance, magnesium (Mg) and calcium (Ca). Elements in this column have two electrons in their outer shells and a strong tendency to lose them both in chemical reactions.

Toward the right side of the table, the two columns headed by oxygen (O), the most abundant element in the Earth, and fluorine (F), a highly reactive toxic gas, group the elements that tend to gain electrons in their outer shells. The elements in the column headed by oxygen tend to gain two electrons; those in the column headed by fluorine tend to gain one electron.

The columns between the two on the left and the two headed by oxygen and fluorine have varying tendencies to gain, lose, or share electrons. The column toward the right side of the table headed by carbon (C) includes silicon (Si), of major abundance in the Earth; both silicon and carbon tend to share electrons.

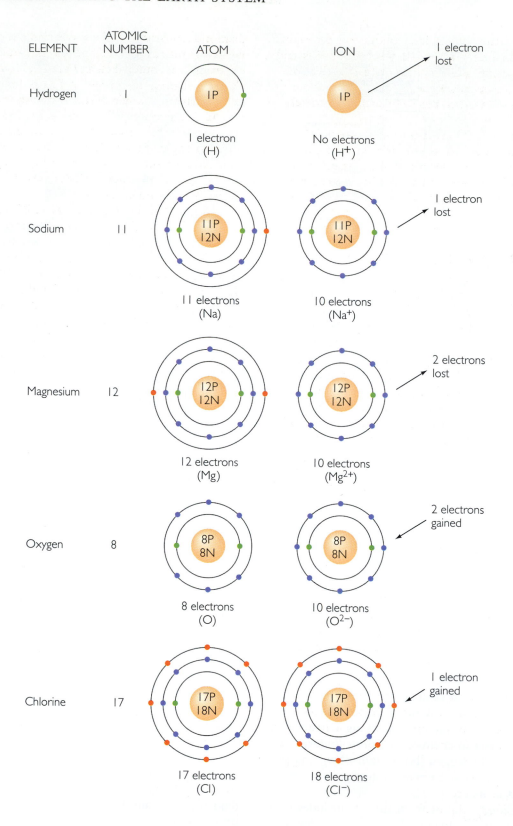

**FIGURE 2.7** Electron shells of atoms and ions of several common elements. Ions are formed as electrons are gained or lost from outer shells. P stands for proton and N for neutron.

The elements in the last column on the right, headed by helium (He), have full outer shells and thus no tendency either to gain or to lose electrons. As a result, these elements, in contrast to those in other columns, do not react chemically with other elements.

In summary, the groupings of the chemical elements by columns and rows in the periodic table reflect the elements' tendencies to gain, lose, or share electrons, which allows us to predict a great many chemical reactions.

# CHEMICAL BONDS

The ions or atoms of elements that make up compounds are held together by electrical forces of attraction between electrons and protons, which we call chemical bonds. The electrical attractions of either shared electrons or gained or lost electrons may be strong or weak, and the bonds created by these attractions are correspondingly strong or weak. Strong chemical bonds keep a substance from chemically decomposing into its elements or into other compounds. They also make minerals hard and keep them from cracking or splitting. Two major types of bonds are found in most rock-forming minerals: ionic bonds and covalent bonds.

## Ionic Bonds

The simplest form of chemical bond is the **ionic bond.** Bonds of this type are formed by electrical attraction between ions of the opposite charge, such as $Na^+$ and $Cl^-$ in sodium chloride (see Figure 2.4). This attraction is of exactly the same nature as the static electricity that can make clothing of nylon or silk cling to the body. The strength of an ionic bond decreases greatly as the distance between ions increases. Bond strength increases as the electrical charges of the ions increase. Ionic bonds are the dominant type of chemical bonds in mineral structures; *about 90 percent of all minerals are essentially ionic compounds.*

## Covalent Bonds

Elements that do not readily gain or lose electrons to form ions and instead form compounds by sharing electrons are held together by **covalent bonds.** One mineral with a covalently bound crystal structure is diamond, consisting of the single element carbon. As we saw in the case of methane ($CH_4$), carbon has four electrons in its outer shell and acquires four more by electron sharing to achieve a full outer shell with eight electrons. In diamond, every carbon atom (not an ion) is surrounded by four others arranged in a regular *tetrahedron,* a four-sided pyramidal form, each side a triangle (Figure 2.8a). In this configuration, each carbon atom shares an electron with each of its four neighbors and thus achieves a stable set of eight electrons in its outer shell.

Atoms of metallic elements, which have strong tendencies to lose electrons, pack together as cations, while freely mobile electrons are shared and dispersed among the ions. This free electron sharing results in a kind of covalent bond that we call a **metallic bond.** It is found in a small number of minerals, among them the metal copper and some sulfides.

The chemical bonds of some minerals are intermediate between pure ionic and pure covalent bonds because some electrons are exchanged and others are shared.

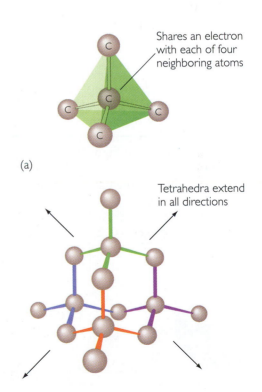

Shares an electron with each of four neighboring atoms

(a)

Tetrahedra extend in all directions

(b)

**FIGURE 2.8** The carbon tetrahedron of diamond. (a) A single tetrahedron formed by a carbon atom bonded to four other carbon atoms. (b) A network of carbon tetrahedra linked to one another.

# THE ATOMIC STRUCTURE OF MINERALS

Recall that to be a mineral, a substance must possess a *specific chemical composition* and a *crystalline structure.* Now that we understand how substances are formed by chemical bonding between atoms and ions, we can see better how the substances we call minerals form in such an orderly way.

## Crystals and Crystallization

Minerals are formed by the process of **crystallization,** the growth of a solid from a material whose constituent atoms can come together in the proper chemical proportions and crystalline arrangement. (Remember that the atoms in a mineral are arranged in an ordered, three-dimensional array.) The bonding of carbon atoms in diamond, a covalently bonded mineral, is one example of crystallization and crystal structure. In satisfying the requirements of electron sharing, carbon atoms come together in tetrahedra, each tetrahedron attaching to another and building up a regular three-dimensional structure from a great many atoms (see Figure 2.8b). As a diamond crystal grows, it extends its tetrahedral structure in all directions, always adding new atoms in the proper geometric arrangement.

The sodium and chlorine ions that make up sodium chloride, an ionically bonded mineral, also crystallize in an orderly three-dimensional array. In Figure 2.9 we can see the geometry of their arrangement, with each ion of one kind surrounded by six ions of the other in a series of cubes extending in three directions. This is called a cubic structure.

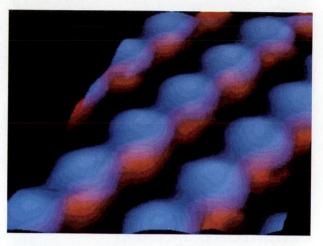

**FIGURE 2.10** A scanning transmission electron microscope representation of the atoms in a crystal of gallium arsenide, a substance used for computer components. The alignment of atoms of gallium (blue) and of arsenic (red) reveals the atomic structure of the crystal. *Randall M. Feenstra, IBM Thomas J. Watson Research Center.*

Atoms and ions are so small, most of them only a few ten-millionths of a centimeter, that we cannot see the crystalline arrangement of a mineral directly even with the most powerful ordinary microscope. With especially high-powered electron and other, newer kinds of microscopes, however, we can now image the atomic arrangements of crystals (Figure 2.10).

Crystallization starts with formation of microscopic single **crystals,** bodies whose boundaries are natural, flat (plane) surfaces (Figure 2.11). These sur-

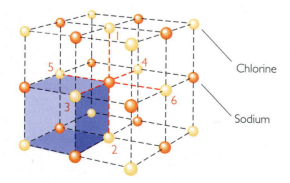

**FIGURE 2.9** Structure of sodium chloride. Ions are not drawn to scale. The lines between ions have been added to show the cubic geometry of this mineral. In this example it is easy to see one sodium ion surrounded by six chlorine ions.

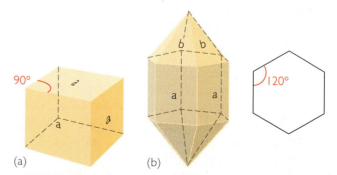

**FIGURE 2.11** Perfect crystals. Although a perfect crystal is rare, no matter how irregular the shapes of the faces, the angles are always exactly the same. (a) Halite, a cubic crystal. (b) Quartz, a hexagonal crystal; the cross section at right angles to the long dimension of the crystal shows a regular hexagon with "a" faces at 120° angles.

**FIGURE 2.12** Snowflake crystals. *W. Bentley; courtesy of D. Blanchard, State University of New York at Albany.*

faces, called *crystal faces,* are the defining external characteristic of a crystal. The crystal faces of a mineral are the external expression of the mineral's internal atomic structure. The simple geometric cubes of sodium chloride crystals (the mineral halite, or rock salt; see Figure 2.20) correspond to the cubic arrangement of its ions. The six-sided (hexagonal) shape of a quartz crystal (as seen in the photo at the beginning of this chapter) corresponds to its hexagonal internal atomic structure. The geometry of the crystals may be beautifully complex, as in snowflakes, the crystals of ice (Figure 2.12).

During crystallization, the initially microscopic crystals grow larger, maintaining their crystal faces as long as they are free to grow. Large crystals with well-formed faces are made when growth is slow and steady and there is space to grow without interference from other crystals nearby. Thus most large mineral crystals form in open spaces in rocks, such as open fractures or cavities (Figure 2.13).

If the spaces between growing crystals are filled in or crystallization proceeds too rapidly, crystal faces are grown over, and the former crystals coalesce to become a solid mass of crystalline particles, which we call *grains.* In the crystallized mass few or no grains show crystal faces (see Figure 2.13). Large crystals that can be seen with the naked eye are relatively unusual, but many microscopic minerals in rocks display crystal faces.

Glassy materials, which solidify from liquids so quickly that they lack any internal atomic order, do not form crystals with plane faces. Instead they are found as masses with curved, irregular surfaces, such as volcanic glass (see Figure 4.5).

When do minerals crystallize? Lowering the temperature of a liquid below its freezing point is one way to start the process. In the case of water, 0°C is the temperature below which crystals of ice start to form. Similarly, magma, a hot, molten liquid rock, crystallizes solid minerals when it cools below their melting points, which may be over 1000°C. (Geolo-

gists usually refer to melting points of magmas rather than freezing points, since freezing normally implies cold.)

Crystallization can also result from evaporation of a solution. A solution is formed when one chemical substance dissolves in another, such as salt in water. When we evaporate a salt solution we lose water but not salt, and the solution becomes more concentrated. Eventually the concentration gets so high that the solution is saturated, which means that it can hold no more salt and thus cannot become more concentrated. As evaporation continues beyond this point, the solution starts to precipitate, or

**FIGURE 2.13** Sample displaying both macro- and microcrystallinity. Large, well-formed crystals lining the central cavity are rimmed by a whitish layer of smaller crystal grains displaying few crystal faces. *Chip Clark.*

drop out of solution, crystals of salt. When seawater evaporates to the point of saturation, as it does in some hot, arid bays or arms of the ocean, the materials dissolved in it crystallize. Some of them remain behind as deposits of rock salt.

Crystals also form by rearrangement of solid materials at high temperatures—for most minerals, at least 250°C. As temperatures increase, ions and atoms in solids become more mobile and rearrange themselves to become new minerals with different crystal structures.

## The Sizes of Ions

Two major factors control the arrangement of atoms and ions in a crystal structure: the number of neighboring atoms or ions and their sizes. We can think of ions as if they were solid spheres, packed together in close-fitting structural units. Figure 2.14 shows the relative sizes of the ions in NaCl. There are six neighboring ions in NaCl's basic structural unit. The relative sizes of the sodium (smaller) and chlorine ions allow them to fit together in a closely packed arrangement.

Ion size is related to the atomic structures of the elements (Figure 2.15). The sizes of ions increase with the number of electrons and electron shells. An ion's charge also affects its size. The more electrons an element loses to become a cation, the stronger its positive charge and the stronger the electrical attraction of its nucleus for the remaining electrons. Many of the cations of abundant minerals are relatively small; most anions are large. This is the case with the

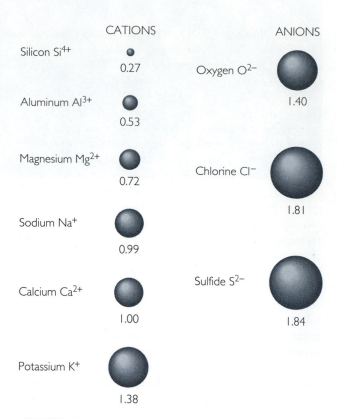

| CATIONS | | ANIONS |
|---|---|---|
| Silicon $Si^{4+}$ 0.27 | | |
| | | Oxygen $O^{2-}$ 1.40 |
| Aluminum $Al^{3+}$ 0.53 | | |
| Magnesium $Mg^{2+}$ 0.72 | | Chlorine $Cl^-$ 1.81 |
| Sodium $Na^+$ 0.99 | | |
| Calcium $Ca^{2+}$ 1.00 | | Sulfide $S^{2-}$ 1.84 |
| Potassium $K^+$ 1.38 | | |

**FIGURE 2.15** Sizes of ions as they are commonly found in rock-forming minerals. Ionic radii are given in $10^{-8}$ cm. (After L. G. Berry, B. Mason, and R. V. Dietrich, *Mineralogy,* San Francisco, W. H. Freeman, 1983.)

most common Earth anion, oxygen. Because anions tend to be larger than cations, it is apparent that most of the space of a crystal is occupied by the anions and that cations fit into the spaces between them. As a result, crystal structures are determined largely by the way the anions are arranged and the way the cations fit between them.

## Cation Substitution

Cations that have similar chemical properties and ions of similar sizes tend to substitute for one another and to make compounds of variable composition. For example, when iron (Fe) and magnesium (Mg) are united with silicon and oxygen (combined as the complex ion silicate, $[SiO_4]^{4-}$) to form natural olivine, abundant in many volcanic rocks, the amounts of iron and magnesium vary but their combined total does not. Iron and magnesium ions are similar in size and both have two positive charges, so they easily substitute for each other in the olivine structure. The

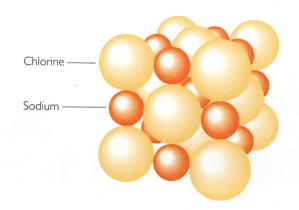

Chlorine ——

Sodium ——

**FIGURE 2.14** The relative sizes of sodium and chlorine ions allow them to pack together in a cubic structure. Ions here are shown in their correct relative sizes.

pure magnesium olivine is $Mg_2SiO_4$; the pure iron olivine is $Fe_2SiO_4$. The composition of natural olivine is given by the formula $(Mg, Fe)_2SiO_4$, which simply means that there are two $Mg^{2+}$ or $Fe^{2+}$ ions, in whatever combination, for every $(SiO_4)^{4-}$ ion. Cation substitution is common in silicate minerals, which are cations combined with silicate ions. In these minerals, some of the most abundant in Earth's crust, aluminum (Al) substitutes for silicon. Aluminum and silicon ions are so similar in size that aluminum can take the place of silicon in many crystal structures. The difference in charge between aluminum (3+) and silicon (4+) ions is balanced by an increase in one of the other cations, such as sodium (1+).

As a consequence of cation substitution, several minerals that have the same crystal structures may have somewhat different chemical compositions. As one cation substitutes for another, the mineral's chemical composition is altered with respect to the ratio of the two cations, but it retains the same structure. One example is the mica group of minerals, all varieties of which have the same general crystal structure but contain various combinations of the cations sodium, potassium, aluminum, magnesium, and iron with silicon and oxygen.

## Polymorphs

Some chemical substances with exactly the same combinations of elements have more than one kind of crystal structure and therefore form different minerals. The different minerals are called **polymorphs** ("many forms"). Because the formation of any particular structure depends strongly on pressure and temperature, and therefore on the depth within the Earth, the polymorph that is found reflects the geological conditions at the time and place it was formed.

Carbon, for example, forms both diamond and graphite (the material that is used as the "lead" in pencils), two minerals with different crystal structures and very different appearances (Figure 2.16). The packing of the carbon atoms is different in the two forms, diamond being more closely packed and therefore of higher density (mass per unit volume), 3.5 $g/cm^3$ (see Figure 2.8b). The less compact graphite has a density of only 2.1 $g/cm^3$. From experimentation and geological observation we know that the denser diamond forms and remains stable at the very high pressures and temperatures of Earth's mantle. Graphite, by contrast, forms and is stable at relatively moderate pressures and temperatures, such as those in the crust. The high pressure in the mantle forces a closer packing of the atoms in diamond.

Low temperatures also favor more compact structures. Quartz, for example, is a relatively dense (2.7 $g/cm^3$), low-temperature polymorph of silica ($SiO_2$), whereas cristobalite, a higher temperature polymorph, has a more open structure and is therefore less dense (2.3 $g/cm^3$).

## THE ROCK-FORMING MINERALS

Although many thousands of minerals are known, geologists commonly encounter only a relatively small number, about 30. These are the minerals that make up most crustal rocks, and thus they are called

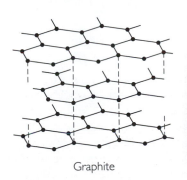

Graphite

Diamond

**FIGURE 2.16** Structures of graphite and of diamond. (*Left*) In graphite, sheets of carbon atoms arranged in hexagons are stacked above one another with weak bonds between the sheets. (*Right*) In diamond, carbon atoms are arranged in a tetrahedral network (compare with Figure 2.8b). *Photo by Chip Clark.*

**TABLE 2.1**

## Relative Abundance of the Ten Most Common Elements in Earth's Crust

| ELEMENT | SYMBOL | ABUNDANCE (PERCENT) |
|---|---|---|
| Oxygen | O | 46.6 |
| Silicon | Si | 27.7 |
| Aluminum | Al | 8.1 |
| Iron | Fe | 5.0 |
| Calcium | Ca | 3.6 |
| Sodium | Na | 2.8 |
| Potassium | K | 2.6 |
| Magnesium | Mg | 2.1 |
| Titanium | Ti | 0.4 |
| Hydrogen | H | 0.1 |

*rock-forming minerals.* The small number of common minerals reflects the small number of elements found in major abundance in Earth's crust: only 10 elements account for 99 percent of the crust (Table 2.1). The most common of the rock-forming minerals are the silicates, composed of the two most abundant elements in Earth's crust, oxygen (O) and silicon (Si), mostly in combination with other abundant elements. Other major rock-forming minerals are oxides, usually made up of oxygen combined with a metallic element, such as hematite ($Fe_2O_3$); carbonates, made up of calcium and magnesium in combination with carbon and oxygen, such as calcite ($CaCO_3$); sulfides and sulfates, both based on the element sulfur, such as pyrite ($FeS_2$); and a few other combinations of elements.

## Silicates

**SILICATE STRUCTURES**   The basis of all silicate mineral structures is a tetrahedron formed by four oxygen ions surrounding and sharing electrons with a silicon ion (Figure 2.17). This tetrahedron is a *silicate ion,* made up of one $Si^{4+}$ and four $O^{2-}$ ions, which has the formula $(SiO_4)^{4-}$. Each silicate tetrahedron is an anion with four negative charges, which

must be balanced by four positive charges to make an electrically neutral mineral in one of two ways: by bonding with cations or by sharing oxygens with other tetrahedra. All silicate minerals are made up of silicate tetrahedra as basic units linked in combinations of these two ways.

Silicates, such as those shown in Figure 2.18, are classified and named according to the way their tetrahedra are linked, as shown in Figure 2.19 and Table 2.2. *Isolated* tetrahedra are linked by the bonding of each oxygen ion of the tetrahedron to a cation (Figure 2.19a), the cations in turn bonding to the oxygens of other tetrahedra. The tetrahedra are thus isolated from one another by cations on all sides. Olivine is a rock-forming mineral with this structure.

*Rings* of tetrahedra are formed by the bonding of two oxygens of each tetrahedron to adjacent tetrahedra in closed rings (Figure 2.19b). This amounts to each tetrahedron's sharing two of its oxygens with other tetrahedra, one on each side. Beryl is a mineral of this type.

*Single chains* form by the same linkage as rings—that is, by the bonding of each tetrahedron to two others by shared oxygens—but in an open-ended chain instead of a closed ring (Figure 2.19c). Single chains are linked to other chains by cations. Pyroxenes are single-chain silicate minerals.

Two single chains may be combined to form *double chains,* which are linked to each other by shared oxygens (Figure 2.19d). Adjacent double chains linked by cations form the structure of the amphibole minerals.

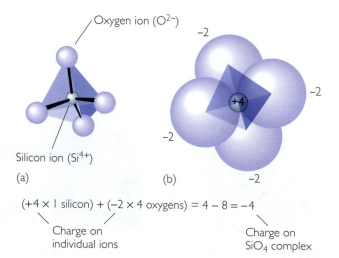

$$(+4 \times 1 \text{ silicon}) + (-2 \times 4 \text{ oxygens}) = 4 - 8 = -4$$

Charge on individual ions

Charge on $SiO_4$ complex

**FIGURE 2.17** The silica tetrahedron. (a) Drawing showing the structure of the tetrahedron. (b) View of a tetrahedron showing atoms drawn to scale.

**FIGURE 2.18** Silicate minerals (clockwise from upper left): feldspar, mica, pyroxene, quartz, and olivine. *Chip Clark.*

*Sheets* are structures in which each tetrahedron shares three of its oxygens with adjacent tetrahedra to build stacked sheets of tetrahedra (Figure 2.19e). Cations may be interlayered with tetrahedral sheets. The micas and clay minerals are the most abundant sheet silicates.

In three-dimensional *frameworks,* the tetrahedra link together by sharing all their oxygens. Feldspar, the most abundant mineral in Earth's crust, is a framework silicate, as is another of the most common minerals, quartz (see Figure 2.18).

**COMPOSITION OF SILICATES**   Chemically, the simplest silicate is silicon dioxide, also called silica ($SiO_2$), which is found most frequently as the mineral quartz. The tendency of silicon to bond with oxygen is so strong that silicon is never found in nature as the pure element; it is always combined with oxygen. (The pure silicon used in computer chips is artificially prepared by advanced chemical techniques.) When the silicate tetrahedra of quartz are linked, sharing two O's for each Si, the total formula adds up to $SiO_2$.

In other silicate minerals, the basic units—rings, chains, sheets, and frameworks—are bonded to such cations as sodium ($Na^+$), potassium ($K^+$), calcium ($Ca^{2+}$), magnesium ($Mg^{2+}$), and iron ($Fe^{2+}$). As we noted in our discussion of cation substitution, alumi-

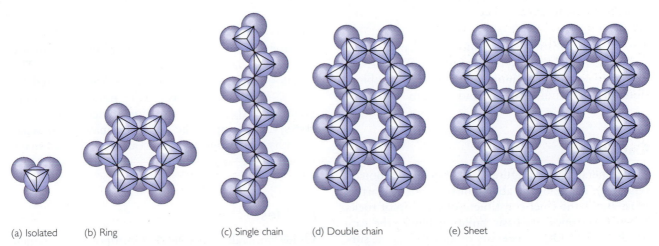

(a) Isolated    (b) Ring    (c) Single chain    (d) Double chain    (e) Sheet

**FIGURE 2.19** The crystal structures of silicate minerals, which are classified according to the different ways in which silica tetrahedra can be linked.

**TABLE 2.2**

## Major Silicate Structures

| GEOMETRY OF LINKAGE OF SiO₄ TETRAHEDRA | | EXAMPLE MINERAL | CHEMICAL COMPOSITION |
|---|---|---|---|
| *Isolated tetrahedra:* No sharing of oxygens between tetrahedra; individual tetrahedra linked to each other by bonding to cation between them | | Olivine | Magnesium-iron silicate |
| *Rings of tetrahedra:* Joined by shared oxygens in three-, four-, or six-membered rings | | Cordierite | Magnesium-iron-aluminum silicate |
| *Single chains:* Each tetrahedron linked to two others by shared oxygens; chains bonded by cations | | Pyroxene | Magnesium-iron silicate |
| *Double chains:* Two parallel chains joined by shared oxygens between every other pair of tetrahedra; the other pairs of tetrahedra bond to cations that lie between the chains | | Amphibole | Calcium-magnesium-iron silicate |
| *Sheets:* Each tetrahedron linked to three others by shared oxygens; sheets bonded by cations | | Kaolinite | Aluminum silicate |
| | | Mica (muscovite) | Potassium-aluminum silicate |
| *Frameworks:* Each tetrahedron shares all its oxygens with other SiO₄ tetrahedra (in quartz) or AlO₄ tetrahedra | | Feldspar (orthoclase) | Potassium-aluminum silicate |
| | | Quartz | Silicon dioxide |

num ($Al^{3+}$) substitutes for silicon in many silicate minerals. All of the several varieties of feldspar, for example, contain aluminum, with various combinations of potassium, sodium, and calcium.

## Carbonates

The mineral calcite (Figure 2.20), calcium carbonate ($CaCO_3$), is one of the abundant minerals of Earth's crust and is the chief constituent of a group of rocks called limestones. Its basic building block, the carbonate ion, $(CO_3)^{2-}$, consists of a carbon atom surrounded by three oxygen atoms in a triangle, as in Figure 2.21a. Groups of carbonate ions are arranged in sheets somewhat like the sheet silicates and are bonded by layers of cations (Figure 2.21b). The sheets of carbonate ions in calcite are separated by layers of calcium ions. The mineral dolomite ($CaMg[CO_3]_2$), another major mineral of crustal rocks, is made up of the same carbonate sheets separated by alternating layers of calcium ions and magnesium ions.

## Oxides

Oxide minerals are defined as compounds in which oxygen is bonded to atoms or cations of other elements, usually metals such as iron (Fe). This group is

**FIGURE 2.20** Nonsilicate minerals (clockwise from upper left): halite, spinel, gypsum, hematite, calcite, pyrite, and galena. *Chip Clark.*

of great economic importance because it includes the ores of most of the metals, such as chromium and titanium, necessary for industrial and technological manufacture. Hematite (see Figure 2.20), one of the most abundant iron oxide ($Fe_2O_3$) minerals, is a chief ore of iron. Most oxide minerals are ionically bonded, their structures varying with the size of the metallic cations. One of the abundant minerals in this group, spinel (also shown in Figure 2.20), is an oxide of two metals, magnesium and aluminum ($MgAl_2O_4$). Spinel has a closely packed cubic structure and high density (3.6 g/cm$^3$), reflecting its formation under high pressure and temperature. Trans-parent, gem-quality spinel resembles ruby and sapphire and is found among the crown jewels of England and Russia.

## Sulfides

The sulfide group of minerals includes compounds of sulfur with metallic elements. The group is named for the sulfide ion, $S^{2-}$. The structures of these minerals are diverse, depending on the way the larger sulfide anions combine with the smaller metallic cat-

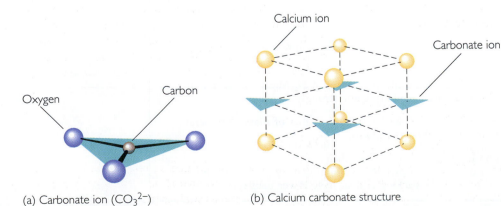

(a) Carbonate ion ($CO_3^{2-}$)

Oxygen  Carbon

Calcium ion

Carbonate ion

(b) Calcium carbonate structure

**FIGURE 2.21** Carbonate minerals, such as calcite (calcium carbonate [$CaCO_3$]), have a layered structure. (a) Top view of the carbonate building block, a carbon atom surrounded in a triangle by three oxygen ions, with a net charge of −2. (b) A view of the alternating layers of calcium and carbonate ions.

ions. The most common mineral of the group is pyrite ($FeS_2$), frequently called "fool's gold" because of its yellowish metallic appearance (see Figure 2.20). Most sulfide minerals look like metals and almost all are opaque. Sulfide minerals are the chief ores of a great many valuable metals, such as copper, zinc, and nickel.

## Sulfates

In sulfates, sulfur is present as the sulfate ion, a tetrahedron made up of one sulfur atom that has lost six electrons joined with four oxygens, giving it the formula $SO_4^{2-}$. The sulfate ion is the basis for a variety of structures. One of the most abundant minerals of this group is gypsum (see Figure 2.20), a calcium sulfate ($CaSO_4 \cdot 2H_2O$). (The dot in this formula signifies that two water molecules are bonded to the calcium and sulfate ions.) Gypsum is the primary component of plaster. Gypsum is formed when $Ca^{2+}$ and $SO_4^{2-}$, two ions abundant in seawater, combine and precipitate as layers of sediment when seawater evaporates. Anhydrite ($CaSO_4$), another calcium sulfate, differs from gypsum in containing no water (as indicated by its name, derived from the word *anhydrous,* for "no water").

# PHYSICAL PROPERTIES OF MINERALS

A diamond—geologically speaking—may not be forever, as the advertisers claim, but it is special. Its glitter is unique. The sparkle and play of colors in diamond are the products of its physical properties, primarily the way it refracts (bends) light. The sparkle is enhanced by diamond's remarkable ability to split perfectly along certain directions of the crystal, which diamond cutters use to advantage in carefully splitting and shaping gems, and by the way multiple facets (faces superficially similar to crystal faces) can be polished. Diamond is the hardest mineral known, so hard that it can scratch any other mineral and remain undamaged. Diamond polishers can grind facets only with other diamonds (in the form of powder or paste). The physical properties that make diamond so special allow it to be identified with certainty by mineralogists and jewelers.

Diamond has a tightly packed crystal structure with strong covalent bonds between the carbon atoms, and these bonds give diamond its characteristic hardness and other physical properties. Diamond is a good example of how important chemical composition and crystal structure are in determining various physical properties of a mineral (Table 2.3). In

---

**TABLE 2.3**

## Physical Properties of Minerals

| PROPERTY | RELATION TO COMPOSITION AND CRYSTAL STRUCTURE |
|---|---|
| Hardness | Strong chemical bonds give high hardness. Covalently bonded minerals are generally harder than ionically bonded minerals. |
| Cleavage | Cleavage is poor if bond strength in crystal structure is high and is good if bond strength is low. Covalent bonds generally give poor or no cleavage; ionic bonds are weak, so give excellent cleavage. |
| Fracture | Type is related to distribution of bond strengths across irregular surfaces other than cleavage planes. |
| Luster | Tends to be glassy for ionically bonded crystals, more variable for covalently bonded crystals. |
| Color | Determined by kinds of atoms and trace impurities. Many ionically bonded crystals are colorless. Iron tends to color strongly. |
| Streak | Color of fine powder is more characteristic than that of massive mineral because of uniform small size of grains. |
| Density | Depends on atomic weight of atoms and their closeness of packing in crystal structure. Iron minerals and metals have high density; covalently bonded minerals have more open packing and so have lower density. |

general, the atoms or ions of a mineral, the types and strength of bonds between them, and the mineral's crystal structure are responsible for such properties as hardness, translucency or opaqueness, color, and density. Geologists now have sophisticated laboratory instruments for determining crystal structures and identifying minerals by their structures or chemical compositions, but the simple time-tested ways of identifying minerals by their easily observable physical properties remain the field geologist's standbys.

In the remainder of this chapter we will look at the range of properties displayed by minerals, many of them valuable for the geologist seeking to identify an unknown mineral quickly.

## Hardness

Just as a diamond scratches glass, a quartz crystal scratches a feldspar crystal because quartz is harder than feldspar. In 1822 Friedrich Mohs, an Austrian mineralogist, devised a scale of hardness based on the ability of one mineral to scratch another. The **Mohs scale of hardness,** covering the range from the softest mineral (talc) to the hardest (diamond), is still one of the best practical means to identify an unknown mineral (Table 2.4). A knife blade and a few of the minerals on the hardness scale are all you need to place an unknown mineral between two points on the Mohs scale. If the unknown mineral is scratched by a piece of quartz but not by the knife, for example, it lies between 5 and 7 on the scale.

The hardness of any mineral depends on the strength of its chemical bonds; the stronger the bonds, the harder the mineral. Among the silicates hardness varies from 1 in talc to 8 in topaz. Most silicates are in the range of 5 to 7; only sheet silicates are relatively soft, with hardnesses between 2 and 3. Within groups of minerals having similar crystal structures, the hardness is related to factors that increase bond strength: size, charge, and packing. The smaller the atom or ion, the smaller the distance between the nucleus and the electrons, and thus the greater the electrical attraction. The larger the electrical charge of cations, the greater the electrical attraction and thus the strength of the bond. The closer the packing, the smaller the distance between atoms and the stronger the bond. Because of these factors, primarily size, most oxides and sulfides of metals with high atomic numbers—such as gold, silver, copper, and lead—are soft, with hardnesses of less than 3. Carbonates and sulfates, with relatively less densely packed structures, are also soft, with hardnesses of less than 5.

| TABLE 2.4 | | |
|---|---|---|
| **Mohs Scale of Hardness** | | |
| **MINERAL** | **SCALE NUMBER** | **COMMON OBJECTS** |
| Talc | 1 | |
| Gypsum | 2 | Fingernail |
| Calcite | 3 | Copper coin |
| Fluorite | 4 | |
| Apatite | 5 | Knife blade |
| Orthoclase | 6 | Window glass |
| | | Steel file |
| Quartz | 7 | |
| Topaz | 8 | |
| Corundum | 9 | |
| Diamond | 10 | |

## Cleavage

**Cleavage** is the splitting, or breakage, of a mineral along planar (flat) surfaces determined by the crystal structure, typified by that of mica. Although both cleavage planes and crystal faces are determined by crystal structure, there are generally fewer cleavage planes than possible crystal faces. The planes of both cleavage and crystal faces reflect planes of similar atoms in a mineral's crystal structure, but crystal faces result from a complex process of crystal growth along many possible planes, whereas cleavage breaks bonds along relatively few planes of weakness.

Cleavage is the result of differences in the strengths of bonds between planes of atoms or ions in a crystal. If the bonds between some of the planes are very weak, the mineral can be made to split along the planes of weakness. Muscovite, a mica, is a sheet silicate that displays excellent cleavage; that is, it breaks along smooth, lustrous, flat, parallel surfaces. Muscovite can be cleaved into thin transparent sheets less than a millimeter thick. Mica's excellent cleavage is the result of weakness of the bonds between the stacks of mica's tetrahedral silica sheets and between interlayer cations and sheets (Figure 2.22).

The number of planes of cleavage varies among minerals. Muscovite has only one, but calcite and dolomite have three excellent cleavage directions that

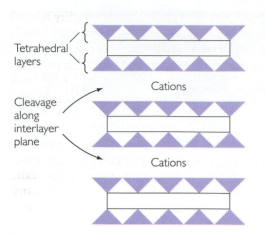

Tetrahedral layers

Cations

Cleavage along interlayer plane

Cations

**FIGURE 2.22** Cleavage of mica. The diagram shows the cleavage planes in the mineral structure, oriented perpendicular to the plane of the page. Horizontal lines mark the surfaces of silica tetrahedral sheets. The photograph shows how thin sheets break along the cleavage planes. *Chip Clark.*

**FIGURE 2.23** Rhomboidal calcite cleavage fragments. *Chip Clark.*

give a rhomboidal shape (Figure 2.23). Calcite can be cleaved by a light hammer blow on a sharp chisel oriented parallel to one of these plane directions. (Practicing with calcite can give you some idea of the skill of diamond cutters, the most expert cleavers in the world.)

Distinctive patterns of cleavage are identifying hallmarks of many common rock-forming minerals in addition to calcite and dolomite. Galena (lead sulfide) and halite cleave along three planes into perfect cubes. Two important groups of silicates that otherwise often look alike, the pyroxenes and amphiboles, can be distinguished on the basis of their different angles of cleavage (Figure 2.24). Pyroxenes, the single chains, are bonded so that their cleavages are almost at right angles (93°) to each other, giving nearly a square cross section. In contrast, amphiboles, the double chains, bond to give two cleavage directions, at 56° and 124° to each other, resulting in a diamond-shaped cross section.

Cleavage can be classified on the basis of the quality of the surfaces it produces and the ease of cleaving. Because of the ease with which muscovite can be cleaved and the extremely high quality of the smooth surfaces that result, this cleavage is called *perfect*. The single- and double-chain minerals (pyroxenes and amphiboles, respectively) show *good* cleavage; these minerals break easily along the cleavage plane but also break across it, and the cleavage surfaces are not so smooth as those of mica. At the low end of the scale, *fair* cleavage is shown by a ring silicate, beryl (emerald and aquamarine are gem-quality

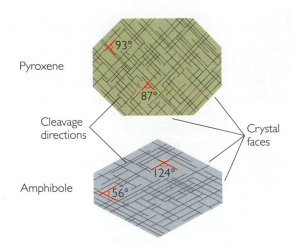

**FIGURE 2.24** Comparison of cleavage directions in pyroxene and amphibole.

beryls). This cleavage is more irregular and the mineral breaks along directions other than cleavage planes relatively easily.

Many minerals lack even fair cleavage. Quartz, one of the commonest, is a framework silicate so strongly bonded in all directions that it breaks only along irregular surfaces. Garnet is an isolated tetrahedral silicate, also bonded strongly in all directions, without cleavage. These two minerals illustrate the absence of a tendency to cleave in silicates with isolated tetrahedra and in most framework silicates.

## Fracture

**Fracture,** the way in which minerals break along irregular surfaces other than cleavage planes, also serves to group minerals and help in their identification. Fracture may be *conchoidal,* showing smooth, curved surfaces like those of a thick piece of broken glass (see Figure 4.5). A common fracture surface with an appearance like that of split wood is described as *fibrous* or *splintery.* The shape and appearance of many kinds of irregular fracture depend on the particular structure and composition of the mineral. The type and irregularity of fracture are related in a complex way to the breaking of bonds in directions that cut across crystal planes.

## Luster

The way the surface of a mineral reflects light gives it a characteristic **luster.** Mineral lusters are described

by the terms listed in Table 2.5. Luster quality is controlled by the kinds of atoms present and their bonding, both of which affect the way light passes through or is reflected by the mineral. Many covalently bonded minerals tend to have an adamantine luster, like that of diamond, whereas many ionically bonded minerals are more vitreous, like glass. Metallic luster is shown by pure metals, such as gold, and sulfides, such as galena (lead sulfide, PbS). Pearly luster is the result of multiple reflections of light from planes beneath the surfaces of translucent minerals, such as the mother-of-pearl inner surfaces of many clam shells, which are made of the mineral aragonite.

## Color

In July 1976 we found out from the *Viking* landing on Mars that the Red Planet really is red. The surface materials have a strong reddish hue that probably comes from iron oxide minerals. The same minerals, blown as very fine particles into the Martian atmosphere by strong winds, give the Martian sky a pinkish color. The **color** of a mineral may be distinctive, but it is not the most reliable clue to its identity. Many minerals show a characteristic color only on freshly broken surfaces, or the color is visible only on

**TABLE 2.5**

## Mineral Luster

| | |
|---|---|
| *Metallic* | Strong reflections produced by opaque substances |
| *Vitreous* | Bright, as in glass |
| *Resinous* | Characteristic of resins, such as amber |
| *Greasy* | The appearance of being coated with an oily substance |
| *Pearly* | The whitish iridescence of such materials as pearl |
| *Silky* | The sheen of fibrous materials such as silk |
| *Adamantine* | The brilliant luster of diamond and similar minerals |

weathered surfaces. Some minerals—precious opals, for example—show a stunning display of colors on reflecting surfaces. Others change color slightly with a change in the angle of the light shining on the surface. Some minerals always show the same color; others may have a range of colors.

Color may be a property of the pure mineral or it may be the result of impurities. The color of pure substances depends on the presence of certain ions, such as iron or chromium, which strongly absorb certain colors of light. Olivine containing iron is green, for example, while pure magnesium olivine is white. Most ionically bonded minerals whose ions have full, stable outer electron shells, such as halite, are colorless. Impurities, often too small to be seen except by the most powerful microscope, such as small dispersed flakes of hematite that color a feldspar crystal brownish or reddish, give a general color to an otherwise colorless mineral. Many of the gem varieties of minerals, such as amethyst (purple quartz) and sapphire (blue corundum), get their color

from trace impurities dissolved in the solid crystal (see the photo at the beginning of this chapter).

**Streak** is the name given to the color of the fine powder made when a mineral is scraped across a tile of unglazed porcelain, called a streak plate. Hematite, for example, always gives a reddish-brown streak, regardless of the color of the mass being streaked, which may be black, red, or brown.

## Specific Gravity and Density

The obvious difference in weight between a piece of hematite iron ore and a piece of sulfur of the same size is easily felt when the two pieces are hefted. A great many common rock-forming minerals, however, are too similar in **density,** which we earlier defined as mass per unit volume, for such simple tests as hefting. As a result, methods were needed that would make it easy to measure this property of minerals accurately. A standard measure of density is

Crystals of asbestos consist of narrow fibers. *Chip Clark.*

people exposed for long periods to moderate amounts of chrysotile show no lung disease. And there is little evidence that general exposure of the public causes lung disease. The lack of specific mineralogical analyses correlated with specific diagnoses of disease has contributed to confusion about nonoccupational exposures to asbestos in buildings. Many medical scientists and mineralogists doubt that we need to spend the $50 to $150 billion it would cost to clean up relatively harmless chrysotile and the four nonasbestiform types of amphibole. The combination of incomplete medical information and general lack of precise identification of the asbestos mineral has led to disagreement on the importance of remedial actions. We are likely to need more mineralogical evaluation with additional medical studies before all of these factors can be sorted out.

**specific gravity,** which is the weight of a mineral in air divided by the weight of an equal volume of pure water at 4°C.

Density depends on the atomic weight of the constituents and the closeness with which the atoms are packed in a mineral's crystal structure. The iron oxide mineral hematite has high density, $5.3 \text{ g/cm}^3$, because of the high atomic weight of iron. The density of the iron silicate olivine, $4.4 \text{ g/cm}^3$, is lower than that of hematite because of the low atomic weight of silicon. The density of the magnesium olivine is even lower, $3.32 \text{ g/cm}^3$, because magnesium is much lower in atomic weight than iron.

As we noted earlier in our comparison of the carbon polymorphs diamond and graphite, density is affected by pressure and temperature. Increases of density caused by pressure affect the way minerals transmit light, heat, and even earthquake waves (see Chapter 18). Temperature also affects density; the higher the temperature, the more open and expanded the structure, and thus the lower the density.

## Crystal Habit

The shape in which individual crystals or aggregates of crystals grow is defined as a mineral's **crystal habit.** Crystal habits are often named after common geometric shapes that they resemble, such as blades, plates, and needles. Some minerals can be recognized by their characteristic crystal habit, such as quartz's six-sided column topped by a pyramidlike set of faces (see Figure 2.11b). These shapes reflect not only the planes of atoms or ions in the mineral's crystal structure but also the typical speed and direction of crystal growth. Thus a needlelike crystal is one that grows very fast in one direction and very slowly in all other directions. A plate-shaped crystal, in contrast, grows fast in all directions perpendicular to a single direction of slow growth. Fibrous crystals are those that crystallize into multiple long, narrow fibers, essentially aggregates of long needles. *Asbestos* is a generic name for a group of silicates with a more or less fibrous habit (see Box 2.1).

**TABLE 2.6**

Some Chemical Classes of Minerals

| CLASS | DEFINING ANIONS | EXAMPLE |
|---|---|---|
| Native elements | None: no charged ions | Copper metal (Cu) |
| Oxides and hydroxides | Oxygen ion ($O^{2-}$)<br>Hydroxyl ion ($OH^-$) | Hematite ($Fe_2O_3$)<br>Brucite ($Mg[OH]_2$) |
| Halides | $Cl^-$, $F^-$, $Br^-$, $I^-$ | Halite (NaCl) |
| Carbonates | Carbonate ion ($CO_3^{2-}$) | Calcite ($CaCO_3$) |
| Sulfates | Sulfate ion ($SO_4^{2-}$) | Anhydrite ($CaSO_4$) |
| Silicates | Silicate ion ($SiO_4^{4-}$) | Olivine ($Mg_2SiO_4$) |

# CHEMICAL PROPERTIES OF MINERALS

Chemical composition is the basis for the main classification of the mineral kingdom. Most minerals are classified by their anions (negative ions). Halite (NaCl), for example, is classed as a chloride from its anion, $Cl^-$, as is its close relative sylvite, potassium chloride (KCl). The anion of the carbonate minerals is $(CO_3)^{2-}$ and of silicates is $(SiO_4)^{4-}$. Such minerals as copper, which occur naturally as the un-ionized pure element, are classified as *native* elements. In this way, all minerals have been grouped into eight classes, some of which are shown in Table 2.6.

Most of what we know about the chemical composition of minerals was learned through the use of ordinary chemical methods to dissolve minerals, separate them into their constituent elements, and then measure their weights or volumes. In the past few decades new instruments have made it possible to measure even very small quantities of some elements—as little as a billionth of a gram in some cases. All natural minerals contain impurities, enough of them so that it is often hard to decide whether an extremely small amount of an element is an essential part of the mineral or merely a contaminant. Elements that make up much less than 0.1 percent of the mineral are reported as "traces," and many of these are called **trace elements.** Some trace elements are useful in interpreting the origins of the minerals in which they are found. Others, such as the trace amounts of uranium in some granites, may contribute to local natural radioactivity.

**FIGURE 2.25**
Hydrochloric acid dropped on a limestone made of calcite causes bubbles to form. *Chip Clark.*

Mineralogists have also relied for years on quick, simple chemical tests that can be carried out in the field. One such test is the "acid test," in which dilute hydrochloric acid (HCl) is dropped on a mineral to see if it fizzes (Figure 2.25). If it does, the mineral is likely to be calcite, a carbonate mineral.

In summary, minerals exhibit a variety of physical and chemical properties that result from their chemical compositions and atomic structures. Many of these properties are useful to the mineralogist or geologist for purposes of identification or classification. The same properties give minerals their practical and decorative values. Geologists use the compositions and structures of minerals to understand the origin of the rocks they make up and thus the nature of the geological processes that operate on and in the Earth. In the rest of this book, we will refer to various minerals in one geological context or another.

# SUMMARY

**What is a mineral?** Minerals, the building blocks of rocks, are naturally occurring inorganic solids with specific crystal structures and chemical compositions that are fixed or vary within a defined range. A mineral is constructed of atoms, the small units of matter that combine in chemical reactions. An atom is composed of a nucleus of protons and neutrons, surrounded by electrons traveling in orbitals around the nucleus. The atomic number of an element is the number of protons in its nucleus, and its atomic weight is the sum of the masses of protons and neutrons.

**How are atoms combined to form the crystal structures of minerals?** Chemical substances react with each other to form compounds either by gaining or losing electrons to become ions or by sharing electrons. Either way, the atoms achieve stable configurations of electron shells. The atoms or ions of a substance are held together by ionic and covalent bonds formed by electrostatic attraction between nuclei and electrons of the constituent elements. When a mineral crystallizes, atoms or ions come together in the proper proportions to form a crystal structure, which is an orderly, three-dimensional geometric array in which the basic arrangement is repeated in all directions.

**What are the major rock-forming minerals?** Silicates, the most abundant minerals in Earth's crust, are crystal structures built of silicate tetrahedra linked in various ways. Tetrahedra may be isolated (olivines) or in rings (beryl), single chains (pyroxenes), double chains (amphiboles), sheets (micas), or frameworks (feldspars). Carbonate minerals are made of carbonate ions bonded to calcium and/or magnesium. Oxide minerals are compounds of oxygen and metallic elements. Sulfide and sulfate minerals are structures made up of sulfur atoms in combination with metallic elements.

**What are the physical properties of minerals?** Physical properties, which reflect the compositions and structures of minerals, include hardness, or the ease with which a mineral surface is scratched; cleavage, or the ability of a mineral to split or break along flat surfaces; fracture, or the way in which minerals break along irregular surfaces; luster, or the nature of a mineral's reflection of light; color, imparted by either transmitted or reflected light to crystals, irregular masses, or a streak (the color of a fine powder); density, or the mass per unit volume; and crystal habit, or the shapes of individual crystals or aggregates.

# KEY TERMS AND CONCEPTS

mineralogy (p. 24)
glassy (p. 25)
atom (p. 25)
nucleus (p. 25)
proton (p. 25)
neutron (p. 25)
electron (p. 25)
atomic number (p. 25)
atomic weight (p. 26)
isotope (p. 26)
chemical reaction (p. 27)

gain or loss of electrons (p. 27)
ion (p. 27)
cation (p. 27
anion (p. 27)
electron sharing (p. 28)
ionic bond (p. 31)
covalent bond (p. 31)
metallic bond (p. 31)
crystallization (p. 32)
crystal (p. 32)
polymorph (p. 35)

Mohs scale of hardness
    (p. 41)
cleavage (p. 41)
fracture (p. 43)
luster (p. 43)
color (p. 43)
streak (p. 44)
density (p. 44)
specific gravity (p. 45)
crystal habit (p. 45)
trace elements (p. 46)

# MINERAL NAMES TO REMEMBER

(See Appendix 3 for specific mineral properties.)

| | | |
|---|---|---|
| amphibole | feldspar | mica |
| anhydrite | galena | olivine |
| beryl | garnet | pyrite |
| calcite | graphite | pyroxene |
| clay mineral | gypsum | quartz |
| diamond | halite | spinel |
| dolomite | hematite | |

# EXERCISES

1. Define a mineral.

2. What is the difference between an atom and an ion?

3. Describe the atomic structure of sodium chloride.

4. What are the two polymorphs of carbon?

5. Give the basic structure types of silicate minerals.

6. How does the cleavage of mica reflect its atomic structure?

7. Name three groups of minerals, other than silicates, based on their chemical composition.

# THOUGHT QUESTIONS

(Consult Appendix 3 for specific mineral properties.)

1. What physical properties would make calcite a poor choice for a good gemstone?

2. Choose two minerals from Appendix 3 that you think might make good abrasive or grinding stones for sharpening steel, and point out which physical property might make them suitable for this purpose.

3. Aragonite, with a density of 2.9 g/cm$^3$, has exactly the same chemical composition as calcite, with a density of 2.7 g/cm$^3$. Other things being equal, which of these two minerals is more likely to have formed under high pressure?

4. Two people are comparing rubies, one natural and the other synthetic. Are either or both minerals? Why?

5. Oxygen exists as three isotopes with atomic weights of exactly 16, 17, and 18. The atomic weight of oxygen found in nature is approximately 16. What does this information tell you about the relative abundance of the three isotopes in nature?

6. What properties of talc make it suitable for face and body powder?

7. Hydrogen (H), the lightest element, has an atomic number of 1 and an atomic weight of 1.008 in nature. What does this information tell you about possible isotopes of hydrogen?

8. In some places in bodies of granite, we can find very large crystals, some as much as a meter across, yet many of them show few crystal faces. What would you surmise about the conditions under which these large crystals grew?

9. What physical properties of sheet silicates are related to their crystal structure and bond strength?

10. How might you identify and differentiate between a single-chain and a double-chain silicate?

11. Diopside is a pyroxene whose formula is $(Ca, Mg)_2Si_2O_6$. What does this tell you about its crystal structure and cation substitution?

# SUGGESTED READINGS

Berry, L. G., B. Mason, and R. V. Dietrich. 1983. *Mineralogy,* 2d ed. San Francisco: W. H. Freeman.

Ernst, W. G. 1969. *Earth Materials.* Englewood Cliffs, N.J.: Prentice-Hall.

Hurlbut, C. S., Jr. 1969. *Minerals and Man.* New York: Random House.

McQuarrie, D. A., and P. A. Rock. 1991. *General Chemistry,* 3d ed. New York: W. H. Freeman.

Prinz, M., G. Harlow, and J. Peters. 1978. *Simon and Schuster's Guide to Rocks and Minerals.* New York: Simon and Schuster.

# ROCKS: RECORDS OF GEOLOGIC PROCESSES

As the tangible record of the geologic processes that made them, rocks are the key to both the geologic past and the nature of processes that are impossible to observe directly, such as the melting of rock deep in the interior. Rocks are made of minerals and fall into three great groups: igneous, sedimentary, and metamorphic. Igneous rocks are formed by the solidification of molten rock. Sedimentary rocks form by the hardening and cementation of layers of sediment deposited at Earth's surface. Metamorphic rocks are formed from preexisting rocks by transformations in the solid state under intense heat and pressure. The rock cycle, driven by plate tectonics, relates the three groups of rocks by the geologic processes that convert one to another: weathering, sedimentation, metamorphism, melting, and tectonic uplift.

3

Sedimentary rock, up-
lifted and somewhat
deformed, is eroded by
the Hunza River, Paki-
stan, which is carrying
away sand and gravel.
The sedimentary rock
represents a former
cycle of uplift, erosion,
and sedimentation; the
sand and gravel repre-
sent the current cycle.
*Art Wolfe.*

What determines the appearance of the rocks we en-counter in road cuts, cliffs, or canyon walls? They vary widely in color, in the sizes of their crystals or grains, and in the kinds of minerals that make them up. Along a road cut, for example, we may find a remnant of an ancient volcanic lava, a black, homo-geneous rock whose constituent particles are too small to be seen with the naked eye. Near it may be a material transformed by heat and pressure deep in the Earth, a brownish rock with abundant large glitter-ing crystals of mica and some grains of quartz and feldspar. Overlying both may be the remains of a former beach, horizontal layers of light-brown rock that appear to be made up of sand grains cemented together.

The appearance of these rocks is determined partly by their **mineralogy,** or the relative propor-tions of their constituent minerals, and partly by their texture, or the sizes and shapes of the mineral grains and crystals and the way they are put together. The grains or crystals, a few millimeters in diameter in most rocks, are commonly categorized as either coarse (if they are large enough to be seen with the naked eye) or fine (if they are not). Mineral grains or crystals may be needle-shaped, flat, platy, or equant (about the same dimension in all directions, like a sphere or cube). The arrangement of the different sizes and shapes and the variations in mineralogy produce larger features, such as alternate layers of coarse and fine rocks.

The major factor that determines the mineralogy and texture of a rock is its geologic origin; that is, where and how it was made (Figure 3.1). The dark rock in our road cut, called *basalt,* was formed by a volcanic eruption; its mineralogy and texture depend on the chemical composition of rocks that were

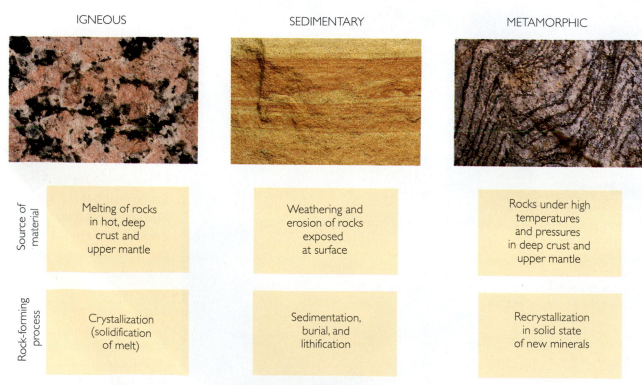

| IGNEOUS | SEDIMENTARY | METAMORPHIC |
|---|---|---|
| **Source of material:** Melting of rocks in hot, deep crust and upper mantle | Weathering and erosion of rocks exposed at surface | Rocks under high temperatures and pressures in deep crust and upper mantle |
| **Rock-forming process:** Crystallization (solidification of melt) | Sedimentation, burial, and lithification | Recrystallization in solid state of new minerals |

**FIGURE 3.1** The minerals and textures of the three great rock groups are formed in differ-ent places in the Earth by different geologic processes, and so they are a guide to the geologic processes that made them.

melted deep in Earth's interior and on the nature of the eruption—whether it was explosive or a quieter lava flow. All rocks that were formed by the solidification of molten rock are called **igneous rocks.**

The light-brown layered rock of the road cut, called *sandstone,* was formed as sand particles accumulated on a beach and eventually were covered over, buried, and cemented together to form a rock. All rocks that were formed as the burial products of layers of sediments such as sand, mud, and calcium carbonate shells, whether they were laid down on the land or under the sea, are called **sedimentary rocks.**

The brownish rock of our road cut, called a *schist,* contains crystals of mica, quartz, and feldspar and was formed deep in Earth's crust as high temperatures and pressures transformed the mineralogy and texture of a buried sedimentary rock. All rocks that are formed by transformations of preexisting rocks in the solid state under the influence of high pressure and temperature are called **metamorphic rocks.**

Understanding rock properties and reasoning from them to deduce their geologic origins is the primary aim of a geologist. Such deductions are also essential for the discovery of economically valuable mineral and energy resources. By studying rocks, geologists have learned how to find oil and gas, coal, metal ores, and many other useful materials. We know that oil, for example, is formed in certain kinds of sedimentary rocks that are rich in organic remains of biological origin. Armed with knowledge of where and how such rocks are formed, we can explore for new oil reserves more intelligently.

If rocks are clues to many of the things we want to know about the Earth, how do we go about interpreting them? We need a key, just as historians needed the Rosetta stone to crack the "code" of Egyptian hieroglyphics before they could read the inscriptions on temples and tombs. The first step in forging the key is to learn to recognize the various kinds of rocks—igneous, sedimentary, and metamorphic—using as main guides their mineralogies and textures. The second step is to understand how the varying compositions and textures of igneous, sedimentary, and metamorphic rocks result from the conditions under which they were formed on Earth's surface and in its interior.

# IGNEOUS ROCKS

Igneous (from Latin *ignis,* fire) rocks form by crystallization from a magma, a mass of melted rock that originates deep in the crust or upper mantle, where temperatures reach the 700°C or more needed to melt most rocks. Geologists distinguish two major types of igneous rocks on the basis of the sizes of their crystals. When magmas cool slowly in the interior, some of the microscopic crystals that start to form as the magma cools below the melting point have time to grow to several millimeters or larger before the whole mass is crystallized as a coarse-grained igneous rock. When a magma flows or erupts from a volcano onto Earth's surface, in contrast, it cools and solidifies rapidly, allowing individual crystals no time for gradual growth. Instead, many tiny crystals form simultaneously. The result is fine-grained igneous rock.

Rocks such as *granite,* formed by slowly crystallizing melts in the interior, are called **intrusive igneous rocks;** they are recognized by their interlocking large crystals, which grew slowly as the magma gradually cooled (Figure 3.2). Magmas cool slowly in the interior because they invade rock masses that conduct heat slowly; in addition, some of their temperatures may not be a great deal cooler than the magma.

Rocks such as basalt, which form from rapidly cooled magmas erupting at the surface, are called

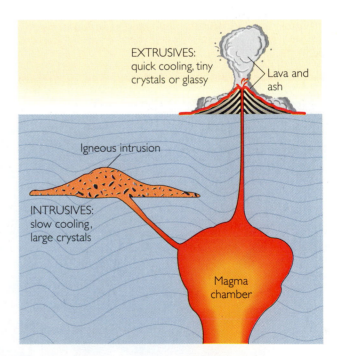

**FIGURE 3.2** The two major classes of igneous rocks are intrusive rocks, formed by slow cooling and crystallization of larger crystals deep in the Earth's crust, and extrusive or volcanic rocks, formed by rapid cooling to glassy and fine-grained lavas and ash.

**extrusive igneous rocks.** They are recognized by their glassy or fine-grained texture (see Figure 3.2). Extrusive igneous rocks range from almost instantaneously crystallized ash particles blown high into the atmosphere to lavas, which do not crystallize so quickly and flow as liquids for some distance on the surface before they solidify.

Most of the minerals of igneous rocks are silicates, partly because silicon is so abundant in the Earth and partly because silicate minerals melt at the temperatures and pressures reached in some parts of the crust and mantle. Some of the common silicate minerals of igneous rocks are quartz, feldspar, mica, pyroxene, amphibole, and olivine, the minerals that typify the various crystal structures we described in Chapter 2. Table 3.1 lists some common minerals found in the three rock types.

## SEDIMENTARY ROCKS

**Sediments,** the precursors of sedimentary rocks, are found on Earth's surface as layers of loose particles, such as sand, silt, and shells of organisms. Particles such as sand grains and pebbles are formed at the surface of the Earth as rocks undergo **weathering**— that is, are broken up into fragments of various sizes (Figure 3.3). Physically deposited sedimentary parti-

cles, such as grains of quartz and feldspar derived from a weathered granite, are called **clastic** (from Greek *klastos,* to break) **sediments.** Clastic sediments are laid down by running water, wind, and ice, forming layers of sand, silt, and gravel in the process.

As a rock weathers, some of its components may dissolve and be carried in river waters to the sea,

---

**TABLE 3.1**

### Some Common Minerals of Igneous, Sedimentary, and Metamorphic Rocks

| IGNEOUS ROCKS | SEDIMENTARY ROCKS | METAMORPHIC ROCKS |
|---|---|---|
| Quartz★ | Quartz★ | Quartz★ |
| Feldspar★ | Clay minerals★ | Feldspar★ |
| Mica★ | Feldspar★ | Mica★ |
| Pyroxene★ | Calcite | Garnet★ |
| Amphibole★ | Dolomite | Pyroxene★ |
| Olivine★ | Gypsum | Staurolite★ |
|  | Halite | Kyanite★ |

Asterisks indicate that a mineral is a silicate.

---

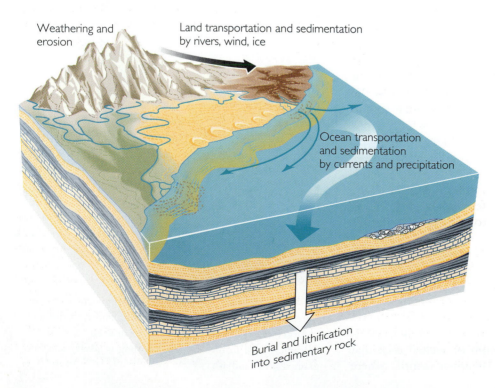

Weathering and erosion

Land transportation and sedimentation by rivers, wind, ice

Ocean transportation and sedimentation by currents and precipitation

Burial and lithification into sedimentary rock

**FIGURE 3.3** Weathering and erosion break down rocks into sediment. The sediment is then transported, deposited, and lithified (hardened into rock).

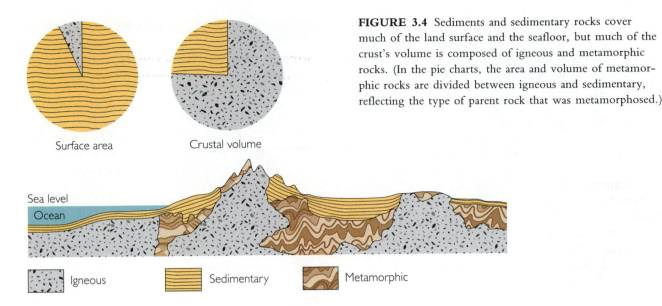

where new chemical substances may be precipitated. These **chemical and biochemical sediments,** as they are called, include layers of minerals such as halite (sodium chloride) and calcite (calcium carbonate, most frequently found in the form of shells).

Sedimentary rocks are formed as sediments undergo **lithification**—that is, are hardened into solid rock. Sediments lithify both by compaction, as the grains are squeezed together into a denser mass than the original, and by cementation, as minerals precipitate around the grains after deposition and bind the particles together. Sediments are compacted and cemented after burial under additional layers of sediment. Thus sandstone forms by the lithification of sand particles and limestone by the lithification of shells and other particles of calcium carbonate.

Sediments and sedimentary rocks are characterized by **bedding,** the formation of parallel layers by the settling of particles to the bottom of the sea, a river, or a land surface. Bedding may reflect variations in mineralogy, as when sandstone is interbedded with limestone, or differences in texture, as when a coarse-grained sandstone becomes interbedded with a fine-grained one.

The common minerals of clastic sediments are silicates, reflecting the dominance of silicate minerals in rocks that weather to form sedimentary particles (see Table 3.1). Thus the most abundant minerals in clastic sedimentary rocks are quartz, feldspar, and clay minerals. The most abundant mineral chemically or biochemically precipitated in the oceans is calcite, most of it the shelly remains of organisms and

the main constituent of limestone. Many limestones also contain dolomite, a calcium-magnesium carbonate precipitated during lithification. Gypsum and halite are formed by chemical precipitation during the evaporation of seawater.

Most rocks found at the Earth's surface are sedimentary, though they account for only a small fraction of all the rocks that make up the crust. Formed as they are by surface processes, over much of the Earth they are a thin covering atop the igneous and metamorphic rocks of the main part of the crust (Figure 3.4).

# METAMORPHIC ROCKS

Metamorphic (from Greek *meta,* change, and *morphē,* form) rocks are produced by mineralogical and textural transformations of all kinds of rocks—igneous, sedimentary, and metamorphic—under the influence of high temperatures and pressures deep in the Earth. Rocks are metamorphosed by temperatures below their melting points (less than about 700°C) but high enough (above 250°C) for the rocks to change by recrystallization and chemical reactions while remaining solid. Metamorphosed rocks may have changed in mineralogy, texture, and chemical composition.

Where high pressures and temperatures extend over large regions, rocks are subject to **regional metamorphism.** Regional metamorphism accompanies plate collisions that result in mountain build-

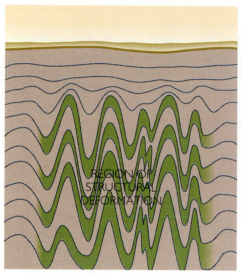

(a) Regional metamorphism

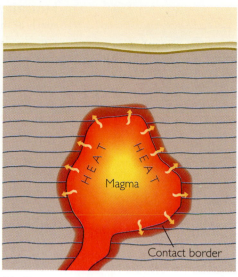

(b) Contact metamorphism

**FIGURE 3.5** Regional and contact metamorphism result from different processes of change caused by heat and pressure. (a) Deep in the crust, regions that are strongly deformed are metamorphosed by accompanying heat and pressure. (b) Contact zones surrounding igneous intrusions are metamorphosed by heat from the intrusion.

ing and structural deformation (Figure 3.5a). Where high temperatures are restricted to the borders of igneous intrusions, the rocks around and in contact with the intrusion are transformed by **contact metamorphism** (Figure 3.5b). Many regionally metamorphosed rocks, such as schists, are recognized by their **foliation,** wavy or flat planes produced in the rock by structural deformation into folds (see Figure 3.1). Granular textures are more typical of most contact metamorphic rocks as well as of some regional metamorphic rocks formed by very high pressure and temperature.

Silicates are the most abundant minerals of metamorphic rocks because they are transformations of other rocks that are rich in silicates (see Table 3.1). Typical minerals of metamorphic rocks are quartz, feldspar, mica, pyroxene, and amphibole, the same kinds of silicates characteristic of igneous rocks. Several other silicates—garnet, staurolite, and kyanite— are more characteristic of metamorphic rocks alone. Calcite is the main mineral of marbles, which are metamorphosed limestones.

# THE CHEMICAL COMPOSITION OF ROCKS

Geologists frequently make chemical analyses of rocks and look for similarities and differences in the proportions of chemical elements from which they can infer geologic origin. The chemical composition of an igneous rock, for example, can reveal what kinds of rock melted to form the magma from which the rock crystallized. Chemical analysis complements mineralogical studies, because the mineralogy alone does not reveal how the chemical elements are divided among the minerals present. Chemical analysis is particularly important for the very fine-grained or glassy rocks, such as volcanic lavas, in which few individual minerals can be made out, even with a microscope.

A chemical analysis of a rock gives the relative proportion of the chemical elements in it. By convention the elements are stated in terms of oxides, the compounds they form by combining with the most abundant element in the Earth, oxygen. The element silicon, for example, is given as the percentage of its oxide, silica, $SiO_2$ (Table 3.2). An ideal chemical analysis consists of the percentages of each element's oxide, adding up to a total of 100 percent.

About two-thirds of a typical igneous rock—a basalt, for example—consists of silica (48 percent) and alumina, the oxide of aluminum (16 percent). There are smaller amounts of iron oxides (14.7 percent) and even smaller amounts of calcium, magnesium, sodium, and potassium oxides (see Table 3.2). These seven *major elements,* along with oxygen, make up the great bulk of all rocks in the Earth.

Differences of a few percentage points in the proportions of some elements may indicate, for example, whether a basalt was formed at a mid-ocean ridge, where plates diverge, or at a subduction zone, where plates converge. One of these differences is the amount of water (the oxide of hydrogen). Chemically bound in minerals, water is a constituent of many rocks, but rarely in abundance. Many igneous rocks are only about 1 percent water. Some sedimentary rocks may contain more water, as much as 5 percent or more, not counting the water that may be present as liquid in the pores of the rock. The larger amount of water in sedimentary rocks is traceable to the abundance of clay minerals in them.

As we will see in Chapter 4, chemical differences among igneous rocks are clues to the origin of the magmas from which the rocks came and how and where they crystallized. The chemical compositions of sedimentary rocks indicate the kinds of rocks that were weathered to provide the sediment particles and the chemical conditions under which precipitated minerals were formed. Chemical analyses of metamorphic rocks are guides to the preexisting rocks that were transformed by heat and pressure.

## WHERE WE SEE ROCKS

Rocks are not found in nature conveniently divided into separate bodies—igneous rocks here, sedimentary rocks there, metamorphic rocks in another place. Instead they are found jumbled together in patterns determined by the geologic history of a region. The geologist maps those patterns and tries to deduce from the kinds and distributions of the rocks as they are now what happened at various times in the geologic past.

If we were to drill a hole into any spot on Earth we would find rocks that carry the geologic history of that place. In the top few kilometers of most regions we would probably find sedimentary rock. Drilling deeper, perhaps 6 to 10 km down, we would eventually penetrate older igneous and metamorphic rocks that underlie the thin cover of sedimentary rocks. The thousands of holes drilled on the continents in the search for rocks that contain oil, water, and mineral resources are major sources of information. In the quest for more data on the deep continental crust, several countries, including the United States, Germany, and Russia, have drilled to depths on the continents deeper than any commercial drills have gone. The deepest hole, in Russia, went deeper than 12 km.

A large part of our knowledge about the rocks of the ocean floor comes from the hundreds of holes

**TABLE 3.2**

### Chemical Analysis of Basalt, an Igneous Rock

| ELEMENT | FORMULA[1] | PERCENT OF WEIGHT |
|---|---|---|
| Silicon | $SiO_2$ | 48.0 |
| Aluminum | $Al_2O_3$ | 16.0 |
| Iron | $Fe_2O_3$, $FeO$ | 14.7 |
| Calcium | $CaO$ | 10.0 |
| Magnesium | $MgO$ | 3.9 |
| Sodium | $Na_2O$ | 3.5 |
| Potassium | $K_2O$ | 1.5 |
| | Total | 97.6[2] |

[1] Formulas are conventionally shown as oxides, though the elements are not actually present in the rock in this form.
[2] Percentages do not add up to 100 because of the omission of minor elements and small errors in analysis.

punched down by the Deep Sea Drilling Program, an ongoing plan for drilling the world's seafloor for geological information. Started by the United States in the late 1960s, at the same time that plate tectonics swept the geological community, it is now an international program carried on with the cooperation of the major maritime countries of the world.

Even with all these sources of information on what lies beneath Earth's surface, geologists continue to rely on the rocks exposed in **outcrops,** places where the bedrock is laid bare of soil (Figure 3.6). On a trip across North America we might run across many kinds of outcrop (Figure 3.7). Starting from the Pacific, we would encounter sea cliffs from Mexico to Canada. From the West Coast to the Rocky Mountain front, which stretches from New Mexico in the south to Alberta in Canada, outcrops are abundant in the canyons, moutainsides, and cliffs of the relatively dry mountainous regions of the western third of the continent.

From the Rockies eastward to the Appalachian Mountains, the landscape is dominated by the plains and prairies of the American Midwest and the Canadian Plains provinces. Here outcrops are scarce be-

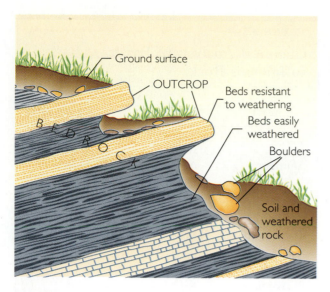

**FIGURE 3.6** A cross section through the surface of the ground shows how an outcrop is related to the bedrock and soil.

cause soils and the sediments deposited by such rivers as the Missouri and their tributaries cover most of the sedimentary bedrock. In the low hills and gentle valleys geologists hunt for dry creeks and interstate highway cuts. Once we reach the eastern mountains,

the Appalachians, the outcrops become more numerous. In this more humid climate, most of the rocks of the low, mountainous ridges are covered by abundant vegetation and soil, but there are many rocky cliffs and ledges, especially on the higher ridges and mountains.

Low coastal plains cover the region from southeastern New Jersey to the Carolinas and Georgia in the east and Texas, Louisiana, Mississippi, and Alabama in the south. Here outcrops of barely lithified, still relatively soft sedimentary rocks are similar to those of the Great Plains. Good exposures can be found in the occasional bluff along the shoreline. To the north, in the hilly, rugged landscape of New England and the Maritime provinces of Canada, we can find good outcrops, but the best exposures are the rocks displayed along rocky coastlines.

This travelogue makes it clear that the presence or absence of different kinds of outcrops depends on the nature of the landscape, which in turn depends on the geologic structure of the region and its history. Later on we will look in more detail at the way rock types relate to geologic structures (Chapter 10) and to landscape (Chapter 16). Here let us turn to the general interrelationships among the three groups of rocks and see how they can be interpreted to reveal geologic structure and history.

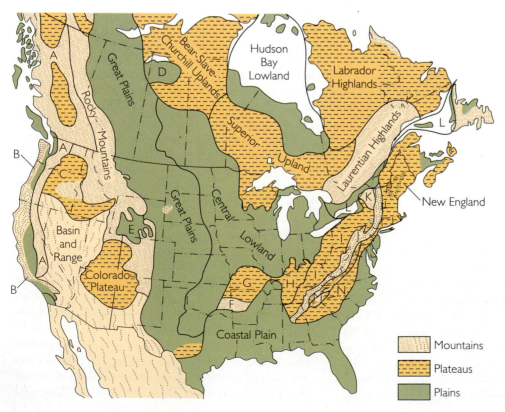

**FIGURE 3.7** Regions of North America where different kinds of outcrop are found. (After C. B. Hunt, *Natural Regions of the United States and Canada,* San Francisco, W. H. Freeman, 1974, p. 63.)

A  Pacific Mountain System; includes
B  Pacific Border
C  Columbia–Snake River Plateau
D  Athabasca Plain
E  Wyoming Basin
F  Ouachita
G  Ozark Plateaus
H  Interior Low Plateaus
I  Appalachian Plateaus
J  Valley and Ridge
K  Adirondack
L  St. Lawrence Lowland
M  Blue Ridge
N  Piedmont Plateau

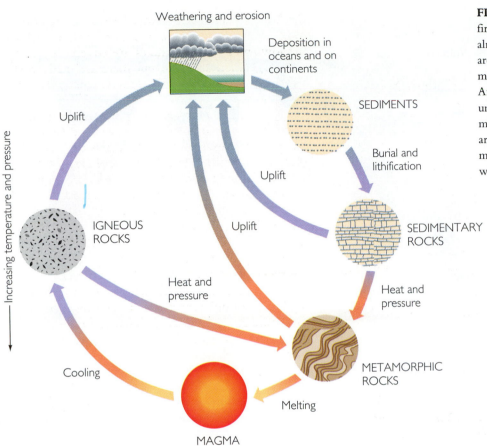

**FIGURE 3.8** The rock cycle, first proposed by James Hutton almost 200 years ago. Rocks are weathered to form sediment, which is then buried. After deep burial, the rocks undergo metamorphism or melting, or both. Later they are deformed and uplifted into mountain chains, only to be weathered again and recycled.

# THE ROCK CYCLE

The three great groups of rocks are related by the **rock cycle,** the circular process by which each is formed from the others (Figure 3.8). The Scotsman James Hutton, frequently called the father of modern geology, played a large role in the understanding of the rock cycle with his book *Theory of the Earth with Proof and Illustrations* (1785). Hutton recognized the cyclic nature of geological changes. Here we give an account of one particular cycle, recognizing that such cycles vary with time and place.

We can start the cycle with igneous rocks, which form by crystallizing from a magma deep in Earth's interior. The magma forms by the melting of preexisting rock of any kind: other igneous rocks, metamorphic rocks, or sedimentary rocks. Hutton called this event the **plutonic episode,** for Pluto, the Roman god of the underworld. Melting destroys all the minerals of the preexisting rocks, homogenizing

their chemical elements in the resulting hot liquid. A new igneous rock is created when the magma cools and new minerals form by crystallization of the melt. As we will see in later chapters, most melting and formation of igneous rock take place along the boundaries of tectonic plates as they collide.

The igneous rocks formed during plate collisions are then uplifted as a high mountain chain. Uplift at plate boundaries, accompanied by crumpling and deformation of the crust, is part of the deformational and mountain-building process that geologists call **orogeny.** After uplift, the rocks of the crust that overlay the uplifted igneous rock are gradually weathered, eroded, and stripped away, until the igneous rock is exposed at the surface. Now, in cool, wet surroundings far from its birthplace in the hot interior, the igneous rock is also weathered, some of its minerals changing to others; iron minerals, for

**FIGURE 3.9** Elements of the rock cycle (shown in boldface) are found in many plate-tectonic settings.

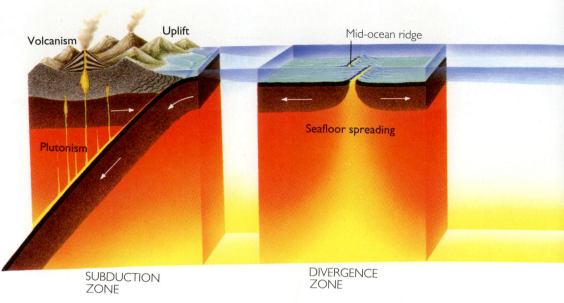

instance, may "rust" to form iron oxides, and feldspars may become clays. The rock debris, consisting of altered and unaltered minerals, and the dissolved substances produced by weathering are transported by streams to the ocean, there to be deposited as layers of sand, mud, and other sediments, such as the calcium carbonate sediments formed of shells.

As these sediments laid down in the sea, as well as those formed by rivers and wind on the land, are buried by successive layers of sediment, they gradually lithify into sedimentary rock. Burial is accompanied by **subsidence;** that is, a depression of Earth's crust caused in part by the weight of the sediment deposited. As subsidence continues, additional sedimentary layers accumulate.

As the lithified sedimentary rock is buried more deeply, it gets hotter. As temperatures climb to over 300°C, the rock's minerals start to change to new kinds of minerals that are more stable at the higher temperatures and pressures of the deeper parts of the crust. This is the process of metamorphism, which transforms sedimentary rocks into metamorphic rocks. With further heating, the rocks may melt, forming a new magma from which igneous rocks crystallize, starting the cycle all over again.

This cycle is only one variation among the many that may occur. Any type of rock, whether meta-

morphic, sedimentary, or igneous, can be uplifted during an orogeny and weathered and eroded to form new sediments. Some of the stages in the cycle may be omitted; as a sedimentary rock is uplifted and eroded, for example, metamorphism and melting are skipped. An igneous rock may become metamorphosed before it is uplifted. And we know from deep drilling that there are igneous rocks many kilometers down in the crust that have never been uplifted or exposed to weathering and erosion. The rock cycle never ends; it is always operating, at different stages in different parts of the world. At any given moment, mountains are forming and eroding in one place and sediments are being laid down and buried in another. The rocks that make up the solid Earth are being recycled continuously.

The rock cycle is driven by plate tectonics (Figure 3.9). Rocks melt and igneous rocks form as plates subduct into the mantle. Plate collisions uplift mountains and create high pressures and temperatures that metamorphose rocks in the interior. Weathering of the mountains produces sediment that gets laid down on continents and ocean floors as plates slowly subside. The sediments become buried and eventually the rock cycle begins again.

With this introduction to the rock world, we are ready to begin the study of igneous rocks. In Chap-

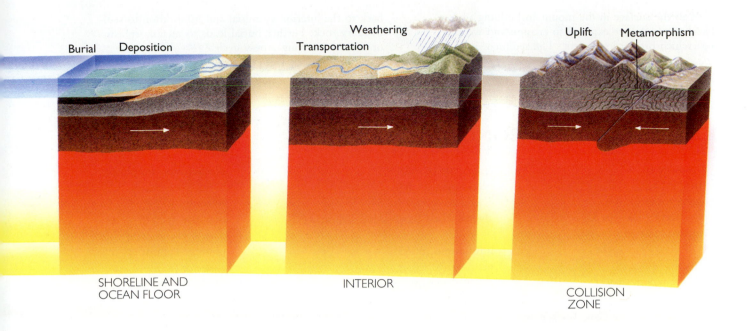

ters 4 and 5 we look at the geologic origin of magmas, the types of igneous rocks that form by their crystallization, the larger picture of plate-tectonic control of igneous processes, and the dynamics of volcanoes and their eruptions. In Chapters 6 and 7 we explore weathering, the nature of sedimentary particles, and the ways various sediments and sedimentary rocks are produced. We complete our discussion of rocks in Chapter 8 by seeing how heat and pressure affect preexisting rocks to form the new minerals and textures of metamorphic rocks and how metamorphism is related to plate tectonics and orogeny.

## SUMMARY

**What determines the properties of the various kinds of rocks that are formed in and on the surface of the Earth?** Mineralogy (the kinds and proportions of minerals that make up a rock) and texture (the sizes, shapes, and spatial arrangement of its crystals or grains) define a rock. The mineralogy and texture of a rock are determined by the geological conditions under which a rock formed, either in the interior under various conditions of high temperature and pressure or at the surface, where temperatures and pressures are low.

**What are the three types of rock and how are they formed?** Igneous rocks are formed by the crystallization of magmas as they cool either in the interior or at the surface of the Earth where lavas and ash erupt from volcanoes. Sedimentary rocks are formed by the lithification of sediments after burial. Sediments are derived from the weathering and erosion of preexisting rocks exposed at the Earth's surface. Metamorphic rocks are formed by alteration in the solid state of any type of rock subjected to high temperatures and pressures in the interior.

**How does the rock cycle describe the formation of rocks as the products of geologic processes?** The rock cycle relates rock-forming geologic processes to the formation of the three types of rocks from one another. We can start the cycle with the formation of igneous rocks by crystallization of a magma in the interior. The igneous rock then is uplifted to the surface in the mountain-building process. There the rocks are exposed to erosion and weathering, which produce sediment. The sediment is cycled back to the interior by burial and lithification to sedimentary rock. Further burial leads to metamorphism or melting, starting a new cycle.

# KEY TERMS AND CONCEPTS

mineralogy (p. 52)
texture (p. 52)
igneous rocks (p. 53)
sedimentary rocks (p. 53)
metamorphic rocks (p. 53)
intrusive igneous rocks
   (p. 53)
extrusive igneous rocks
   (p. 54)

sediments (p. 54)
weathering (p. 54)
clastic sediments (p. 54)
chemical and biochemical sediments
   (p. 55)
lithification (p. 55)
bedding (p. 55)
regional metamorphism
   (p. 55)

contact metamorphism
   (p. 56)
foliation (p. 56)
outcrops (p. 57)
rock cycle (p. 59)
plutonic episode (p. 59)
orogeny (p. 59)
subsidence (p. 60)

# EXERCISES

1. What are the differences between extrusive and intrusive igneous rocks?

2. What are the differences between regional and contact metamorphism?

3. What are the differences between clastic and chemical or biochemical sedimentary rocks?

4. Name three silicate minerals that are common in each group of rocks: igneous, sedimentary, and metamorphic.

5. Which kinds of rocks are formed at Earth's surface and which in the interior of the crust?

# THOUGHT QUESTIONS

1. What geologic processes accompany the transformation of a metamorphic rock to a sedimentary rock?

2. What minerals might you use to distinguish between an extrusive igneous rock and a fine-grained sedimentary rock formed from lithified mud?

3. What differences in origin might cause differences in the sizes of the crystals of two granites, one

with crystals about 1 cm in diameter, the other with crystals about 2 mm in diameter?

4. Which igneous intrusion would you expect to have a wider contact metamorphic zone: one intruded very deep in the crust or one intruded very near the surface?

5. Describe the geologic processes by which an igneous rock is first transformed to a metamorphic rock and then exposed to erosion.

## SUGGESTED READINGS

Dietrich, Robert V., and Brian J. Skinner. 1980. *Rocks and Rock Minerals*. New York: Wiley.

Ehlers, Ernest G., and Harvey Blatt. 1982. *Petrology. Igneous, Sedimentary, and Metamorphic*. San Francisco: W. H. Freeman.

Ernst, W. Gary. 1969. *Earth Materials*. Englewood Cliffs, N.J.: Prentice-Hall.

Prinz, Martin, George Harlow, and J. Peters. 1978. *Simon and Schuster's Guide to Rocks and Minerals*. New York: Simon and Schuster.

# IGNEOUS ROCKS: SOLIDS FROM MELTS

Igneous rocks formed early in Earth's history when a magma ocean crystallized to form the crust. Geologists 200 years ago divided igneous rocks into the coarsely crystalline intrusives, products of slow cooling deep in the crust, and the finely crystalline extrusives, products of fast cooling where volcanoes bring melted rock to the surface. Igneous rocks are also classified on the basis of mineral composition into a series ranging from light-colored rocks rich in silica, such as granite, to dark-colored rocks poor in silica, such as basalt. Two general paths of crystallization—a continuous series of feldspars and a discontinuous series of mafic minerals—account for the differentiation of the melt as newly formed crystals selectively remove some of the chemical components of the magma.

Columnar basalts;
Isle of Staffa,
Scotland. Masses of this
kind of rock fracture
along more or less
symmetrical columnar
joints when they cool.
*Colin Prior/Tony Stone
Worldwide.*

For thousands of years, people who lived near volcanoes could see that the hot liquid lava that spilled onto Earth's surface cooled and solidified to hard rock in a few hours. It was not until the eighteenth century that geologists began to understand that bands or sheets of rock seen cutting across other formations had also been made by the cooling and solidifying of a hot liquid, but in this case the process had proceeded slowly, while the liquid rock was still buried in the Earth's crust.

How early geologists came to make this connection is one of the milestones in the development of geology into a modern observational and experimental science. Today we know that deep in the hot crust and mantle of the Earth, rocks melt and rise toward the surface, and they sometimes break through. Igneous rocks are the products of the cooling and solidification of these melts. Knowing where and how rocks melt and resolidify is a key to understanding how the Earth's crust is formed, since much of the crust, as we saw in Chapter 3, is composed of igneous rock, some metamorphosed and some not.

In seeking to discover where these processes take place, modern geologists have learned that the origins of a wide variety of igneous rocks are connected with plate-tectonic movements, especially the sinking of one converging plate below another and the spreading apart of two divergent plates. By studying igneous rocks, then, geologists can gain more insight into the tectonic history of the Earth. Although much is still to be learned about the exact mechanisms of melting and solidification, geologists now have good answers to questions that long perplexed them: How do igneous rocks differ from one another? Where do igneous rocks form? How do rocks solidify from a melt? Where do melts form?

In the late eighteenth century, long before plate-tectonic theory was introduced, geologists sought

answers to these questions in field observations. By the late nineteenth century they were collecting all kinds of igneous rocks and taking them back to the laboratory to determine their mineral and chemical compositions. They classified their rock samples in the same general way we do today, pigeonholing each by the geological setting in which it was found as well as by its texture and its mineral and chemical composition. We start our investigation of igneous rocks, then, as those early geologists did, with a descriptive classification of the rocks whose origins we want to explain.

# MAJOR TYPES OF IGNEOUS ROCKS

The first division of the igneous rocks was made on the basis of texture: either coarsely or finely crystalline (Figure 4.1). This was a distinction that was simple and practical to make in the field. It was easy to see the separate crystals of a coarse-grained rock such as granite with the naked eye. The crystals of fine-grained rocks, in contrast, were too small to be seen, even with the aid of the small magnifying lens that no field geologist would be without. The difference was clear, but what did it mean?

## The Importance of Texture

The first clue to the meaning of texture came from observations of rocks being formed during volcanic eruptions. Lava (the term we apply to magma flowing out on the surface) cooled rapidly to a finely crystalline rock or to a glassy one in which no crystals at all could be distinguished. Yet in the middle of a

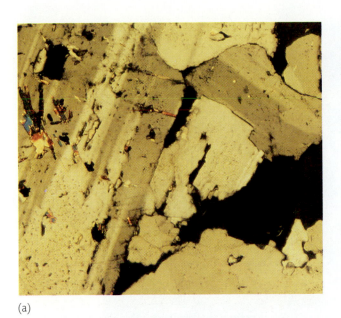

(a)

(b)

**FIGURE 4.1** Photomicrographs of coarsely and finely crystalline igneous rocks: (a) granite; (b) basalt. *Raymond Siever.*

thick flow many meters high, where the lava cooled more slowly, some larger crystals had formed.

The second clue to the implications of texture came in the nineteenth century, as experimental scientists came to understand the nature of solidification from studies of familiar liquids. Anyone who has frozen an ice cube knows that water solidifies to ice in a few hours as its temperature drops below the freezing point. If you pull a tray of ice cubes from the freezer before they are completely solid, you may be able to see thin ice crystals forming at the surface and along the sides of the compartments. During the crystallization process the water molecules line up into fixed positions in a solid crystal structure so that they are no longer free to move around as they were when the water was liquid. Magmas, like all other liquids, "freeze" in the same way. (Commonly, we use the terms "freezing," "solidification," and "crystallization" interchangeably.)

The first tiny crystals that form from a crystallizing liquid serve as a pattern: new atoms or ions attach themselves in such a way that the tiny crystals grow larger. It takes some time for the atoms or ions to "find" their correct places on a growing crystal. Thus if a liquid freezes very quickly, as a magma does when it erupts onto the cool surface of the Earth, there is no time for the crystals to grow larger, and all of the liquid freezes into tiny crystals. For large crystals to grow from a magma, they have to grow slowly.

## The Origin of Granite

The study of volcanoes pointed to the origin of finely crystalline igneous rocks. Early geologists could observe the process at work as lavas poured out on the surface and solidified to volcanic rocks in a matter of hours. These observations allowed early geologists to link finely crystalline textures with extrusive rocks and to see old extrusives as evidence of former volcanism. But how did they come to realize without direct observation that the coarse-grained intrusive rocks were formed by slow cooling deep in the interior? And what geological evidence supports that conclusion? Granite, one of the commonest rocks of the continents, turned out to be a crucial clue to the origin of intrusive rocks (Figure 4.2). The evidence on the origin of granite remains important today as one of the major underpinnings of the basic theory of igneous rock formation.

Geologists first had to demonstrate that granite was formed by the crystallization of hot molten rock deep in the crust. The evidence that proved crucial consisted of the details of granite's relationships to other rocks. James Hutton, whom we noted in Chapter 3 as the father of modern geology and the rock cycle, was one of the first to appreciate this evidence. As he worked in the field in Scotland near the end of the eighteenth century, he saw granites cutting across and disrupting the bedding of sedimentary rocks. The sedimentary rocks were somehow frac-

**FIGURE 4.2** Granite, Yosemite National Park. Weathering has revealed the coarsely crystalline nature of this massive rock, which makes up large parts of the Sierra Nevada. *Art Wolfe.*

tured and invaded by the granite. To Hutton it looked as if the granite had been forced into the fractures as a liquid.

As Hutton looked at more and more granites and focused on the sedimentary rocks bordering them, he noticed that those along the contact line were changed. The minerals of the contacting sedimentary rocks were different from those of sedimentary rocks some distance from the granite. His recognition of what we now call contact metamorphism led him to conclude that the changes in the sedimentary rocks had resulted from great heat emanating from the granite.

Hutton also noted the texture of granite, a pattern of interlocked crystals held together like the pieces of a jigsaw puzzle (see Figure 4.1a). This is the texture expected from a slow crystallization process. Chemists had long known that such textures form when a salty solution gradually evaporates; as the liq-

uid disappears, it leaves behind a mass of intergrown salt crystals.

Hutton put together the evidence of liquid behavior, heat, and crystal size and proposed that the granite was formed from a hot molten material that solidified deep in the Earth. The evidence was conclusive. No other explanation fitted all the facts. As other geologists saw the same characteristics of granites in widely separated places in the world, they came to recognize that granite and many similar coarsely crystalline rocks were the products of magma that had crystallized slowly. The origin of the intrusive rocks was no longer in doubt.

Now we can see the full significance of the textural distinction between coarse and fine crystals. Texture is linked to the rapidity and thus the place of cooling: slow cooling in the interior, fast cooling at the surface. Slow cooling produced the intrusive igneous rocks; fast cooling gave the extrusives.

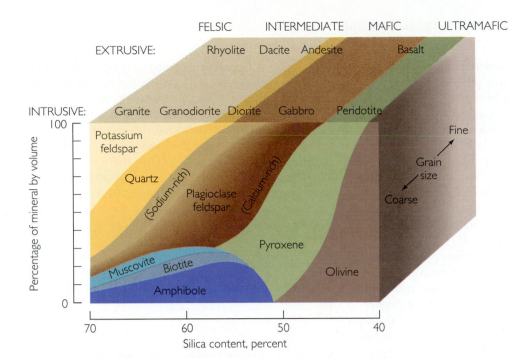

**FIGURE 4.3** Classification of igneous rocks. The mineral composition is plotted as percentage by volume of a given mineral (vertical axis) for a rock of given silica content (horizontal axis). Thus a granodiorite of about 60 percent silica content (as determined by chemical analysis) would contain about 15 percent amphibole, 12 percent biotite, 50 percent plagioclase feldspar, 19 percent quartz, and 5 percent potassium feldspar.

## Classifying Igneous Rocks

**CHEMICAL AND MINERAL COMPOSITION** In addition to the broad categories of extrusive and intrusive rocks, the igneous rocks are subdivided on the basis of their chemical and mineral compositions. One of the first criteria to be used when igneous rocks began to be studied in the last century was a simple chemical one: the amount of silica ($SiO_2$) in a rock. Silica is abundant in most igneous rocks, accounting for 40 to 70 percent of their total weight. Today we still refer to rocks rich in silica, such as granite, as *silicic*.

Modern classifications are based primarily on mineral composition, which, as we will see, gives much the same general groupings as chemical composition. The minerals used to classify common igneous rocks are all silicates: quartz, feldspar (both potassium and plagioclase), pyroxene, olivine, biotite and muscovite micas, and amphibole (Table 4.1). Igneous rocks are classified and identified on the basis of the relative contents of these silicate minerals.

As Figure 4.3 shows, differences in mineral composition correspond to a systematic series of silica contents. This figure graphically represents the classification of igneous rocks on the basis of their mineral compositions and the names that by common agreement are applied to them. The proportions of the minerals were determined by thousands of mineralogical analyses performed by geologists all over the world.

Rocks at the high-silica end of the series contain abundant quartz and potassium and sodium feldspar,

| TABLE 4.1 | | |
|---|---|---|
| **Common Minerals of Igneous Rocks** | | |
| **MINERAL** | **CHEMICAL COMPOSITION** | **SILICATE STRUCTURE** |
| Olivine | $(Mg, Fe)_2SiO_4$ | Isolated tetrahedra |
| Pyroxene | $\left.\begin{array}{l} Mg \\ Fe \\ Ca \\ Al \end{array}\right\} SiO_3$ | Single chains |
| Amphibole | $\left.\begin{array}{l} Mg \\ Fe \\ Ca \\ Na \end{array}\right\} Si_8O_{22}(OH)_2$ | Double chains |
| Biotite (mica) | $\left.\begin{array}{l} K \\ Mg \\ Fe \\ Al \end{array}\right\} Si_3O_{10}(OH)_2$ | Sheets |
| Muscovite (mica) | $KAl_3Si_3O_{10}(OH)_2$ | |
| Plagioclase feldspar | $\left\{\begin{array}{l} NaAlSi_3O_8 \\ CaAl_2Si_2O_8 \end{array}\right.$ | Frameworks |
| Potassium feldspar | $KAlSi_3O_8$ | |
| Quartz | $SiO_2$ | |

(MAFIC applies to rows Olivine through Biotite; FELSIC applies to rows Muscovite through Quartz.)

minerals rich in silica. The dominant minerals at the low-silica end of the series are pyroxenes and olivines, minerals poor in silica but rich in magnesium and iron. The rocks richer in silica are commonly termed **felsic** (from *fel*dspar and *si*lica) and the rocks poorer in silica are called **mafic** (from *ma*gnesium and *f*errous, from Latin *ferrum,* iron). Peridotite, a rock made entirely of pyroxene and olivine, is termed an **ultramafic** rock. The terms are applied to both the minerals and the rocks they constitute. Felsic minerals and rocks tend to be light in color; mafic minerals and rocks tend to be dark.

The division of igneous rocks into compositional groups is explained geologically by a strong correlation between mineralogy and the temperatures of crystallization. Mafic minerals crystallize at higher temperatures than felsic minerals. Thus mineralogy is a clue to the temperature of origin; the temperature tells us much about where and how the magma formed and crystallized, as we shall see.

As the mineral and chemical compositions of igneous rocks became known, it was soon discovered that rocks of the same composition could be either extrusive or intrusive, the only difference being textural. A common mafic igneous rock, basalt, for example, is found extruded as a lava from a volcano. Its corresponding intrusive, called gabbro, with exactly the same composition as the basalt, is formed deep in the Earth's crust (Figure 4.4). Thus there are two parallel sets of rock groups, the extrusive and the intrusive. Most compositions have extrusive and intrusive representatives; exceptions are some very highly mafic rocks that rarely or never appear as extrusives.

We look first at the intrusives—the plutonic rocks, as Hutton termed them. They were the first group to be studied intensively, partly because their minerals were coarsely crystalline and thus easily identified.

**GRANITE TO GABBRO: THE INTRUSIVE SERIES** Using silica levels and characteristic silicate minerals as a basis for classification, we can now define **granite,** the best known and one of the most

IGNEOUS

**FIGURE 4.4** Basic igneous rocks: basalt (*upper left*), gabbro (*upper right*), rhyolite (*lower left*), and granite (*lower right*). *Chip Clark.*

abundant of the intrusive igneous rocks. With about 70 percent silica, granite contains abundant quartz and potassium feldspar and a lesser amount of sodium-rich plagioclase feldspar (see the left side of Figure 4.3). These light-colored felsic minerals give granite its pink or gray color.

Moving to the right of granite on the diagram, with lower levels of silica we find **granodiorite,** a light-colored felsic rock that looks something like granite. It has abundant quartz, but plagioclase is the predominant feldspar. Still lower in silica is **diorite,** a rock dominated by plagioclase feldspar with little or no quartz. Diorites contain a moderate amount of the mafic minerals biotite, amphibole, and pyroxene and tend to be darker than granite or granodiorite. At even lower amounts of silica we find **gabbro,** a dark-gray rock abundant in mafic minerals, especially pyroxene, but containing no quartz and only moderate amounts of plagioclase feldspar. At very low silica levels, only about 45 percent, is **peridotite,** a dark greenish-gray rock made up primarily of pyroxene and olivine.

The plagioclase feldspars, which may contain more or less calcium and sodium, are richer in sodium near the granite end and richer in calcium near the gabbro end. This finding is consistent with the high-temperature origin of the mafic intrusives, since the calcium-rich plagioclases crystallize at higher temperatures than the sodium-rich plagioclases.

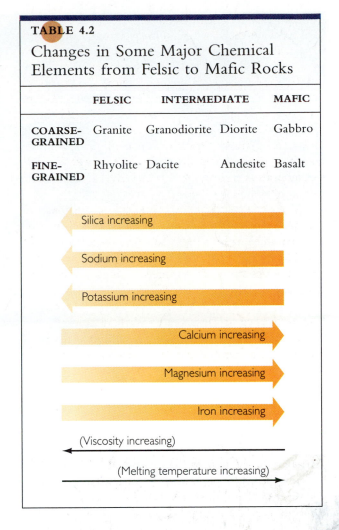

**TABLE 4.2**

## Changes in Some Major Chemical Elements from Felsic to Mafic Rocks

| | FELSIC | INTERMEDIATE | | MAFIC |
|---|---|---|---|---|
| COARSE-GRAINED | Granite | Granodiorite | Diorite | Gabbro |
| FINE-GRAINED | Rhyolite | Dacite | Andesite | Basalt |

Silica increasing →

Sodium increasing →

Potassium increasing →

Calcium increasing →

Magnesium increasing →

Iron increasing →

← (Viscosity increasing)

(Melting temperature increasing) →

**BASALT TO RHYOLITE: THE EXTRUSIVE SERIES** Extrusive or volcanic (either term can be used) counterparts to the intrusive rocks are classified in the same way as the intrusives. At the silicic end is **rhyolite,** the volcanic equivalent of granite. It shares granite's felsic composition and light coloration but is much finer grained. **Basalt,** at the mafic end of the series, is dark gray to black, the fine-grained equivalent of gabbro. Intermediate between the felsic and mafic ends of the series are **andesite,** the volcanic equivalent of diorite, and **dacite,** the extrusive that corresponds to granodiorite.

It is relatively easy to make out and identify the minerals of the coarse-grained intrusives. The extrusives, in contrast, have such small crystals that individual minerals are hard to discern, even with a magnifying lens. Some extrusives may also contain much glass, which is formless even under the microscope. These glassy rocks, as well as very fine-grained ones, are frequently classified by chemical analysis (Table 4.2). Some of them are classified and named according to special textural characteristics as well.

**SPECIAL TEXTURAL EXTRUSIVE TYPES** The special textural properties of volcanic rocks are related to the many ways in which they are extruded. Many kinds of lavas are solidified from flows of magma, ranging in appearance from smooth and ropy to sharp, spiky, and jagged, depending on their composition and the fluidity of the magma.

Other volcanic rocks are fragmented by more violent eruptions. These **pyroclastic rocks** (Figure 4.5), made up of broken pieces of lava and glass thrown high in the air, result from volcanic explosions such as the eruptions of Mount Pinatubo in the Philippine Islands in June 1991. These eruptions buried nearby towns and two U.S. airfields under a blanket of volcanic debris several meters thick. The finest fragments of this debris make up **volcanic ash,** extremely small particles that accumulate as layers of

**FIGURE 4.5** Pyroclastics: obsidian (*left*), volcanic ash (*center*), and pumice (*right*). *Chip Clark.*

loose and uncemented material. Larger pyroclastics are made up of several kinds of material: crystals that started to form before the explosion, fragments of previously solidified lava, and pieces of glass that cooled and then fractured during the eruption. All ...canic rocks that lithify from ash layers and other ...ieties of pyroclastic material are lumped under the ...rm **tuff.**

Volcanic glass, which can be a component of both lavas and pyroclastics, comes in a variety of forms when it makes up all of an extrusive. One common glassy rock type is **pumice,** a frothy mass with a great number of bubble holes, called *vesicles,* formed when gas escapes from the melt. Sharp, spiky glass fragments of vesicle walls are called *shards.* Another wholly glassy volcanic rock is **obsidian,** which, unlike pumice, has no vesicles and so is solid and dense. The glassy, sharp, broken edges of obsidian made it perfect for Native American arrowheads.

What can we make of a volcanic rock of mixed texture, one that seems to be mostly finely crystalline but has a number of large crystals "floating" in the fine matrix? This kind of rock, called a **porphyry** (Figure 4.6), is formed when a magma starts to cool slowly in the interior and then suddenly finds its way to the surface through a volcanic eruption. The large crystals, called *phenocrysts,* were the first to form

while the magma was still in the crust. Then, before other crystals could grow, the eruption quickly cooled the rest of the magma to a finely crystalline mass. We shall see in Chapter 5 how all these volcanic rocks are formed during volcanism.

Thus we classify igneous rocks by their textural, mineral, and chemical characteristics. To see how such characteristics arise, we must investigate igneous processes, such as the creation of magmas, that are not directly observable in nature but can be geologically inferred or simulated by experiments.

## THE ORIGINS OF MAGMAS

Where does magma come from? To put it another way, where does melting take place? That is the start of the whole process of intrusion and extrusion of igneous rocks. We know from the way the Earth transmits earthquake waves that the bulk of it is solid for thousands of kilometers down to the boundary of the core. But obviously there are some liquid regions where magmas originate. Volcanoes on land and under the sea—everywhere that molten rock erupts—give us information about where magmas are located.

Some **magma chambers**—pockets of magma in an otherwise solid interior—encompass a volume as large as several cubic kilometers. While we cannot say exactly what magma chambers look like in three dimensions, we can determine their depths and sizes from earthquake waves. We envision these chambers as large liquid-filled cavities in solid rock, which expand as more of the surrounding rock melts or liquid migrates in through cracks and other small openings between crystals. Magma chambers contract as they expel magma to the surface in eruptions. The magmas themselves form by the melting of rocks in either the crust or the upper mantle.

## Where Do Rocks Melt?

To answer that question we have to know where temperatures get high enough to melt various rocks. Temperatures recorded in mines and drill holes tell us that the temperature of the interior increases with depth. We can calculate from the rate at which temperature rises in various places around the world that in some tectonically and volcanically active areas the increase is very rapid. In these places at a depth of 40 km, not far below the base of the crust, the temperature is 1000°C, almost at the melting point of basalt. In tectonically stable regions the temperature rises much more slowly, reaching only 500°C at the same depth. So we know that some parts of the mantle and crust are hotter than others.

Rocks do not melt everywhere at the very high temperatures deep in the mantle because the very high pressures at these depths raise the melting temperatures. Only where temperatures are very high in relation to pressure does melting occur.

We know from many laboratory experiments the temperatures and pressures at which different kinds of rock melt. The temperatures give us some idea of where melting may take place. Mixtures of sedimentary rocks, for example, melt at temperatures several hundred degrees lower than the melting point of basalt. Thus both tectonics and rock composition determine if and where different kinds of rock melt. This information leads us to expect that basalt may start to melt near the base of the crust in tectonically active regions of the upper mantle and that sedimentary rocks melt at shallower depths than basalt.

The geometry of plate motions is the link we need to tie tectonic activity and rock compositions to melting (Figure 4.7). Two types of plate boundaries are associated with magma formation: mid-ocean ridges, where the divergence of two plates causes the seafloor to spread, and subduction zones, where one plate dives beneath another. At mid-ocean ridges, rising convection currents in the mantle cause the formation of one kind of magma, basalt. (Geologists label magmas according to the name of their corresponding intrusive or extrusive rock; it is more common to use the extrusive names for the mafic members basalt and andesite.) Melted rock forms in the hot upper mantle below mid-ocean ridges and rises

**FIGURE 4.6** Porphyry. *Peter Kresan.*

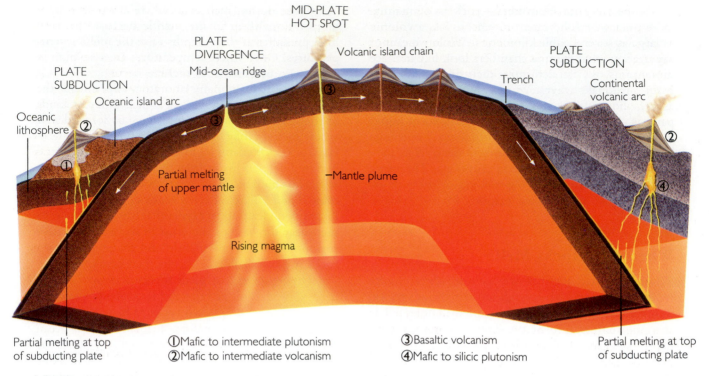

PLATE
SUBDUCTION

Oceanic island arc

PLATE
DIVERGENCE

Mid-ocean ridge

MID-PLATE
HOT SPOT

Volcanic island chain

PLATE
SUBDUCTION

Trench

Continental
volcanic arc

Oceanic
lithosphere

Partial melting
of upper mantle

—Mantle plume

Rising magma

Partial melting at top
of subducting plate

①Mafic to intermediate plutonism
②Mafic to intermediate volcanism

③Basaltic volcanism
④Mafic to silicic plutonism

Partial melting at top
of subducting plate

**FIGURE 4.7** Plate tectonics controls where rocks of the crust and upper mantle melt
and are intruded or extruded.

to collect in shallow magma chambers near the crest of the ridge. Tremendous quantities of basaltic magma flow intermittently from the rifts and fissures of mid-ocean ridges.

Other kinds of magmas underlie volcanic belts, such as the Andes Mountains of South America, and the volcanoes of island arcs, such as the Aleutian Islands of Alaska. Both kinds of volcanic belt are generated by the subduction of one plate under another. The magmas of subduction zones are formed from the melting of a mixture of seafloor sediments and basaltic crust as the subducting plate gets carried down to deep regions where temperatures are high enough. The compositions of these source materials are reflected in the types of igneous rock formed. The extrusive igneous rocks of these subduction zones are more silicic than the basalts of mid-ocean ridges and include much andesite and lesser amounts of more felsic volcanics. Deep in the crust, beneath the volcanoes, intrusives—from diorite to granite—are formed at the same time.

Basalts similar to those of mid-ocean ridges are found in thick accumulations over some parts of continents distant from plate boundaries. The Columbia and Snake rivers flow over a great area in the states of Washington, Oregon, and Idaho that is covered by

this kind of basalt, which flowed out as lavas millions of years ago. Large quantities of basalt are also erupted in isolated volcanic islands far from plate boundaries, such as the Hawaiian Islands. At these two kinds of places mantle materials are melted into basaltic magmas as slender, pencillike *plumes* of hot rock rise from deep in the mantle, perhaps as deep as the core-mantle boundary. Mantle plumes, most of them far from plate boundaries, are the "hot spots" of the Earth and are responsible for the outpouring of huge quantities of basalt.

We now know what kinds of rock solidify from a magma and where magmas form, but to get a complete picture we must examine more closely some strange behavior of melting rocks that geologists discovered when they looked at the details of the melting process.

## How Do Rocks Melt?

Early in the twentieth century, pioneering geologists turned to laboratory experimentation to determine the melting temperatures of rocks of different compositions. As they melted rocks they discovered that the whole rock did not melt all at once at one temperature. Instead, the rock gradually melted over a range of temperatures as the heat increased. They also noticed that the chemical composition of the melt changed as temperatures rose. When they examined any rock that remained unmelted at some temperature in the melting range, they found that some minerals had melted and were missing and others had remained solid. If at any temperature in the melting range they stopped increasing the temperature and kept it stabilized, melting ceased and a constant mixture of solid rock and melt was maintained.

These experiments showed that at any given temperature in the melting range of a rock, a **partial melt** is formed. A partial melt is the fraction of the rock that has melted. At the lower end of its melting range much of the hot rock is still solid, but there are appreciable amounts of liquid in small droplets in the tiny spaces between crystals throughout the mass. To visualize a partial melt you might think of how a chocolate chip cookie would look if you heated it to the point where the chocolate chips had melted while the main part of the cookie stayed solid.

The proportion of liquid to solid in a partial melt depends on the composition and melting temperatures of the original rocks and on the temperature at the depth in the crust or mantle where melting takes place. If the temperature were near the beginning of the melting range, the partial melt might be less than 1 percent of the volume of the original rock. Many partial melts of basaltic magma in the upper mantle, for example, are estimated to be only 1 to 2 percent melted. At the high end of the melting temperature range, most of the rock would be liquid, with only a small amount of unmelted crystals in it.

Geologists seized on this new knowledge of partial melts to help them determine how different kinds of magma form at different temperatures and regions in the interior. As you can imagine, the composition of a partial melt where only the minerals with the lowest melting points have melted may be significantly different from the composition of a completely melted rock. Therefore, if most basaltic magmas formed in the upper mantle, for instance, had about the same composition, then most basaltic magmas must come from about the same proportion of partial melt.

## The Formation of Magma Chambers

One more piece of the magma puzzle remained. How do the small droplets of a partial melt become the large magma chambers that feed intrusives and extrusives? A possible explanation presented itself when geologists thought about what would happen to partial melts under the great pressures of the interior. Liquid rock, like most substances, has a lower density than solid rock; that is, a given volume of melt would weigh less than the same volume of solid rock. Therefore, if the less dense melt were given a chance to move, it would move upward, just as oil, which is less dense than water, floats to the surface of a mixture of oil and water. Being liquid, the partial melt is able to move slowly upward through pores and along the boundaries between crystals of the overlying rocks. As the hot drops move upward, they coalesce with other drops and gradually form larger pools of magma, finally melting and pushing aside surrounding solid rock to form magma chambers. This is only a general description of the way the process might work. Exactly how magma chambers form from ascending drops of partial melt is still not known and is the subject of much research.

## The Effects of Water on Melting

The many experiments on melting temperatures and partial melting paid other dividends. Geologists knew from analyses of natural lavas that there was water in some magmas, so they added small amounts of water to the rocks they were melting. They discovered that the compositions of partial and complete melts vary not only with temperature but also with the amount of water present.

At the low pressures of the Earth's surface, if only a small amount of water is present, pure albite (the sodium feldspar), for example, will remain solid up to temperatures a little over 1000°C. At these high temperatures, hundreds of degrees above the boiling point of water, the water is present as a vapor (gas). If more water is present, the melting temperature of the albite is lower. In the same way, the melting temperatures of all the feldspars and other silicate minerals drop considerably in the presence of large amounts of water.

The water contained in sedimentary rocks is an important factor in lowering the melting temperatures of mixtures of sedimentary and other rocks. Sedimentary rocks contain much more water in their pore spaces than igneous or metamorphic rocks. In addition, the clay minerals that are so abundant in shales, the most abundant sedimentary rocks, contain much water chemically bound in their crystal structure. As sediments become very deeply buried—as they do in subduction zones, for example, as the subducting lithospheric plate moves down into the lower crust—the increase in heat releases the water from the clay minerals as well as some water remaining in pore spaces. Much of this water is released by chemical reactions as the temperature increases to about 150°C at moderate depths of about 5 km. At greater depths, from 10 to 20 km, more water is released and contributes to the melting of sedimentary and other rocks.

# MAGMATIC DIFFERENTIATION

By the early twentieth century, accumulating geological data on igneous rocks were posing new questions. What accounts for the variety of igneous rocks? Do they arise from magmas of different chemical compositions made by the melting of different kinds of rocks? Or do some processes produce variety from an originally uniform parent material? The answers started coming from experiments similar to those used for studying melting. Geologists mixed chemical elements in proportions that simulated those of natural igneous rocks and melted them in high-temperature furnaces. Then they allowed the melts to cool and solidify, carefully observing the temperatures at which crystals formed and their chemical compositions. The results of these experiments gave rise to the theory of **magmatic differentiation, a process by which a variety of minerals could be crystallized from a single homogeneous magma at different temperatures.** During the crystallization process the composition of the magma changed as it was depleted of the chemical elements withdrawn to make the crystallized minerals.

Crystallization experiments showed solidification behavior that was the reverse of partial melting. As the magma began to cool, the first minerals to crystallize turned out to be the same as the last minerals to melt in partial-melting experiments. As the crystallization started, the chemical elements that made up the minerals were withdrawn from the melt, and the magma began to change composition. As cooling continued, the next minerals to crystallize were those that melted in the same temperature range in melting experiments. Again the magma changed in composition as different chemical elements were withdrawn from the liquid. Finally, at the point of complete solidification, at the lowest temperature, the last minerals to crystallize were the same as those that melted first when the rock was heated to the beginning of its melting temperature range.

In the course of many experiments, two patterns of crystallization emerged. In one pattern, illustrated

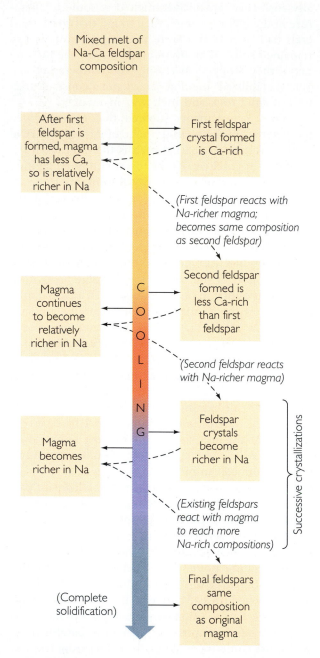

**FIGURE 4.8** At successive stages in the crystallization of a molten plagioclase feldspar, the liquid and crystals become richer in sodium, but the crystals formed are always richer in calcium than the liquid. The crystals already formed continue to react with the liquid so that at any given time all existing crystals, old and new, are of the same composition. When complete solidification is reached, all the crystals have reacted to attain the same composition as the original liquid.

by the plagioclase feldspars, the composition of the successively formed feldspars changed continuously and gradually as crystallization proceeded. In the other pattern, characteristic of the mafic minerals such as olivine and pyroxene, the composition of the crystals changed discontinuously during cooling, one mineral abruptly changing to another at a particular temperature. Because these crystallization patterns are so basic to an understanding of magmatic differentiation, the details are important.

## The Continuous Reaction Series

When melts of various plagioclase feldspar compositions were cooled, the first crystals to form were always richer in calcium than the melt. Their formation partially depleted the melt of calcium, so that the remaining melt became richer in sodium. As a result, when the melt continued to cool, the next crystals to form were more sodium-rich. At the same time, the calcium-rich crystals formed earlier reacted chemically with the now more sodium-rich melt. In this reaction calcium ions in the crystal were replaced by sodium ions from the melt so that the calcium-rich crystals formed earlier became richer in sodium. All crystals, both earlier and new, now had the same composition (Figure 4.8). As the process continued, both melt and crystals gradually became richer in sodium and poorer in calcium, until, when crystallization was complete, the final homogeneous solid mass of crystals had arrived at the same composition as the original melt. At all times the mineral being crystallized was a plagioclase feldspar.

The key to this process is the continuous reaction of crystals with the melt as both change by small amounts, so that at any point in the course of crystallization, all of the crystals have the same composition. Crystals and melt change continuously through a series of compositions, in earlier stages being richer in calcium, in later stages richer in sodium. With continued cooling this **continuous reaction series** proceeds until crystallization is complete.

## The Discontinuous Reaction Series

A somewhat different process is involved in the crystallization of mafic minerals, such as olivine, pyroxene, amphibole, and biotite mica. Experiments like those on plagioclase feldspars, in which melts made up of the components of mafic minerals were allowed to cool slowly so that crystals could react with the liquid, also showed a systematic order of crystallization. Olivine crystallized first, starting at 1800°C, and continued to crystallize until the melt cooled to 1557°C. Below that temperature, a completely different mineral, pyroxene, abruptly started to form, and all of the earlier olivine crystals were converted to pyroxene (Figure 4.9). At 1543°C cristobalite, a high-temperature silica mineral, began to form, and pyroxene crystallization continued until solidification was complete. In other experiments, similar transformations, first to amphibole, then to biotite mica, took place at successively lower temperatures. In this **discontinuous reaction series,** reactions take place between the melt and minerals of two definite compositions only at particular temperatures. This is a different process from the gradual evolution of plagi-

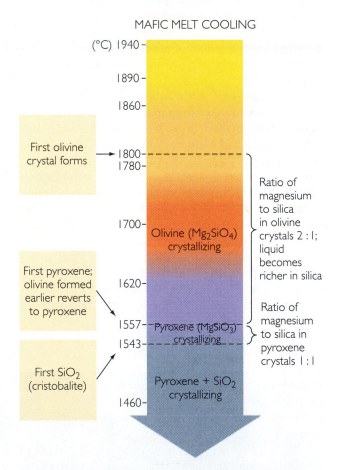

FIGURE 4.9 The sequence of events in the crystallization of a mafic melt, a cooling liquid of magnesium and silica in which silica is about 50 percent by weight. This sequence forms one part of the discontinuous reaction series by which mafic minerals are successively crystallized from a melt. The exact temperature at which olivine first starts to crystallize from a cooling melt depends on the composition of the melt.

oclase feldspars and parent melt over a continuous range of compositions and temperatures.

The crystal structures of the minerals of the two reaction series are part of the differences in crystallization patterns (see Table 4.1). Throughout the changes in the continuous reaction series, the basic feldspar crystal structure stays the same; only the proportions of calcium and sodium change. The crystal structures of the discontinuous reaction series change as one mineral gives way to another. At the highest temperatures, olivines are made of isolated silica tetrahedra, the basic building blocks of silicate minerals (see Chapter 2). In the next stage, the pyroxenes are single chains of tetrahedra. Then come the amphiboles, double chains of tetrahedra, followed by the micas, sheets of tetrahedra. At the end stages, quartz and the feldspars are three-dimensional frameworks of silica tetrahedra. Thus as temperatures fall, the structures change discontinuously to different arrangements of silica tetrahedra.

In the cooling of a natural magma, which normally contains the chemical elements of both plagioclase feldspars and mafic minerals, both patterns of crystallization go on simultaneously. As the temperature of such a magma drops below 1550°C, for example, pyroxene and a pure calcium plagioclase feldspar crystallize together.

The two reaction series explain the composition of many igneous rocks but fall short of accounting for many others. For example, the crystallization products of a particular natural magma, if all crystals reacted completely with the liquid at all stages of crystallization, would include only the final products, a single plagioclase feldspar corresponding to the composition of the original melt and a pyroxene. No traces of the original calcium-rich feldspar and

the original olivine that were the first crystallization products would be left. Yet geologists did find many igneous rocks with calcium-rich plagioclases and with olivine. Some part of the theory was missing.

## Fractional Crystallization

What was missing from the theory of magmatic differentiation was a way to account for the preservation of minerals formed earlier as the composition of the melt changed. A mechanism for doing so was first proposed about 75 years ago by N. L. Bowen, a Canadian geologist who had become interested in the chemical basis for igneous rock formation even as an undergraduate. Because Bowen's proposal is one of the most important parts of magmatic differentiation theory, we will explain it in some detail by following his researches.

Bowen had done much work on the continuous and discontinuous crystallization series. He focused on the course of crystallization in situations in which the plagioclase feldspars—or mafic minerals—did *not* constantly react with the liquid to change composition. Such might be the case, for example, if a magma cooled more rapidly than usual. In such a magma, crystals had time to grow, but only the outer surfaces of the crystals formed earlier had time enough to react with the changing liquid. As a result only the outer layer of each crystal would change composition, each successive layer being covered by a layer richer in sodium as crystallization proceeded.

The end product of this process of limited reaction is a mass of **zoned crystals,** whose compositions change gradually from calcium-rich interiors to sodium-rich exteriors (Figure 4.10). There is more to

**FIGURE 4.10** A zoned crystal. *Chip Clark.*

limited reaction than this, however. If the calcium-rich centers of the growing crystals cannot react with the liquid, the liquid remains richer in sodium than it would be in a slow continuous reaction, because the calcium from the crystal interiors cannot replace the sodium in the melt.

Bowen proposed a new theory of magmatic differentiation based on both experimentation and field observation. Though the mechanisms he suggested are no longer accepted for the differentiation of most igneous rocks, his ideas served as a foundation for

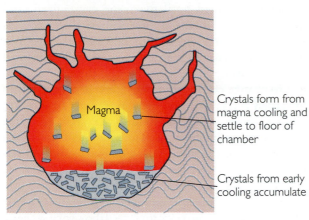

(a) Early crystallization

Crystals form from magma cooling and settle to floor of chamber

Crystals from early cooling accumulate

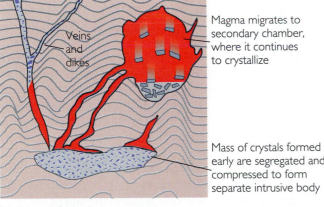

(b) Later deformation squeezes remaining liquid from crystal mush

Magma migrates to secondary chamber, where it continues to crystallize

Mass of crystals formed early are segregated and compressed to form separate intrusive body

**FIGURE 4.11** Two stages in the evolution of a magma differentiated by fractional crystallization. In the first stage (a), crystals formed early settle to the floor of the magma chamber. As cooling proceeds, structural deformation may squeeze the remaining liquid from the chamber and separate those crystals as a distinct intrusive body (b), while the liquid migrates elsewhere to form veins, dikes, and other magma chambers, where it continues to crystallize.

most later work and still teach us much today. Bowen reasoned what the course of crystallization would be if the first-formed crystals were to be removed from reaction. He suggested that early crystals formed in a magma chamber might settle to the bottom and thus be removed from further reaction with the remaining liquid (Figure 4.11a).

Another possibility is that structural deformation occurring midway through the crystallization process might squeeze the liquid away from the crystals (Figure 4.11b). In either scenario, crystals formed early would be segregated from the remaining melt, which would then behave as though it had just begun to crystallize. In the continuous reaction series, for example, the melt, already enriched in sodium at that point, would start to crystallize a feldspar much richer in sodium than any that would have crystallized from an unsegregated magma. The end result of continued crystallization would be a mass of feldspars that would be much richer in sodium than the rock that the original melt would have produced. Meanwhile, the first crystals that formed, the ones removed from reaction with the melt, would form a mass of feldspar much richer in calcium than the original melt. However it happened, **fractional crystallization** of the melt—that is, separation and removal of successive fractions of crystals formed from a cooling magma—could account for the preservation of early-formed calcium-rich feldspars and the crystallization of sodium-rich plagioclases from an originally calcium-rich magma.

Fractional crystallization could work with the discontinuous mafic mineral series as well. Just as the first plagioclase feldspars to crystallize may be removed, the first-formed crystals of olivine in a discontinuous reaction series may settle out and so be removed from further reaction. These mafic minerals would be found with their corresponding plagioclase feldspars. While the first-formed olivine was removed, the magma went on to crystallize pyroxene. Thus both continuous and discontinuous crystallization paths might produce a range of products similar to those found in natural igneous rocks. But before the theory of fractional crystallization became accepted, it had to be tested.

## The Palisades Intrusion

Any good theory must be checked against the facts. Where could geologists find an outcrop in which they could see the products of fractional crystallization? What came to mind was a somewhat strange outcrop of basaltic igneous rocks that thousands of geology students had visited on field trips. It contained abundant olivine near the bottom, pyroxene and plagioclase in the middle, and mostly plagioclase near the top.

Facing the city of New York on the west bank of the Hudson River is the Palisades, a massive cliff about 80 km long and in places more than 300 m from top to bottom. The Palisades is an igneous formation that was intruded as a melt of basaltic composition into almost horizontal sedimentary rocks. The variation in mineral composition from top to bottom of this formation made it a perfect test case for the theory of fractional crystallization and showed how laboratory experiments could help explain field observations. From experiments on the melting of rocks with about the same proportions of the various minerals found in the Palisades intrusion, we know that the melt had to have been at about 1200°C. The

parts of the magma within a few meters of the relatively cold upper and lower contacts of the surrounding sedimentary rocks cooled quickly to become a fine-grained basalt, preserving the chemical composition of the original melt. But the hot interior of the intrusion cooled more slowly, so that slightly larger crystals could form.

Experiments on fractional crystallization lead us to think that the first mineral to crystallize from the slowly cooling interior was olivine, which is heavy and sank through the melt to the bottom of the intrusion. There it can be found today in a coarse-grained, olivine-rich layer just above the chilled, fine-grained basaltic layer along the bottom contact (Figure 4.12). Cooling continued until pyroxene crystallized, and then almost immediately calcium-rich plagioclase feldspar as well. These minerals, too, settled out through the magma to accumulate in the lower third of the intrusion. After these stages, the melt continued to change composition until, as successive layers of settled crystals neared the top, mostly plagioclase feldspar, now richer in sodium, crystallized. This explains the abundance of plagioclase feldspar in the upper parts of the intrusion. The explanation of the layering of the Palisades intrusive as the result of fractional crystallization was an early success of the first version of the theory of magmatic differentiation. It firmly tied field observations to laboratory experiments and was solidly based on chemical knowledge.

## Bowen's Theory of Magmatic Differentiation

How could fractional crystallization and magmatic differentiation explain the differences among igneous rocks? Two seemingly contradictory facts needed to be explained: (1) the widespread distribution and abundance of granite and (2) the equally widespread distribution and abundance of basalt. Granite, an intrusive rock, is at the silicic end of the igneous rocks, containing abundant sodium-rich plagioclases and other minerals with low melting temperatures. Basalt, an extrusive rock, is at the mafic end of the spectrum, containing calcium-rich plagioclases and other minerals with high melting temperatures. Studies of the lavas of volcanoes showed that basaltic magmas are common, far more common than the rhyolitic magmas that correspond in composition to granites. How could the abundant granites have been derived from basaltic magmas?

Bowen's idea was that an originally basaltic magma would gradually cool and differentiate to a more silicic, lower temperature melt by fractional

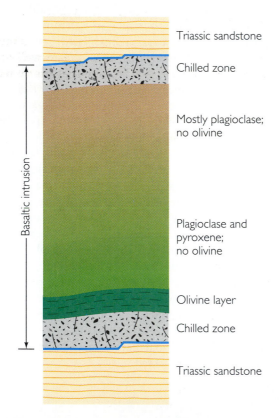

Triassic sandstone

Chilled zone

Mostly plagioclase; no olivine

Basaltic intrusion

Plagioclase and pyroxene; no olivine

Olivine layer

Chilled zone

Triassic sandstone

**FIGURE 4.12** The Palisades is a basaltic formation that intruded into sedimentary rocks as a melt some 200 million years ago. The interpretation of the vertical variation in texture and mineral composition of this formation is a classic example of the application of ideas of fractional crystallization to rocks in the field.

crystallization. Early stages of differentiation of a basaltic magma by fractional crystallization would produce andesitic magma, which might erupt to form andesite lavas or solidify by slow crystallization to diorite intrusives. Intermediate stages would make magmas of granodiorite composition. If this process were carried far enough, its late stages would form rhyolite lavas and granite intrusions.

## Bowen's Reaction Series

In 1928 Bowen capped more than 10 years of active experimentation by proposing a simplified general scheme for magmatic differentiation that combined the continuous and discontinuous fractional crystallizations of the major minerals of igneous rocks. The **Bowen reaction series,** as it is called, starts with the cooling of a high-temperature basaltic magma that gradually differentiates by fractional crystallization along two simultaneous paths (Figure 4.13). One path is the continuous path of the plagioclase feldspars, starting at high temperatures with calcium-rich feldspar and proceeding to the lower temperature sodium-rich feldspar. The other path is the discontinuous path of the mafic minerals, starting at the high-temperature end with olivine, then progress-

ing to pyroxene, amphibole, and biotite mica as the magma cools.

The paths of the two series converge at a final, low-temperature (about 600°C) magma crystallizing the minerals of granite: albite (sodium-rich plagioclase feldspar) and orthoclase (potassium feldspar), muscovite mica, and quartz.

To sum up, we have seen a train of observations and experiments that led to the Bowen reaction series as an explanation for the great variety of igneous rocks. First, the melting and crystallization behavior revealed the continuous and discontinuous reaction series. Then, the idea of fractional crystallization explained how magmas could evolve through a series of compositions by the separation of the crystals formed earlier from the magma. Finally, the continuous and discontinuous series were put together as an explanation for the evolution of granites and other intrusive and extrusive rocks from an originally basaltic magma.

## Modern Theories since Bowen

At first Bowen's theory of magmatic differentiation seemed to be a great success. It explained well how different types of igneous rocks could form by frac-

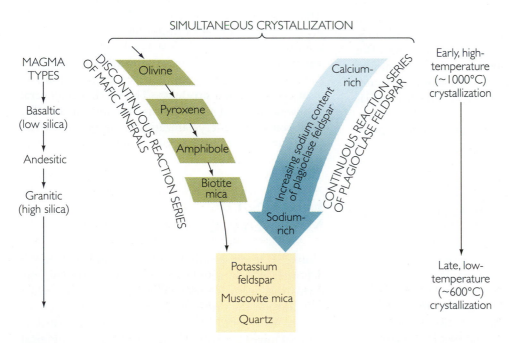

**FIGURE 4.13** Bowen's reaction series, showing how the sequence of fractional crystallization of a melt could lead to the formation of differentiated magmas.

tional crystallization, and it provided an understanding of the kinds of rocks seen in the field, such as the Palisades mafic intrusion. It also explained how rhyolite, the extrusive equivalent of granite, formed toward the end of a series of eruptions that started with basaltic lavas. Zoned plagioclase crystals, common in granites and granodiorites, were explained by fractional crystallization. Temperatures of basaltic lavas, measured in such places as the Volcano Observatory on Kilauea volcano, on the island of Hawaii, corresponded to melting temperatures measured in laboratory experiments and were much higher than those of more silicic melts that solidified to rhyolitic lavas.

More than half a century has passed since Bowen proposed his theory of magmatic differentiation. As often happens when a scientific theory quickly comes to dominate an area of science, further work showed the need to modify and add to Bowen's original version. Some researchers showed that such great lengths of time would be needed for small crystals of olivine to settle through a dense, viscous magma that they might never reach the bottom. Many layered intrusives, similar to but much larger than the Palisades, did not show the simple progression of layers predicted by Bowen's theory.

A major problem was the great volume of granite. Because much liquid volume is lost by crystallization during successive stages of differentiation, an initial volume of basaltic magma on the order of 10 times the size of a granitic intrusion would be required to produce the amount of granite found. That abundance would imply the crystallization of huge quantities of basalt underlying granite intrusions. But geologists could not find anything like that amount of basalt. Where great volumes of basalts are found, at mid-ocean ridges, there is no wholesale differentiation to granite.

Most in question is Bowen's original idea that all granitic rocks evolve from the differentiation of a single type of magma, a basaltic melt. Melting of varied source rocks is responsible for much variation in magma composition. Rocks in the upper mantle may partially melt to produce basaltic magma, but a mixture of sedimentary rocks and basaltic oceanic rocks might melt to form andesitic magma. A melt of sedimentary, igneous, and metamorphic continental rocks might produce granitic magma.

Geologists now recognize that differentiation does exist but that its mechanisms are more complex than Bowen recognized. Partial melting is of great importance in producing magmas of varying composition. Magmas do not cool uniformly but may en-

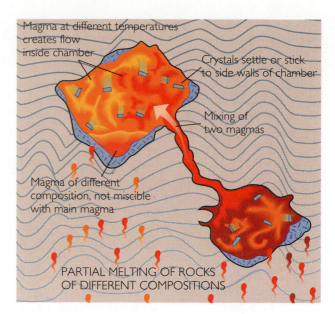

**FIGURE 4.14** Modern ideas of magmatic differentiation. Melting is usually partial. Some magmas derived from rocks of varying compositions may mix, while other magmas are immiscible. Crystals may be transported to various parts of the magma chamber by currents in the liquid.

compass a wide range of temperatures within a magma chamber. These differences in temperature may cause the chemical composition of the magma to vary from one region to another (Figure 4.14). A few melt compositions are immiscible—that is, they do not mix with each other, just as oil and water remain separate liquids when they are mixed. Such magmas can coexist in one magma chamber, each forming its own crystallization products. The mixing of several of the majority of magmas that are indeed miscible may give rise to a crystallization path different from that followed by any one of the magmas alone.

# FORMS OF MAGMATIC INTRUSIONS

Unlike volcanic eruptions of lavas and pyroclastic deposits, which geologists can study directly, the forms that intrusive igneous rocks take when magmas intrude the crust are impossible to observe directly. We can only deduce the shapes and distributions of intrusives from evidence gained in geological fieldwork done millions of years after the rocks were formed, long after the magma cooled and the rocks were uplifted and exposed to erosion.

To be sure, we do have indirect evidence of current magmatic activity. Earthquake waves, for example, show us the general outlines of magma chambers that underlie some active volcanoes, but they cannot tell us the shape or size of intrusives supplied from those magma chambers. In some nonvolcanic but tectonically active regions, such as an area near the Salton Sea in southern California, high temperatures in deep drill holes reveal a crust much hotter than normal, which may be evidence of an intrusion at depth.

But in the end, it is geologists in the field who have mapped and compared a wide variety of outcrops of intrusive igneous rock and have reconstructed how these intrusives are emplaced. Their studies have resulted in the description and classification of the many irregular and variable forms of intrusive bodies.

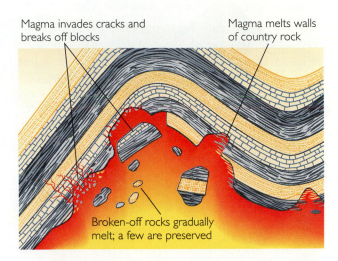

Magma invades cracks and breaks off blocks

Magma melts walls of country rock

Broken-off rocks gradually melt; a few are preserved

**FIGURE 4.15** Mechanisms by which a pluton intrudes surrounding rock and makes room for itself.

## Plutons

Large igneous bodies formed at depth in the crust are called **plutons.** They range in size from a cubic kilometer to hundreds of cubic kilometers. When uplift and erosion uncover them, or when mines or drill holes cut into them, geologists have found them to be highly variable not only in size but also in shape and in their relation to the **country rock**—that is, the invaded rocks surrounding them. Plutons exclude dikes and sills, which are generally smaller bodies.

Magma rising through the crust makes space for itself in several ways: by cracking the overlying rock, by breaking off large blocks of rock, and by melting its way along. There are few holes or openings in rocks at the depths where most magmas intrude—deeper than 8 to 10 km—because the pressure of the overlying rock is so great that it tends to close them. But even that high pressure is overcome by the presence of the upwelling magma. As the magma lifts that great weight, it fractures the rock, then penetrates the cracks, wedges them open, and so flows into the rock (Figure 4.15).

Magma can push its way upward by breaking off blocks of the invaded crust. These blocks sink into the magma, melt, and blend into the liquid, in some places changing the magma composition. Intrusion can also split rocks apart or bow them up. Melting of some of the surrounding country rock also makes room for the magma.

Although most plutons show sharp contacts with country rock and other evidence of intrusion of a liquid magma into solid rock, some plutons grade into country rock and show structures that vaguely resemble those of sedimentary rocks. Such features suggest that preexisting sedimentary rocks underwent **granitization**—that is, they were converted to granite by partial melting and invasions by hot solutions and gases percolating up from great depths with the magma.

**Batholiths,** the largest plutons, are huge intrusives that by definition cover at least 100 km². Similar but smaller plutons are called **stocks.** Both batholiths and stocks, are **discordant intrusives;** that is, they cut across the layers of the country rock they intrude. Geological field evidence is accumulating to show that batholiths are horizontal sheetlike or lobate thick bodies that extend from a funnel-shaped central region. Their bottoms may extend 10 to 15 km deep, and a few are estimated to go even deeper. Batholiths are generally coarse-grained—a consequence of the slow cooling of deeply buried intrusives. Batholiths are found in the cores of tectonically deformed mountain belts.

## Sills and Dikes

Smaller tabular intrusions, although similar in many ways to the larger plutons, are distinguished by their shape and by their relationship to the layering of in-

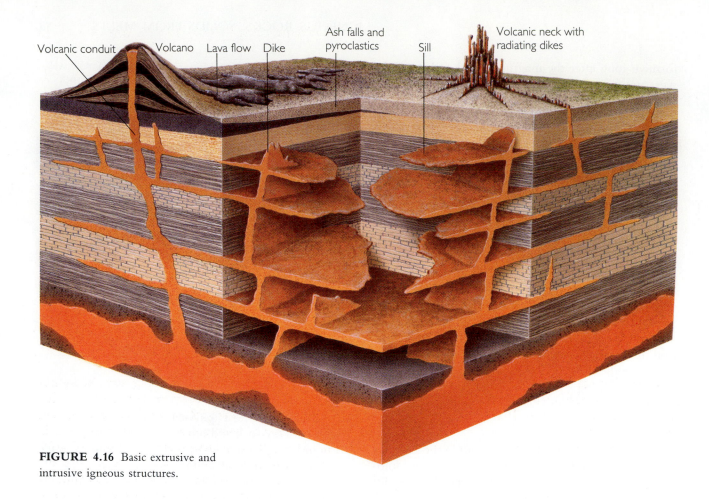

Volcanic conduit    Volcano    Lava flow    Dike    Ash falls and pyroclastics    Sill    Volcanic neck with radiating dikes

**FIGURE 4.16** Basic extrusive and intrusive igneous structures.

truded rocks. A **sill,** which is a tabular body that has been formed by injection of magma between beds of layered, or bedded, rock, is a **concordant intrusive;** that is, its boundaries are parallel to the layering, whether or not that layering is horizontal (Figure 4.16). Sills range in thickness from only a centimeter to hundreds of meters and can extend over considerable areas. Figure 4.17 shows a large sill at Big Bend National Park, Texas. The 300-m-thick Palisades intrusion is another sill.

Sills can be distinguished from layers of lava flows and pyroclastics, which they may superficially resemble, by the fact that sills lack the ropy, blocky, and vesicle-filled structures so common in volcanic rocks. Sills also are coarser-grained than volcanics because they have cooled more slowly. The rocks above and below sills show the effects of heating: they have been bleached or mineralogically altered by contact metamorphism. And many lava flows, unlike sills, overlie weathered older flows or soils formed between successive flows.

**Dikes,** like sills, are tabular igneous bodies. Unlike sills, they are discordant. Dikes can force open fractures formed earlier, but more often their channels follow new cracks opened by the pressure of magmatic injection. Some individual dikes can be followed across country for tens of kilometers. Their widths vary from centimeters to many meters. Some dikes contain fragments of country rock completely surrounded by the intrusive material; these fragments were once floating in the intruding magma (Figure 4.18). Such fragments provide good evidence of disruption of the surrounding rock during the intrusion process. Dikes rarely occur alone; more typically large numbers, or swarms, of hundreds or thousands of dikes are found in a region that has been deformed by a large intrusion.

Many dikes and sills are coarsely crystalline, with an appearance typical of intrusive rocks (Figure 4.19). Many others are finer-grained and look much more like extrusives. Remembering that this textural difference is caused by different cooling rates, we can

**FIGURE 4.17** A sill intruded into sedimentary rocks at Big Bend National Park, Texas. *Tom Bean.*

understand that the fine-grained ones were emplaced nearer the Earth's surface, where rocks are cold in comparison with intrusives, whereas the coarse-grained ones were intruded at depths of many kilometers, where the rocks are warmer, much closer in temperature to the intrusives.

## Hydrothermal Veins

Numerous **veins,** irregular pencil- or sheet-shaped intrusions, branch off from the tops and sides of many intrusive bodies. Most veins are very different in mineralogy from the surrounding rock. Veins may be a few millimeters to several meters across and tend to be tens of meters to kilometers long or wide. The famous Mother Lode of the 1849 gold rush in California is a vein of quartz bearing crystals of gold.

Veins may be filled not only with metal ores but also with minerals that contain much chemically bound water and are known from experimental evidence to crystallize from hot-water solutions. Though the temperatures at which crystallization occurs are high, they are much lower than the temperatures of magmas; they typically range from 250

**FIGURE 4.18** A dike with floating fragments of country rock. *Martin Miller.*

**FIGURE 4.19** Granite pegmatite dike. The center of the dike displays coarse crystallinity associated with slow cooling. The finer crystals along the boundaries of the dike cooled more rapidly. *Martin Miller.*

# BOX 4.1 JAPAN: A GROWING ISLAND ARC

The terrain of the Japanese Islands is a prime example of the complex of intrusives and extrusives that evolves over many millions of years at a subduction zone. Everywhere in this small country are all kinds of extrusive igneous rocks of various ages intercalated with mafic and intermediate intrusives, metamorphosed volcanic rocks, and sedimentary rocks derived from erosion of the igneous rocks. The erosion of these kinds of rocks has contributed to the distinctive landscapes portrayed in so many classical and modern Japanese paintings.

Japan lies at the intersection of three converging oceanic plates, the giant Pacific and Eurasian plates and the small Philippine Plate. Just east of the Japanese Islands are deep trenches marking the lines along which the Pacific and Philippine plates subduct beneath the Eurasian Plate. As these plates slip downward, they provoke the earthquakes that are so prevalent throughout the islands. Japan is dotted with active and dormant volcanoes, the most famous of which, Fujiyama, is a traditional object of reverence.

Japanese geologists have worked out the history of this subduction complex by mapping the intrusive and extrusive rocks of various ages to show that originally the archipelago was a narrow arc of small islands more like the Marianas of the western Pacific. As subduction continued, the volcanism, accompanied by intrusions at depth, built up a widening belt of land, while some of the igneous rocks emplaced earlier underwent structural deformation. In the course of this igneous and tectonic activity, mountains were elevated, one chain of them spectacular enough to be called the Japanese Alps.

Thus the consequence of continued subduction has been the growth of sizable islands that through magmatic differentiation, structural deformation, and sedimentation have come to resemble tiny continents.

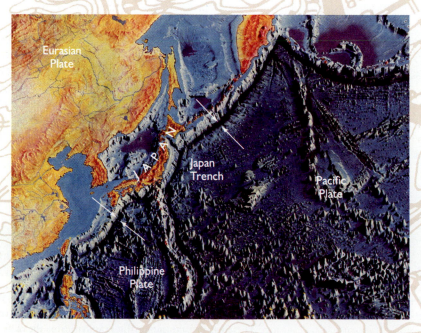

Seafloor topography of the Western Pacific shows the Japan Trench, part of the system of subduction zones that bounds the Pacific, Philippine, and Eurasian plates. (*World Ocean Floor*, based on bathymetric studies by Bruce C. Heezen and Marie Tharp. Painting by Heinrich C. Berann. Copyright © Marie Tharp, 1977.)

to 350°C. These **hydrothermal** (from Greek *hydōr,* water, and *thermē,* heat) **veins** show that abundant water was present as the veins formed. Some of the water may have come from the magma itself, but some may be from underground water in the cracks and pore spaces of the intruded rocks. Some of these groundwaters in the intruded rocks originated by the slow downward movement of rainwaters that seeped into the soil and surface rocks. (The geology of hydrothermal veins and the valuable ores they contain are discussed in detail in Chapter 23.)

# PLUTONISM AND PLATE TECTONICS

Since the advent of the theory of plate tectonics in the 1960s, geologists have been trying to fit the facts and theories of igneous rock formation into its framework (see Figure 4.7). We noted that batholiths, for example, are found in the cores of many mountain ranges. These mountain ranges were formed by the convergence of two plates. This observation implies a connection between plutonism and the mountain-making process and between both of them and the major dynamics responsible for Earth movements, plate tectonics.

Subduction zones, where one plate dives below another, are major sites of rock melting. The top of a subducting lithospheric plate includes oceanic crust, which is largely basalt originally formed at a mid-ocean ridge. The lithosphere also carries water and still-soft oceanic sediment laid down on it during the plate's travels from mid-ocean ridge to subduction zone. As the oceanic plate moves down, the increase in temperature and pressure converts the sediments first to sedimentary rocks and then at greater depths to metamorphic rocks. These materials, because of the relatively large amounts of water they contain, have lower melting temperatures than dry crustal or mantle rock. As the lithosphere moves deeper, it reaches the melting temperatures of the sedimentary or metamorphic rocks. Continuing downward, it finally reaches temperatures at which the top parts of the basalt melt. Subduction thus creates magma, or perhaps magmas of several kinds.

As the magmas work their way upward from the top of the melting subducting slab, they may melt portions of the overriding plate and change their composition. At the same time, the magmas may differentiate by fractional crystallization. The result is a range of igneous rocks, both intrusives and extrusives. Volcanoes over the deeper parts of the subduction zone, where melting is going on, extrude rocks from basalts to andesites and more silicic extrusives. These volcanoes and their extrusives form the islands of oceanic volcanic arcs, such as the Aleutian Islands of Alaska. Where subduction takes place beneath a continent, the many volcanoes and their extrusives coalesce to form a mountainous arc on land. This is the case in northern California, Oregon, and Washington, where subduction of an offshore oceanic plate has generated the Cascade Range with its active volcanoes, such as Mount St. Helens.

As mountains are forming above, intrusives are crystallizing deep below, forming bodies of mafic to felsic igneous rocks, depending on the composition of the magma and on the degree of its differentiation. Working backward from the compositions of the igneous rocks, we can estimate the composition of the magma and the depth of the descending slab. Thus, by mapping patterns of igneous rocks, geologists can reconstruct and make sense of the structure of the subduction zone (see Box 4.1).

In all of these ways, igneous rocks reflect the major forces shaping the Earth. Each plate-tectonic setting produces its own pattern of igneous rocks: the lava flows and pyroclastics extruded from volcanoes; the batholiths, dikes, and sills intruded at depth; and the wide variety of rocks that come from magmas of distinctive compositions following their own routes of differentiation.

# SUMMARY

**How are igneous rocks classified?** All igneous rocks can be divided into two broad textural classes: the coarsely crystalline rocks, which are intrusive and therefore cooled slowly; and the finely crystalline ones, which are extrusive and cooled rapidly. Within each of these broad categories the rocks are classified on the chemical basis of their silica content or by the mineralogical equivalent, the proportion of lighter, felsic minerals to darker, mafic minerals. Felsic rocks, such as granite and its corresponding extrusive, rhyolite, are rich in silica and dominated by quartz, potassium feldspar, and sodium-rich plagioclase feldspar. Mafic rocks, such as gabbro and its corresponding extrusive, basalt, are poor in silica and consist primarily of pyroxene, olivine, and calcium-rich feldspar. Intermediate rocks are granodiorite and diorite and their corresponding extrusives, dacite and andesite.

**How and where do magmas form?** Magmas form at places in the lower crust and mantle where temperatures and pressures are high enough for at least partial melting. Basalt can partially melt in the upper mantle where convection currents bring hot rock upward at mid-ocean ridges. Mixtures of basalt and other igneous rocks with sedimentary rocks, which contain significant quantities of water, have lower melting points than dry igneous rocks and therefore melt in subduction zones, where the subducting plate heats up as it moves down.

**How does magmatic differentiation account for the variety of igneous rocks?** Minerals crystallize from magmas along two paths: (1) a continuous reaction series of the plagioclase feldspars and (2) a discontinuous reaction series of the mafic minerals. In these series, crystals continuously react with the melt through successive stages of crystallization and magma composition until they solidify completely, at which point the final product has the same composition as the original magma. If there is fractional crystallization, so that the crystals do not react with the melt, either because they grow very rapidly or because they settle from the liquid, the final product may be more silicic than the earlier, more mafic crystals.

Bowen's continuous and discontinuous reaction series explain how fractional crystallization can produce mafic igneous rocks from earlier stages of crystallization and differentiation and felsic rocks from later stages. Different kinds of igneous rocks may also be produced by variations in the compositions of magmas caused by the melting of different mixtures of sedimentary and other rocks and by mixing of magmas.

**What are the forms of intrusive and extrusive igneous rocks?** Igneous bodies of large size are plutons. The largest plutons are batholiths, which are thick tabular masses with a central funnel. Stocks are smaller plutons. Less massive than plutons are sills, which are concordant, following the layering of the intruded rock, and dikes, which are discordant, cutting across the layering. Hydrothermal veins are formed where water is abundant, either in the magma or in surrounding country rock.

**How are igneous rocks related to plate tectonics?** The two major sites of magmatic activity are mid-ocean ridges, where basalt wells up from the upper mantle, and subduction zones, where a series of differentiated magmas produces both extrusives and intrusives in island or continental volcanic arcs as the subducting oceanic lithosphere moves down into the deep crust and upper mantle. Large volumes of basalt are produced at oceanic islands and on landmasses that overlie mantle plumes.

# KEY TERMS AND CONCEPTS

felsic rocks (p. 70)
mafic rocks (p. 70)
ultramafic rocks (p. 70)
granite (p. 70)
granodiorite (p. 71)
diorite (p. 71)
gabbro (p. 71)
peridotite (p. 71)
rhyolite (p. 71)
basalt (p. 71)
andesite (p. 71)
dacite (p. 71)
pyroclastic rocks (p. 71)

volcanic ash (p. 71)
tuff (p. 72)
pumice (p. 72)
obsidian (p. 72)
porphyry (p. 72)
magma chamber (p. 73)
partial melt (p. 75)
magmatic differentiation (p. 76)
continuous reaction series
    (p. 77)
discontinuous reaction series
    (p. 77)
zoned crystal (p. 78)

fractional crystallization (p. 79)
Bowen reaction series (p. 81)
pluton (p. 83)
country rock (p. 83)
granitization (p. 83)
batholith (p. 83)
stock (p. 83)
discordant intrusive (p. 83)
sill (p. 84)
concordant intrusive (p. 84)
dike (p. 84)
vein (p. 85)
hydrothermal vein (p. 87)

# EXERCISES

1. Why are intrusive igneous rocks coarsely crystalline and extrusive rocks finely crystalline?

2. What kinds of minerals would you find in a mafic igneous rock?

3. What kinds of igneous rock contain quartz?

4. Name two intrusive igneous rocks that are more silicic than gabbro.

5. In which plate-tectonic settings would you expect magmas to form?

6. What is the difference between the continuous and discontinuous reaction series?

7. How does fractional crystallization lead to magmatic differentiation?

8. Where in the crust, mantle, or core would you expect to find a partial melt of basaltic composition?

9. How do you distinguish between a sill and a dike?

10. How are dikes different from lavas?

# THOUGHT QUESTIONS

1. How would you classify a coarse-grained igneous rock that contained about 12 percent quartz, 10 percent potassium feldspar, 35 percent plagioclase feldspar, and small amounts of biotite and amphibole?

2. What kind of rock would contain some plagioclase feldspar crystals about 5 mm long "floating" in a dark-gray matrix of crystals of less than 1 mm?

3. What differences in texture might you expect to find between two sills, one intruded at a depth of about 12 km and the other at a depth of 0.5 km?

4. If you were to drill a hole through the crust of a volcanic island arc, what igneous rocks would you expect to encounter from the surface to the base of the crust?

5. If a magma had a certain ratio of calcium to sodium, would the same ratio characterize the plagioclase feldspars formed at complete solidification if there were or were not any fractional crystallization during the solidification process?

6. Why would you not expect to find a magma crystallizing olivine at the same stage of crystallization as a sodium-rich plagioclase feldspar?

7. How does a zoned crystal indicate fractional crystallization?

8. Why are plutons more likely than dikes to show the effects of magmatic crystallization?

9. What might be the origin of a rock composed almost entirely of olivine?

10. Are porphyries more likely to occur in rapidly cooled extrusives or very slowly cooled intrusives?

# SUGGESTED READINGS

Barker, D. S. 1983. *Igneous Rocks*. Englewood Cliffs, N.J.: Prentice-Hall.

Best, M. G. 1982. *Igneous and Metamorphic Petrology*. San Francisco: W. H. Freeman.

Cox, K. G., J. D. Bell, and R. J. Pankhurst. 1979. *The Interpretation of Igneous Rocks*. London: Allen and Unwin.

Ehlers, Ernest G., and Harvey Blatt. 1982. *Petrology. Igneous, Sedimentary, and Metamorphic*. San Francisco: W. H. Freeman.

Ernst, W. Gary. 1969. *Earth Materials*. Englewood Cliffs, N.J.: Prentice-Hall.

Hyndman, D. W. 1985. *Petrology of Igneous and Metamorphic Rocks*, 2d ed. New York: McGraw-Hill.

# VOLCANISM

About 80 percent of Earth's surface, the seafloor as well as the land, began as molten rock that rose from deep inside the Earth. When it emerged onto the surface, it cooled and hardened into volcanic rock. This is the process called volcanism, which occurs principally near plate boundaries. Volcanic rock preserves information about its origins and in a sense serves as a window through which we can dimly perceive the Earth's interior. Volcanoes can be beneficial as sources of fertile soil, chemicals, and minerals. They can be troublesome—the dust and gas thrown high into the atmosphere in a severe eruption can partially block out solar radiation and lower Earth's mean annual temperature by several degrees for a few years. And they can be deadly, destroying cities and even bringing civilizations to an end.

Lava stream flowing from a vent near the summit of Mount Etna, Sicily, in the eruption of January 1992. *Roger Ressmeyer/Starlight.*

Imagine a volcanic eruption in which the ground covering an area the size of New York City collapses, a region larger than Vermont is buried by hot ash that snuffs out all life, and farms 1000 or 2000 km away are covered by ash and rendered infertile. Imagine that the volcanic dust thrown into the atmosphere dims the sun for a year or two, so that there are no summers. Unbelievable? It happened at least twice in what is now the United States: at Yellowstone in Wyoming 600,000 years ago and at Long Valley, California, 700,000 years ago. This was long before humans first reached North America, only 30,000 years ago, but not so long ago on the 4.6-billion-year geologic time scale. We know that these events took place because the volcanic rocks that were formed by these eruptions have been identified.

Ancient philosophers were awed by volcanoes and their fearsome eruptions of molten rock. In their efforts to explain volcanoes they spun myths about a hellish, hot underworld below Earth's surface. They had the right idea. Modern scientists, who also seek an explanation, see in volcanoes evidence of Earth's internal heat. Measurements of the temperatures of lava as it erupts and of rocks as far down as humans have drilled (about 10 km) show that the Earth does get hotter with depth. From these measurements, geologists can infer temperatures at even greater depths. As we saw in Chapter 4, geologists now believe that at the depths of the asthenosphere, which extends from about 75 to 250 km, the temperature reaches 1100 to 1200°C, high enough for the rocks there to begin to melt. This is why they identify this region as a main source of magma, the molten rock below Earth's surface, which we know as lava after it erupts. Remelting of sections of the solid lithosphere, which is above the asthenosphere, may provide another source of magma. In some places, perhaps where the lithosphere is fractured, the magma floats up because it is less dense than the surrounding rock. Some of it eventually reaches the surface and erupts as lava. Figure 5.1 is a simplified sketch of the plumbing system of a volcano such as Kilauea on Hawaii,

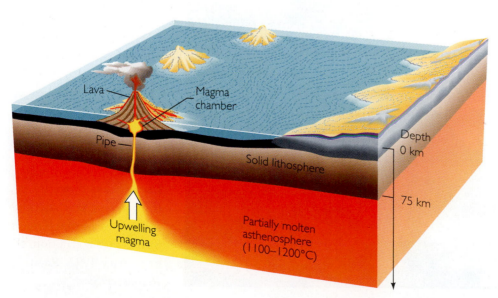

**FIGURE 5.1** The plumbing system of a volcano. Magma, which originates in the partially molten asthenosphere, rises through the lithosphere to erupt as basaltic lava.

which taps a pool of molten rock at depth and gives vent to it at the surface. Note the pipe or conduit through which the magma rises and the magma chamber, a shallow reservoir in the crust below the summit, which periodically fills with magma rising from below and empties to the surface in cycles of eruptions.

Because lava is a sample of the Earth's interior, it is interesting to geologists. Unfortunately, it is not a perfect sample. Lava differs from magma in that it loses some gaseous constituents to the atmosphere or ocean during an eruption and may have gained or lost other chemical components on its way to the surface. Despite these differences, lava and other materials that erupt from volcanoes still provide important information about the chemical composition and physical state of the upper mantle. Volcanic rocks are important not only for what they reveal about the interior, but also because they tell us something about the eruptions that created them so many thousands or millions of years ago. The chemical and mineralogical compositions of lava have much to do with the way it erupts and the kind of landform it leaves behind when it solidifies. To understand the variety of shapes of volcanoes and why some eruptions are explosive and others gentle, we must begin by looking at the available evidence: the rocks that were left behind.

# VOLCANIC DEPOSITS

The major types of lavas are placed within the categories of igneous rocks, as we saw in Chapter 4. Table 4.2 describes the three major groups of igneous rocks and their lavas: felsic, intermediate, and mafic. The rocks are further classified according to whether they are intrusive (they cooled below the surface and are coarse-grained as a result) or extrusive (they cooled on the surface and are finer-grained). The major intrusive rocks are granite (felsic), diorite (intermediate), and gabbro (mafic). The major extrusive counterparts are the lavas rhyolite (felsic) and the more common andesite (intermediate) and basalt (mafic). With this classification in mind, let us examine the volcanic process.

Before magma erupts, it floats up from the asthenosphere and enters a shallow chamber within the volcano. In time the magma is forced to move within the volcano by the pressure that builds up when the magma chamber is filled. It may erupt onto the surface as molten lava or be violently ejected into the air as a spray of solidified fragments called pyroclasts.

## Lava Flows

The several types of lavas leave behind different landforms—volcanic mountains that vary in shape and solidified lava flows that vary in character. The major differences among lavas reflect their chemical composition, their gas content, and their temperature. The higher the silica content and the lower the temperature, the more viscous the lava and the more slowly it flows; the more gas the lava contains, the more violent the eruption is likely to be.

Like all liquids, lavas flow downhill. Basaltic lava erupts at temperatures of 1000 to 1200°C—close to the temperature of the upper mantle. Because of the high temperature and low silica content, basaltic lava is so fluid that it can flow fast and far. Streams of lavas have been observed to flow as fast as 100 km per hour, although velocities of a few kilometers per hour are more common. In 1938 two daring Russian volcanologists measured lava temperatures and collected gas while they floated 2 km down a river of lava on a raft of colder solidified lava. The surface temperature of their raft was 300°C, and the lava river had a temperature of 870°C. Lava streams that extend more than 50 km have been witnessed in historical times. Highly fluid basaltic lava that erupts on flat terrain can spread out in thin sheets, and successive flows often pile up into immense lava plateaus, as at the great Columbia Plateau of Oregon and Washington (Figure 5.2).

Rhyolite, the most felsic lava, has a lower melting point than basalt and erupts at temperatures of 800 to 1000°C. It is much more viscous because of its lower temperature and higher silica content; it resists flow, moves 10 or more times more slowly than basalt, and tends to pile up in thick, bulbous deposits. Andesite, with a silica content between that of basalt and rhyolite, has properties that fall between those of basalts and rhyolites. Geologists can recognize the type of lava that flowed from the Earth even millions of years ago by the distinguishing features of the solidified rock it left behind.

**BASALTIC FLOWS**  Basaltic flows fall into two categories, according to their surface forms: **pahoehoe** (pronounced pa-ho-ee-ho-ee) and **aa** (ah-ah).

Pahoehoe (the word is Hawaiian for "ropy") forms when a highly fluid lava spreads in sheets and a thin, glassy elastic skin congeals on its surface as it cools. The skin is dragged into ropy, filamented folds as the molten liquid continues to flow below the surface (Figure 5.3).

"Aa" is what one exclaims if one walks barefoot on lava that looks to the unwary like clumps of

**FIGURE 5.2** Flood basalts form the layered rocks in this eroded part of the Columbia Plateau in Oregon. *Ellen Morris Bishop.*

moist, freshly plowed earth. Aa is lava that has lost its gases and consequently has become more viscous than pahoehoe and moves more slowly, allowing a thick skin to form. As the layer continues to move, the thick skin breaks into rough, jagged blocks (see Figure 5.3). The blocks ride on the viscous, massive interior, advancing as a steep front of angular boulders like a tractor tread. A single lava flow commonly has the features of pahoehoe near its source, where the lava is still fluid and hot, and of aa farther downstream, where the flow's surface, having been exposed to cool air for some time, has solidified into a thicker layer, with still hot, viscous lava flowing below. Aa is truly treacherous to cross. A good pair of boots has an average lifetime of about a week on it, and the traveler or geologist can count on cut knees and elbows. Many a mainland haole (nonnative) has paid dearly for an acre of Hawaiian aa. His shoes would be cut to ribbons if he tried to walk on his own land.

**PILLOW LAVAS** A geologist who comes across **pillow lavas**—piles of ellipsoidal, sacklike blocks about a meter wide—knows they formed in an underwater eruption (Figure 5.4) even if they are now on dry land. In fact, pillow lavas are an important indicator that a region was once under water. Geologist-divers have actually observed pillow lavas form on the ocean floor. Tongues of molten lava develop a tough, plastic skin on contact with the ocean water because the surface chills quickly. The lava inside the skin cools more slowly, and the pillow's interior becomes crystalline while the skin remains glassy.

Lavas have many other features that reflect the conditions under which they formed. They can be glassy or fine-grained if they cool quickly or coarsely crystalline if they cool slowly beneath the surface. They can have little bubbles, created when pressure falls suddenly as the lava cools. Lava is typically charged with gas. When lava rises, the pressure on it decreases, just as it does in a bottle of soda when the cap is removed. And just as the soda's carbon dioxide

**FIGURE 5.3** Ropy pahoehoe lava and jagged blocks of aa lava on Kilauea, Hawaii. *Peter Kresan.*

FIGURE 5.4 Pillow lava, characteristic of underwater volcanic eruptions, on the seafloor near the Galápagos Islands. *Woods Hole Oceanographic Institute.*

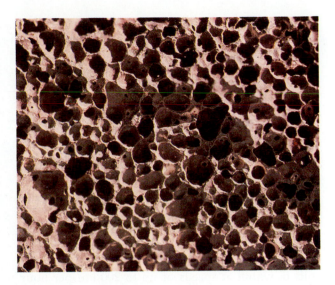

FIGURE 5.5 Vesicular lava. *Peter Kresan.*

creates bubbles as it is released, water vapor and other gases escaping from lava create gas cavities, or *vesicles*. The presence of bubbles left behind in the solidified lava provides geologists with details of the rock's volcanic origins (Figure 5.5). One extremely vesicular, generally rhyolitic lava is pumice. Some pumice has so much void space that it is light enough to float.

## Pyroclastic Deposits

Water and dissolved gases in magmas can have even more dramatic effects on eruptive styles. Before eruption, the confining pressure of the overlying rock keeps these volatiles from escaping. When the magma rises close to the surface and the pressure drops, the volatiles may be released with explosive force, shattering the lava and any overlying solidified rock into fragments of various sizes, shapes, and textures. These pyroclasts are particularly likely to be formed by the gas-rich viscous rhyolitic and andesitic lavas.

Pyroclasts, as we saw in Chapter 4, are any fragmentary volcanic rock materials that are ejected into the air. These rocks, minerals, and glasses are classified according to size. The finest, less than 2 mm in diameter, are called **ash.** Fragments ejected as blobs of magma and then rounded and cooled in flight and chunks torn loose from previously solidified volcanic rock can be much larger (Figure 5.6). Some volcanic

FIGURE 5.6 A field of volcanic ejecta at Anak Krakatoa, Indonesia. *Katia Krafft/Explorer.*

ejecta as big as a house are known to have been thrown more than 10 km in violent eruptions. Volcanic ash fine enough to stay aloft can be carried great distances. Within two weeks of the 1991 eruption of Mount Pinatubo in the Philippines, its volcanic dust was traced all the way around the world by orbiting Earth satellites. Sooner or later pyroclasts fall, usually building deposits near their source. As they cool, hot sticky fragments become welded together (or lithified). The rocks created from the smaller fragments are called **volcanic tuffs;** those formed from the larger fragments are called **volcanic breccias** (Figure 5.7).

One particularly spectacular and often devastating form of eruption occurs when hot ash, dust fragments, and gases are ejected in a glowing cloud that rolls downhill at speeds of up to 200 km per hour. The solid particles are actually buoyed up by hot gases and entrained air, so that there is little frictional resistance to this incandescent **pyroclastic flow** (Figure 5.8). In 1902 a pyroclastic flow with an internal temperature of 800°C exploded from the side of Mont Pelée, on the Caribbean island of Martinique, with very little warning. The avalanche of choking hot gas and glowing volcanic ash plunged down the slopes at a hurricane speed of 160 km per hour. In one minute and with hardly a sound, the searing emulsion of gas, ash, and dust enveloped the town of

**FIGURE 5.7** Volcanic breccia in the Grapevine Mountains, Death Valley, California. *Tom Bean.*

**FIGURE 5.8** A pyroclastic flow plunges down the slopes of Mount Unzen, in Japan, in June 1991. Note the fireman and fire engine in the foreground, trying to outrun the hot ash cloud descending on them. Three scientists who were studying this volcano were killed when they were engulfed by a similar flow. *AP/Wide World Photos.*

**FIGURE 5.9** Mont Pelée steams above St. Pierre, the town it destroyed by a pyroclastic flow on May 8, 1902. (Copyright © 1902 by by Underwood and Underwood. Library of Congress.)

St. Pierre and killed 29,000 people (Figure 5.9). It is sobering to scientists who render advice to others to recall the statement of one Professor Landes, issued the day before the cataclysm: "The Montagne Pelée presents no more danger to the inhabitants of Saint Pierre than does Vesuvius to those of Naples." Professor Landes perished with the others. Volcanists Maurice and Katia Krafft, whose photographs are used in this chapter, were killed by a pyroclastic flow at Mount Unzen, Japan, in 1991.

# ERUPTIVE STYLES

Now that we have examined various types of volcanic materials that pour or blow out of Earth's interior, we can look more closely at styles of eruptions and the characteristic formations they leave behind. Eruptions do not always create the majestically symmetrical cone of a Fujiyama. The hundreds of thousands of square kilometers of monotonous layers of basalt that make up the Columbia Plateau of Washington and Oregon represent another variant.

## Central Eruptions

Central eruptions create the most familiar of all volcanic features—the volcanic mountain shaped like a cone. These eruptions discharge lava or pyroclastic materials from a **central vent,** an opening atop a pipe through which the material rises to the Earth's surface.

**LAVA ERUPTIONS**   A lava cone is built by successive flows of lava from a central vent. If the lava is basaltic, it flows easily and spreads widely. If it is copious, the flows create a broad, shield-shaped volcano many tens of kilometers in circumference and more than 2 km high. The slopes are relatively gentle. Mauna Loa, on Hawaii, is the classic example of a

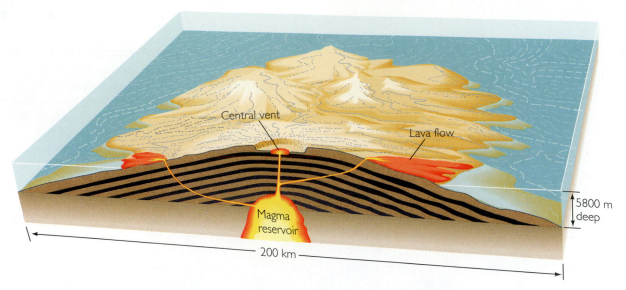

**FIGURE 5.10** A shield volcano is built up by the accumulation of thousands of thin basaltic lava flows that spread widely and cool as gently sloping sheets. Each layer shown represents an accumulation of many hundreds of thin lava flows. Modeled after Mauna Loa, Hawaii.

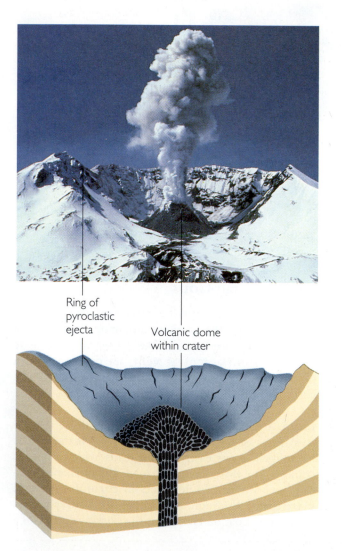

**shield volcano** (Figure 5.10). Although it rises only 4 km above sea level, it is the world's tallest structure: measured from its base on the seafloor, the volcano is 10 km high. It has a base diameter of 120 km—an area roughly three times that of Rhode Island. It grew to this enormous size by the accumulation of thousands of lava flows, each only a few meters thick, over a period of a few million years. In fact, the island of Hawaii actually consists of the tops of a series of overlapping shield volcanoes emerging through the ocean surface (see Box 5.1).

In contrast to basaltic lavas, felsic lavas are so viscous that they can just barely flow. They usually produce a **volcanic dome,** a rounded, steep-sided mass of rock (Figure 5.11). Domes look as though lava had been squeezed out of a vent like toothpaste, with very little lateral spreading. Domes often plug vents, trapping gases. Pressures increase until an explosion occurs, destroying the dome. This occurred in the eruptions of Mount St. Helens in 1980.

**FIGURE 5.11** Volcanic domes are bulbous masses of felsic lava, which are so viscous that instead of flowing they pile up over the vent. Shown here is a growing dome within the crater of Mount St. Helens. *Lyn Topinka/USGS Cascades Volcano Observatory.*

## BOX 5.1  KILAUEA: AN INSTRUMENTED VOLCANO

The giant shield volcano Mauna Loa and the smaller Kilauea on its eastern flank make up the southern half of the island of Hawaii. Because the U.S. Geological Survey operates a volcano observatory on the rim of Kilauea Caldera, this volcano is perhaps the best studied in the world, and what has been learned from it has profoundly influenced our notions of volcanic processes. A modern network of instruments and laboratory facilities is used to track the movement of magma within the volcano and the changing chemistry of the erupting lavas and gases. Seismographs, which measure movements within the Earth, detect and locate the small earthquakes that are often correlated with movements of magma. These instruments can locate earthquakes as deep as 55 km beneath Kilauea. Such quakes often mark the entrance of magma into the channels leading from the asthenosphere through the lithosphere to the Earth's surface. The upward migration of the magma can be followed over a period of months by seismic disturbances progressively nearer the surface as the magma rises. Tilt-

meters (instruments that measure tilting of the ground) indicate when the volcano begins to swell as the rising magma fills a shallow magma chamber below the summit. The first sign that an outbreak of lava is imminent is a swarm of small earthquakes, thousands of them, signifying that the magma is splitting its way to the surface.

Very often, geologists know where the eruption will occur from the location of earthquakes and changes in their pattern. In January 1960, for example, Geological Survey scientists detected a swarm of earthquakes not far from the village of Kapoho, on the flank of Kilauea. As they expected, an eruption broke out, destroying Kapoho but causing no casualties. A new landscape was created as the lava flowed to the sea. Twenty-foot walls were built in a futile attempt to divert the lava and save a seashore community. When it was all over, the tilt-meters showed that the volcano had deflated, signifying that the magma chamber below had been drained in the Kapoho eruption. The cycle is repeated every few years.

Wahaula Visitor Center, Kilauea, engulfed in lava. *J. D. Griggs/USGS.*

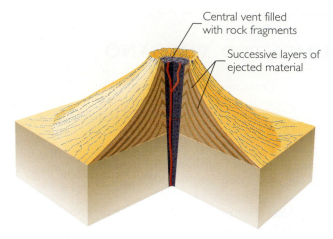

Central vent filled with rock fragments

Successive layers of ejected material

**FIGURE 5.12** In a cinder cone, ejected material is deposited as layers that dip away from the crater at the summit. The vent beneath the crater is filled with fragmental debris.

**PYROCLASTIC ERUPTIONS**  When volcanic vents discharge pyroclasts, the solid fragments build up and form **cinder cones** (Figures 5.12, 5.13, and 5.14). The profile of a cone is determined by the maximum angle at which the debris remains stable instead of sliding downhill (see Figure 11.1). The larger fragments, which fall near the summit, can form very steep but stable slopes. Finer particles are carried farther from the vent and form gentle slopes at the base of the cone. The classic concave-shaped volcanic cone with its summit vent reflects this variation in slope.

**COMPOSITE ERUPTIONS**  When a volcano emits lava as well as pyroclasts, a concave-shaped **composite volcano** or **stratovolcano** is built of alternating lava flows and beds of pyroclasts (Figure 5.15). This is the most common form of such large volcanoes as Fujiyama (Figure 5.16), Mount Vesuvius, Mount Etna, and Mount St. Helens.

**OTHER FEATURES OF CENTRAL ERUPTIONS**  A bowl-shaped pit or crater is found at the summit of most volcanoes, centered on the vent. During the eruption of a lava volcano, the upwelling lava overflows the crater walls; when eruption ceases, the lava that remains in the crater often sinks back into the vent. When pyroclasts erupt, the material is literally blasted out of the crater, which later becomes partially filled by the debris that falls back into it. Because a crater's walls are steep, they may cave in or become eroded in time, until the diameter of the crater is several times that of the vent and hundreds of meters deep. The crater of Mount Etna in Sicily, for example, has grown to 300 m (more than three football fields) in diameter and at least 850 m deep.

**FIGURE 5.13** Hot pyroclastic material ejected from Parícutin in Mexico builds a cinder cone. *R. E. Wilcox/ USGS Photo Library, Denver.*

**FIGURE 5.14** Cerro Negro in 1968. This volcano, near Managua, Nicaragua, is a cinder cone built on an older terrain of lava flows. *Mark Hurd Aerial Surveys*.

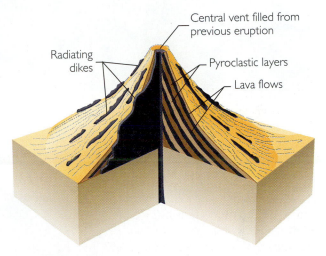

**FIGURE 5.15** A composite volcano is built up of alternating layers of pyroclastic material and lava flows. Lava that has solidified in fissures forms riblike dikes that strengthen the cone. (After R. G. Schmidt, USGS.)

**FIGURE 5.16** Fujiyama, a composite volcano in Japan. *Shizuo Ijima/Tony Stone Worldwide*.

**FIGURE 5.17** Aerial view of snow-covered summit caldera of Mauna Loa, Hawaii. *D. W. Peterson/ USGS.*

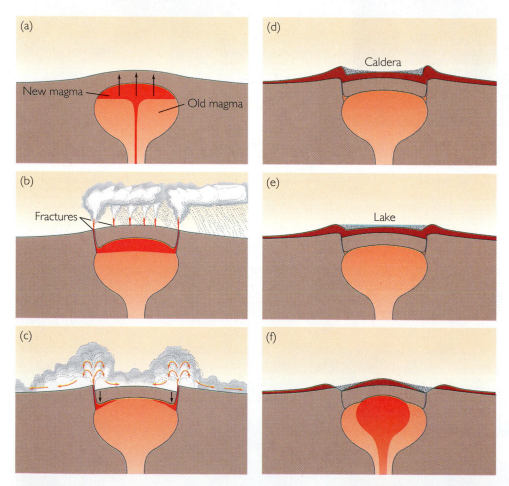

**FIGURE 5.18** Stages in the evolution of a resurgent caldera. (a) Fresh magma fills a magma chamber a few kilometers below the surface, doming the surface. (b) A ring of vertical fractures forms around the dome, and the gas-charged magma at the top (bright red) explodes, erupting columns of incandescent pumice and ash into the atmosphere. (c) Now that the magma chamber is empty, the roof collapses along the fractures and pyroclastic flows are ejected. (d) The caldera and surrounding area are blanketed with pyroclastic debris. (e) The caldera wall begins to erode, and a lake may form in the depression. (f) Many thousands of years later, fresh magma enters the chamber and the caldera floor begins to dome again. Minor volcanic activity may persist along the ring fracture for millions of years. (After P. Francis, "Giant Volcanic Calderas," *Scientific American,* June 1983, p. 60.)

**FIGURE 5.19** This phreatic explosion occurred when lava encountered seawater, generating steam; Kilauea, Hawaii. *J. D. Griggs/USGS.*

After a violent eruption in which large volumes of magma are discharged from a magma chamber a few kilometers below the vent, the empty chamber may no longer be able to support its roof. In such cases the overlying volcanic structure can collapse catastrophically, leaving a large, steep-walled, basin-shaped depression much larger than the crater, called a **caldera** (Figure 5.17). Calderas are impressive features, ranging in size from a few kilometers to as much as 50 km or more in diameter, or about the area of greater New York City. After some hundreds of thousands of years, fresh magma can reenter the collapsed magma chamber and reinflate it, forcing the caldera floor to dome upward again—perhaps to repeat the cycle of eruption, collapse, resurgence, eruption, and so on (Figure 5.18). This phenomenon is known as a *resurgent caldera.* The collapse of large resurgent calderas causes some of the most destructive natural phenomena on Earth. Fortunately, no catastrophic ones have occurred during human history. Yellowstone Caldera in Wyoming, marked today by a leftover relic, Old Faithful geyser, ejected some 1000 km³ of pyroclastic debris during its eruptive stages about 600,000 years ago, more than a thousand times the amount of material ejected by Mount St. Helens in 1980. Ash deposits fell over much of what is now the United States. Other resurgent calderas are Valles Caldera in New Mexico, Long Valley Caldera in California, the still-active Kilauea, and the dormant Crater Lake in Oregon.

Caldera watching is very important to geologists today. They have found that Yellowstone and Long Valley calderas are slowly beginning to inflate again.

At one time it was thought that a caldera was formed by a huge explosion of material when a volcano literally blew its top. However, geologic mapping of the debris around volcanoes and of the pattern of faulting produces a picture more consistent with the collapse of the roof than with its ejection upward. This is not to say that volcanic explosions have not left their marks on Earth's surface. We have already mentioned the explosive "unplugging" of a felsic dome. Another instance occurs when hot, gas-charged magma encounters groundwater or seawater. The vast quantities of superheated steam generated cause **phreatic,** or steam, **explosions** (Figure 5.19). One of the most destructive volcanic eruptions in history, that of Krakatoa in Indonesia, was a phreatic explosion (see Box 5.2).

Sometimes when hot matter from the deep interior escapes explosively, the vent is left filled with breccia. The resulting structure is called a **diatreme.** Ship Rock, which towers over the surrounding plain in New Mexico, is a diatreme exposed by the erosion of the sedimentary rocks through which it originally burst. To transcontinental air travelers, Ship Rock looks like a gigantic black skyscraper in the red desert (Figure 5.20). The eruptive mechanism that produces diatremes has been pieced together in great detail from the geologic record. The kinds of minerals and

## BOX 5.2    THE EXPLOSION OF KRAKATOA

The 1883 explosion of the volcano Krakatoa, in the strait between Java and Sumatra, was one of the greatest ever witnessed. Now almost completely submerged, Krakatoa was then a small island formed from a group of volcanic cones in an ancient caldera. The caldera, 6 km across, was a remnant of a collapsed prehistoric andesitic stratovolcano. On August 27, after many smaller explosions, Krakatoa blew its top in a phreatic explosion with the energy of 100 million tons of TNT (5000 times greater than the nuclear explosion that destroyed Hiroshima). It is believed that much of the energy was provided by the violent expansion of hot steam after the walls of the volcano first ruptured, letting seawater into the magma chamber. The result can be viewed as the biggest steam-boiler explosion and the loudest noise in recorded history.

The explosion was heard in Australia, nearly 2000 km away. Volcanic ash fell over an area of some 700,000 km². Almost total darkness settled on Jakarta, 150 km away, when the dust blotted out the Sun's rays. Fine dust rose to the stratosphere and drifted around the Earth, lowering Earth's mean annual temperature a few degrees for the next year or so by blocking 13 percent of the Sun's light from reaching Earth. The explosion also generated a *tsunami*, or giant sea wave, that reached a height of almost 40 m, destroying 295 coastal towns as far as 80 km away and drowning 36,000 people. The tsunami was recorded on tide gauges as far away as the English Channel. After the eruption, most of Krakatoa disappeared, leaving in its place the current 300-m-deep, water-covered basin.

**FIGURE 5.20** Ship Rock, towering 515 m above the surrounding flat-lying sediments of New Mexico, is a diatreme, or volcanic pipe, exposed by erosion of its enclosing sedimentary rock. *Fred Padula.*

Vent of Anak
Krakatoa

Outline of
1883 caldera

KRAKATOA
TODAY

5 km

Anak Krakatoa ("Child of Krakatoa") is a new volcano that is building up in the caldera left by the 1883 eruption. The new cone rose above sea level in 1928. Since then successive eruptions have largely filled in the 1883 caldera and may eventually build a new island in roughly the same position as the original Krakatoa. *Katia Krafft/ Explorer.*

rocks found in some diatremes could have been formed only at great depths—100 km or so, well within the upper mantle. This observation indicates that diatremes are formed when gas-charged magmas melt their way upward, finally ejecting gases, lava fragments from the vent walls, and fragments from the deep crust and mantle, all with explosive energy and sometimes at supersonic speed. Such an eruption would probably look like the exhaust jet of a giant rocket upside down in the ground blowing rocks and gases into the air.

Another diatreme is encountered in the underground workings of the fabled Kimberley mines of South Africa, one of the world's richest sources of diamonds. This diatreme is a peridotite, a rock composed mostly of the mineral olivine. It also contains diamonds, which form from carbon under the great pressures found in the mantle, and other fragments of mantle rock the magma picked up en route to the surface. Geologists view this diatreme much as they

would a 300-km drill core into the mantle in which the rocks were scrambled by eruption. The fragments provide the only direct evidence on the materials of the mantle and so have been studied extensively. This is an example of the evidence that enables geologists to conclude that peridotite is a major constituent of the upper mantle.

## Fissure Eruptions

Imagine basaltic lava flowing out of a crack in the Earth's surface tens of kilometers long, flooding large areas. Such **fissure eruptions** have occurred on Earth countless times in the last four billion years, but in recorded history humans have witnessed such an eruption only once, on Iceland in 1783. One-fifth of the Icelandic population perished as a result. A fissure 32 km long opened and spewed out some 12 km$^3$ of basalt, enough to cover Manhattan about

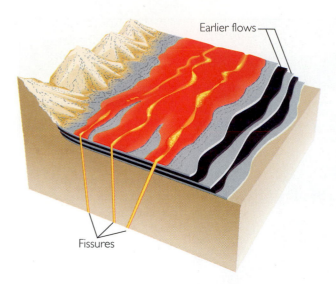

**FIGURE 5.21** In a fissure eruption of highly fluid basalt, lava rapidly flows away from fissures and forms widespread layers, rather than building up into a volcanic mountain. (After R. S. Fiske, USGS.)

**FIGURE 5.22** The area covered by the Columbia River flood basalts. (After R. S. Fiske, USGS.)

halfway up the Empire State Building. The geologic record contains ample evidence of basalt flooding from great fissures. When **flood basalts** erupt from fissures, the lavas build a plain or accumulate as a plateau, rather than piling up as a volcanic mountain as they do when they erupt from a vent (Figure 5.21). The flood basalts (see Figure 5.2) that made the Columbia Plateau buried 200,000 km² of the preexisting topography (Figure 5.22). Some individual flows were more than 100 m thick, and some were so fluid that they spread more than 60 km from their source. An entirely new landscape with new river valleys has since evolved atop the lava that buried the old surface. Plateaus made by flood basalts are found on every continent.

Fissure eruptions of pyroclastic materials are more likely when the parent magma is felsic. Such eruptions have produced extensive sheets of hard volcanic tuffs called **ash-flow deposits** (Figures 5.23, 5.24). As far as is known, humans have never witnessed one of these spectacular events. The early Tertiary ash-flow deposits of the Great Basin in Nevada and adjacent states, formed in this way, cover an area of about 200,000 km² and are as much as 2500 m thick in some places. Yellowstone National Park, in Wyoming, has been covered by a number of ash-flow sheets. A succession of forests there have been buried by these deposits. Among the geologically important fissure eruptions are those that occur along mid-ocean ridges.

## Other Volcanic Phenomena

**LAHARS**     Among the most dangerous volcanic events are the torrential mudflows of wet volcanic debris that are called **lahars.** They can form when a pyroclastic flow meets a river or a snowbank; when the wall of a crater lake breaks, suddenly releasing

**FIGURE 5.23** A hand-sized hard volcanic tuff from an ash-flow sheet. The dark lenses are volcanic fragments. The "glue" that binds the fragments together is glass formed when droplets of lava freeze rapidly. Laboratory experiments indicate that tuffs welded together by glass form at temperatures in the range of 700 to 900°C. *R. G. Schmidt and R. L. Smith/USGS.*

**FIGURE 5.24** These ash-flow sheets on O-shima Island, Japan, were formed when intensely hot, gas-charged volcanic dust, ash, and pumice spread swiftly over the surface to settle, cool, and harden. *S. Aramaki.*

water; when glacial ice is melted by a lava flow; or when heavy rainfall transforms new ash deposits into mudflows. One formation in the Sierra Nevada of California contains 8000 km³ of material of lahar origin, enough to cover Delaware with a deposit more than a kilometer thick. Lahars have been known to carry huge boulders for tens of kilometers. When Nevada del Ruiz in Colombia erupted in 1985, lahars triggered by the melting of glacial ice near the summit plunged down the slopes and buried the town of Armero, killing 25,000 people.

**VOLCANIC GASES**  The nature and origin of volcanic gases are of considerable interest and importance. It is thought that over geologic time these gases have created the oceans and the atmosphere and may even affect today's climate. Volcanic gases have been collected by courageous volcanologists and studied to determine their composition (Figure 5.25). Water vapor is the main constituent of volcanic gas (70 to 95 percent), followed by carbon dioxide, sulfur dioxide, and traces of nitrogen, hydrogen, carbon monoxide, sulfur, and chlorine. Every eruption re-

**FIGURE 5.25** Scientist Maurice Krafft took this photo of his wife Katia measuring properties of a lava flow. *Maurice Krafft.*

leases enormous amounts of these gases. Some volcanic gas may come from deep within the Earth, making its way to the surface for the first time; some may be recycled groundwater and ocean water, recycled atmospheric gas, or gas that has been trapped in earlier generations of rocks.

The correlation between volcanic eruptions and changes in weather and climate is receiving increasing attention. For instance, the 1982 eruption of El Chichón in southern Mexico and the 1991 eruption of Mount Pinatubo injected sulfurous gases into the stratosphere, 10 km above the Earth. Through various chemical reactions the gases formed an aerosol (a collection of small droplets suspended in air) of tens of millions of metric tons of sulfuric acid droplets. The aerosol can absorb solar radiation and cool the surface of the Earth for as long as a year. Scientists observed that the eruption of Mount Pinatubo, one of the largest explosive eruptions of the century, led to a global cooling of as much as 0.5°C in 1992. For similar reasons the debris lofted into the stratosphere during the 1815 eruption of Mount Tambora in Indonesia cooled the Northern Hemisphere and resulted in a very cold summer in 1816. Crop losses and food shortages caused great suffering in that "year without a summer."

**HOT SPRINGS AND GEYSERS**   The late stages of volcanic activity are often marked by emissions of gas and vapor unaccompanied by lava or pyroclastic matter. Circulating groundwater that has reached and been heated by buried magma (which retains heat for a long time) produces hot springs and geysers (Figure 5.26). A geyser is a hot-water fountain that spouts intermittently with great force, frequently accompanied by a thunderous roar. The best known geyser in the United States is Old Faithful in Yellowstone Park, which erupts about every 65 minutes, sending a jet of hot water as high as 60 m.

For decades after a major eruption, volcanoes continue to emit gas fumes, steam, and hot water through small vents. All of these volcanic emanations contain dissolved materials that precipitate as the water evaporates and cools, forming various sorts of encrusting deposits (such as travertine), some of which contain valuable minerals.

## THE GLOBAL PATTERN OF VOLCANISM

Before the advent of plate-tectonics theory, geologists noted the concentration of volcanoes around the boundaries of the Pacific Ocean and called it "the Ring of Fire." We now know that the Ring of Fire coincides with plate boundaries; this correlation has been an important clue to the nature of volcanoes. The 500 to 600 active volcanoes of the world are not randomly distributed, but show a definite pattern. About 80 percent are found at boundaries where plates converge, 15 percent where plates separate, and the remaining few within plates (Figure 5.27).

**FIGURE 5.26** Geyser Hot Springs, in the Black Rock Desert of northwestern Nevada, periodically spews forth hot water and steam. Between eruptions, hot water presumably fills underground cavities. Further heating converts some of the water into steam, generating the pressure that forces the discharge. *Stephen Trimble.*

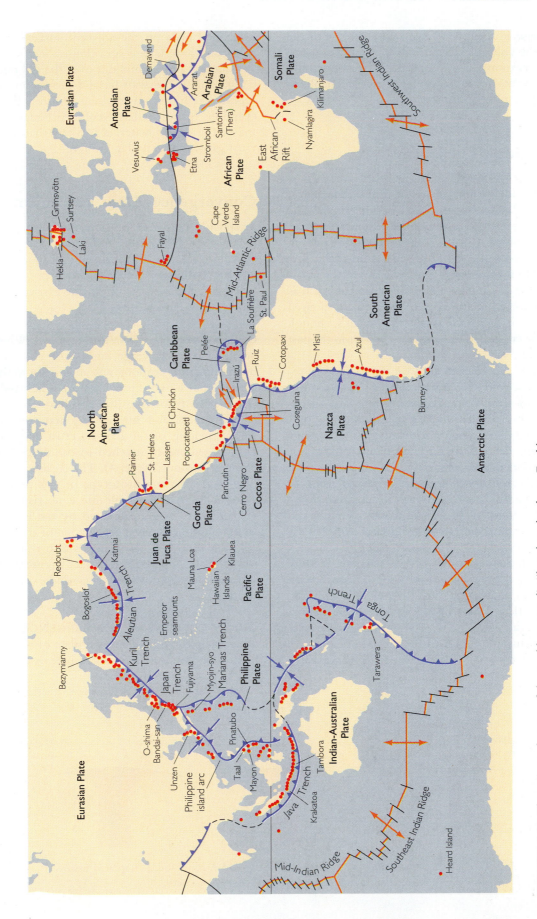

**FIGURE 5.27** The active volcanoes of the world are not distributed randomly on Earth's surface; they tend to be associated with the boundaries where plates collide or separate. Convergent boundaries are shown in blue, divergent boundaries in orange. Black lines are transform faults. Active volcanoes are marked by red dots.

We also saw in Chapter 4 that lavas vary with the locations of volcanoes (Figure 4.7). Can we weave these observations into a hypothesis that not only describes events but explains them as well? Thanks to the theory of plate tectonics, we can.

## Ocean-Ridge Volcanism

As we saw in Chapters 1 and 4, the seafloor is broken by a worldwide system of rifts or mid-ocean ridges, along which plates separate and basalt erupts. The fissure between the separating plates extends down to the asthenosphere. Basaltic magmas rise buoyantly in the gap between the separating plates and overflow the fissure to form ocean ridges, volcanoes, the basaltic seafloor crust, and basaltic island plateaus, such as Iceland. Enormous amounts of basalt have poured out of this world-encircling system of cracks, enough to build the crust of all the present seafloor in the past 200 million years.

Much of the volcanic heat on the seafloor is removed when cold seawater circulates in the fissures of the ocean-ridge volcanic system. Seawater that is heated and rich in dissolved minerals after contact with magmas forms extremely hot (350°C) springs and smoking vents along the cracks, and it is a major source of minerals (see Box 17.2).

Iceland, an exposed segment of an extension of the Mid-Atlantic Ridge, provides an unmatched opportunity to view the process of fissure eruption and seafloor spreading directly (see Figure 1.16). The island is composed mostly of basalt. Repeated surveys show that Iceland is in a state of tension, literally being pulled apart—one half moving eastward with the Eurasian Plate, the other westward with the North American Plate. Cracks develop, and magma flows in from below and overflows onto the surface. At the conclusion of each episode, the lava solidifies to form a nearly vertical dike in the fissure and nearly horizontal beds on the adjacent surface. With each new episode of lateral spreading, a new crack opens and another flow pours out over the old one (Figure 5.28). This is the way Iceland grows, primarily by repeated eruptions from long fissures, but also by eruptions from localized vents. Although the details may differ under water, it is likely that the seafloor crust grows in a similar fashion.

## Convergence-Zone Volcanism

Scientists are now sorting out the many phenomena that occur where plates converge. One of the most striking features is the chain of volcanoes that parallels any boundary between two plates that collide, whether both are oceanic or one carries a continent on its edge (Figure 5.29).

**OCEAN-OCEAN CONVERGENCE**   Where two oceanic plates converge, an arc of volcanic islands builds up from seafloor volcanism, typically by the extrusion of basalts and andesites. The basalts probably derive from the asthenosphere above the descending plate, and the andesites may come from the partial melting of the basaltic crust and the ocean-bottom sediments attached to the descending plate (as described in Chapter 4). The creation of the Japanese island arc is a prototype of this process. Fujiyama is an andesitic volcanic cone rising within a growing arc. The arc formed by the Philippine Islands is another convergence zone of this type. The collision of two ocean plates is also responsible for Mount Pelée on Martinique (see Figure 5.9), which squeezes out viscous felsic lavas and explosive pyroclastic flows.

**OCEAN-CONTINENT CONVERGENCE**   When an ocean plate is overridden by a plate carrying a continent on its leading edge, an arcuate volcanic moun-

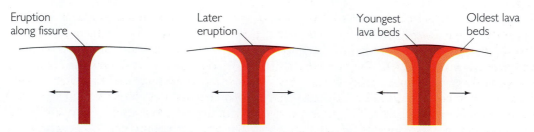

**FIGURE 5.28** Iceland is an exposed part of the Mid-Atlantic Ridge. Repeated fissure eruptions and lateral spreading are the mechanisms of its growth.

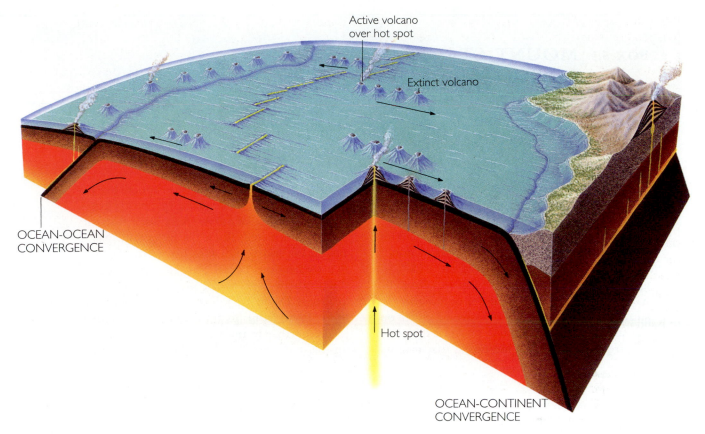

Active volcano
over hot spot

Extinct volcano

OCEAN-OCEAN
CONVERGENCE

Hot spot

OCEAN-CONTINENT
CONVERGENCE

**FIGURE 5.29** Volcanism associated with plate separation at a mid-ocean ridge, with an
ocean-ocean plate collision, with an ocean-continent plate collision, and with plate motion
over a hot spot. Hot spots account for the creation of mid-plate chains of volcanic islands.
Magma rises from a hot spot to create a volcano. As the plate moves, it carries the volcano
off the hot spot and the volcano becomes extinct. A new volcano develops over the hot spot.
In this way, a chain of volcanoes develops; these volcanoes get progressively older with distance from the hot spot.

tain chain grows in the zone of collision near the continental margin (see Figure 5.29). The Andes mark
the convergence boundary between the oceanic
Nazca Plate and South America. Farther north, the
subduction of the small Juan de Fuca Plate under the
North American Plate gives rise to the volcanoes of
the Cascade Range, which stretch from northern
California to British Columbia; one of them is
Mount St. Helens (see Box 5.3). During a typical
volcanic eruption, large quantities of ash and of andesitic and basaltic lavas are ejected. The lava is basaltic when magma from the asthenosphere reaches the
surface without being contaminated by sediments or
the felsic crust through which it rises. The more felsic
eruptives seem to indicate remelting of the subducted
plate, contamination by sediments, or actual melting
of continental crust. Mount St. Helens, Parícutin

(Mexico), Coseguina (Nicaragua), Irazú (Costa
Rica), and Cotopaxi (Ecuador) are all felsic volcanoes
(see Figure 5.27).

## Intraplate Volcanism

For many years, volcanism far from plate boundaries
posed a problem for the theory of plate tectonics: it
seemed to be an exception to the neat correlation of
volcanism and plate boundaries. Take the Hawaiian
Islands, in the middle of the Pacific Plate. This island
chain begins with the active volcanoes on Hawaii and
continues as a string of progressively older, extinct,
eroded, and submerged volcanic ridges and mountains. Frequent large earthquakes do not occur along
the Hawaiian chain—that is, it is essentially aseismic

## BOX 5.3 MOUNT ST. HELENS: DANGEROUS BUT PREDICTABLE

Long before Mount St. Helens erupted in 1980, geologists knew it to be the most active and explosive volcano in the contiguous United States. They could piece together a 4500-year history of destructive lava flows, hot pyroclastic flows, lahars, and distant ash falls by examining the geologic record. Beginning on March 20, 1980, a series of earthquakes under the volcano signaled the start of a new eruptive phase after 123 years of dormancy. The earthquakes moved the U.S. Geological Survey (USGS) to issue a formal hazard alert. The first outburst of ash and steam erupted from a newly opened crater on the summit one week later. In April the seismic tremors increased, indicating that magma

was moving beneath Mount St. Helens, and an ominous swelling of the northeastern flank was noticed. The USGS issued a more serious warning, and people were ordered out of the vicinity.

On May 18 the climactic eruption began abruptly. A large earthquake apparently triggered a massive landslide. As a huge flow of debris plummeted down the mountain, gas and steam under high pressure were released in a tremendous lateral blast that blew out the northern flank of the mountain. USGS geologist David A. Johnston was monitoring the volcano from his observation post 8 km to the north. He must have seen the advancing blast wave before he radioed his last mes-

Mount St. Helens before and after the cataclysmic eruptions of May 1980. *USGS Cascades Volcano Observatory.*

sage: "Vancouver, Vancouver, this is it!" A northward-directed jet of superheated (500°C) ash, gas, and steam roared out of the breach with hurricane force, devastating a zone 20 km outward from the volcano and 30 km wide. A vertical eruption sent an ash plume 25 km high, twice as high as a commercial jet flies. The ash cloud drifted to the east and northeast with the prevailing winds, bringing darkness at noon to an area 250 km to the east and depositing ash up to 10 cm deep on much of Washington, northern Idaho, and western Montana. The energy of the blast was equivalent to about 25 million tons of TNT. The volcano's summit was destroyed, its elevation reduced by 350 m.

The local devastation was spectacular. Within an inner blast zone extending 10 km, the thick forest was denuded and buried under several meters of pyroclastic debris. Beyond this zone, out to 20 km, trees were stripped of their branches and blown over like broken matchsticks aligned radially away from the volcano. As far as 26 km away, the hot blast was so intense that it overturned a truck and melted its plastic parts. Some fishermen were severely burned and survived only by jumping into a river. More than 60 other people were killed by the blast and its effects.

The lahar formed when the landslide and pyroclastic debris—fluidized by groundwater, melted snow, and glacial ice—flowed 28 km down the valley of the Toutle River, filling the valley bottom to a depth of 60 m. Beyond this debris pile, muddy water flowed into the Columbia River, where sediments clogged the ship channel and stranded many vessels in Portland. Mount St. Helens may go on erupting for 20 years or more in its present episode of activity.

Although much of the devastated area is still barren, almost 20 percent of the surface is showing evidence of revegetation: native species of grass, legumes, and young trees are beginning to come back.

(without earthquakes)—so it is called an **aseismic ridge.**

Aseismic ridges of volcanic origin, which also occur elsewhere in the Pacific and in the other large oceans, were difficult to incorporate into the framework of plate-tectonics theory until the concept of **hot spots** was introduced. Hot spots have also been proposed to explain some forms of volcanism within continents, far from plate boundaries. According to this hypothesis, illustrated in Figure 5.29, hot spots are the volcanic manifestations of jets or plumes of hot material that rise from deep within the mantle (perhaps even from the core-mantle boundary), penetrate the lithosphere, and erupt at the surface. These columnar currents are supposedly fixed in the mantle and do not move with the lithospheric plates. As a result, the hot spot leaves a trail of extinct, progressively older volcanoes as the plate moves over it. The tracks of the extinct volcanoes that constitute the Hawaiian Islands and the Emperor seamount chain shown in Figure 5.27 trace the motion of the Pacific Plate over the hot spot marked by the active volcanoes on Hawaii. The bend in the chain records a change in the direction of plate motion.

If hot spots are indeed fixed in the mantle, the trail of volcanoes carried away from the hot spot provides a powerful method of measuring the velocity as well as the direction of plate motion. We could calculate the velocity, for example, by dividing the distance an extinct volcano has traveled from the hot spot by the age of that volcano's youngest lava. Results of deep-sea drilling along the Hawaiian and Emperor seamount chains are consistent with the hypothesis that they originated over hot spots. Drilling reveals that the farther the lavas along these island chains are from an active hot spot, the older they are. It also reveals that the plates are moving several centimeters a year and that the change in direction of plate motion occurred 40 million years ago.

The origin of fissure eruptions of basalt on continents, such as those that formed the Columbia River Plateau, is a subject of debate. Some geologists suggest that fractures (of unknown origin) penetrated the continental lithosphere and that basaltic lavas, which represent partial melts of the underlying mantle, spurted rapidly to the surface without much contamination from the felsic crust. Eruptions of basalt that mark the initial stages of continental rifting and the opening of a new ocean can be documented in several parts of the world. For example, basalt is found in association with the rift valleys of East Africa—a feature that some geologists interpret as representing a breakup of Africa that never was completed (see Figure 5.27).

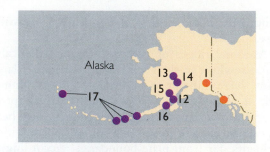

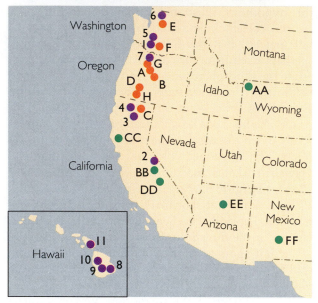

● Volcanoes that have short-term eruption periodicities (100–200 years or less), or have erupted in the past 200–300 years, or both:

| Cascades | | Hawaii | | Alaska | |
|---|---|---|---|---|---|
| 1 | Mount St. Helens | 8 | Kilauea | 12 | Augustine volcano |
| 2 | Mono-Inyo craters | 9 | Mauna Loa | 13 | Redoubt volcano |
| 3 | Lassen Peak | 10 | Hualalai | 14 | Mount Spurr |
| 4 | Mount Shasta | 11 | Haleakala | 15 | Iliamna volcano |
| 5 | Mount Rainier | | | 16 | Katmai volcano |
| 6 | Mount Baker | | | 17 | Aleutian volcanoes |
| 7 | Mount Hood | | | | |

● Volcanoes that appear to have eruption periodicities of 1000 years or greater and last erupted 1000 years or more ago:

| Cascades | | Alaska | |
|---|---|---|---|
| A | Three Sisters | I | Mount Wrangell |
| B | Newberry volcano | J | Mount Edgecumbe |
| C | Medicine Lake volcano | | |
| D | Crater Lake (Mount Mazama) | | |
| E | Glacier Peak | | |
| F | Mount Adams | | |
| G | Mount Jefferson | | |
| H | Mount McLoughlin | | |

● Volcanoes that last erupted more than 10,000 years ago, but beneath which exist large, shallow bodies of magma that are capable of producing exceedingly destructive eruptions:

| | | | |
|---|---|---|---|
| AA | Yellowstone Caldera | DD | Coso volcanoes |
| BB | Long Valley Caldera | EE | San Francisco Peak |
| CC | Clear Lake volcanoes | FF | Socorro |

# VOLCANISM AND HUMAN AFFAIRS

Teams of archeologists and marine geologists have pieced together the story of the volcano Thera (at one time called Santorini), in the Aegean Sea. It appears that the eruption of Thera about 3500 years ago was far more violent than that of Krakatoa. The volcanic debris and sea waves that were produced destroyed dozens of coastal settlements over a large part of the eastern Mediterranean. Some scientists have attributed the mysterious disappearance of the Minoan civilization to this cataclysm. And it is thought that the legend of the lost continent of Atlantis may have its origin in the destruction of land that accompanied the eruption. The course of history was probably changed by this one volcanic event. It could happen again.

Can volcanic eruptions be predicted? We have already seen in Boxes 5.1 and 5.3 that to some extent they can—fortunately for us all, because there are about 100 high-risk volcanoes in the world, and some 50 erupt each year. Certainly with our new understanding of volcanism we can improve the terrible record of the past 500 years. Over that period some 200,000 people have been killed by volcanic eruptions (see Box 5.4).

Can volcanic eruptions be controlled? Not likely, although in special circumstances and on a small scale the damage can be reduced. Perhaps the most successful attempt to control volcanic activity was made on the Icelandic island of Heimaey in January 1973. By spraying the advancing lava with seawater, Icelanders cooled and slowed the flow, preventing the lava from blocking the port entrance and saving some homes from destruction. In the years ahead, the best policy for protecting the public will be the establishment of warning and evacuation systems and the restriction of settlements in potentially dangerous locations. But even these precautions may not help. Dormant or long-extinct volcanoes can come to life suddenly—as Vesuvius and St. Helens did after hundreds of years. Some potentially dangerous volcanoes in the United States are Mount St. Helens and

**FIGURE 5.30** Locations of potentially hazardous volcanoes in the United States. Volcanoes within each group are listed in the order of declining probable cause for concern, subject to revision as studies progress. (After R. A. Bailey, P. R. Beauchemin, F. P. Kapinos, and D. W. Klick, USGS.)

## BOX 5.4  REDUCING THE RISKS OF HAZARDOUS VOLCANOES

Of Earth's 500 to 600 active volcanoes, one out of six has claimed human lives. Volcanoes kill people and damage property by explosive blasts, ash falls, rapid outpouring of hot lethal gas, lava flows, and mudflows or lahars. Let us contrast the consequences of several destructive volcanic eruptions.

Scientists monitored Mount St. Helens (discussed in Box 5.3) and Mount Pinatubo and issued warnings of imminent major eruptions. Government infrastructures were in place to evaluate the warnings and to issue and enforce evacuation orders. In the case of Pinatubo, the warning was issued a few days before the cataclysmic eruption. A quarter of a million people were evacuated, including some 16,000 residents of the nearby U.S. Clark Air Force Base (since permanently abandoned). Tens of thousands of lives were saved from the lahars that destroyed everything in their path. The casualties were limited to the few who disregarded the order.

Heat from a minor eruption of Nevada del Ruiz in Colombia in 1985 melted the snowcap, and the water triggered lahars that buried an entire town and killed more than 25,000 people. Scientists knew this volcano to be dangerous and were prepared to issue warnings, but unfortunately, no evacuation procedure was in place.

The science of volcanoes has progressed to the point that the world's dangerous volcanoes can be identified and their potential hazards characterized from the volcanic deposits laid down in earlier eruptions. These hazard assessments can be used to introduce zoning regulations to restrict land use—the most effective measure to reduce casualties. Volcanoes can be monitored by instruments (as described in Box 5.1). Such signals as earthquakes, swelling of the volcano, and emissions warn of impending eruptions. People at risk can be evacuated if the authorities are organized and prepared. Volcanic eruptions cannot be prevented, but their catastrophic effects can be significantly reduced by a combination of science and enlightened public policy.

---

Mount Rainier (Washington); Mono-Inyo craters, Lassen Peak, and Mount Shasta (California); and several other peaks in the Cascade Range of the Pacific Northwest. (More are listed in Figure 5.30.) Mount Rainier is not far from the Seattle-Tacoma metropolitan area. An even more difficult problem of prediction is posed by such eruptions as that of the volcano Parícutin, which rose up with little warning from a small hole in a Mexican cornfield in 1943. The farm and whole towns were buried by ash and lava as the new volcano grew by repeated eruptions. Learning how to sense the movements of deep lava in relation to possible new outlets to the surface is a real challenge for geologists.

We have seen something of the beauty of volcanoes and also something of their destructiveness. Volcanoes contribute to our well-being in many ways. In Chapter 1 we mentioned that the atmosphere and the oceans may have originated in volcanic episodes of the distant past. Soils derived from volcanic materials are exceptionally fertile because of the mineral nutrients they contain. Emissions of volcanic rock, gases, and steam are also sources of important industrial materials and chemicals, such as pumice, boric acid, ammonia, carbon dioxide, and some metals. Seawater circulating through fissures in the ocean-ridge volcanic system is a major factor in the formation of ores. The thermal energy of volcanism is being harnessed in more and more places. Most of the houses in Reykjavík, Iceland, are heated by hot water tapped from volcanic springs. Geothermal steam, originating in water heated by contact with hot volcanic rocks below the surface, is exploited as a source of energy for the production of electricity in Italy, New Zealand, the United States, Mexico, Japan, and the former Soviet Union.

# SUMMARY

**Why does volcanism occur?** Volcanism occurs because molten rock inside the Earth rises to the surface, squeezed up by the weight of the overlying layers.

**What are the three major categories of lava?** Lavas are classified as felsic (rhyolite), intermediate (andesite), or mafic (basalt), on the basis of the decreasing amounts of silica and the increasing amounts of magnesium and iron they contain. The chemical composition and gas content of lava are important factors in the form an eruption takes.

**How are the structure and terrain of a volcano related to the kind of lava it emits and the type of its eruption?** Basalt can be highly fluid and can erupt in sheets from fissures to build a lava plateau. A shield volcano grows from repeated eruptions of basalt from vents. Silicic magma is more viscous and, when charged with gas, tends to erupt explosively. The pyroclastic debris piles up into a cinder cone or covers an extensive area with sheets of ash flow. A stratovolcano is built of alternating layers of lava flows and pyroclastic deposits. The rapid ejection of magma from a magma chamber a few kilometers below the surface, followed by collapse of the chamber's roof, results in a large surface depression, or caldera. Giant resurgent calderas are among the most destructive natural cataclysms.

**How is volcanism related to plate tectonics?** The ocean crust forms from basaltic magma that rises from the asthenosphere into fissures of the ocean ridge–rift system where plates separate. Intermediate and felsic materials tend to erupt on islands and continental margins, in regions where plates collide. They may originate by remelting of the sinking lithosphere or by assimilation of felsic seafloor sediments or felsic continental crust in rising magma. Within plates volcanism may occur above hot spots, which are manifestations of plumes of hot material that rise from deep in the mantle.

**What are some beneficial effects of volcanism?** Over the course of Earth's evolution, volcanic eruptions released the water and gases that formed the oceans and atmosphere. Geothermal heat drawn from areas of recent volcanism is of growing importance as a source of energy. An important ore-forming process occurs when water circulates in fissures between the surface (or seafloor) and underlying magma.

# KEY TERMS AND CONCEPTS

pahoehoe (p. 93)
aa (p. 93)
pillow lava (p. 94)
ash (p. 95)
volcanic tuff (p. 96)
volcanic breccia
    (p. 96)
pyroclastic flow (p. 96)

central vent (p. 97)
shield volcano (p. 98)
volcanic dome (p. 98)
cinder cone (p. 100)
composite volcano/stratovolcano
    (p. 100)
caldera (p. 103)
phreatic explosion (p. 103)

diatreme (p. 103)
fissure eruption (p. 105)
flood basalts (p. 106)
ash-flow deposits
    (p. 106)
lahar (p. 106)
aseismic ridge (p. 113)
hot spot (p. 113)

# EXERCISES

1. The asthenosphere has been identified as a major source of magma. Why? What forces magma to rise to the surface?

2. What is the difference between magma and lava? Give some examples of types of lavas and their coarse-grained, intrusive counterparts.

3. Describe the principal styles of volcanic eruptions and the deposits and landforms each style produces.

4. The accompanying photograph shows the remains of a building in a village 5 km south of El Chichón volcano, in southeastern Mexico. The

village was destroyed when El Chichón erupted in April 1982. From the debris and the bent reinforcing rods evident in the photograph, what can you conclude about the nature of the flow, its force, and its direction?

5. What is the association between plate boundaries and volcanism? Can the eruptive style and composition of volcanic deposits be correlated with plate boundaries?

6. Under what circumstances do lahars occur? Hot springs? Ash-flow deposits?

7. What features of volcanoes are the most dangerous?

8. What observations warn of impending eruptions?

R. I. Tilling/USGS.

# THOUGHT QUESTIONS

1. What public policy initiatives do the eruptions of Mount St. Helens, Mount Pinutabo, and Nevada del Ruiz suggest should be undertaken in such areas as zoning, land use, insurance, warning systems, and public education?

2. Do a risk-benefit analysis of volcanoes—that is, tabulate their dangers and their contributions to humankind—and decide whether you would prefer an Earth with or without them.

3. What have we learned about the Earth's interior from volcanoes?

# SUGGESTED READINGS

Decker, R. W., and B. Decker. 1989. *Volcanoes,* rev. and updated ed. New York: W. H. Freeman.

Dvorak, John J., Carl Johnson, and Robert I. Tilling. 1982. Dynamics of Kilauea volcano. *Scientific American* (August):46–53.

Edmond, John M., and Karen L. Von Damm. 1992. Hydrothermal activity in the deep sea. *Oceanus* (Spring):74–81.

Francis, Peter. 1983. Giant volcanic calderas. *Scientific American* (June):60–70.

Heiken, G. 1979. Pyroclastic flow deposits. *American Scientist* 67:564–571.

Simkin, T., L. Siebert, L. McClelland, D. Bridge, C. Newhall, and J. H. Latter. 1981. *Volcanoes of the World.* New York: Academic Press.

Tilling, Robert I. 1989. Volcanic hazards and their mitigation: progress and problems. *Reviews of Geophysics* 27(no. 2):237–269.

Vink, Gregory E., and W. Jason Morgan. 1985. The Earth's hot spots. *Scientific American* (April):50–57.

White, Robert S., and Dan P. McKenzie. 1989. Volcanism at rifts. *Scientific American* (July):62–72.

# WEATHERING AND EROSION

Weathering and erosion are the main geologic processes affecting the surface of the land. The rate of weathering depends on the climate, rock type, and length of time a rock has been exposed to the atmosphere. Weathering produces sedimentary particles and soils; erosion takes away weathering products, transporting them ultimately to the oceans. Chemical weathering breaks down feldspars to clay minerals and iron-rich silicates to iron oxides; dissolved ions and silica are carried away by waters in the ground and rivers on the surface. Carbonates weather by completely dissolving. Physical weathering breaks rocks into fragments and aids chemical weathering by exposing fresh rock to the atmosphere. The three great soil groups originate as the products of weathering in wet, temperate, and dry climates.

Weathering of easily
erodible sedimentary
rocks in Bryce Canyon
National Park, Utah.
*Peter Kresan.*

Unprotected, a newspaper left outside for a few months and exposed to rain, sun, ice, and wind will become tattered and illegible and finally will rot away. An old automobile left in a junkyard for years will rust and fall apart. We keep painting bridges and houses to preserve them and look for materials more durable than wood or steel. Solid as rock may seem, even the hardest rocks eventually weaken and crumble as they are exposed to water and the gases of the atmosphere over thousands of years.

The general process by which rocks are broken down by conditions at the Earth's surface is called **weathering.** We recognize two mechanisms by which rocks weather. **Chemical weathering** occurs when the minerals in a rock are chemically altered or dissolved. **Physical weathering** occurs when solid rock becomes fragmented by physical processes that do not change its chemical composition. The blurring or disappearance of lettering on old gravestones and monuments is attributable mainly to chemical weathering. The rubble of broken stone blocks and columns that were once stately temples in ancient Greece is primarily the result of mechanical (physical) weathering.

After rocks are weathered, their broken-up and altered products can be moved. Loose fragments on a slope will slide or roll downhill, pulled by gravity. Eventually the fragments will be carried away by a river, wind, or glacial ice. Rivers and glaciers are also able to cut into and break up solid bedrock. The removal of the debris of weathering, the cutting into bedrock, and the consequent sculpturing of landscape are components of the process of erosion, which we defined in Chapter 3 as the wearing away of solid rock.

Weathering and erosion are closely intertwined geological processes in nature. As a rock weathers, it becomes susceptible to erosion. As erosion takes away weathered solid material, it exposes fresh, unaltered rock to weathering. When a rock such as limestone or rock salt weathers by dissolving in rainwater, no solid residue is left. All of the material is simultaneously dissolved and carried away in the water as ions in solution. In this instance, weathering and erosion are inseparable.

Weathering and erosion are major geological processes in the rock cycle, as we saw in Chapter 3. Working together with the other elements of the rock cycle, tectonics and volcanism, weathering and erosion change the form of Earth's surface and alter rock materials. After tectonics and volcanism have made mountains, weathering and erosion wear them away. Weathering and erosion convert igneous and other rocks into sediment and form soil. The material dissolved during chemical weathering contributes most of the dissolved material in the oceans.

**Soil,** as the basis for agriculture, is one of our most valuable natural resources. Soil is composed of fragments of bedrock, clay minerals formed by the alteration of bedrock minerals, and organic matter produced by the organisms that live in it. The weathering processes that make fertile soils are of intense interest to agricultural scientists. So are the thinning and stripping of soils by erosional processes, which are now major environmental and economic concerns in many countries.

## GEOLOGIC FACTORS THAT CONTROL WEATHERING

All rocks weather, but not all in the same way or at the same rate. After a hundred years, a limestone headstone in New England may appear to have melted away, much as the name on a bar of soap disappears after a few washes. Yet a monument of alabaster, a much more easily weathered rock, appears untouched after several thousand years in the desert of North Africa. Everyday observation of rock weathering under different conditions points to four key factors that control the disintegration of rocks and minerals: rock type, climate, soil, and time.

## Rock Type

The carved letters on a recently erected gravestone, whether the stone is an easily dissolved rock, such as limestone, or a rock resistant to chemical alteration or dissolution, such as granite, stand out in sharp relief from the stone's polished surface. When we look at inscriptions on gravestones and learn that they were erected a century ago, we see that different rocks weather chemically at very different rates (Figure 6.1). Limestone weathers rapidly in a moderately rainy climate; granite lasts longer. After a hundred years, limestone's surface dulls and the outlines of the letters become rounded and blurred, while granite shows only minor changes. But after several hundred years, a granite monument will have weathered appreciably, its surface and letters also somewhat dulled and blurred. Thus even a resistant rock, given enough time, will ultimately decay.

If we look more closely at the weathered granite, perhaps with a hand lens, we see that its mineral grains have not all weathered at the same rate. The feldspars show signs of corrosion, and their surfaces are covered with a thin layer of soft clay. The outer layers of the grains have undergone a change in chemical composition and turned into a new mineral, but the quartz crystals are fresh and unaltered. A comparison of many kinds of weathered rocks reveals that different minerals, and the rocks they make up, have a wide range of weathering rates.

Rock type also affects physical weathering. Shale, a sedimentary rock that splits easily across thin bedding planes, breaks up into small pieces so quickly that only a few years after a new road is cut through a shale, the rock will become rubble. But granite monuments may remain unbroken and uncracked even after centuries, though they may show evidence of some chemical weathering.

## Climate

A tour of graveyards across the continent, from the lower United States to northern Canada and Alaska, reveals that different climates produce differences in rates of chemical and physical weathering. Old gravestones in hot, humid Florida are badly chemically weathered, but those of the same age in the equally hot but arid Southwest are hardly touched. And gravestones in cold, dry arctic regions are even less chemically weathered than those in the Southwest. We conclude that **climate**—the amount of rainfall and the temperature—exerts strong control over the rate of chemical weathering, speeding it up in warmer and more humid regions and slowing it down in colder and dryer ones.

Physical weathering, in contrast, may be active in climates in which chemical weathering is minimal. In cold climates, moisture chemically weathers rocks only slightly, because it is frozen and thus chemically unreactive. But water that freezes in cracks can physically push rock apart. Lack of water causes chemical weathering to be slow in arid regions, but physical weathering can still proceed from causes unrelated to climate, such as the heat of forest fires and the weakening of rock by even small chemical changes. Ancient tombs and monuments in the deserts of Egypt are cracked and broken, while the figures carved in their stone surfaces remain relatively sharp.

## Effects of Soil

The presence or absence of soil, itself a product of weathering, is another factor in chemical and physical weathering. An old nail that has been buried in soil usually will be so badly rusted that you can snap it like a matchstick. Yet a nail pried from the wood of a centuries-old house may still be strong, covered with only a thin layer of rust. Similarly, a mineral in soil in a lowland valley may be badly altered and corroded, while the same mineral exposed in a nearby cliff of bedrock will be much less affected by weathering. Soil promotes weathering by providing a moist and acidic environment that alters or dissolves minerals. The vegetation, bacteria, and animals living on and in the soil are responsible for much of the acidity, while plant roots aid physical weathering.

**FIGURE 6.1** The rate at which a gravestone weathers is affected by the properties of the rock. *Ric Ergenbright Photography.*

Though the bare rock of a cliff is exposed to occasional rain, most of the time the rock is dry and weathering proceeds very slowly. No soil forms on the cliff because any particles that are created by weathering are quickly washed down by rain to the valley, where they can accumulate.

Soil production is a *positive-feedback process;* that is, the product of the process works to advance that process. Once soil starts to form, it works as a geological agent to weather rock more rapidly, and more soil is formed.

## The Time Factor

The longer the time available for weathering, the greater the alteration, dissolution, and physical breakup of rock. Rocks that have been exposed at the surface for many thousands of years have thick rinds of weathered material surrounding the fresh, unaltered rock. The lavas and ash deposits newly extruded from volcanoes, in contrast, are completely unweathered. Because we know the dates of modern eruptions, such as those of Mount St. Helens in 1980, we can measure the times required for various degrees of weathering to occur. In the years that have passed since those eruptions, the ash deposits have become appreciably weathered and altered to other minerals. In the same period of time, lava, a different rock type, is still relatively fresh. The difference occurs mainly because the ash is made up of very small particles, which weather faster than more massive lavas.

These simple kinds of observations make it possible to summarize the important factors (Table 6.1) in chemical and physical weathering:

1. Minerals and rocks of different compositions weather at different rates.
2. Climate, which includes both rainfall and temperature, strongly affects chemical weathering: warmth and heavy rainfall accelerate chemical weathering, whereas cold and aridity hinder it. Physical weathering, which is also aided by water, may be appreciable in cold or arid climates, even though chemical weathering proceeds very slowly.
3. The presence of soil, itself a product of weathering, furthers weathering.
4. The longer a rock is exposed at the surface, the more weathered it becomes.

Although chemical and physical weathering are not completely separable, it is simpler to discuss them separately first and show their relationships afterward.

## CHEMICAL WEATHERING

Chemical weathering results from chemical reactions between minerals in rocks and air and water. Some minerals dissolve and others combine with water and components of the atmosphere, such as oxygen and carbon dioxide, to form new chemical compounds: minerals formed by weathering. As we have seen, the factors that control these chemical reactions can be deduced in part from observations in the field. We can get a better picture of the mechanisms of chemical weathering if we combine observations with laboratory experiments that simulate the natural process. We begin our investigation with feldspar, the most abundant mineral in Earth's crust.

---

**TABLE 6.1**

## Major Factors Controlling Rates of Weathering

| | WEATHERING RATE | | |
|---|---|---|---|
| | **SLOW** | | **FAST** |
| *Mineral solubility* | Low (e.g., quartz) | Moderate (e.g., pyroxene, feldspar) | High (e.g., calcite) |
| *Rainfall* | Low | Moderate | Heavy |
| *Temperature* | Cold | Temperate | Hot |
| *Vegetation and other life* | Sparse | Moderate | Lush |
| *Soil cover* | Bare rock | Thin to moderate soil | Thick soil |

# Chemical Weathering of Feldspar

Feldspar is a key mineral in a great many igneous, sedimentary, and metamorphic rocks. Feldspar is also representative of the many other kinds of rock-forming silicate minerals. Thus an understanding of feldspar's behavior during weathering contributes much to our grasp of the weathering process in general, both because of the overwhelming abundance of silicate minerals in the Earth and because the same chemical processes of dissolution and alteration that characterize feldspar weathering affect other kinds of minerals.

We observed that feldspars in weathered granite gravestones have been corroded and altered. A more extreme example of feldspar weathering can be found in granite boulders in soils of the humid tropics. Here all the factors that promote weathering—heavy rainfall, high temperature, the presence of soil, and abundant organic activity—are at a maximum. These boulders are so weakened that they can easily be kicked or pounded into a heap of loose mineral grains. Most of the feldspar particles in these boulders have been altered to clay. Greatly magnified under an electron microscope, any remaining feldspar grains can be seen to be corroded and coated with a clay rind, in contrast to clear and unaltered quartz crystals (Figure 6.2).

In a sample of unweathered granite, the rock is hard and solid because the interlocking network of quartz, feldspar, and other crystals holds it tightly together with strong cohesive forces. But when the feldspar is altered to a loosely adhering clay, the network is weakened and the mineral grains separate (Figure 6.3). In this instance, chemical weathering,

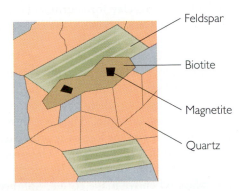

(a) Fresh granite

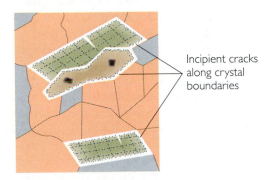

(b) Feldspar, biotite, and magnetite start to decay

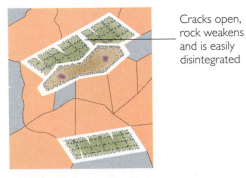

(c) Feldspar, biotite, and magnetite decay extensively

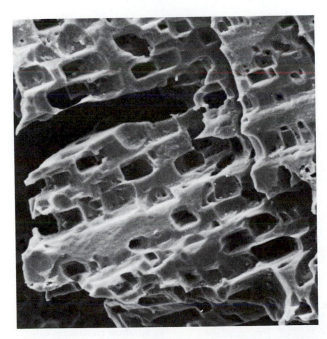

**FIGURE 6.2** A scanning electron micrograph of feldspar etched and corroded by chemical weathering in soil. (From R. A. Berner and G. R. Holden, Jr., "Mechanism of Feldspar Weathering: Some Observational Evidence," *Geology,* vol. 5, 1977, p. 369.)

**FIGURE 6.3** Microscopic views of stages in the disintegration of granite. As weathering proceeds, grain boundaries weaken and the rock disintegrates into fragments.

by producing the clay, also promotes physical weathering, the disintegration of the rock.

The white to cream-colored clay produced by the weathering of feldspar is **kaolinite,** a hydrous (that is, containing water in the crystal structure) aluminum silicate. Pure kaolinite—named for Kao-ling, a hill in southwestern China where it was first obtained—is the raw material of pottery and china. The Chinese had used this clay for centuries before a European visitor took some home in the eighteenth century.

Only in the severely arid climates of some deserts and polar regions does feldspar remain relatively unweathered. This observation points to water as an essential component of the chemical reaction by which feldspar turns to kaolinite. That reaction results in the gain of water and the loss of several chemical components from the solid feldspar. The process can be understood by analogy to the chemical reaction that takes place when we make coffee. Solid coffee, analogous to unweathered feldspar, chemically reacts with hot water to make a solution—the liquid coffee. Caffeine and other components of the bean are extracted from the solid by the reaction, leaving behind spent coffee grounds, analogous to the kaolinite, as a residue (Figure 6.4).

The finer the coffee beans are ground, the more coffee is extracted from the beans and the stronger the brew becomes. The rate of a chemical reaction increases with an increase in the surface area of the solid (the total of the surface areas of all the grains used) that is exposed to the fluid reacting with it. Because the only part of a solid available for reaction with a fluid is its surface, the more we increase the surface area, the more we speed up the reaction. As we grind coffee finer, we increase the ratio of surface area to volume of the particles. The same holds true for minerals and rocks. Smaller fragments yield more surface area. This ratio increases greatly as the average particle size decreases, as shown in Figure 6.5.

We can now write the beginnings of an unbalanced chemical reaction for the weathering of the common feldspar of granite, orthoclase ($KAlSi_3O_8$), which is made up of potassium (K), aluminum (Al), silicon (Si), and oxygen (O):

$$\text{feldspar} + \text{water} \rightarrow \text{kaolinite}$$
$$KAlSi_3O_8 \quad\quad H_2O \quad\quad Al_2Si_2O_5(OH)_4$$

## DISSOLVING FELDSPAR IN THE LABORATORY

To learn more about the loss of components as feldspar weathers, we can perform a simple chemical experiment: immerse feldspar in pure water and analyze the solution for the kinds of material that have dissolved. First we grind the feldspar to a fine powder, to speed its dissolution by exposing more surface area to the water. The feldspar dissolves very slowly in the water, and after some time we take samples from the solution and analyze them. We find small amounts of dissolved potassium and silica ($SiO_2$) in the water. We can now write a qualitatively more complete but still unbalanced chemical reaction:

$$\text{feldspar} + \text{water} \rightarrow$$
$$KAlSi_3O_8 \quad\quad H_2O$$

$$\text{kaolinite} + \underset{\text{silica}}{\text{dissolved}} + \underset{\substack{\text{potassium}\\\text{ion}}}{\text{dissolved}}$$
$$Al_2Si_2O_5(OH)_4 \quad\quad SiO_2 \quad\quad K^+$$

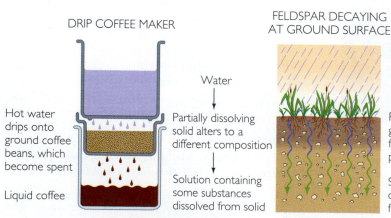

DRIP COFFEE MAKER

Water

Hot water drips onto ground coffee beans, which become spent

Liquid coffee

Partially dissolving solid alters to a different composition

Solution containing some substances dissolved from solid

FELDSPAR DECAYING AT GROUND SURFACE

Rainfall filters into ground, altering feldspar in rock particles to kaolinite

Soil water containing dissolved substances from feldspar

**FIGURE 6.4** The process by which feldspar decays is analogous to the making of coffee. In both processes, water dissolves some of the solid, leaves behind an altered material, and produces a solution containing substances drawn from the original solid.

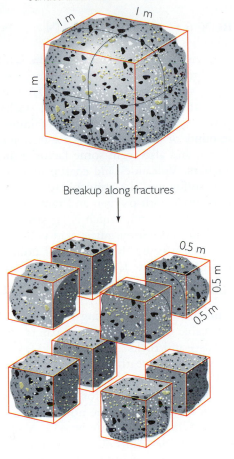

Single boulder, approximately 1 m on a side
Volume = 1 m³
Surface area = 6 m²

1 m   1 m

1 m

Breakup along fractures

0.5 m

0.5 m

0.5 m

8 fragments, each approximately 0.5 m on a side
Volume = (0.5)³ × 8 = 1 m³
Surface area = 12 m²

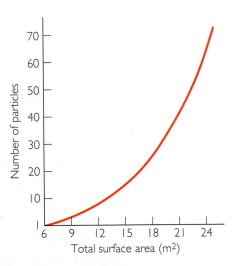

**FIGURE 6.5** As a rock mass breaks into smaller pieces, much more surface becomes available for the chemical reactions of weathering. The graph shows the increase of total surface area as the number of roughly equal-sized particles increases with no change in the total volume (1 m³).

Two major points about this reaction give us information on the gains and losses of material that occur as feldspar weathers: *(1) The potassium and silica dissolved from the feldspar appear as dissolved materials in the water solution. (2) Water is used up in the reaction; it is absorbed into the kaolinite crystal structure.* The absorption of water, or **hydration,** is one of the major processes of weathering.

**CARBON DIOXIDE SPEEDS FELDSPAR WEATHERING**  The reaction of feldspar with pure water in a laboratory is a very, very slow process. It would take thousands of years to weather even a small amount of feldspar under such conditions, so this reaction cannot account for the more rapid weathering we observe widely in nature. If we wanted to speed up weathering in the laboratory, we could add a strong acid, such as hydrochloric acid, and dissolve the feldspar in a few days. An acid is a substance that releases hydrogen ions ($H^+$) to a solution; a strong acid produces abundant hydrogen ions, a weak one relatively few. Since hydrogen ions have a strong tendency to combine chemically with other substances, acids make excellent solvents.

On the Earth's surface, the most common natural acid—and the one responsible for increasing weathering rates—is carbonic acid ($H_2CO_3$). This weak acid is formed by the solution in rainwater of a small amount of carbon dioxide ($CO_2$) gas from the atmosphere:

$$\text{carbon dioxide} + \text{water} \rightarrow \text{carbonic acid}$$
$$CO_2 \qquad\qquad H_2O \qquad\qquad H_2CO_3$$

We are familiar with everyday solutions of carbon dioxide in water: soft drinks become carbonated when pressurized carbon dioxide is pumped into the liquid. In this process a large quantity of carbon dioxide is dissolved in the beverage. As pressure is lowered, the dissolved gas bubbles out of solution, both the bubbles and the carbonic acid contributing to the "bite" on the tongue. The amount of carbon dioxide dissolved in water decreases as the amount of gaseous carbon dioxide in contact with the water decreases. Thus the amount of carbon dioxide dissolved in rainwater is small because the amount of carbon dioxide gas in the atmosphere is small. As carbon dioxide from the burning of oil, gas, and coal increases in the atmosphere, the amount of carbonic acid in rain increases slightly. Most of the acidity of acid rain, however, comes from sulfur and nitrogen gases (see Box 6.1).

About 0.03 percent of the molecules in the atmosphere are carbon dioxide. That number may seem small, but it makes carbon dioxide the fourth most abundant gas, just behind argon (0.9 percent), oxygen (21 percent), and nitrogen (78 percent). The amount of carbonic acid formed in rainwater is very small, only about 0.0006 gram per liter, but that is enough to weather feldspars and dissolve great quantities of rock over a long time. Adding carbon dioxide, in the form of carbonic acid, to our equation, we can now write the full balanced form of the weathering reaction:

$$\underset{2KAlSi_3O_8}{feldspar} + \underset{2H_2CO_3}{\underset{acid}{carbonic}} + \underset{H_2O}{water} \rightarrow$$

$$\underset{Al_2Si_2O_5(OH)_4}{kaolinite} + \underset{4SiO_2}{\underset{silica}{dissolved}} + \underset{2K^+}{\underset{potassium}{dissolved}} + \underset{2HCO_3^-}{\underset{ion}{\underset{bicarbonate}{dissolved}}}$$

The bicarbonate ions are the residue of the carbonic acid. They are left behind as the hydrogen ions from the acid combine with the oxygens of the feldspar to form the water in the kaolinite structure (Figure 6.6). Because hydrogen ions are used up in the reaction, the solution becomes less acidic. The dissolved potassium and silica are carried away by rain and river waters and ultimately transported to the ocean. The solid alteration product, clay, becomes part of the soil or is carried away as sediment.

**HOW FELDSPAR WEATHERS IN NATURE**  Now that we understand the chemical reaction by which rainwater weathers feldspar, let us return to the field to see how the chemical reaction explains how weathering works on outcrops and in the soil. We have noted that feldspars on bare rock surfaces are much better preserved than those buried in damp soils. The equation for feldspar weathering gives us a clue to the reason. First, the reaction requires water. The feldspars on a bare rock weather only while it is moist with rainwater; during all the dry periods, the only moisture that touches the bare rock is dew. In moist soil, however, the feldspar is constantly in contact with small amounts of water in the pores, the spaces between grains in the soil. Thus in moist soil feldspar weathers continuously.

Second, there is more acid in the pore water of the soil than in rainwater. As rainwater sinks into the soil, it carries not only its original carbonic acid but additional carbonic acid and other acids produced by the roots of plants, by the many insects and other animals that live in the soil, and by the bacteria that degrade plant and animal remains.

## BOX 6.1 ACID RAIN

In many industrialized areas of the world, the air is greatly polluted with sulfur-containing gases, such as sulfur dioxide ($SO_2$). These gases are emitted from the smokestacks of power plants that burn coal containing large amounts of the mineral pyrite, iron sulfide ($FeS_2$), and also from some factories and smelters. Volcanoes and coastal marshes also emit sulfur into the atmosphere. The sulfur gases react with oxygen and rainwater to form sulfuric acid, which is far stronger than carbonic acid. Some nitric acid is formed in the same way by the reaction of polluting nitrogen gases with the atmosphere. Small amounts of these strong acids turn harmless rainwater into acid rain. Although acid rain is much too weak to sting the skin, it does noticeable damage to fabrics, paints, and metals—and rocks.

Acid rain weathers our stone monuments and outdoor sculptures at a rapid rate. Acid rain is also responsible for massive kills of fish in many lakes in the northeastern United States, Canada, and Scandinavia. A study of 40 lakes in the Adirondack Mountains of New York, for example, showed that as early as 1975, half of them no longer contained any fish, whereas in 1930 their fish populations had been abundant and varied. The number of fish in these lakes is declining steadily as the acidification continues.

The relationship between the burning of coal and acid rain has been firmly established. Agencies of the U.S. and Canadian governments and independent scientific panels have tracked sulfur gases emitted by smokestacks to downwind locations where they are precipitated as acid rain. Careful tracing of pollutants from their sources to the sites of acid rain is necessary to demonstrate that it is coal-burning power plants that are responsible for much of the acid rain. Concerned scientists have recommended restriction of sulfur emissions from power plants and smelters, but so far political resistance has prevented effective control. In 1991, after many years of wrangling, the U.S. Congress

A monument before and after deterioration caused by acid rain. *Westfälisches Amt für Denkmalpflege.*

passed the Clean Air Act, which requires a reduction of 10 million tons by the year 2000 in the amount of sulfur emitted annually by coal-burning power plants. The problem is equally critical in Europe, where scientists have noted damage to lakes and forests caused by acid rain.

Emissions of carbon dioxide from the burning of coal and oil are another cause for concern. As the amount of carbon dioxide in the atmosphere increases, as it is expected to do (see Chapter 22), the amount of carbon dioxide in rainwater will increase too, and so will the acidity of the water.

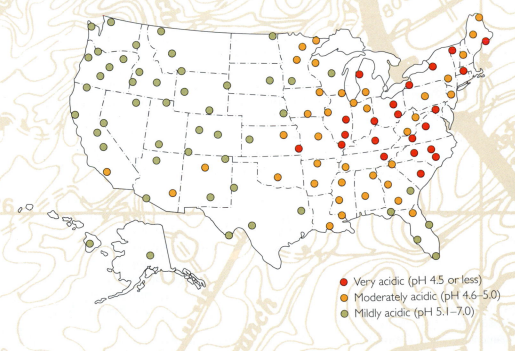

Acidity of precipitation for December 24, 1990, to January 20, 1991.

- ● Very acidic (pH 4.5 or less)
- ● Moderately acidic (pH 4.6–5.0)
- ● Mildly acidic (pH 5.1–7.0)

Rock weathers more rapidly in the tropics than in temperate and cold climates because the weathering rate increases as the temperature rises. The main reason is that plants and bacteria grow quickly in warm, humid climates, contributing the acids that promote weathering. In addition, most chemical reactions, weathering included, speed up with an increase in temperature.

## Chemical Weathering of Other Silicates

Just as feldspar weathers to form kaolinite, silicate minerals other than feldspar, such as amphibole, pyroxene, and olivine, may also weather to form clays. The weathering reactions of these other silicates follow the general course of feldspar weathering. As the mineral is weathered, water is absorbed and silica and ions such as sodium, potassium, calcium, and magnesium are lost to the solution. In general, clay minerals are formed by these reactions. Clay minerals are water-containing silicates derived from alteration of other silicates. The kinds of clays formed depend on the composition of the parent silicates and the climate. For example, the clay mineral montmorillonite, which swells when it absorbs large quantities of water, forms from the weathering of volcanic ash. This clay mineral is also common as a weathering product in semiarid environments, such as the high plains of the southwestern United States.

Because the rock-forming silicates constitute so large a fraction of the Earth's crust and weathering at the surface is so widespread, clay minerals are a principal component of soils and sediments everywhere. Deposits pure enough to be used as the raw material for pottery, chinaware, and industrial ceramics are found in some uncommon soils and sedimentary rocks. Brick, clay tiles, and other structural ceramics require less pure material.

Not all silicates weather to form clay minerals. Some rapidly weathering silicates, such as some pyroxenes and olivines, may dissolve completely in humid climates, leaving no residue of clay mineral. Quartz, one of the slowest of the abundant silicate minerals to weather, also dissolves without forming any clay mineral. And materials other than clay minerals may form from silicate weathering. **Bauxite,** for example, a clayey ore composed of aluminum hydroxide that is the major source of aluminum metal, is formed when clay minerals derived from the weathered silicates weather further, losing all their silica and ions other than aluminum. It is found in tropical regions with heavy rainfall, where weathering is intense.

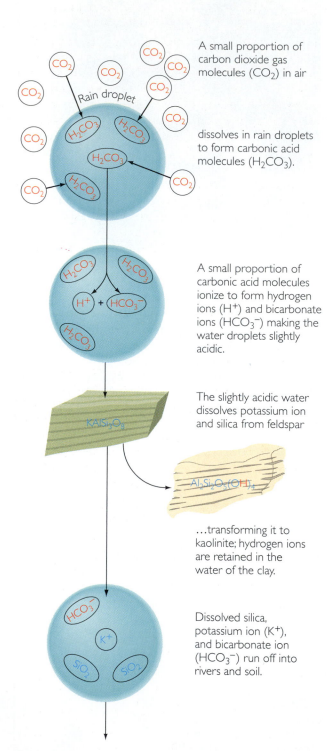

A small proportion of carbon dioxide gas molecules ($CO_2$) in air

dissolves in rain droplets to form carbonic acid molecules ($H_2CO_3$).

A small proportion of carbonic acid molecules ionize to form hydrogen ions ($H^+$) and bicarbonate ions ($HCO_3^-$) making the water droplets slightly acidic.

The slightly acidic water dissolves potassium ion and silica from feldspar

...transforming it to kaolinite; hydrogen ions are retained in the water of the clay.

Dissolved silica, potassium ion ($K^+$), and bicarbonate ion ($HCO_3^-$) run off into rivers and soil.

**FIGURE 6.6** The weathering of feldspar is a reaction with rainwater containing carbon dioxide. Two products are formed: kaolinite clay and a solution containing dissolved silica, potassium ion, and bicarbonate ion.

# Chemical Weathering of Iron Silicates

Weathering is responsible for the formation of some of our most valuable mineral resources, iron ores. Iron is one of the eight most abundant elements in the Earth's crust, and iron metal, the chemical element in its pure form, is the main component of the steel used in industry. Metallic iron is rarely found in nature; it is found only in certain kinds of meteorites that fall to Earth from other places in the solar system. Most of the iron ores used for the production of iron and steel are composed of iron oxide minerals that were formed originally as the weathering products of iron-rich silicate minerals, such as pyroxene and olivine. The iron released by dissolution of these minerals combines with oxygen from the atmosphere to form iron oxide minerals by a chemical reaction called **oxidation.** Thus oxidation is a chemical combination of an element with oxygen. Oxidation is defined more generally as a chemical reaction in which an ion or element loses electrons (see Chapter 2). Like hydration, oxidation is one of the important chemical weathering processes.

The iron in minerals may be present in one of several forms. In the metallic iron of meteorites, the iron atoms are uncharged; that is, they have neither gained nor lost electrons by reaction with another element. In silicate minerals such as pyroxene, the iron is **ferrous,** $Fe^{2+}$; that is, the iron atoms have lost two of the number of electrons they would have in the metallic form and thus have become ions. The iron in the most abundant iron oxide at the Earth's surface, **hematite,** $Fe_2O_3$, is **ferric,** $Fe^{3+}$; that is,

these iron atoms have lost three electrons of the number in the metallic form. Iron ions oxidize by losing an additional electron, going from 2+ (ferrous) to 3+ (ferric). All of the several kinds of iron oxides formed at the surface of the Earth are ferric. The electrons lost by the iron are gained by oxygen atoms as they become oxygen ions ($O^{2-}$). Thus oxygen atoms from the atmosphere oxidize ferrous iron to ferric iron. As we know well from our experience with rusting iron, iron metal is also oxidized by contact with the atmosphere, forming ferric iron oxide.

When an iron-rich mineral such as pyroxene dissolves in water, the silicate structure dissolves and its ferrous iron is oxidized by oxygen to the ferric form. Because of the strength of the chemical bonds between ferric iron and oxygen, ferric ion is very insoluble in most natural surface waters, precipitating to form ferric iron oxide.

We can show the overall weathering reaction by the equation:

$$\text{iron pyroxene} + \text{oxygen} \rightarrow \text{hematite} + \text{dissolved silica}$$
$$4FeSiO_3 \qquad O_2 \qquad 2Fe_2O_3 \qquad 4SiO_2$$

Although the equation does not show it explicitly, water is required for this reaction to proceed.

Iron minerals, which are widespread, weather to the characteristic red and brown colors of oxidized iron (Figure 6.7). Iron oxides are found as coatings and encrustations that color soils and weathered surfaces of iron-containing rocks. The red soils of Georgia and other warm, humid regions are colored by

**FIGURE 6.7** Red and brown iron oxides color weathering rocks in Arches National Park, Utah. *Peter Kresan.*

iron oxides. Iron minerals weather so slowly in frigid regions that iron meteorites frozen in the ice of Antarctica are almost entirely unweathered.

## Chemical Weathering of Carbonates

Even olivine, the silicate mineral that weathers most rapidly, is relatively slow to dissolve in comparison with some other rock-forming minerals. Limestone, made of the calcium and magnesium carbonate minerals calcite and dolomite, is one of the rocks that weather most quickly in humid regions. Caves are formed in limestone formations as water in the ground dissolves away great quantities of the carbonate minerals below the surface. Farmers and gardeners add ground limestone to soil to counterbalance acidity because it dissolves rapidly over the growing season. Old limestone buildings also show the effects of dissolution by rainwater (Figure 6.8). When limestone dissolves, no clay minerals are formed. The solid dissolves completely, and its components are carried off in water solution.

**FIGURE 6.8** Weathered limestone blocks of 2500-year-old Greek ruins at Segesta, Sicily, show pitted, etched surfaces caused by chemical solution. *Ric Ergenbright Photography*.

Carbonic acid promotes the dissolution of limestone as well as the weathering of silicates. The overall reaction by which calcite, the major mineral of limestones, dissolves in rain or other water containing carbon dioxide is

$$\text{calcite} + \text{carbonic acid} \rightarrow \text{calcium ion} + \text{bicarbonate ion}$$
$$CaCO_3 \qquad H_2CO_3 \qquad\qquad Ca^{2+} \qquad\qquad 2HCO_3^-$$

This reaction proceeds only in the presence of water, which contains the carbonic acid and dissolved ions. When calcite dissolves, the calcium and bicarbonate ions are carried away in solution. When dolomite ($CaMg[CO_3]_2$), another abundant carbonate mineral, dissolves, equal amounts of magnesium and calcium ions are produced.

Because carbonate minerals dissolve faster and in greater amounts than any silicates, the weathering of limestone accounts for more of the total chemical weathering of the land surface each year than that of any other rock, even though much larger areas are covered by silicate rocks.

## Chemical Stability and Relative Weathering Rates

We can compare the weathering rates of many minerals and see that they cover a great range, from the rapid rates of carbonates to the slow rate of quartz. Differences in weathering rate account for a variety of sediment compositions (discussed in Chapter 7) and many differences in landscape (discussed in Chapter 16). The rates at which minerals weather reflect the chemical stability of the minerals under weathering conditions. If we know the chemical stabilities of minerals in the presence of water at given surface temperatures, we can predict the intensity of weathering in any given area.

Chemical stability can be described in much the same way as mechanical stability. A book lying flat on a table is stable; that is, it will remain in the same position unless it is moved. A book balanced on its edge is unstable; that is, a bump or a push will cause it to fall to the flat, stable position. **Chemical stability** is a measure of the tendency for a chemical substance to remain in a given chemical form rather than to react spontaneously to become a different chemical form. Thus iron metal in a meteorite in outer space, where it is exposed to no oxygen or water, is chemically stable and remains unaltered for billions of years. If that meteorite falls to Earth and is exposed to oxygen and water, it is chemically unstable and spontaneously reacts to form iron oxide.

**TABLE 6.2**

Stability of Common Minerals under Weathering Conditions
Compared with Bowen's Reaction Series

| STABILITY OF MINERALS | BOWEN'S REACTION SERIES[1] |
|---|---|
| **Most stable** | |
| Iron oxides (hematite) | |
| Aluminum hydroxides (gibbsite) | |
| Quartz | **Last to crystallize** |
| Clay minerals | Quartz |
| Muscovite mica | Muscovite |
| Potassium feldspar (orthoclase) | Orthoclase          Albite |
| Biotite mica | Biotite |
| Sodium-rich feldspar (albite) | |
| Amphibole | Amphibole |
| Pyroxene | Pyroxene |
| Calcium-rich feldspar (anorthite) | Anorthite |
| Olivine | Olivine |
| Calcite | |
| Halite | |
| **Least stable** | **First to crystallize** |

INCREASING STABILITY — INCREASING RATES OF WEATHERING

Mafic minerals / Plagioclase feldspars

[1] The arrow labeled "Mafic minerals" denotes the discontinuous crystallization series; the arrow labeled "Plagioclase feldspars" denotes the continuous crystallization series.

Whereas the book on the table is said to be simply stable or unstable, the stability of chemical substances is variable: some substances are more stable in one environment than in another. Feldspar, for example, is stable at the conditions found deep in the Earth's crust (high temperatures and small amounts of water) but unstable at the conditions of the surface (lower temperatures and abundant water).

The *solubility* of a mineral in water, measured by the amount dissolved when the water solution is saturated—that is, when the water cannot hold any more of the dissolved substance—partially determines its relative stability under weathering. The higher its solubility, the less its stability. For example, rock salt, which is highly soluble in water, is unstable under weathering conditions and is leached from a soil by even small amounts of water. Quartz, in contrast, is fairly stable under most weathering conditions because its solubility in water is very low, only about 0.008 gram per liter of water.

The rate of *dissolution,* measured by the amount of a chemical substance dissolved in an unsaturated solution in a given length of time, also contributes to the stability. The faster a mineral dissolves, the less stable it is. Thus quartz dissolves at a much slower rate than feldspar, and primarily for that reason it is more stable than feldspar under weathering conditions.

In regions where chemical weathering is intense, only the most stable minerals will be left on an outcrop or in the soil. Where weathering is at a minimum, many unstable minerals remain. When we compare the stabilities of all the common rock-forming minerals, we get a series from salt and carbonate minerals at the least stable end to iron oxides at the most stable end (Table 6.2). In a way, the position of the silicate minerals in this weathering order based on chemical stability is roughly the reverse of the order of crystallization of silicate minerals from a basalt magma given in the Bowen reaction series, which we discussed in Chapter 4. Whereas olivine and calcium plagioclase are the first minerals to crystallize as the melt cools, indicating their stability under high temperatures and pressures, they are the least stable and the first to disappear when they are exposed at the surface, where temperature and pressure are greatly reduced.

The relative stability revealed by both weathering order and reaction series is determined by the nature of the chemical bonds that characterize the crystal structures of the silicate minerals (see Chapter 2). The least stable under weathering conditions are isolated tetrahedra, as in the mineral olivine. Somewhat more stable are the single-chain silicates, the pyroxenes, and the double chains, the amphiboles. Next in order of stability are the sheet silicates, the micas and clay minerals; the framework silicate, quartz; and the iron oxides.

# PHYSICAL WEATHERING

Weathered outcrops in arid regions are covered by a rubble of various-sized fragments, from individual mineral grains only a few millimeters in diameter to boulders more than a meter across. The differences in size reflect varying degrees of physical weathering and patterns of breakage of the parent rock. As physical weathering continues, the larger particles are cracked and broken into smaller ones. Some of these fragments are broken along planes of weakness in the parent rock. The smaller grains of sand are individual crystals of different minerals broken apart from one another. The finest materials are clays, formed from the chemical weathering of silicates.

Though there are signs of chemical weathering, such as clays and altered feldspars, physical weathering dominates. Yet chemical weathering has prepared the way for physical weathering, as noted earlier. Even slight alteration of feldspar and other minerals in a rock weakens the cohesive forces that hold the crystals together. As small cracks form and widen, individual quartz or feldspar crystals are freed by a combination of physical and chemical weathering and fall to the ground. Fractures enlarge and large blocks of rock are separated from the outcrop.

While chemical weathering promotes physical weathering, fragmentation in turn promotes chemical weathering by opening channels where water and air can penetrate and react with minerals inside the rock. Breakup into smaller pieces exposes more surface area to weathering and so speeds the chemical reactions. Both chemical and physical weathering are affected by the activity of organisms, from bacteria to tree roots, all working in ways that destroy the rock. Once a crack has been widened and a tree takes root in it, for example, the physical force of the growing root system helps pry the crack apart.

Physical weathering is not always so dependent on chemical weathering. There are processes by which unweathered rock masses are broken up, such as the freezing of water in cracks. And some rocks are made particularly susceptible to physical weathering by fracturing produced by tectonic forces as rocks are bent and broken during mountain building. On the Moon, physical fragmentation works alone, for there is no water to make chemical weathering possible. On that lifeless terrain, rocks are broken into boulders and fine dust by the impacts of large and small meteorites.

**FIGURE 6.9** Weathered, enlarged joint patterns developed in several directions in rocks at Joshua Tree National Monument, California, in a desert climate. *Travis Amos.*

**FIGURE 6.10** Tree roots in fractured rock. *Peter Kresan.*

## Rock Breakage

Rocks have natural zones of weakness along which they tend to crack. Such sedimentary rocks as sandstone and shale, formed originally by the deposition of successive layers of sediment, tend to break between the layers. The parallel planes of fractures developed in such metamorphic rocks as slate make them split easily to form roofing tiles. Granites and other rocks that are massive—that is, large masses that show no changes in rock type or structure—tend to fracture along regularly spaced planes called **joints** (see Chapter 10) at intervals of one to several meters (Figure 6.9).

These joints and other more irregular fractures form while the rock is still deeply buried in the crust. They open slightly when the weight of tons of overlying rock is removed as the rocks are gradually brought to the surface by uplift and erosion. The process is not unlike the way in which poorly glued joints in a wooden chair separate as the clamps that held them together while the glue dried are released. Once the cracks are open a little, both chemical and physical weathering work to widen the fracture. Bacteria and plant roots promote chemical weathering, while the growth of plant roots exerts pressure that helps wedge the openings apart (Figure 6.10).

As we noted earlier, one of the most efficient mechanisms for widening cracks is the freezing of water. Water expands as it freezes, and the outward force it exerts is strong enough to wedge open the crack and split the rock, just as it can crack open the engine block of a car that is not protected by antifreeze (Figure 6.11).

Minerals crystallizing from solutions in cracks can exert great expansive forces that split rocks. This phenomenon is most common in arid regions, where dissolved substances derived from chemical weathering may crystallize as a solution evaporates. The minerals are commonly calcium carbonate, occasionally gypsum, and rarely, rock salt.

Are there other ways in which rock can be cracked and fragmented? One recurring idea is that rocks break from the daily alternation of hot days and cold nights in a desert. At twilight on the desert the temperature may drop from 43 to 15°C in an hour. Part of the breakage process might be weakening of the rock caused by its expansion in the heat and contraction during the cold. We know that a fire built over a rock surface can crack the rock. Various attempts to simulate in the laboratory the natural breakage process presumed to be caused by temperature extremes have failed to confirm the idea. But supporters of the mechanism point out that it is difficult to match in a few months of laboratory experiments the thousands of years it takes for natural weathering to occur. The matter remains unsettled.

Breakage not directly related to earlier fractures or joints sometimes follows large curved outcrop surfaces as well as planes. **Exfoliation** is the peeling off of large planar or curved sheets or slabs of rock (Figure 6.12). Exfoliated outcrops may look like the layers of a large onion. **Spheroidal weathering** is also a cracking and splitting off of curved layers from a generally spherical boulder, but usually on a much smaller scale (Figure 6.13). As common as exfoliation and spheroidal weathering are, no generally accepted explanation of their origin has yet emerged.

**FIGURE 6.11** A rock split by ice. *Travis Amos.*

FIGURE 6.12 Exfoliation on Half Dome, Yosemite National Park, California. *Peter Kresan.*

FIGURE 6.13 Spheroidal weathering. *Michael Follo.*

## Physical Weathering and Erosion

As we noted earlier, weathering is closely related to erosion: the two processes help each other. Erosion is especially closely tied to physical weathering. As a rock breaks into small pieces, it becomes more easily transported and therefore more easily eroded. The first steps in this process are downhill movements of masses of weathered rock, such as landslides, and transportation of individual particles by flows of rainwater down a slope. Wind can blow away the finer particles, and glacial ice can carry away large blocks torn from bedrock. Thus the sizes of materials formed by physical weathering are related to erosional processes.

The intertwined processes of physical weathering and erosion are closely tied to the ways in which wind, water, and ice work to transport weathered material. As weathered material is transported, it may change in size, shape, and composition and undergo additional weathering. When transportation stops, deposition of the sediment formed by weathering begins. All of these processes contribute to the formation of different kinds of landscape; they are discussed in Part 2.

## SOIL: THE RESIDUE OF WEATHERING

Not all weathering products are eroded and immediately carried away by streams or other transport agents. On moderate and gentle slopes, plains, and lowlands, a layer of loose, heterogeneous, weathered material overlying bedrock remains. It may include particles of weathered and unweathered parent rock, clay minerals, iron and other metal oxides, and other products of weathering. Engineers and construction workers refer to this entire layer as "soil." Geologists, however, prefer to call this material **regolith,** reserving "soil" for the topmost layers, which contain organic matter and can support plant life. The organic matter in soil, **humus,** comes from the remains and waste products of the many plants, animals, and bacteria living in it.

Soils vary in color, from the brilliant reds and browns of iron-rich soils to the black of soils rich in organic matter. Soils also vary in texture. Some are full of pebbles and sand; others are entirely clay. Because soils are easily eroded, they do not form on very steep slopes or where high altitude or frigid climate prevents plant growth.

Soil is such an essential part of our environment and economy that a separate science, soil science, has developed in the twentieth century. Soil scientists, agronomists, geologists, and engineers study the composition and origin of soils, their suitability for agriculture and construction, their value as a guide to climatic conditions in the past, and many other practical aspects of soil (see Box 6.2).

## Soil Profiles

A road or trench cut through a soil reveals its vertical structure, the soil profile (Figure 6.14). Usually not much more than a meter or two thick, its topmost layer is usually the darkest, containing the highest concentration of organic matter. This layer is called the **A-horizon** (a particular level in a rock section is commonly called a "horizon"). In a thick soil that has

formed over a long period of time, the inorganic components of this top layer are mostly clay and insoluble minerals such as quartz. Soluble minerals have been leached from this layer. Beneath the topmost section is the **B-horizon,** where organic matter is sparse. In this layer soluble minerals and iron oxides have accumulated in small pods, lenses, and coatings. The lowest layer, the **C-horizon,** is slightly altered bedrock, broken and decayed, mixed with clay from chemical weathering.

Most soils are residual; that is, they evolve in one place from bedrock to regolith to well-developed soil horizons. Residual soil forms faster and becomes thicker where weathering is intense. Even there it may take thousands of years for the A-horizon to develop to the point where it can support crops. Soil forms so slowly because the most active chemical weathering goes on only during the short time when fresh rain enters it. During dry intervals the reactions continue, but very slowly and only when some moisture remains in the soil. Where soil dries out completely between rains, chemical weathering is almost completely suspended.

Soils may accumulate in some lowlands when they are eroded from surrounding slopes and transported downhill. These transported soils owe their thickness to deposition rather than to weathering in place. Transported soils are not common and should not be confused with ordinary sediment laid down by rivers, wind, and ice.

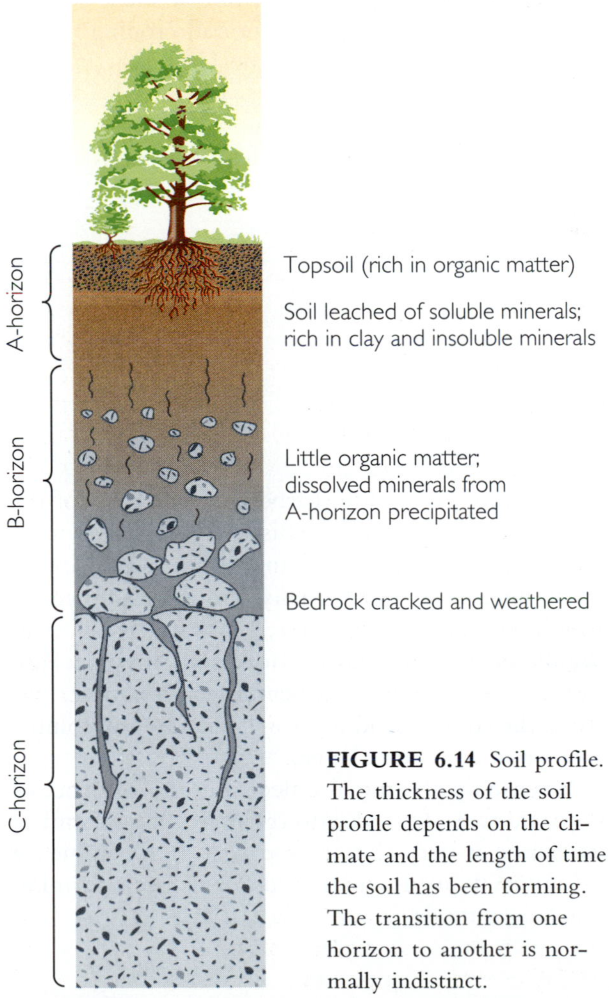

**FIGURE 6.14** Soil profile. The thickness of the soil profile depends on the climate and the length of time the soil has been forming. The transition from one horizon to another is normally indistinct.

Labels in figure:
A-horizon
B-horizon
C-horizon

Topsoil (rich in organic matter)

Soil leached of soluble minerals; rich in clay and insoluble minerals

Little organic matter; dissolved minerals from A-horizon precipitated

Bedrock cracked and weathered

## The Effects of Climate and Time on Soil Formation

Because climate so strongly affects weathering, it has a great influence on the nature of the soil formed on any given parent rock. In warm, humid climates, weathering is fast and intense and soils become thick. The higher the temperature and humidity, the lusher the vegetation. Abundant vegetation and moisture and warm temperatures so greatly speed up chemical weathering that the uppermost layer of the soil is leached of all soluble and easily weatherable minerals. Feldspars, other silicates, and even quartz are altered and dissolved. The residue of this fast weathering is a soil containing clay minerals and reddish iron oxides.

Lack of water and absence of vegetation slow weathering, so soils are thin in arid regions. Their A-horizons contain much unweathered parent rock mineral and fragments of the parent rock. When rainfall is too sparse to dissolve significant amounts of soluble minerals, more of them may remain in the A-horizons.

## BOX 6.2 SOIL EROSION

Because soils take such a long time to form, they cannot be renewed quickly after they become eroded. There is some balance between the moderate natural erosion of soils by streams and the wind and the slow formation of new soil. If soil is formed and eroded at roughly the same rate, its thickness remains constant. If soil is eroded more slowly than it is formed, it grows thick. If it erodes much more rapidly than it forms, new soil has no opportunity to form and the existing soil is quickly lost.

Before settlers began to farm the prairies of the United States and Canada in the nineteenth century, the soils were moderately thick. Soil formed relatively quickly in this area because the glaciation that ended about 10,000 years ago had deposited an abundance of easily weathered ground-up bedrock material.

With the advent of agriculture, soil erosion accelerated. Plowing breaks up the soil and eliminates the erosion-resistant natural plant cover. Soil erosion has been especially severe on the North American prairies, where plows have bit deep and conservation practices were long ignored. The loosened soils have been thinned and carried away by the region's rivers. The loosening of soil has also made it vulnerable to wind erosion and dust storms during long periods of drought (see Chapter 14).

One soil conservation practice that can lessen such serious consequences is contour plowing; that is, cultivating along a level path, following contour lines (lines of equal elevation). This practice covers the field with curved furrows instead of straight tracks that follow property lines up and down slopes. Contour plowing inhibits erosion because much of the rainwater that falls on the soil is prevented from running off downhill by the ridges of the contoured furrows.

Despite widespread promotion of such agricultural techniques, soils are still eroding at very high rates in parts of the United States and Canada; over 10 tons of topsoil per acre of cropland are lost annually. Over the whole United States, 2 billion tons of topsoil are lost to erosion each year, twice

---

The characteristics of soils in regions with moderate rainfall and temperatures depend on the climate, the type of parent rock, and the length of time the soil has had to develop and thicken. The longer the time soil has to develop, the less it will reflect the influence of its parent rock. For example, the soil developed after a relatively short time on a granite bedrock in a climate with moderate temperatures and humidity may be very different from a soil formed on limestone under the same conditions. The soil on the granite may still contain remnants of the silicate minerals and be dominated by the clay minerals forming from feldspar, a chief constituent of the parent rock. The soil on the limestone may still have a few remnants of calcium carbonate, but most of the limestone fragments will be dissolved. The clay minerals will be mainly those found as impurities in the parent limestone. After many thousands of years, though, the differences between the two soils may dwindle or even disappear. Both soils may develop the same clay minerals, depending on the exact nature of the climate, and both will have lost all soluble minerals in the upper layers.

Intense weathering also decreases the influence of the parent rock. In cold, arid regions, where chemical weathering is very slow, the parent rock's influence is dominant, even where soils have been forming over long periods.

Contour plowing, as on these farms in Oregon, minimizes soil erosion. *William Garnett.*

the amount of soil that is formed in the same period. The loss is equivalent to 780,000 acres of cropland. Comparable losses are estimated for other agricultural regions of the world. If these losses persist, the next century will see decreasing agricultural yields from thinned soils and the inevitable abandonment of agriculture in the most seriously affected regions. And once gone, a soil takes thousands of years to form again.

## Major Soil Groups

Because of soil's importance to agriculture, different kinds of soils have been mapped over much of the world. Some kinds of soil are excellent for cropland; others are poor. We can distinguish three major soil groups on the basis of their mineralogy and chemical composition, both of which correlate well with climate (Figure 6.15). **Pedalfers** are soils rich in aluminum and iron oxides and hydroxides (the name is derived from *pedon,* the Greek word for "ground" or "soil," and *al* and *fer,* from the chemical symbols for aluminum and iron). They are characteristic of much of the areas of moderate to high rainfall in the eastern United States, most of Canada, and much of Europe. The upper and middle layers of these pedalfers contain abundant insoluble minerals such as quartz, clay minerals, and iron alteration products. Carbonates and other more soluble minerals are absent. Pedalfers make good agricultural soils.

The **pedocals** are soils rich in calcium from the calcium carbonate and other soluble minerals they contain. These soils are found in warm, dry areas such as the southwestern United States. In such climates, much of the soil water is drawn up near the surface and evaporates between rainfalls, leaving precipitated nodules and pellets of calcium carbonate,

**FIGURE 6.15** Major soil types. (a) Pedalfer soil profile developed on granite in a region of high rainfall. The only mineral materials in the upper parts of the soil profile are iron and aluminum oxides and silicates such as quartz and clay minerals, all of which are very insoluble. Calcium carbonate is absent. (b) Pedocal soil profile developed on sedimentary bedrock in a region of low rainfall. The A-horizon is leached; the B-horizon is enriched in calcium carbonate precipitated by evaporating soil waters. (c) Laterite soil profile developed on a mafic igneous rock in a tropical region. In the upper zone, only the most insoluble precipitated iron and similar oxides remain, plus occasional quartz. All soluble materials, including even relatively insoluble silica, are leached; thus the whole soil profile may be considered to be an A-horizon directly overlying a C-horizon.

mostly in the middle layer of the soil. Pedocals are not as fertile as the pedalfers because the combination of mineralogy and dryness is less favorable for a large population of organisms in the soil. Hence these soils are low in organic matter.

**Laterite,** found in the tropics, is a deep-red soil in which all silicates have been completely altered, leaving behind mostly aluminum and iron oxides and hydroxides. Silica as well as calcium carbonate has been leached from the soil. Although laterites may support lush vegetation in equatorial jungles, they

are not very productive soils for crop plants. Most of the organic matter is constantly recycled from the surface to the vegetation, with a very thin humus layer at the top of the soil. The clearing of trees and tilling of the soil allow its superficial layer of rich humus to oxidize quickly and disappear, uncovering the infertile layer below. For this reason many laterites can be farmed intensively for only a few years after they have been cleared before they become barren and have to be abandoned. Large regions of India are now in this condition. As sections of the Amazon

**FIGURE 6.16** Sand grains of various minerals. *Rex Elliott.*

rain forest of Brazil are cleared, they too become barren after only a few years. Under natural conditions it takes a very long time, many thousands of years, to restore a forest on laterite soil.

## WEATHERING MAKES THE RAW MATERIAL OF SEDIMENT

Soil is only one of many products of weathering. Processes that weather rocks and break them apart produce fragments that vary greatly in size and shape, from huge boulders 5 m across to small pebbles, sand grains, and clay particles too fine to see without a microscope. Pieces larger than a large sand grain (2 mm in diameter) tend to be rock fragments containing mineral grains of the parent rock. Sand and silt grains are usually individual crystalline grains of any of the various minerals that make up the rock (Figure 6.16). The smallest pieces of weathered material are made of clay minerals, which are the products of the chemical alteration of silicate minerals.

Collectively, all the atoms and ions that are products of chemical and physical weathering are equal to all the atoms and ions of the original rock that weathered. A certain mass of granite, for example, is broken down by weathering into the following classes: (1) mineral and rock fragments of the parent rock that are still identifiable as such; (2) solid products of chemical alteration, such as clay minerals and iron oxides; (3) ions dissolved in rainwater and soil water. The mass of all these products, minus the water and carbon dioxide derived from the atmosphere, equals the original mass of the granite that was weathered.

In this way a granite is transformed into the raw material of sediment. The weathered products are eventually carried away from the weathering site by wind, water, and ice, ultimately to be deposited as various kinds of sediment, such as the sand, silt, and mud found in river valleys. This sediment is buried by additional deposits and gradually turns into sedimentary rock, the subject of Chapter 7.

## SUMMARY

**What is weathering and how is it geologically controlled?** Rocks are broken down at the surface of the Earth by chemical weathering, the chemical alteration or dissolution of a mineral, and by physical weathering, the fragmentation of rocks by physical processes. Erosion, the wearing away of the land, carries away the products of weathering, the raw material of sediment. The nature of the parent rock affects weathering because different minerals weather at different rates. Climate affects weathering: warmth and heavy rainfall speed weathering; cold and dryness slow it down. The presence of soil promotes weathering, and the longer a rock has been weathered, the more it breaks down.

**How does weathering work?** Potassium feldspar ($KAlSi_3O_8$), in a way typical of silicate minerals, weathers in the presence of water by a chemical reaction in which potassium (K) and silica ($SiO_2$) are lost to the water solution and the solid feldspar alters to a clay mineral, kaolinite ($Al_2Si_2O_5[OH]_4$). Carbon dioxide dissolved in water promotes chemical weathering by providing acid. Carbonate minerals weather by simple dissolution, leaving no residue. Iron (Fe), which is found in ferrous form in many silicates, weathers by oxidation, producing ferric oxides in the process. The rates at which all of these processes operate vary, reflecting the chemical stabilities of minerals under weathering conditions.

**What are the processes of physical weathering?** Rocks are fragmented as they break along crystal boundaries or along joints in rock masses. Physical weathering is promoted by chemical weathering, which weakens grain boundaries; by the freezing of water and crystallization of minerals in cracks; by fires; by organisms, such as expanding tree roots; and perhaps by alternating extremes of heat and cold. Patterns of breakage such as exfoliation and spheroidal weathering result from interactions between chemical and physical weathering processes.

**How do soils form as products of weathering?** Soil is a mixture of clay minerals, weathered rock particles, and organic matter, which forms as organisms interact with weathering rock and water in the soil. Soils form faster in warm, humid climates than in cold, dry ones because weathering is controlled by climate and the activity of organisms. Young soils reflect the composition of the parent rock, but older soils primarily reflect climate. The three major soil groups are pedalfers, found in temperate climates; pedocals, which form in warm, dry climates; and laterites, found in the humid tropics.

# KEY TERMS AND CONCEPTS

weathering (p. 120)
chemical weathering (p. 120)
physical weathering (p. 120)
soil (p. 120)
climate (p. 121)
kaolinite (p. 124)
hydration (p. 125)
bauxite (p. 128)

oxidation (p. 129)
ferrous iron (p. 129)
hematite (p. 129)
ferric iron (p. 129)
chemical stability (p. 130)
joints (p. 133)
exfoliation (p. 133)
spheroidal weathering (p. 133)

regolith (p. 134)
humus (p. 134)
A-horizon (p. 135)
B-horizon (p. 135)
C-horizon (p. 135)
pedalfer (p. 137)
pedocal (p. 137)
laterite (p. 138)

# EXERCISES

1. What do the different kinds of rocks used for monuments tell us about weathering?

2. What rock-forming minerals found in igneous rocks weather to clay minerals?

3. How does abundant rainfall affect weathering?

4. Which weathers faster, a granite or a limestone?

5. How does physical weathering affect chemical weathering?

6. What are the main factors controlling the development of different soil types?

# THOUGHT QUESTIONS

1. You are planning to use polished decorative limestone facings for monuments in Tucson, Arizona (a hot, arid region), and Seattle, Washington (a cool, rainy region). How do you think the monuments might look after a hundred years?

2. In northern Illinois you can find two soils developed on the same kind of bedrock; one is 10,000 years old and the other is 40,000 years old. What differences in soil composition or profile would you expect?

3. Which igneous rock would you expect to weather faster, a granite or a basalt?

4. Given a granite with crystals about 4 mm across and a rectangular system of joints spaced about 0.5 to 1 m apart, what would be the size of the largest weathered particle you would ordinarily expect?

5. Compare the speeds of weathering of (a) a basalt rich in volcanic glass and crystals of pyroxene and calcium-rich feldspar that are about 0.5 mm across, and (b) a gabbro of exactly the same mineral composition but with crystals about 3 mm across.

6. Why do you think a road built of concrete, an artificial rock, tends to crack and develop a rough, uneven surface in a cold, wet region even when it is not subjected to heavy traffic?

7. Pyrite is a mineral in which ferrous iron is combined with sulfide ion. What do you think is the major chemical process that weathers pyrite?

8. Rank the following rocks in the order of their rapidity of weathering in a warm, humid climate: granite; a sandstone made of pure quartz; a limestone made of pure calcite; a deposit of rock salt (halite, NaCl).

9. What differences do you expect to find in the weathering of a pure magnesium olivine ($Mg_2SiO_4$) and a pure iron olivine ($Fe_2SiO_4$)?

# SUGGESTED READINGS

Blatt, Harvey. 1992. *Sedimentary Petrology,* 2d ed. New York: W. H. Freeman.

Carroll, Dorothy. 1970. *Rock Weathering.* New York: Plenum Press.

Colman, S. M., and D. P. Dethier. 1986. *Rates of Chemical Weathering of Rocks and Minerals.* New York: Academic Press.

Gauri, K. L. 1978. The preservation of stone. *Scientific American* (June):126.

Loughnan, F. C. 1969. *Chemical Weathering of the Silicate Minerals.* New York: Elsevier.

# SEDIMENTS AND SEDIMENTARY ROCKS

Much of Earth's surface is covered with sediments, materials that originated either as solid debris or as dissolved substances from the weathering of the continents. The solid materials are laid down by rivers, the wind, glaciers, and ocean currents. The dissolved substances are deposited chemically or biochemically in the oceans and in lakes. The deposition of sediments by the settling of layers of particles results in bedding or stratification, a hallmark of sediments and sedimentary rocks. Sediments may be preserved by burial under more layers of sediment. As a result of burial over long periods of time, soft, loose sands, gravels, and muds harden to sandstones, conglomerates, and mudstones; the carbonate sands and muds of coral reefs in the tropical oceans are transformed to limestone.

Marble Canyon, a section of the Grand Canyon, Colorado River, Arizona. An immense length of geologic time is represented by these layers of sedimentary rock. *Peter Kresan.*

ediments and sedimentary rocks are important elements of the rock cycle, described in Chapter 3. Placed as they are in the cycle, between rocks brought up from the interior by tectonics and rocks returned to the interior of the crust by burial, sedimentary rocks are controlled by the processes that occur in the surface part of the rock cycle: weathering, erosion, transport, deposition, sedimentation, burial, and diagenesis (alteration in the properties of sediments after they are deposited) (Figure 7.1).

Because sedimentary rocks were once sediments, they are records of the conditions at the surface when and where they were deposited. In the Grand Canyon, for example, we may see beds of limestone containing the fossils of marine organisms; so we know that at one time, long before the Grand Canyon was formed, this region, now high on the continent, was the floor of an ocean.

Even a single quartz grain in a sandstone has a story to tell. It may originally have been a crystal in a granite. Perhaps weathering freed it from the other minerals of the granite, and it became a loose sand grain. It was then carried by a stream to the ocean, where currents transported it to its ultimate resting place. There it joined other grains to form a layer of sediment. Finally, it may have become cemented to its neighboring grains by the precipitation of additional quartz around the original grain.

Geologists work backward from the sedimentary rocks of a particular time exposed over a broad region to infer the sources of sediment and the various kinds of places in which the sediments were originally deposited. In one part of a region, for example, the sediments may have been sands laid down under the sea, later to become sandstone. In a bordering region, carbonate reefs may have been laid down

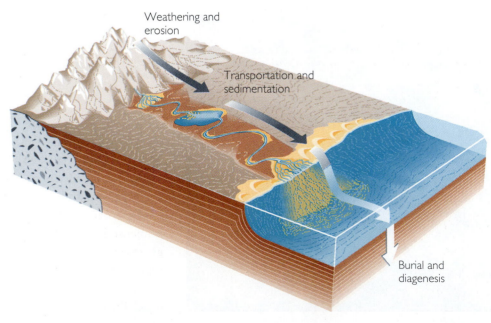

Weathering and erosion

Transportation and sedimentation

Burial and diagenesis

**FIGURE 7.1** The sedimentary stages of the rock cycle embrace several overlapping processes: *physical and chemical weathering; erosion; transportation;* and *burial.* Burial converts the sediment to rock through *diagenesis.*

along the edge of a continental shelf. Beyond, there may have been a near-shore area in which the sediments were shallow ocean carbonate muds that later became thin-bedded limestones. In this way we can infer shorelines, mountains, plains, deserts, and swamps, and we can draw a map of the continents and oceans of long ago. At the same time, we can infer former plate-tectonic movements from the compositions of clastic sedimentary rocks, which reflect their origin in volcanic arcs, rift valleys, or collisional mountains.

The study of sediments and sedimentary rocks has enormous practical value, for oil, gas, and coal, our most valuable sources of energy, are found in these rocks. So is much of the uranium that is used for nuclear power. Phosphate rock used for fertilizer is sedimentary, as is much of the world's iron ore. Knowing how these kinds of sediments are formed helps us to explore for additional resources more intelligently.

# THE RAW MATERIAL OF SEDIMENT: PARTICLES AND DISSOLVED SUBSTANCES

As we saw in Chapter 6, chemical weathering and mechanical fragmentation of rock at the surface make both solid and dissolved products. The solid fragments produced by weathering—from pebbles and boulders to particles of sand, silt, and clay—are called **clastic particles.** Accumulations of such ma-

terials are **clastic sediments.** The dissolved products of weathering are ions or molecules in the waters of soils, rivers, lakes, and oceans. These dissolved substances are precipitated from water by chemical and biochemical reactions and accumulate as **chemical and biochemical sediments.** Clastic sediments are about 10 times more abundant in the Earth's crust than chemical and biochemical sediments, because in most of the world much more of the rock becomes fragmented by physical weathering than dissolved by chemical weathering. Thus clastic particles are formed and accumulate much more rapidly than chemical or biochemical precipitates.

Clastic sediment may be a mixture of minerals of the parent weathered rock that are resistant to weathering and thus stable, such as quartz; partially altered fragments of minerals less resistant to weathering and so less stable, such as feldspar; and newly formed minerals, such as clay minerals. Where weathering is intense, the sediment contains only clastic particles made of chemically stable minerals, mixed with clay minerals. Where weathering is slight, many minerals that are unstable under surface conditions will survive as clastic particles (Table 7.1).

Clastic particles vary widely in size and shape. The shapes of boulders, cobbles, and pebbles are determined by natural breakage along joints, bedding planes, and other fractures in the parent rock. Sand grains tend to inherit their shapes from the individual crystals formerly interlocked in the parent rock.

Layers of chemical or biochemical sediment commonly consist of precipitated sedimentary particles that are made of just one or two kinds of miner-

**TABLE 7.1**

Varying Intensities of Weathering Produce Different Sets of Minerals in Clastic Sediments

| | INTENSITY OF WEATHERING | | |
| --- | --- | --- | --- |
| | LOW | MEDIUM | HIGH |
| Minerals remaining | Quartz | Quartz | Quartz |
| | Feldspar | Feldspar | Clay minerals |
| | Mica | Mica | |
| | Pyroxene | Clay minerals | |
| | Amphibole | | |

als or a few closely related types. Calcium carbonate sediments of the deep sea, for example, are made of the shells of only a few kinds of organisms and are entirely calcite. Shallow-water calcium carbonate sediments may be mixtures of two calcium carbonate minerals, calcite and aragonite. Sediments formed by the evaporation of seawater may consist of layers composed only of gypsum or of halite.

# TRANSPORTATION OF SEDIMENT

Once clastic particles and dissolved ions have been formed by weathering, they start a journey to a sedimentation area. Chemically or biochemically produced particles may also be transported from one area to another, usually nearby. With the exception of wind and some ocean currents, all agents of transportation carry material downhill. The fall of a rock from a cliff, the transport of sand in a river flowing to the sea, and the slow movement of glacial ice are all responses to gravity. Although winds may blow material from a low elevation to a height and back again, in the long run gravity is relentless, and wind-blown sand and dust settle in response to its pull. And once a particle drops to the ocean and settles through the water, it is trapped. It can be picked up again only by an ocean current, which may transport it only to a depositional site on the seafloor.

## Transportation by Currents

Much of the transportation of sediment is accomplished by currents, which are movements of fluids such as air and water. (Scientists include gases among fluids because, like liquids, they take the shapes of their containers, whereas solids maintain their own shapes.) Rivers, like ocean currents, are currents of water. Currents in air move material too, but in far smaller quantities than rivers or ocean currents. As the particles are lifted into the fluid, the current carries them downriver or downwind. The stronger the current—that is, the faster it flows—the larger the particles it transports.

Working against the ability of a current to carry a particle is the tendency of a particle to settle to the bottom of the flow because of the effect of gravity. Although it is a basic law of physics that particles of all sizes fall to Earth at the same speed in a *vacuum*, large grains settle faster than small ones in a *fluid*. The settling velocity is proportional to the density of the particle as well as to its size. But because the most common minerals in sediments have roughly the same density (about 2.6 to 2.9 g/cm$^2$) and size is more conveniently measured than density, we use size as an indicator of settling velocity. As a current carrying particles of various sizes begins to slow, it can no longer keep the largest particles suspended, and they settle. As the current slows even more, smaller particles settle. When the current stops completely, even the smallest particles settle.

Thus particles become separated in accordance with their size; gravels may be deposited by very strong currents, while sands and muds are kept in suspension and settle only when the current weakens. A fast current may lay down a bed of gravel; as it slows, it will lay down a bed of sand on top of the gravel. If the current stops, a layer of mud may be deposited on top of the sand bed. This tendency for variations in current velocity to segregate sediments according to size is called **sorting**. A well-sorted sediment consists mostly of particles of a uniform size; a poorly sorted sediment contains particles of many sizes (Figure 7.2).

Particles are generally transported intermittently rather than steadily. Fast currents give way to weak flows or stop entirely. A river may transport large quantities of sand and gravel when it floods but will drop them as the flood recedes, only to pick up the materials and carry them farther in the next flood. Strong winds may carry large amounts of dust for a few days, then die down and deposit the dust as a layer of sediment.

## Transportation of Dissolved Material

Chemical substances that are dissolved in water during weathering are carried along with the water as a homogeneous solution. Because materials such as dissolved calcium ions are part of the water solution itself, they can never settle out of it. They appear as solid particles only if a chemical reaction takes place, as when calcium reacts with carbonate ions to precipitate as calcium carbonate. Most of the material dissolved by weathering on the continents eventually arrives in the ocean, which is a huge reservoir of these substances.

## Transportation by Glaciers

As glaciers move downhill in response to gravity, they incorporate large quantities of solid particles eroded from soil and bedrock into their ice and carry

**FIGURE 7.2** Well-sorted and poorly sorted sand grains. *Rex Elliott.*

this clastic material along with them. Because the particles cannot settle through the solid ice, they have no tendency to become sorted, and the ice carries a heterogeneous mixture of sizes. Where the ice melts at the edge of a glacier, the particles are dropped and then carried away by rivers of meltwater, which then sort them by size. Although the speed of a glacier's movement varies, it is generally much steadier than the intermittent flow of a river, the wind, or an ocean current. (Transportation and sedimentation by glaciers are discussed in detail in Chapter 15.)

## Transportation and Weathering

Weathering processes continue during transportation, for clastic material is still in contact with the chief agents of chemical weathering, water and the oxygen and carbon dioxide of the atmosphere. The time during which the material is actually being transported by a current is too short for slow weathering reactions to have much effect. Most weathering is accomplished during the long intermittent periods when the sediment is temporarily deposited before being picked up again by the current. When a river floods its valley for a few days, for example, it deposits sand, silt, and clay. After the flood recedes,

weathering of the deposits resumes and continues until the next flood. That flood may pick up the sediment of the earlier flood and redeposit it farther downstream, where the sediment begins to weather again. In this way, episodes of transportation can alternate with episodes of weathering.

## The Effects of Transportation on Clastic Particles

Because sedimentation may be intermittent, the total time between the formation of clastic debris and final sedimentation may be many hundreds or thousands of years, depending on the distance to the final depositional area and the number of stop-offs along the way. Clastic particles eroded by the headwaters of the Missouri River in the mountains of western Montana, for example, take hundreds of years to travel the 2000 miles down the Missouri and Mississippi rivers to the Gulf of Mexico. During that long time, the particles carried by these rivers may be affected by the transportation process as well as by intermittent weathering.

Transportation by water and wind currents affects particles in two ways. It reduces the sizes of the particles and it rounds the originally angular frag-

Distance of transport

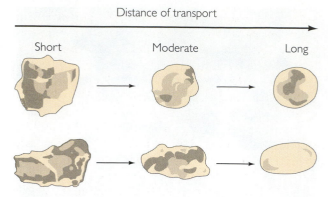

Short    Moderate    Long

**FIGURE 7.3** The effects of transportation on the size and angularity of detritus. Grains become rounded and slightly smaller as they are transported, though the general shape of the grain may not change significantly.

ments (Figure 7.3). As particles are transported, they tumble and strike one another or rub against bedrock. Pebbles or large grains that collide forcefully may break into two or more smaller pieces. Weaker hits may chip small pieces from edges and corners. Abrasion by bedrock, together with grain impacts, also rounds the particles, wearing off and smoothing sharp edges and corners.

Larger particles lose a greater proportion of their size and become more rounded than smaller ones. Boulders break into smaller pieces after they have traveled short distances; cobbles and pebbles travel somewhat farther before they break up and become rounded. Sand grains are much less likely to break into several pieces and become rounded over long distances.

Glacial transport affects transported particles by slowly pressing them against one another, splitting them, and breaking up the bedrock at its bottom and sides. The particles become smaller, but they are not rounded.

# SEDIMENTATION: THE END OF THE LINE

We have mentioned before that sedimentation starts where transportation stops. Clastic particles are laid down as beds of sediment where currents die down and particles settle. Chemical or biochemical sedimentary particles are formed after river waters merge with ocean or lake waters and conditions for precipitation are met.

Most of the clastic particles produced by weathering and erosion of the land end up at the bottom of the ocean. The seafloor is blanketed by the many kinds of sediment brought there by rivers, the wind, and glaciers. Most chemical and biochemical sediments, too, are deposited on the floor of the ocean.

Smaller amounts of clastic sediments are laid down on the way to the oceans. Chemical or biochemical sediments may be deposited in lakes or swamps. Only a small fraction of these land-deposited sediments are buried and preserved because erosion eventually removes much of the sediment on the land surface. A much larger fraction of sediment deposited on the ocean floor is buried and preserved, primarily because most currents at the ocean floor are not strong enough to erode the sediment once it has settled.

## Clastic Sedimentation

The types of clastic sediments laid down are determined by the nature of the transporting currents and the abundance and kinds of sediment supplied by weathering and erosion.

Gravels are laid down by currents strong enough to roll and slide pebbles, cobbles, and boulders. When these currents wane, the gravel is deposited while finer materials such as sand and mud continue to be carried along. An abundant supply of coarse detritus is common in rapidly flowing rivers in mountainous terrains, where erosion is rapid. Beach gravels are deposited where ocean waves actively erode rocky shores. A glacier produces and carries much coarse clastic debris, which it deposits as gravels at the edge of the ice as it melts.

Sands are laid down by moderately strong currents. Most rivers have currents strong enough to carry and deposit sand beds in their channels. Sands may be spread by floodwaters over river valleys. Sands are also blown and deposited by winds, especially in deserts. Sand grains are transported and deposited on beaches and in the ocean by waves and currents.

Muds, the finest clastic particles, can be carried even by very weak currents. They settle to the bottom only when the current slows almost to a stop, as on the floor of a valley when floodwaters recede slowly or stop flowing entirely. Muds are deposited in the ocean generally some distance from shore, where waves and currents are too weak to keep fine particles in suspension. Much of the floor of the open ocean is covered by mud particles that originally were transported by surface waves and currents or by the wind. These particles gradually settle to depths where currents and waves are stilled. Once in the quieter depths, they settle undisturbed all the way to the bottom.

## Chemical and Biochemical Sedimentation

The driving force of chemical and biochemical sedimentation is chemistry rather than gravity. Huge quantities of the chemical substances dissolved from rocks in the course of weathering are carried down to the sea, where they enter saline surroundings that are chemically very different from the more or less fresh waters of soils and rivers. Much smaller quantities of dissolved materials may be precipitated in saline or alkaline lakes, which also may differ from river waters in composition.

The ocean may be thought of as a huge chemical mixing tank. Dissolved materials are constantly being brought in by rivers, rain, wind, and glaciers. Smaller quantities of dissolved materials enter the ocean by hydrothermal chemical reactions between seawater and hot basalt at mid-ocean ridges. Currents and waves constantly mix these materials with ocean water. The ocean continuously loses water by evaporation at the surface. The inflow and outflow of water to and from the oceans are so exactly balanced that the amount of water in the oceans remains constant.

The entry and exit of dissolved materials, too, are balanced. Materials that enter the ocean as dissolved substances exit as chemical and biochemical precipitates. As a result of this balance, the salinity— that is, the total amount of dissolved substances in a given volume of seawater—remains constant.

We can see some of the mechanisms by which this chemical balance is maintained by focusing on the element calcium. Calcium is an important component of the most abundant biochemical precipitate formed in the oceans, calcium carbonate ($CaCO_3$). Calcium is dissolved when limestone and silicates containing calcium, such as some feldspars and pyroxenes, are weathered on land and brought as ions ($Ca^{2+}$) to the oceans. There a wide variety of marine organisms biochemically combine the calcium ions with bicarbonate ions ($HCO_3^-$), also present in seawater, to form their calcium carbonate shells. The calcium that entered the ocean as dissolved ions leaves it as a solid sediment when the organisms die and their shells settle and accumulate as calcium carbonate sediment on the seafloor, ultimately to be transformed by diagenesis into limestone. The chemical balance that keeps constant the levels of calcium dissolved in the ocean is thus regulated in part by the activities of organisms.

Chemical balance is also maintained by nonbiological mechanisms. Sodium ions ($Na^+$) brought into the oceans, for example, combine in a chemical reaction with chloride ion ($Cl^-$) to form the precipitate sodium chloride ($NaCl$) when evaporation raises the amounts of sodium and chloride ions to the point of supersaturation. As we saw in Chapter 2, supersaturated solutions crystallize minerals when they become so rich in dissolved materials that they react spontaneously to form precipitates. The intense evaporation required to crystallize salt takes place in warm, shallow arms of the sea.

Each of the many dissolved components of seawater participates in some chemical or biochemical reaction that precipitates it out of the water and onto the seafloor. Totaled over all the oceans of the world, this precipitation balances the total inflow of dissolved material from continental weathering and hydrothermal activity at mid-ocean ridges.

Organic sedimentation is yet another type of biochemical precipitation. Vegetation may be preserved from decay in swamps and accumulate as a rich organic material, **peat,** which is ultimately buried and transformed by diagenesis to **coal.** In both lake and ocean waters, the remains of algae, bacteria, and other microscopic organisms may accumulate in sediments as organic matter that is transformed to **oil** and **gas.**

## Sedimentary Environments

Of the many ways in which sedimentation can be classified, geologists have found the concept of sedimentary environments most useful. A **sedimentary environment** is a geographic location characterized by a particular combination of environmental conditions and geological processes (Figure 7.4). Environmental conditions include the kind and amounts of water (ocean, lake, river, arid land), the topography (lowland, mountain, coastal plain, shallow ocean, deep ocean), and biological activity. Geological processes include the nature of currents that transport and deposit sediment (water, wind, ice). Thus a beach environment brings together the dynamics of waves approaching and breaking on the shore and the

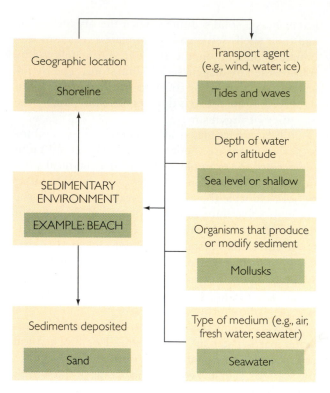

**FIGURE 7.4** A sedimentary environment is characterized by particular environmental conditions and geological processes that continue in a particular way. Properties of one specific example—a beach—are given in small boxes.

distribution of sediments on the beach. Sedimentary environments are related to plate-tectonic settings. The deep trenches of the oceans, for example, are found at subduction zones, whereas thick alluvial deposits are typically associated with mountains formed by the collision of continents. Figure 7.5 shows some common sedimentary environments, grouped by their locations on the continents, near shorelines, or in the oceans.

Sedimentary environments are distinguished by the kinds of sediment found in them. All of the sediments in an alluvial environment, for instance, are clastic; only chemical and biochemical sediments are produced in a coral reef. Thus geologists group sedimentary environments into two broad classes, depending on the dominance of clastic sedimentation or chemical and biochemical sedimentation.

**CLASTIC SEDIMENTARY ENVIRONMENTS**
**Clastic sedimentary environments** are those dominated by clastic sediments (Table 7.2). They include the continental alluvial (river), desert, lake, and glacial environments, as well as the shoreline environments transitional between continental and marine: deltas, beaches, and tidal flats. They also include the oceanic environments of the continental shelf, continental margin, and deep ocean floor. The sediments

**TABLE 7.2**
## Clastic Sedimentary Environments

| ENVIRONMENT | AGENT OF TRANSPORTATION, DEPOSITION | SEDIMENTS |
|---|---|---|
| Alluvial | Rivers | Sand, gravel, mud |
| Lake | Lake currents, waves | Sand, mud |
| Desert | Wind | Sand, dust |
| Glacial | Ice | Sand, gravel, mud |
| Delta | River + waves, tides | Sand, mud |
| Beach | Waves, tides | Sand, gravel |
| Shallow shelf | Waves, tides | Sand, mud |
| Deep sea | Ocean currents, settling | Mud |

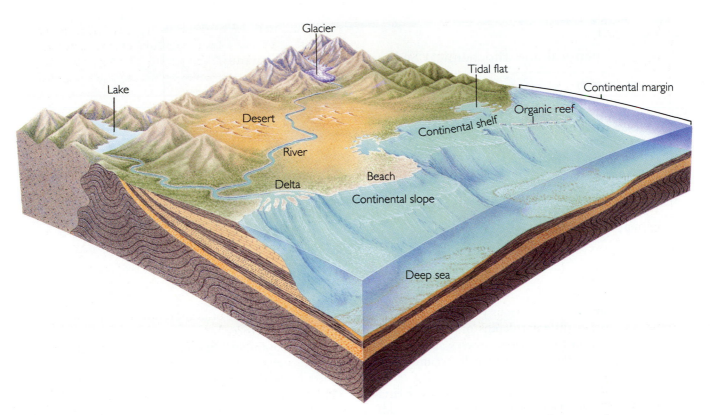

**FIGURE 7.5** Common sedimentary environments. **Continental:** The *alluvial environment* encompasses the river channel itself, the borders of the channel, and the flat valley floor on either side of the channel that is covered by water when the river floods. A *desert environment* is arid; sediment there is formed by a combination of wind action and the work of rivers that flow (mostly intermittently) through it. A *lake environment* is controlled by the relatively small waves and moderate currents of inland bodies of fresh or saline water. A *glacial environment* is controlled by the dynamics of moving masses of ice. **Shoreline:** Shoreline environments are dominated by the dynamics of waves, tides, and currents on rocky or sandy shores. Shoreline environments include *deltaic environments,* where rivers enter lakes or the sea; *tidal flat environments,* where extensive areas exposed at low tide are dominated by tidal currents; and *beach environments,* strips of sand or gravel laid down by wave action. **Marine:** *Continental shelf environments* are located in the shallow waters off continental shores, where sedimentation is controlled by relatively gentle currents. *Continental margin environments* are found in the deeper waters at the edges of the continents, where sediment is deposited by a special type of deep-water current. *Organic reefs* are composed of carbonate structures built up on continental shelves or on oceanic volcanic islands. *Deep-sea environments* include all the floors of the deep ocean, far from the continents, where the quiet waters are disturbed only occasionally by ocean currents.

of these environments are often called **terrigenous,** to indicate their origin on land.

### CHEMICAL AND BIOCHEMICAL SEDIMENTARY ENVIRONMENTS

**Chemical and biochemical sedimentary environments** are those characterized principally by chemical and biochemical precipitation. In most places they contain little clastic sediment, which in abundance may dilute chemical sediment or modify chemical sedimentary processes (Table 7.3). By far the most abundant are **carbonate environments,** marine settings where calcium carbonate, principally of biochemical origin, is the main sediment. Carbonate shell materials are secreted by literally hundreds of species of mollusks and other invertebrate organisms as well as by calcareous (cal-

**TABLE 7.3**

Major Chemical and Biochemical Sedimentary Environments

| ENVIRONMENT | AGENT OF PRECIPITATION | SEDIMENTS |
|---|---|---|
| Carbonate (includes reef, bank, deep sea, etc.) | Shelled organisms, inorganic precipitation from seawater | Carbonate sands and muds, reefs |
| Evaporite | Evaporation of seawater | Gypsum, halite, other salts |
| Deep sea | Shelled organisms | Silica sediment |
| Swamp | Vegetation | Peat |

cium-containing) algae. Various populations of these organisms live at different depths of water, both in quiet areas and in places where waves and currents are strong, and as they die, their shells accumulate to form sediment.

Except for those of the deep sea, carbonate environments are found in the warmer tropical or subtropical regions of the oceans, where chemical conditions are favorable for the precipitation of calcium carbonate. These regions include organic reefs, carbonate sand beaches, tidal flats, and shallow carbonate banks. The organic reef environment is created by corals and other organisms that secrete calcium carbonate and live in warm, shallow seas. The carbonate of the corals, intergrown with calcareous algae and the shells of other organisms, forms a hard, cemented rock structure that is resistant to waves. Reefs may be small islands or long ridges that extend over many hundreds of kilometers, such as the Great Barrier Reef of Australia. Organic reefs may rim shallow carbonate banks such as the Bahama Islands, off the southeastern coast of the United States. Carbonate banks form along broad areas of continental shelves where warm ocean waters and a lack of clastic material provide a hospitable environment for carbonate-secreting organisms. Carbonate sands derived mainly from the breakup of shell materials are deposited on these banks. Carbonate muds are derived both from minute crystals secreted by algae and from the precipitation of inorganic chemicals. Deep-sea carbonate environments are dominated by the calcium carbonate remains of surface-dwelling, single-celled shelled organisms that settle to the bottom after death.

**Marine evaporite environments** are dominated by the salts crystallized from seawater by evap-

oration. The degree of evaporation and the length of time it has proceeded control the kinds of salt formed. Commonly, the first mineral to precipitate in marine evaporites is calcium carbonate; as evaporation continues, gypsum (calcium sulfate, $CaSO_4 \cdot 2H_2O$) precipitates, followed by crystallization of halite (NaCl). These are the most common evaporite sedimentary minerals. In some evaporites, continued evaporation beyond the halite stage produces various magnesium and potassium chlorides and sulfates.

**Siliceous environments** are deep-sea environments named for the remains of silica shells deposited in them. The organisms that secrete silica grow in surface waters where nutrients are abundant, and their shells settle to the ocean floor to accumulate as layers of siliceous sediment.

# DIAGENESIS AND LITHIFICATION

After sediments are deposited and buried, they are subject to many physical and chemical changes. The term **diagenesis** refers to all postdepositional alterations in the properties of either sediment or sedimentary rock, such as its mineral composition and the amount of space between the grains. Diagenesis begins as soon as deposition stops and continues until the sediment or sedimentary rock is either exposed to weathering or metamorphosed by heat and pressure. The temperature at which diagenesis gives way to metamorphism is typically around 300 to 350°C, corresponding to a depth of about 10 to 12 km. Geological processes that result in diagenesis are termed *diagenetic*.

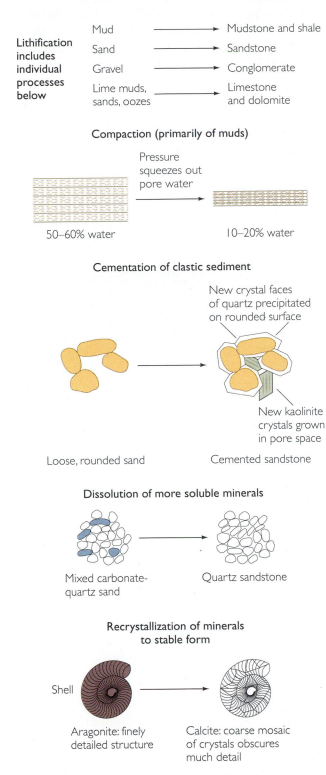

Lithification includes individual processes below

Mud → Mudstone and shale

Sand → Sandstone

Gravel → Conglomerate

Lime muds, sands, oozes → Limestone and dolomite

Compaction (primarily of muds)

Pressure squeezes out pore water

50–60% water → 10–20% water

Cementation of clastic sediment

New crystal faces of quartz precipitated on rounded surface

New kaolinite crystals grown in pore space

Loose, rounded sand → Cemented sandstone

Dissolution of more soluble minerals

Mixed carbonate-quartz sand → Quartz sandstone

Recrystallization of minerals to stable form

Shell

Aragonite: finely detailed structure → Calcite: coarse mosaic of crystals obscures much detail

**FIGURE 7.6** Diagenetic processes produce changes in composition and texture. Most of the changes tend to transform a loose, soft sediment into a hard, lithified sedimentary rock.

A major *chemical* diagenetic change is the addition of mineral cements that bind clastic sediments and rocks. **Cementation** results in a decrease in **porosity,** the relative proportion of a rock's volume occupied by open pores between grains. Cementation also results in **lithification,** the diagenetic processes by which soft sediment is hardened into rock (Figure 7.6). In some sands, for example, calcium carbonate is precipitated and acts as a cement, binding the grains and hardening the resulting mass into sandstone (Figure 7.7). Other minerals, such as quartz, may cement sands, muds, and gravels into sandstone, mudstone, and conglomerate.

Another chemical diagenetic change in clastic sediments and rocks is the chemical alteration of clay minerals originally deposited as clastic particles.

The major *physical* diagenetic change is **compaction,** a decrease in porosity that results when the grains are squeezed closer together by the weight of overlying sediment. Sands are fairly well packed during deposition, so they do not compact much. Newly deposited muds, however, are highly porous; often over 60 percent of the sediment is water in pore space. As a result, muds compact greatly after burial, losing more than half of their water. Compaction is frequently, but not always, accompanied by the precipitation of cementing materials. Many of the muddy sediments of the deep sea, even after burial to hundreds of meters, are compacted but uncemented and therefore not lithified into hard rock.

Carbonate sediments are strongly prone to chemical and textural alteration. For example, arago-

**FIGURE 7.7** Photomicrograph of sandstone showing quartz grains (white and gray) cemented by calcite cement (brightly colored and variegated) introduced after deposition. *Peter Kresan.*

nite, the less stable form of calcium carbonate, is the main constituent of many of the shell materials that make up carbonate sediments. During diagenesis soon after the beginning of burial, the aragonite tends to recrystallize to the more stable form of calcium carbonate, calcite, which is the most common mineral of limestones. Carbonate sediments are commonly lithified to limestone soon after deposition and burial.

The chemical and physical changes produced during diagenesis are promoted by burial. As a sediment is buried, it is subjected to increasingly high temperatures in the Earth's interior. In the crust, the increase in temperature averages about 1°C for each 30 m, so that sediments buried to a depth of 4 km may reach temperatures of more than 120°C. Many chemical reactions among the minerals and pore waters of sedimentary rocks will take place only at these higher temperatures. Another factor causing diagenesis is the increase in pressure with burial, on the average about 1 atmosphere for each 4.4 m of depth. This pressure is responsible for the compaction of sediments.

An understanding of diagenetic processes is of great practical value because coal, oil, and gas are formed by the diagenesis of buried organic matter.

# CLASSIFICATION OF SEDIMENTS AND SEDIMENTARY ROCKS

We can now use our knowledge of sedimentation to classify sediments and their lithified counterparts, sedimentary rocks. The major divisions are the clastic and the chemical and biochemical (Table 7.4). Clastic sediments and sedimentary rocks are classified according to the sizes of their particles, because the size of a particle affects the ability of a current to transport and deposit it. Chemical and biochemical sediments and sedimentary rocks are classified by the chemicals in the parent solutions.

## Clastic Sediments and Sedimentary Rocks

Clastic sediments and sedimentary rocks are classified on the basis of their textures, primarily the sizes of the grains. They are divided into (1) coarse-grained clastics: **gravels** and their lithified equivalents, **conglomerates;** (2) medium-grained clastics: **sands** and their lithified equivalents, **sandstones;** and (3) fine-grained clastics: **silts** and their lithified equivalents, **siltstones; clays** and **muds** and their lithified equivalents, **mudstones** and **shales** (Table 7.5).

Within each textural category, clastics are further subdivided by mineralogy, which reflects the parent rocks. Thus there are quartz-rich and feldspar-rich sandstones and calcareous, siliceous, and organic-rich shales.

## Chemical and Biochemical Sediments and Sedimentary Rocks

Chemical and biochemical sediments and sedimentary rocks are classified by their chemical composition (Table 7.6), which in the case of marine sediments reflects the chemical elements dissolved in seawater. The most abundant such elements are chlo-

---

**TABLE 7.4**

Major Classes of Sediments and Sedimentary Rocks

|  | CLASTIC | CHEMICAL AND BIOCHEMICAL |
|---|---|---|
| Raw material | Broken solid fragments | Dissolved constituents |
| Sediment/ sedimentary rock types | Gravel/conglomerate | Carbonate sediment/limestone, dolostone |
|  | Sand/sandstone | Evaporite sediment/halite, gypsum |
|  | Mud/mudstone, shale | Siliceous sediment/chert |

**TABLE 7.5**

Particle Size Classification of Clastic Sediments and Sedimentary Rocks

| SEDIMENT | | PARTICLE SIZE | ROCK |
|---|---|---|---|
| COARSE | | | |
| Gravel | Boulder | 256 mm | Conglomerate |
| | Cobble | 64 mm | |
| | Pebble | 2 mm | |
| Sand | | 0.062 mm | Sandstone |
| Mud | Silt | 0.0039 mm | Siltstone |
| | Clay | | Mudstone (blocky fracture) Shale (breaks along bedding) |
| FINE | | | |

**TABLE 7.6**

Classification of Chemical and Biochemical Sediments and Sedimentary Rocks

| SEDIMENT | ROCK | CHEMICAL COMPOSITION | MINERALS |
|---|---|---|---|
| Carbonate sand and mud | Limestone | Calcium carbonate $CaCO_3$ | Calcite (aragonite) |
| No primary sediment | Dolostone | Calcium-magnesium carbonate $CaMg(CO_3)_2$ | Dolomite |
| Iron oxide sediment | Iron formation | Iron silicate; oxide; carbonate $FeO_3$; $FeCO_3$ | Hematite Limonite Siderite |
| Evaporite sediment | Evaporite | Sodium chloride; calcium sulfate $NaCl$; $CaSO_4$ | Gypsum Anhydrite Halite Other salts |
| Siliceous sediment | Chert | Silica $SiO_2$ | Opal Chalcedony Quartz |
| Peat, organic matter | Organics | Carbon C | (Coal) (Oil) (Gas) |
| No primary sediment | Phosphorite | Calcium phosphate $Ca_3(PO_4)_2$ | Apatite |

rine (as chloride, Cl⁻), sodium (Na), potassium (K), calcium (Ca), magnesium (Mg), sulfur (as sulfate, $SO_4$), and carbonate ($CO_3$). Silica ($SiO_2$) and phosphorus, major components of some sedimentary rocks, are found in minor quantities in seawater.

Most abundant are the **carbonate sediments and sedimentary rocks,** composed of either calcium or calcium-magnesium carbonates. Most of them were originally secreted biochemically as shells by organisms living near the surface or on the bottom of the sea. The abundance of carbonate rocks is due to the large amounts of calcium and carbonate present in seawater—carbonate derived from the carbon dioxide in the atmosphere and calcium and carbonate from the easily weathered limestone on the continents.

**Evaporite sediments and sedimentary rocks** are precipitated inorganically from evaporating seawater and contain minerals formed by the crystallization of sodium chloride (halite), calcium sulfate (gypsum and anhydrite), and other combinations of the ions commonly found in seawater. Other chemical sediments and sedimentary rocks originally precipitated inorganically include **iron formations** (iron

oxides, silicates, and carbonates) and **phosphorites** (sediments and sedimentary rocks rich in the element phosphorus).

Coal, a sedimentary rock rich in organic carbon, is classified as an **organic sedimentary rock.** Although oil and gas are fluids, not normally classed with sedimentary rocks, they too can be considered organic sediments because they are formed by the diagenesis of organic materials in the pores of sedimentary rocks.

# BEDDING AND SEDIMENTARY STRUCTURES

Sediments and sedimentary rocks display several types of **bedding** or **stratification.** Most bedding is horizontal, but there are some exceptions. One kind of nonhorizontal bedding, called **cross-bedding,** consists of sets of bedded material inclined at angles up to 35° from the horizontal (Figure 7.8). Crossbeds are formed by deposition of the grains on the

**FIGURE 7.8** Cross-bedding in Navaho sandstone, Zion National Park, Utah. The different directions of cross-bedding reflect different wind directions at the time of deposition of sand dunes. *Peter Kresan.*

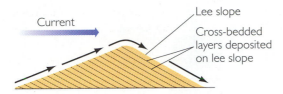

**FIGURE 7.9** Formation of cross-bedding on the lee or downcurrent slope of a dune or ripple.

steeper, downcurrent (lee) slopes of sand dunes on land or of sandbars in rivers and under the sea (Figure 7.9). Cross-bedding is common in sandstones and also occurs in gravels and some carbonate sediments.

**Graded bedding** is a form of bedding in which beds grade upward, from coarse grains at the base to fine grains at the top. A graded bed, comprising one set of coarse to fine beds, normally ranges from a few centimeters to several meters thick. The grading reflects a waning of the current. Accumulations of many graded beds, the total thickness reaching many hundreds of meters, are deposited in deep ocean wa-

ters by distinctive currents flowing along the bottom (see Chapter 17).

All kinds of bedding and a variety of other forms produced in the process of sedimentation are called **sedimentary structures. Ripples** are sedimentary structures common in both modern sands and ancient sandstones. Ripples are low, narrow ridges or corrugations, most only a centimeter or two high, separated by wider troughs. Ripples can be seen on the surfaces of windswept dunes, on underwater sandbars in shallow streams, and under the waves at beaches (Figure 7.10). Geologists can distinguish the symmetrical ripples made by waves on a beach from the asymmetrical ripples formed on sandbars in rivers or by the wind on dunes (Figure 7.11).

Bedding in many sedimentary rocks is broken or disrupted, sometimes crossed by roughly cylindrical tubes a few centimeters in diameter that extend vertically through several beds. These sedimentary structures are remnants of burrows and tunnels excavated by clams, worms, and many other marine organisms that live on the bottom of the sea. In a process called **bioturbation,** these organisms burrow through muds and sands, ingesting sediment for the bits of

**FIGURE 7.10** Ripples in modern sand on a beach (*left*) and ancient ripple-marked sandstone (*right*). *Left: Raymond Siever. Right: Reg Morrison/Auscape.*

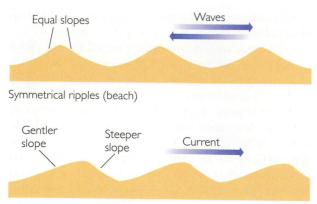

FIGURE 7.11 The shapes of ripples on beach sand, produced by the back-and-forth movements of waves, are symmetrical. Ripples on dunes and river bars, produced by the movement of a current in one direction, are asymmetrical.

organic matter it contains and leaving the reworked sediment behind, which fills the burrow (Figure 7.12). From bioturbation structures geologists can deduce the kinds of organisms that burrowed the sediment and reconstruct the sedimentary environment.

Another important characteristic of sedimentary rocks is **bedding sequences,** which are patterns of interbedding of sandstone, shale, and other sedimentary rock types. The nature of these bedding sequences helps geologists reconstruct how all of the sediments were deposited. Rivers frequently deposit repeated cycles of bedding sequences in which sediments grade from coarse at the base to fine at the top

FIGURE 7.12 Bioturbation. Tracks thought to have been made by trilobites in middle Cambrian mudstone of Montana. *Chip Clark.*

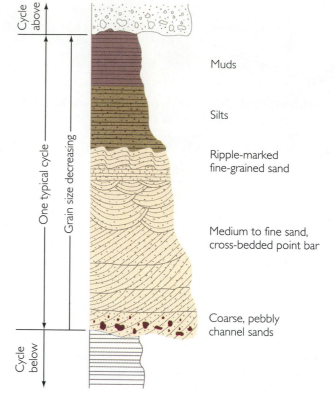

FIGURE 7.13 A typical alluvial cycle. The widths of the sections drawn are proportional to the sizes of the grains in the sediment. The thicknesses of cycles range from a few meters for small streams to 20 m or more for large ones.

(Figure 7.13). Other characteristic bedding sequences can be used to recognize deposition at the shoreline and in the deep sea.

# CLASTIC SEDIMENTS AND SEDIMENTARY ROCKS

Clastic sediments and sedimentary rocks—muds and shales, sands and sandstones, gravels and conglomerates—constitute more than three-quarters of the total mass of all types of sedimentary rocks in the Earth's crust (Figure 7.14). Of the various types of clastic sediments and sedimentary rocks, silt, siltstone, mud, mudstone, and shale are by far the most abundant—about seven times more common than the coarser clastics. The abundance of the fine-grained clastics, which contain large amounts of clay minerals, reflects the chemical weathering to clay minerals of the large amounts of feldspar and other silicate minerals in the Earth's crust.

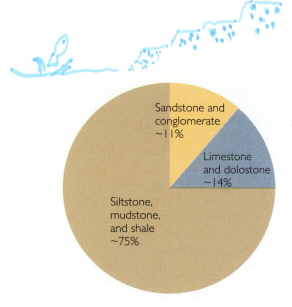

FIGURE 7.14 The relative abundance of the major sedimentary rock types. In comparison with these three, all other sedimentary rock types—including evaporites, cherts, and other chemical sediments—occur in only minor amounts.

Classifying the various clastic sediments and rocks on the basis of the sizes of their particles in effect also distinguishes them by the conditions of sedimentation. As we have seen, the larger the particle, the stronger the current needed to move and deposit it. For that reason, like-sized particles tend to accumulate in sorted beds; that is, most sand beds do not contain pebbles or mud, and most muds consist only of finer particles.

*Gravels and conglomerates* consist mostly of pebbles and cobbles, all of which are defined as particles larger than 2 mm in diameter (see Table 7.5). Such large particles indicate sedimentation by strong river currents in mountains or by high waves on a rocky beach. *Silts, siltstones, muds, mudstones, and shales* are made up of particles less than 0.062 mm in diameter. These particles indicate sedimentation in quiet waters that allow the finest particles to settle to the bottom as layers of silt and mud. Between gravels and muds are the medium-sized particles, 0.062 to 2 mm in diameter, that make up *sands and sandstones*. These sediments are moved by moderate currents, such as those of rivers, waves at shorelines, and the winds that blow sand into dunes. The wide range of particle sizes that are found in clastic sedimentary rocks can be easily observed in the samples shown in Figure 7.15.

## Sand and Sandstone

Sandstone has been studied more than other clastic sedimentary rocks both because it is abundant and widespread and because the information it contains about its origin is easy to read. Sand particles are large enough to be seen with the naked eye, and many of their features are easily discerned with a low-power magnifying glass.

FIGURE 7.15 Clastic sedimentary rocks (clockwise from top left): breccia, conglomerate, sandstone, shale. *Chip Clark.*

CALCITE

**SIZES OF SAND GRAINS**    Sand particles may be fine, medium, or coarse. Although the average size of the grains in any one sandstone reflects both the strength of the current that carried them and the sizes of the crystals eroded from the parent rock, the size does not tell the whole story. The range and relative abundance of the various sizes are also significant. If all the grains are close to the average size, the sand is well sorted; if many grains are much larger or smaller than the average, the sand is poorly sorted. The degree of sorting can help distinguish, for example, between sands of beaches (well sorted) and muddy sands deposited by glaciers (poorly sorted).

**SHAPES OF SAND GRAINS**    The shapes of sand grains can also be important clues to their origin. The abrasion that results when grains are knocked together as they are transported gives them rounded contours. Angular grains imply short distances of transportation; rounded ones indicate long journeys down a large river system. Sand grains also become rounded as they are moved constantly back and forth by the waves on beaches. Most sand grains inherit their spheroidal, elongate, or disc shapes from the shapes of the original crystals in the parent rock.

**MINERALOGY**    From the mineralogy of sands and sandstones geologists can deduce the nature of the source areas that were eroded to produce the sand grains. The presence of sodium- and potassium-rich feldspars with abundant quartz, for example, might indicate that the sediments were eroded from a granitic terrain. Other minerals, as we will see in Chapter 8, would be indicative of metamorphic parent rocks. The mineralogy of a sand or sandstone may not match the mineralogy of the parent rocks exactly. Chemical weathering at the source area can alter and dissolve—and thus remove—much of the feldspar from a granite. Under these conditions, what remains might be grains mainly of quartz. Thus we can infer, from an analysis of the mineralogy of a sand or sandstone, its parent rocks and something about the factors, such as climate, that influence weathering in the source area. The mineralogical composition of parent rocks can also be correlated with plate-tectonic settings. Sandstones containing abundant fragments of mafic volcanic rocks, for example, are derived from the volcanic arcs of subduction zones.

**MAJOR KINDS OF SANDSTONES**    Sandstones fall into several major groups on the basis of their mineralogy and texture. Some, called quartz arenites, are made up almost entirely of quartz grains, usually well sorted and rounded (Figure 7.16a). These pure quartz sands result from extensive weathering before and during transport that removed everything but quartz, the most stable mineral. Another group of sandstones, called arkoses, contain more than 25 percent feldspar; the grains tend to be poorly rounded and less well sorted than those of pure quartz sandstones (Figure 7.16b). These feldspar-rich sandstones come from rapidly eroding granitic and metamorphic terrains where chemical weathering is subordinate to physical weathering.

Lithic sandstones contain many fragments derived from fine-grained rocks, mostly shales, volcanic rocks, and fine-grained metamorphic rocks (Figure 7.16c). Graywacke, another kind of sandstone, is a heterogeneous mixture of rock fragments and angular grains of quartz and feldspar, the sand grains being surrounded by a fine-grained clay matrix (Figure 7.16d). Much of this matrix is formed by the chemical alteration and mechanical compaction and deformation of relatively soft rock fragments, such as those of shale and some volcanic rocks, after deep burial of the sandstone formation.

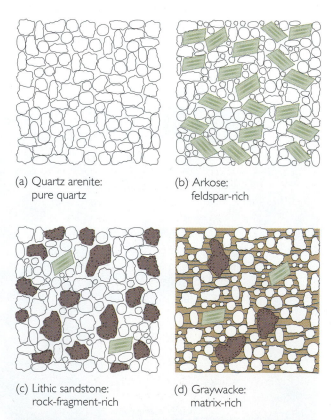

(a) Quartz arenite: pure quartz

(b) Arkose: feldspar-rich

(c) Lithic sandstone: rock-fragment-rich

(d) Graywacke: matrix-rich

**FIGURE 7.16** The mineralogy of four major groups of sandstones.

Knowing the origins of sandstones and being able to predict their characteristics have practical and economic benefits. Much of the oil and gas discovered in the last 150 years has been found in buried sandstones. Both petroleum geologists and groundwater geologists, who seek supplies of water in porous sandstones, must know about the porosity and cementation of these rocks.

## Gravel and Conglomerate

Pebbles, cobbles, and boulders are the main constituents of the coarsest-grained sediments—gravels—and their lithified equivalents—conglomerates (see Figure 7.15). Because of their large size, pebbles are easy to study and identify. Granite pebbles in a river-deposited conglomerate provide firm evidence of a mass of exposed granite in the drainage areas feeding the rivers that transported the gravel. There are relatively few environments in which currents are strong enough to transport pebbles: mountain streams, beaches with strong waves, and the meltwaters of glaciers. Strong currents also carry sand, and we almost always find sand between the pebbles, some of it deposited with the gravel and some of it infiltrated into pores between fragments after the larger ones have been deposited.

Pebbles and cobbles abrade and become rounded very quickly in the course of transport. Travel of 100 km can make pebbles smooth and round. Beach gravels moved back and forth constantly by strong waves also become rounded.

In contrast to conglomerate, some coarse-grained clastic rocks contain sharp, angular fragments that show little or no signs of abrasion. These are **sedimentary breccias,** characterized by the angularity of the rock fragments that make them up (see Figure 7.15). Sedimentary breccias are found in deposits close to their source, where sediments were deposited before they had been transported very far. Breccias not of sedimentary origin may form by the breaking up of volcanic materials during eruptions or by the fragmentation of rocks along faults.

## Mud, Silt, Mudstone, Siltstone, and Shale

The finest-grained clastic sediments and sedimentary rocks—the muds and mudstones, shales, silts, and siltstones—cover a wide range of grain sizes and mineral compositions. *Mud* is any clastic sediment in which most of the particles are less than 0.062 mm in diameter. *Silt* is a clastic sediment in which most of the grains are between 0.0039 and 0.062 mm in diameter. Clay-sized particles (we are referring only to the sizes of the particles, not to clay minerals) are less than 0.0039 mm in diameter. The rock equivalents of muds are shales (see Figure 7.15), which break readily along bedding planes, and mudstones, which are blocky and show poor or no bedding. Bedding may have been well marked in many mudstones when the sediments were deposited, but it may have been lost by bioturbation (see Figure 7.12). Rocks made up exclusively of clay-sized particles are called claystones. The lithified equivalent of silt is siltstone, which commonly looks similar to mudstone but may appear more like very fine-grained sandstone.

The most abundant component of fine-grained sediment and sedimentary rock is clay-sized material that consists largely of clay minerals. Many muds, mudstones, and shales contain more than 10 percent carbonate, forming deposits of calcareous shales. Black, or organic, shales contain abundant diagenetically altered organic matter; some, called oil shales, contain large quantities of oily organic material, which makes them a potentially important source of oil.

Fine-grained sediments are deposited by the gentlest currents, which allow slow settling of the finest particles. After a river has flooded its lowlands and the flood recedes, the current slows and mud settles; this mud contributes to the fertility of river bottomlands. Muds are left behind by ebbing tides along many tidal flats where wave action is mild. Much of the deep ocean floor, where currents are weak or absent, is blanketed by mud. Dust, which includes clay- and silt-sized particles, is deposited by the wind after dust storms on arid plains (see Chapter 14).

Certain clay minerals in fine-grained sediments, especially kaolinite, are economically valuable for making ceramic products. Potentially most valuable are the oil shales, which we discuss in detail in Chapter 22.

# CHEMICAL AND BIOCHEMICAL SEDIMENTS AND SEDIMENTARY ROCKS

Whereas clastic sediments and sedimentary rocks give us information about continental parent rocks and weathering, the chemical and biochemical sediments tell us about chemical conditions in the environment of sedimentation, predominantly the ocean.

FIGURE 7.17 Chemical and biochemical sedimentary rocks (clockwise from top left): gypsum, halite, limestone, chert. *Chip Clark.*

Although chemical sedimentation takes place in some lakes, particularly those of arid regions where evaporation is intense, such as the Great Salt Lake of Utah, such sediments account for only a very small fraction of the amounts deposited along the ocean's shorelines, on continental shelves, and in the deep ocean.

spectacular vacation spots is the deep ocean floor, the site of most of the carbonate deposited today. The calcite shells of **foraminifera,** tiny single-celled organisms that live in surface waters, settle to the seafloor when the organisms die and accumulate there as sediment (Figure 7.18).

## Limestone and Dolostone

The dominant biochemical sedimentary rock is **limestone,** composed mainly of calcium carbonate ($CaCO_3$) in the form of the mineral calcite (Figure 7.17). Limestones are lithified carbonate sediments, which consist of shells ranging in size from many centimeters to microscopic and some fine-grained carbonate muds of mixed origin. Carbonate sands are abundant in many environments; mud dominates in others. In addition to calcite, most carbonate sediments contain aragonite, a less stable form of calcium carbonate. Some organisms precipitate calcite, others aragonite, and some both. Another abundant carbonate rock is **dolostone,** made up of the mineral dolomite, which is composed of calcium-magnesium carbonate ($CaMg[CO_3]_2$). As we will see later in this chapter, dolomite is not a constituent of shells or newly deposited carbonate sediment.

The oceans of the world are the scenes of most carbonate sedimentation, from the coral reefs of the Pacific and Caribbean to the shallow banks of the Bahama Islands. Less accessible for study than these

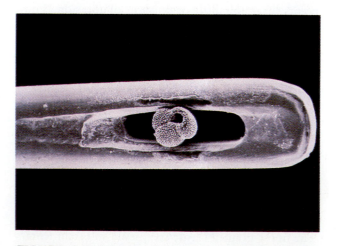

FIGURE 7.18 Scanning electron micrograph of a foraminifer in the eye of a needle. These single-celled organisms secrete shells of calcium carbonate, which they extract from seawater. *Chevron Corporation.*

**FIGURE 7.19** A coral reef surrounding a central volcanic island. In the foreground is the reef, which shelters the shallow lagoon behind it. *Jean-Marc Truchet/Tony Stone Worldwide.*

Most of the carbonate sediments of the ocean are derived from the shells and skeletons of foraminifera and other organisms that secrete calcium carbonate extracted from seawater. Coral reefs of warm seas are constructed of the carbonate skeletons of millions of corals. The calcium carbonate of the coral and other organisms, in contrast to the soft, loose sediment produced in other environments, forms a rigid, wave-resistant structure—a cemented buttress of solid limestone—that is built up to and slightly above sea level (Figure 7.19). The solid limestone of the reef is produced directly by the action of organisms; there is no soft sediment stage. On and around these reefs live hundreds of species of other carbonate-precipitating organisms, many of them similar to the familiar clams, oysters, and mussels of our shorelines. Carbonate is also precipitated by marine algae, single-celled organisms similar to primitive plants, which grow on reef tracts and in other carbonate environments. Carbonate muds, consisting partly of microscopic fragments of shells and calcareous algae and partly of inorganic—that is, purely chemical rather than biochemical—precipitation, form in the lagoons behind reefs and on extensive shallow banks such as those of the Bahamas. Reef tracts are found on passive continental margins and fringing volcanic island arcs in the warmer parts of the oceans where they are not flooded with clastic sediment.

Until recently many geologists thought that there was no such thing as direct inorganic precipitation of calcium carbonate from seawater, but research in lagoons and banks in the Bahama Islands has shown that a significant fraction of the carbonate mud on shallow banks is inorganically precipitated directly from seawater.

The chemical basis for inorganic carbonate precipitation is the relative abundance of calcium ($Ca^{2+}$) and bicarbonate ($HCO_3^-$) ions in seawater. The warm, tropical parts of the ocean are supersaturated with calcium carbonate and precipitate carbonate by the following chemical reaction:

$$\underset{Ca^{2+}}{\text{calcium ion}} + \underset{2HCO_3^-}{\underset{\text{(dissolved)}}{\text{bicarbonate ion}}} \rightarrow$$

$$\underset{CaCO_3}{\underset{\text{(precipitated)}}{\text{calcium carbonate}}} + \underset{H_2CO_3}{\underset{\text{(dissolved)}}{\text{carbonic acid}}}$$

When living organisms secrete their carbonate shells, they are effectively producing the same chemical reaction by biochemical means. The chemical reaction that precipitates calcium carbonate is the reverse of the reaction that takes place during the chemical weathering of limestone (see Chapter 6):

$$CaCO_3 + H_2CO_3 \rightarrow Ca^{2+} + 2HCO_3^-$$

By comparing the lithologies and textures of carbonate sediments being deposited today and those of limestones and dolostones of the past, geologists can tell how the older rocks were formed. Many limestones and dolostones were originally deposited as limestone reefs; they contain as evidence the fossilized remains of the organisms that made them (Figure 7.20). At present corals are the main constructors of reefs, but at earlier times in Earth's history other organisms were builders of wave-resistant structures, some of them cemented buttresses of solid limestone.

The precursors of ancient limestones made up of sand-sized carbonate particles, some of them crossbedded, can be found in the carbonate sands of today's beaches and underwater banks and bars in the Caribbean and the Bahamas.

Carbonate muds are laid down in deeper waters or in protected lagoons or bays, where waves and currents do not have the power to carry away the fine particles. These muds become compacted and cemented after burial, forming dense, fine-grained limestone.

Dolostones are diagenetically altered carbonate sediments and limestones. The mineral dolomite does not form as a primary precipitate from ordinary seawater, and no organisms secrete shells of dolomite. Instead, the original calcite or aragonite of a carbonate sediment is converted to dolomite soon after deposition by the addition of magnesium ions from seawater slowly passing through the pores of the sediment. Many dolostones form in this way in shallow bays and flats where seawater is concentrated. As the seawater becomes concentrated, the proportion of magnesium ions increases in relation to calcium ions, leading to a reaction between water and sediment by which some calcium ions in calcite or aragonite are exchanged for magnesium ions, converting these calcium carbonate minerals ($CaCO_3$) to dolomite ($CaMg[CO_3]_2$). Other dolostones form after moderate or deep burial of limestones by reaction with groundwaters containing a large proportion of magnesium ions.

The marine environments where carbonate sedimentation produces rigid limestone structures, including reefs, carbonate banks, and deep-water deposits in the open ocean, are discussed further in Chapter 17.

## Evaporites

One of the commonest chemical sediments formed from evaporating seawater is table salt, the mineral halite ($NaCl$) (see Figure 7.17). Deep under the city of Detroit, Michigan, for example, beds of salt laid down by an evaporating arm of an ancient ocean are mined in extensive workings. Another mineral formed by the evaporation of seawater is **gypsum,** calcium sulfate ($CaSO_4 \cdot 2H_2O$), the principal component of plaster (see Figure 7.17). Halite and gypsum are major components of **marine evaporites,** the chemical sediments and sedimentary rocks formed by the evaporation of seawater. As seawater evaporates and becomes more concentrated, minerals are crystallized in sequence, some of them primary precipitates, such as halite, others formed by diagenetic reaction, such as dolomite. As dissolved ions precipitate to form each of these minerals, the evaporating seawater changes composition.

The first precipitates to form as seawater starts to evaporate are the carbonates, first calcite alone, then dolomite by diagenetic reaction. Continued evaporation leads to the precipitation of gypsum. By this time no carbonate ions are left in the water. After further evaporation, halite starts to form. In the final stages of evaporation, after the sodium chloride is gone, magnesium and potassium chlorides and sulfates are precipitated. This sequence of precipitation has been studied in the laboratory and is matched by the bedding sequences found in certain natural salt formations.

Most of the evaporites of the world consist of thick sequences of dolomite, gypsum, and halite and do not contain the final-stage precipitates. Many do not even go so far as halite. The absence of the final stages indicates that the water did not evaporate completely but was replenished by normal seawater as evaporation continued. The great volume of many

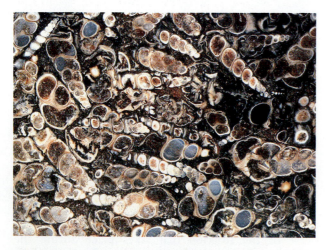

**FIGURE 7.20** Fossiliferous limestone made up of fossil shells of gastropods (snails) in the Eocene Green River Formation of Wyoming. *Peter Kresan.*

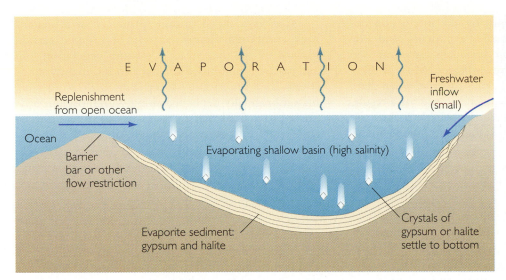

**FIGURE 7.21** Gypsum and halite form as evaporite sediments when seawater evaporates in a shallow basin with a restricted connection to the open ocean. Evaporation removes much more water than the fresh water flowing in can replace. As the evaporating basin gets appreciably more saline than the water of the open sea, gypsum is precipitated. A further increase in salinity leads to the crystallization of halite.

evaporites, some hundreds of meters thick, shows that they could not have formed from the small amount of water that could be held in a shallow bay or pond. A huge amount of seawater had to have evaporated.

The water of a bay or arm of the sea evaporates steadily where its freshwater supply from rivers is small, connections to the open sea are constricted, and the climate is arid (Figure 7.21). Here thick evaporites form. The openings allow seawater to flow in to replenish the evaporating waters of the bay, which stay at constant volume but are more saline than the open ocean. The evaporating bay waters stay more or less constantly supersaturated with halite, gypsum, or other minerals and steadily deposit evaporite minerals on the floor of the evaporite basin.

Evaporite sediments also form in lakes that have little or no river outlet, where evaporation controls the lake level and incoming salts derived from chemical weathering accumulate. The Great Salt Lake of Utah is one of the best known such lakes. River waters enter the lake, bringing salts dissolved in the course of weathering. In the dry climate of Utah, evaporation has more than balanced the inflow of fresh water from rivers and rain, concentrating dissolved ions in the lake to make it one of the saltiest bodies of water in the world, eight times saltier than seawater. Other small lakes in arid regions may collect unusual salts, such as borates (compounds of the element boron), and some become alkaline. The water in this kind of lake is poisonous. Economically valuable resources of borates and nitrates (minerals containing the element nitrogen) are found in the sediments beneath some of these lakes.

## Chert

One of the first sedimentary rocks to be used for practical purposes by our prehistoric ancestors was **chert,** a sedimentary rock made up of chemically or biochemically precipitated silica ($SiO_2$) (see Figure 7.17). Early hunters used it for arrowheads and other tools because it could be chipped and shaped to form hard, sharp implements. A common name for chert is **flint,** and the terms are practically interchangeable. The silica in most cherts is in the form of extremely finely crystalline quartz; some geologically young cherts consist of a less well crystallized form of silica, opal.

Like calcium carbonate, much silica sediment is precipitated biochemically, secreted as shells by ocean-dwelling organisms. When these organisms die, they sink to the deep ocean floor, where their shells accumulate as layers of silica sediment. After these silica sediments are buried by later sediments, they are diagenetically cemented into chert.

## Other Chemical and Biochemical Sediments

Many other kinds of chemical and biochemical sediments are deposited in the sea and in lakes. *Phosphorite,* sometimes called phosphate rock, is composed of calcium phosphate precipitated from phosphate-rich seawater in places where currents of deep, cold water containing phosphate and other nutrients rise along continental margins. The phosphorite forms diagenetically by the interaction between muddy or carbonate sediments and the phosphate-rich water. *Iron formations* are composed of iron oxides and some iron silicates. Most of these materials formed early in Earth's history, when there was less oxygen in the atmosphere, and, as a result, iron was more easily soluble. In soluble form, iron was transported to the sea and precipitated there.

*Coal* is a biochemically produced sedimentary rock composed almost entirely of organic carbon formed by the diagenesis of swamp vegetation. *Oil* and *gas* are found in sedimentary rocks, mainly sandstones and limestones. Deep burial alters organic matter originally deposited along with inorganic sediments to a fluid that then escapes to other, porous formations and becomes trapped there.

# SUMMARY

**What are the major processes involved in the formation of sedimentary rock?** Weathering and erosion produce the clastic particles and dissolved ions that compose sediment. Currents of water, wind, and ice transport the sediment to its ultimate resting place, the site of sedimentation. Sedimentation, or deposition, a settling of particles from the transporting agent, produces bedded sediments in river channels and valleys, on sand dunes, and at the edges and floors of the oceans. Lithification and diagenesis harden the sediment to sedimentary rock after it has been buried by additional sediment.

**What are the two major divisions of sediments and sedimentary rocks?** Clastic sediments are formed from the fragments of parent rock produced by physical weathering and the clay minerals produced by chemical weathering. These solid products are carried by water and wind currents and ice to the oceans, sometimes being sedimented along the way. Chemical and biochemical sediments originate from the ions dissolved in water during chemical weathering. These ions are transported in solution to

the oceans, where they are mixed into seawater. Through chemical and biochemical reactions, the ions are precipitated from solution and the precipitated particles settle to the ocean floor.

**How do we classify the major kinds of clastic and chemical and biochemical sedimentary rocks?** Clastic sediments and sedimentary rocks are classified as gravels and conglomerates; sands and sandstones; and muds, mudstones, and shales; in accordance with the sizes of their particles. This method of classifying sediments reflects the importance of the strength of the current as it transports and deposits solid materials. The chemical and biochemical sediments and sedimentary rocks are classified on the basis of their chemical compositions. The most abundant of these rocks are the carbonate rocks, limestone and dolostone. Limestone is made up largely of biochemically precipitated shells. Dolostone is formed by the diagenetic alteration of limestones. Other chemical and biochemical sediments include evaporites, siliceous sediments, phosphorites, iron formations, and coal.

# KEY TERMS AND CONCEPTS

clastic particles (p. 145)
clastic sediments (p. 145)
chemical and biochemical sediments (p. 145)
sorting (p. 146)
peat (p. 149)
coal (p. 149)
oil (p. 149)
gas (p. 149)
sedimentary environment (p. 149)
clastic sedimentary environment (p. 150)

terrigenous sediment (p. 151)
chemical and biochemical sedimentary environments (p. 151)
carbonate environment (p. 151)
marine evaporite environment (p. 152)
siliceous environment (p. 152)
diagenesis (p. 152)
cementation (p. 153)
porosity (p. 153)
lithification (p. 153)
compaction (p. 153)

gravel (p. 154)
conglomerate (p. 154)
sand (p. 154)
sandstone (p. 154)
silt (p. 154)
siltstone (p. 154)
clay (p. 154)
mud (p. 154)
mudstone (p. 154)
shale (p. 154)
carbonate sediments and sedimentary rocks (p. 156)

evaporite sediments and
   sedimentary rocks (p. 156)
iron formation (p. 156)
phosphorite (p. 156)
organic sedimentary rock (p. 156)
bedding (p. 156)
stratification (p. 156)

cross-bedding (p. 156)
graded bedding (p. 157)
sedimentary structures (p. 157)
ripples (p. 157)
bioturbation (p. 157)
bedding sequence (p. 158)
sedimentary breccia (p. 161)

limestone (p. 162)
dolostone (p. 162)
foraminifera (p. 162)
gypsum (p. 164)
marine evaporites (p. 164)
chert (p. 165)
flint (p. 165)

# EXERCISES

1. How do clastic sedimentary rocks differ from chemical and biochemical sedimentary rocks?

2. How and on what basis are the clastic sedimentary rocks subdivided?

3. What kind of sedimentary rocks were originally formed by the evaporation of seawater?

4. What processes change sediment into sedimentary rock?

5. Name three clastic environments.

6. How do organisms produce or modify sediments?

# THOUGHT QUESTIONS

1. Weathering of the continents has been much more widespread and intense in the past 10 million years than it was in earlier times. How might this observation be reflected in the sediments that now cover the Earth's surface?

2. In what respects might you consider a volcanic ash fall a sediment?

3. A geologist is heard to say that a particular sandstone was derived from a granite. What information could she have gleaned from the sandstone to lead her to that conclusion?

4. Name a sedimentary rock that is essentially the product of diagenesis and that has no exact equivalent as a sediment.

5. You discover a bedding sequence that has a conglomerate at the base, grades upward to a sandstone and then to a shale, and finally, at the top, grades to a limestone of cemented carbonate sand. What changes in the area of the sediment's source or in the sedimentary environment would have been responsible for this sequence?

6. How can you use the sizes and sorting of sediments to distinguish between sediments deposited in a glacial environment and those deposited on a desert?

7. Describe the beach sands that you would expect to be produced by the beating of waves on a coastal mountain range consisting largely of basalt.

8. How are chert and limestone similar in origin?

9. What role do transporting currents play in the origin of some kinds of limestone?

# SUGGESTED READINGS

Blatt, Harvey. 1992. *Sedimentary Petrology,* 2d ed. New York: W. H. Freeman.

Goreau, T. F., N. I. Goreau, and T. J. Goreau. 1979. Corals and coral reefs. *Scientific American* (August):124–136.

Laporte, L. F. 1979. *Ancient Environments,* 2d ed. Englewood Cliffs, N.J.: Prentice-Hall.

Leeder, M. R. 1982. *Sedimentology.* London: Allen and Unwin.

Siever, Raymond. 1988. *Sand.* New York: Scientific American Library.

# METAMORPHIC ROCKS

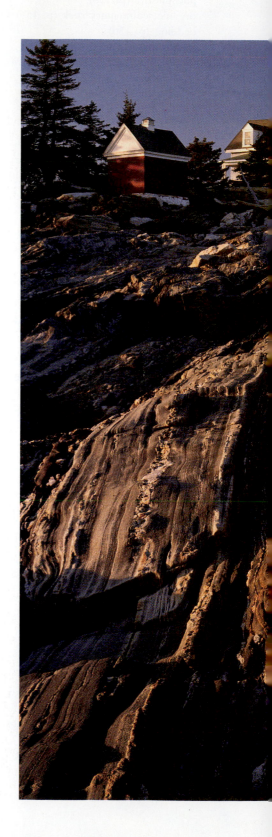

Rocks change as the conditions they encounter change. Igneous rocks are formed by melting and crystallization, whereas sedimentary rocks are produced by weathering. Rocks subjected in the solid state to high temperatures and pressures change dramatically to metamorphic rocks. The minerals and textures of metamorphic rocks, which form deep in the crust and mantle, are guides to the patterns of pressure and temperature that accompany mountain building and the intrusion of igneous bodies. As geologists deduce the conditions for metamorphism, they are able to place the origins of rocks in their plate-tectonic settings. Metamorphism, then, can be related to the plate movements that make mountains and deform rocks and to the igneous activity that bakes the surrounding rock.

## 8

Metamorphic rock
showing bands of
different compositions
produced by heat and
pressure. These bands
are not to be confused
with the layering of
sedimentary rocks.
Pemaquid Point
lighthouse, Maine.
*Larry Ulrich.*

Deep in Earth's crust, tens of kilometers below the surface, temperatures and pressures are high, but not high enough to melt rocks. This is the principal domain of metamorphism, where increases in heat and pressure and changes in the chemical environment result in changes in the mineral compositions and crystalline textures of sedimentary and igneous rocks, *which remain solid all the while*. The result is the third large class of rocks, the **metamorphic** or "changed form" **rocks.** A limestone filled with fossils, for example, may become transformed to a white marble in which no trace of fossils remains. The rock, originally made of calcite, may be unchanged in mineral and chemical composition while its texture changes drastically. Shale, a well-bedded rock so fine-grained that no individual mineral grain can be seen, may be changed to a form in which bedding is obscured and large crystals of mica glitter in the sun. In this metamorphic transformation, both mineral composition and texture have changed while the overall chemical composition of the rock remains the same. Other rocks change in mineralogy, texture, *and* chemical composition.

While sediments and sedimentary rocks belong to the surface environments of Earth and igneous rocks to the melts of the lower crust and mantle, most metamorphic rocks exposed at the surface are the products of processes acting on rocks at depths ranging from the upper to the lower crust. Most have formed at depths of 10 to 30 km, the middle to lower half of the crust. Large parts of the mantle may be metamorphic, but we see such rocks only in the exposed cores of some deeply eroded mountain belts. Although most metamorphism occurs at depth in the crust and mantle, metamorphism can take place at the Earth's surface; we can see metamorphic changes in the baked surfaces of soils and sediments just beneath volcanic lava flows.

The pressure and heat that drive metamorphism are consequences of the internal heat of the Earth, the weight of overlying rock, and the horizontal pressures developed as rocks become deformed. Long ago, miners and geologists learned through experience that the deeper we go into the Earth, the hotter it gets and the heavier are the supports needed to keep mining tunnels from collapsing under the pressure. (Geothermal gradients are discussed in Chapter 19.) Temperatures in most of the continental crust increase by about 30°C for each kilometer of depth. At a depth of 15 km, for example, the temperature will be about 450°C, much higher than the average temperature of the surface, which ranges from 10 to 20°C in most regions. The pressure at that depth would be equivalent to the weight of all the overlying rock, about 4000 times the pressure at the surface. High as these temperatures and pressures may seem, they are only in the middle range of metamorphism, as Figure 8.1 shows. We refer to the metamorphic rocks formed under the lower temperatures and pressures of shallower crustal regions as **low-grade rocks** and the ones formed at deeper zones of higher temperature and pressure as **high-grade rocks.**

Some metamorphic rocks may have been subjected early in their history to high pressures and temperatures, which produced a high-grade rock, then much later to lower temperatures and pressures. In that case, the new conditions produce a low-grade rock, a process called **retrograde metamorphism.**

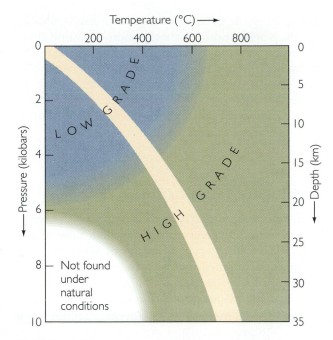

**FIGURE 8.1** Temperatures, pressures, and depths at which low- and high-grade metamorphic rocks are formed. The wide band shows common rates at which temperature and pressure increase with depth over much of the continents.

# PHYSICAL AND CHEMICAL FACTORS CONTROLLING METAMORPHISM

Metamorphic changes bring a preexisting rock into equilibrium with new surroundings. A sedimentary rock formed by diagenesis, for example, is in equilibrium with the moderate pressures and temperatures corresponding to burial a few kilometers deep. Later this rock may be caught up in an orogeny that buries it much deeper and subjects it to a temperature of more than 500°C. The rock is changed mineralogically and texturally so that it is brought into equilibrium with the new temperatures and pressures.

## Temperature

Heat has a profound effect on a rock's mineralogy and texture. As we saw in Chapter 4, heat can break chemical bonds and alter the existing crystal structures of igneous rocks. As the rock adjusts to its new temperature, its atoms and ions link up (recrystallize) in new arrangements, creating new mineral assemblages. Many new crystals will grow larger than they were in the original rock, and the rock may become banded as minerals are segregated into separate planes. Because different minerals are known to crystallize and remain stable at different temperatures, the metamorphic geologist, like the igneous geologist, uses a rock's composition to gauge the temperature at which the rock formed. Heat reduces the strength of rocks, and so metamorphic rocks are likely to be severely folded and deformed in orogenic belts.

## Pressure

Pressure changes a rock's texture as well as its mineralogy. Solid rock is subjected to two basic kinds of pressure, also called **stress.** One type is general pressure in all directions, like the pressure the atmosphere exerts at the surface of the Earth. High levels of this kind of pressure, called *confining pressure,* alter mineralogy by squeezing atoms together to form new minerals with denser crystal structures. The other kind of pressure, called *directed pressure,* is exerted in a particular direction, as when a ball of clay is squeezed between thumb and forefinger. The compressive action of converging plates is a form of directed pressure, which results in deformation of the rock.

Depending on the kind of stress applied to the rocks, metamorphic minerals may be compressed, elongated, or rotated to line up in a particular direction. Thus directed pressure guides the shape and orientation of the new metamorphic crystals formed as the minerals recrystallize under the influence of both heat and pressure. During recrystallization of micas, for example, the crystals grow with the planes of their sheet-silicate structures aligned perpendicular to the directed stress. Elongate minerals, such as amphiboles, will also line up in planes perpendicular to the directed stress.

## Chemical Metamorphic Changes

A rock's chemical composition can be altered significantly during metamorphism by the introduction or removal of chemical components. Chemical changes in surrounding rocks, such as shale or limestone, commonly follow the intrusion of a magma. Hydrothermal fluids rise from the magma, carrying dissolved sodium, potassium, silica, copper, zinc, and other chemical elements soluble in hot water under pressure. These elements may be derived from both the magma and the intruded rock. As hydrothermal solutions percolate up to the shallower parts of the crust, they react with the rocks they penetrate, changing their chemical and mineral compositions and sometimes completely replacing one mineral by another without changing the rock's texture. This kind of change in a rock's bulk composition is called **metasomatism.** Many valuable deposits of copper, zinc, lead, and other metal ores are formed by this kind of chemical substitution.

## Fluids in Metamorphism

Many of the chemical and mineralogical changes that occur during metamorphism take place in fluids that permeate the solid rock. Although metamorphic rocks as we see them in outcrops appear to be completely dry and of extremely low porosity, most contain fluid in their minute, thin pores (the spaces between grains). This intergranular fluid—typically water containing dissolved gases, salts, and traces of the components of the rock's minerals—acts as a medium that accelerates metamorphic chemical reactions. As changes in temperature and pressure break up crystal structures, atoms and ions move back and forth between the rock and its fluid. When in the fluid, the atoms and ions are able to migrate through the rock more rapidly and react with the solids to form new minerals.

As metamorphism proceeds, the water itself reacts with the rock as chemical bonds between minerals and water molecules form or break. The minerals of mafic volcanic rocks, for example, which contain almost no water in their crystal structures, take up some water molecules from pore fluids during early stages of metamorphism and form micas, chlorite, and other minerals with chemical bonds between water and other mineral components. The clay minerals in sedimentary rocks, by contrast, initially contain much chemically bound water and the rocks contain additional water in the pores, both of which are largely lost during metamorphism. As metamorphic grade increases, most of both chemically bound and pore water is lost.

## KINDS OF METAMORPHISM

Thanks to modern technology, geologists working in laboratories can now duplicate metamorphic conditions and determine the precise combinations of pressure, temperature, and chemical composition under which transformations might take place. But to understand how any particular such combination is related to the geology of metamorphism—that is, when, where, and how these conditions came about in the Earth—geologists must go into the field. Observations in the field have allowed metamorphic rocks to be grouped into several categories on the basis of the geological circumstances of their origin. These categories are described below and summarized in Figure 8.2.

## Regional Metamorphism

The most widespread type of metamorphism occurs where both high temperature and high pressure are imposed over large belts of the crust. We refer to this as **regional metamorphism,** to distinguish it from more localized changes near igneous intrusions or faults. Regional metamorphism destroys some or all of the original igneous or sedimentary textures as it causes the growth of new minerals. Some regional metamorphic belts are created by high temperatures and moderate to high pressures near the volcanic arcs formed where subducted plates dive deep into the mantle. Other belts are formed by high pressures and moderate temperatures near oceanic trenches, where subduction drags down relatively cold oceanic crust. Regional metamorphism under very high pressures and temperatures occurs in deeper levels of the crust along boundaries where converging continental tectonic plates deform rock and raise high mountain belts.

## Contact Metamorphism

Igneous intrusions metamorphose the immediately surrounding rock by their heat and pressure, subjecting the minerals of the preexisting rock to new conditions. This type of localized transformation, called **contact metamorphism,** normally affects only a thin region of intruded rock along the contact. In many contact metamorphic rocks, especially along shallow intrusions, the mineral transformations are largely related to the high temperature of the magma. Pressure effects are important where the magma was intruded at great depths. Contact metamorphism by extrusives is limited to very thin zones because lavas cool quickly at the surface and their heat has little time to penetrate deep into the surrounding rocks and cause metamorphic changes. Contact metamorphism, like the igneous activity with which it is linked, is found along plate convergences and oceanic and continental hot spots. Because igneous activity

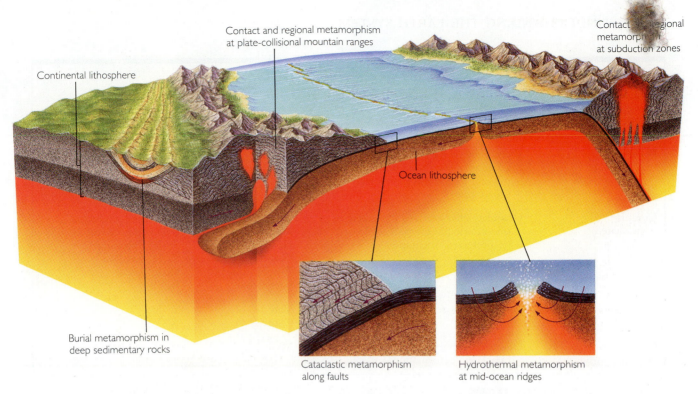

Continental lithosphere

Contact and regional metamorphism
at plate-collisional mountain ranges

Contact and regional
metamorphism
at subduction zones

Ocean lithosphere

Burial metamorphism in
deep sedimentary rocks

Cataclastic metamorphism
along faults

Hydrothermal metamorphism
at mid-ocean ridges

**FIGURE 8.2** Metamorphic rocks are formed in four main plate-tectonic settings: subduction zones, continental collisions, mid-ocean ridges, and deeply subsiding regions on continents.

may also occur in regional metamorphic belts, contact metamorphism is also found in deformed mountain belts.

## Cataclastic Metamorphism

Metamorphism may be found along faults, where tectonic movements cause the crust to crack and slip. As the rocks along the fault plane shear past each other, they grind and mechanically fragment solid rock into a pasty mass. This **cataclastic metamorphism** produces a broken, pulverized texture that is found in strongly deformed mountain belts where faulting is extensive. Thus cataclastic rocks are frequently found together with regionally metamorphosed rocks.

## Hydrothermal Metamorphism

Another form of metamorphism, called **hydrothermal metamorphism,** is frequently associated with mid-ocean ridges, where plates spread apart and rising basalt magmas form new oceanic crust. Seawater percolates through the hot, fractured basalts of the ridge flanks and itself becomes heated. The increase in temperature promotes chemical reactions between the seawater and the rock, forming altered basalts whose chemical compositions differ distinctively from that of the original basalt. Hydrothermal metamorphism also occurs on the continents where fluids rising from igneous intrusions metamorphose overlying rocks.

## Burial Metamorphism

Recall from Chapter 7 that as sedimentary rocks are gradually buried by the sinking of the crust, they slowly heat up as they come into equilibrium with the crustal temperatures surrounding them. In this process diagenesis alters the mineralogy and texture of the rock. Diagenesis grades into **burial metamorphism,** a low grade of metamorphism caused by the heat and pressure exerted by overlying sediments and sedimentary rocks. While temperatures and pressures are not so great as those that accompany regional metamorphism, they are high enough to produce partial alteration of the mineralogy and texture of the sedimentary rock. Bedding and other sedimentary structures are preserved.

**TABLE 8.1**

**TABLE 8.1**

## Classification of Metamorphic Rocks Based on Texture

| CLASSIFICATION | CHARACTERISTICS | ROCK NAME | TYPICAL PARENT ROCK |
|---|---|---|---|
| Foliated | Distinguished by slaty cleavage, schistosity, or gneissic foliation; mineral grains show preferred orientation | Slate Phyllite Schist Gneiss | Shale, sandstone |
| Granular (nonfoliated) | Granular, characterized by coarse or fine interlocking grains; little or no preferred orientation | Quartzite Marble Hornfels Granulite[1] Amphibolite[2] Greenstone | Quartzose sandstone Limestone, dolomite Shale, volcanics Shale, basalt Shale, basalt Basalt |
| Porphyroblastic | Large crystals set in fine matrix | Slate to gneiss | Shale |
| Crushed | Grains pulverized and streaked by cataclastic deformation | Mylonite | Shale, sandstone |

[1] High-temperature, high-pressure rock.
[2] Typically contains much amphibole, which may show alignment of long, narrow crystals.

# METAMORPHIC TEXTURES

All the various kinds of metamorphism imprint new textures on the rocks they alter. The texture of a metamorphic rock is determined by the sizes, shapes, and arrangement of its constituent crystals (Table 8.1). Some metamorphic textures are dependent on the particular kinds of minerals formed, such as the micas, which have a platy habit. Metamorphic textures may be inherited in part from the parent; thus the sizes of grains in a sedimentary rock may be reflected in the sizes of crystals formed during metamorphism. Each textural variety tells us something about the metamorphic process that created it.

## Foliation and Cleavage

Many of the best known metamorphic rocks show a set of parallel planes that generally cut the rocks at an angle to the bedding of the original sediment, although they may coincide with the bedding in some

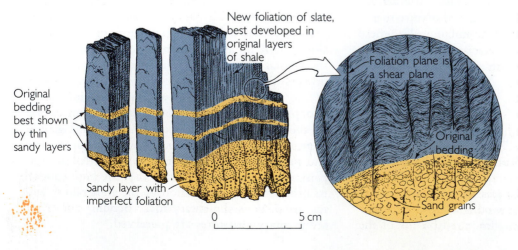

Original bedding best shown by thin sandy layers

Sandy layer with imperfect foliation

New foliation of slate, best developed in original layers of shale

Foliation plane is a shear plane

Original bedding

Sand grains

0    5 cm

**FIGURE 8.3** Fragments of slate *(left)* show foliation (vertical lines) and relics of original bedding. The enlargement shows small faultlike offsets of the bedding along the surfaces of foliation. (After J. Gilluly, A. C. Waters, and A. O. Woodford, *Principles of Geology,* 4th ed., San Francisco, W. H. Freeman, 1975.)

FIGURE 8.4  An outcrop of intensely foliated and deformed metamorphic rock produced by very high pressures and temperatures and strong tectonic forces. Blue Ridge Parkway, North Carolina. *Gary Meszaros.*

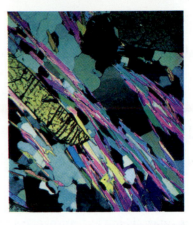

FIGURE 8.5  Preferred orientation of platy and elongate crystals in foliated rocks *(top)*. Arrows indicate direction of compressive forces. A photomicrograph of schist *(bottom)* shows the preferred orientation of mica (blue, green, purple) and staurolite (yellow) crystals. *S. Dobos.*

places (Figure 8.3). This texture, called **foliation,** is the most prominent textural feature of regionally metamorphosed rocks (Figure 8.4). The form that foliation takes depends on the mineral composition of the rock.

One of the main causes of foliation is the presence of platy minerals, chiefly the micas and chlorite. Platy minerals tend to crystallize as thin, platelike or sheety crystals. The planes of all the platy crystals are aligned parallel to the foliation. The parallel planes are called the *preferred orientation* of the minerals (Figure 8.5). As platy minerals crystallize, their planes take a preferred orientation that is usually perpendicular to the main direction of the forces squeezing the rock during the deformation that accompanies metamorphism. Preexisting minerals may acquire a preferred orientation and thus produce foliation by rotating crystals until they lie parallel to the developing plane. Plastic deformation, or the softening and

bending of the hot rock without breakage, also may produce crystals with a preferred orientation.

Minerals whose crystals have an elongate, pencil-like shape, such as the amphiboles, also tend to assume a preferred orientation during metamorphism, the crystals normally lining up parallel to the foliation plane (Figure 8.6). Rocks that contain abundant amphiboles, typically metamorphosed mafic volcanics, show this kind of texture.

The most familiar form of foliation is seen in slate, a common metamorphic rock, which is easily split along smooth, parallel surfaces into thin sheets. This *fracture cleavage* (as distinguished from mineral cleavage) develops along moderately thin, regular intervals in the rock. Slate splitters learned long ago to recognize this property and use it to make thick or thin slates for roofing tiles and blackboards. We still use flat slabs of slate for flagstone walks in parts of the country where slate is abundant.

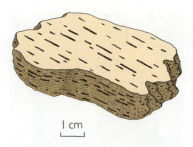

**FIGURE 8.6** Preferred orientation of elongate minerals such as pyroxenes and amphiboles. Viewed end on, the elongate minerals appear as dots.

## Foliated Rocks

The foliated rocks are classified according to the nature of their foliation, the size of their crystals, the degree of segregation of their minerals into lighter and darker bands, and their metamorphic grade (Figure 8.7).

Examples of the major types of foliated rocks are shown in Figure 8.8.

**SLATE**  Slates are the lowest grade of foliated rocks. These rocks, with their excellent planar partings, are so fine-grained that their individual minerals cannot be seen easily without a microscope. They are commonly produced by the metamorphism of shales or, less frequently, of volcanic ash deposits. Slates are usually dark gray to black, colored by small amounts of organic material originally present in the parent shale. Slates tinged red and purple get their color from iron oxide minerals, and greenish slates are colored by chlorite, a sheety iron silicate mineral closely related to the micas. Whereas shales may split along

bedding planes, the foliation along which slates split is normally at an angle to the bedding—good evidence that foliation is unrelated to the texture of the parent rock. It is the product of metamorphic rather than sedimentary processes.

**PHYLLITE**  Of slightly higher grade than slates but of similar character and origin are the **phyllites,** which tend to have a more or less glossy sheen from crystals of mica and chlorite that have grown a little larger than those of slates.

**SCHIST**  At low grades of metamorphism, platy minerals are generally too small to be seen, the foliation is closely spaced, and the layers are very thin. As metamorphic rocks are more intensely metamorphosed to higher grades, the foliation becomes more conspicuous and pervasive throughout the rock. At the same time, the platy crystals grow to sizes visible to the naked eye, and the minerals may tend to segregate in lighter and darker bands. The result is a texture called *schistosity,* the coarse, wavy, pervasive foliation that characterizes **schists.** Schists are among the most abundant metamorphic rock types. They contain more than 50 percent platy minerals, mainly the micas muscovite and biotite. Depending on the quartz content of the original shale, schists may contain thin layers of quartz, feldspar, or both. Schists are frequently named for their most abundant mineral constituent. Thus there are mica schists, chlorite schists, and quartz schists.

**GNEISS**  Even coarser foliation is shown by high-grade **gneisses,** light-colored rocks with coarse bands of segregated light and dark minerals throughout the rock. Gneisses do not split along the foliation and there are few sheetlike minerals along the folia-

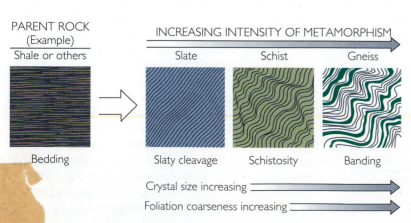

PARENT ROCK
(Example)

INCREASING INTENSITY OF METAMORPHISM

| Shale or others | Slate | Schist | Gneiss |

Bedding    Slaty cleavage    Schistosity    Banding

Crystal size increasing

Foliation coarseness increasing

**FIGURE 8.7** Classification of foliated rocks. Cleavage and schistosity do not generally correspond to the original bedding direction of sedimentary rocks.

**FIGURE 8.8** Foliated and nonfoliated metamorphic rocks. Clockwise from top left: mica schist, gneiss, slate, marble, quartzite. All except the marble and quartzite are foliated. Though most gneisses are light-colored, this dark gneiss shows well the characteristic banding. *Chip Clark.*

tion planes. Gneisses are coarse-grained, and the ratio of granular to platy minerals is higher than in slate or schist. The result is poor foliation and thus little tendency to split. The relative lack of micas, which changes the character of the foliation, is the consequence of metamorphism at very high temperatures and pressures. Under these conditions, the mineral assemblages of the lower grade rocks containing micas and chlorite alter to new assemblages dominated by quartz and feldspars, with lesser amounts of micas and amphiboles.

The banding of gneisses into light and dark layers results from the segregation of lighter-colored quartz and feldspar and darker amphiboles and other mafic minerals. In some places gneisses are the high-grade equivalents of schists; in others they are the metamorphic equivalents of granites.

## Nonfoliated Rocks

Not all metamorphic rocks are foliated. Some show a very weak preferred orientation of crystals, which results in little or no foliation. Other rocks show no preferred orientation and therefore no foliation. **Nonfoliated rocks** fall into two groups: (1) contact metamorphic rocks; and (2) those regional, hydrothermal, and burial metamorphic rocks composed mainly of crystals that grow in equant (equidimensional) shapes, such as cubes and spheres, rather than platy or elongate shapes.

**NONFOLIATED CONTACT METAMORPHIC ROCKS**  Many contact metamorphic rocks have undergone little or no deformation. Their platy or elongate crystals are oriented randomly and foliated texture is absent. Contact metamorphic rocks such as these are called **hornfels.** They have a granular texture overall, even though they commonly contain pyroxene, which makes elongate crystals, and some micas.

**NONFOLIATED REGIONAL AND OTHER METAMORPHIC ROCKS**  Regionally metamorphosed nonfoliated rocks include quartzite, marble, argillite, greenstone, and granulite. **Quartzites** are derived from quartz-rich sandstones. Quartzites may be massive—that is, unbroken by preserved bedding or foliation. Other quartzites may contain thin bands of slate or schist, relics of former interbedded layers of clay or shale (see Figure 8.8).

**Marbles** are the metamorphic products of heat and pressure acting on limestones and dolomites. They may result from either contact or regional metamorphism. Some white, pure marbles, such as the famous Italian Carrara marbles prized by sculptors, show an even, smooth texture of intergrown calcite crystals of uniform size. Other marbles show irregular banding or mottling from silicate and other mineral impurities in the original limestone (see Figure 8.8).

**Argillite** is a low-grade metamorphic rock made from a shaly sedimentary rock. Some shadowy relic bedding may be preserved. In contrast to the well-foliated slates, phyllites, and schists that come from similar parents, this rock breaks with an irregular or conchoidal fracture. The lack of foliation may be attributable partly to a lower grade of deformation and partly to an abundance of quartz silt or other minerals that are neither platy nor elongate in the parent shale.

**Greenstones** are metamorphosed mafic volcanics. Many of these low-grade rocks are formed when mafic lavas and ash deposits react with percolating seawaters or other solutions. Large areas of the seafloor are covered with basalts slightly or extensively altered in this way at mid-ocean ridges. On the continents, buried volcanic and plutonic mafic igneous rocks react with groundwaters at temperatures of 150 to 300°C and form similar greenstones. An abundance of chlorite gives these rocks their greenish cast.

**Granulites** are high-grade, medium- to coarse-grained rocks in which the crystals are equant and show only faint or no foliation. Containing chiefly feldspar, pyroxene, and garnet, they are formed by the metamorphism of shale, impure sandstone, and many kinds of igneous rock.

## Large Crystal Textures

New minerals may grow into large crystals surrounded by a much finer-grained matrix of other minerals. These large crystals, termed **porphyroblasts,** are found in both contact and regionally metamorphosed rocks (Figure 8.9). They grow by reorganization of the chemical components of the matrix and thus replace portions of the matrix. Porphyroblasts form where a strong contrast between the chemical and crystallographic properties of the matrix and those of the porphyroblast minerals causes the porphyroblast crystals to grow faster than the slow-growing minerals of the matrix, at the expense of the matrix. Porphyroblasts vary in size, ranging from a few millimeters to several centimeters in diameter. Their composition also varies. Garnet and staurolite are two common minerals that form porphyroblasts, but many others are also found. The precise composition and distribution of porphyroblasts of these two minerals can be used to infer the

**FIGURE 8.9** Garnet porphyroblasts in schist matrix. *Chip Clark.*

pressures and temperatures of metamorphism. Pure garnets, transparent and beautifully colored in shades of red, green, and black, are valued as semiprecious gems.

## Deformational Textures

We have already seen that mechanical deformation along fault planes produces cataclastic metamorphism. As the movement of two rock surfaces against each other pulverizes minerals and strings them out in bands or streaks, metamorphic rocks called **mylonites** are formed. These fine-grained rocks can be foliated when they are formed very deep in the crust, where rocks under very high pressures have been plastically deformed.

Structural deformation accompanies most metamorphism. The textural effects of deformation are most obvious in mylonites, but they are also prominent in the foliated rocks. Metamorphism may precede the deformation, be contemporaneous with it, or follow it. Many metamorphic rocks have complex geological histories. A sedimentary rock may become a low-grade metamorphic greenschist as a result of deep burial without significant deformation. Later it may be caught up in deformation associated with mountain building and become strongly foliated during a higher grade of metamorphism. Finally it may be altered to an even higher grade nonfoliated metamorphic rock from contact metamorphism adjacent to igneous intrusions injected into the orogenic belt.

# REGIONAL METAMORPHISM AND METAMORPHIC GRADE

Because metamorphic rocks form under a wide range of conditions, their minerals and textures are clues to the pressures and temperatures in the crust where and when they were formed. Geologists who study the geologic formation of metamorphic rocks constantly seek to determine the intensity and character of metamorphism more precisely than the designation of low or high grade indicates. Minerals are used as pressure gauges and thermometers to make these finer distinctions. The techniques are best illustrated by their application to regional metamorphism.

## Mineral Isograds

When geologists study broad belts of regionally metamorphosed rocks, they can see many outcrops, some showing one set of minerals, some showing others. Different parts of these belts may be distinguished by the characteristic minerals that define them. For example, one may cross from a region of unmetamorphosed shales to a belt of weakly metamorphosed slates and then to a belt of high-grade schists. The sedimentary minerals are unchanged in the shales, but at the slate belt a new mineral, chlorite, appears. Moving in the direction of increasing metamorphism, the geologist may successively encounter biotite, then garnet, then staurolite, the schists becoming progressively more foliated. These minerals, called *index minerals,* define metamorphic zones formed under a restricted range of pressures and temperatures.

Lines called *isograds* are drawn on a map connecting places where index minerals first appear, to define where one zone of metamorphic grade changes to another. Figure 8.10 shows a series of regionally metamorphosed rocks produced by the metamorphism of a shale. The pattern of isograds tends to follow the structural grain of a region as folds and faults reveal it. An isograd based on a single index mineral, such as the biotite isograd, is a good approximate measure of metamorphic pressure and temperature. Pressure and temperature can be determined more precisely by examining a group of two or three minerals whose textures indicate that they crystallized together. These groups of minerals have been carefully studied in the laboratory to determine more exactly the pressures and temperatures at which they

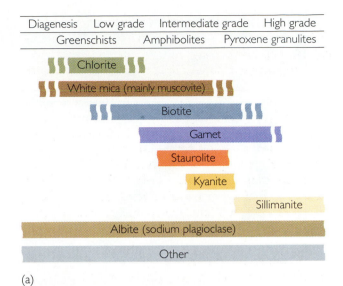

(a)

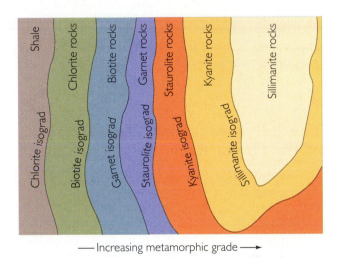

— Increasing metamorphic grade ⟶

(b)

**FIGURE 8.10** (a) Changes in the mineral composition of shales metamorphosed under the same conditions. (b) A regionally metamorphosed terrain in which shales have been metamorphosed under conditions of intermediate pressure and temperature. The isograd lines mark the first appearance of the index mineral and correspond to diagram (a).

formed. The results are used to calibrate the field mapping of isograds.

The isograd sequence in one metamorphic belt may differ from that in another belt, because pressure and temperature do not increase at the same rate in all geologic settings. Pressure increases more rapidly than temperature in some places, more slowly in others (Figure 8.11).

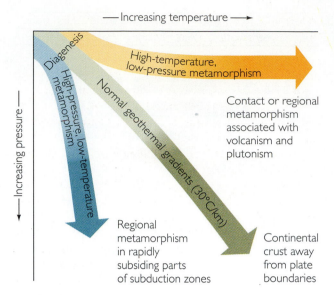

**FIGURE 8.11** The routes taken by increases in pressure and temperature as depth increases depend on tectonic and igneous activity. The various routes give different sequences of metamorphic rock types.

## The Influence of the Composition of the Parent Rock

The kind of metamorphic rock that results from a given grade of metamorphism depends partly on the mineral composition of the parent rock. The shale metamorphism shown in Figure 8.10 reveals the effects of metamorphic conditions on rocks rich in clay minerals, quartz, and perhaps some carbonate minerals. Metamorphism of mafic volcanics, composed predominantly of feldspars and pyroxene, follows a different course (Figure 8.12).

In the regional metamorphism of a basalt, for example, the lowest grade rocks characteristically contain various **zeolite** minerals, complex hydrous aluminosilicates formed by the alteration of mafic volcanic rocks and sedimentary rocks derived from volcanics at very low temperatures and pressures. Rocks that include this group of minerals are thus identified as zeolite grade. The next higher grade of metamorphosed mafic volcanic rocks is the **greenschist** grade, whose abundant minerals include chlorite and epidote (an aluminosilicate). Next are the **amphibolites,** which contain large amounts of the key minerals hornblende (an amphibole mineral), plagioclase feldspar, and garnet. The highest grade of metamorphosed mafic volcanics are the **pyroxene granulites,** rocks containing pyroxene and calcium plagioclase.

Pyroxene granulites are the products of high-grade metamorphism in which the temperature is high and the pressure moderate. In the opposite situation, where the pressure is high and the temperature moderate, rocks of a variety of compositions—from mafic volcanic rocks to shaly sedimentary rocks—form rocks of **blueschist** grade. The name comes from the abundance in these rocks of glaucophane, a blue amphibole. Still another metamorphic rock, formed at extremely high pressures and moderate to high temperatures, is **eclogite,** a rock rich in garnet and pyroxene.

## Metamorphic Facies

We can put all this information together in a "map" of the way these various metamorphic grades derived from parent rocks of different composition fit on a graph of temperature and pressure (Figure 8.13). The groupings of rocks of various compositions formed under different grades of metamorphism are called **metamorphic facies.** The essentials of the concept of metamorphic facies are that (1) different kinds of metamorphic rocks are formed from parent rocks of different composition at the same grades of metamorphism, and (2) different kinds of metamorphic rocks are formed under different grades of metamorphism from parent rocks of the same composition (Table 8.2). Because parent rocks vary so greatly in composition, there are no sharp boundaries between the groups shown in Figure 8.13.

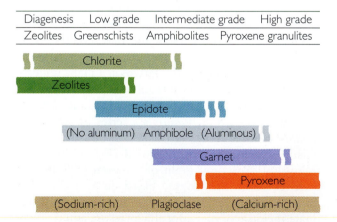

**FIGURE 8.12** Changes in the mineral composition of basalts and other mafic rocks metamorphosed under conditions of intermediate pressure and temperature. Compare the mineral assemblages of shales metamorphosed under the same conditions in Figure 8.10 to see the effect of original composition on metamorphic mineralogy.

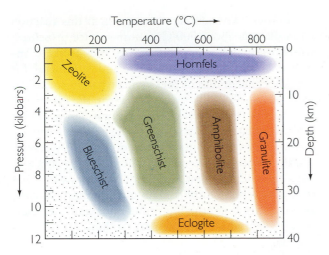

**FIGURE 8.13** The various types of metamorphic rocks may be grouped in accordance with the temperature and pressure fields in which they were formed. There are no sharp boundaries between any of these groupings, or facies.

At the high-grade end of this series, metamorphic rocks become partly melted and thus transitional to igneous rocks. These rocks are badly deformed and contorted and are penetrated by many veins, blebs, and lenses of melted rock. This kind of very high grade veined gneiss is called **migmatite,** a term that is applied to rocks of mixed igneous and metamorphic origin. Some migmatites are mainly metamorphic, with only a small proportion of igneous material. Others have been so affected by melting that they are considered almost entirely igneous.

# CONTACT METAMORPHIC ZONES

The best place to see the metamorphic effects of an igneous body on the rocks it intruded is at an outcrop where a shale has been intruded by a dike or sill. At the border, where the shale is in contact with the dike, the shale has lost all of its original texture: bedding is gone, fossils are obliterated, and the shale's mineralogy has completely changed. Instead of the fine-grained clay minerals of the shale, the rock immediately adjacent to the dike now consists of large crystals of pyroxenes and aluminosilicate minerals such as andalusite, which are not found in sedimentary rocks. Farther from the contact, from several centimeters to a meter away, faint outlines of the shale's bedding can still be seen, but the clay minerals have been altered to crystalline micas. At some greater distance from the contact, the shale is unaltered. Thus contact metamorphic zones, like regional metamorphic ones, are characterized by index minerals that reflect different grades of metamorphism.

The thickness and character of the rim of metamorphically altered rock around an igneous intrusion, called the *contact aureole,* depend on the temperature of the magma and the depth in the crust where the intrusion took place. As one might expect, the contact aureole is most prominent where a mafic intrusion at about 1000°C invaded shallow crustal rocks only a few kilometers deep, where normal temperatures are only about 60 to 90°C. In this kind of intrusion, the temperature is very high at the con-

**TABLE 8.2**

## Major Minerals of Metamorphic Facies Produced from Parent Rocks of Different Composition

| FACIES | MINERALS PRODUCED FROM SHALE PARENT | MINERALS PRODUCED FROM BASALT PARENT |
|---|---|---|
| Greenschist | Muscovite, chlorite, quartz, sodium-rich plagioclase feldspar | Albite, epidote, chlorite |
| Amphibolite | Muscovite, biotite, garnet, quartz, plagioclase feldspar | Amphibole, plagioclase feldspar |
| Granulite | Garnet, sillimanite, plagioclase feldspar, quartz | Calcium-rich pyroxene, calcium-rich plagioclase feldspar |
| Eclogite | Garnet, sodium-rich pyroxene, quartz | Sodium-rich pyroxene, garnet |

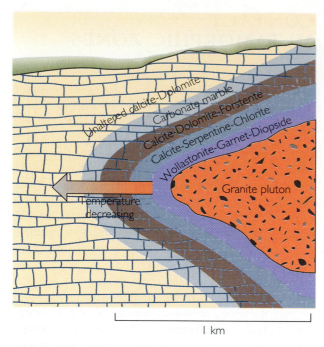

**FIGURE 8.14** The contact metamorphism of a limestone composed of calcite and dolomite produces a contact aureole of several zones of minerals. Grading from unaltered rock to the contact, they change from pure carbonate marble to bands of various calcium-magnesium silicate minerals and finally to a carbonate-free silicate rock.

in the rock. The temperature at which this reaction takes place is about 500°C at near-surface pressures, rising to higher temperatures as the pressure increases. Thus wollastonite is an index mineral for metamorphic grades of rocks whose parent rocks had this composition (Figure 8.14). Near the contact is the mineral produced at the highest temperature, wollastonite, together with garnet. Farther from the contact, the zone contains serpentine, a magnesium silicate containing chemically bound water, together with chlorite. Farthest from the intrusion, the zones of lowest temperature contain a magnesium olivine. Beyond this zone, the limestone mineralogy is unaffected. The whole aureole may be hundreds of meters wide.

Contact metamorphism of silicate rocks such as shale also produces progressive mineral zones, but the groups of minerals in them differ from those seen in metamorphosed limestones (Figure 8.15). The quartz, clay, and carbonate of the shale are transformed in the cooler outer zones to micas, amphi-

tact but drops rapidly a short distance away from it. Intrusions at lower temperatures, such as granites at about 600°C, invade deeper parts of the crust where temperatures are already high, do not heat the surrounding rocks so much, and thus cause less metamorphic change.

Contact metamorphic zones vary with the kinds of parent rock that contact a hot intrusive. Although they are not ordinarily included in metamorphic facies diagrams such as Figure 8.13, contact metamorphic rocks show the same relationships between grade and composition of parent rock. The patterns of mineral grades shown by limestones, for example, which are composed of carbonate minerals, differ from those shown by shales, which are made of silicate minerals. In a typical contact metamorphic change in limestones, the carbonate minerals react with silica impurities in the rock to form wollastonite, a light-colored calcium pyroxene:

calcite + silica → wollastonite + carbon dioxide

The carbon dioxide produced by this reaction generally escapes as a gas through fractures and pores

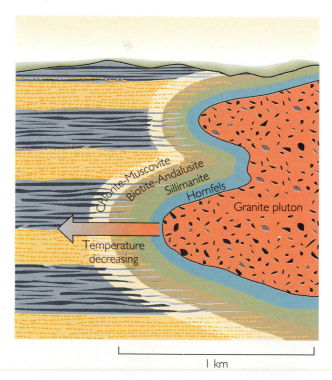

**FIGURE 8.15** A contact aureole in sandstones and shales intruded by a graded pluton is a series of zones characterized by minerals ranging from the unaltered quartz and clay minerals of the sandstones and shales to bands of aluminum, iron, and magnesium silicates. The rock at the contact is a pyroxene hornfels.

boles, and calcite. In inner zones, closer to the heat of the intrusion, pyroxenes and pure aluminum silicates, such as andalusite, are characteristic.

# PLATE TECTONICS AND METAMORPHISM

Soon after the theory of plate tectonics was proposed, geologists started to see how patterns of metamorphism fitted into the larger framework of plate-tectonic movements that cause volcanism and orogeny. At the beginning of this chapter, we explored the link between plate-tectonic settings and the geologic processes that cause the various types of metamorphism (see Figure 8.2). We can also often deduce a rock's site of metamorphism on the basis of its grade and composition.

Greenstones—metamorphosed basalts—are associated with metamorphism at mid-ocean ridges, where the seafloor spreads and hot basaltic magma wells up from the mantle. The heat of the magma transforms newly extruded basalts into low-grade metamorphic rocks of the greenschist facies. Hydrothermal circulation through the basalt also plays a role in the alteration of basalts.

Blueschists—the metamorphosed volcanic and sedimentary rocks whose minerals indicate that they were formed under very high pressures but at relatively low temperatures—form in the forearc region of a subduction zone, the area between the seafloor trench and the volcanic arc. There sediments are carried down the subduction zone along the surface of a cool subducting lithospheric slab. The subducted plate moves down at such a high rate of speed that it heats up very slowly while the pressure increases rapidly.

The opposite conditions of high temperature and low pressure are found along the volcanic island arcs on the overriding plate at a subduction zone. Here the magmas formed by melting of the deeply subducted plate rise to shallow depths in the overriding plate and by their heat transform shallowly buried volcanics and sediments into greenschist and higher grade rocks. Both kinds of metamorphism, high pressure–low temperature and low pressure–high temperature, are found as paired belts along plate convergences. On the oceanic side, close to the trench, where the descending slab is still cool, we find the high-pressure, low-temperature metamorphics. On the landward or volcanic arc side we find

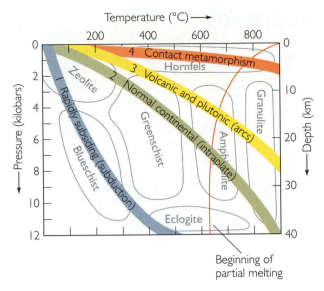

**FIGURE 8.16** The routes taken by increases in pressure and temperature in various tectonic settings superimposed on metamorphic facies. Curve 1 shows rapidly increasing pressure with a slow rise in temperature, characteristic of the forearcs of subduction zones. Curve 2 is the path followed in continental regions away from plate boundaries, places where burial metamorphism occurs. Curve 3 is the higher temperature route followed in regions of volcanic arcs, typical of regional metamorphism. Curve 4 is the high-temperature, low-pressure path of contact metamorphism at shallow and moderate depths.

the high-temperature, low-pressure rocks. Figure 8.16 shows these pressure-temperature relationships superimposed on the metamorphic facies diagram.

Regional metamorphic belts are associated with continental collisions that cause orogenies. In the cores of the major mountain belts of the world, from the Appalachians to the Alps, we find long belts of regionally metamorphosed and deformed sedimentary and volcanic rocks that parallel the lines of folds and faults of the mountains. As continents collide and the lithosphere thickens, the deeper parts of the continental crust heat up and metamorphose to different grades, while in deeper zones melting may begin. In this way the complex mixture of metamorphic and igneous rocks forms the cores of orogenic belts that evolve during mountain building. Millions of years afterward, when erosion has stripped off the surface layers, the cores are exposed at the surface, providing the geologist with a rock record of the metamorphic processes that formed the schists, gneisses, and other metamorphic rocks.

# SUMMARY

**What factors cause metamorphism?** Metamorphism—alteration in the solid state of preexisting rocks—is caused by increases in pressure and temperature and reaction with chemical components introduced by migrating fluids. As pressures and temperatures deep within the crust increase as a result of tectonic or igneous activity, the chemical components of the parent rock rearrange themselves into a new set of minerals that are stable under the new conditions. Rocks metamorphosed at relatively low pressures and temperatures are referred to as low-grade rocks and those metamorphosed at high temperatures and pressures as high-grade rocks. Chemical components of a rock may be added or removed during metamorphism, most commonly by the influence of fluids migrating from nearby intrusions.

**What are the various kinds of metamorphism?** The two major types of metamorphism are (1) regional metamorphism, during which large areas are metamorphosed by high pressures and temperatures generated during orogenies; and (2) contact metamorphism, during which rocks surrounding magmas are metamorphosed primarily by the heat of the igneous body. Three additional metamorphic types are (3) cataclastic metamorphism, during which rocks along fault planes are pulverized and mineralogically altered; (4) hydrothermal metamorphism, during which hot fluids percolate through and metamorphose various crustal rocks; and (5) burial metamorphism, a variety of regional metamorphism, during which deeply buried sedimentary rocks are altered by the more or less normal increase of pressure and temperature in the crust.

**What are the chief types of metamorphic rocks?** Metamorphic rocks fall into two major textural classes: the foliated (displaying fracture cleavage, schistosity, or other forms of preferred orientation of minerals) and the nonfoliated. The kinds of rocks produced by metamorphism depend on the composition of the parent rock and the grade of metamorphism. The regional metamorphism of a shale leads to zones of foliated rocks of progressively higher grade, from slate to phyllite, schist, and gneiss. Regional metamorphism of mafic volcanic rocks progresses from zeolite grade to greenschist, then to amphibolite and pyroxene granulite. Among nonfoliated rocks, marble is derived from the metamorphism of limestone, quartzite from quartz-rich sandstone, argillite from mudstone, and greenstone from basalt. Hornfels is the product of contact metamorphism of fine-grained sedimentary rocks and other types of rock containing an abundance of silicate minerals. Mylonites are produced by cataclastic metamorphism. According to the concept of metamorphic facies, rocks of the same grade may differ because of variations in the chemical composition of the parent rocks, while rocks of the same composition may vary because of different grades of metamorphism.

# KEY TERMS AND CONCEPTS

metamorphic rocks (p. 170)
low-grade rocks (p. 170)
high-grade rocks (p. 170)
retrograde metamorphism (p. 170)
stress (p. 171)
metasomatism (p. 172)
regional metamorphism (p. 172)
contact metamorphism (p. 172)
cataclastic metamorphism (p. 173)
hydrothermal metamorphism (p. 173)

burial metamorphism (p. 173)
foliation (p. 175)
phyllite (p. 176)
schist (p. 176)
gneiss (p. 176)
nonfoliated rocks (p. 177)
hornfels (p. 177)
quartzite (p. 177)
marble (p. 177)
argillite (p. 178)
greenstone (p. 178)

granulite (p. 178)
porphyroblast (p. 178)
mylonite (p. 178)
zeolite (p. 180)
greenschist (p. 180)
amphibolite (p. 180)
pyroxene granulite (p. 180)
blueschist (p. 180)
eclogite (p. 180)
metamorphic facies (p. 180)
migmatite (p. 181)

# EXERCISES

**1.** What is the difference between a granite and a slate?

**2.** Name a mineral commonly found in a schist that shows preferred orientation.

3. What kinds of metamorphism are related to igneous intrusions?

4. What is an isograd?

5. In which plate-tectonic settings would you expect to find regional metamorphism?

6. Name two nonfoliated metamorphic rocks.

7. What is a porphyroblast?

8. How are metamorphic facies related to temperatures and pressures?

# THOUGHT QUESTIONS

1. Why would you not expect to find burial metamorphosed rocks at a mid-ocean ridge?

2. You have mapped metamorphic rocks in which north-south isograd lines run from kyanite in the east to chlorite in the west. In which direction were metamorphic temperatures higher?

3. What kinds of preferred orientation of minerals would you expect to find in an amphibolite?

4. Are cataclastic rocks more likely to be found in a continental rift valley or in a volcanic arc?

5. Contrast the minerals found in the contact metamorphism of a pure limestone and a limestone containing appreciable layers of shale.

6. Which kind of pluton would produce the highest grade metamorphism, a granite intrusion 20 km deep or a gabbro intrusion at a depth of 5 km?

7. Would you choose to rely on chemical composition or type of foliation to determine metamorphic grade?

# SUGGESTED READINGS

Best, Myron G. 1982. *Igneous and Metamorphic Petrology*. New York: W. H. Freeman.

Ehlers, Ernest G., and Harvey Blatt. 1982. *Petrology. Igneous, Sedimentary, and Metamorphic*. New York: W. H. Freeman.

Ernst, W. Gary (ed.). 1975. *Metamorphism and Plate Tectonic Regimes*. New York: Dowden, Hutchinson, and Ross.

Hyndman, Donald W. 1985. *Petrology of Igneous and Metamorphic Rocks*, 2d ed. New York: Wiley.

Winkler, Helmut G. 1979. *Petrogenesis of Metamorphic Rocks*, 5th ed. New York: Springer-Verlag.

# THE ROCK RECORD AND THE GEOLOGIC TIME SCALE

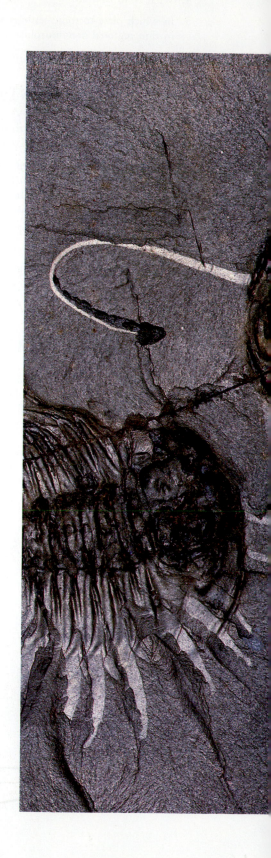

The geological processes that shape Earth's surface and give structure to its interior work over millions and billions of years. Geologists learn to deal with these extraordinarily long intervals in order to understand both the nature of such slow processes and the geologic history of the planet. Geologists of the nineteenth century built a geologic time scale from the time and space relationships of rocks exposed at the surface or in drill holes. That time scale enables us to place the geological events in Earth's history in sequence, from the oldest to the most recent. In the twentieth century, we have been able to measure the actual number of years spanned by the phases of geologic history by correlating geologic periods with the dates derived from the decay of radioactive atoms in rocks.

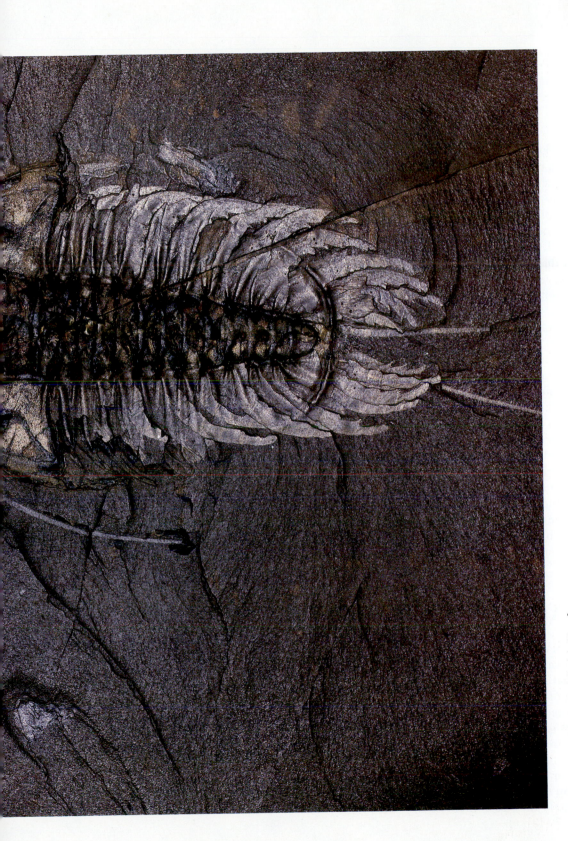

## 9

Trilobites in Burgess Shale from Canadian Rockies. These particular strange creatures, now extinct, are found only in rocks about 530 million years old, illustrating the correspondence of geologic time and fossils. *Chip Clark*.

Despite an occasional earthquake or volcanic eruption, the Earth seems to provide a reasonably stable foundation upon which to build a civilization. But organized human society is only a few thousand years old. Over the past millions of years, Earth has been far less stable: continents, oceans, and mountains have moved great distances. Even at this instant there is hardly a place on Earth that is not moving vertically and horizontally, however slowly. Learning the patterns and rates of these movements is a key job of the geologist.

One of the most important reasons for placing geological processes in their proper sequence is to learn about the evolution of the planet we see today. When were the Rocky Mountains formed? What was happening in East Africa when early humans were evolving? To answer these kinds of questions, we need a tool for organizing and dating the rock record: a geologic calendar to determine the sequence in which rock layers formed and their ages, and a way to compare the ages of rocks situated continents apart. Two centuries of modern geologic research have resulted in such a tool: the geologic time scale. It enables scientists to determine the age of the Earth, to unravel complex geologic histories, and even to study the origin and evolution of life.

## TIMING THE EARTH

Geologists (and astronomers) differ from most other scientists in their attitude toward time. Physicists and chemists study processes that last only fractions of a second, such as the splitting of an atomic nucleus or a fast chemical reaction. Other scientists do experiments that last from minutes to hours. Geologists, by contrast, deal with Earth processes that unfold over a great range of time periods (Figure 9.1). Earthquake tremors may last seconds or minutes, whereas the building of a mountain chain takes many millions of years. We can measure directly the rise and fall of a

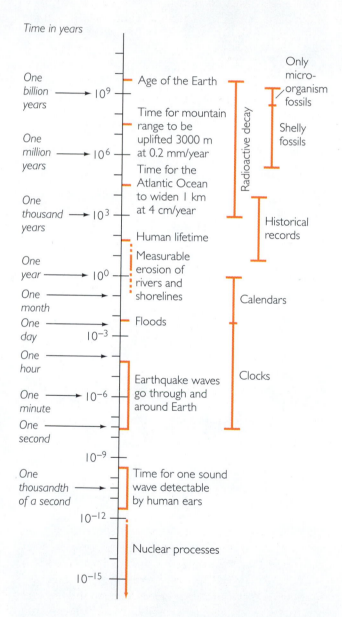

**FIGURE 9.1** Orders of magnitude of times required for some common processes and events. To the right are the timekeepers we use for time scales. The scale is logarithmic; that is, it has equal divisions between successive powers of 10.

188

river's floodwaters over a few days and even the much slower movements of glaciers, which may take a year to move 50 m.

Many geological processes, however, such as the erosion of a hillside, are too slow to be measured directly. We may rely on historical records to determine the amounts of time required for some of these processes (Figure 9.2). But even the oldest historical records extend back only a few thousand years, a time span that is still inadequate to study the great many very slow geological processes that shape the planet. Our only resource for timing such processes is the rock record. Rocks formed in the past and preserved from erosion serve as Earth's memory, recording geological events, such as glaciations, that lasted many thousands or millions of years.

Geologists of the nineteenth century used the principles of stratigraphy and an understanding of fossils to determine the **relative ages** of sedimentary rock layers; that is, how old they are in relation to one another. Geologists could then put the geological events that created these rock formations into chronological order. Today geologists use the physics of radioactive decay to pinpoint a rock's **absolute age;** that is, how many years ago it formed.

The geologists who worked out the geologic time scale did more than merely date rocks. They precipitated a revolution in the way we think about time, our planet, and even ourselves. They discovered that Earth is far older than anyone had previously imagined and, contrary to the common beliefs of earlier times, that its surface and interior have been changed and shaped repeatedly by the same geological processes that operate now. They found, too, that not only the planet but its inhabitants have evolved over time. And humans, they discovered, account for only the briefest moment of Earth's long history.

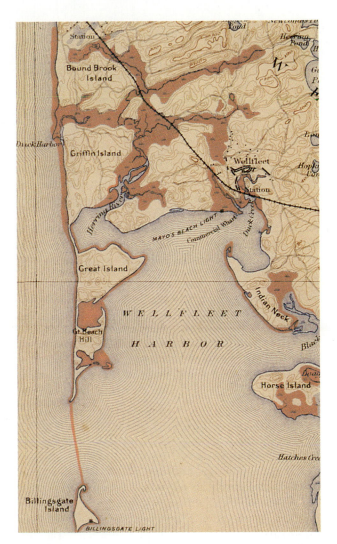

**FIGURE 9.2** Historical records such as old maps and land surveys are useful for measurement of some geological processes. Tidelands near Wellfleet Bay on Cape Cod have been filled in by sand, silt, and mud (brown) brought in by waves and tides since 1887, when this map was being prepared.

## Relative Dating and the Stratigraphic Record

Stratification—the layering that is the hallmark of sedimentary rocks—is basic to two simple principles used to interpret geologic events from the sedimentary rock record. The first, called the **principle of original horizontality,** states that sediments are deposited as essentially horizontal beds. Observation of modern marine and nonmarine sediments in a wide variety of environments supports this generalization. (Although cross-bedding, discussed in Chapter 7, is inclined, the overall orientation of cross-bedded units is horizontal.) If we find a sequence of sedimentary rock layers that are folded or tilted, we know that the rocks were deformed by tectonic stresses after their sediments were deposited.

The second principle, called the **principle of superposition,** states that each layer of sedimentary rock in a tectonically undisturbed sequence is younger than the one beneath it and older than the one above it. Geological common sense tells us that a younger layer cannot slip beneath one that has already been deposited. This principle enables us to view a series of layers as a kind of vertical time line; that is, a partial or complete record of the time

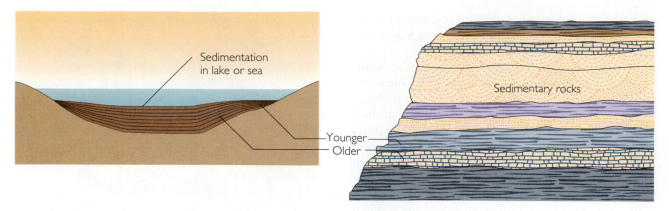

**FIGURE 9.3** Sediments are deposited in horizontal layers. The sediments gradually change into sedimentary rocks, and if they are undisturbed by tectonic processes they remain horizontal, their relative ages the same as those of their predecessor sediments: oldest on the bottom, youngest on the top.

elapsed from the deposition of the lowest bed to the deposition of the uppermost bed (Figure 9.3).

The two principles of the study of strata, or **stratigraphy,** allow us to view a vertical set of strata, called a **stratigraphic sequence,** as a chronological record of the geologic history of a region. The corresponding time line based on the sequence is called the **geologic time** spanned by the sequence.

With a geologic timekeeper, or "stratigraphic clock," geologists can tell whether one rock layer is older than another, although they cannot necessarily tell how much older. If sediments accumulated continuously at a steady rate, if they compacted a constant amount as they lithified, and if they did not erode, then a stratigraphic sequence might also provide a *measure* of absolute time. If we knew that muddy sediments accumulated at a rate of 10 m per million years, for instance, then 100 m of mudstone would represent 10 million years of deposition.

In practice, however, there are complications that make it impossible to gauge absolute time from stratigraphy with any accuracy. First, sediments do not accumulate at a constant rate in any sedimentary environment. During a flood, a river may deposit several meters of sand in its channel in just a few days, whereas in the years between floods it will deposit only a few centimeters of sand. Even in the deep ocean, where it may take 1000 years to deposit 1 mm of mud, sedimentation is unsteady, and the thickness of sediment cannot be used for precise timekeeping. In addition, the rate at which sediment is deposited varies widely in different sedimentary environments.

Second, the rock record does not tell us how many years may have passed between periods of dep-

osition. Many places on the floor of a river valley receive sediment only during times of flood. The times between floods are not represented by any sediment. Over the course of Earth's history in various places, there have been long intervals, some lasting *millions* of years, in which no sediments were deposited at all. In other places and at other times, sedimentary rocks may have been removed by erosion. Although we often can tell where a gap in the record occurs, we rarely can say how long an interval it represents.

Most important to geologists who wish to compare the geologic histories of different parts of the Earth, stratigraphy alone cannot be used to determine the relative ages of two widely separated beds. A geologist might be able to follow one bed or a series of beds by walking along an outcrop over a limited distance, but there is no way of knowing if a rock layer in Arizona, say, is older or younger than one in northern Canada.

The key to detecting missing time intervals and to correlating the ages of rocks at different geographic locations lay in the discovery of fossils. Fossils became the single most important tool for constructing a geologic time scale that is accurate for the entire planet.

## Fossils as Timepieces

Fossils are the remains of ancient organisms. Some look very much like animals living today; others are the remains of extinct life forms. Fossils can be shells, teeth, bones, impressions of plants, or tracks of animals. The most common fossils in rocks of the last half-billion years are shells of invertebrates, such as clams and oysters (Figure 9.4a). Much less common are the bones of vertebrates, such as mammals, reptiles, and dinosaurs. Plant fossils are abundant in some rocks, particularly those associated with coal beds, where leaves, twigs, branches, and even whole tree trunks can be recognized (Figure 9.4b).

The ancient Greeks were probably the first to surmise that fossils are records of ancient life, but it was not until modern times that the concept took hold and its consequences were explored. One of the first modern thinkers to make the connection between fossils and once-living organisms was Leonardo da Vinci, in the fifteenth century. In the seventeenth century Nicolaus Steno compared what he called "tongue-stones" found in the Mediterranean region with similarly shaped teeth of modern sharks and concluded that the stones were the remains of ancient life. Many people scoffed at Steno's conclu-

sions; the idea seemed preposterous. To others it was blasphemous, for it ran counter to the belief that the stones were divine creations. But by the end of the eighteenth century, after hundreds of fossils and their relationships to modern organisms had been described and catalogued, the evidence that fossils are the remains of formerly living creatures had become overwhelming. Thus **paleontology,** the study of the history of ancient life from the fossil record, took its place beside geology, the study of Earth's history from the rock record.

The dividends from the study of fossils did not accrue to geology alone. Young Charles Darwin's famous voyage as naturalist on the *Beagle* (1831–1836) vastly extended his knowledge of the great variety of fossil organisms and what their presence in rocks portended. He also had an opportunity to see a host of unfamiliar animal and plant species in their

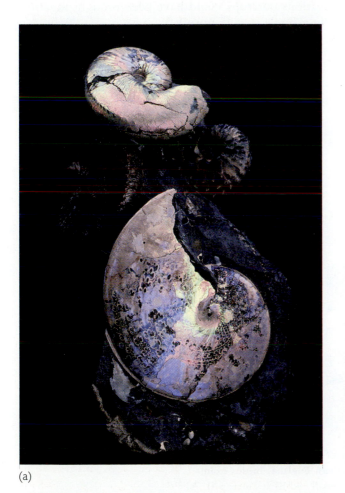

(a)

(b)

**FIGURE 9.4** (a) Ammonite fossils, ancient examples of a large group of organisms that is now largely extinct, represented in the modern world by a single species, the chambered nautilus. *Chip Clark.* (b) Petrified Forest, Arizona: ancient logs, millions of years old, whose substance was completely replaced by silica, which preserved all the original details of form. *Tom Bean.*

native habitats. In 1859, putting his and others' knowledge of both ancient and modern organisms together with an understanding of Earth's history that came from the geologic time scale, Darwin proposed the theory of evolution. It revolutionized scientific thinking about the origins of the millions of species of animal and plant life and provided a sound theoretical framework for paleontology.

Well before Darwin, in 1793, a surveyor working in southern England recognized that fossils could be used to date the relative ages of sedimentary rocks. William Smith, who was fascinated by the variety of fossils, collected them in the rock strata he saw exposed along canals and outcrops. Smith knew nothing of the idea of organic evolution that Charles Darwin was to enunciate over half a century later. He did note, however, that different layers contained different kinds of fossils, and he was able to tell one layer from another by the characteristic fossils each layer contained. He established a general order for the sequence of fossils and strata, from lowermost (oldest) to uppermost (youngest) rock layers. Regardless of the location of any new outcrop he came across, he could predict the stratigraphic position of any particular strata from the fossil assemblages he found in them. This stratigraphic ordering of the fossils is known as a *faunal succession.*

Smith was the first person to use faunal succession to correlate rocks from different outcrops. In each outcrop he identified distinct formations. A **formation** is a series of rock layers that everywhere has about the same physical properties (lithology) and contains the same assemblage of fossils. Some formations may consist of a single rock type, such as limestone. Others may be made up of thin, interlayered beds of different kinds, such as sandstone and shale. However they vary, each formation comprises a distinctive set of rock layers that can be recognized and mapped as a unit.

Using his knowledge of faunal successions, Smith matched up the similarly aged formations found in different outcrops. By noting the vertical order in which the formations were found in each place, he compiled a composite stratigraphic sequence for the entire region. His composite series showed how the complete sequence would have looked if the formations at different levels in all the various outcrops could have been brought together in a single spot (Figure 9.5).

Using this same approach of combining faunal successions with stratigraphic sequences, geologists of the last two centuries have carefully correlated formations around the world. The result, as we shall see, is a geologic time scale for the entire Earth.

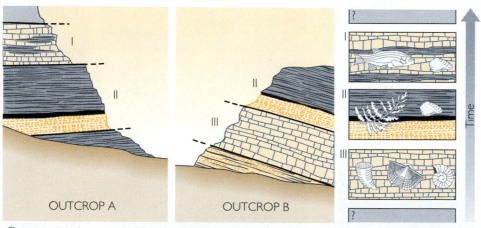

Outcrops may be separated by a long distance

**FIGURE 9.5** William Smith could piece together the sequence of rock layers of different ages containing different fossils by correlating outcrops found in southern England. In this example, formations I and II were exposed at outcrop A and formations II and III at outcrop B. A composite of the two would show formations I and II both overlying formation III and therefore younger than formation III.

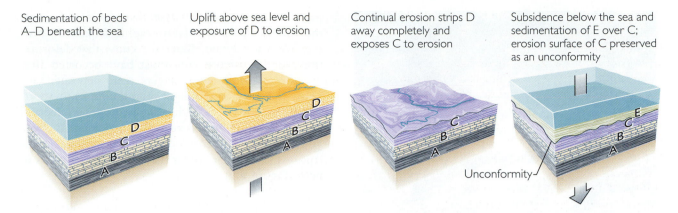

Sedimentation of beds A–D beneath the sea

Uplift above sea level and exposure of D to erosion

Continual erosion strips D away completely and exposes C to erosion

Subsidence below the sea and sedimentation of E over C; erosion surface of C preserved as an unconformity

Unconformity

**FIGURE 9.6** The sequence of events in the making of an unconformity.

## The Mark of Missing Time

In putting together sequences of formations, geologists often find places in which a formation is missing; either it was never deposited or it was eroded away before the next strata were laid down. The boundary along which the two existing formations meet is called an **unconformity;** that is, a surface between two layers that were not laid down in an unbroken sequence (Figure 9.6).

In many sequences, an unconformity arises after the lower beds have been folded by tectonic processes and then eroded to a more or less even plane. The next series is then laid down in horizontal layers. The resulting **angular unconformity** is an erosion surface that separates two sets of layers having bedding planes that are not parallel (Figures 9.7 and 9.8).

## Cross-Cutting Relationships

In addition to being folded and otherwise deformed, the layering of sedimentary rocks can be interrupted by discordant dikes or other magmatic intrusions (see Chapter 4) that cut through sedimentary layers. Faults displace bedding planes as they break apart blocks of rock (see Chapter 10). Faults can also dis-

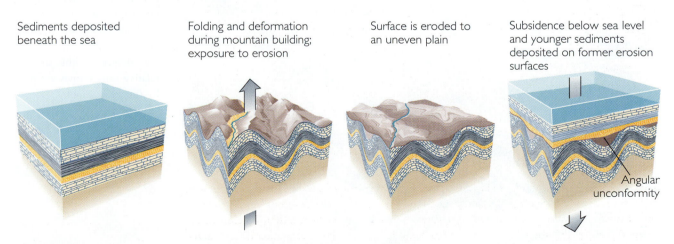

Sediments deposited beneath the sea

Folding and deformation during mountain building; exposure to erosion

Surface is eroded to an uneven plain

Subsidence below sea level and younger sediments deposited on former erosion surfaces

Angular unconformity

**FIGURE 9.7** The sequence of events in the making of an angular unconformity.

**FIGURE 9.8** The Great Unconformity in Grand Canyon, with the Tapeats sandstone (Cambrian) above and the Vishnu schist (Precambrian) below. *Peter Kresan.*

place dikes and sills. These disturbances, because they can be fitted into stratigraphic sequences, provide clues for dating. Since we know that deformational or intrusive events must have occurred after the affected sedimentary layers were deposited, we know that they are younger than the rocks they cut. If the intrusions or fault displacements are eroded and planed off at an unconformity, then overlain by a younger series of formations, we know that the intrusions or faults are older than the younger series of beds (Figure 9.9).

## Ordering the Geologic Record

We have now seen several ways of ordering rock strata and correlating them with a time sequence of geologic events. First, we can determine the relative ages of sedimentary rocks both by the simple rule of superposition and by the local and global fossil record. Second, we can use deformation and angular unconformities to date tectonic episodes in relation to the stratigraphic sequence. And third, we can use cross-cutting relationships to establish the relative ages of igneous bodies or faults cutting through sedimentary rocks. Combining all three, we can decipher the history of geologically complicated regions (Figure 9.10).

By using these relative dating principles and by piecing together information from outcrops all over the world, geologists worked out the entire **geologic time scale,** a relative-age calendar of Earth's

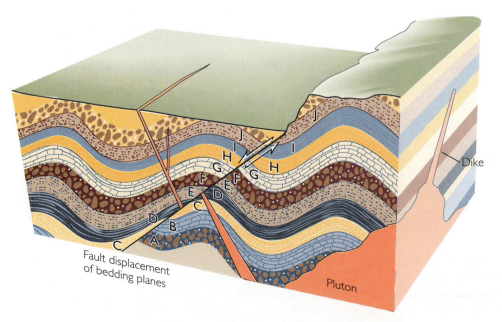

**FIGURE 9.9** Cross-cutting relationships of igneous intrusions and deformation (such as folding and faulting) allow us to place geological events within the relative time frames given by the stratigraphic sequence.

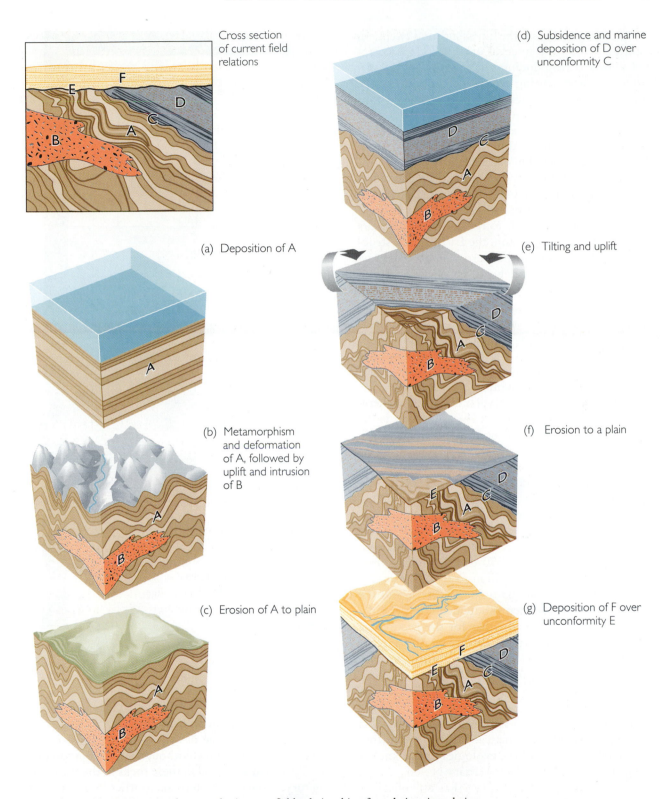

Cross section of current field relations

(a) Deposition of A

(b) Metamorphism and deformation of A, followed by uplift and intrusion of B

(c) Erosion of A to plain

(d) Subsidence and marine deposition of D over unconformity C

(e) Tilting and uplift

(f) Erosion to a plain

(g) Deposition of F over unconformity E

**FIGURE 9.10** Here we see how geologists use field relationships for relative time dating. From field mapping, a geologist makes a cross section of four formations: A, deformed metamorphic rocks; B, a granite pluton; D, sandstones, limestones, and shales containing marine fossils; F, sandstones containing land fossils. A and D are separated by an unconformity (C). D and F are separated by an angular unconformity (E). From these relationships the stages of this area's geologic history are reconstructed in diagrams (a) through (g).

geologic history, during the nineteenth and twentieth centuries. Each time interval on this scale is correlated with a corresponding set of rocks and fossils. Although it is still being refined here and there, the main divisions of geologic time have remained constant for the last century.

The geologic time scale, shown in Figure 9.11, is divided into four major time units: eons, eras, periods, and epochs, listed in order of decreasing length. An **eon** is the largest division of history. The oldest eon is the Archean (from Greek *archaios,* "ancient"). Archean rocks range from the earliest rocks known, almost 4 billion years old, to rocks 2.5 billion years old. During the Archean eon, the Earth's basic structure and dynamics, from the core, mantle, and crust to the surface, were still being formed. Fossils of primitive unicellular microorganisms are found in some sedimentary rocks of this age. The next younger set of rocks was formed during the Proterozoic eon (2.5 billion to 570 million years ago). During Proterozoic times, the surface and interior were close to the state they attained during later geologic times, with some significant exceptions. One exception was the level of oxygen in the atmosphere, which did not approach present levels until late in the Proterozoic. Life forms of the Proterozoic were still unicellular, but by its end more advanced forms started to evolve and be preserved as fossils.

The most recent and best understood eon, covering the last 570 million years, is known as the Phanerozoic; many rock formations of this age contain an abundance of shells and other fossils, such as vertebrate bones. The Phanerozoic is subdivided into three **eras:** the Paleozoic (570 to 245 million years ago), the Mesozoic (245 million to 65 million years ago), and the Cenozoic (65 million years ago to the present). The eras are then subdivided into **periods,** most of which are named either for the geographic locality in which the formations are best displayed or were first described or from some distinguishing characteristic of the formations. The Jurassic period, for example, is named for the Jura Mountains of France and Switzerland, and the Carboniferous period is named for the coal-bearing sedimentary rocks of Europe and North America. Periods are further subdivided into **epochs,** the geologically best known being the subdivisions of the Tertiary period, such as the Pliocene, the youngest (5 to 1.6 million years ago). In Box 9.1 the geologic time scale is used to interpret one of the world's most spectacular outcrops, the Grand Canyon.

The names of the eons and eras originally were derived from early ideas about life forms preserved as

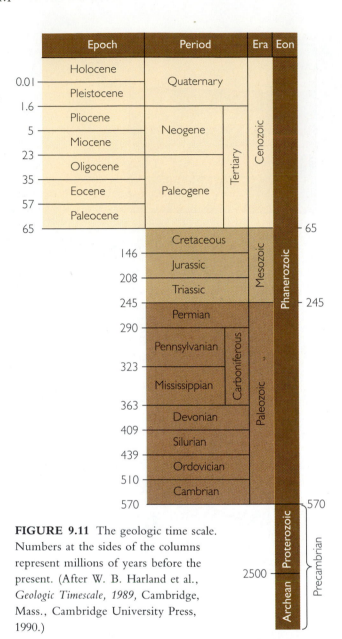

**FIGURE 9.11** The geologic time scale. Numbers at the sides of the columns represent millions of years before the present. (After W. B. Harland et al., *Geologic Timescale, 1989,* Cambridge, Mass., Cambridge University Press, 1990.)

fossils. The Archean and the Proterozoic eons were first considered to be devoid of fossils and so were called Azoic, meaning without life. As microscopic fossils were discovered in these rocks in the twentieth century, the name Proterozoic (earlier life) came into use. Phanerozoic means visible life; Paleozoic, ancient life; Mesozoic, middle life; and Cenozoic, recent life. We now know that the start of the Paleozoic era (the Cambrian period) roughly marked the appearance of multicellular animals and that other major evolutionary developments happened at various times during the Proterozoic and Phanerozoic eons.

In the process of working out this geologic time scale, geologists had to change their way of thinking about the Earth. Led by James Hutton in the late eighteenth century and, in the nineteenth century, by Charles Lyell, author of one of the first and most influential geology textbooks (*Principles of Geology*), they came to understand that the planet was not shaped by a series of catastrophic events over a mere few thousand years, as many people believed, especially those who accepted the first chapter of Genesis as a literal account of the Earth's creation. Rather, Earth was the product of ordinary geological processes operating steadily over much longer time intervals. As we noted in Chapter 3, Hutton was the first to grasp the cyclical nature of the geologic changes that result from erosion, weathering, sedimentation, burial, igneous and tectonic activity, and finally the renewal of the cycle by mountain building and erosion again.

Also inherent in Hutton's thinking and enunciated most forcefully in Lyell's textbook was the principle of uniformitarianism, which (we recall from Chapter 1) states that the processes we see shaping the Earth today are the same as those that have been operating during all of Earth's history. While different kinds of sediments may have been deposited at different rates in different places throughout Earth's history, we can be sure that the depositional processes that laid down sediments millions and billions of years ago were working then the same way they do today.

## ABSOLUTE TIME AND THE GEOLOGIC TIME SCALE

The geologic time scale based on studies of stratigraphy and fossils is a relative one; with it, geologists can say whether one formation is older than another, but they cannot pinpoint precisely when a rock formed. It's like knowing that World War I preceded World War II but not knowing the specific years in which each conflict began and ended. Nineteenth-century geologists could only estimate that it might take millions of years for one fossil assemblage to change to another. Although they had estimated times for various geological processes, such as the laying down of sediments on lake bottoms or the

erosion of a river valley, they could not devise a way to measure the duration of those processes accurately. Some nineteenth-century physicists had estimated the age of the Earth and the solar system from astronomical and physical principles and calculated it to be many millions of years old. But these were educated estimates based on the physical principles of the day, some now outmoded, and varied a good deal, from 25 million to 75 million years. It wasn't until the early twentieth century that scientists found a way to make reliable and accurate absolute measurements of geologic time.

An advance in modern physics paved the way for these measurements. In 1896 radioactivity in uranium was discovered by Henri Becquerel, a French physicist. Within less than a year of Becquerel's find, the French chemist Marie Sklodowska-Curie discovered and isolated a different and highly radioactive element, radium. It soon became clear that rays emitted from radioactive materials are indications that these materials transform at constant rates from one chemical element to another. In 1905 the British physicist Ernest Rutherford suggested that radioactivity could be used to measure the exact age of a rock. He was able for the first time to tell the absolute age of a uranium mineral from measurements in his laboratory. In the next few years, the ages of many more rocks were determined as dating methods were refined and more radioactive elements were found. This was the start of what is called **radiometric dating;** that is, the use of naturally occurring radioactive elements to determine the ages of rocks. When the results of Rutherford's first measurements were announced, it became clear that the Earth is billions of years old and that the Phanerozoic eon alone was a little over half a billion years long.

## Radioactive Atoms: The Clocks in Rocks

How do geologists use radioactivity to determine the ages of rocks? What the pioneers of nuclear physics discovered at the turn of the century was that atoms of uranium, radium, and several other elements that exhibit **radioactivity** are unstable. The nucleus of a radioactive atom spontaneously disintegrates, forming an atom of a different element and emitting radiation, a form of energy, in the process. We call the original atom the parent; its decay product is known as the daughter. The parent isotope rubidium-87, for instance, decays by emitting an electron from the

# BOX 9.1 INTERPRETING THE GRAND CANYON SEQUENCE

The rocks of the Grand Canyon have many stories to tell. They record a long history of sedimentation in a variety of environments, sometimes on land and sometimes under the sea. Unconformities mark erosion intervals, and angular unconformities also record ancient tectonic movements that built mountains. The rocks contain a succession of fossils, which reveals the evolution of new organisms and the extinction of old ones. But one of the most important stories that can be read from the rocks of the Grand Canyon is a geologic history billions of years long that can be reconstructed from the rock sequences exposed along the canyon wall.

The lowermost—and therefore oldest—rocks exposed at the Grand Canyon are dark igneous and metamorphic rocks that form a complex Precambrian basement formation, called the Vishnu schist (after Vishnu's Temple, an erosion-sculpted mass on the Colorado River named for a Hindu god). Because no fossils are preserved in schists, there is no quick way to tell the Vishnu's geologic age, only that it was formed by metamorphism of sedimentary and igneous rocks a long time ago.

Above the Vishnu is the younger Precambrian Grand Canyon series of formations, a group of interlayered sandstones, shales, and limestones originally deposited as sands and muds along rivers, in lakes, and in shallow seas. An angular unconformity separates the Vishnu and Grand Canyon, signifying a period of structural deformation accompanying metamorphism of the Vishnu before the deposition of the Grand Canyon. The tilting of the Grand Canyon beds at an angle from their originally horizontal position shows that they too were folded after deposition and burial. None of the rocks in the Grand Canyon series contains fossils of shelled organisms, but some do include fossils of single-celled microorganisms, an indication that these rocks are about a billion years old.

Yet another angular unconformity divides the Grand Canyon beds from the overlying horizontal Tapeats sandstone. This unconformity indicates a long period of erosion after the lower rocks had been tilted. The Tapeats sandstone contains no fossils, but it can be dated by reference to the overlying Bright Angel shale. The Bright Angel shale can be dated as Cambrian by its fossils, many of which are trilobites, extinct relatives of modern crayfish. (The fossils shown in the photograph at the beginning of this chapter are trilobites.) These fossils also indicate that the sequence was laid down under a sea.

Above the Bright Angel shale is a group of horizontal limestone and shale formations (Muav limestone, Temple Butte limestone, Redwall limestone) that represent about 200 million years from the late Cambrian period to the end of the Mississippian period. We know that there are unconformities between these formations because there are gaps in the faunal succession.

The next formation, high up on the canyon wall, is the Supai group of formations (Pennsylvanian and Permian), which contains fossils of land plants like those found in coal beds of North America and other continents. Of even greater interest are fossil footprints of primitive land reptiles. Overlying the Supai is the Hermit, a sandy red shale.

Continuing up the canyon wall, we find another continental deposit, the Coconino sandstone, which contains more vertebrate animal tracks and is extensively cross-bedded. The animal tracks and cross-bedding suggest that the Coconino was formed by wind-blown sand in an arid environment during Permian times. All of these features not only help geologists date the formation as Permian and understand its origin but also tell something of the paleoclimate, the climate of the past. Learning about past climates is especially vital today because we are becoming increasingly concerned about the economic and social impact of climate changes in the near future.

At the top of the cliffs at the canyon rim are two more formations of Permian age: the Toroweap, made mostly of limestone, overlain by the Kaibab, a massive layer of sandy and cherty limestone. These two formations record subsidence below sea level and the deposition of marine sediments. Still higher, younger formations are exposed at some distance from the canyon rim. Some of these formations contain dinosaur bones and fossilized tree trunks of the famous Petrified Forest (see Figure 9.4b).

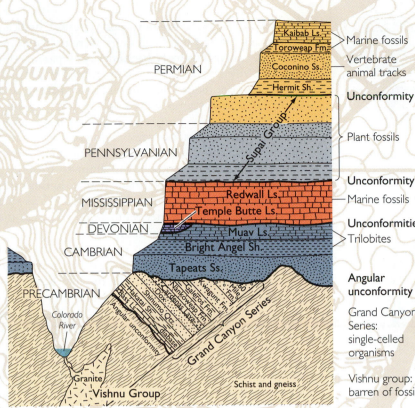

PERMIAN

Kaibab Ls. — Marine fossils
Toroweap Fm. — Vertebrate animal tracks
Coconino Ss.
Hermit Sh. — **Unconformity**

Supai Group

PENNSYLVANIAN

Plant fossils

**Unconformity**

MISSISSIPPIAN

Redwall Ls.
Temple Butte Ls. — Marine fossils

DEVONIAN — **Unconformities**
Muav Ls. — Trilobites

CAMBRIAN

Bright Angel Sh.

Tapeats Ss.

PRECAMBRIAN — **Angular unconformity**

Colorado River

Hakatai Sh., Bass Ls., Shinumo Qz., Dox, Nankoweap Fm., Cardenas Lava, Galeros Fm., Kwagunt Fm., Sixtymile Fm.

Angular unconformity

Grand Canyon Series

Granite
**Vishnu Group**
Schist and gneiss

**Grand Canyon Series:** single-celled organisms

**Vishnu group:** barren of fossils

*Left:* Generalized stratigraphic section of the rock units in the Grand Canyon Sequence. (After S. S. Beus and M. Morales (eds.), *Grand Canyon Geology,* New York, Oxford University Press/Museum of Northern Arizona Press, 1990, p. 463.) *Below:* View of the Grand Canyon from South Kaibab Trail. *Peter Kresan.*

Kaibab Ls.
Toroweap Fm., Coconino Ss., Hermit Sh.
Supai Group
Redwall Ls.
Redwall Ls.
Temple Butte Ls.
Muav Ls.
Bright Angel Sh.
Tapeats Ss.
Vishnu Schist
Grand Canyon Series
Vishnu Schist

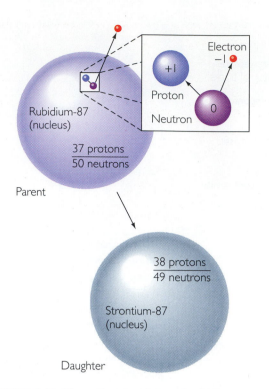

**FIGURE 9.12** The radioactive decay of rubidium to strontium. A neutron in a rubidium-87 atom ejects an electron, leaving an additional proton, which changes the atom to strontium-87.

nucleus to form a stable daughter, strontium-87 (Figure 9.12). (Recall from Chapter 2 that an atomic nucleus consists of protons and neutrons and that the isotopes of a given element contain the same number of protons but different numbers of neutrons.) The radioactive decay of rubidium-87 is essentially the emission of an electron from one of the neutrons in the nucleus, converting that neutron into a proton. With the additional proton, the former rubidium atom, with 37 protons, becomes a strontium atom with 38 protons.

The reason radioactive decay offers a dependable means of keeping time is that the average rate of nuclear disintegration is fixed: the decay rate does not vary with any of the changes in temperature, chemistry, or pressure that typically accompany geological processes in the Earth or other planets. This means that once atoms of a radioactive isotope are created anywhere in the universe, they start to act like a ticking clock, steadily altering from one type of atom to another at a fixed rate. If we know the decay rate and can count the number of newly formed daughter isotope atoms as well as the remaining parent atoms, then we can calculate the time that has elapsed since

the radioactive clock began to tick. In effect, we can work back to the time when there were no daughter isotopes, only those of the undecayed parent element.

Geologists count the number of parent and daughter isotopes with a mass spectrometer, a very precise and sensitive instrument that can detect even minute quantities of isotopes. Suppose we counted 950 rubidium-87 atoms and 50 strontium-87 atoms in a rock. Using the known rubidium-to-strontium decay rate, we would then calculate that, in this case, 4 billion years had passed since the rubidium in that rock began to disintegrate. (Four billion years is the age of the oldest rocks on Earth.)

To geologists, this is the age of a rock, or more exactly, the time since rubidium was first trapped in newly formed minerals. Rubidium, like other elements, is incorporated into minerals as they crystallize from a magma or recrystallize in a metamorphic rock. In the process, the rubidium is separated chemically from any strontium daughters produced before a magma solidified or a parent rock metamorphosed, because the two elements crystallize in different kinds of minerals. This action sets the radioactive clock back to zero. Inside the solid mineral the radioactive decay of rubidium-87 continues, and new strontium atoms, which cannot escape from the rock, start to accumulate. Thus rubidium-87 and other radioactive isotopes in igneous rocks provide a way of measuring when a magma was emplaced and cooled, and those in metamorphic rocks enable us to measure the time that has elapsed since they were metamorphosed. Rubidium decay is not used for dating sedimentary rocks because the minerals in clastic sediments are, in general, older than the sediment. Chemically and biochemically precipitated minerals in sediments, such as carbonates, do not usually contain enough newly precipitated rubidium to allow accurate isotopic analysis.

In addition to rubidium-87, geologists use a number of other naturally occurring radioactive elements to determine the ages of rocks (Table 9.1). Among these elements are uranium-238, which decays to lead-206 by a complex series of transformations; potassium-40, which decays to argon-40; and carbon-14, which decays to nitrogen-14. One intermediate daughter product of the uranium decay series is radon, a radioactive gas that poses a hazard to health in some places (see Box 9.2).

Each radioactive element has its own decay rate. Those that decay slowly over billions of years, like rubidium-87, are used to measure the ages of old rocks. Those that decay rapidly, so that most of the radioactive parent isotope disappears over only a few

**TABLE 9.1**

## Major Radioactive Elements Used in Radiometric Dating

| ISOTOPES | | HALF-LIFE OF PARENT (YEARS) | EFFECTIVE DATING RANGE (YEARS) | MINERALS AND OTHER MATERIALS THAT CAN BE DATED |
|---|---|---|---|---|
| PARENT | DAUGHTER | | | |
| Uranium-238 | Lead-206 | 4.5 billion | 10 million–4.6 billion | Zircon<br>Uraninite |
| Potassium-40 | Argon-40<br>Calcium-40 | 1.3 billion | 50,000–4.6 billion | Muscovite<br>Biotite<br>Hornblende<br>Whole volcanic rock |
| Rubidium-87 | Strontium-87 | 47 billion | 10 million–4.6 billion | Muscovite<br>Biotite<br>Potassium feldspar<br>Whole metamorphic or igneous rock |
| Carbon-14 | Nitrogen-14 | 5730 | 100–70,000 | Wood, charcoal, peat<br>Bone and tissue<br>Shell and other calcium carbonate<br>Groundwater, ocean water, and glacier ice containing dissolved carbon dioxide |

tens of thousands of years—carbon-14 is one—are useful for determining the ages of very young rocks. There must be a measurable amount of parent and daughter atoms in the rock to make radiometric dating feasible. If, for example, the rock is very old and the decay rate is fast, almost all the parent atoms may already have been transformed. In that case we know that the radiometric clock has run down, but we have no way of knowing how long ago.

Radioactive decay rates are commonly stated in terms of an element's **half-life**—the time required for one-half of the original number of radioactive atoms to decay. The half-lives of geologically useful radioactive elements range from thousands to billions of years. At the end of the first half-life after a radioactive isotope is incorporated into a new mineral, one-half of the parent atoms remains. At the end of a second half-life, half of that half, or one-quarter of the original number, is left; at the end of a third half-life, an eighth is left; and so on (Figure 9.13). Like rubidium-87, uranium-238 and potassium-40 have half-lives of billions of years and therefore are also excellent choices for dating old rocks (see Table 9.1).

Carbon-14, which rapidly decays to nitrogen-14, has a half-life of 5730 years. In a rock that is 30,000 years old, then, more than five half-lives have passed

and all but less than $\frac{1}{32}$ of the original amount of carbon-14 is gone. By the time 70,000 years have passed, too few carbon-14 atoms are left to count accurately. Carbon-14 is thus most suitable for measuring times in the relatively recent geologic past.

Carbon-14 is an especially important tool for dating fossil, bone, shell, wood, and other organic

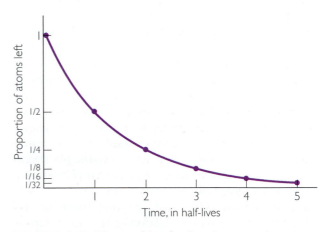

**FIGURE 9.13** The number of radioactive atoms in any mineral declines over time.

## BOX 9.2 RADON: AN ENVIRONMENTAL THREAT

Radon is an invisible, odorless, radioactive gas, possibly the most lethal of natural hazards. It is one of the products of the radioactive decay of uranium. Minerals containing uranium are unevenly distributed in rocks and soils and are usually present in amounts so small as to be negligible, although concentrations occur in certain rocks, such as shales formed from organic-rich muds and some granites. Under certain conditions uranium is easily dissolved by groundwater, transported over considerable distances, and reprecipitated in other rocks and soils.

As a gas, radon can move freely from its source in rock or soil into the atmosphere, where it can be inhaled. In the open air it is dispersed by winds and presents no danger. If it seeps into homes, however, it can build up to hazardous levels. Radon works its damage by releasing its own radioactive decay products, which can lodge in lung tissues. (Radon itself is inhaled and exhaled without significant buildup.) In a small percentage of individuals exposed to radon for many years the result is lung cancer; smokers are particularly vulnerable. Some estimate that there may be as many as 6000 to 25,000

radon-induced cancer deaths each year in the United States. This is the same order of risk as dying by an accidental fall or a fire in the home.

A recent survey of several thousand homes in seven states by the Environmental Protection Agency revealed that nearly one in three homes exceeds federal health guidelines for indoor levels of radon. In most cases the radon originates in the underlying soil or rocks and seeps into homes through cracks and openings in basement walls and floors. Fortunately, it is possible to identify homes at risk and take preventive measures. Elevated concentrations of radon often occur in "hot belts" corresponding to specific geological formations. Because radon escapes into the atmosphere so readily, it is hard to measure directly over a region. A map of the abundance of its parent, uranium, however, reveals the geographic distribution of areas in which the levels of uranium in soil and rock are high enough to produce radon exceeding safe levels.

It is wise to follow the advice of the U.S. Health Service and invest in an inexpensive device (costing $25 or less) available

---

materials in very young sediments because all of these materials contain carbon. Unlike radioactive elements such as rubidium and uranium, which are incorporated into rocks during metamorphism or crystallization from a magma, carbon, including a small amount of carbon-14, is an essential element of the living cells of all organisms. As plants grow they continuously incorporate into their tissues a small amount of carbon-14, along with other (stable) carbon isotopes, from carbon dioxide in the atmosphere. (The process by which this occurs, photosynthesis, is discussed in Box 9.3.) When a plant dies, it stops absorbing carbon dioxide and so no new carbon of any kind is added. At that moment, the amount of carbon-14 in relation to the stable carbon isotopes is identical to that in the atmosphere. But the

amount of carbon-14 in the dead plant tissue steadily decreases as radioactive atoms decay. Because the nitrogen-14 daughter atoms of carbon-14 are gaseous and thus leak from the sediment, we cannot measure the daughters accurately. Instead, we can calculate the time elapsed since the plant died by comparing the amount of carbon-14 left in the plant material with the original amount that would have been in equilibrium with the atmosphere, which is assumed to be approximately constant over the periods of time involved.

Radiometric dating cannot be used for every rock a geologist samples. If a rock containing uranium were to lose some of its lead through weathering, for example, we would get an erroneously young age. Or if an igneous rock were metamor-

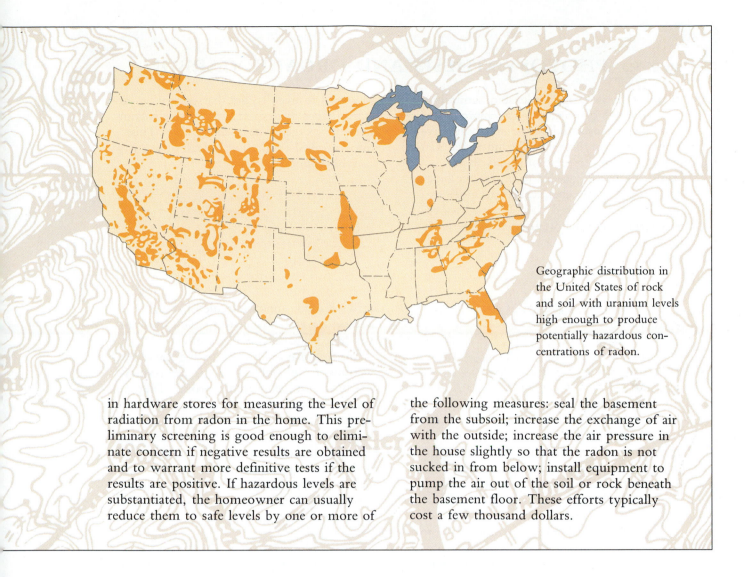

Geographic distribution in the United States of rock and soil with uranium levels high enough to produce potentially hazardous concentrations of radon.

in hardware stores for measuring the level of radiation from radon in the home. This preliminary screening is good enough to eliminate concern if negative results are obtained and to warrant more definitive tests if the results are positive. If hazardous levels are substantiated, the homeowner can usually reduce them to safe levels by one or more of the following measures: seal the basement from the subsoil; increase the exchange of air with the outside; increase the air pressure in the house slightly so that the radon is not sucked in from below; install equipment to pump the air out of the soil or rock beneath the basement floor. These efforts typically cost a few thousand dollars.

phosed, the daughter isotopes that had accumulated since the crystallization of the magma might be lost, resetting the clock to the time of metamorphism rather than the time of initial formation.

## Other Geologic Clocks

Because time is so central to the study of the Earth, geologists continue to search for additional ways to gauge geologic time. Paleomagnetic stratigraphy, for example, is under steady development as an adjunct to radiometric dating. (This method is discussed in detail in Chapter 19.) Briefly, periodic reversals of Earth's magnetic field about every half-million years are recorded in the orientation of magnetic minerals in rocks, especially those on the seafloor. The magnetic time scale has been calibrated both by radioactive age determinations and by the stratigraphic ages of overlying and underlying fossiliferous formations.

## The Geologic Time Scale

Once geologists could determine radiometric ages and then link them to their earlier studies of fossils and stratigraphy, they added absolute dates to the geologic time scale (see Figure 9.11). With these three sources of information, geologists can deduce approximate ages of rock formations, even those that contain no materials amenable to radiometric analysis. If, for example, we know from radiometric dat-

## BOX 9.3  PHOTOSYNTHESIS AND ORGANIC CARBON

Photosynthesis is a complex biochemical process in which green plants use chlorophyll and energy from sunlight to make carbohydrates out of carbon dioxide and water. Carbohydrate, in the form of sugars, is the source of energy needed to power the biological world. In the chemical reaction below, the simple compound $CH_2O$ is used to represent all of the many different carbohydrates that actually form.

$$\text{carbon dioxide} + \text{water} \xrightarrow[\text{and chlorophyll}]{\substack{\text{in the presence} \\ \text{of sunlight}}}$$
$$CO_2 \qquad H_2O$$
$$\text{carbohydrate} + \text{oxygen}$$
$$CH_2O \qquad O_2$$

For each 30 g of carbohydrate produced in this reaction, approximately 112,000 calories from sunlight are converted to chemical energy and stored in sugars. At the same time, for each atom of carbon incorporated into organic matter, one molecule of oxygen is produced.

Because animals cannot synthesize carbohydrates for themselves, they depend on food from plants or other animals for the energy they need to live. To get that energy, an organism takes in oxygen and in its cells combines the oxygen with carbohydrate. This process, called respiration, releases the energy stored by photosynthesis:

$$\text{carbohydrate} + \text{oxygen} \xrightarrow{\text{chemical energy}}$$
$$CH_2O \qquad O_2$$
$$\text{carbon dioxide} + \text{water}$$
$$CO_2 \qquad H_2O$$

In this reaction, one molecule of oxygen is used up for each atom of organic carbon incorporated into carbon dioxide.

As you can see by comparing the chemical reactions for photosynthesis and respiration, these processes are reciprocal. They link carbon dioxide, oxygen, and carbohydrate (or organic carbon, a term we will use for all the great number of organic compounds biologically synthesized from the building blocks supplied by photosynthesis). Photosynthesis and respiration are the vehicles for cycling carbon dioxide and oxygen between the atmosphere and oceans and the biological world.

Some of the organic carbon produced by photosynthesis is preserved in sediments. Bits

---

ing that an igneous intrusion is 500 million years old, then sedimentary layers cross-cut by the intrusions must be older than 500 million years. And if these same layers overlie metamorphic rocks radiometrically dated at 550 million years, then we know that the sedimentary layers were formed between 550 and 500 million years ago. If these sedimentary rocks contain fossils indicating Cambrian and late Ordovician stratigraphic ages, we know the absolute ages of parts of these geological periods. With this kind of "bracketing" of ages, geologists worked out the entire geologic time scale. After almost a century of radioactive dating and continued work on the stratigraphy of the world, this time scale remains undisputed in all major respects.

# ESTIMATING THE RATES OF VERY SLOW EARTH PROCESSES

Now that we have a way of dating rocks, let us see what the time scale can tell us about the rates of some slow geological processes. Consider the opening of an ocean, in which seafloor plates spread away from

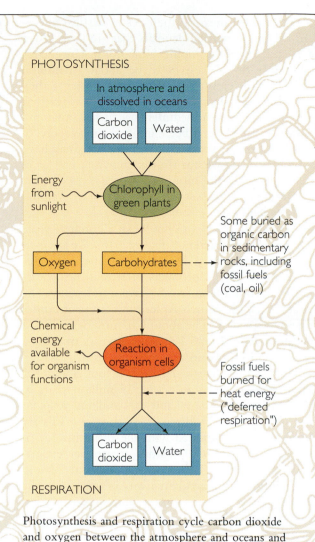

PHOTOSYNTHESIS

In atmosphere and dissolved in oceans

Carbon dioxide | Water

Energy from sunlight

Chlorophyll in green plants

Oxygen | Carbohydrates

Some buried as organic carbon in sedimentary rocks, including fossil fuels (coal, oil)

Chemical energy available for organism functions

Reaction in organism cells

Fossil fuels burned for heat energy ("deferred respiration")

Carbon dioxide | Water

RESPIRATION

Photosynthesis and respiration cycle carbon dioxide and oxygen between the atmosphere and oceans and the biological world.

and pieces of the organic matter of dead organisms and waste products of living ones are incorporated into sediments and buried in the crust. If photosynthesis and respiration were in perfect balance, all organic matter would be used up in respiration, and the carbon dioxide and oxygen from the two processes would be in balance. The burial of organic matter creates an imbalance, an excess of photosynthesis (and the production of $O_2$) over respiration (and the production of $CO_2$). For each molecule of organic matter buried there is a molecule of oxygen left over that cannot be used up by respiration. Thus the oxygen in our atmosphere today is the legacy of organic matter buried over billions of years of geologic time.

When the organic carbon buried in sedimentary rocks is exposed to weathering by uplift and erosion, it oxidizes to form carbon dioxide and water, just as in respiration. This "deferred respiration" reverses the imbalance between photosynthesis and respiration created by the burial of organic matter. For the past 100 years, humans have greatly accelerated this process by burning fossil fuels—coal, oil, and gas—at an increasing rate. As a result, carbon dioxide is now accumulating in the atmosphere at a faster rate than it leaves via photosynthesis.

each other at mid-ocean ridges. The southern Atlantic Ocean from South America to Africa is a little over 5000 km wide. The seafloor at the edges of these continents is known from fossils in sediments to be about 100 million years old (middle Cretaceous age). Thus the average rate of spreading of this part of the ocean is 5000 km every 100 million years, or about 5 cm per year. In other parts of the oceans, spreading rates may be as low as 1 or 2 cm per year or as high as 10 or 12 cm per year.

This kind of rough calculation gives a surprisingly good estimate of the rate of seafloor spreading. In 1987, using the technology of long-ranging laser

beams and satellites, scientists were first able to measure the spreading rate of the Atlantic seafloor directly. Their results agreed with the spreading rates calculated by marine geologists from the age and position of the seafloor.

Another slow process, and one of direct concern to the people of California, is the movement of crustal blocks along the San Andreas fault, which has been responsible for many major earthquakes. Figure 9.14 shows how the Pacific Plate slides past the North American Plate along this transform fault. We can determine the rate of movement by matching up distinctive geological formations of various ages that

**FIGURE 9.14** The North American Plate is moving southeast along the San Andreas fault at a rate of about 5 cm per year relative to the Northern Pacific Plate, which is moving northwest.

have been split by the fault and have moved away from each other. Dividing the distance that now separates any pair of these formations by the elapsed time—the geological time since they were formed and split—we get the rate of movement. In this way we can estimate that the average movement over the past several million years was about 5 cm per year in northern and central California. At that rate, about 25 million years ago, the part of the Pacific Plate block containing the San Francisco coast would have been at the present latitude of Los Angeles.

We can also measure the rates of vertical movement by dating marine deposits that are now above sea level. Parts of the Alpine range, for instance, contain marine fossils known to be about 15 million years old. Now elevated 3000 m above sea level, they were originally deposited on the shallow seafloor close to sea level. The sedimentary rocks containing the fossils must have been uplifted an average of about 0.2 mm per year, although the rates may have been higher or lower for shorter time intervals or in different parts of the mountain range.

Erosional processes are continuously wearing down the land surface, but they are so slow that two photographs of a river valley taken a hundred years apart show little difference (Figure 9.15). We can estimate erosion rates by adding up all of the disintegrated and dissolved products of erosion being carried away from a land area by rivers and wind. The rate at which the North American continent erodes has been estimated to be about 0.03 mm per year. At this rate it would take 100 million years to wear a 3000-m-high mountain down to sea level.

Thus in these particular cases it took about 100 million years to open an ocean, 15 million years to raise a mountain range, and 100 million years to erode it. But as we will see, these time intervals are relatively short compared with the entire history of the planet. During that history, Earth has experienced many cycles of mountain building and erosion.

In addition to providing a way of measuring the rates of geological processes and reconstructing Earth's history, the time scale and the ideas on which

**FIGURE 9.15** Two photos of Bowknot Bend on the Green River in Utah taken nearly 100 years apart show that little has changed in the configuration of rocks and formations in that time interval.

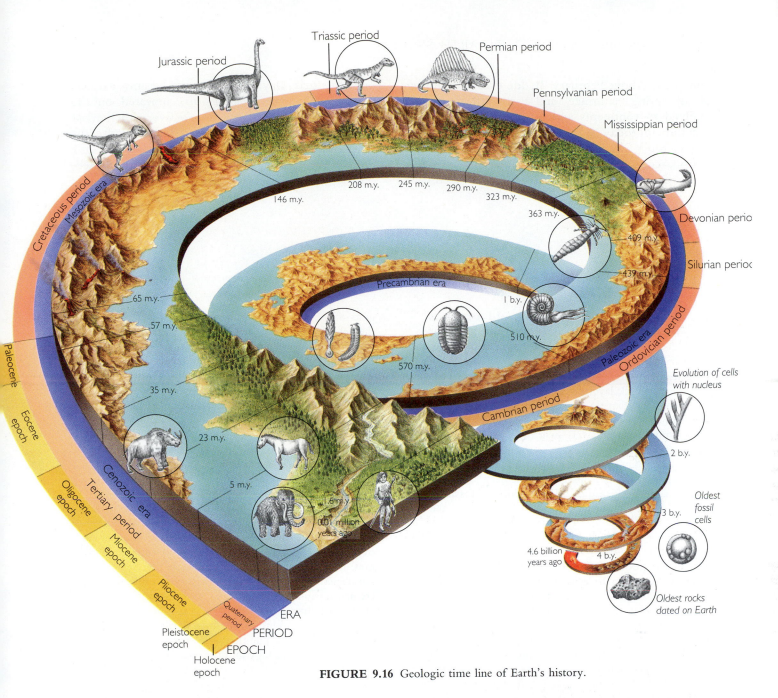

**FIGURE 9.16** Geologic time line of Earth's history.

# AN OVERVIEW OF GEOLOGIC TIME

it is based offer a tool for validating plate tectonics. Without a way to measure the ages of rocks on the seafloor and the continents, geologists would not have been able to deduce the spreading of the seafloor and other plate motions.

We can now combine the geologic time scale with absolute dates from the analysis of radioactive decay and the evolution of organisms to construct a time line for the whole history of the Earth, beginning at its birth 4.6 billion years ago (see the inside back cover). Figure 9.16 shows the whole sweep of geologic time as a spiral pathway, each revolution of the spiral corresponding to one billion years. From a look at this illustration we can see how short a proportion of the total of Earth's history is taken up by the eras of the Phanerozoic eon and what a tiny amount of time has elapsed since human beings evolved.

As another way of comprehending this extraordinarily long period of time, think of Earth's age as being only one calendar year. On January 1 the Earth was formed. During January and part of early February the Earth became organized into core, mantle, and crust. About February 21, life evolved. During all of spring, summer, and early fall the Earth evolved to continents and ocean basins something like those of today and plate tectonics became active. On October 25, at the beginning of the Cambrian period, complex organisms, including those with shells, arrived. On December 7 reptiles evolved, and on Christmas Day the dinosaurs became extinct. Modern humans, *Homo sapiens,* appeared on the scene at 11 P.M. on New Year's Eve, and the last glacial age ended at 11:58:45 P.M. Three-hundredths of a second before midnight, Columbus landed on a West Indian island. And a few thousandths of a second ago, you were born.

# SUMMARY

### How do geologists know how old a rock is and whether one rock is older than another?

Geologists determine the order in which rocks were formed by studying their stratigraphy, fossils, and arrangement in the field. An undeformed sequence of sedimentary rock layers will be horizontal, with each layer younger than the layers beneath it and older than the ones above it. In addition, since animals and plants have evolved progressively over time, their fossil remains change in known ways in the stratigraphic sequence. Knowing the faunal succession makes it easier for geologists to spot missing sedimentary layers, known as unconformities. Most important, fossils enable geologists to correlate rocks located all over the world.

### How did geologists create a geologic time scale that is applicable around the world?

Using fossils to correlate rocks of the same geological age and piecing together the sequences exposed in hundreds of thousands of outcrops around the world, geologists compiled a stratigraphic sequence applicable everywhere in the world. The composite sequence represents the geologic time scale. The use of radiometric dating allowed scientists to assign absolute dates to the units of the time scale. Radiometric dating is based on the behavior of radioactive elements, in which unstable parent atoms are transformed into daughter isotopes at a constant rate. When radioactive elements are locked into minerals as rocks are formed, their daughters accumulate and the number of parents decreases; by measuring parents and daughters we can calculate absolute ages.

### Why is the geologic time scale important to geology?

The geologic time scale enables geologists to reconstruct the chronology of events that have shaped the planet. The time scale has been instrumental in validating and studying plate tectonics and in estimating the rates of geological processes too slow to be monitored directly, such as the opening of an ocean over millions to hundreds of millions of years. In addition, the development of the time scale revealed that the planet is much older than early geologists and others had imagined and that since its beginning it has undergone almost constant change as a result of gradual processes working throughout Earth's history. The creation of the geologic time scale paralleled the development of paleontology and the theory of evolution, one of the most revolutionary and powerful ideas in science.

# KEY TERMS AND CONCEPTS

relative age (p. 189)
absolute age (p. 189)
principle of original horizontality
    (p. 189)
principle of superposition
    (p. 189)
stratigraphy (p. 190)

stratigraphic sequence (p. 190)
geologic time (p. 190)
paleontology (p. 191)
formation (p. 192)
unconformity (p. 193)
angular unconformity (p. 193)
geologic time scale (p. 194)

eon (p. 196)
era (p. 196)
period (p. 196)
epoch (p. 196)
radiometric dating (p. 197)
radioactivity (p. 197)
half-life (p. 201)

# EXERCISES

1. Name the geological periods, from youngest to oldest.

2. Give the absolute times of the beginnings of the Paleozoic, the Mesozoic, and the Cenozoic eras.

3. To what element does rubidium-87 decay radioactively?

4. Give the age of a sediment that might be dated by carbon-14.

5. What geologic events are implied by an angular unconformity?

6. What is the principle of superposition?

7. What is the principle of original horizontality?

# THOUGHT QUESTIONS

1. As you pass by an excavation in the street, you see a cross section showing paving at the top, soil below that, and bedrock at the base. You also notice that a vertical water pipe extends from a drain in the street into a sewer in the soil. What can you say about the relative ages of the various layers and the water pipe?

2. How would you be able to ascertain the relative ages of several volcanic ash falls exposed in an outcrop?

3. What evidence could you give to a friend to support the idea that a particular formation was many millions of years old?

4. Construct a diagram similar to the one in Figure 9.10 to show the following series of geological events: (a) sedimentation of a limestone formation; (b) uplift and folding of the limestone; (c) erosion of the folded terrain; (d) subsidence of the terrain and sedimentation of a sandstone formation.

5. Many fine-grained muds are deposited at a rate of about 1 cm per 1000 years. At this rate, how long would it take to accumulate a stratigraphic sequence 0.5 km thick?

6. What radioactive elements might you use to date a schist that is approximately one billion years old?

7. What geological event is dated by radioactive decay of a mineral in a schist?

8. What geological event is dated by radioactive decay of a mineral in a basalt?

# SUGGESTED READINGS

Eicher, Don L. 1976. *Geologic Time,* 2d ed. Englewood Cliffs, N.J.: Prentice-Hall.

Faure, Gunter. 1986. *Principles of Isotope Geology,* 2d ed. New York: Wiley.

Palmer, Allison R. 1984. *Decade of North American Geologic Time Scale*. Geological Society of America, Map and Chart Series MC-50.

Stanley, Stephen. 1989. *Earth and Life Through Time,* 2d ed. New York: W. H. Freeman.

Simpson, George G. 1983. *Fossils*. New York: Scientific American Books.

# FOLDS, FAULTS, AND OTHER RECORDS OF ROCK DEFORMATION

The eighteenth- and nineteenth-century scientists who laid the foundations of modern geology concluded that most sedimentary rocks were originally deposited as soft horizontal layers at the bottom of the sea and hardened over time. But geologists were puzzled that so many hardened rocks were tilted, bent, or fractured (like those shown on the facing page). They wondered: What forces could have deformed these hard rocks in this way? Can we reconstruct the history of the rocks from the patterns of deformation found in the field? Today's geologists would add: How does the deformation relate to plate tectonics? We will answer these questions in this and later chapters. But we must learn the features of deformed rocks before we can see order and then explain the patterns of deformation.

Intricately folded sediments on Nuptse–Lhotse Wall, Mount Everest. Folding, faulting, and uplift in the Himalayas reflect the collision of India with Asia. *Glen Rowell.*

Scientists are trained to be keen observers so that they can gather accurate information on which to base their theories. Geologists rely heavily on the detailed information they collect from field observations to reconstruct the geologic history of a region. They might ask: What kinds of rocks were originally deposited? What happened to them then? If they were deformed by tectonic forces, what was the nature of the deformation and what kinds of forces were involved? Fortunately, in their search for answers geologists found similar patterns of deformation around the globe. A few concepts can explain in general why and how rocks were deformed.

**Folding** and **faulting** are the most common forms of deformation in the rocks that make up Earth's crust. Folds in rocks are like folds in clothing (Figure 10.1). Tectonic forces can also cause a rock formation to break and slip on both sides of a fracture, parallel to the fracture (Figure 10.2). This is a fault. Geologic folds and faults can range in size from centimeters to tens of kilometers. Many mountain ranges are actually a series of large folds or faults, or both, that have been weathered and eroded. Geologists now believe that the forces that move the large plates are ultimately responsible for the deformations found in most local areas.

## INTERPRETING FIELD DATA

To figure out how rock formations are deformed, geologists need accurate information about the geometry of the beds they can see. A basic source of information is the outcrop, where the bedrock that underlies the surface everywhere is exposed, not obscured by soil or loose boulders. A sedimentary bed that has been bent into a fold is shown in Figure 10.3. Often folded rocks, unlike those in the photo, may be partly exposed in an outcrop and show only as an

inclined layer. The orientation of the layer is important information for the geologist to use in piecing together a picture of the overall deformed structure. It takes only two measurements to describe the orientation of a layer of rock at a given location: the strike and the dip. Figure 10.4 shows how the strike and the

**FIGURE 10.1** Small-scale folds in a Precambrian iron formation, Beresford Lake area, Manitoba. *Geological Survey of Canada.*

212

**FIGURE 10.2** Small-scale faults offset once-continuous rock layers; Kabab Canyon, Utah. *Tom Bean.*

**FIGURE 10.3** An outcrop of originally horizontal rock layers bent into folds by compressional tectonic forces. *Phil Dombrowski.*

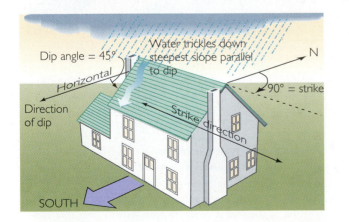

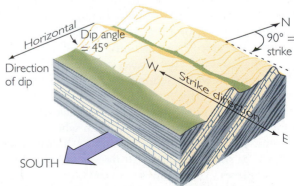

**FIGURE 10.4** Geologists use the strike and dip of a formation to define its orientation at a particular place. (After A. Maltman, *Geological Maps: An Introduction,* New York, Van Nostrand Reinhold, 1990, p. 37.)

**FIGURE 10.5** Dipping limestone and shale beds; coast of Somerset, United Kingdom. Children are walking along the strike of beds that dip to the left. Dip is the angle of steepest descent of the bed from the horizontal; strike is at right angles from the dip direction. *Chris Pellant.*

dip are observed and measured in the field. The **strike** is the direction of a rock layer as it intersects a horizontal surface. The **dip,** which is measured at right angles to the strike, is simply the amount of tilting—that is, the angle at which the bed inclines from the horizontal (Figure 10.5). A geologist might describe a particular outcrop, for example, as "beds of coarse-grained sandstone striking north and dipping 30 degrees west."

A convenient means of organizing information is a geological map, on which geologists record the locations of outcrops, the nature of the rocks, and the dips and strikes of inclined layers. (Appendix 4 gives more information on geological maps.) Also helpful in piecing together a geological story is a geological cross section, a diagram showing the features along a vertical slice through part of the crust. A natural cross section can often be observed in the vertical face of a cliff, a quarry, or a road cut. A cross section can also be constructed from the information on a geological map. Figure 10.6 shows a simple geological map of an area where sedimentary rocks, originally horizontal, were bent into a fold.

How does the geologist reconstruct the deformed shapes of the rock layers, even when erosion has removed parts of the section? It's like putting together a three-dimensional jigsaw puzzle with missing pieces. Common sense and intuition play important roles. The geologist may first look for such features as sedimentary layers and surmise that they were originally deposited as horizontal beds at

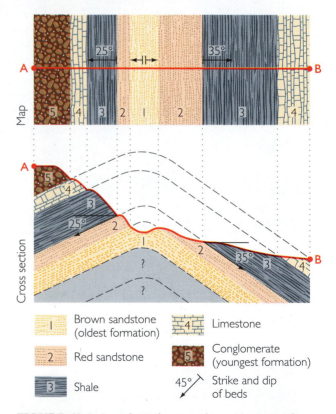

Brown sandstone (oldest formation) — 1
Red sandstone — 2
Shale — 3
Limestone — 4
Conglomerate (youngest formation) — 5
45° Strike and dip of beds

Map

Cross section

**FIGURE 10.6** A geological map and a cross section derived from it. The orientation of the layers indicates that formation 1 is at the bottom and therefore is the oldest; flanking formations are successively younger. The eroded portion of the fold is enclosed with dashed lines that connect identical formations, shown by the same colors and numbers, and match the observed dip.

the bottom of the sea. If the layers are now found to be inclined, as they are in Figure 10.6, this is taken as evidence that later events forced the rocks to be tilted or bent. The law of superposition (younger beds are laid down over older beds, as we saw in Chapter 9) tells the geologist that the oldest bed is the one numbered 1, and the flanking, overlying beds are successively younger. Using the dip of the formations, the geologist constructs a cross section—a vertical cut as it would exist along the traverse A-B on the map. The geologist finds identical formations on both sides of the oldest formation and links them up by drawing dashed lines in the map space representing the air to match the observed dips. These dashed lines represent the boundaries of the portions of the beds removed by erosion. The geologist projects the trend of the beds below the ground, even though the rocks cannot be seen. The map in Figure 10.6 tells the story of an ancient ocean, now gone, in which a succession of sedimentary rocks was deposited on the seafloor. The once-horizontal beds were subjected to crustal forces of compression, bent into folds, and raised above sea level. Erosion could then remove a major portion of the section, leaving the present-day remnant portrayed by the map and cross section. The folded rocks in Figure 10.5 have had this history.

# HOW ROCKS BECOME DEFORMED

For years geologists were baffled by the problem of how rocks, which seem strong and rigid, could be distorted into folds by tectonic forces or broken along faults. The forces can be of three types: **compressive forces,** which squeeze and shorten a body; **tensional forces,** which stretch a body and tend to pull it apart; and **shearing forces,** which deform a body so that one side slides past the portion on the opposite side (think of what happens to a deck of cards between your palms when your hands move parallel to each other but in opposite directions; the cards slide past one another, deforming the deck by shear). Figure 10.7 shows how rocks generally become deformed under these three types of forces. Recall from Chapter 1 that the same kinds of forces are active at the boundaries between plates: compressive forces dominate at convergent boundaries, where plates collide; tensional forces dominate at divergent boundaries, where plates are pulled apart; and shearing forces dominate at transform fault plate boundaries, where plates slide horizontally past each other.

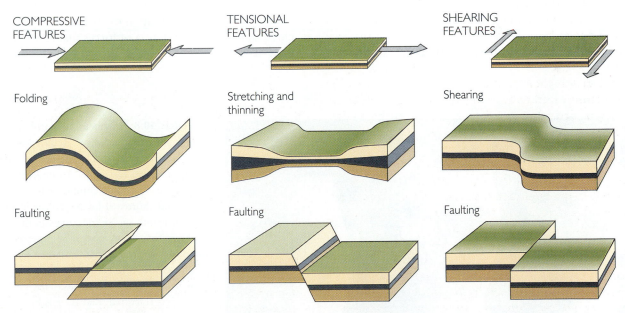

**FIGURE 10.7** Rocks are deformed by folding or by faulting when they are subjected to different kinds of tectonic forces. Geologists see the pattern of deformation in the field and infer the nature of the forces that caused it.

Although geology depends heavily on field observations, many geologists conduct laboratory experiments to discover why rock formations are folded in one place and fractured in another. Experiments have been performed in which rocks are squeezed at the low pressure and temperature characteristic of conditions near the surface. Pressures that correspond to the weight of 30 km of rock and temperatures of several hundred degrees Celsius have also been used to simulate conditions at a depth of about 30 km in the Earth. Figure 10.8 shows what happens to a small cylinder of marble when it is squeezed by compressive forces applied at the ends. The experimenter applies force by pushing down on one end of a rock sample with a piston while applying pressure to all sides of the sample. These pressures simulate both the force of compression and the confining pressure to which materials deep in the crust are subjected by the weight of the overlying rock. (Confining pressure is like the pressure you feel all over your body when you dive deep underwater.) Under low confining pressures like those at shallow depths in the crust, the middle sample deformed by fracturing; under high confining pressures like those at greater depths in the crust, the third sample deformed slowly and steadily without fracturing. It behaved as a pliable or moldable material; that is, it deformed plastically into its shortened, bulging shape. The investigators concluded that if a bed of this particular marble were subjected to tectonic forces near the surface, it would tend to deform by fracturing and faulting; but if it were deeper than a few kilometers, it would behave as a plastic material and change shape by gradually folding. Experiments also show that if rocks are hot when forces are applied, this kind of smooth, plastic deformation occurs more readily.

In this way rocks, like most solids, can be classed as **brittle** or **ductile,** on the basis of how they are deformed by forces. As forces are increased, a brittle material experiences little change until it breaks suddenly; ductile substances experience smooth and continuous plastic deformation. Glass that is near room temperature is a familiar brittle material, and modeling clay is ductile. Marble is brittle at shallow depths but ductile deeper in the crust.

Natural conditions are more complex, of course, than those under which such simple experiments are performed. Tectonic forces are applied over millions of years, whereas the laboratory experiment is performed in a few hours. Nevertheless, the experiments shed some light on how rocks respond to forces, and they give us more confidence in our interpretations of field evidence. When we see folds and fractures in the field, we can remember that some rocks are brittle, others are ductile, and the same rock can be brittle at shallow depths and ductile deep in the crust. The details vary from one type of rock to another. Laboratory experiments also teach us that we should expect most igneous rocks to be less deformable than most sedimentary rocks and **basement rocks** (old, underlying igneous or metamorphic rocks) to be more brittle than the ductile young sediments that may cover them.

Folds and fractures are the signatures of deformation that geologists map in the field. They provide clues to the larger panorama of forces that result from plate tectonics.

(a)  (b)  (c)

**FIGURE 10.8** Results of laboratory experiments conducted to discover how rocks, in this case marble, are deformed by compressive forces. (a) An undeformed sample. (b) A sample compressed, fractured, and shortened by 20 percent of the original length under conditions in the shallow crust. (c) A sample compressed, smoothly deformed, and shortened by 20 percent of the original length under conditions in the deeper crust. Note that sample (b) has fractured; that is, it is brittle at the laboratory equivalent of shallow depth. It is pliable or ductile at greater depth. *M. S. Patterson, Australian National University.*

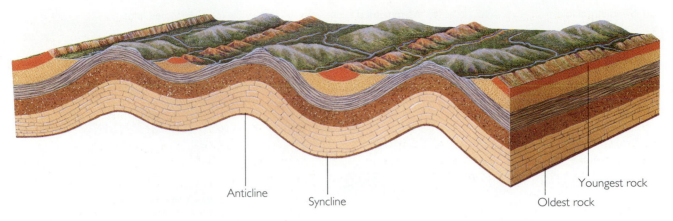

**FIGURE 10.9** Anticlines fold upward; synclines fold downward.

# FOLDS

The term *fold* implies that a structure that originally was planar, such as a sedimentary bed, has been bent. The deformation may be produced by either horizontal or vertical forces in the crust, just as pushing in on opposite sides of a piece of paper or up from below may fold it. Folding is a common form of deformation of layered rocks and is most typically manifested in mountain belts. As we mentioned earlier, in many young mountain systems where erosion has not yet erased them, majestic, sweeping folds can be traced, some of them with dimensions of many kilometers (as in the photograph at the beginning of this chapter). On a much smaller scale, very thin beds

can be crumpled into folds a few centimeters long (see Figure 10.1). The bending can be gentle or severe, depending on the magnitude of the applied forces, the length of time they are applied, and the ability of the beds to resist deformation.

Layered rocks can fold in several basic ways in response to compressional forces, depending on the properties of the rocks and the details of the forces. To describe their observations, geologists have given these folds and their parts the names shown in Figures 10.9 and 10.10. Upfolds, or arches, of layered rocks are called **anticlines** (Figure 10.11). Downfolds, or troughs, are called **synclines.** The two

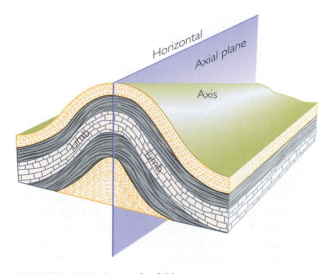

**FIGURE 10.10** Parts of a fold.

**FIGURE 10.11** A sharply folded anticline in sandstone, Anza-Borrego Desert, California. *Bill Evarts.*

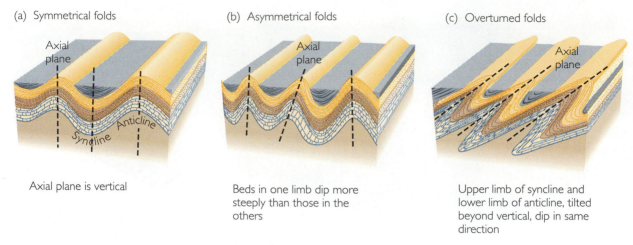

(a) Symmetrical folds

Axial plane

Syncline     Anticline

Axial plane is vertical

(b) Asymmetrical folds

Axial plane

Beds in one limb dip more
steeply than those in the
others

(c) Overturned folds

Axial plane

Upper limb of syncline and
lower limb of anticline, tilted
beyond vertical, dip in same
direction

**FIGURE 10.12** Symmetrical, asymmetrical, and overturned folds.

**FIGURE 10.13** Overturned fold. *Geological Survey of Israel.*

sides of a fold are its **limbs.** The **axial plane** is a surface that divides a fold as symmetrically as possible, with one limb on either side of the plane. The line made by the intersection of the axial plane with the beds is the **axis** of the fold. If the axis of a fold is not horizontal, as in Figure 10.10, the fold is called a **plunging fold;** the plunge is given by the angle the axis makes with the horizontal.

Not every fold has a vertical axial plane with limbs dipping symmetrically away from the axis, as in Figure 10.12a. With increasing horizontal force the folds can be thrown into **asymmetrical** shapes, with one limb dipping more steeply than the other (Figure 10.12b). This is a common situation. When the deformation is intense and one limb has been tilted beyond the vertical, the fold may be **overturned.** Both limbs of an overturned fold dip in the same direction, as in Figure 10.12c, but the order of the layers in the bottom limb is precisely the reverse of their original sequence; that is, older rocks are on top of younger rocks. The folds in Figures 10.3 and 10.13 have been overturned to such an extent that the axial plane is nearly horizontal. One limb has been rotated into a completely upside-down sequence, with older beds on top of younger beds.

Follow the axis of any fold in the field and sooner or later the fold dies out, just as wrinkles in a tablecloth do. The fold appears to plunge into the ground as it disappears. Figures 10.14 and 10.15 show the geometry of **plunging anticlines** and **plunging synclines** and the zigzag pattern of outcrops that may appear in the field after erosion has removed much of the surface rock.

A **dome** is an anticlinal structure, a broad circular or oval upward bulge of rock layers. The flanking

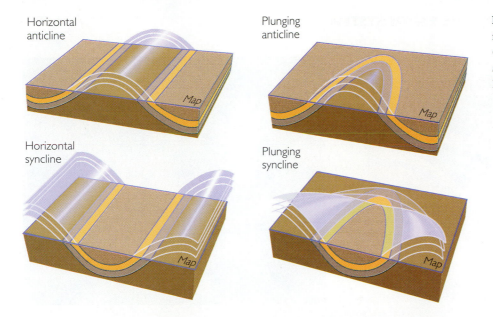

Horizontal anticline

Plunging anticline

Horizontal syncline

Plunging syncline

**FIGURE 10.14** Plunging folds. (After A. Maltman, *Geological Maps: An Introduction,* New York, Van Nostrand Reinhold, 1990, p. 84.)

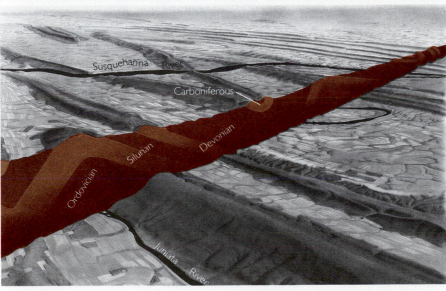

Susquehanna River

Carboniferous

Devonian

Silurian

Ordovician

Juniata River

**FIGURE 10.15** The erosional remnants of plunging folds show a characteristic zigzag pattern in this view of the Valley and Ridge belt of the Appalachian Mountains 30 miles northwest of Harrisburg, Pennsylvania. In the drawing below, the imaginary trench reveals the subsurface structure. (From J. S. Shelton, *Geology Illustrated,* San Francisco, W. H. Freeman, 1966.)

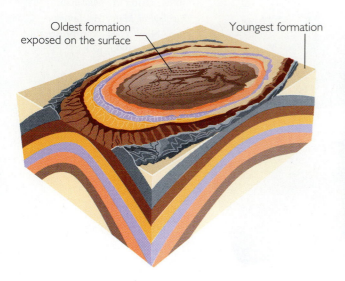

Oldest formation
exposed on the surface

Youngest formation

**FIGURE 10.16** The characteristic circular or elliptical out-crop pattern of a dome. The oldest bed is in the core; the flanking formations are successively younger and dip away from the core.

**FIGURE 10.17** An eroded dome in strata 6 miles east of Rawlins, Wyoming. The highway and railroad at the lower right suggest its dimensions. (From J. S. Shelton, *Geology Illustrated,* San Francisco, W. H. Freeman, 1966.)

beds of a dome encircle a central point and dip radially away from it (Figures 10.16 and 10.17). A **basin** is a synclinal structure, a bowl-shaped depression of rock layers in which the beds dip radially toward a central point. Domes and basins are typically many kilometers in diameter. They are recognized in the field by outcrops with the characteristic circular or oval shapes seen in Figures 10.16 and 10.17. Domes are very important in oil geology because oil is buoyant and tends to migrate upward through permeable rocks. If the rocks at the high point of a dome are not easily penetrated, the oil becomes trapped against them. It is not entirely clear why domes and basins form. Some domes can be attributed to igneous rock that intrudes the crust, pushing the overlying sediments upward. Some basins are formed when a heated portion of the crust cools and contracts, causing the overlying sediments to subside. Others result when the crust is stretched by tectonic forces. The weight of sediments deposited in a shallow sea can depress the crust, forming a basin. There are many domes and basins in the central portion of the United States. The Black Hills of South Dakota form an eroded dome; much of the lower peninsula of Michigan is a sedimentary basin.

Observations in the field seldom provide geologists with complete information. Either bedrock is obscured by overlying soils or erosion has removed much of the evidence of earlier structures. So geologists search for clues they can use to work out the relationship of one bed to another. For example, in the field or on a map, an eroded anticline would be recognized by a strip of older rocks forming a core bordered on both sides by younger rocks dipping away, as in Figure 10.6. A syncline would show as a core of younger rocks bordered on both sides by older rocks dipping toward the core (Figure 10.18).

Folds typically occur in elongated groups. A strip of country in which the rock layers are folded—that is, a **fold belt**—suggests to a geologist that the region was compressed at one time by horizontal tectonic forces. The Valley and Ridge province of the Appalachians is a folded mountain belt (Figure 10.19). We will see in Chapter 21 that an ancient plate collision accounts for the wrinkling of the once-flat layers of sedimentary rocks in this region.

# HOW A ROCK FRACTURES: JOINTS AND FAULTS

We have seen that the way rocks deform depends on the kinds of forces to which they are subjected and the conditions that prevail. Some layers crumple into folds, and some fracture. There are two kinds of fractures, joints and faults. A **joint** is a crack along which

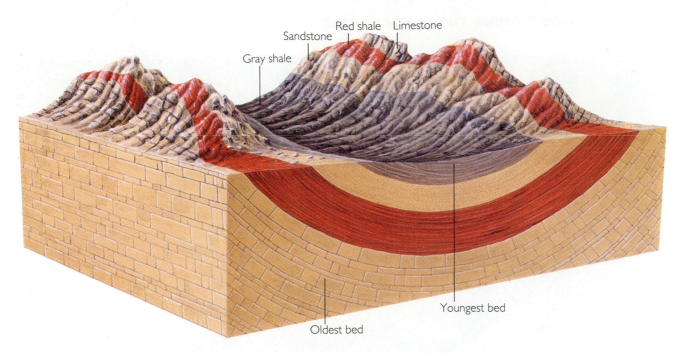

**FIGURE 10.18** Geologists typically work from available surface outcrops of rock formations to reconstruct subsurface structures. This diagram shows the surface expression of eroded remnants of a syncline and the characteristic core of younger rocks flanked on both sides by older rocks dipping toward the core.

no appreciable movement has occurred. A **fault** is a fracture with relative movement of the rocks on both sides of it, parallel to the fracture. Just like folds, joints and faults tell geologists something about the forces a region has experienced in the past.

## Joints

Joints, which can be caused by tectonic forces, are found in almost every outcrop. Like any other brittle material, brittle rocks break more easily at flaws or weak spots when they are subjected to pressure.

These flaws can be tiny cracks, fragments of other material, or even fossils. Regional forces—compressional, tensional, or shearing—that have long since vanished may leave their imprint in the form of a set of joints. Joints can also form when erosion has stripped away surface layers. The removal of those layers releases the confining pressure on underlying formations and causes the rocks to part at flaws. Joints can form in lava as a result of the contraction of the lava as it cools.

When a formation fractures at many places and develops joints, the joints are usually only the beginning of a series of changes that will significantly alter

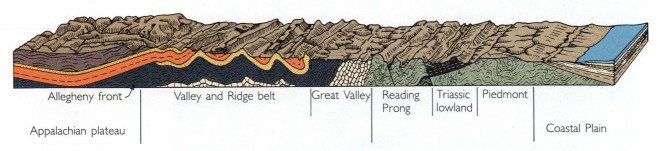

**FIGURE 10.19** The Valley and Ridge province of the Appalachian Mountains is the eroded remnant of a folded mountain belt. (After D. Johnson, *Stream Sculpture on the Atlantic Slope,* New York, Columbia University Press, 1931.)

**FIGURE 10.20** Eroded joints; Arches National Park, Utah. *Grant C. Willis/Utah Geological Survey.*

**FIGURE 10.21** The San Andreas fault runs from top to bottom in the middle of the photo. Note the offset of the stream as it crosses the fault, which is caused by the northward movement of the Pacific Plate on the left with respect to the North American Plate on the right. *Gudmundar E. Sigvaldason/Nordic Volcanological Institute.*

the formations. For example, joints provide channels through which water and air can reach deep into the formation and speed the weathering and weakening of the structure internally. If two or more sets of joints intersect, weathering may cause a formation to break into large columns or blocks (Figure 10.20).

## Faults

Unlike folds, which usually signify that compressive forces were at work, faults can be caused by all three types of forces: compressive, tensional, and shearing. These forces are particularly intense at plate boundaries. Some transform faults that occur where plates slide past each other, such as the San Andreas fault of California (Figure 10.21), show such large displacements that the relative movement of the two plates may amount to hundreds of kilometers. Faults are common features of mountain belts, which are asso-

ciated with plate collisions, and of rift valleys, where plates are being pulled apart. Crustal forces can also be strong within plates and cause faulting in rocks far from plate boundaries.

Geologists, like other scientists, develop a logical terminology to describe the features they observe. They define faults by the direction of relative movement, or slip, at the fracture. The surface along which the formation fractures and slips is the *fault plane* (Figure 10.22). Two terms defined earlier, *dip* and *strike,* are also used here to describe the orientation of the fault plane. A *dip-slip fault* involves relative movement of the formations up or down the dip of the fault plane. A *strike-slip fault* is one in which the movement is horizontal, parallel to the strike of the fault plane. (A transform is a strike-slip fault that forms a plate boundary.) An oblique movement along the strike and simultaneously up or down the dip is described as an *oblique-slip fault*. Dip-slip faults are associated with compression or tension, and

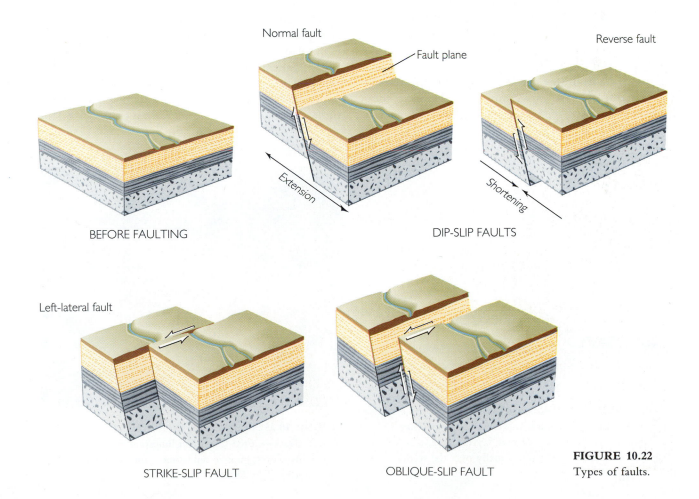

BEFORE FAULTING

Normal fault

Fault plane

Extension

DIP-SLIP FAULTS

Reverse fault

Shortening

Left-lateral fault

STRIKE-SLIP FAULT

OBLIQUE-SLIP FAULT

**FIGURE 10.22**
Types of faults.

**FIGURE 10.23** View looking north along the Hurricane fault at the Arizona-Utah boundary. This normal fault follows the base of the 300-m cliff. The rocks on the left of the fault have downdropped 1400 m relative to those on the right. *John Shelton.*

strike-slip faults indicate that shearing forces were at work. An oblique-slip fault suggests a combination of the two.

Faults need to be characterized further, because the movement can be up or down or right or left, as Figure 10.22 indicates. In a *normal fault,* the rocks above the fault plane move down in relation to the rocks above, causing an extension. Figure 10.2 shows

a small-scale normal fault in an outcrop; Figure 10.23 is an aerial view of a large-scale normal fault. A *reverse fault,* then, is one in which the rocks above the fault plane move upward in relation to the rocks below, causing a shortening. Movement of this sort results from compression. If, as we face a strike-slip fault, the block on the other side is displaced to the right, the fault is a *right-lateral fault;* if the block on the

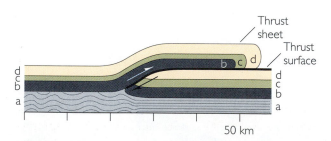

**FIGURE 10.24** A large-scale overthrust sheet of a kind found in California and southern Nevada. Compressive forces have detached a sheet of rock layers—d, c, b— thrusting it a great distance horizontally over the section d, c, b, a.

**FIGURE 10.25** The Keystone thrust fault of southern Nevada. Dark-colored Cambrian limestone has been thrust over light-colored Jurassic sandstone, younger by some 350 million years. *John Shelton.*

other side of the fault is displaced to the left, it is a *left-lateral fault*. These movements result from shearing forces.

Finally, a reverse fault at which the dip of the fault plane is small, so that the overlying block is pushed mainly horizontally, is a *thrust fault* (Figure 10.24). Thrust faults at which one block has been pushed a great distance horizontally over the other are often found in intensely deformed mountain belts. These *overthrusts* are expressions of compressive forces (Figure 10.25). In effect the crust accommodates these forces by shortening, in this case by breaking, and one sheet overrides the other. Often the shortening may cover many tens of kilometers and involve multiple thrust faults. Think of a closed venetian blind. When the blind is raised so that the slats override one another, the blind is effectively shortened.

Tensional forces, such as those that split a plate apart, leave normal faults behind as evidence of their action. This splitting can result in the development of a **rift valley**—a depression where one block looks as though it had dropped between two flanking blocks that have been pulled apart. The tensional forces create a long, narrow trough bounded by one or more parallel normal faults (Figure 10.26). The East African rift valleys, the rifts of mid-ocean ridges, the Rhine River valley, and the Red Sea Rift (Figure 10.27) are famous rift valleys.

Geologists recognize faults in the field in several ways. If the relative movement is large, as in many transform faults, the formations separated by the fault probably differ in lithology and age, and often the offset bed is so far away that it cannot be found.

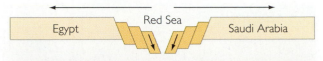

**FIGURE 10.27** The African Plate, on which Egypt rides, and the Arabian Plate, bearing Saudi Arabia, are drifting apart. The tensional forces have created a rift valley, filled by the Red Sea. *NASA.* The diagram shows parallel normal faults bounding the rift valley in the crust beneath the sea.

When movements are smaller, offset features can be observed and measured, as in Figure 10.2. In establishing the time of faulting, geologists use a simple rule: A fault must be younger than the youngest rocks it cuts (the rocks had to be there before they could break) and older than the oldest undisrupted formation that covers it.

## UNRAVELING GEOLOGIC HISTORY

Usually the geologic history of a region is a succession of episodes of deformation and other geological processes. Let us take what appears to be a complicated example and see how some of the concepts introduced in this chapter lead to a simple interpretation. The cross sections in Figure 10.28 represent a few tens of kilometers in the Basin and Range provinces of Nevada and Utah. Figure 10.28e shows the end result of a succession of events: First came the

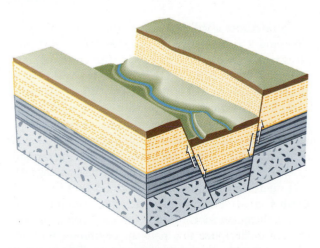

**FIGURE 10.26** A downfaulted block, or rift valley.

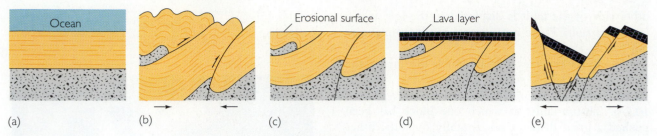

**FIGURE 10.28** Stages in the development of the Basin and Range provinces of Nevada and Utah. (a) Horizontally stratified sediments are deposited on the seafloor. (b) Horizontal compression (indicated by arrows) causes folding and faulting. (c) Uplift above sea level is evidenced by the development of a new horizontal erosional surface. (d) Volcanic eruptions deposit sheets of lava over the eroded surface. (e) Stretching forces (indicated by arrows) cause new normal faults, breaking up the earlier features into blocks. A geologist sees only the last stage and attempts to reconstruct from the structural features all of the earlier stages in the history of a region. (After P. B. King, *The Evolution of North America,* Princeton, Princeton University Press, 1977.)

deposition of horizontal layers of sediment, which then became tilted and folded by the horizontal forces of compression. They were then uplifted above sea level. There erosion gave them a new horizontal surface, which was covered by lava when forces deep in the Earth's interior caused a volcanic eruption. In the final stage, horizontal stretching resulted in normal faulting, which broke the crust into blocks. The geologist sees only the last stage (Figure 10.28e) but visualizes the entire sequence. Once the sedimentary beds are identified, the geologist starts with the knowledge that the beds must originally have been horizontal and undeformed at the bottom of an ancient ocean (Figure 10.28a). The succeeding events can then be reconstructed.

# EXPRESSION OF DEFORMATION IN THE LANDFORM

Present-day surface relief, such as we find in the Alps, the Rocky Mountains, the Pacific Coast Ranges, and the Himalayas, can be traced in large part to deformation that occurred over the past few tens of millions of years. These younger mountain systems still contain much of the information the geologist needs to piece together the history of deformation. Deformation that occurred hundreds of mil-

lions of years ago, however, no longer shows as prominent mountains. Erosion has left behind only the remnants of folds and faults in the old basement rocks of the continental interior (see Figure 10.15).

As we have seen, deformation—in the form of mountain belts with their structures of folds and faults, rift valleys, and strike-slip faults—leaves its unmistakable mark on the landscape. These topographic expressions are often guides to the deformation structures that shaped them. Even such relatively small-scale features as the shapes of hills and valleys and the courses of streams may be controlled by the complex interaction between underlying structures and erosion (Figure 10.29; see also Figure 10.21).

Sometimes a valley forms in the trough of a syncline and a ridge forms at the crest of an anticline. However, we should not expect the crests of anticlines always to form ridges and the troughs of synclines always to become valleys. When stratified rocks are deformed, an important factor in the shaping of landforms is the amount of resistance the individual beds offer to weathering and erosion, as well as whether the layers are tilted, folded, or faulted.

We have seen that there is a pattern in the way rocks deform that relates to the forces ever present in Earth's crust. Plate movements play an important role in the generation of these forces. Geologists have learned to decipher this pattern, beginning with the formation of rocks and reconstructing their subsequent deformation and erosion.

**FIGURE 10.29** This scarp is a fresh surface feature that formed when a new reverse fault broke and caused a devastating earthquake in Armenia in 1988. *Armando Cisternas, Université Louis Pasteur.*

# SUMMARY

**What do laboratory experiments tell us about the way rocks deform when they are subjected to crustal forces?** Laboratory studies show that rocks vary in strength and in the way they respond to the forces to which they are subjected. Some are ductile, others brittle. These qualities depend on the kind of rock, the temperature, the surrounding pressure, the magnitude of the force, and the speed with which it is applied.

**What are some of the deformation structures that show up in rocks in the field?** Among the geologic structures in rock formations that result from deformation are folds, domes, basins, joints, and faults.

**What kinds of forces are involved in the formation of these structures?** Folds are usually formed by compressive forces, as at the boundaries where plates collide. Domes and basins are not fully understood. Some domes are formed by intrusion of magma deep in the crust. Basins can be formed when tensional forces stretch the crust or when a heated por-

tion of the crust cools and contracts. The weight of sediments deposited in a basin can contribute to its deepening. Joints are fractures in rock caused by regional stresses or by the cooling and contraction of the rocks. Normal faults can be caused by tensional or stretching forces, such as those that occur at boundaries where plates diverge. Reverse faults and thrust faults can be produced by compressive forces, such as those that occur at boundaries where plates converge. Shearing forces can produce strike-slip faults.

**How do geologists reconstruct the history of a region?** Geologists see the end results of a succession of events: deposition, deformation, erosion, volcanism, and so forth. They deduce the deformational history of a region by identifying and fixing the ages of the rock layers, recording the geometric orientation of the beds on maps, mapping folds and faults, and reconstructing cross sections of the subsurface consistent with the surface observations. They can ascertain the relative age of a deformation by finding a younger undeformed formation lying unconformably on an older deformed bed.

# KEY TERMS AND CONCEPTS

folding (p. 212)
faulting (p. 212)
strike (p. 214)
dip (p. 214)
compressive forces (p. 215)
tensional forces (p. 215)
shearing forces (p. 215)
brittle rocks (p. 216)

ductile rocks (p. 216)
basement rocks (p. 216)
anticline (p. 217)
syncline (p. 217)
limb (p. 218)
axial plane (p. 218)
axis (p. 218)
plunging fold (p. 218)
asymmetrical fold (p. 218)

overturned fold (p. 218)
plunging anticline (p. 218)
plunging syncline (p. 218)
dome (p. 218)
basin (p. 220)
fold belt (p. 220)
joint (p. 220)
fault (p. 221)
rift valley (p. 225)

# EXERCISES

1. Why do some rock layers fold and others break into faults when they are subjected to crustal forces?

2. What types of deformation structures would be expected at the three types of plate boundaries?

3. How would you identify a fault in the field? How would you tell whether it was a normal, reverse, or strike-slip fault?

4. If you found tilted beds in the field, how would you tell if they were part of an anticline or a syncline?

5. Draw a cross section of a rift valley and indicate by arrows the nature of the forces that produced it. Do the same for a thrust fault.

6. Although anticlines are upfolds and synclines are downfolds, we often find synclinal ridges and anticlinal valleys. Explain why.

7. What was the direction of the crustal forces that deformed the Appalachian block depicted in Figure 10.19?

8. Draw a geological cross section that tells the following story: A series of marine sediments are deposited and subsequently deformed into folds and thrust faults. These events are followed by erosion. Volcanic activity ensues, and lava flows over the eroded surface. A final stage of high-angle faulting breaks the crust into several upheaved and downdropped rocks.

# THOUGHT QUESTIONS

1. If you were asked to describe the geologic history of a region that had not yet been explored, how would you proceed?

2. Other than crustal forces, what might cause rocks to deform?

# SUGGESTED READINGS

Hills, E. S. 1972. *Elements of Structural Geology,* 2d ed. New York: Wiley.

Ramsay, J. F. 1967. *Folding and Fracturing of Rocks.* New York: McGraw-Hill.

Twiss, R. J., and E. Moores. 1992. *Structural Geology.* New York: W. H. Freeman.

## PART 2

# SURFACE PROCESSES

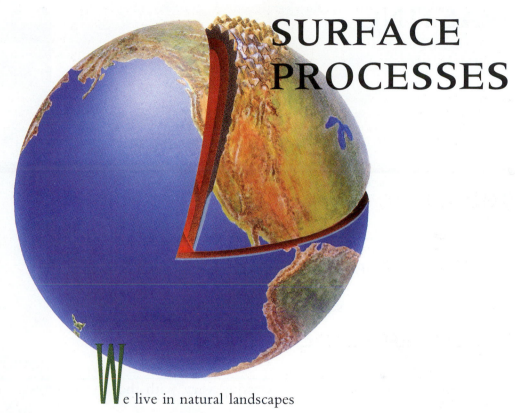

We live in natural landscapes shaped by rivers, glacial ice, the wind, and groundwater. Although humans may alter our environment by building cities, draining wetlands, and pumping carbon dioxide into the atmosphere, our existence depends on the basic geological processes that govern the dynamics of the land surface and on the vast bodies of water that cover much of the solid planet. The dynamics of Earth's surface are controlled by the Sun, whose radiant energy drives the atmosphere and oceans in a complex circulation pattern that ultimately produces our climate and transports water over the globe. Surface processes result from the interaction of the external solar heat engine with Earth's internal heat engine.

# MASS
# WASTING

Great quantities of weathered material are eroded from the land surface and begin their long journey to the ocean by falls, slides, slumps, and flows down hilly and mountainous slopes. Mass wasting includes all the processes by which masses of rock and soil move downhill, eventually to be carried away by other transporting agents. The mechanisms of mass wasting vary with the material, but the presence of water as a lubricant is usually an important factor. Movements of rock and unconsolidated material are classified according to whether the material slides or flows and by the speed, which ranges from very slow, imperceptible displacements of soil to high-velocity mudflows racing down steep slopes. Although the vast bulk of mass wasting is natural, human activities may promote or trigger landslides in vulnerable areas.

## 11

Landslide following the undercutting of soft sedimentary rocks by a flash flood. Rancho Mirage, California, July 1979. *Joel Sternfeld, Pace/MacGill Gallery.*

The snout of the debris flow was twenty feet high, tapering behind. Debris flows sometimes ooze along, and sometimes move as fast as the fastest river rapids. The huge dark snout was moving nearly five hundred feet a minute and the rest of the flow behind was coming twice as fast, making roll waves as it piled forward against itself—this great slug, as geologists would describe it, this discrete slug, this heaving violence of wet cement. Already included in the debris were propane tanks, outbuildings, picnic tables, canyon live oaks, alders, sycamores, cottonwoods, a Lincoln Continental, an Oldsmobile, and countless boulders five feet thick.

John McPhee, *The Control of Nature,*
New York, Farrar Straus Giroux, 1989, p. 219.

John McPhee was writing about the rain-soaked, stormy night of February 9, 1978, in the San Gabriel Mountains, which tower over the northern part of the Los Angeles area. It was then that the disastrous debris flow ran down Shields Canyon, burying people, houses, and everything else in its path. As terrible as this event was for many people, it was neither the biggest nor the most damaging debris flow to inundate the area, and most people had forgotten it just a few years later.

The debris flow that sped down Shields Canyon was just one of many kinds of downhill movement of masses of soil, rock, mud, or other unconsolidated (loose and uncemented) materials that we call **mass movements.** The masses move by themselves, pulled down by the force of gravity, not by the action of any other erosional agent, such as running water, the wind, or glacial ice. Movement takes place by various combinations of falling, sliding, and flowing. Mass movements range from small, almost imperceptible displacements of soil down a gentle hillside to huge landslides that dump tons of earth and rock on valley floors below steep mountain slopes.

Every year mass movements take their toll of lives and property around the world. In 1985 more than 20,000 people lost their lives in a giant mudflow of unconsolidated volcanic ash in the Andes Mountains of Colombia. Because mass movements are responsible for so much destruction, we want to be able to predict them. We certainly want to avoid provoking them by unwise interference with natural processes. We cannot prevent most natural mass movements, but we can control construction and land development to minimize our losses.

We use the term **mass wasting** for the general process by which mass movements occur and erode the land surface. Mass wasting is one of the consequences of weathering and rock fragmentation. It is an important part of the general erosion of the land, especially in hilly and mountainous regions. Mass movements change the landscape by scarring mountainsides as great masses of material fall or slide away from the slopes. The material that moves ends up as tongues or wedges of debris on the valley floor, sometimes piling up and damming the stream running through the valley. The scars and debris deposits are clues to mass movements in the past; by reading those clues geologists can predict whether new movements are likely to occur in the future and can issue timely warnings.

## WHAT MAKES MASSES MOVE?

Field observations of many kinds of mass movements have led geologists to identify three primary factors that affect them: (1) the steepness and instability of slopes; (2) the nature of the slope materials; and (3) the amount of water in the material (Table 11.1). The steepness and instability of slopes contribute to the tendency of materials to fall, slide, or flow under various conditions. Slope materials may be solid masses of bedrock, regolith (the surface debris, including soil, formed by weathering), or sediment. These slope materials may be **unconsolidated**—that is, loose and uncemented—or **consolidated,** mean-

**TABLE 11.1**

Factors That Influence Mass Movements

| MATERIAL | SLOPE | MOISTURE | CONSEQUENCES |
|---|---|---|---|
| Loose sand or sandy silt | Angle of repose | Dry | Stable unless oversteepened by excavation |
| | | Wet | May flow if sand is water-saturated |
| Unconsolidated mixture of sand, silt, and soil | Moderate | Dry | Stable unless oversteepened |
| | | Wet | Prone to slump, slide, or flow |
| | Steep | Dry | Temporarily stable |
| | | Wet | Very likely to slide or flow |
| Rock, jointed and deformed | Moderate to steep | Dry or wet | Rockfall or slide possible |
| Rock, massive | Moderate | Dry or wet | Stable |
| | Steep | Dry or wet | Rockfall or slide possible |

ing compacted or cemented. The amount of water in the materials depends on how porous they are and on how much rain or other water they have been exposed to. All of these factors operate in nature, but slope stability and water content are most strongly influenced by human activity, such as excavation for building and highway construction.

## Unconsolidated Materials

The role that the steepness and instability of slopes play in mass movements can be seen in the behavior of loose sand. Children's sandboxes have made nearly everyone familiar with the characteristic slope of a pile of dry sand: the angle between the slope of any particular pile of sand and the horizontal is the same whether the pile is only a few centimeters high or 3 m high. For most sands the angle is about 35°. If some sand is scooped from the base of the pile very slowly and carefully, the angle of slope will steepen a little and hold temporarily. But if someone jumps on the ground near it, the sand cascades down the side of the pile, which again assumes its original angle with the horizontal. Although the cascading sand appears to move as a unit, much of the movement is by indi-

vidual grains over and around one another. We can see this behavior of sand on the steepest slopes of sand dunes. The original and resumed angle of the sand pile is the **angle of repose,** the maximum angle at which a slope of loose material is stable (Figure 11.1). A slope that is steeper than the angle of repose is unstable and will tend to collapse to the stable angle.

The angle of repose varies significantly with a number of factors, including the size and shape of the particles and the amount of moisture between them. Larger, flatter, and more angular pieces of loose material remain stable on steeper slopes. Damp sand has a higher angle of repose than dry sand because the small amount of moisture between the grains tends to hold them together by surface tension. **Surface tension** is the attractive force between molecules at a surface (Figure 11.2a) that makes water drops spherical and allows a thin razor blade or paper clip to float on a smooth water surface (Figure 11.2b). Too much water has the opposite effect. If the sand is saturated—that is, if all the pore space between grains is occupied by water—the sand will run like a fluid and collapse to a flat pancake shape. The water separates the grains and allows them to move freely over one another (Figure 11.2c).

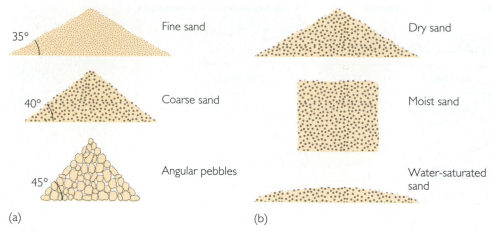

(a)                                    (b)

**FIGURE 11.1** (a) The angle of repose of a mound of particles increases as the size of the particles increases and their shapes become more angular. (b) The angle of repose depends on the amount of moisture between the particles. Moist sand sticks together, so it can have vertical sides, whereas water-saturated sand flows to a thin lens.

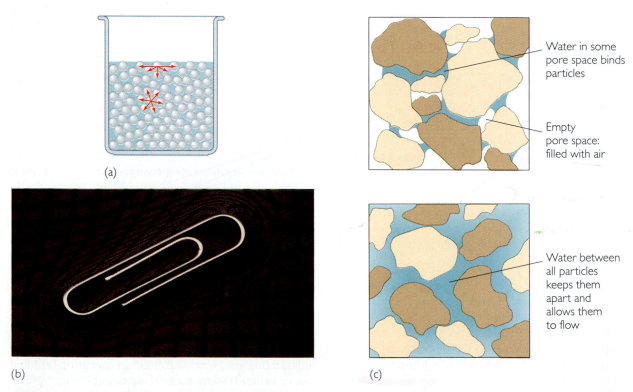

**FIGURE 11.2** (a) Molecules in the interior of a liquid are attracted in all directions, but molecules at the surface have a net inward attraction that results in surface tension. (b) Surface tension causes a paper clip to float on water; the water surface behaves as if it were an elastic membrane, stopping the clip from falling into the liquid. (c) In an unsaturated soil, the surface tension of a thin film of water adhering to particles binds them so that they resist movement. When the soil is saturated, water between all the particles keeps them apart and allows them to flow.

## Consolidated Materials

Consolidated dry materials, such as compacted and cemented sediments and vegetated soils, do not have the simple angles of repose characteristic of loose materials. The slopes of consolidated materials may be steeper and irregular. The particles of consolidated sediments are bound together by cohesive forces associated with tightly packed particles (the particles stick together), and some are weakly cemented by minerals precipitated in pore spaces. The intricately branching system of plant roots binds the particles of vegetated soil together. The resulting resistance to movement is sometimes called "internal friction" because it is like the friction that resists the movement of any piece of matter against another. In a material with high internal friction, the particles are not so free to move as are loose particles such as sand. When these materials move, they tend to do so as a unit.

Mass movement of consolidated materials usually can be traced to the effects of moisture, often in combination with other factors, such as loss of vegetation or oversteepening of the slope. When the ground becomes saturated with water, the material is lubricated, the internal friction is lowered, and the particles or larger aggregates can move past one another more easily. Water may seep into the bedding planes of muddy or sandy sediments, for example, and promote the slippage of beds past one another. When large amounts of water are absorbed, the pressure of the water in the pores of the material may be great enough to separate the grains and distend the mass. Then the material may start to flow like a fluid.

When vegetation is stripped from the soil by burning or deforestation, the soil is no longer bound by root systems and thus becomes more susceptible to mass movement (Figure 11.3). Oversteepening of a slope of consolidated material, such as by a stream that undercuts part of a valley wall, acts to make the slope unstable in the same way as does steepening the angle of a sand pile. Sooner or later the unstable valley wall will move down to assume a more stable angle.

## Rock Slopes

Rock slopes range from relatively gentle slopes of easily weathered shales and volcanic ash beds to vertical cliffs of hard rocks such as granite. The stability of these slopes depends on the weathering and fragmentation of the rock. Shales, for example, tend to weather and fragment into small pieces that form a

**FIGURE 11.3** Stripping of vegetation and root systems in Yellowstone National Park after a fire caused a weakening of the soil that made it susceptible to erosion and mass movement. *Grant Meyer.*

thin layer of loose rubble covering the bedrock. The slope angle of the bedrock is similar to the angle of repose of loose coarse sand. As weathered rubble gradually builds up to an unstable slope, eventually some of the loose material will slide down.

Hard, cemented sandstones, in contrast, which resist erosion and break into large blocks, may show steep, bare bedrock slopes above and less steep slopes covered with broken rock below. The bedrock cliffs are fairly stable, except for the occasional mass of rock that falls and rolls down to the rock-covered slope below. That slope has a steep angle of repose but otherwise shows the same property of building to an unstable slope and then sliding down. Where such sandstones are interbedded with shale, slopes may be stepped. As shale slides from under the sandstone beds, the harder beds are undercut, become less stable, and eventually fall as large blocks. If much water seeps into such an interbedded valley wall, the shale may be lubricated enough to enable movement of whole masses of shale and interbedded sandstone.

The structure of the beds influences their stability, especially when the dip of the beds parallels the angle of the slope. In this position, bedding planes that are zones of potential weakness because of differences in lithology or ability to absorb water may become unstable, allowing masses of rock to slide along the weak bedding planes.

## Triggering Mass Movements

When the right combination of materials, moisture, and steepened slope angle makes a hillside unstable, a slide or flow is inevitable. All that is needed is a trigger. Sometimes a slide or debris flow, like the one in the San Gabriel Mountains described by John McPhee, is provoked by a heavy rainstorm. Many slides are set off by vibrations, such as those of earthquakes. Others may not be precipitated by a single detectable event but just by gradual steepening until the slope suddenly gives (see Box 11.1).

Most of the damage in the great Alaskan earthquake of March 27, 1964, was caused by the slides it triggered in the city of Anchorage. Mass movements of rock, earth, and snow wreaked havoc in residential areas, and there were major submarine slides along lakeshores and the seacoast. Huge landslides took place along flat plains below bluffs 30 to 35 m high along the coast. The bluffs were composed of interbedded clays and silts. During the earthquake the ground shook so hard that unstable, water-saturated sandy layers in the clay were transformed into fluid slurries, a process called *liquefaction*. Enormous blocks of clay and silt were shaken down from the bluffs and slid along the flat ground with the liquefied sediments, leaving a completely disrupted terrain of jumbled blocks and broken buildings (Figure 11.4). Houses and roads were carried along by the slides and destroyed. The whole process took only five minutes, beginning about two minutes after the first shock of the earthquake. At one locality three people were killed and 75 homes destroyed.

Studies of the instability of those slopes and the likelihood of earthquakes had indicated that this area was a prime candidate for landslides. A geological report issued more than a decade earlier had warned of the hazards of development, but the great scenic beauty of the area overwhelmed people's judgment.

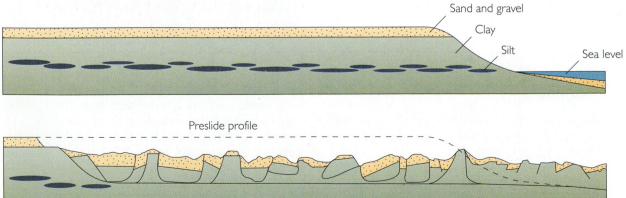

**FIGURE 11.4** *Left:* Landslide at Turnagain Heights, Alaska, triggered by the earthquake of 1964. *Steve McCutcheon/Alaska Pictorial Service. Below:* Cross sections of the bluffs at Anchorage, Alaska, after the earthquake.

## BOX 11.1  PREVENTING LANDSLIDES

Many natural mass movements are so large or inevitable that we must learn to live with them. But some smaller ones and others provoked by human activity can be prevented or their effects minimized. We need to take three main steps to prevent or decrease loss of life and damage to property. First, we should avoid construction in areas that are prone to mass movements. Second, where slopes are stable naturally, we should build in a way that does not make them unstable. Third, because saturation is critical, we must engineer water drainage so that slope materials will not become waterlogged and likely to slide or flow.

Avoidance of unwise building development requires good zoning regulations based on adequate geological assessment of the terrain. A geological survey of an area will quickly reveal the conditions that make construction unsafe, such as poorly vegetated slopes covered with unconsolidated material that takes up much water. It would take such great effort and expense to engineer safe structures in some areas where slopes and materials are unstable that they would be bet-

ter off undeveloped. In other areas, no amount of engineering will counter nature's tendency for mass movements.

Areas that are relatively stable should not be made unstable by stripping the vegetation that binds the soil or by artificially oversteepening the slopes. Avoidance of these errors requires architectural and landscape design that takes the natural situation into account. If vegetation must be eliminated and slopes steepened during construction, the area should be properly graded with a low slope and replanted as soon as possible.

Preventing a buildup of water in poorly drained soil or other unconsolidated material is essential. Slopes that are otherwise stable may creep or slump after heavy rains if retaining walls or the construction itself prevents drainage of the water. In areas where heavy rains tend to continue over long periods, storm drainage of potentially unstable slopes is vital. A frequent culprit when soil becomes waterlogged is poorly designed drainage from septic tanks in a hillside home development. Good sewer systems are part of slide prevention.

Weathered, rubbly rock and soil

Thin concrete retaining wall

Impermeable bedrock

Clay

(a) After construction activities

(b) Slide after prolonged rain

To build a house on some slopes is to court disaster. This slope is unstable, since it parallels the dip of the underlying beds and rests on a clay layer that would act as a lubricant if it became waterlogged. The slope in back of the house has been oversteepened, and the concrete retaining wall is too thin to hold it.

Many other equally hazardous slopes in earthquake-prone regions are being developed by people who willfully or mistakenly deny the high potential for dangerous mass movements.

# CLASSIFICATION OF MASS MOVEMENTS

Most people call any mass movement a landslide. But there are many kinds of mass movements, each with its own characteristics. We will use the term *landslide* only to refer to mass movements in general. Geologists classify mass movements in accordance with several characteristics: (1) the nature of the material (for example, whether it is rock or unconsolidated debris); (2) the speed of the movement (from a few centimeters per year to many kilometers per hour); and (3) the nature of the movement: whether it is *sliding* (the bulk of the material moves more or less as a unit) or *flowing* (the material moves as if it were a fluid). Table 11.2 summarizes the various kinds of mass movements geologists distinguish by means of these criteria.

Some movements have characteristics that are intermediate between sliding and flowing; most of the mass may be moving by sliding, for example, but parts of it along the base may be moving as a fluid. In this range of intermediate motions, the mass movements are named for the dominant mechanism. Some of the movements in this classification overlap because of the range of transitional intermediate motions. It is not always easy to tell the exact mechanism of a movement, since the nature of the movement must be reconstructed from the debris deposited after the event is over. Geologists are not often present and prepared to observe and measure a mass movement as it takes place, and no one is about to experiment with large masses.

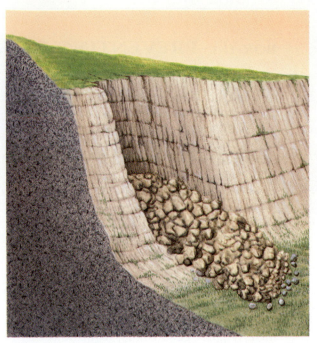

Rockfall

**FIGURE 11.5** Rock mass movements.

**TABLE 11.2**

Classification of Mass Movements

| DOMINANT MATERIAL | NATURE OF MOTION | VELOCITY | | |
| | | SLOW (1 cm/year or less) | MODERATE (1 km/hr or more) | FAST (5 km/hr or more) |
|---|---|---|---|---|
| Rock | Flow | | | Rock avalanche |
| | Slide or fall | | Rockslide | Rockfall |
| Unconsolidated | Flow | Creep | Earthflow Debris flow | Mudflow Debris avalanche |
| | Slide or fall | | Slump | Debris slide |

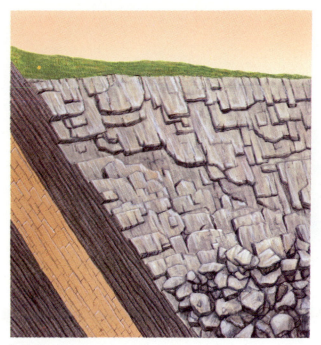

Rockslide

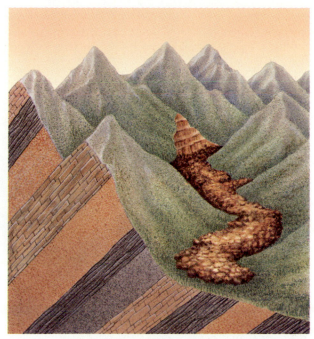

Rock avalanche

## Rock Mass Movements

Rock movements encompass several kinds of rapid downhill movements of small blocks or larger masses of bedrock (Figure 11.5). During a **rockfall,** individual blocks plummet in free fall from a cliff or steep mountainside. Sometimes large slabs fall and are broken up into blocks when they hit the ground. These angular slabs are broken from the outcrop by chemical and physical weathering and erosion. Weathering weakens the bedrock along joints until the slightest pressure, often caused by the expansion of water as it freezes in a crack, is enough to dislodge it.

People are rarely around to see rockfalls, but the evidence for their origin is clear. The lithologies of blocks in a rocky accumulation at the foot of a steep bedrock cliff can be matched with the lithologies of outcrops on the cliff. The accumulations of blocks are called **talus.** Individual rocks fall suddenly but very infrequently. Thus talus accumulates slowly, building up blocky slopes along the base of a cliff over long periods of time.

In many places rocks do not fall freely but slide down a slope. These are **rockslides,** the fast movements of large masses of bedrock sliding more or less as a unit, commonly along downward-sloping bedding or joint planes (Figure 11.6). **Rock avalanches,** like the more common snow avalanches, are flows

**FIGURE 11.6** Rockslide along the Icefields Parkway in the Canadian Rockies. Massive rock, weakened by weathering, collapsed and slid down the steep slope. *Ben Gadd.*

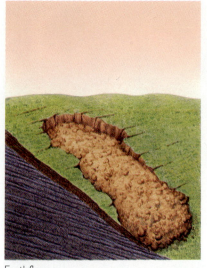

Earthflow

Debris flow

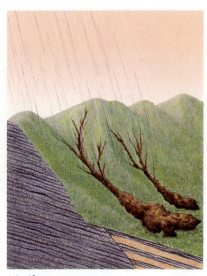

Mudflow

Creep

**FIGURE 11.7** Unconsolidated mass movements.

rather than slides. They are composed of large masses of rocky materials that have broken up into smaller pieces by falling and sliding and then flow farther downhill at high velocity, many tens of kilometers per hour.

Most rock mass movements occur in higher mountainous regions; they are rare in lower hilly areas. These movements tend to happen where weathering and fragmentation have attacked rocks already predisposed to breakage by structural deformation or relatively weak bedding or metamorphic cleavage planes. In many of these regions, extensive talus accumulations have been built by infrequent but large-scale rockfalls and rockslides. The talus slopes themselves are eroded at their lower edges as the rocks there continue to weather. These rocks are bro-

ken down into smaller pieces and eventually are carried away by streams or mass movements of unconsolidated material.

## Unconsolidated Mass Movements

Most unconsolidated mass movements are slower than most rock movements. While some unconsolidated materials move as coherent units, many flow like very viscous fluids. (Recall that viscosity is a measure of a fluid's resistance to flow.) The slower speed results largely from the lower slope angles at which these materials move. Unconsolidated material, frequently called "debris," includes various mixtures of soil, broken-up bedrock, trees and shrubs,

Debris avalanche

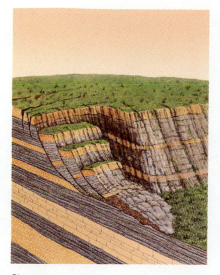

Slump

Debris slide

and materials of human construction, from fences to cars and houses. Figure 11.7 shows the full range of unconsolidated mass movements.

The slowest unconsolidated mass movement is **creep,** the downhill movement of soil or other debris at a rate of about 1 to 10 mm per year, depending on the kind of soil, the climate, the steepness of the slope, and the density of the vegetation. Accumulations of debris from creep along gentle slopes build up so slowly that the rate of movement is difficult to measure over short time periods. The movement is a very slow deformation of the regolith, the upper layers of the regolith moving down the slope faster than the lower layers. Such slow movements are why trees, telephone poles, and fences, which apparently are fixed firmly in the soil, tend to lean or to be moved slightly downslope (Figure 11.8). The heavy weight of masses of soil creeping downhill may break poorly supported retaining walls and crack the walls and foundations of buildings.

**Earthflows** and **debris flows** are fluid mass movements that travel slightly faster than creep, up

**FIGURE 11.8** A fence offset by creep in Marin County, California. *Travis Amos.*

**FIGURE 11.9** An earthquake in January 1989 in Tadzhikistan produced 15-m-high mudflows on slopes weakened by rain. *Vlastimir Shone/Gamma.*

to a few kilometers per hour, primarily because their resistance to flow is lower. Earthflows are movements of relatively fine-grained materials, such as soils, weathered shales, and clay. Debris flows contain much material coarser than sand and tend to move more rapidly.

**Mudflows** are flowing masses of mixed mud, soil, rock, and water (Figure 11.9). Because many mudflows have absorbed large quantities of water, they offer less resistance to flow and thus tend to move faster than earthflows or debris flows. Many move at several kilometers per hour. Most common in hilly and semiarid regions, mudflows start after infrequent, sometimes prolonged, rains. The previously dry, cracked mud keeps absorbing water as the rain continues. In the process, its physical properties change; the internal friction drops and the mass becomes much less resistant to movement. The slopes, which are stable when dry, become unstable and any disturbance triggers movement of waterlogged masses of mud. The mudflows travel down the upper valley slopes and merge on the valley floor. Where mudflows exit from confined valleys into broader, lower valley slopes and flats, they may splay out to cover large areas with wet debris. Mudflows can carry large boulders, trees, and even houses.

Fastest of all unconsolidated flows are **debris avalanches,** which usually occur in humid mountainous regions. Their speed is due to the combination of high water content and steep slopes. Water-saturated debris may move as fast as 70 km per hour, a speed comparable to that of water flowing down a moderately steep slope. These flows carry everything in their paths with them. In 1962 a debris avalanche in the high Peruvian Andes traveled almost 15 km in about 7 minutes, engulfing most parts of eight towns and killing 3500 people.

On May 31, 1970, in the same region, an earthquake toppled a large mass of glacial ice at the top of one of the highest mountains, Nevado Huascarán. Breaking up, the ice mixed with the debris of the high slopes and became an ice-debris avalanche. As it flowed down it picked up more debris and increased its speed to an almost unbelievable 200 to 435 km per hour. More than 50 million cubic meters of muddy debris roared down into the valleys, killing 17,000 people and wiping out scores of villages.

Almost exactly 20 years later, on May 30, 1990, an earthquake shook another mountainous area in northern Peru, again setting off mudflows and debris avalanches. It was just two days before a memorial ceremony scheduled to commemorate the earlier disasters. In a region such as this, where unstable slopes build up and earthquakes are frequent, there can be no doubt about the necessity of learning how to predict both earthquakes and the dangerous mass movements that follow.

Mudflows and debris avalanches are easily triggered on slopes of volcanic cinder cones when the accumulations of unconsolidated ash and other

**FIGURE 11.10** Ashfall at Spirit Lake, Mount St. Helens. Following the violent eruption of 1980, huge volumes of ash mixed with a catastrophic landslide and filled in the former lake. *USGS*.

erupted materials become saturated with rainwater. Some flows of volcanic debris are set off by events associated with the volcanic eruption itself, such as earthquakes, downpours of rain and ash from the eruption cloud, and sudden falls of volcanic debris. More than 55 such flows, many of them associated with eruptions, have occurred on the slopes of Mount Rainier, Washington, in the past 10,000 years, according to geologists of the U.S. Geological Survey. In 1980 a giant volcanic debris avalanche triggered by the eruption of Mount St. Helens, Washington, roared down the north slope of the mountain at a speed of about 200 km per hour and covered an area of over 600 km² below the mountain (Figure 11.10).

A **slump** is a slow slide of unconsolidated material that travels as a unit. In most places the slump slips along a basal surface that is concave upward, like a spoon. Faster than slumps are **debris slides,** in which the debris slides largely as one or more units along planes of weakness, either within or at the base of the debris. During the slide, some of the debris may behave like a chaotic, jumbled flow. Such a slide may turn to mostly flow as it moves rapidly downhill and most of the material mixes as if it were a fluid.

**Solifluction** is a type of movement that occurs only in cold regions when water in the surface layers of the soil alternately freezes and thaws (Figure 11.11). When the surface zones thaw, the soil there becomes waterlogged. The water cannot seep downward into the deeper layers of the soil, as it might in more temperate regions, because the lower layers are still frozen. Because the water cannot escape to the icy, impermeable deeper parts of the regolith or bedrock, water continues to accumulate and saturates the upper layers of soil so thoroughly that they ooze downhill, carrying broken rocks and other debris with them.

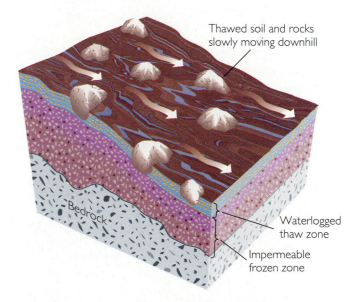

Thawed soil and rocks slowly moving downhill

Bedrock

Waterlogged thaw zone

Impermeable frozen zone

**FIGURE 11.11** Solifluction operates in cold regions when surface layers of soil thaw while deeper layers are still frozen.

**REDUCING LOSS FROM LANDSLIDES**

Landslides take a heavy toll in human lives and property damage. The San Francisco Bay area alone sustained over $25 million in land–slide costs in the winter of 1968–1969. The Bay region was hit hard again in the winter of 1992–1993, when prolonged and torrential rains saturated the ground, breaking a six-year drought. In some countries the losses exceed those of all other natural hazards combined. In March 1987, for example, the vibrations of an earthquake in the mountains of Ecuador after a month of heavy rains triggered a landslide that killed at least 1000 people and ruptured the trans-Ecuadoran oil pipeline—the nation's most important source of income—resulting in a loss of approximately $1.5 billion.

In the United States, 25 to 50 people die each year in landslides, and economic losses range from $1 billion to $2 billion. Here, as in many other countries, losses in life and property will continue to increase as newly constructed homes, roads, commercial buildings, and public works push into the hilly and unstable terrains that are most susceptible to landslides. The pressure for development stems from growing populations, the flight of city dwellers to the countryside, and ignorance or denial of the hazards and costs.

Japan's landslide mitigation program is one model for other countries to follow.

Much of the terrain of the Japanese Islands is a combination of hilly and mountainous topography, volcanic deposits, and deformed metamorphic rocks, a mixture that makes many slopes potential landslide hazards. The Japanese government dictates how land can be used and what types of building construction will be permitted. Land-use management controls are established to minimize the possibility of potentially dangerous mass movements. The government conducts research and puts into practice advanced engineering techniques, such as soil drainage networks, to avoid landslides. It forces adherence to strong building and grading regulations.

In addition, the Japanese government sponsors a strong program of public education and broadcasts warnings of landslide danger to local populations. Experience and education have prepared people to evacuate hazardous areas at the first warning, and they know where to go. The government's investment in these landslide control programs, which began in 1958, has been successful. As the accompanying chart indicates, 500 lives were lost and some 130,000 homes were destroyed or badly damaged by major landslide disasters in 1938. In 1976, the worst year for landslides since the control program started, fewer than 125 lives were lost and only 2000 homes were destroyed or damaged.

# CATASTROPHIC MASS MOVEMENTS

We can best understand how the steepness of a slope, the nature of the materials, and their water content interact to create mass movements by looking at both natural mass movements and those provoked by human activities. There are more than enough to study. Every year thousands of landslides occur throughout the world, many causing great loss of life and property (see Box 11.2). To find the causes of modern slides, geologists dovetail eyewitness reports with geological investigations of the source of the slide and the distribution and nature of the debris dropped in the valley below. Causes of prehistoric slides can be inferred from geological evidence alone where the debris is still present and can be analyzed for size, shape, and composition.

## Natural Landslides

In April 1983 a huge slide of mud and debris, estimated to be 4 million cubic meters, came down from a wall of Spanish Fork Canyon in Utah (Figure 11.12). The slide of unconsolidated material buried a major transcontinental railroad and blocked the major east-west highway through central Utah. It also dammed the river in the canyon and created a large lake that submerged the small town of Thistle,

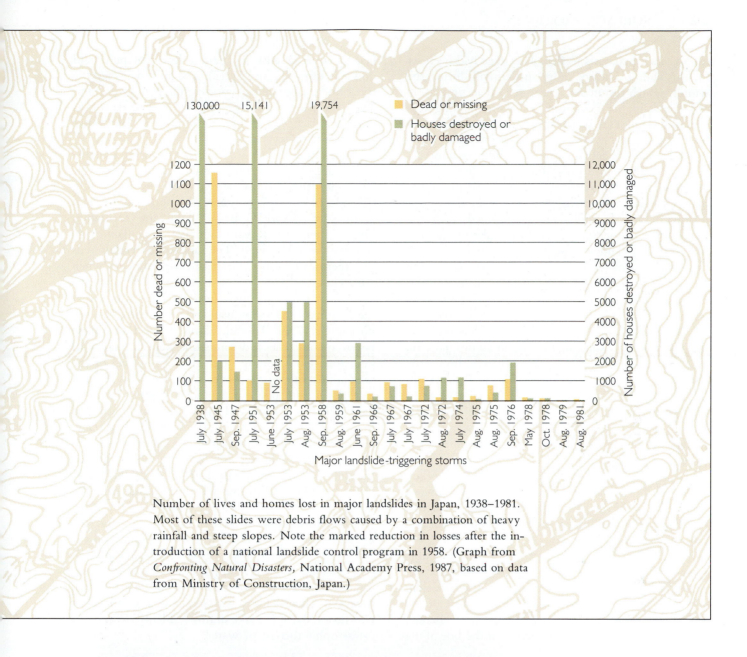

Number of lives and homes lost in major landslides in Japan, 1938–1981. Most of these slides were debris flows caused by a combination of heavy rainfall and steep slopes. Note the marked reduction in losses after the introduction of a national landslide control program in 1958. (Graph from *Confronting Natural Disasters,* National Academy Press, 1987, based on data from Ministry of Construction, Japan.)

**FIGURE 11.12** Mudslide at Spanish Fork Canyon, Thistle, Utah, April 1983. The slide followed melting of a record snow pack, which weakened the steep valley walls. *R. Morgan/Utah State Engineering Office.*

forcing its inhabitants to flee. The stage had been set for this slide by heavy snow and rainfall during the winter, which resulted in a near-record accumulation of snow at higher elevations in the surrounding Wasatch Mountains. A warm spring caused the snow to melt rapidly, and heavy rainstorms joined with meltwaters to set off debris flows and slides in unprecedented numbers in Spanish Fork and other canyons of the Wasatch Range. In this case, the nature of the materials—soil and rock that was structurally deformed, cracked, and weak—combined with the steep slopes of the Spanish Fork to make the canyon susceptible to slides. The rain and melting snow caused a mass of regolith and weathered rock on the wall of the canyon to become saturated, lubricating both the mass and the bedrock surface of the slide below it. Sooner or later the mass had to give.

If we look at a classic landslide of this century, the Gros Ventre slide in the mountainous Jackson Hole area of western Wyoming, we can see how various factors interact to produce slides in bedrock. In the spring of 1925, melting snow and heavy rains made water abundant in the Gros Ventre River valley, both in the stream and in groundwater, ultimately producing a great slide down one side of the valley. One local rancher barely escaped the slide on horseback. Warned by the roar of the slide, he looked up to see the whole side of the valley racing toward his ranch. From the gate to his property he watched the slide hurtle past him at about 80 km per hour and bury everything he owned.

It has been estimated that about 37 million cubic meters of rock and soil slid down one side of the valley, surged more than 30 m up the opposite side, and then fell back to the valley floor. Most of the slide was a confused mass of blocks of sandstone, shale, and soil, but one large section of the side of the valley, covered with soil and a forest of pine, slid down as a unit (Figure 11.13). The slide dammed the river, and a large lake grew over the next two years. Then the lake overflowed the dam. The dam was quickly breached, and the rapidly draining lake water flooded the valley below.

The causes of the Gros Ventre slide were all natural; in fact, the stratigraphy and structure of the valley made a slide almost inevitable (Figure 11.14). On the side of the valley where the slide occurred, a permeable, erosion-resistant sandstone formation dipped about 20° toward the river, paralleling the slope of the valley wall. The sandstone overlay beds of impermeable soft shale that became slippery when wet. The conditions became ideal for a slide when the river channel cut through most of the sandstone at the bottom of the valley wall and left it with virtually no support. Only friction along the bedding plane between the shale and sandstone kept the layer of sandstone from sliding. The river's removal of the sandstone's support was equivalent to the oversteepening of a sand pile when sand is scooped from its base. The heavy rains and snow meltwater saturated the sandstone and the surface of the underlying shale, creating a slippery surface along the bedding planes at the top of the shale. We do not know what triggered the slide, but at some point the force of gravity overcame friction and almost all of the sandstone slid down along the water-lubricated surface of the shale.

The formation of a dam on a river and the growth of a lake, as at both Thistle, Utah, and the Gros Ventre River valley, are common consequences of a landslide (Figure 11.15). Because most slide materials are permeable and weak, such a dam is soon breached when the lake water reaches a high level or overflows. Then the lake drains suddenly, releasing a catastrophic torrent of water.

**FIGURE 11.13** Landslide at Gros Ventre, Jackson Hole, Wyoming, 1925. This photo was taken after the slide had dammed the river and created a lake that drowned houses. *Jackson Hole Museum.*

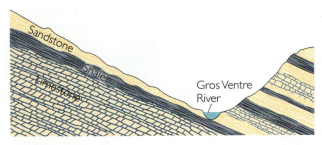

(a) Before slide

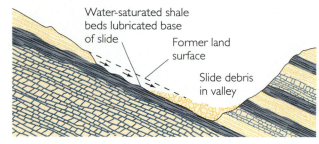

Water-saturated shale beds lubricated base of slide

Former land surface

Slide debris in valley

(b) After slide

**FIGURE 11.14** Cross sections of the 1925 Gros Ventre slide. (After W. C. Alden, "Landslide and Flood at Gros Ventre, Wyoming," *Transactions of the American Institute of Mining, Metallurgical, and Petroleum Engineers,* 1928, pp. 345–361.)

## Slides Caused by Human Activities

Slides are often provoked when human activities change natural slopes. When a highway was being widened in Quebec in 1955, the side of a road cut was oversteepened by excavation at the base of the road-cut slope. The cut was made in soft formations of silt and clay that were susceptible to weakening by water. After a period of heavy rain, the steepened slope of the road cut became saturated with water and suddenly gave way in a debris flow that carried away buildings, roads, and people. Three deaths can be traced to this case of poor highway engineering.

Some geological settings are so susceptible to landslides that engineers may have to avoid construction projects entirely. A landslide in one such place ultimately killed 3000 people. The place was Vaiont, an Italian alpine valley, and the time was the night of October 9, 1963. A large reservoir in the valley was impounded by a concrete dam (the second highest in the world at 265 m) and bordered by steep walls of interbedded limestone and shale. A great debris slide of 240 million cubic meters, 2 km long, 1.6 km wide, and over 150 m thick, plunged into the deep water of the reservoir behind the dam (Figure 11.16). The debris filled the reservoir for a distance of 2 km upstream of the dam and created a giant spillover. The 3000 people died in the violent torrent that hurtled downstream as a flood wave 70 m high.

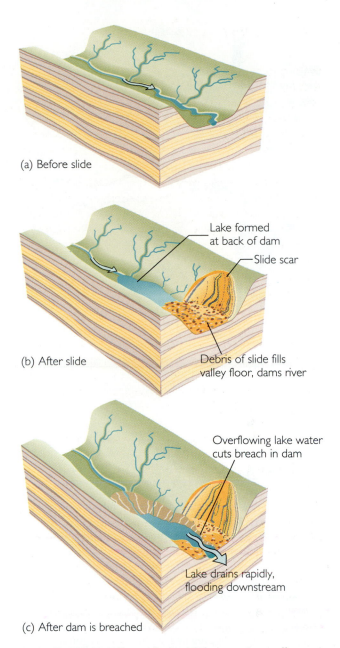

(a) Before slide

Lake formed at back of dam

Slide scar

(b) After slide

Debris of slide fills valley floor, dams river

Overflowing lake water cuts breach in dam

Lake drains rapidly, flooding downstream

(c) After dam is breached

**FIGURE 11.15** Effects of a landslide on a river valley.

Engineers had underestimated three warning signs at Vaiont: (1) the weakness of the cracked and deformed layers of limestone and shale that made up the steep walls of the reservoir; (2) the scar of an ancient slide on the valley walls above the reservoir; and (3) a forewarning of danger signaled by a small rockslide in 1960, just three years before. Although the landslide was natural and could not have been prevented, its consequences could have been much less severe. It might have caused only some damage to roads and buildings and perhaps a few injuries if the reservoir had been located in a geologically safer place, where the water was less likely to spill over its walls.

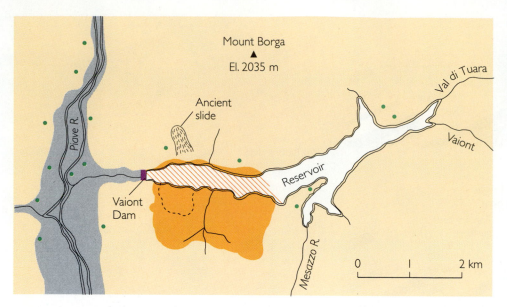

**Legend:**

▮ Limit of landslide, October 9, 1963

▨ Area filled by slide, 1963

▮ Limit of flood, downstream from slide

- - - - Limit of 1960 slide

• Cities and towns

# SUMMARY

**What are mass movements and what kinds of material are involved?** Mass movements are slides, flows, or falls of large masses of material down slopes in response to the pull of gravity. Mass wasting is the erosion and sculpturing of the land surface by mass movements. Mass movements may be imperceptibly slow or too fast to outrun. The masses consist of bedrock; consolidated material, including compacted sediment or regolith; or unconsolidated material, such as loose, easily disaggregated sediment or regolith. Rock movements include rockfalls, rockslides, and rock avalanches. Unconsolidated material moves by creep, slump, debris slide, debris avalanche, earthflow, mudflow, and debris flow.

**What are the factors responsible for mass movements, and how are such movements triggered?** The three factors that have the greatest bearing on the predisposition of material to move down a slope are (1) the steepness and instability of the slope, (2) the nature of the material, and (3) the water content of the material. Slopes become unstable when they become steeper than the angle of repose, the maximum slope angle that unconsolidated material will assume. Slopes in consolidated material may also become unstable when they are oversteepened or denuded of vegetation. Water absorbed by the material contributes to instability by lowering internal friction and thus resistance to flow or by lubricating planes of weakness in

the material. Mass movements can be triggered by earthquakes or sudden absorption of large quantities of water after a torrential rain. In many places slopes build to a point of instability at which the slightest vibration will set off a slide, flow, or fall.

**What factors are responsible for catastrophic mass movements, and how can such movements be prevented or minimized?** Analysis of both natural mass movements and those induced by human activity shows that one of the main contributory factors is the oversteepening of slopes, either by natural erosional processes or by human construction or excavation. Because water content has such a strong effect on stability, absorption of water from prolonged or torrential rains is often an important factor. The structural attitude of the beds, especially when bedding dips parallel to the slope, can promote mass movements. Volcanic eruptions may produce a tremendous fallout of ash and other materials that build to unstable slopes. Slides or flows of the volcanic materials are triggered by earthquakes accompanying eruptions. Loss of life and damage to property from catastrophic mass movements can be prevented or minimized by avoidance of steepening or undercutting of slopes. Careful engineering can keep water from making material more unstable. In some areas that are extremely prone to mass movements, development may have to be restricted.

# KEY TERMS AND CONCEPTS

mass movement (p. 232)
mass wasting (p. 232)
unconsolidated materials (p. 232)
consolidated materials (p. 232)
angle of repose (p. 233)
surface tension (p. 233)

rockfall (p. 239)
talus (p. 239)
rockslide (p. 239)
rock avalanche (p. 239)
creep (p. 241)
earthflow (p. 241)

debris flow (p. 241)
mudflow (p. 242)
debris avalanche (p. 242)
slump (p. 243)
debris slide (p. 243)
solifluction (p. 243)

# EXERCISES

1. What role do earthquakes play in the occurrence of landslides?

2. What kinds of mass movement advance so rapidly that a person could not outrun them?

3. How does absorption of water weaken unconsolidated material?

4. What is the difference between a slide and a flow?

5. What is the angle of repose and how does it vary with water content?

6. How does steepness of a slope affect mass wasting?

7. What is a debris flow and how does it differ from a debris avalanche?

# THOUGHT QUESTIONS

1. You are planning a highway through hills made of unlithified sands and gravels. What construction practices would you avoid to minimize mass movements?

2. Would a prolonged drought affect the potential for landslides?

3. What geological conditions might you want to investigate before you bought a house at the base of a steep hill of bedrock covered by a thick mantle of regolith?

4. While you are excavating the base of a slope prone to landsliding, taking care not to oversteepen it, heavy rain falls for three days. Why might you temporarily stop heavy construction trucks from driving near the slope?

5. Would you expect a talus slope of large blocks of granite to be prone to debris flows?

6. What evidence would you look for to indicate that a mountainous area had undergone a great many prehistoric landslides?

7. What factors would make a mountainous terrain in the rainy tropics have a greater or lesser potential for mass movements than a similar terrain in a desert?

8. What kind(s) of mass movements would you expect from a steep hillside with a thick layer of soil overlying unconsolidated sands and muds after a prolonged period of heavy rain?

# SUGGESTED READINGS

Bloom, Arthur L. 1991. *Geomorphology*, 2d ed. Englewood Cliffs, N.J.: Prentice Hall.

Costa, John E., and Victor R. Baker. 1981. *Surficial Geology*. New York: Wiley.

Eckel, E. B. (ed.). 1958. *Landslides and Engineering Practice*. Highway Research Board Special Report 29. National Academy of Sciences, National Research Council Publication 544.

Lundgren, Lawrence. 1986. *Environmental Geology*. Englewood Cliffs, N.J.: Prentice-Hall.

Porter, Stephen C., and G. Orombelli. 1981. Alpine rockfall hazards. *American Scientist* 69:67–75.

Voight, Barry (ed.). 1978. *Rockslides and Avalanches. 1. Natural Phenomena.* New York: Elsevier.

# THE HYDROLOGIC CYCLE AND GROUNDWATER

Water plays a central role in the dynamics of the

Earth's surface: it powers weathering and erosion,

transports material on land and in the oceans, and

sinks into surface materials to form large reservoirs of

groundwater. Civilization and life itself depend on pure,

pollution-free waters. Water travels over the globe in the

atmosphere and on and below Earth's surface. The

amounts of water cycled from one place to another are

strongly affected by climate and the geology of the land.

The transport of water by clouds, rivers, and ice is readily

observable, but we can deduce the movement of water

beneath the surface only indirectly. Groundwaters are

major sources of water in many parts of the world, and

they can be replenished, although it may take many years.

## 12

Lush vegetation flourishes in wetlands, such as those shown here in the Everglades, Florida. Wetlands store immense quantities of water and moderate large changes in water supply. *Planet Earth Pictures*.

The study of water, the science of **hydrology,** is an important part of geology. Water moving in rivers and frozen in glacial ice is a major agent of erosion, which helps shape the landscape of the continents. Water is essential to weathering, both as a solvent of minerals in rock and soil and as a transport agent that carries away the dissolved material. Water provides the lubrication for many landslides and other forms of mass movement. Hydrothermal ore deposits are the products of hot water circulating over igneous bodies or through mid-ocean ridges.

In addition, water is vital to all life on this planet. Humans cannot survive more than a few days without it, and even the hardiest desert plants and animals could not exist without some water. The water we require for modern civilization is far greater than we need for simple physical survival. Water is used in immense amounts for industry, agriculture, and such urban needs as sewage systems. The United States, one of the heaviest users of water in the world, has been steadily increasing its consumption since the nineteenth century. Between 1950 and 1985 alone we nearly tripled our water use, from 34 billion gallons a day to about 90 billion gallons a day. As we will see, this kind of growth cannot continue indefinitely.

Hydrology is becoming more important to all of us as the demand on limited water supplies increases. To protect those supplies while we satisfy our needs, we must understand not only where to find water but also how water supplies are renewed. With that knowledge, we can use and dispose of water in ways that do not endanger future supplies.

## FLOWS AND RESERVOIRS

We can see water moving from one place to another in the rivers on Earth's surface, and we can see water stored on the surface in lakes and oceans. What is harder to see are the massive amounts of water stored in the atmosphere and underground and the flows into and out of these storage places. As water evapo-

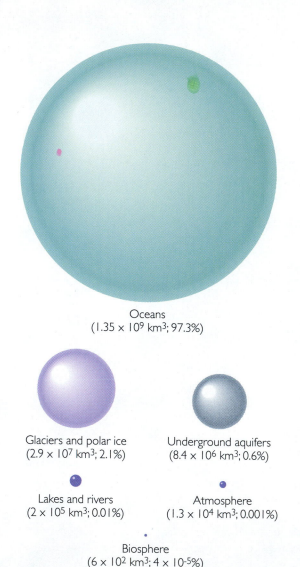

Oceans
($1.35 \times 10^9$ km$^3$; 97.3%)

Glaciers and polar ice
($2.9 \times 10^7$ km$^3$; 2.1%)

Underground aquifers
($8.4 \times 10^6$ km$^3$; 0.6%)

Lakes and rivers
($2 \times 10^5$ km$^3$; 0.01%)

Atmosphere
($1.3 \times 10^4$ km$^3$; 0.001%)

Biosphere
($6 \times 10^2$ km$^3$; $4 \times 10^{-5}$%)

**FIGURE 12.1** Distribution of water in the Earth. The amounts of water present in various natural reservoirs are shown as spheres of comparative volumes. The content of each reservoir is given in cubic kilometers and as a percentage of the whole. (After J. P. Peixoto and M. Ali Kettani, "The Control of the Water Cycle," *Scientific American,* April 1973, p. 46.)

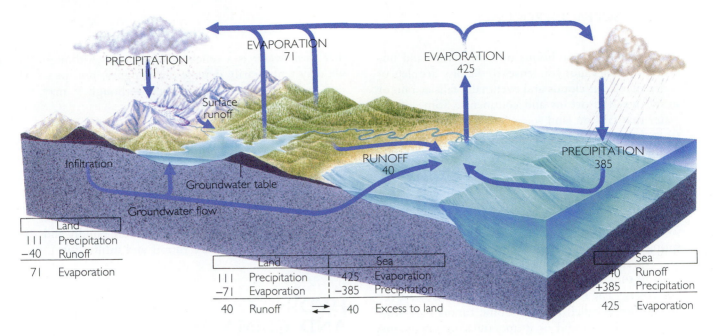

**FIGURE 12.2** The hydrologic cycle. The movement of water into the atmosphere by evaporation from the oceans and continents is matched by precipitation as rain and snow. Evaporation from the oceans is balanced by surface runoff from the continents and rainfall over the oceans. The water-flow budgets of the oceans, the land, and the atmosphere are calculated at the bottom of the diagram. All figures are given in thousands of cubic kilometers per year.

rates, it vanishes into the atmosphere as vapor. As rain sinks into the ground, it flows beneath the surface to become **groundwater.**

Each of the environments in which water is stored is referred to as a **reservoir.** Reservoirs have inflows by which they gain water, such as rain and river inflow, and outflows by which they lose water, such as evaporation and river outflow. If the inflow and outflow are equal, the reservoir stays the same size, even though water is constantly entering and leaving. Earth's main natural reservoirs and the distribution of water among them are shown in Figure 12.1. The oceans are by far the largest reservoirs. Although the total amount of water in rivers and lakes is relatively small, these reservoirs are important to human populations because they contain fresh water, ready for use. The amount of groundwater is over 40 times the amount in rivers and lakes, but much of it is unusable because it contains large amounts of dissolved material.

## How Much Water Is There?

The world's total water supply is enormous, about 1.36 billion cubic kilometers, distributed among the various reservoirs. If it covered the land area of the United States, it would submerge the 50 states under a layer about 145 km deep. This total is constant, even though the flows from one reservoir to another may vary from day to day, year to year, and century to century. Over these geologically short time intervals, there is no net gain or loss of water to or from Earth's interior, nor any significant loss of water from the atmosphere to outer space.

## The Hydrologic Cycle

Water at or beneath the Earth's surface moves or "cycles" among the main reservoirs: the oceans, the atmosphere, and the land. The **hydrologic cycle** is a simplified description of this endless circulation and the amounts of water moved (Figure 12.2).

Water may move from one reservoir to another by changing its state. Water is able to shift among the three states of matter—liquid (water), gas (water vapor), and solid (ice)—within the range of temperatures found at the Earth's surface, and these transformations power some of the main flows in the hydrologic cycle. The external heat engine of Earth, powered by the Sun, drives the hydrologic cycle, mainly by evaporating water from the oceans and transporting it as water vapor in the atmosphere.

Under the right conditions of temperature and humidity, water vapor condenses to the tiny droplets of water that form clouds and eventually falls as rain or snow over the oceans and continents. Some of the water that falls on land soaks into the ground by **infiltration,** a process by which water enters the small pore spaces between particles of soil or rock. Some groundwater evaporates through the soil surface. Another part is absorbed by plant roots, carried up to the leaves, and returned to the atmosphere by **transpiration,** the release of water vapor from plants. Other groundwaters may return to the surface via springs that empty into rivers and lakes.

The rainwater that does not infiltrate the ground runs off the surface, gradually collecting into streams and rivers. The sum of all rainwater that flows over the surface is called **runoff.** Some runoff may later seep into the ground or evaporate from rivers and lakes, but most of it flows into the oceans.

Snowfall may be converted to ice in glaciers, which return the water to the oceans by melting and runoff or to the atmosphere by **sublimation,** the transformation from a solid (ice) directly to a gas (water vapor). The largest part of the water that evaporates from the oceans returns to them as rain and snow, commonly grouped together as **precipitation.** The remainder falls over the land and either evaporates or returns to the ocean as runoff.

Figure 12.2 shows how the total flows among reservoirs balance one another. The land surface, for example, gains water by precipitation and loses the same amount of water by evaporation and runoff. The ocean gains water from runoff and precipitation and loses the same amount of water by evaporation. Thus each reservoir stays constant. As we noted earlier, more water evaporates from the oceans than falls on them as rain. This loss is balanced by the water returned as runoff from the continents. About one-third of the precipitation on land comes from evaporation from the oceans, and that one-third returns to the ocean as runoff.

## How Much Water Can We Use?

The global hydrologic cycle wields ultimate control over water supplies. Almost all of the water we use is fresh water—that is, water that is not salty, like the oceans. Rain, rivers, lakes, and some groundwaters are fresh, as is water melted from snow or ice on land. Since all of these waters are ultimately supplied by precipitation, the limit to the amount of natural fresh water that we can ever envision using is the amount steadily supplied to the continents by precipitation. At the same time, the steady supply means

that fresh water is a renewable resource. Although we may temporarily deplete our supplies, precipitation will eventually rebuild them—though it may take thousands of years to do so.

Our water supplies are strongly affected by the division of precipitation into runoff, evaporation, and infiltration. We cannot use the fraction of precipitation that evaporates. The fraction of precipitation that infiltrates to become groundwater can be used, but only if we can recover it by digging wells. Runoff is the most readily available fraction. The most desirable water supplies are those that are rapidly and continually replenished by runoff and infiltration.

# HYDROLOGY AND CLIMATE

For most practical purposes, the local hydrology—that is, the amount of water there is in a region and the way it flows from one reservoir to another—is more important than the global picture. The strongest influence on local hydrology is the climate, which includes both temperature and precipitation. Some of us live in warm areas where rain falls frequently throughout the year and water supplies are abundant, both at the surface and underground. Others live in warm arid or semiarid regions where it rarely rains and water is a precious resource. People who live in icy climates can usually rely on meltwaters from snow and ice. In some parts of the world, seasons of heavy rains, called monsoons, alternate with long dry seasons during which water supplies shrink, the ground dries out, and vegetation shrivels.

## Humidity and Rainfall

Many of these differences in climate are related to the temperature of the air and the amount of water vapor it contains. The **relative humidity** is the amount of water vapor in the air as a proportion of the maximum amount, or saturation value, that the air could hold at the same temperature. The amount of moisture in air at 15°C when the relative humidity is 50 percent, for example, is one-half the maximum amount that the air could hold at 15°C.

At any given relative humidity, warm air holds much more water vapor than cold air. When warm air at a certain relative humidity cools enough, it becomes supersaturated and some of the vapor condenses to form water droplets. When warm air in the atmosphere cools, the condensed water droplets form clouds. We can see clouds because they are formed of visible water droplets rather than invisible water vapor. When enough moisture has condensed in clouds and the droplets have grown so large that they are too heavy to stay suspended by air currents, they fall as rain.

Most of the world's rain falls in warm humid regions near the equator. In these tropical regions, both the surface waters of the oceans and the air are warm, so a great deal of the water evaporates and humidity is high. When water-laden winds from these oceanic regions rise over nearby continents, the air cools and becomes supersaturated. The result is heavy rainfall over the land, even at great distances from the coast.

Polar climates, in contrast, tend to be very dry. The polar oceans are cold and so is the air, so evaporation from the sea surface is minimized and the air can hold little moisture. Between the tropical and polar extremes are the temperate regions, where rainfall and temperatures are moderate.

Landscape can affect precipitation. Mountain ranges form "rain shadows," areas of low rainfall on the leeward (downwind) slopes. As moisture-filled air rises over high mountains, it cools and precipitates much rain on the windward slopes (Figure 12.3). By the time the air reaches the opposite, leeward slope, it has lost much of its moisture. As it drops to the lower slopes, it warms and relative humidity declines, further decreasing the moisture available for rain. A rain shadow can be seen on the eastern side of the Cascade Mountains of Oregon. Moist winds blowing over the Pacific hit the mountain's western slopes, where rainfall is heavy. The eastern slopes, on the other side of the range, are dry and barren.

## Droughts

All climates can include droughts, periods stretching from months to years during which precipitation is much lower than normal, but drier regions are especially vulnerable to decreases in their water supplies from this cause. During prolonged droughts, rivers, lacking replenishment from precipitation, shrink and dry up and reservoirs evaporate. The soil dries and cracks and vegetation dies. As populations grow, demands on reservoirs increase; a drought will deplete water supplies that may already be inadequate.

The severest droughts in the last few decades have affected lands along the southern border of the Sahara Desert, where tens of thousands of lives have been lost to famine caused by breakdowns in agriculture and cattle grazing. This long drought has expanded the desert and effectively destroyed farming and grazing in the area (as we shall see in Chapter 14).

Another prolonged but less severe drought affected most of California from 1987 until February 1993, when torrential rains broke the drought, at least temporarily. During the drought, groundwater and reservoirs dropped to their lowest levels in 15 years. Some control measures were instituted, but a move to reduce the extensive use of water supplies for irrigation encountered strong political resistance from farmers and the agricultural industry (see Box 12.1).

The midwestern United States and parts of Canada experienced a severe but short-lived drought in 1988, when surface water supplies shrank and the Mississippi River was lowered by many feet and closed to traffic. The next year precipitation over the region was back to normal.

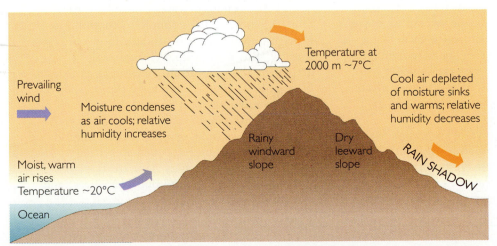

**FIGURE 12.3** A mountain range may produce a rain shadow on its leeward slope by forcing warm, moist air to rise with a prevailing wind from the ocean. The rising air cools, causing precipitation on the windward slope and leaving the leeward slope dry.

Prevailing wind

Moisture condenses as air cools; relative humidity increases

Temperature at 2000 m ~7°C

Cool air depleted of moisture sinks and warms; relative humidity decreases

Rainy windward slope

Dry leeward slope

RAIN SHADOW

Moist, warm air rises Temperature ~20°C

Ocean

# BOX 12.1   WATER, A PRECIOUS RESOURCE: WHO SHOULD GET IT?

For years most people in the United States took their water supply for granted. The cost was generally low, and local, regional, and national governments assumed responsibility for quality and safety and managed supplies by drilling wells and building dams and reservoirs when they were needed. Scientific analysis of available supplies and user needs, however, indicates that in many areas of the country water shortages will occur with increasing frequency and conflict will grow among the several sectors of consumers—residential, industrial, agricultural, and recreational. In recent years droughts and mandatory restrictions on water use, such as have occurred in California, were widely publicized and alerted the public that the nation faces major water problems. Public concern waxes and wanes, however, as periods of the hydrologic cycle of droughts and abundant rainfall come and go, and governments are not pursuing long-term solutions with the urgency they deserve. Here are some facts to ponder:

- A human can survive with about 2 liters of water per day. In the United States the per capita use for all purposes is about 6000 liters per day.
- Industry uses about 38 percent and agriculture about 43 percent of water withdrawals.
- Per capita domestic water use in the United States is two to four times more than in Western Europe, where users pay up to 350 percent more for their water.
- Although the western states receive one-fourth the rainfall, per capita water use (mostly for irrigation) is 10 times greater than that of the eastern states, and at much lower prices.
- The traditional ways of increasing water supply, such as building dams and reservoirs and drilling, have become extremely costly because most of the good (and therefore cheaper) sites have been

used. Also, the negative environmental impact and opponents such as fishing interests and Native American tribes have led to costly delays or rejection of proposals for new facilities.
- Almost all the fresh water used in the United States eventually returns to the hydrologic cycle. But it may return to a reservoir that is not well located for human use, and the quality is often decreased. Recycled irrigation water often has increased salinity and is loaded with pesticides. Polluted urban waste water ends up in the oceans.
- Global climatic change may lead to reduced rainfall in western states, exacerbating the problems there and making long-term solutions even more urgent.

Many analysts believe that the problem is not one of supply but of allocation. By shifting from inefficient users to efficient users who contribute more to the economy, water can be transferred to where it will do the most good. For example, 16 percent of California's water, enough to supply the needs of 30 million people, is used for subsidized irrigation of alfalfa, a feed for horses and cattle. With better allocation policies, more efficient users would have higher priority and would pay water suppliers the more realistic market prices. Most scarce resources are sold at market prices, and charging market prices for water is a natural extension.

There is enough water available in the United States for the country to continue to grow economically and produce food cheaply. But institutional changes will be needed to alter allocation and pricing policies. New policies should encourage the transfer of water to more efficient users, with pricing that promotes conservation and efficiency. A voice in the decision-making process should also be given to interests such as Native American tribes that may have treaty rights and wish to preserve traditional practices, and recreational users.

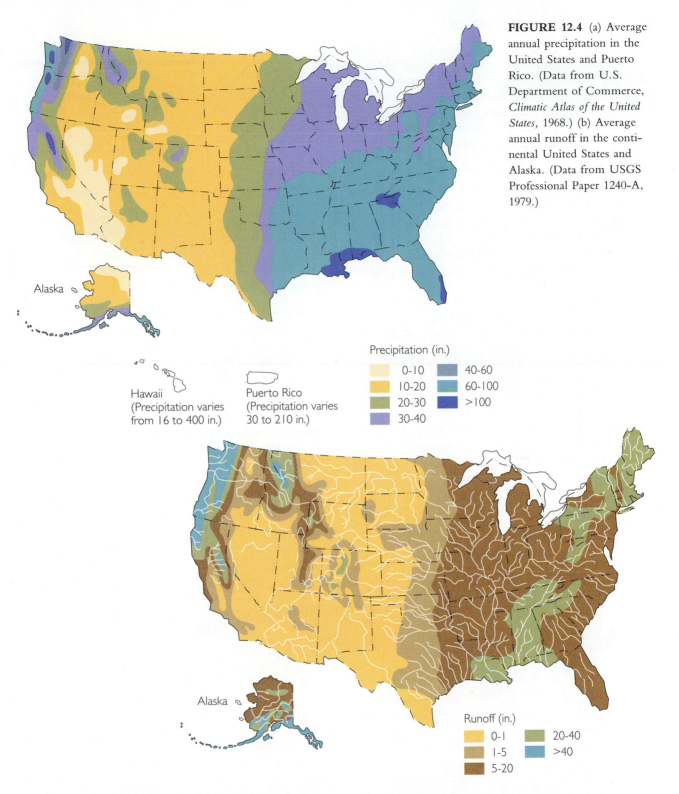

**FIGURE 12.4** (a) Average annual precipitation in the United States and Puerto Rico. (Data from U.S. Department of Commerce, *Climatic Atlas of the United States,* 1968.) (b) Average annual runoff in the continental United States and Alaska. (Data from USGS Professional Paper 1240-A, 1979.)

Alaska

Hawaii
(Precipitation varies
from 16 to 400 in.)

Puerto Rico
(Precipitation varies
30 to 210 in.)

Precipitation (in.)

| | |
|---|---|
| 0-10 | 40-60 |
| 10-20 | 60-100 |
| 20-30 | >100 |
| 30-40 | |

Alaska

Runoff (in.)

| | |
|---|---|
| 0-1 | 20-40 |
| 1-5 | >40 |
| 5-20 | |

# THE HYDROLOGY OF RUNOFF

In a few days a heavy rainstorm can force sudden changes in local stream and river runoff. The connection between precipitation and runoff can be seen when weather forecasters predict local flash flooding after torrential rains. When levels of precipitation and runoff are measured over a large area, such as all of the states drained by a major river, and over a long period, such as a year, the relationship is less extreme but still strong, as the maps of precipitation and runoff shown in Figure 12.4 illustrate. When we compare them we see that areas of low precipitation have low runoff.

The relationship between precipitation and run-off from a large river flowing over a great distance is less obvious. Such a river may carry large amounts of water from an area with high rainfall to an area with low rainfall. The Colorado River, for example, begins in an area of moderate rainfall in Colorado and then runs through an arid region in western Arizona and southern California.

In regions of low rainfall, such as southern California, Arizona, and New Mexico, only a small fraction of precipitation ends up as runoff. In such dry regions, much of the precipitation is lost by evaporation and infiltration. In humid regions such as the southeastern United States, a much higher proportion of the precipitation runs off in rivers.

Most surface runoff is carried by major rivers. The millions of small and intermediate-sized rivers carry about half of the world's entire runoff; the other half is carried by only about 70 major rivers. And half of that is carried by the Mississippi, the largest river of North America, and the Amazon of South America, which carries about 10 times more water than the Mississippi (Table 12.1).

Surface runoff collects and is stored in natural lakes and artificial reservoirs created by the damming of rivers. Areas of swamp and marshland, often called wetlands, also act as storage depots for runoff (Figure 12.5). All of these reservoirs, if their volumes

| TABLE 12.1 Water Flows of Some Great Rivers | |
|---|---|
| | WATER FLOW (m³/s) |
| Amazon, South America | 175,000 |
| La Plata, South America | 79,300 |
| Congo, Africa | 39,600 |
| Yangtze, Asia | 21,800 |
| Brahmaputra, Asia | 19,800 |
| Ganges, Asia | 18,700 |
| Mississippi, North America | 17,500 |

are large enough, can absorb short-term inflows of major rainfalls without flooding. During dry seasons or droughts, the reservoirs continue to release water, either to streams or to water systems for human use. In these ways reservoirs smooth out seasonal or yearly variations in runoff and release steady flows

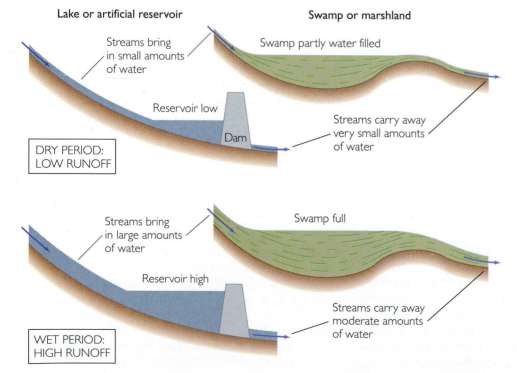

**FIGURE 12.5** A swamp or marshland, like a natural lake or an artificial reservoir behind a dam, stores water during times of rapid runoff and slowly releases it during periods of little runoff.

**FIGURE 12.6** Groundwater exiting from a cliff; Vasey's Paradise, Marble Canyon, Grand Canyon National Park, Arizona. This is a dramatic example of a spring, formed where hilly topography allows water in the ground to flow out onto the surface. *Larry Ulrich.*

downstream. Thus they are important in flood control, holding some of the water that would otherwise spill over the riverbanks. Because natural wetlands are so important in flood control, some geologists have worked to stop artificial draining of wetlands for real estate development. If unchecked, draining of wetlands could seriously damage a region's natural flood regulation.

## GROUNDWATER

Groundwaters are an enormous reservoir of water in the Earth. The amount of water stored beneath the Earth's surface equals about 22 percent of all the fresh water stored in lakes and rivers, glaciers and polar ice, and the atmosphere. For thousands of years people have used this resource, either by digging shallow wells or by storing water that flows out on the surface at springs. Springs are direct evidence of water moving below the surface (Figure 12.6).

Groundwater is formed as raindrops infiltrate soil and other unconsolidated surface materials, even sinking into cracks and crevices of bedrock. We tap groundwater by drilling wells and pumping the water to the surface. Well drillers in temperate climates know that they are most likely to find a good supply of water if they drill into porous sand or sandstone beds not far below the surface. Beds that carry water are called **aquifers.**

## How Water Flows through Soil and Rock

If water moves into and through the ground, what determines where and how fast it flows? With the exception of caves, there are no large open spaces for

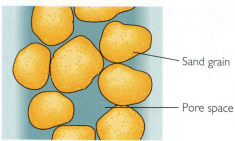

(a)  Porous sandstone

— Sand grain

— Pore space

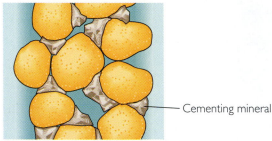

(b)  Cemented sandstone

— Cementing mineral

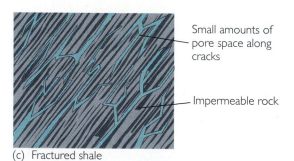

(c)  Fractured shale

Small amounts of pore space along cracks

Impermeable rock

(d)  Unfractured shale

Very small amounts of pore space between clays and silt grains

Silt grains

**FIGURE 12.7**  Pores in rocks are normally filled partly or entirely by water. (Pores in oil- or gas-bearing sandstones and limestones are filled with oil or gas.) A highly porous sandstone (a) has large amounts of pore space between grains. A cemented sandstone (b) has lower porosity because a cementing mineral has been precipitated in some of the pore space by fluids moving through the pores. A fractured shale (c) has higher porosity than a fine-grained, unfractured shale (d).

pools or rivers of water underground. The only space available for water is the pore space between grains of sand and other particles that make up the soil and bedrock, and in fractures. Some pores, however small and few, are found in every kind of rock and soil, but large amounts of pore space are most often found in sandstones and limestones.

The amount of pore space in rock, soil, or sediment is its **porosity**—the percentage of its total volume that is taken up by pores. Porosity depends on the size and shape of the grains and how they are packed together. As Figure 12.7 shows, the more tightly packed the particles, the less the pore space between the grains. The smaller the particles and the more they vary in shape, the more tightly they fit together. Sediments and sedimentary rocks are higher in porosity than igneous or metamorphic rocks. Porosities are highest—over 40 percent of volume—in loose sand and gravel layers found below the soil (Table 12.2). The porosities of many sandstones are as high as 30 percent, but those of most shales are less than 10 percent. Pore space varies in limestones, depending on how many pores were created by dissolution by groundwater or during weathering. The porosities of massive, unfractured igneous and metamorphic rocks are as low as 1 or 2 percent. Extensively fractured and jointed rocks, such as structurally deformed shales, granites, and metamorphic rocks, may contain appreciable pore space in their many cracks, up to 10 percent of the volume.

While porosity tells us how much water the rock can hold if all the pores are filled, it gives us no information on how rapidly water can flow through the pores. Water travels through a porous material by winding between grains and through cracks. The smaller the pore spaces and the more tortuous the path, the more slowly the water travels. The ability of a solid to allow fluids to pass through is its **permeability.** Generally, the greater the porosity, the greater the permeability, but permeability also depends on the sizes of the pores, how well they are connected, and how tortuous a path the water must travel to pass through.

Both porosity and permeability are important factors when one is searching for a groundwater supply. In general, a good groundwater reservoir will be a body of rock or sediment with both high porosity, so it can hold large amounts of water, and high permeability, so the water can be pumped from it easily. A rock with high porosity but low permeability may contain a great deal of water, but because the water flows so slowly, it is hard to pump it out of the rock.

**TABLE 12.2**

## Porosity and Permeability of Aquifer Rock Types

| ROCK TYPE | POROSITY | PERMEABILITY |
|---|---|---|
| Gravel | Very high | Very high |
| Coarse- to medium-grained sand | High | High |
| Fine-grained sand and silt | Moderate | Moderate to low |
| Sandstone, moderately cemented | Moderate to low | Low |
| Fractured shale or metamorphic rocks | Low | Very low |
| Unfractured shale | Very low | Very low |

## The Groundwater Table

As a well driller brings up samples of the soil and rock being drilled, the material varies in wetness. At shallower depths, the material is unsaturated; that is, the pores are not completely filled with water but also contain some air. This level is called the **unsaturated zone** (Figure 12.8). Below it, in the **saturated zone,** the pores of the soil or rock are com-

pletely filled with water. The boundary between the two zones is the **groundwater table,** usually called simply the "water table." When a hole is drilled below the water table, water from the saturated zone flows into the hole and fills it to the level of the water table.

Groundwater moves under the force of gravity, and so some of the water in the unsaturated zone may be on its way down to the water table. A fraction of the water, however, will remain in the unsaturated zone, held in small pore spaces by surface tension, the attraction between the water molecules and the surfaces of the particles. Surface tension keeps the sand in a sandbox or at the beach moist even though there are spaces below to which water could travel by gravity. Evaporation of water in pore spaces in the unsaturated zone is slowed both by the effect of surface tension and by the relative humidity of the air in the pore spaces—possibly close to 100 percent.

If we were to drill wells at several sites and measure the elevations of the water levels in the wells, we would get a map of the water table. A cross section of the landscape might look like the one in Figure 12.9. The water table follows the general shape of the surface topography, but the slopes are gentler. The water table is at the surface in river and lake beds and at springs. Under the influence of gravity, groundwater moves downhill from an area where the water-table elevation is high—under a hill, for example—to places where the water-table elevation is low, such as a spring where groundwater exits to the surface.

Water enters the saturated zone by infiltration of rain or snow meltwater from the surface and leaves the saturated zone by springs, lake and river beds,

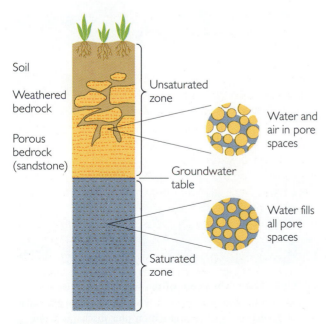

**FIGURE 12.8** The groundwater table is the top of the saturated zone. The saturated and unsaturated zones can be in either unconsolidated material or bedrock.

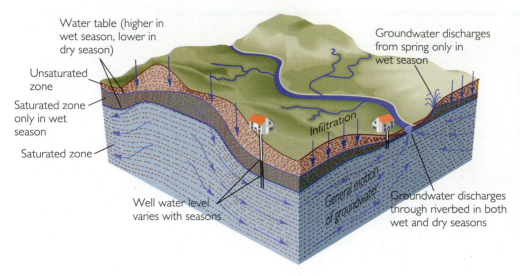

FIGURE 12.9 Dynamics of the groundwater table in permeable shallow formations in a temperate climate. Water enters the ground by infiltration of rain and melted snow and discharges at springs and rivers. The unsaturated zone varies in thickness as the water table rises in wet seasons and falls in dry seasons.

and pumped wells. The infiltration of water into any subsurface formation is called **recharge,** and its exit to the surface is called **discharge.** In addition to infiltration, recharge may take place through the bottom of a stream where the stream channel lies above the water table (Figure 12.10). This kind of stream, called an **influent stream,** is most characteristic of arid regions, where the water table is deep. In the reverse case, when the channel intersects the water table, water discharges from the groundwater to the stream. Such an **effluent stream** is typical of humid areas. Effluent streams continue to flow long after runoff has stopped because they are fed by groundwater. Thus the reservoir of groundwater may be increased by influent streams and depleted by effluent streams.

## Artesian Flows

Many permeable aquifers, typically sandstones, are bounded above and below by shale beds of low permeability. Groundwater cannot flow through these relatively impermeable beds, or **aquicludes,** or it flows through them very slowly. When aquicludes lie both over and under an aquifer, they confine the water flow to that aquifer (Figure 12.11). These flows in **confined aquifers** are called **artesian flows.** Many other flows near the surface are **unconfined aquifers**—that is, the water travels through beds that extend with more or less uniform permeability to the surface in both discharge and recharge areas. In an unconfined aquifer, the level of the reservoir is given by the height of the water table.

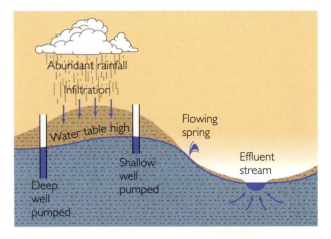

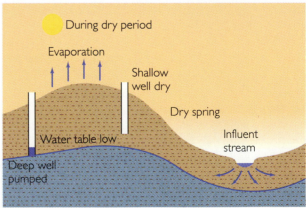

**FIGURE 12.10** The depth of the water table fluctuates in response to the balance between what is added by precipitation and what is lost by evaporation plus discharge from wells, springs, and streams. Streams become influent mainly in arid climates but may also do so after prolonged dry periods in temperate climates.

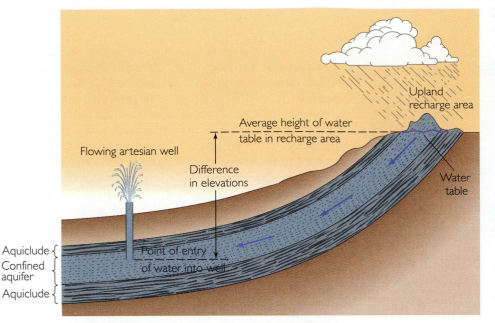

FIGURE 12.11 A confined aquifer is created where water enters an aquifer situated between two aquicludes (beds of low permeability). The artesian well flows in response to the difference in natural pressure (before the well was drilled) between the height of the water table in the recharge area and the bottom of the well. The actual pressure difference that governs the flow from the top of the well is the difference between the elevation of the water table and the top of the well. If the wellhead were as high as the water table in the recharge area, there would be no pressure difference and thus no flow.

The impermeable beds above an artesian aquifer prevent rainwater from infiltrating downward into the aquifer. Instead, the confined aquifer is recharged by rainwater that enters the ground where the formation outcrops. Precipitation over the outcrop area enters the ground and travels down the aquifer (see Figure 12.11). Water in a confined aquifer is under pressure. At any point in the aquifer the pressure is equivalent to the weight of the water in the aquifer above that point.

If a well is drilled into a confined aquifer at a point where the elevation of the ground surface is lower than the water table of the aquifer in the recharge area, the water will flow out of the well spontaneously. Any well that flows to the surface spontaneously in this way is called an **artesian well.** Artesian wells are extremely desirable because it doesn't take any energy to pump the water to the surface; instead, the water is brought up by its own pressure.

An aquiclude may lie below the water table in a shallower aquifer and above the water table in a deeper aquifer (Figure 12.12). The water table in the shallower aquifer is called a **perched water table**

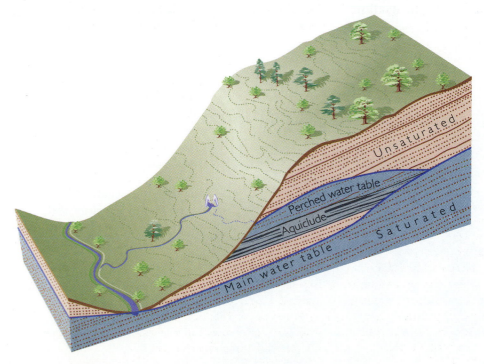

FIGURE 12.12 A perched water table formed by a shale aquiclude located above the main water table in a sandstone aquifer.

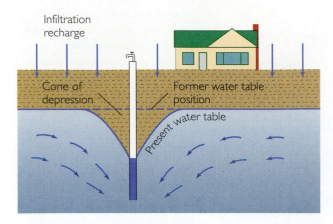

Infiltration recharge

Cone of depression

Former water table position

Present water table

**FIGURE 12.13** Excessive pumping in relation to recharge draws down the water table into a cone-shaped depression around a well. The water level in the well is lowered to the depressed level of the water table.

because it is above the main water table in the lower aquifer. Many perched water tables are small lenses, but some extend for hundreds of square kilometers.

## Balancing Recharge and Discharge

In natural situations, when recharge and discharge are balanced, the reservoir of groundwater and the water table remain constant, even though water is continually flowing through the aquifer. Discharge balances recharge when rainfall is frequent enough to match the sum of runoff from rivers and outflow from springs and wells.

Because of seasonal variation in rainfall, recharge and discharge will not always be equal. Typically the water table drops in drier seasons and rises in wet periods. If there is a decrease in recharge, such as during a prolonged drought, a longer term imbalance and lowering of the water table follow. The same imbalance can result from an increase in discharge, usually from increased well pumping. As the water table drops, the bottoms of some shallow wells end up in the unsaturated zone and become impossible to pump. The wells have "gone dry."

When a well pumps water out of an aquifer faster than recharge can replenish it, the water level is lowered in a cone-shaped area around the well, called a cone of depression (Figures 12.13 and 12.14). If the cone of depression extends below the bottom of a well, that well goes dry. If, however, the bottom of the well is above the base of the aquifer, deepening the well farther into the aquifer may allow more water to be withdrawn, even at high pumping rates. If the well is deepened so much that the entire aquifer is tapped and the rate of pumping is kept high, the aquifer may be depleted as the cone of depression reaches the bottom of the aquifer. The aquifer can recover only if the pumping rate is reduced enough to give it time for recharge.

People who live near the ocean's edge may face a different problem when pumping rates are high in relation to recharge: the incursion of salt water into the well. Underground near shorelines or a little offshore, there is a boundary between salt water under the sea and fresh water under the land. This boundary slopes down and inland from the shoreline in such a way that salt water underlies the fresh water of

**FIGURE 12.14** In Antelope Valley, California, overpumping of groundwater has led to fissures and sinklike depressions on Rogers Lakebed at Edwards Air Force Base, the landing site for the space shuttle. This fissure, formed in January 1991, is about 625 m long. *Edwards Air Force Base.*

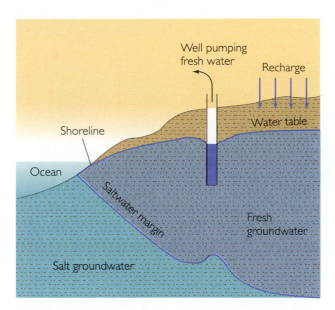

(a) Before extensive pumping

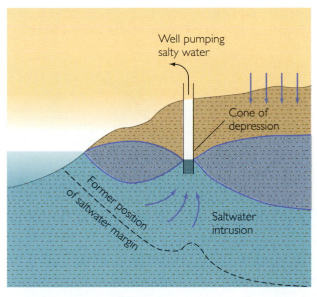

(b) After extensive pumping by many wells

**FIGURE 12.15** The boundary between fresh groundwater and salty groundwater along shorelines is determined by the balance between recharge and discharge in the freshwater aquifers. (a) Normally the pressure of fresh water keeps the saltwater margin slightly offshore. (b) Extensive pumping lowers the pressure of the fresh water, allowing the saltwater margin to move inland. This movement creates not only a cone of depression but an inverted cone of depression that brings salty water into the well. A well that formerly pumped fresh water now pumps salty water.

the aquifer (Figure 12.15a). A fresh groundwater lens under many ocean islands floats on a base of seawater. The fresh water floats because it is less dense than seawater (1.00 g/cm$^3$ compared with 1.02 g/cm$^3$, a small but significant difference).

As long as recharge of the freshwater part of the aquifer by rainwater is at least equal to discharge, fresh water can continue to be pumped. But if water is withdrawn faster than it is recharged, a cone of depression develops at the top of the aquifer. The cone of depression is mirrored by an inverted cone rising from the freshwater-seawater boundary below. The cone of depression at the upper part of the aquifer makes it more difficult to pump fresh water, and the inverted cone below leads to an intake of salt water at the bottom of the well (Figure 12.15b). The people living closest to the shore are the first affected. Towns on Cape Cod, on Long Island, and in many other nearshore areas suffer from this problem; some have had to post notices that town drinking water contains more salt than is considered healthful by environmental agencies. There is no ready solution to this difficulty other than to slow the pumping or, in some places, to recharge the aquifer artificially by funneling runoff into the ground.

## The Speed of Groundwater Flows

The balance between discharge and recharge is a balance of the *speeds* at which water moves in the ground. Most groundwaters move slowly, a fact of nature responsible for our groundwater supplies. If groundwater moved as rapidly as rivers, aquifers would run dry after a period of time without rain, as many small streams do. At the same time, the slowness of groundwater flows makes rapid recharge impossible once groundwater levels have been lowered by excessive pumping. Though all groundwaters flow through aquifers slowly, some flow more slowly than others.

The reason was first worked out in the middle of the nineteenth century by Henry Darcy, town engineer of Dijon, France. While studying the town's water supply, he measured the elevations of water in various wells and mapped the varying heights of the water table in the district. He calculated the distances traveled by the water from well to well and measured the permeability of the aquifers. (Remember that permeability is the ease with which water passes through the pore spaces of the aquifer.)

Darcy found that for a given aquifer and distance of travel, the rate at which water flows from one place to another is directly proportional to the drop in elevation of the water table in the two places; that

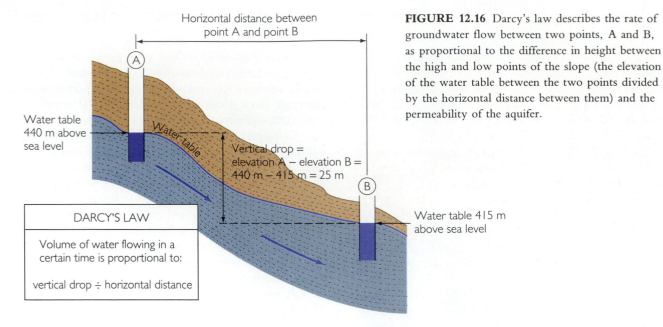

Water table
440 m above
sea level

Water table

Horizontal distance between
point A and point B

A

Vertical drop =
elevation A − elevation B =
440 m − 415 m = 25 m

B

DARCY'S LAW

Volume of water flowing in a
certain time is proportional to:

vertical drop ÷ horizontal distance

Water table 415 m
above sea level

**FIGURE 12.16** Darcy's law describes the rate of groundwater flow between two points, A and B, as proportional to the difference in height between the high and low points of the slope (the elevation of the water table between the two points divided by the horizontal distance between them) and the permeability of the aquifer.

is, as the difference in elevation increases, the rate of flow increases. He also found that the rate of flow for a given aquifer and given difference in elevation is inversely proportional to the horizontal distance the water travels; that is, as the distance increases, the rate decreases. The ratio between the elevation difference (a vertical distance) and the horizontal distance is the slope of the water table, defined in the same way as is the slope of a ground surface. Just as a ball runs faster down a steeper slope, groundwater runs more quickly down a steeper water-table slope (Figure 12.16).

Darcy's law, as this discovery is called, can be expressed in a simple relationship: the volume of water flowing in a certain time is proportional to the vertical drop divided by the horizontal distance. Darcy reasoned that this relationship should hold whether the water is moving through a porous sandstone aquifer or an open pipe. You might guess (correctly) that the water would move more quickly through a pipe than through the tortuous turns of pore spaces in an aquifer. Darcy recognized this factor and included permeability in his final equation, so that, other things being equal, the greater the permeability and thus the greater the ease of flow, the faster the flow.

Velocities calculated by Darcy's law have been confirmed experimentally by measuring how long it takes a harmless dye introduced into one well to reach another. In most aquifers, groundwater moves at a rate of a few centimeters per day. In very permeable gravel beds near the surface, groundwater may travel as much as 15 cm per day. This is still much slower than the speeds of 20 to 50 cm per second typical of river flows.

# WATER RESOURCES FROM MAJOR AQUIFERS

Large parts of North America rely on groundwater for all their water needs. The demand on groundwater resources has grown as populations have increased and uses such as irrigation have expanded (Figure 12.17). Many areas of the Great Plains and

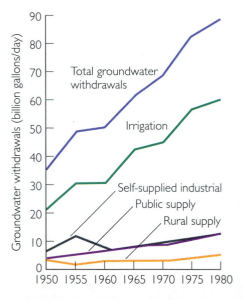

**FIGURE 12.17** Billions of gallons of groundwater withdrawn in the United States, 1950–1980 (1 billion gallons per day = $3.785 \times 10^6$ m³ per day). Public, industrial, and rural withdrawals have risen moderately over this period; the major increase in total withdrawal comes from increases in irrigation. These trends are continuing to the present. (Data from USGS Water-Supply Paper 2250, 1984.)

other parts of the Midwest are underlain by sandstones that transport waters over hundreds of kilometers and constitute a major resource. Thousands of wells have been drilled into these formations, most of which are confined aquifers like the one shown in Figure 12.11. The aquifers are recharged from outcrops in the western high plains, some very close to the foothills of the Rocky Mountains. From there the water runs downhill in an easterly direction.

Darcy's law tells us that water flows at rates proportional to the slopes of the aquifers between their recharge areas and the areas of discharge from wells. In the western plains the slopes are gentle and waters move slowly through the aquifers, recharging them at low rates. At first many of these wells were artesian and the water flowed freely. As more wells were drilled, the water levels were lowered and the wells began to need pumping. Because extensive pumping from some of these aquifers has withdrawn the water faster than the slow recharge from far away can fill them, the reservoirs have been gradually depleted (see Box 12.2).

Some aquifers can be replenished by artificially increasing their recharge. In Long Island, New York, for example, the water authority drilled a large system of recharge wells—wells used to put water into the aquifer from the surface. Used water was first treated to purify it and then was pumped back into the ground. In addition, large shallow basins were constructed over natural recharge areas to catch run-off and divert it to augment infiltration from surface waters, including storm and industrial waste drainage. In this way, the Long Island aquifer was rebuilt, although not to its original level.

Urban development may decrease recharge by interfering with infiltration. As urbanization progresses, water is prevented from infiltrating the ground by the impermeable materials used to pave large areas for streets, sidewalks, and parking lots. This practice increases the amount of rainwater that runs off and may decrease the natural infiltration into the ground enough to deprive the aquifers of much of their recharge. One remedy is to catch and use the storm runoff in a systematic program of artificial recharge.

# EROSION BY GROUNDWATER

Every year thousands of people visit caves, either on tours of well-advertised attractions such as Mammoth Cave, Kentucky, or in adventurous explorations of little-known caves. These underground open spaces are produced by the dissolution of limestone —or, rarely, other soluble rocks such as evaporites— by groundwater (Figure 12.18). The amounts of limestone that have dissolved to make caves may be

**FIGURE 12.18** Powderhouse Cave, West Virginia. Numerous stalactites can be seen at top. At right center, one stalactite has joined a large stalagmite. *Chip Clark.*

## BOX 12.2 DEPLETED GROUNDWATERS AND WATER RESOURCES

For over a hundred years, wells have provided water for the people of the cities, towns, ranches, and farms of western Texas and New Mexico. The water comes from the Ogallala aquifer, a formation of sands and gravels. The population of the region has climbed from a few thousand late in the nineteenth century to about a million now. The Ogallala continues to provide the irrigation water needed to support the agriculture that serves as the area's economic base, but the water pressure in the wells has declined steadily and the water table has dropped by 30 m or more.

The Ogallala aquifer of the southern plains is very slow to recharge naturally because rainfall is sparse, the degree of evaporation is high, and the recharge area is small. Water has been pumped from the reservoir so extensively—about 6 billion cubic meters per year—that recharge cannot keep up. At the current rates of recharge, if all pumping were to stop, the water table would take several thousand years to recover its original position, with pressure restored. Some artificial recharge experiments have injected water from shallow lakes that form in wet seasons on the high plains into the aquifer and have managed to increase recharge, but the aquifer is still in danger over the long term.

It has been estimated that the remaining supplies in the Ogallala will last only until the year 2000. As this valuable underground reservoir is drained, about 5.1 million acres of irrigated land in western Texas and eastern New Mexico will dry up. This region currently supplies 12 percent of the country's production of cotton, corn, sorghum, and wheat and a significant fraction of the feedlots for the nation's cattle.

Other aquifers in the northern plains and elsewhere in North America are in a similar condition. In three major areas of the United States—Arizona, the high plains, and California—groundwater supplies are being significantly depleted. We are essentially mining the groundwater as a nonrenewable resource in places where recharge is so slow. The amount left in the ground steadily diminishes as we pump, just as coal does when it is mined.

It is likely that these water-hungry regions will have to start serious water conservation measures, ranging from restrictions on water sprinklers and swimming pools to decreases in the huge amounts used for irrigation. As water use all over the country increases in relation to supply, all of us will eventually have to adopt sensible conservation practices.

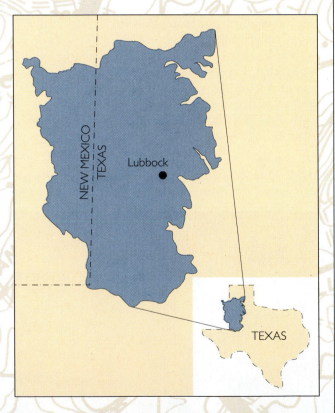

The southwestern high plains of Texas and New Mexico, underlain by the Ogallala aquifer. The blue region represents the aquifer. The general recharge area is located along the western margin of the aquifer. (From USGS.)

huge. Mammoth Cave, for example, has tens of kilometers of large and small interconnected chambers, and the large room at Carlsbad Caverns, New Mexico, is more than 1200 m long, 200 m wide, and 100 m high. Limestone formations are widespread in the upper parts of the crust, but caves form only where these relatively soluble rocks are at or near the surface and enough water infiltrates the surface to dissolve extensive amounts of limestone.

As we saw in Chapter 6, the dissolution of limestone is enhanced by the atmospheric carbon dioxide dissolved in rainwater. Waters that infiltrate soils may pick up even more of the gas from the carbon dioxide given off by plant roots, bacteria, and other soil-dwelling organisms. As this carbon dioxide–rich water moves down to the water table, through the unsaturated zone to the saturated zone, it dissolves carbonate minerals. As limestones are dissolved along joints and fractures, these openings are enlarged to form a network of rooms and passages. Much of the dissolution of these networks takes place in the saturated zone; because the caves are filled with water, dissolution takes place over the entire surface of the openings, on their floors, walls, and ceilings.

We can explore caves that were once dissolved below the water table but are now in the unsaturated zone as a result of a drop in the water table. In these caves, now air-filled, water saturated with calcium carbonate may drip from the ceiling. As it drips, some of the dissolved carbon dioxide it gained from the soils through which it passed will escape to the cave atmosphere. The loss of carbon dioxide from the groundwater solution makes calcium carbonate less soluble, and the water precipitates a small amount of calcium carbonate on the ceiling from each drop. Each drop adds a little more and, in the way an icicle grows, a long narrow spike of carbonate, called a **stalactite,** hangs down from the ceiling (see Figure 12.18). As the drop hits the floor, more carbon dioxide escapes and another small amount of calcium carbonate is precipitated on the floor. Below the stalactite, a mass of calcium carbonate grows upward from the floor to form an irregular, cone-shaped **stalagmite.** Eventually a stalactite and stalagmite may grow together to form a column.

In some places, dissolution may thin the roof of a limestone cave so much that it collapses, producing a depression or **sinkhole** in the land surface above (Figure 12.19). Such collapses may be so sudden that they can swallow automobiles whole. Sinkholes are common on the surface of cavernous limestone formations and contribute to a distinctive form of topography known as **karst,** named for a region in northern Yugoslavia with an irregular terrain of hills and many sinkholes. Karst topography lacks a normal surface drainage system of small and large rivers. Streams are short and scarce, and they frequently end in sinkholes, where they detour underground, sometimes reappearing miles away. In North America, karst topography is found in limestone terrains of Indiana and Kentucky and in the Yucatan Peninsula of Mexico.

**FIGURE 12.19** A large sinkhole formed by the collapse of a shallow underground cavern. Such collapses may occur so suddenly that moving cars can be buried. Winter Park, Florida. *Leif Skoogfors/Woodfin Camp.*

# WATER QUALITY

Most residents of Canada and the United States take a supply of fresh, pure water for granted. But a growing number of people are getting worried about contaminants in their water and are beginning to buy bottled spring water; some even install home purifying systems. Almost all water supplies in North America are free of bacterial contamination, and the vast majority are chemically pure enough to drink safely. But in some places toxic wastes have infiltrated aquifers from surface dumps. We have noted the problem faced by some shoreline communities, where heavy pumping has caused an incursion of seawater and unacceptable levels of sodium in the water.

Lead, a well-known pollutant derived from industrial processes, is routinely eliminated from public water supplies by chemical treatment before the water is distributed through the water mains. Yet in many older neighborhoods, where lead pipes are still common, some lead contamination occurs from the water mains themselves or from lead pipes in old houses. Even in newer construction, the lead solder used to connect copper pipes is a small source. Replacing old lead mains and substituting durable plastic pipe can reduce lead contamination.

Some groundwaters, though perfectly healthful to drink, may have a slightly disagreeable taste. Some taste of "iron" or are slightly sour. Other waters may taste good but are called "hard" because it is difficult to make lather or soapsuds with them. Hard waters contain relatively large amounts of dissolved calcium carbonate and usually some magnesium car-

bonate, both of which are harmless to drink but do interfere with soaping and laundering. Many of the highest quality, best tasting public water supplies come from lakes and artificial surface reservoirs, many of which are simply collecting places for rainwater. But there are some groundwaters that taste just as good. How do these differences in taste and quality arise?

Some so-called impurities that enter groundwater are actually from natural sources, such as the ions dissolved from minerals weathering at the surface and small amounts of organic compounds dissolved from plants and animals (Figure 12.20). It is this load of dissolved substances that gives groundwaters their chemical composition and thus their taste.

Waters that pass through rocks that weather only slightly, such as sandstones made up largely of quartz, gain little in dissolved substances and thus taste fresh. Some waters that pass through waterlogged soils containing aromatic organic compounds and hydrogen sulfide, in contrast, may dissolve enough of these compounds to have a disagreeable taste. In greater quantities some of these dissolved substances are toxic.

## Is the Water Drinkable?

Water that is good enough to drink is called **potable water.** The amounts of dissolved substances in potable waters are very small, usually measured by weight in parts per million (ppm). Just as 1 percent is defined as one part per hundred, 1 ppm is one part

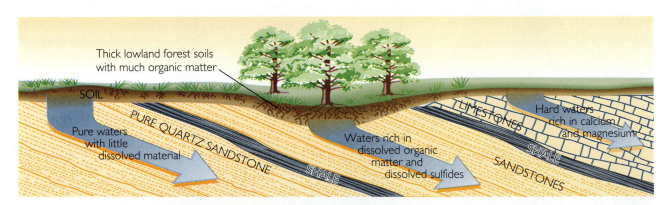

**FIGURE 12.20** Groundwater picks up natural dissolved materials as it passes through rock and unconsolidated soils and sediments. Groundwaters passing through limestone dissolve carbonate minerals and carry away calcium, magnesium, and bicarbonate ions, making the water "hard." Pure quartz sandstones resist weathering and contribute almost no dissolved materials. Water coming through waterlogged forest or swampy soils may contain dissolved organic compounds and hydrogen sulfide.

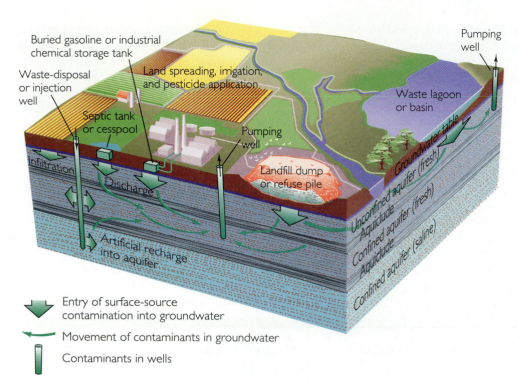

Buried gasoline or industrial chemical storage tank

Waste-disposal or injection well

Septic tank or cesspool

Land spreading, irrigation, and pesticide application

Pumping well

Waste lagoon or basin

Pumping well

Infiltration

Discharge

Landfill dump or refuse pile

Groundwater table

Unconfined aquifer (fresh)

Aquiclude

Confined aquifer (fresh)

Aquiclude

Confined aquifer (saline)

Confined aquifer (fresh)

Artificial recharge into aquifer

Entry of surface-source contamination into groundwater

Movement of contaminants in groundwater

Contaminants in wells

**FIGURE 12.21** Sources of groundwater contamination. Normal groundwater flow carries contaminants from surface sources such as dumps and subsurface sources such as septic tanks into aquifers. Pumping wells may incorporate the contaminants into water supplies. Waste-disposal wells are designed to pump contaminants into deep saline aquifers, but they may accidentally leak into freshwater aquifers above. (Modified from U.S. Environmental Protection Agency.)

per million. Potable groundwaters of good quality typically contain about 150 ppm total dissolved materials, since even the purest natural waters contain some dissolved substances derived from weathering. Only distilled water contains less than 1 ppm dissolved substances.

Groundwater is almost always free of solid particles when it seeps into a well from a sand or sandstone aquifer. The tortuous passageways of the rock or sand act as a fine filter that removes small particles of clay and any other solids. They even strain out bacteria and large viruses. Limestone aquifers may have larger pores and so be less efficient as filters. If there are bacteria at the bottom of a well, it is usually because of contamination, introduced either from the surface by the pump materials or from nearby underground sewage disposal, such as a septic tank located too close to the well. Sewage put underground will be clear of suspended solid material after it has traveled through hundreds of meters of sand. The dissolved matter that may be in the groundwater cannot be so easily removed because it cannot be physically strained out by the aquifer.

## Contamination of the Water Supply

Groundwaters that are naturally clean can be contaminated by human activities (Figure 12.21). Chemical industry waste lagoons, leaking chemical storage barrels, sanitary landfill operations, and city garbage dumps may introduce hazardous or foul-tasting contaminants. Buried gasoline storage tanks may leak, and road salt inevitably drains into the soil and ultimately into aquifers. Pesticides, herbicides, and fertilizers used in agriculture may be washed by rain into the soil and percolate downward into aquifers. Keeping groundwaters free of these contaminants is becoming steadily more important as the demand for water increases.

The widespread use of septic tanks in rapidly growing urban and suburban areas without sewer networks has multiplied the sources of contamination. Septic tanks are settling tanks buried at shallow depths in which the solid wastes from house sewage are decomposed by bacteria. In a properly designed system, as the sewage water flows through and into the soil, any harmful bacteria and bits of solid waste and sludge are filtered out. Septic tank waters may also contain dissolved materials such as phosphate, nitrate, and toxic metals that cannot be filtered. Avoiding contamination of the water requires proper placement and drainage of septic tanks at sufficient distance from water wells in shallow aquifers.

Can we reverse contamination of water supplies? The answer is a qualified yes, but the process is costly and very slow. The faster an aquifer recharges, the easier it will be to clean. If the recharge is fast, once we close off the sources of contamination, fresh water moves into the aquifer, and in a short time the water recovers its quality. Even fast recoveries, however, may take a few years. Contamination of slowly recharging reservoirs is more serious, because the rate of groundwater movement may be so slow that contamination from some distance may take a long time to show up. By the time it does, it is too late for rapid recovery. Some contaminated deep reservoirs in which the water has traveled hundreds of kilometers may take many decades to respond to cleaned-up recharge.

## WATER DEEP IN THE CRUST

All rocks below the groundwater table are saturated with water. Even in the deepest wells drilled for oil, some 8 or 9 km deep, geologists always find water in permeable formations. At these depths waters move so slowly, probably less than a centimeter per year, that they have plenty of time to dissolve even very insoluble minerals from the rocks they pass through. Thus dissolved materials become more concentrated in them than in near-surface waters. Groundwaters that pass through salt beds, which are quick to dissolve, tend to become greatly enriched in sodium chloride.

At depths greater than 12 to 15 km, the basement igneous and metamorphic rocks that everywhere underlie the sedimentary formations of the upper part of the crust have extremely low porosities and permeabilities. Even these rocks are saturated, although the total amounts of water are extremely small because the porosity, distributed along small cracks and the boundaries between crystals, is so low (Figure 12.22). In some deeper regions of the crust that are undergoing active metamorphism, such as along subduction zones, hot concentrated waters play an important role in the chemical reactions by which metamorphic rocks are made, helping to dissolve some minerals and precipitating others (see Chapter 8). Even some mantle rocks are presumed to have very minute quantities of water.

### Hydrothermal Waters

Natural hot springs occur in many places in the world. Hot Springs, Arkansas; Banff Sulfur Springs, Alberta; and Reykjavik, Iceland, are only a few of the better known places where hot waters deep in the crust migrate rapidly upward without losing much

heat and emerge at the surface, sometimes at boiling temperatures. Such hot waters in the crust are termed hydrothermal waters.

These waters are loaded with chemical substances dissolved from rocks at high temperatures. The dissolved material can remain in solution as long as the water remains hot. But as hydrothermal waters coming to the surface quickly cool, they may precipitate various minerals, such as opal (a form of silica) and calcite or aragonite (forms of calcium carbonate). Crusts of calcium carbonate that form at some hot springs build up to form the rock travertine, prized for its beauty as a polished stone used for buildings

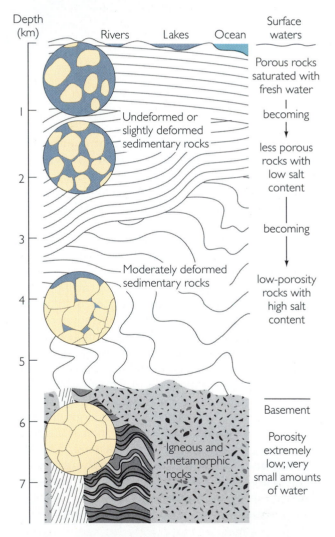

**FIGURE 12.22** The distribution of water in a typical section of continental crust. Most water is at the surface or in sedimentary rocks buried at shallow depths. Porosity and water content generally decrease with increasing depth and greater structural deformation.

**FIGURE 12.23** Travertine ($CaCO_3$) deposits form as hot waters emerge at springs on the surface and dissolved material is deposited. Mammoth Hot Springs, Yellowstone National Park. *Peter Kresan.*

and tables (Figure 12.23). Hydrothermal waters, while still in the crust, are also responsible for depositing some of the world's richest metallic ores as they cool after migration (see Chapter 23).

Most of the hydrothermal waters of the continents derive from surface waters that percolated downward to deeper regions of the crust. Such waters, which derived originally from rain or snow, are termed **meteoric waters** (from the Greek *meteōron*, "phenomenon in the sky," which also gives us the word *meteorology*). They may be very old; it has been determined that the water at Hot Springs, Arkansas, was derived from rain and snow that fell over 4000 years ago and slowly infiltrated the ground.

The other source of hydrothermal waters is water that escapes from a magma. In areas of igneous activity, sinking meteoric waters encounter hot masses of rocks, become heated, and then mix with water released from the nearby magma. The mixture of hydrothermal water then returns to the surface as hot springs or geysers (Figure 12.24).

Other hot springs come from meteoric waters that move downward into deep sedimentary rock formations, where they are heated by the normal increase in temperature with depth, and then return as hydrothermal waters to the surface. Many metallic ores and other mineral deposits in sedimentary rocks far from any igneous activity originated in this way.

In the search for new and clean sources of energy, geologists have turned to hydrothermal waters. The steam generated by hydrothermal activity in areas where hot springs and geysers are found can be used to drive electricity-generating turbines. In

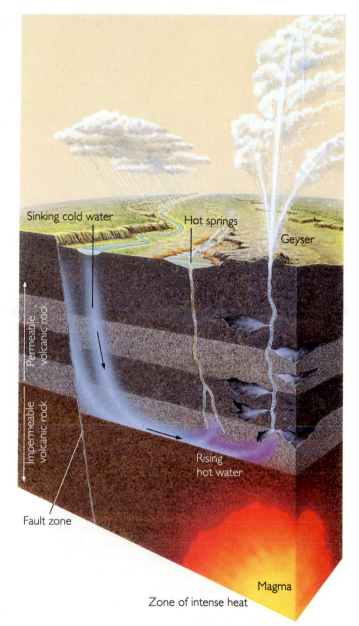

**FIGURE 12.24** Circulation of water over a magma body produces geysers or hot springs. Cold rainwater soaks into the soil and filters down through permeable rocks. As it approaches the magma, it heats up and becomes less dense, thus setting up a circulation system that returns it to the surface.

northern California, Iceland, Italy, and New Zealand, hydrothermal waters are already in practical use for power (see Chapter 22).

Hydrothermal waters, however important they may be for power generation, ore deposits, and their supposed healing qualities, do not contribute to surface water supplies, primarily because they contain so much dissolved material.

## The Usable Waters of the Earth

This survey of water in and on the Earth leads us to an inescapable conclusion: the accessible and usable waters of the Earth are limited to the surface and the near-surface parts of the crust. Most of the water is in surface reservoirs: oceans, rivers, lakes, and glacial ice. The rest is stored in rocks, most of it at depths of not much more than 10 to 15 km. Although there are enormous quantities of fresh water stored in glacial ice, it has not proved practical so far to melt and transport it. Thus the overwhelming majority of the world's *usable* fresh waters are in the ground rather than on the surface. Over the long term, we must depend on the total precipitation that falls on the continents to replenish the water in all reservoirs. The rain is both the ceiling for our water supply and our standard for purity.

# SUMMARY

**How does water move around and in the Earth in the hydrologic cycle?** Water moves in such a way that a constant balance is maintained among the major reservoirs of water at or near the Earth's surface: oceans, lakes, rivers, glaciers, and groundwater. Water is transferred to the atmosphere by evaporation from the oceans, evaporation and transpiration from the continents, and sublimation from glaciers. It is returned from the atmosphere to the oceans and continents by precipitation as rain and snow. Part of the precipitation that falls on land is returned to the ocean by runoff in rivers, and the remainder infiltrates the ground to become groundwater. Local variations in the evaporation-precipitation-runoff-infiltration balance arise because of differences in climate.

**How does water move below the ground?** Groundwater forms as rain infiltrates the surface of the ground and travels through pore spaces in the soil, sediment, or rock that serves as an aquifer. Water moves through the upper, unsaturated zone through the water table into the saturated zone. Groundwater moves downhill under the influence of gravity, eventually emerging at springs, where the water table intersects the ground surface. Over the long term, a groundwater aquifer is in dynamic balance between recharge and discharge. Groundwater may flow in unconfined aquifers, which are continuous to the surface, or in confined aquifers, which are bounded by aquicludes. Confined aquifers produce artesian flows and spontaneously flowing artesian wells. Darcy's law describes the groundwater flow rate in relation to the slope of the water table and the permeability of the aquifer.

**What factors govern our use of groundwater resources?** As population has grown in many areas, demand for groundwaters has greatly increased, particularly where irrigation is widespread. Many aquifers, such as those of the western plains of North America, have such slow recharge rates that continued pumping over the last century has reduced the pressure in artesian wells. As pumping discharge continues to be out of balance with recharge, such aquifers are being depleted, and there is no prospect of renewal for many years. Artificial recharge may help renew some aquifers, but conservation will be required to preserve others. Contamination of groundwater by sewage and industrial effluents reduces the potability of some waters and limits our resources.

**What geological processes are affected by groundwater?** Erosion by groundwater in humid limestone terrains leads to karst topography, caves, and sinkholes. Heating of downward-percolating meteoric waters by magma bodies leads to a circulation that brings hydrothermal waters to the surface as geysers and hot springs. At great depths in the crust, more than 12 to 15 km, dense rocks have low porosities and hence contain extremely small quantities of water.

# KEY TERMS AND CONCEPTS

hydrology (p. 252)
groundwater (p. 253)
reservoir (p. 253)
hydrologic cycle (p. 253)
infiltration (p. 254)
transpiration (p. 254)

runoff (p. 254)
sublimation (p. 254)
precipitation (p. 254)
relative humidity (p. 254)
aquifer (p. 259)
porosity (p. 260)

permeability (p. 260)
unsaturated zone (p. 261)
saturated zone (p. 261)
groundwater table (p. 261)
recharge (p. 262)
discharge (p. 262)

influent stream (p. 262)

effluent stream (p. 262)

aquiclude (p. 262)

confined aquifer (p. 262)

artesian flow (p. 262)

unconfined aquifer (p. 262)

artesian well (p. 263)

perched water table
    (p. 263)

stalactite (p. 269)

stalagmite (p. 269)

sinkhole (p. 269)

karst (p. 269)

potable water (p. 270)

meteoric waters (p. 273)

# EXERCISES

1. What are the main reservoirs of water at or near the surface of the Earth?

2. How do mountains form rain shadows?

3. What is an aquifer?

4. What is the difference between the saturated and unsaturated zones of groundwater?

5. How do aquicludes make a confined aquifer?

6. How are recharge and discharge balanced to make a groundwater table stable?

7. How does Darcy's law relate groundwater movement to permeability?

# THOUGHT QUESTIONS

1. If the Earth warmed so that evaporation from the oceans greatly increased, how would the hydrologic cycle of today be altered?

2. If you lived near the seashore and started to notice a slight salty taste to your well water, how would you explain the change in water quality?

3. Why would you recommend against extensive development and urbanization of the recharge area of an aquifer that serves your community?

4. If it were discovered that radioactive waste had seeped into groundwater from a nuclear processing plant, what kind of information would you need to predict the length of time before the radioactivity appeared in well water 10 km from the plant?

5. What geological processes would you infer are occurring at depth at Yellowstone National Park, which has many hot springs and geysers?

6. Why should septic tanks be maintained in good condition?

7. Why are more and more communities in cold climates restricting the use of salt to melt snow and ice on highways?

# SUGGESTED READINGS

Dolan, R., and H. G. Goodell. 1986. Sinking cities. *American Scientist* 74:38–47.

Dunne, Thomas, and Luna B. Leopold. 1978. *Water in Environmental Planning.* San Francisco: W. H. Freeman.

Frederick, K. D. 1986. *Scarce Water and Institutional Change.* Washington, D.C.: Resources for the Future.

Freeze, R. Allen, and John A. Cherry. 1979. *Groundwater.* Englewood Cliffs, N.J.: Prentice-Hall.

Heath, R. C. 1983. *Basic Groundwater Hydrology.* U.S. Geological Survey Water-Supply Paper 2220.

Jennings, J. N. 1983. Karst landforms. *American Scientist* 71:578–586.

Leopold, Luna B. 1974. *Water: A Primer.* San Francisco: W. H. Freeman.

National Research Council. 1993. *Solid-Earth Sciences and Society.* Washington, D.C.: National Academy Press.

U.S. Geological Survey. 1990. *Hydrologic Events and Water Supply and Use.* National Water Summary 1987, U.S. Geological Survey Water-Supply Paper 2350.

# RIVERS: TRANSPORT TO THE OCEANS

Rivers are the major geological agents operating on the surface of the land. As they erode bedrock and transport and deposit sand, gravel, and mud, streams of all sizes, from tiny rills to major rivers, are preeminent carvers of the landscape. To understand how rivers accomplish their geological work, we need to know how water flows in currents, how currents carry sediment, and how streams break up and erode solid rock. On a larger scale, streams carve valleys and assume a variety of forms as they channel water downstream. Knowing the form and size of the channel and the quantity of water carried is important for flood prediction and control. We will follow a river's course from its headwaters to its delta to see how the entire length of a river acts as a large system that adjusts itself to geological changes.

## 13

Tanner Creek, Oregon. The current in this fast-flowing stream is rapid enough to carry large boulders at flood levels. *Gary Ellis Nature Photography*.

Rivers are deeply embedded in the imagery of language. We picture rivers when we speak of flowing waters, babbling brooks, raging torrents. In ordinary language many different words are used to describe waterways, but geologists tend to use the word **stream** for any flowing body of water, large or small, and to reserve **river** for the major branches of a large stream system.

Almost every town or city in most parts of the world has a river, creek, stream, or brook running through it. Rivers are major factors in the economies of many countries, serving as commercial waterways for barges and steamers and as water resources for large populations and industries. The river Nile was vital to the agricultural economy of ancient Egypt and remains important to Egypt today. Living near a river also entails risk. When rivers flood, they destroy lives and property, sometimes on a huge scale.

Streams are pivotal to the dynamics of the Earth's surface, for they play the major role in shaping the face of the continental landscape. Glaciers may produce dramatic landscapes, but they cover only a small portion of the land surface, whereas streams cover by far the greatest part. Streams erode mountains, carry the products of weathering down to the oceans, and deposit billions of tons of sediment along the way in bars and flood deposits. At their mouths, at the edges of the continents, they dump even greater quantities of sediment, building new land out into the oceans. They carry the bulk of the rainwater that falls on land back to the sea, completing the hydrologic cycle.

As they flow, streams assume many forms. Some stretches are straight; others curve gently or loop around sharp bends. At their headwaters streams may run through deep, narrow mountain valleys. Farther downstream they flow in wider channels along broad, low depressions that may be difficult to recognize as valleys. In periods of low water, streams may flow slowly; in floods, their flow is great enough to carry away buildings. How these changing forms and flows arise and what effects they have on the landscape are the subject of this chapter. We begin by looking at the ways in which running water moves in currents and how that movement enables streams to carry various kinds of sediment.

## HOW STREAM WATERS FLOW

All flows of fluid, large and small, share some basic characteristics. Two kinds of fluid flow can be pictured by the lines of motion we call streamlines. In the simplest kind of movement along streamlines, termed **laminar flow,** the streamlines run parallel to one another without mixing or crossing between layers (Figure 13.1a). A common instance of this kind of flow is the slow movement of thick syrup over a pancake, with strands of unmixed melted butter flowing in parallel but separate paths.

A more complex type of movement, **turbulent flow,** shows a confused pattern of streamlines mixing, crossing, interfering, and forming swirls and eddies (Figure 13.1b). Fast-moving river waters typically show this kind of motion. Turbulence—that is, the degree to which there are irregularities and eddies in the flow—may be low or high.

Whether a flow is laminar or turbulent depends on the velocity of the flow, the geometry of the flow (primarily its depth), and the physical properties of the particular fluid. The slower and shallower the flow, the more likely it is to be laminar.

The physical property of a given fluid that promotes laminar flow is viscosity. **Viscosity** is a measure of a fluid's resistance to flow. The more viscous (the "thicker") a fluid is, the more it resists flow.

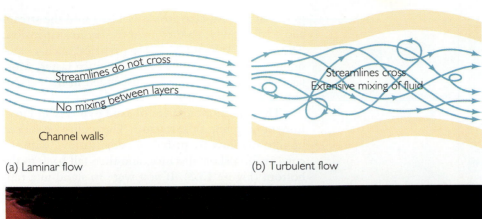

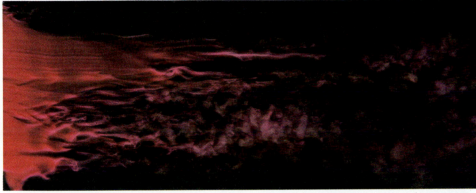

(c) Laminar (left) to turbulent (right)

**FIGURE 13.1** (a) Laminar flow of a fluid between two solid channel walls. (b) Turbulent flow of a fluid between two solid channel walls. (c) The transition from laminar to turbulent flow in water along a flat plate, revealed by injection of a dye. *ONERA.*

Viscosity arises from the attractive forces between the molecules of a fluid; these forces tend to impede the slipping and sliding of molecules past one another. The greater the attractive forces, the greater the resistance to mixing with neighboring molecules and the higher the viscosity, which leads to a greater tendency for laminar flow. A cold syrup or a high-viscosity cooking oil, for example, is sluggish and laminar when it is poured. The viscosity of most fluids, including water, decreases as the temperature rises.

A flow may change from laminar to turbulent as the velocity or depth increases and return to laminar again when either factor decreases. If a fluid heats up, its viscosity may decrease enough to change a laminar flow to a turbulent one.

Water has low viscosity in the common range of temperatures at the Earth's surface, and for this reason alone most watercourses in nature tend to turbulent flow. In addition, the rapid movement of water in most streams makes them turbulent. In nature, we are likely to see laminar flows of water only in thin sheets of rain runoff flowing slowly down nearly level slopes and, in cities, in thin, small flows in gutters. Because most streams and rivers are broad and

deep and their velocities are high, their flows are almost always turbulent.

A stream may show turbulent flow over much of its width but be in laminar flow along its edge, where the flow is shallow and slow. The flow velocity is highest near the center of the stream, and we commonly refer to a rapid flow as a strong current.

## STREAM LOADS AND SEDIMENT MOVEMENT

### Erosion and Transport

Different kinds of fluid flow have different abilities to erode and carry sand grains and other sediment. Laminar flows of water can lift and carry only the smallest, lightest, clay-sized particles. Turbulent flows, depending on their speed, can move particles from clay size up to pebbles and cobbles. As turbulence lifts particles from the bed into the flow, it carries them downstream. It also rolls and slides larger particles along the bottom. The **suspended load** of a stream includes all the material being carried sus-

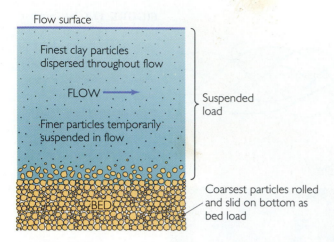

Flow surface

Finest clay particles
dispersed throughout flow

FLOW ⟶

Finer particles temporarily
suspended in flow

Suspended
load

BED

Coarsest particles rolled
and slid on bottom as
bed load

**FIGURE 13.2** A current flowing over a bed of sand, silt, and clay transports particles in two ways: as bed load, the material sliding and rolling along the bottom; and as suspended load, the material temporarily or permanently suspended in the flow itself.

pended in the flow. The **bed load** of a stream is the material carried along the bottom by sliding and rolling (Figure 13.2).

The faster the current, the larger the particles carried as suspended and bed load. The ability of a flow to carry material of a given size is its **competence.** As the current increases in velocity and coarser particles are suspended, the suspended load grows. At the same time, more of the bed material is in motion and the bed load increases. As we would expect, the larger the volume of a flow, the more suspended and bed load it can carry. The total sediment load carried by a flow is its **capacity.**

Interactions between the velocity and volume of a flow affect both the competence and capacity of the stream. Along most of its length, the Mississippi River flows at moderate speeds and carries only fine to medium-sized particles, clay to sand, but it carries huge quantities of them. A small, steep, fast-flowing mountain stream, in contrast, may carry boulders, but only a few of them.

## Settling from Suspension

A stream's ability to carry sediment depends on a balance between the uplifting forces that turbulence exerts on particles and the competing downward pull of gravity, which makes grains settle out of the current and become part of the bed. The speed with which suspended particles of various weights settle to the bottom is called the **settling velocity.** Small

grains of silt and clay are easily lifted into the stream and settle slowly, so they tend to stay in suspension. The settling velocity of larger particles, such as medium- and coarse-grained sand, is much faster. Most larger grains therefore stay suspended in the current only a short time before they settle.

The typical movement of sand grains is an intermittent jumping, or **saltation.** The grains are sucked up into the flow by turbulent eddies, move with the current for a short distance, and then fall back to the bottom (Figure 13.3). If you were to stand in a rapidly flowing sandy stream, you might see a cloud of saltating sand grains moving around your ankles. The bigger the grain, the longer it will tend to remain on the bed before it is picked up. Once it is in the current, it will quickly settle. The smaller the grain, the more frequently it will be picked up and the longer it will take to settle.

To study how rivers carry sediment, geologists and hydraulic engineers measure the relationships between the force the flow exerts on particles in the suspended and bed loads and the sizes of the grains. Engineers use these data to calculate how much sediment a particular flow can move, and how rapidly. With this information, they can judge how to design dams and bridges or estimate how quickly artificial reservoirs behind dams will fill with sediment. Geologists also can infer the velocities of ancient currents from the sizes of grains in sedimentary rocks.

The relationship between grain size and current velocity is shown in Figure 13.4. The lower line bounding the yellow area is the velocity at which all particles of a given size settle to the bed. The upper line is the velocity at which particles are eroded from the bed. The gray area is a transition zone between erosion and settling that depends on the depth of

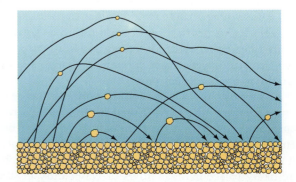

**FIGURE 13.3** Saltation is an intermittent jumping motion of grains. In general, the smaller the particle, the higher it jumps and the farther it travels.

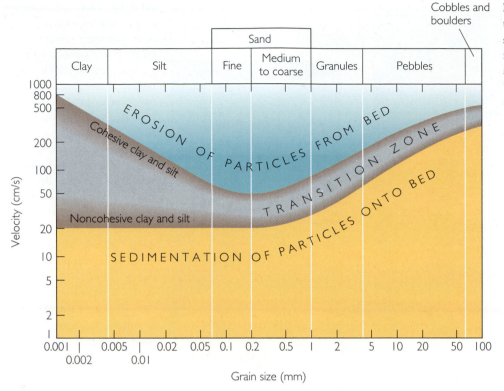

**FIGURE 13.4** The relationship between particle size and velocity of the flow. (After F. Hjulstrom, as modified by A. Sundborg, "The River Klaralven," *Geografisk Annaler,* 1956.)

water in the flow and on factors other than grain size and current velocity. The top line is for depths of 10 m and the bottom line for depths of 0.1 m or less. The right side of the graph shows the steady increase of velocity required to erode and transport larger and larger grains. On the left side of the graph, for grains about 0.1 mm in diameter and smaller, the lower line and transition zone are flat. For these small grains, settling velocities are so slow that even a gentle current, about 20 cm per second, is able to keep the particles in suspension and transport sediment. Fine sediment particles that are cohesive—that is, they stick together, as many clay minerals do—are harder for the flow to lift from the streambed than noncohesive ones. The finer the cohesive particles, the greater the velocity required to erode them, as shown by the uppermost line on the left side of the graph.

## Bedforms: Ripples and Dunes

When sand grains on a streambed are transported by saltation, they tend to form cross-bedded ripples and dunes (see Chapter 7). **Ripples** are low, narrow ridges separated by somewhat wider troughs. The ridges have a gentle slope upstream and a steeper slope downstream. They range in height from less than a centimeter to several centimeters. **Dunes** have the same general form as ripples but are larger, ranging up to many meters high in large rivers. Though harder to observe, underwater ripples and dunes form in the same way, and just as commonly, as those formed by air currents on land. As sand grains move by saltation, they are eroded from the upstream side of ripples and dunes and deposited on the downstream side. The steady downstream transfer of grains across the ridges causes the ripple and dune forms to migrate downstream at speeds slower than individual grain speeds and much slower than the water. (We will look at ripple and dune migration in more detail in Chapter 14.)

As ripples and dunes migrate, the grains are deposited at a characteristic angle of 30° to 45° on the downstream slope, forming cross-bedding. The size of the cross-bedding is proportional to the size of the ripples or dunes. From the cross-bedding alone, even if the ripple or dune form is not exposed, geologists can estimate the relative velocity of the current (see Figure 14.13).

The shape and migration speed of ripples and dunes change as the velocity of the current increases (Figure 13.5). At the lowest velocities, with few

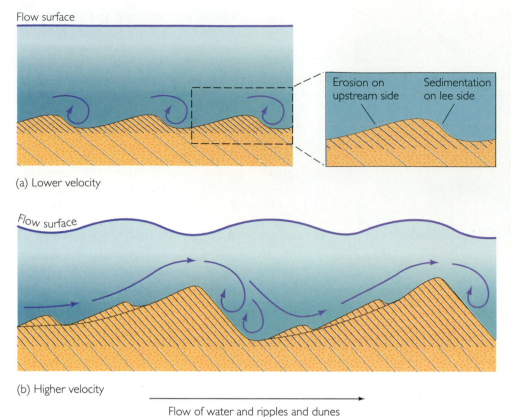

Flow surface

(a) Lower velocity

Erosion on upstream side

Sedimentation on lee side

Flow surface

(b) Higher velocity

Flow of water and ripples and dunes

**FIGURE 13.5** The change in the form of a sandy bed with increasing velocity. (a) At lower velocity a rippled bed forms. Ripples migrate downstream and have a cross-bedded structure. (b) At higher velocity rippled dunes form and migrate downstream. Dunes have the same cross-bedded structure as ripples. Because ripples migrate downstream faster than dunes, they tend to climb over the backs of dunes. Reverse eddies form in the lee of ripples and dunes. (After D. A. Simons and E. V. Richardson, "Forms of Bed Roughness in Alluvial Channels," *American Society of Civil Engineers Proceedings,* vol. 87, 1961, pp. 87–105.)

grains saltating, the sand bed of a stream is flat. At slightly higher velocities, the number of grains saltating increases and small ripples begin to form. As the velocity increases further, the ripples grow larger and migrate faster until, at a certain point, ripples are replaced by dunes. As these dunes grow larger, small ripples form and migrate over the dunes. Very high velocities will wipe out the dunes and form a high-velocity flat bed below a dense cloud of rapidly saltating sand grains. Most of these grains hardly settle to the bottom before they are picked up again. Some are in permanent suspension.

## HOW RUNNING WATER ERODES SOLID ROCK

We can easily see the rapid process of a current picking up loose sand from its bed and carrying it away, thus eroding the bed. At high water levels and during floods, streams can even scour and cut into unconsolidated banks, which then slump into the flow and are carried away. Gullies—valleys made by small streams eroding soft soils or weak rocks—cut their

way headward into higher land. This headward erosion, which accompanies widening and deepening of the valleys, may be extremely rapid, up to several meters in a few years in easily erodible soils.

We cannot so easily see the erosion of solid rock, which takes much longer than the erosion of unconsolidated sediment. One of the major ways in which a river breaks apart and erodes rock is by slow abrasion of the bottom by the sand and pebbles it carries. This sandblasting action wears away the hardest rock. On some river bottoms, swirling eddies can wear deep **potholes** into the river bottom (Figure 13.6). At low water, we can see the pebbles and sand at the bottom of the exposed potholes, a remnant of the material that constantly flushes through them when water is high.

Chemical weathering of rock, which alters its minerals and weakens it along joints and cracks, helps destroy rocks in streambeds just as it does on the land surface. Violent crashes of boulders and constant smaller impacts of pebbles and sand split the rock along cracks. As a result, rock is broken up much faster in river channels than it is by slow weathering on a gently sloping hillside. Once large blocks of bedrock are loosened by impacts and

**FIGURE 13.6** Potholes in river rock; McDonald River, Glacier National Park, Montana. The pebbles rotate inside the potholes, grinding deep holes in the bedrock. *Ric Ergenbright Photography.*

weathering, strong upward eddies may pull them up and out by a sudden, violent plucking action.

Rock erosion is particularly strong at rapids and waterfalls. Rapids are places in a stream where the flow is extremely fast because the slope of the river-bed suddenly steepens, typically at rocky ledges. Because of the speed of the water and the great turbulence, blocks quickly break into smaller pieces and are carried away by the strong current.

Channels at the bottom of waterfalls are eroded at fast rates by the tremendous impact of huge volumes of plunging water and tumbling boulders. Waterfalls erode the underlying rock and recede upstream as the cliff that forms the falls is undercut and the upper beds collapse (Figure 13.7). Erosion by falls is fastest where the rock layers are horizontal and erosion-resistant rocks at the top are underlain by softer rocks, such as shales. Historical records at Niagara Falls, perhaps the best known falls in North America, show that the main part of the falls has been moving upstream at a rate of a meter per year.

## STREAM VALLEYS, CHANNELS, AND FLOODPLAINS

As streams erode the Earth's surface—in some places bedrock, in others unconsolidated sediment—they create valleys. A stream **valley** comprises the entire

**FIGURE 13.7** A waterfall on the Iguaco River, Brazil, retreats upstream as falling water and sediment pound on the bottom and undercut it. From the center to the upper left, one can see the steep walls of the river channel created by retreat of the falls to the right. *Donald Nausbaum.*

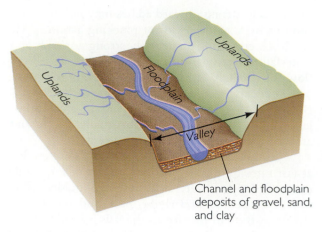

**FIGURE 13.8** A river flows in a channel that moves over a broad, flat floodplain in a wide valley eroded from uplands. Floodplains may be narrow or absent in steep valleys.

area between the tops of the slopes on both sides of the river (Figure 13.8). The cross-sectional profile of many river valleys is V-shaped, but many others show a low and broad profile. At the bottom of the valley is the **channel,** the trough through which the water runs. At low water levels, the stream may run only along the bottom of the channel. At high water levels, the stream occupies most of the channel. In broader valleys a **floodplain,** a flat area about level with the top of the channel, lies on either side of the channel. This is the part of the valley that is flooded when the river spills over its banks.

## River Valleys

In high mountains, stream valleys are narrow and steep-walled, and the channel may occupy most or all of the valley bottom. A small floodplain may be visible only at low water levels. In such valleys, the stream is actively cutting into the bedrock, a characteristic of tectonically active, newly uplifted highlands. In lowlands, where tectonic uplift has long since ceased, stream erosion of valley walls is helped by chemical weathering and mass wasting. With a long time to operate, these processes produce gentle slopes and floodplains many kilometers wide.

## Channel Patterns

Stream channels display a variety of patterns as they run over floodplains. Channels may run straight for some stretches, but for most of their lengths they follow various irregular paths along the bottom of the valley, in some places splitting into multiple channels. The channel may flow along the center of the floodplain in some places and hug one edge of the valley in others. In addition to straight stretches, channel patterns fall into two main types, meandering and braided.

**MEANDERS**    On a great many floodplains, channels follow curves and bends called **meanders** (Figure 13.9). The word comes from the Maiandros (now Menderes) River in Turkey, fabled in ancient times for its winding, twisting course. Meanders are

**FIGURE 13.9** A meandering river west of Anchorage, Alaska. The white areas are point bars formed at the insides of bends. *Peter Kresan.*

**FIGURE 13.10** This section of the San Juan River, Utah, is a good example of incised meanders. *Tom Bean.*

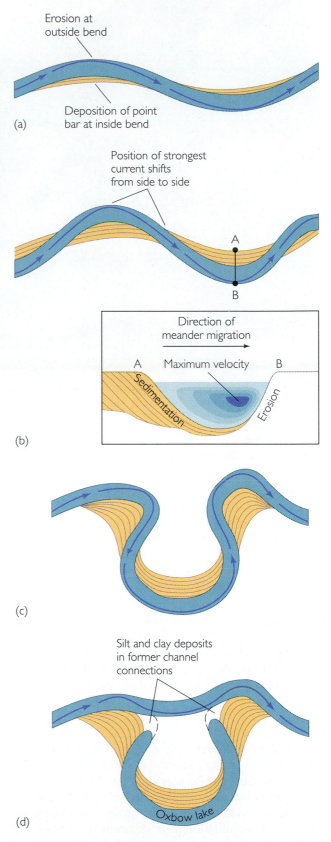

Erosion at outside bend

Deposition of point bar at inside bend

(a)

Position of strongest current shifts from side to side

A
B

(b)

Direction of meander migration →

A   Maximum velocity   B

Sedimentation   Erosion

(c)

Silt and clay deposits in former channel connections

(d)

Oxbow lake

**FIGURE 13.11** (a,b) Meanders migrate over time, eroding the outsides of bends, where the current is strongest, and depositing point bars on the insides of bends, where the current is weakest. (c,d) An oxbow lake evolves as a former meander is cut off.

normal for streams flowing on low slopes in plains or lowlands, where channels typically cut through unconsolidated sediments—fine sand, silt, or mud—or easily eroded bedrock. Meanders are less pronounced but still common where the channel flows on higher slopes and harder bedrock. In the latter terrain, meandering stretches of streams may alternate with long, relatively straight ones.

Some meandering streams have deeply eroded bedrock and completely occupy the valley floor, with no floodplain present (Figure 13.10). Others may meander on somewhat wider floodplains bounded by steep, rocky valley walls. The reasons for these differences are not clear.

Meanders on a floodplain migrate over periods of many years. They shift position from side to side and also downstream, in a snaking motion something like that of a long rope being snapped. This migration is powered by the erosion of the outside bank of a bend, where the current is faster. At the same time, a curved sandbar, called a **point bar,** is laid down along the inside bank of a meander, where the current is slower (Figure 13.11). Migration may be rapid: some meanders on the Mississippi shift as much as 20 m per year. As meanders move, so do the point bars, building up an accumulation of sand and silt over the part of the floodplain across which the channel migrated.

As meanders migrate, sometimes unevenly, the bends may get closer and closer to one another. When they get very close, the river may cut across

**FIGURE 13.12** Itkillik River, Brooks Range, Alaska, a braided stream with numerous splittings and rejoinings of multiple channels. *Tom Bean.*

the neck of the next loop during a major flood, shortening its course. The abandoned, water-filled loop is left behind as an **oxbow lake.**

Engineers sometimes artificially straighten and confine a meandering river, channeling it along a straight path with the aid of concrete abutments. The result is the destruction of wetlands and much of the natural vegetation and animal life of the floodplain. In a change of heart driven by environmental concerns, one such river, the Kissimmee in central Florida, may be about to be restored to its original meandering course, if the U.S. Congress approves funds. If the river is left to its own natural processes, it may take many decades or hundreds of years to restore itself.

**BRAIDED STREAMS** Some streams have not a single channel but many channels, which split apart and then rejoin in a pattern resembling braids of hair. These **braided streams** are found in many settings, from broad valleys in lowlands to stream deposits in wide, downfaulted valleys adjacent to mountain ranges. Braids tend to form wherever rivers have large variations in volume of flow combined with a high sediment load and easily erodible banks. They are well developed, for example, in sediment-choked

streams formed at the edges of melting glaciers (Figure 13.12).

**CHANNEL STABILITY** The three channel patterns—meandering, braided, and straight—blend into one another without sharp divisions. A close look at long straight stretches of river shows some slight curves and bends. Meandering rivers may have some straight stretches. Rivers may be braided in their upper courses and meandering lower down. Whatever the pattern, channels are stable; some keep the same pattern for hundreds of years. The particular set of patterns that a stream shows throughout its length is a response to varying conditions of flow rates and volumes, sediment loads, and erodibility of banks.

## The Stream Floodplain

A stream floodplain is created by the migration of the channel over the floor of the valley. As we have seen, channel migration leaves behind a series of point bars, which build up the surface of the floodplain. As the stream overflows its banks during a flood, parts or all of the floodplain may be covered with sediment deposited by floodwaters. Alternatively, a stream may erode bedrock or unconsolidated sediment as it migrates, producing an erosional floodplain covered with a thin layer of sediment.

When a stream in flood spills over its confining banks, the velocity of the water rapidly decreases as the flow spreads out over the floodplain. As the current loses velocity, it loses its ability to carry sediment, which then settles on the floodplain. Along the immediate borders of the channel, the speed of the flooding waters drops most quickly; as a result, the current deposits much coarse sediment, typically sand and gravel, along a narrow strip at the edge of the channel. Successive floods build up ridges of the coarse material at the edge of the channel. These **natural levees** confine the stream within its banks between floods, even when water levels are high (Figure 13.13). Where levees have built to a height of several meters and the channel is almost filled by the stream, the floodplain level is below the stream level. You can walk the streets of an old river town built on a floodplain, such as Vicksburg, Mississippi, and look up at the levee, knowing that the river waters are rushing by above your head.

During floods, finer sediments—silts and muds—are carried well beyond the channel banks, frequently over the entire floodplain, and deposited there as floodwaters continue to lose velocity. As floodwaters

recede, standing ponds and pools of water are left behind. Here the finest clays are deposited as the water gradually disappears by evaporation and infiltration. Fine-grained floodplain deposits have been a major resource for agriculture since ancient times. The fertility of the floodplains of the Nile and other rivers of the Middle East, which contributed to the evolution of the early cultures that flourished there thousands of years ago, depended on frequent flooding. Today the great, broad floodplain of the Ganges River in northern India continues to play an important role in India's life and agriculture (see Box 13.1).

## STREAMS CHANGE WITH TIME AND DISTANCE

Streams are dynamic systems, constantly changing as they go from low to high waters and floods in a few years. They also change the shapes of their valleys over longer periods. The flow of a stream appears steady when you look at it from a bridge for a few minutes or canoe along it for a few hours. But the volume and velocity of a stream flow at a single place may change appreciably from month to month and season to season. Streams also change their flows and channel dimensions as they move downstream from narrow valleys in their upland headwaters to broader floodplains in their middle and lower courses. Most of these longer term changes in streams are adjustments in the normal (nonflood) volume and velocity of flow as well as the depth and width of the channel.

The changes at any point in a stream are reflections of an imbalance between the input from surface runoff and groundwater and the output by drainage downstream. If it rains heavily for days or weeks in the area drained by the stream, more water will flow for a time, until the rain diminishes and the flow declines to normal. A common seasonal change is the higher discharge that follows spring rains and melting snows in colder temperate regions.

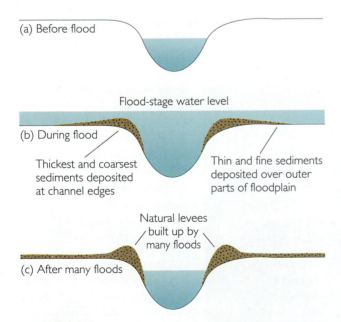

(a) Before flood

Flood-stage water level

(b) During flood

Thickest and coarsest sediments deposited at channel edges

Thin and fine sediments deposited over outer parts of floodplain

Natural levees built up by many floods

(c) After many floods

**FIGURE 13.13** The formation of natural levees by river floods.

### Discharge

We measure the size of a stream's flow by its discharge. The **discharge** is the volume of water that passes a given point in a given time as it flows through a channel of a certain width and depth. Discharge is commonly measured in cubic meters per second or cubic feet per second. A typical small

## BOX 13.1  THE DEVELOPMENT OF CITIES ON FLOODPLAINS

Near the beginnings of historical time, about 4000 years ago, cities began to dot river floodplains in Egypt along the Nile, in the ancient land of Mesopotamia along the Tigris and Euphrates rivers, and in the East, along the Indus River of India and the Yangtze and Huang rivers of China. These were natural sites for urban settlements, for they combined easy transportation along the river with closeness to the best agricultural lands on the fertile soils of the floodplain. Many of the capital cities of Europe are on river floodplains: London on the Thames, Paris on the Seine, Rome on the Tiber. Floodplain cities in North America include St. Louis on the Mississippi, Cincinnati on the Ohio, and Montreal on the St. Lawrence.

Floods periodically destroyed the lower parts of these ancient and modern cities, but each time they were rebuilt. Today most large cities are protected by artificial levees that strengthen and heighten the river's natural levees. Extensive systems of dams can help to control flooding that would affect these cities, but they cannot eliminate the risk entirely.

In 1973 the Mississippi went on a rampage with a flood that continued for 77 consecutive days at St. Louis. It reached a record 4.03 m above flood stage (the height at which the river first overflows the channel banks). In 1993 the Mississippi and its tributaries broke loose again, shattering the old record and causing billions of dollars in damage. Some geologists think that the artificial levees and other engineering works along the river, such as dikes protecting harbors, have confined the Mississippi unnaturally. This confinement prevents the river from eroding its banks and widening its channel to accommodate some of the additional water flowing during times of high discharge and thus contributes to the record high floods.

What are cities and towns in this position to do? Some have urged a halt to all construction and development on the lowest parts of the floodplains. Some have called for the elimination of federally subsidized disaster funds for rebuilding in such areas. Harrisburg, Pennsylvania, hit hard by a flood in 1972, turned some of its devastated riverfront area into a park. Smaller towns have moved to higher ground. Yet some people who have lived all their lives on floodplains want to stay and are prepared to live with the risk. The costs of protecting some river-bottom areas are prohibitive, and these places will continue to pose public policy problems.

In the summer of 1993, the Mississippi swept over its banks in the most extensive flood recorded in its history, inundating towns up and down the river. In Davenport, Iowa (shown here), the river eclipsed all records, cresting on July 9 at 6.9 m, 2.3 m above flood stage. *John Eastcott and Yua Momatiuk/Woodfin Camp.*

stream may vary in discharge from about 0.25 to 300 m³ per second. The discharge of the Mississippi River at times is as low as 1400 m³ per second and in times of flood may be more than 57,000 m³ per second.

We find the discharge by multiplying the velocity of the flow and the cross-sectional area (the width multiplied by the depth of the part of the channel occupied by water):

$$\text{Discharge} = \text{cross section} \times \text{velocity}$$
$$\begin{array}{cc} (\text{width} \times & (\text{distance} \\ \text{depth}) & \text{traveled} \\ & \text{per} \\ & \text{second}) \end{array}$$

This equation leads us to expect that if discharge is to increase, either velocity or cross-sectional area, or both, will have to increase. As you increase the discharge of a garden hose by turning up the water pressure, the cross-sectional area of the hose, measured by its diameter, cannot change, so the water comes out at higher speed. As discharge of a stream increases at a particular point, both the velocity and the cross-sectional area tend to increase. The cross-sectional area increases as the flow occupies more of the channel's width and depth (Figure 13.14).

Normal discharge in most rivers increases downstream as more and more water is collected from tributaries. As we have seen, increased discharge means that width, depth, or velocity must increase too. Velocity does not increase downstream as much as the increase in discharge leads us to expect, because of decreases in slope along the lower courses of a stream (decreasing slope reduces velocity). Where discharge does not increase significantly downstream and slope decreases greatly, a river will flow more slowly.

## Floods

A flood is an extreme case of an increased discharge that results from a short-term imbalance between input and output. As the discharge increases, the flow velocity in the channel increases and the water gradually fills the channel. As the discharge continues to increase, the water floods over the banks. Rivers flood regularly, some at infrequent intervals, others almost every year. Some floods are large, with very high water levels lasting for days; at the other extreme are minor floods that barely break out from the channel before they recede. Small floods are more

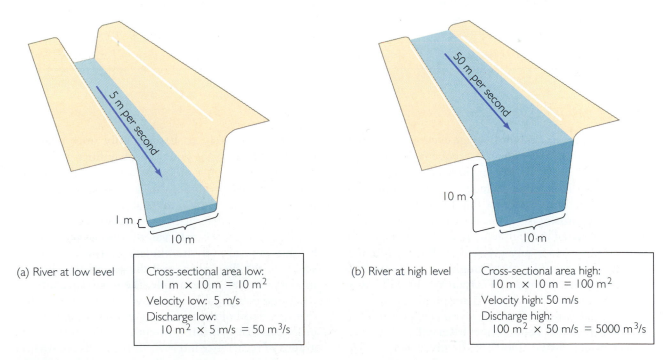

**FIGURE 13.14** Discharge depends on velocity and cross-sectional area. (a) A river at low discharge. (b) A river at high discharge.

## BOX 13.2  HISTORIC FLOODS AND FLOOD CONTROL

A year seldom passes without reports of a major flood somewhere in the world. History records so many disastrous floods that only the greatest are mentioned. One of the largest of modern times was the great flood of the Huang Ho in China in 1931, which killed 4 million people. More recently, floods in Bangladesh near the mouths of the Ganges and Brahmaputra rivers led to the deaths of 300,000 people.

Advance warning of a flood makes the difference in preventing deaths. One of the most famous floods in American history took place in Johnstown, Pennsylvania, in 1889, when a dam on the Conemaugh River failed and 2200 people died. In 1977, Johnstown was hit again when another dam failed, but this time, although more property was lost than in the earlier flood, the death toll was reduced to 49.

Flooding is not restricted to a few dangerous localities. It is estimated that 20 million people in the United States alone live in areas subject to river and coastal flooding. In recent decades, floods have caused property damage running over $1.5 billion annually in addition to the enormous suffering of families flooded out of their homes.

Flood control by dams is moderately effective for smaller floods but may offer little protection from very large ones. Large floods along the Susquehanna River of Pennsylvania were supposed to be contained by a system of 23 dams built after a flood in 1936. In June 1972, Hurricane Agnes dropped 4 to 18 inches of rain on various localities, amounting to 3 trillion gallons of water over six days. Four to six inches of rain had fallen during the three previous weeks and the ground was already saturated. Most of the dams spilled over on the third day of the hurricane. The flood crested soon after at Harrisburg, the capital of Pennsylvania, at a meter higher than the previous record of 1936. Calculations have indicated that 522 new dams would have to be built to protect Harrisburg against another such flood. Clearly, the absurd expense of building so many dams and permanently flooding vast tracts of land behind them make this an unfeasible option.

Flood prediction depends on careful monitoring of weather conditions, such as prolonged storms or hurricanes. Heavy rainfall, particularly when the ground may already be saturated with water from earlier

frequent, occurring on the average every 2 or 3 years. Large floods are generally less frequent, usually occurring only every 10, 20, or 30 years.

Because it is impossible to predict months in advance exactly how high a flood will occur in any given year, geologists measure the frequency of a flood in terms of the *probability* that a flood of a given height will occur in any given year. For a particular stream, for example, there might be a 20 percent probability that a flood of a certain height will occur in any one year. This chance corresponds to an average time interval—in this case 5 years (20 percent = 1 in 5)—called the **recurrence interval,** which we expect between two floods of the given height. We speak of a flood of this height as a 5-year flood. A 50-year flood on the same stream will be much

higher but is likely to happen only once every 50 years. A graph of the annual probabilities and recurrence intervals for a range of flood heights in one river is shown in Figure 13.15. When we speak of recurrence intervals, we must always remember that they express probabilities, not certainties.

The recurrence interval of floods of different heights depends on the climate of the region, the width of the floodplain, and the size of the channel. In a dry climate, for example, the recurrence interval of a flood of a given height may be much longer than that of a flood of the same height on a similar stream in an area that gets intermittent rain. For this reason, graphs of recurrence intervals of major rivers are necessary if river towns are to be prepared to cope with floods of various heights (see Box 13.2).

Houses destroyed by the 1889 flood in Johnstown, Pennsylvania. *Johnstown Area Heritage Association.*

rains or melting snows, predictably leads to increased discharge and eventual flooding. A detailed knowledge of how fast a particular river's discharge increases after rains can be used to determine the probability of flooding so that warnings can be given.

The combination of some dams for partial control and the careful and calculated raising of artificial levees can mitigate many floods, but it appears that we cannot count on such measures to eliminate the danger from large floods everywhere. Our efforts might be better spent on altering the pattern of urban development in the most hazardous areas.

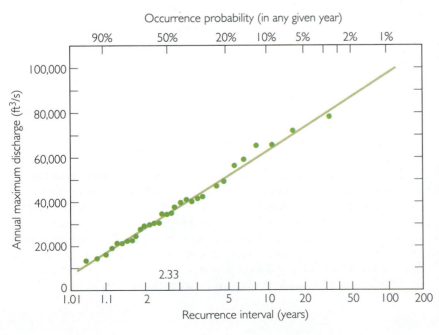

**FIGURE 13.15** The flood frequency curve for annual floods on Skykomish River at Gold Bar, Washington. This curve predicts the probability that a flood of a certain height will occur in any given year. (After T. Dunne and L. B. Leopold, *Water in Environmental Planning,* San Francisco, W. H. Freeman, 1978.)

# The Longitudinal Profile and the Concept of Grade

We have seen that stream flow at any locality balances inputs and outputs, which temporarily get out of balance during floods. There is also a larger scale and longer term balance, revealed by studies of changes in discharge, velocity, channel dimensions, and especially slope along the entire length of a stream, from headwaters to mouth. This long-term balance reflects an equilibrium between erosion of the streambed and sedimentation in the channel and floodplain along the entire course of the stream.

This equilibrium is controlled by several factors: topography, climate, stream flow (including both discharge and velocity), and the resistance of rock to weathering and erosion. A particular combination of factors—such as high topography, humid climate, high discharge and velocity, hard rocks, and low sediment load—would make the stream erode a steep valley into bedrock and carry downstream all sediment derived from that erosion. Conversely, downstream, where topography is lower and the stream might be flowing over easily erodible sediments, the river would deposit bars and floodplain sediments, building up the elevation of the streambed by sedimentation.

We describe the slope of a river from headwaters to mouth by plotting the elevation of its streambed against distances from the headwaters (Figure 13.16). The smooth, concave-upward curve on this graph is the **longitudinal profile** of the river, notably steep near the stream's head and low, almost level, near its mouth. All streams, from small rills to large rivers, show this same general concave-upward profile.

Why do all streams, many of which differ so markedly in detail, follow this profile? The answer lies in the combination of factors that control erosion

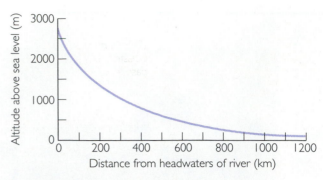

**FIGURE 13.16** The longitudinal profile of the Platte and South Platte rivers from the headwaters of the South Platte in central Colorado to the mouth of the Platte at the Missouri River in Nebraska. (Data from H. Gannett, in "Profiles of Rivers in the United States," USGS Water-Supply Paper 44, 1901.)

and sedimentation. Because streams run downhill, they all start at elevations higher than their lower courses and erode faster upstream than downstream. In their lower courses, with sediments inevitably derived from erosion of the upper courses, sedimentation becomes more significant. Differences in topography and the other factors may make the longitudinal profile steeper or shallower in the upper and lower courses of the stream, but the general shape remains concave upward.

The longitudinal profile is controlled at its lower end by a stream's **base level,** the elevation at which it enters a large standing body of water, such as a lake or the ocean, and so disappears as a stream (Figure 13.17). Streams cannot cut below base level, for base level is the "bottom of the hill"—the lower limit of the longitudinal profile.

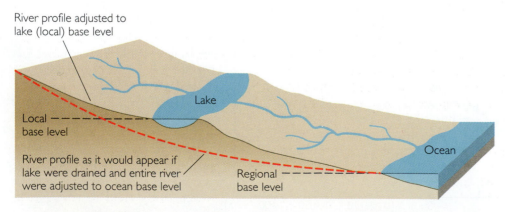

**FIGURE 13.17** Natural regional and local base levels as illustrated by longitudinal profiles of a river flowing into a lake and from the lake into the ocean. In each river segment the profile adjusts to the lowest level it can reach.

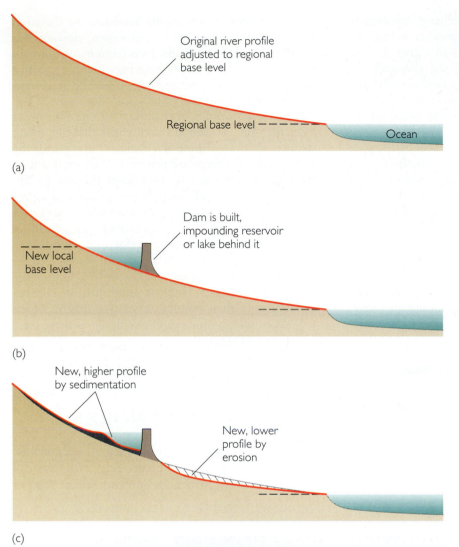

(a)

(b)

(c)

Original river profile
adjusted to regional
base level

Regional base level

Ocean

Dam is built,
impounding reservoir
or lake behind it

New local
base level

New, higher profile
by sedimentation

New, lower
profile by
erosion

**FIGURE 13.18** A change in the base level of a river caused by human intervention and its consequences for the river's profile. (a) The original profile is adjusted to equilibrium with the regional base level. (b) The equilibrium is upset when a dam is built. The dam impounds a lake behind it and raises the local base level. (c) Upstream, the riverbed is adjusted to a new higher profile as sediment is deposited in and in front of the lake; downstream, the bed is eroded to a new lower profile.

Changes in natural base level affect the longitudinal profile in predictable ways. If the regional base level rises, as when sea level rises, the profile shows the effects of sedimentation as the river builds up channel and floodplain deposits to reach the new base-level elevation. Damming a river artificially can create a new local base level, with similar effects on the longitudinal profile (Figure 13.18). The slope of the river upstream from the dam decreases, lowering its velocity and decreasing its ability to transport sediment. This causes the stream to deposit some of the sediment on the bed, which makes the concavity somewhat shallower than it was before the dam was built. Below the dam, the river, now carrying much less sediment, adjusts its profile to the new conditions and typically erodes its channel in the section just below the dam.

As sea level falls, the regional base levels of all streams flowing into the ocean are lowered, and their valleys are cut into former stream deposits. When the drop in sea level is large, as it was during the last glacial period, rivers erode steep valleys into coastal plains and continental shelves. Because fluctuations in sea level are caused by local tectonics, plate tectonics, continental drift, and worldwide glaciations, the courses of ancient rivers reflect many of the Earth's most important geological processes.

As a stream's profile becomes stable over a period of years, a balance is achieved between erosion and sedimentation. That balance is governed by the elevation of the stream's headwaters, its base level, and all of the other factors we have mentioned that control the equilibrium of the stream profile. At equilibrium, the stream is a **graded stream,** one in which the slope, velocity, and discharge combine to transport its sediment load, with neither sedimentation nor erosion. If the conditions that give rise to a particular graded stream profile are changed, the stream's profile changes to reach a new equilibrium. This may involve changes in depositional and erosional patterns and alterations in the shape of the channel.

Over geologic times, where regional base level is constant, the longitudinal profile reflects the balance between tectonic uplift and erosion on the one hand and transport and deposition on the other. If uplift is dominant, typically in the upper courses of a stream, the profile is steep and expresses the dominance of erosion and transport. As uplift slows, the profile is lowered as the headwater region is eroded.

One of the places in which a river must adjust suddenly to changed conditions is at a mountain front, where streams leave narrow mountain valleys for broad, relatively flat valleys. Along such mountain fronts, typically at steep fault scarps, streams drop large amounts of sediment in cone- or fan-shaped accumulations called **alluvial fans** (Figure 13.19). This deposition results from the sudden decrease in velocity as the channel widens greatly. To a minor extent, a lowering of slope below the front also slows the stream velocity. The surface of the alluvial fan normally shows a concave-upward profile connecting the steeper mountain part of the profile with the gentler valley or plains profile. Coarse mate-

rials, from boulders to sand, dominate on the steep upper slopes of the fan. Lower down, deposits are finer sands, silts, and muds. Fans from many adjacent streams along a mountain front may merge to form a long wedge of sediment whose appearance may mask the outlines of the individual fans that make it up.

The role of uplift in changing the equilibrium of a stream valley is seen in the **terraces** that line many streams. Many of these flat benches that climb the valley walls like steps are paired, one on each side of the stream, both at the same level (Figure 13.20). Terraces are made of floodplain deposits and represent former floodplains, constructed when the stream was at a higher level. The sequence of events forming terraces starts when a stream forms a floodplain. Rapid uplift then changes the stream's equilibrium, causing it to cut down into the floodplain. In time the stream reestablishes a new equilibrium at a lower level. It may then build another floodplain, which will also undergo uplift and be sculpted into another, lower pair of terraces.

## DRAINAGE NETWORKS

Every rise between two streams, whether a few meters in height or a mountainous ridge, forms a **divide,** a line along which all rain that falls is shed as

**FIGURE 13.19** An alluvial fan in Copper Canyon, Death Valley, California. The fan starts at the upper center of the photo, where the stream leaves the confined channel in the mountains for the broad lowlands. *Peter Kresan.*

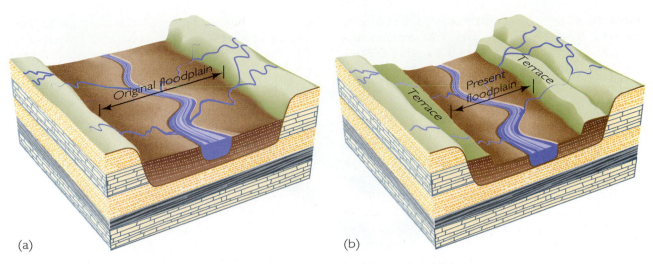

(a)                                                          (b)

**FIGURE 13.20** River terraces form when a river erodes into its floodplain and establishes a new floodplain at a lower level. The terraces are remnants of the former floodplain.

runoff down one side of the rise or the other. All of the divides that separate a stream and its tributaries from their neighbors define its **drainage basin,** the area that funnels all its water into the network of streams draining the area (Figure 13.21). Drainage basins range from a small area, such as a ravine surrounding a small stream, to a great region drained by a major river and its tributaries (Figure 13.22). A continent is divided into major drainage basins separated by a major divide. In North America the continental divide along the Rocky Mountains separates all waters flowing into the Pacific Ocean from those going to the Atlantic.

Many divides tend to keep their places over long times as they are eroded to low ridges. But in some places, divides do change. If a stream on one side of a

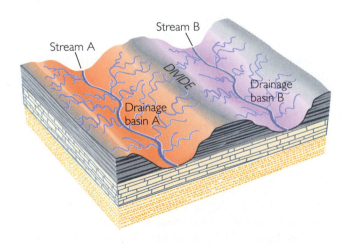

**FIGURE 13.21** Stream valleys and drainage basins are separated by divides, which are ridges, gentle uplands, or mountain ranges.

**FIGURE 13.22** The natural drainage basin of the Colorado River covers about 630,000 km², a large part of the southwestern United States. This basin is surrounded by divides that separate it from the neighboring drainage basins. (After USGS.)

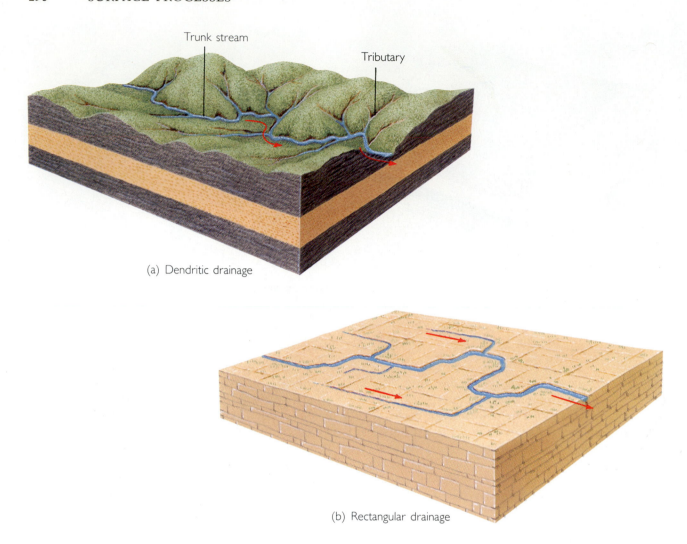

(a) Dendritic drainage

(b) Rectangular drainage

divide is able to erode and transport sediments much more rapidly than a stream on the opposite side, the divide may be eroded unevenly. At some place, the more active stream may break through the divide and "capture" the drainage of its slower neighbor, a case of **stream piracy.** As streams become larger, piracy becomes much less common, and it is rare where rivers are large. Stream piracy explains such odd landscapes as narrow valleys that have no active streams running in them.

## Drainage Patterns

A map of the courses of streams, large and small, tributaries and main rivers, shows a pattern of connections between tributaries called **drainage net-** **works.** As streams are followed upstream, steadily divided by tributaries into smaller and smaller streams, their drainage networks show characteristic branching patterns (Figure 13.23). Branching is a general property of many kinds of networks in which material is collected and distributed. A stream drainage network collects water as it transports it downstream. We will see shortly that at their mouths rivers follow a branching network in reverse that distributes the flow among many channels. We can compare these patterns with those of the human circulatory system, which distributes blood to the body through a branching system of arteries and collects it through a corresponding system of veins.

Perhaps the most familiar branching networks are those of trees and roots. Most rivers follow the same kind of pattern, called **dendritic drainage,** from the Greek word for tree, *dendron*. This fairly

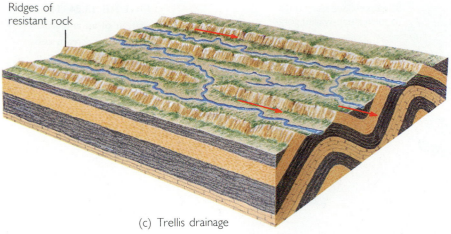

(c) Trellis drainage

**FIGURE 13.23** Typical drainage networks. (a) A river with dendritic drainage is characterized by branches similar to the limbs of a tree. (b) In a typical rectangular drainage pattern developed on a strongly jointed rocky terrain, drainage tends to follow the joint pattern. (c) Trellis drainage develops in valley and ridge terrain, where rocks of varying resistance to erosion are folded into anticlines and synclines. (d) Radial drainage patterns develop on a single large peak, such as a large dormant volcano.

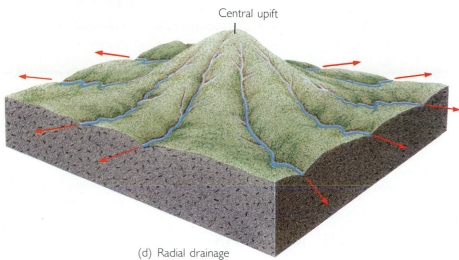

(d) Radial drainage

random drainage pattern is typical of terrains where the bedrock is uniform, such as horizontal sedimentary rocks or massive igneous or metamorphic rocks. Where rapid weathering along fractures or joints in bedrock controls stream courses, a more orderly **rectangular drainage** pattern develops. A special variety of rectangular drainage, the **trellis** pattern, resembles the right-angled geometry of wooden trellises. This pattern evolves where bands of rock resistant to weathering alternate with bands that erode more rapidly, a situation found in terrains where sedimentary rocks have been deformed into parallel folds. The larger streams run along valleys eroded from the more easily eroded rocks and are joined at right angles by short tributaries running down from ridges of more resistant rocks. Drainage from a central high point, such as a volcano or domal uplift, is **radial drainage.**

## Drainage Patterns and Geologic History

We can observe directly or judge from historical records how most stream drainage patterns evolve, but others elude a simple explanation. Some streams, for example, cut through ridges to form steep-walled notches or gorges, in many places through bedrock that is resistant to erosion. We would have expected the stream to run along the lowland on either side of the ridge rather than cutting a narrow valley directly through the ridge.

We can explain this kind of stream pattern with a knowledge of the geologic history of the region. If a ridge is formed by structural deformation while a preexisting stream is flowing over it, the stream may erode its valley as fast as the ridge is elevated (Figure 13.24). Such a stream is called an **antecedent stream** because it was present before the ridge.

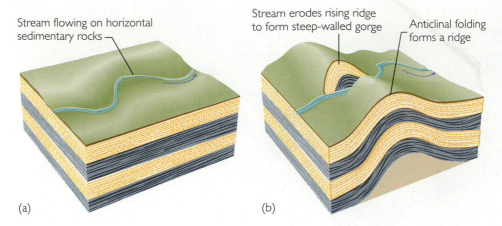

Stream flowing on horizontal sedimentary rocks

(a)

Stream erodes rising ridge to form steep-walled gorge

Anticlinal folding forms a ridge

(b)

**FIGURE 13.24** The formation of an antecedent stream as a stream cuts through a ridge as it is uplifted.

In another geological situation, a stream may be flowing over horizontal sedimentary rocks in a dendritic drainage pattern and cut down into underlying folded and faulted rocks having varying resistance to erosion. Such streams tend to continue the pattern they developed earlier rather than adjusting to their new conditions. They are termed **superposed streams** because they are imposed on the lower set of rocks from above. In this case, a dendritic pattern is forced onto a surface that would otherwise have evolved a rectangular network (Figure 13.25).

# DELTAS: THE MOUTHS OF RIVERS

Sooner or later all rivers end as they flow into large standing bodies of water. There the current gradually comes to a stop as the water can no longer flow downhill. As the river current flows into a lake or the ocean, it mixes with the surrounding water and gradually loses its forward momentum. The largest rivers, such as the Amazon and the Mississippi, can maintain some current many kilometers out to sea (Figure 13.26). Where smaller rivers enter a turbulent, wave-swept coast, the current disappears almost immediately beyond the river's mouth.

## Delta Sedimentation

As its current gradually dies out, a river progressively loses the power to transport sediment. The coarsest material, normally sand, is dropped first, in most rivers right at the mouth. Finer-grained sands are dropped farther out, followed by silt, and still farther by clay. As the floor of the lake or sea slopes to deeper water away from the shore, the dropped

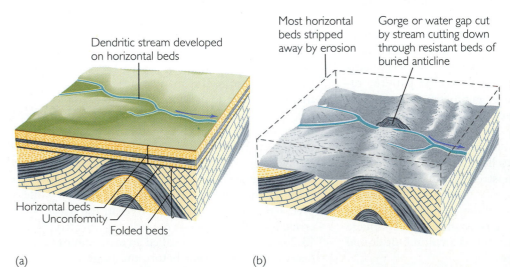

Dendritic stream developed on horizontal beds

Horizontal beds
Unconformity
Folded beds

(a)

Most horizontal beds stripped away by erosion

Gorge or water gap cut by stream cutting down through resistant beds of buried anticline

(b)

**FIGURE 13.25** The development of a superposed stream by erosion of horizontal beds unconformably overlying folded beds of varying resistance to erosion. As the downcutting stream encounters a buried anticline, it erodes a narrow gorge, or water gap, in the resistant beds of the anticline.

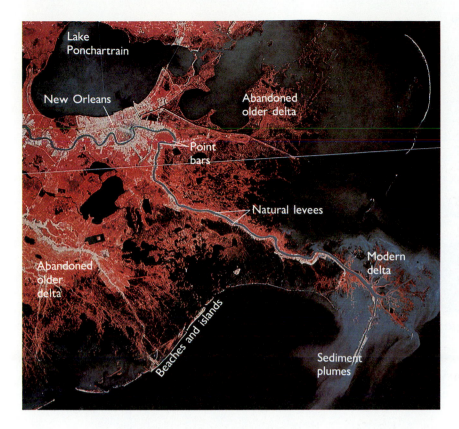

FIGURE 13.26 The infrared-sensitive film used to shoot this satellite image of the Mississippi Delta causes the vegetation to appear red, relatively clear water to appear dark blue, and water with suspended sediment to appear light blue. At the upper left are New Orleans and Lake Ponchartrain. Well-defined natural levees and point bars are at the center. At the lower left are beaches and islands that were formed as sand from the river was transported from the delta by waves and currents. (From G. T. Moore, "Mississippi River Delta from Landsat 2," *Bulletin of the American Association of Petroleum Geologists,* 1979.)

materials build up a depositional platform called a **delta** (Figure 13.27). (We owe the name "delta" to the Greek historian Herodotus, who traveled through Egypt around 450 B.C. The roughly triangular shape of the sediments deposited at the mouth of the Nile prompted him to name it after the Greek letter Δ, delta.)

Materials dropped on top of the delta, normally sand, make up horizontal **topset beds.** Downcurrent, fine-grained sand and silt are deposited to form gently inclined **foreset beds,** which resemble large-scale cross-beds. Spread out on the seafloor seaward of the foreset beds are thin, horizontal **bottomset beds** of mud.

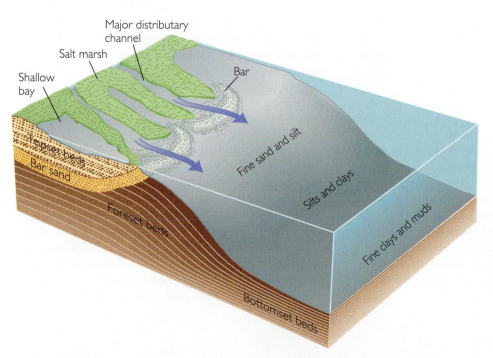

FIGURE 13.27 A typical large marine delta, many kilometers in extent, in which the foreset beds are fine-grained and deposited at a very low angle, normally only 4 to 5° or less. Sandbars form at the mouths of the distributaries, where the currents' velocity suddenly decreases. The delta builds forward by the advance of the bar and topset, foreset, and bottomset beds. Between distributary channels, shallow bays fill with fine-grained sediment and become salt marshes. This general structure is found on the Mississippi Delta.

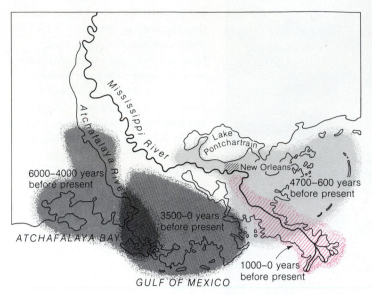

**FIGURE 13.28** The Mississippi Delta. Over the past 6000 years, the river has built its delta first in one direction and then in another as water flow shifted from one major distributary to another. The modern delta was preceded by deltas deposited to the east and west.

## The Growth of Deltas

As the delta builds forward, the mouth of the river advances into the sea, leaving new land in its wake. A major river such as the Mississippi or the Nile forms a large delta thousands of square kilometers in area. The delta of the Mississippi, like many other major river deltas, has been growing for millions of years. It started out, about 150 million years ago, around what is now the junction of the Ohio and the Mississippi rivers, at the southern tip of Illinois. It has advanced about 1600 km since then, creating almost the entire states of Louisiana and Mississippi as well as major parts of adjacent states.

As rivers approach their deltas, where the slope profile is almost level with the sea, they reverse their normal upstream-branching drainage pattern. Instead of collecting more water from **tributaries,** they assume a pattern of **distributaries,** smaller rivers that branch off *downstream,* thus distributing the water and sediment into many channels. As deltas grow they shift the flows from some distributaries to others with shorter routes to the sea. As a result of such shifts, the delta grows in one direction for some hundreds or thousands of years, then breaks out into a new distributary and begins to grow into the sea in another direction (Figure 13.28).

## Effects of Waves and Tides

Strong waves, shoreline currents, and tides affect the growth and shapes of deltas built into the sea. Waves and shoreline currents may move the sediment along the shore almost as rapidly as it is dropped by the river. The delta front then becomes a long beach shoreline with only a slight seaward bulge at the mouth. Where tidal currents move in and out, they redistribute deltaic sediment into elongate bars parallel to the direction of the currents, which in most places are at approximately right angles to the shore.

Where waves and tides are strong enough, deltas cannot form. The sediment brought down to the sea by the river is dispersed along shorelines as beaches and bars and is transported into deeper waters offshore. The east coast of North America lacks deltas for this reason. The Mississippi has been able to build out its delta because neither waves nor tides are very strong in the Gulf of Mexico.

Regardless of the kinds of deltas formed, or whether any are formed at all, the ultimate end of all of a river's water and sediment load is the ocean. Along with 45 trillion cubic meters of water, the world's rivers bring about 7 billion tons of fine sediments and 1 to 2 billion tons of coarse sediments to the oceans each year. But they deposit more than solid sediments. Material dissolved by chemical weathering is also brought by rivers to the sea, there to be mixed with seawater. The nearly 4 billion tons per year of these dissolved materials transported by rivers contribute the major part of the salts dissolved in seawater.

# SUMMARY

**How is flowing water in streams able to erode solid rock and to transport and deposit sediment?** Any fluid can move in either laminar or turbulent flow, depending on its velocity, viscosity, and flow geometry. The turbulence that characterizes most streams is responsible for transporting sediment by suspension (clays), saltation (sands), and rolling and sliding along the bed (sand and gravel). The tendency for particles to be carried in suspension is countered by the gravitational force that leads them to settle to the bottom, measured by the settling velocity. When a stream flow slows, it loses its competence to carry sediment and deposits it, in many places as beds of rippled, cross-bedded sand. Running water erodes solid rock by abrasion, impact breakage, enlargement and opening of cracks induced by weathering, plucking, and undercutting.

**How do stream valleys and their channels and floodplains evolve?** As a stream flows, it carves a valley with steep to gently sloping walls and a more or less broad floodplain on either side of the channel. The channel may be straight, meandering, or braided. Although the channel carries all of the water and sediment during normal, nonflood times, as a stream increases its discharge to flood stage, it overflows its banks and floods. As floodwaters inundate the floodplain, the velocity slows, and the waters drop sediment that builds up natural levees and floodplain deposits. Flood heights are related to the frequency of flooding by recurrence intervals, a measure relating the probability that a flood of a given height will occur in any year to the interval of time between floods of that height.

**How does a stream's longitudinal profile reflect the equilibrium between erosion and sedimentation?** A stream is in dynamic equilibrium between erosion and sedimentation over its entire length. This equilibrium is affected by topography, discharge, velocity, and slope, such that a stream's longitudinal profile, always concave upward, reflects the elevation at its headwaters and the base level at its mouth in a lake or the ocean. Uplift at the upper end of a stream and the rise and fall of sea level at the lower end will change the profile. Alluvial fans form at mountain fronts, in response primarily to an abrupt widening of the valley and secondarily to a change in slope.

**How do drainage networks work as collection systems and deltas as distribution systems for water and sediment?** Rivers and their tributaries constitute a branching-upstream drainage network that collects the water and sediment running off its drainage basin, which is separated from its neighbors by a divide. Drainage networks show various kinds of branching patterns: dendritic, rectangular, trellis, or radial, depending on topography, rock type, and structure. Near its mouth, as it forms its delta, a river tends to branch downstream into distributaries. Deltas are major sites of deposition of sediment as rivers drop their sediment load as topset, foreset, and bottomset beds. Deltas are modified or even absent where waves, tides, and shoreline currents are strong.

# KEY TERMS AND CONCEPTS

stream (p. 278)
river (p. 278)
laminar flow (p. 278)
turbulent flow (p. 278)
viscosity (p. 278)
suspended load (p. 279)
bed load (p. 280)
competence (p. 280)
capacity (p. 280)
settling velocity (p. 280)
saltation (p. 280)
ripple (p. 281)

dune (p. 281)
pothole (p. 282)
valley (p. 283)
channel (p. 284)
floodplain (p. 284)
meander (p. 284)
point bar (p. 285)
oxbow lake (p. 286)
braided stream (p. 286)
natural levee (p. 287)
discharge (p. 287)

recurrence interval (p. 290)
longitudinal profile (p. 292)
base level (p. 292)
graded stream (p. 293)
alluvial fan (p. 294)
terrace (p. 294)
divide (p. 294)
drainage basin (p. 295)
stream piracy (p. 296)
drainage network (p. 296)
dendritic drainage (p. 296)

rectangular drainage (p. 297)
trellis drainage (p. 297)
radial drainage (p. 297)
antecedent stream (p. 297)
superposed stream (p. 298)
delta (p. 299)
topset bed (p. 299)
foreset bed (p. 299)
bottomset bed (p. 299)
tributary (p. 300)
distributary (p. 300)

# EXERCISES

**1.** How does velocity determine whether a given flow is laminar or turbulent?

**2.** How does the size of a sediment grain affect the speed with which it settles to the bottom of a flow?

**3.** What kind of bedding characterizes a ripple or a dune?

**4.** How do braided and meandering river channels differ?

**5.** What is the discharge of a stream and how does it vary with velocity?

**6.** What is the commonest kind of drainage network developed over horizontally bedded sedimentary rocks?

**7.** What is a delta distributary?

# THOUGHT QUESTIONS

**1.** Why might the flow of a very small, shallow stream be laminar in winter and turbulent in summer?

**2.** In some places, engineers have artificially straightened a meandering stream. If such a straightened stream is then left free to adjust its course naturally, what changes would you expect?

**3.** Your hometown, built on a river floodplain, experienced a 50-year flood last year. What are the chances that another flood of that height will occur next year?

**4.** In the first few years after a dam was built on it, a stream severely eroded its channel downstream of the dam. Could this erosion have been predicted?

**5.** If global warming becomes responsible for a significant rise in sea level as polar ice melts, how would the longitudinal profiles of the world's rivers be affected?

**6.** What kind of drainage network do you think is being established on Mount St. Helens since its violent eruption in 1980?

**7.** A major river, which carries a heavy sediment load, has no delta where it enters the ocean. What conditions might be responsible for the lack of a delta?

**8.** The Delaware Water Gap is a steep, narrow valley cut through a structurally deformed high ridge in the Appalachian Mountains. How could it have formed?

# SUGGESTED READINGS

Chorley, Richard J., Stanley A. Schumm, and David E. Sugden. 1984. *Geomorphology*. New York: Methuen.

Dingman, S. L. 1984. *Fluvial Hydrology*. San Francisco: W. H. Freeman.

Dunne, Thomas, and Luna B. Leopold. 1978. *Water in Environmental Planning*. San Francisco: W. H. Freeman.

Leopold, Luna B., M. G. Wolman, and John P. Miller. 1964. *Fluvial Processes in Geomorphology*. San Francisco: W. H. Freeman.

Morisawa, Marie. 1968. *Streams: Their Dynamics and Morphology*. New York: McGraw-Hill.

Ritter, Dale F. 1986. *Process Geomorphology*, 2d ed. Dubuque, Iowa: W. C. Brown.

Schumm, Stanley A. 1977. *The Fluvial System*. New York: Wiley-Interscience.

# WINDS AND DESERTS

The wind is a potent force shaping the surface of the land, particularly in deserts, where strong winds can howl for days on end. The deserts of the world are the scenes of widespread and intensive winds that are far more active geologically than the winds of humid regions. But wherever they blow, winds are major erosional and depositional agents, moving enormous quantities of sand, silt, and dust over large regions of the continents. The power of the wind can be seen in the great heaps of sand accumulating as dunes in a tremendous array of shapes and sizes. No less dramatic are the thick layers of dust that accumulate when dust storms die down. When the wind carries large quantities of sand, it becomes an effective agent of erosion, wearing away rock outcrops, boulders, and pebbles and sculpting the desert landscape.

## 14

Giant sand dunes;
Rub'al-Khali, Saudi
Arabia. These dunes are
formed by the wind in
desert climates where
sand is abundant.
*Edward Mann.*

We all have been caught, at one time or another, in a wind so strong that it could have blown us over if we hadn't leaned into it or held on to something solid. London, England, which rarely gets strong winds, experienced a major windstorm on January 25, 1990. Winds blowing at more than 175 km per hour ripped roofs off buildings, blew trucks over, and made it virtually impossible to walk on the streets. A wind strong enough to move a body weighing more than 50 kg is easily capable of blowing sand grains into the air, as anyone who has ever been in a sandstorm can attest.

The ability of the wind to erode, transport, and deposit sediment is much like that of water, for the same general laws of fluid motion that govern liquids govern gases as well. There are differences, however, that make the wind less powerful than water currents, as we shall see. In contrast to a stream, whose discharge is dependent on rainfall, wind works most effectively when rain is lacking. In this chapter we will discuss the deserts of the Earth in particular detail because so many of the geological processes of the desert are related to the work of the wind. The ancient Greeks called the god of winds Aeolus, and geologists today use the term **eolian** for the geological processes powered by the wind.

# WIND AS A FLOW OF AIR

Wind is a horizontal flow of air—horizontal in relation to the surface of the rotating planet. As with water flows, we can describe air flows by streamlines. Though winds obey all the laws of fluid flow that apply to water in streams, as discussed in Chapter 13, there are some differences. In contrast to the flow of water in river channels, winds are generally unconfined by solid boundaries, except for the ground surface and narrow valleys. Air flows are free to spread out in all directions, including upward into the atmosphere.

Like the water flowing in rivers, air flows are nearly always turbulent. Recall from Chapter 13 that turbulence depends on the density and viscosity of a fluid as well as on its velocity. The extremely low density and viscosity of air—$\frac{1}{1000}$ the density and $\frac{1}{50}$ the viscosity of water—make it turbulent even at the velocity of a light breeze (Table 14.1). As with water turbulence, the turbulence of air increases in proportion to the velocity of the flow. A light breeze will barely move tall grass, but the turbulence of a fresh wind can easily lift a hat. It takes a strong gale with turbulent gusts to rock a moving car. Turbulence also leads to sudden changes in the speed and direc-

---

**TABLE 14.1**

## Wind Speeds

| | WIND SPEED (KM/HR) | DESCRIPTION | EFFECT OF WIND ON SEA SURFACE |
|---|---|---|---|
| INCREASING TURBULENCE | 1 | Calm | Mirror surface |
| | 1–19 | Light to gentle breeze | Ripples and wavelets |
| | 20–49 | Moderate to strong breeze | Moderate to large waves, whitecaps |
| | 50–188 | Moderate to strong gale | High waves, foam, spray |
| | 89–117 | Whole gale to storm | Very high waves, rolling sea |
| | 117 | Hurricane | Sea white with spray and foam, low visibility |

SOURCE: Modified from 1939 International Agreement and N. Bowditch, *American Practical Navigator,* U.S. Navy Hydrographic Office Publication 9, 1958.

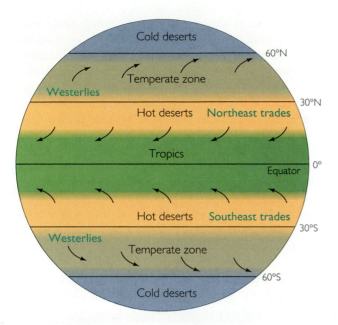

**FIGURE 14.1** Circulation of Earth's atmosphere, with prevailing wind belts.

tion of the wind, which can be strong enough to jolt and buck a large airplane.

Winds vary in speed and direction from day to day, but over the long term prevailing winds tend to come mainly from one direction. In temperate climates, the prevailing winds come from the west and are referred to as the westerlies (Figure 14.1). In the tropics, the trade winds, or "trades" (named for an archaic use of the word *trade* to mean a track or course), blow from the east. Within these belts, winds shift with the weather from one direction to another as storms pass.

# WIND AS A TRANSPORT AGENT

The wind exerts the same kind of force on particles on the land surface as a river current exerts on its bed. Turbulence and forward motion combine to lift particles into the wind and carry them along, at least temporarily. Even the lightest breezes carry dust, the finest-grained material, usually consisting of particles less than 0.01 mm in diameter but often including somewhat larger silt particles. Dust can be carried to heights of many kilometers by the wind. But it takes higher velocities to carry coarser particles, such as sand grains, which are larger than 0.06 mm in diame-

ter. Moderate breezes can roll and slide these grains along a sandy bed, but it takes a fresh wind to lift sand grains into the air flow. The wind usually cannot transport the largest particles, however, because of the low viscosity and density of air. As strong as winds can be, they can only rarely move large pebbles and cobbles the way rapidly flowing rivers do.

## The Motion of Sand Grains in Winds

Sand moves in the wind by a combination of sliding and rolling along the surface and by saltation, the temporary suspension of grains in the air flow. Saltation, the jumping motion of sand grains in a current that we described in Chapter 13, works the same way in air as it does in a river (see Figure 13.3). In air, however, saltation is much more pronounced. Sand grains frequently rise to heights of 50 cm over a sand bed and 2 m over a pebbly surface, much higher than grains of the same size can jump in water. The reason is partly air's lower viscosity, which retards bounding grains less than does more viscous water. High jumps of sand grains are also induced by the impact of falling grains as they hit the surface. These collisions, cushioned hardly at all by the air, kick surface grains into the air in a sort of "splashing" effect (Figure 14.2).

Impacts by saltating grains on the bed can push forward grains too large to be thrown up into the air, causing a creep of the sand bed in the direction of the wind. A sand grain striking the surface at high speed can move forward another grain up to six times its own diameter. Surface creep moves grains less rapidly than saltation. Large and small grains may then be separated as the small ones are gradually blown away, leaving behind a pavementlike surface of coarse sand and gravel.

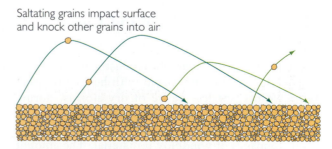

**FIGURE 14.2** As a saltating grain falls to the ground, it may strike another grain lying on the surface with sufficient force to throw the struck grain into a saltation trajectory.

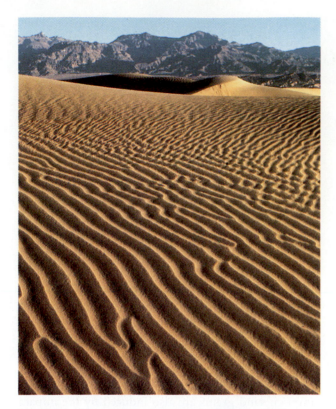

**FIGURE 14.3** Wind ripples in sand; Stovepipe Wells, Death Valley, California. Although complex in form, these ripples are always transverse to the wind direction. *Tom Bean.*

various speeds can erode from a meter-wide strip across a sand dune's surface. A strong wind of 48 km per hour can move half a ton of sand (equivalent in volume to about two large suitcases) in a single day from this small surface area, and the amounts that can be moved at higher speeds increase rapidly. No wonder entire houses can be buried by a sandstorm lasting several days.

Because sand is transported mostly by saltation near the ground and sand grains are likely to be buried in dunes after a relatively short time of travel, most windblown sands are locally derived. Few deposits have been transported more than a few hundred kilometers. The extensive sand dune areas of major deserts such as the Sahara and the wastes of Saudi Arabia are exceptions. In those great sandy regions, sand grains may have traveled more than 1000 km.

Air's capacity to hold dust is staggering. In large dust storms, 1 km$^3$ of air may carry up to 1000 tons, equivalent to the volume of a small house (see Box 14.1). When such storms cover hundreds of square kilometers, they may carry more than 100 million tons and deposit dust layers several meters thick. In dusty regions, even when winds are moderate, large quantities of dust remain suspended in the air, creating a permanent haze.

An almost inevitable consequence of the movement of sand along a bed by the wind is the formation of ripples and dunes much like those formed by water (Figure 14.3). At low to moderate wind speeds, small ripples form. As speed increases, the ripples become larger. Ripples migrate in the direction of the wind over the backs of larger dunes. Because some wind is almost always blowing, a sand bed is almost always rippled to some extent.

## How Much Can the Wind Carry, and How Far?

Most of us are familiar with rainstorms or snow-storms, high winds associated with heavy precipitation. We may be less aware of dry storms, during which high winds blow sand and dust for days on end. These sandstorms and dust storms carry enormous tonnages of windblown material. The amount of sand, silt, and dust that the wind can carry depends on the sizes of the particles, the strength of the wind, and the surface material of the area over which it blows. Figure 14.4 shows how much sand winds of

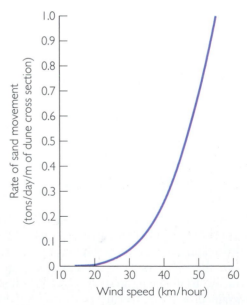

**FIGURE 14.4** The amount of sand moved daily across each meter of width of a dune's surface in relation to wind speed. High-speed winds blowing for several days can move enormous quantities of sand. (After R. A. Bagnold, *The Physics of Blown Sand and Desert Dunes,* London, Methuen, 1941.)

# BOX 14.1 DROUGHTS AND DUST BOWLS

On November 12 and 13, 1933, an enormous dust storm blew up in the southwestern plains of Texas, Oklahoma, and parts of adjacent states. Roads and houses were buried under thousands of tons of dust. The storm extended into the Midwest and as far east as New England, where tons of dust settled, turning snow-covered areas a dark brown.

The immediate cause of this great dust storm was a severe and prolonged drought in the early 1930s, coupled with farming practices that left the soil vulnerable to erosion. In the southern Great Plains, the rainfall is low, the soils are thin, and the winds are strong. But even after decades of intensive agriculture, the land had been fairly stable under these conditions until the long drought came. The killing dryness greatly reduced the vegetation, destroying the root systems that had held the soil together. When the plains were hit by the big windstorm of 1933, the strong winds picked up the parched, loosened dust that the soil had become and swirled it high in the atmosphere.

The dust that choked the plains states gave the region a new name: the Dust Bowl.

During the later stages of the drought, even moderate winds blew clouds of dust. Refugees from the drought-stricken fields packed their belongings and migrated, many to California, in search of a livelihood. Their often tragic story was told by the Nobel Prize–winning novelist John Steinbeck in *The Grapes of Wrath* and sung by the folksinger Woody Guthrie in his "Dust Bowl Ballads."

The former Dust Bowl has been stable for more than 50 years, since the drought broke. Soil conservation methods have reduced erosion of the soil, and irrigation offsets the worst effects of short droughts. Nevertheless, the region remains vulnerable. Wind erosion of the soil in the plains was very high during dry periods in the 1950s and 1970s, though never so high as it was during the 1930s. Even in normal times, some plains areas are enveloped in a brownish dust haze during the plowing season, when the fields are dry and bare of vegetation. We cannot predict exactly how global climate changes will affect this area, only that another severe drought prolonged for several years will spell trouble.

A dust storm in Elkhart, Kansas, May 21, 1937. *Library of Congress.*

Many dust particles are so small that they remain suspended in the atmosphere for a very long time, traveling long distances with the prevailing winds. Volcanic explosions such as that of Mount Pinatubo in the Philippine Islands in 1991 inject huge quantities of dust high into the atmosphere. The volcanic particles from Pinatubo have circled the globe, and the finest-grained ones were still suspended in 1993.

Most nonvolcanic dust does not travel over the whole world, but much of it can travel for thousands of kilometers. The Sahara Desert is one of the chief sources of dust in the world today, and its finer-grained particles have been traced as far as England and across the Atlantic Ocean to Barbados in the Caribbean. Most coarse-grained dust, though, comes from not more than several hundred kilometers away.

## Materials Carried by the Wind

Windblown sand may consist of almost any kind of mineral grain produced by weathering, but quartz grains are by far the most common because quartz is such an abundant constituent of many surface rocks, especially sandstones. In a few places, feldspar grains are abundant in eolian sands. Rock fragments of fine-grained shale or finely crystalline metamorphic and igneous rocks are unusual, because the continuous impacts of saltating grains break down these materials to a very small size.

Many windblown quartz grains have frosted or matte (roughened and dull) surfaces like the inside of a frosted light bulb (Figure 14.5). Some of the grain frosting is produced by wind-driven impacts, but most is the result of slow, long-continued dissolution by dew. Even the tiny amounts of dew found in arid climates are enough to dissolve microscopic pits and hollows, creating the frosted appearance. Because frosting is not found in environments other than eolian, it is good evidence that a sand has been windblown.

Windblown calcium carbonate grains accumulate where there are abundant fragments of shells and coral, as in Bermuda and on many coral islands in the Pacific Ocean. The White Sands National Monument in New Mexico is a prominent example of sand dunes made of gypsum sand grains eroded from evaporite bedrock.

Dust includes microscopic rock and mineral fragments of all kinds, especially silicates, as might be expected from their abundance as rock-forming minerals. Two of the most important sources of silicate minerals in dust are clays blown from soils in dry plains and volcanic dust from eruptions. Organic materials, such as pollen and bacteria, are also com-

**FIGURE 14.5**
Photomicrograph of a frosted, rounded grain of quartz from late Tertiary sand in southern Louisiana. The pits and etched pattern are characteristic of windblown sand in arid climates. *David H. Krinsley.*

mon. Charcoal is abundant downwind of forest fires; when it is found in buried sediments, it is an indication of forest fires in former geologic times. Since the beginning of the Industrial Revolution, we have been pumping new kinds of synthetic dust into the air, from ash from burning coal to the many solid chemical compounds produced by manufacturing processes, incineration of wastes, and auto exhausts.

# WIND AS AN AGENT OF EROSION

By themselves, winds can do little to erode large masses of solid rock exposed at the surface. It is only when the rock is fragmented by chemical and physical weathering that particles can be picked up. But those particles can be lifted by the wind only when they are dry, for wet soils and moist fragmented rock are held together by moisture. Thus winds erode most effectively in arid climates, where strong winds are dry and any moisture is quickly evaporated.

## Deflation

As particles of dust, silt, and sand become loose and dry enough to be lifted by the blowing wind and carried away, the surface of the ground is gradually lowered. This process, called **deflation,** occurs on dry plains and deserts and on temporarily dried-up river floodplains and lake beds. Deflation can scoop out shallow depressions or hollows. Deflation is slow where vegetation is firmly established; even the sparse vegetation of arid and semiarid regions can retard it. The roots bind soil together and the stems and leaves break up the wind and shelter the ground surface. Deflation works fast, however, where the vegetation cover is broken, either naturally by killing drought or artificially by cultivation, construction, or tracks of motor vehicles.

**Desert pavement** is a coarse, gravelly ground surface produced by wind erosion. Deflation can remove the finer-grained particles from a mixture of gravel, sand, and silt in sediments and soils, producing a remnant surface of gravel too large to be transported by the wind (Figure 14.6). Over thousands of years, as deflation removes the finer-grained particles from successive deposits of streams, the gravel accumulates as a layer of desert pavement, which protects the soil or sediments below from further erosion.

## Sandblasting

Wind alone cannot erode solid rock. But when a wind contains blown sand, it becomes an effective natural **sandblasting** agent. The common method of cleaning buildings and monuments with compressed air and sand works on exactly the same principle: solid surfaces are worn away by the impact of high-speed particles. Natural sandblasting mainly works close to the ground, where most sand grains are carried. Sandblasting rounds and erodes rock outcrops, boulders, and pebbles and frosts the occasional glass bottle.

**Ventifacts** are wind-faceted pebbles that show several curved or almost flat surfaces that meet at sharp ridges (Figure 14.7). Each surface or facet is made by sandblasting of the pebble's windward side. Occasional storms roll or rotate the pebbles, exposing a new windward side to be sandblasted to a plane. Many ventifacts are found in deserts and in glacial gravel deposits, places where the necessary combination of gravel, sand, and strong winds is present. Strong cold, dry winds blowing from the edge of a melting glacier over extensive meltwater river flats and strong winds blowing in sandy desert regions are well equipped to blast, roll, and facet loose gravel.

On a much larger scale, eolian processes erode streamlined parallel ridges aligned with the direction of a strong prevailing wind. These ridges, called **yardangs,** are commonly less than 10 m high and 100 m or more long, but very large ones may be 80 m high and more than 100 km long. They are found in the middle of extremely arid regions with little or no sand and so seem to be the products of abrasion by silt and dust. Exactly how the wind forms yardangs remains controversial.

(a)

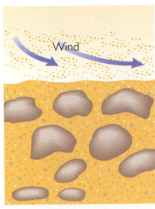

Mixture of coarse and fine particles at surface

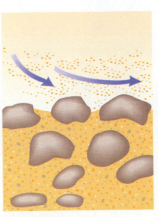

Wind gradually removes finer particles

Desert pavement formed

Desert pavement prevents further wind erosion

(b)

**FIGURE 14.6** (a) Desert pavement; Chuckwalla Mountains, California. *Bill Evarts.* (b) The evolution of desert pavement as the wind removes fine-grained materials from a heterogeneous soil or sediment. Coarse gravel gradually becomes more concentrated into a packed layer as deflation lowers the surface by removing finer sand and silt particles.

**FIGURE 14.7** Ventifacts, pebbles and cobbles pitted and faceted by windblown sand; Sweetwater County, Wyoming. *M. R. Campbell, USGS.*

# WIND AS A DEPOSITIONAL AGENT

When the wind dies down, it can no longer transport the sand, silt, and dust it has carried. The coarser material is deposited in variously shaped sand dunes ranging in size from low knolls to huge hills more than 100 m high. The finer silt and dust fall as a more or less uniform blanket of silt and clay. Geologists have observed these depositional processes working today and have linked them to the sediments' characteristics, primarily bedding and texture, to infer past climates and wind patterns from ancient sandstones and dust falls.

## Where Sand Dunes Form

Sand dunes are found in relatively few environmental settings. Most people are familiar with the dunes formed behind beaches along ocean coasts or large lakes (Figure 14.8). Some dunes are found on the sandy floodplains of large rivers in semiarid and arid regions. Most spectacular are fields of dunes in desert regions (Figure 14.9), which cover large expanses and may reach heights of several hundred meters, truly mountains of sand.

These places all have a ready supply of loose sand: beach sands along coasts, sandy river bars or floodplain deposits in river valleys, and sandy bedrock formations in deserts. Another common factor is wind power. On oceans and lakes, strong winds blow onshore off the water. Strong winds, sometimes of long duration, are common in deserts.

As we noted earlier, wind cannot easily pick up wet materials, so most dunes are found in drier climates. The exception is dune belts along a coast,

**FIGURE 14.8** Back beach dunes. Winds, mainly onshore, blow sand back from the beach in a complex of parabolic dunes. *Loren McIntyre.*

**FIGURE 14.9** Transverse (foreground) and linear (background) dunes in a desert dune field; Death Valley, California. *Peter Kresan.*

where sand is so abundant and dries so quickly in the wind that dunes can form even in humid climates. But in such climates the dunes become covered with soil and vegetation not very far inland from the beaches, and the sand is no longer blown.

## How Sand Dunes Form and Move

How does a dune start? Given enough sand and wind, any obstacle, such as a large rock or clump of vegetation, can start a dune. Streamlines of wind, like those of water, separate around obstacles and rejoin downwind, creating a wind-shadow zone downstream of the obstacle (Figure 14.10). Wind velocity is much lower in the wind-shadow zone than in the main flow around the obstacle, low enough to allow sand grains blown into the shadow to settle there. Because the velocity is low, these grains can no longer be picked up, and they accumulate as a **sand drift,** a small pile of sand in the lee of the obstacle. As this process continues, the sand drift becomes an obstacle itself. If there is enough sand and the wind continues to blow in the same direction long enough, the drift grows into a dune. Dunes may also grow by the enlargement of ripples, just as underwater dunes do.

As a dune grows, the whole mound starts to migrate downwind by the combined movements of a host of individual grains. Sand grains constantly saltate to the top of the low-angle windward slope, then

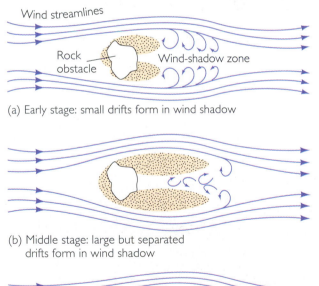

(a) Early stage: small drifts form in wind shadow

(b) Middle stage: large but separated drifts form in wind shadow

(c) Final stage: drifts coalesce into dune

**FIGURE 14.10** Formation of a sand drift in the lee of a rock obstacle. By separating the flow streamlines, the rock creates a wind shadow in which the eddies are weaker than the main flow, allowing the grains to settle and build up a drift. (After R. A. Bagnold, *The Physics of Blown Sand and Desert Dunes,* London, Methuen, 1941.)

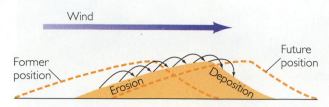

**FIGURE 14.11** A ripple or dune advances by the movements of individual grains. The whole form moves forward slowly as sand erodes from the windward slope and is deposited on the leeward slope.

fall over into the wind shadow on the lee slope, as in Figure 14.11. These grains gradually build up a steep unstable accumulation on the upper part of the lee slope. Periodically the steepened buildup gives way and spontaneously slips or cascades down this **slip face,** as it is called, to a new slope at a lower angle (Figure 14.12). If we overlook the short-term unstable steepenings in slope, the slip face maintains a sta-

ble, constant slope angle, its angle of repose. As we saw in Chapter 11, this angle increases with the size and angularity of the particles.

Successive slip faces deposited at the angle of repose create the cross-bedding that is the hallmark of windblown dunes. As dunes accumulate, interfere with one another, and become buried in a sedimentary sequence, the cross-bedding is preserved even though the original shapes of the dunes are lost. Sets of sandstone cross-bedding many meters thick are one of the lines of evidence that geologists use to infer high windblown dunes. From the directions of these eolian cross-beds geologists can reconstruct wind directions of the past.

As more sand accumulates on the windward slope of a dune than blows off onto the slip face, the dune grows in height. In this way dunes commonly build to heights of 30 m. Huge dunes in Saudi Arabia reach 250 m, but this seems to be the limit. The explanation for limited dune height lies in the relationship of wind streamline behavior, velocity, and topography. Wind streamlines advancing over the back of a dune become more compressed as the dune grows higher, as Figure 14.13 shows. As more air rushes through a smaller space, the wind velocity increases. Ultimately the air speed at the top of the dune gets so great that sand grains blow off the top of the dune as quickly as they are brought up the windward slope, and the height remains constant.

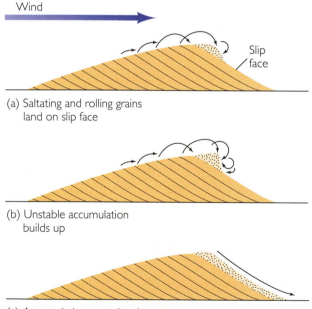

(a) Saltating and rolling grains land on slip face

(b) Unstable accumulation builds up

(c) Accumulation cascades down to base, advancing the dune

**FIGURE 14.12** The formation of the slip face of a dune. In successive positions, (a) the slip face accumulates an unstable slope (b) as a result of the deposition of saltating, rolling, and sliding grains. Intermittently this accumulation becomes so unstable that (c) it spontaneously slips to the base and a new stable slope is formed at a lower angle.

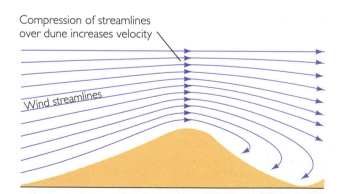

**FIGURE 14.13** Limitation of the height of a dune by a compressed wind stream. As the dune grows higher, the wind streamlines become more compressed and thus the velocity increases. With increased velocity comes increased competence to transport sand grains. Eventually a height is reached at which the wind is so fast that all of the sand is transported, and the dune stops growing vertically.

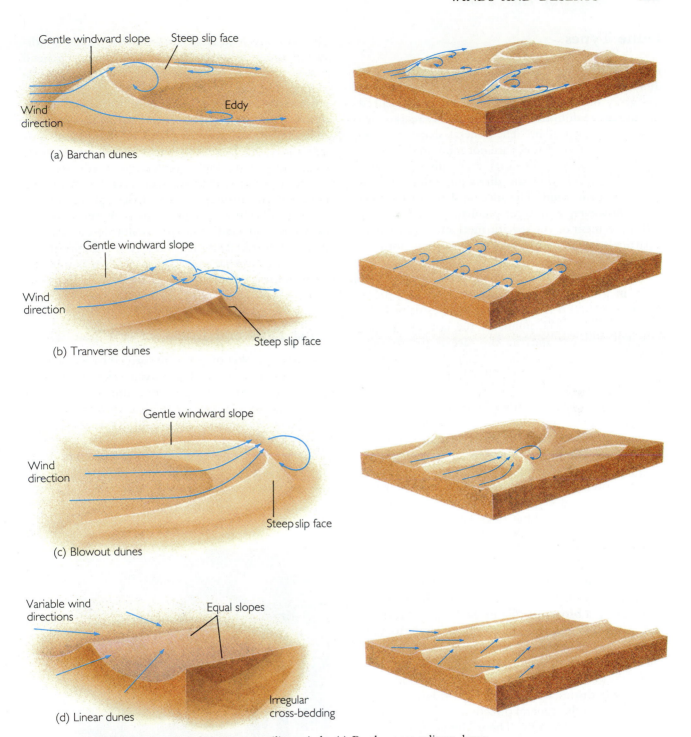

**FIGURE 14.14** Dune types in relation to prevailing winds. (a) Barchans are solitary dunes whose horn shapes point downwind; (b) transverse dunes are long ridges oriented at right angles to the wind; (c) blowout dunes are almost the reverse of barchans, with their slip faces convex downwind as opposed to the barchan's concave downwind slip face; (d) linear dunes are long ridges of sand oriented parallel to the wind.

## Dune Types

A person standing in the middle of a large expanse of dunes might be bewildered by the seemingly orderless array of undulating slopes. It takes a practiced eye to see the dominant pattern and may require observation from the air. The general shapes and arrangements of sand dunes are not random. They depend upon a variety of factors, especially the amount of sand available and the direction, duration, and strength of the wind. The specific shape that a dune takes, however, is not yet predictable, and there is little agreement on the specific mechanisms by which a particular wind regime results in one dune form or another.

A **barchan** is a crescent-shaped dune, found often in groups or occasionally as a solitary dune, that moves over a flat surface of pebbles or bedrock (Figure 14.14a). The points of the crescent are directed downwind and the slip face is the concave curve advancing downwind. Barchans are the products of limited sand supply and unidirectional winds. Several barchans may coalesce to form irregular ridges whose long directions lie transverse (perpendicular) to the wind direction. These merged barchans may be transitional to **transverse dunes,** long, wavy ridges that lie transverse to the prevailing unidirectional wind (Figure 14.14b). Transverse dunes form in arid regions where abundant sand dominates the landscape and vegetation is absent.

Typically, sand dune belts behind beaches are transverse dunes formed by strong onshore winds. In temperate or humid regions, they are stabilized by vegetation at some distance from the beach. If a section of the dune belt is deflated and stabilizing vegetation is overwhelmed by sand, a parabola-shaped dune, called a **blowout** (Figure 14.14c), will form at that point and migrate inland. In contrast to barchans, which they superficially resemble, the arms of the blowout are directed upwind and the slip face forms the convex curve advancing downwind.

**Linear dunes** are long, straight ridges more or less parallel to the general direction of the prevailing winds (Figure 14.14d). These dunes may reach heights of 100 m and may extend many kilometers. The origin of linear dunes remains controversial, but most experts believe that they are depositional forms created by winds of varying direction. Most areas covered by linear dunes have a moderate sand supply, a rough pavement, and winds that may shift but are always in the same general direction. During summer, for example, the winds blow from the southwest across linear dunes oriented east-west, depositing sand on a northeast-facing slip face. In winter, winds blow from the northwest and trans-port sand over the dune to a slip face facing southeast. The combination advances the dune in a general easterly direction while producing two different orientations of slip faces.

Extremely large, high, hilly forms are called **draas.** They are composites of superimposed dunes of several kinds, in some places reaching heights of 400 m. Large dunes like these move much more slowly than small dunes, in places as slowly as 0.5 m per year. Where sand is abundant over wide areas and winds blow strongly, large fields of dunes are formed. The most extensive areas of this type are **ergs,** "seas of sand" found in major deserts such as those of Saudi Arabia (see the photograph at the opening of this chapter). Ergs may cover as much as 500,000 km$^2$, twice the size of the state of Nevada.

## Dust Falls and Loess

**Loess** is a blanket of sediment formed by the settling of fine-grained particles from dust clouds. Beds of loess lack internal stratification, and in compacted deposits more than a meter thick, loess tends to form vertical cracks and to break off along sheer walls during erosion (Figure 14.15). The vertical cracking may be caused by a combination of root penetration and uniform downward percolation of groundwater, but the exact mechanisms are still unknown.

**FIGURE 14.15** Pleistocene loess, showing vertical cracking; Colorado. *H. E. Malde, USGS.*

The best known loess deposit in North America is in the upper Mississippi Valley, a legacy of recent glacial activity in the Pleistocene epoch. Its origin as windblown material deposited during melting of the large glaciers of that time was originally demonstrated in the early part of this century by its pattern of distribution as a blanket of more or less uniform thickness on both hills and valleys, all in or near formerly glaciated areas. Changes in the regional thickness of the loess in relation to the prevailing westerly winds confirm its eolian origin. The loess on the eastern sides of major river floodplains is 8 to 30 m thick, greater than on the western sides, and the thickness decreases downwind rapidly to 1 to 2 m farther east from the floodplains. The source of the dust was the abundant silt and clay deposited on extensive floodplains of rivers draining the edges of melting glaciers. Strong winds dried the floodplains, whose frigid climate and rapid rates of sedimentation inhibited vegetation, and blew up tremendous amounts of dust, which then settled to the east.

Great loess deposits formed in the past 2 million years are found in China, where they lie 30 to 100 m thick over wide areas in the northwest. The winds blowing from the Gobi Desert and the arid regions of central Asia were the source of the dust, which still blows over the Chinese interior.

In China, North America, and elsewhere, soils formed on loess are fertile and highly productive. They also pose environmental problems, for they are easily eroded into gullies by small streams and deflated by the wind when they are poorly cultivated.

Airborne dust has been measured far out to sea by oceanographic research vessels. Comparison of the composition of this dust with many of the components of deep-sea sediments in the same region indicates that windblown dust is an important contributor to oceanic sediment. Much of this dust is of volcanic origin, and there are individual ash beds marking very large eruptions. Volcanic dust is abundant because much of it is very fine-grained and is injected high in the atmosphere, where it travels farther than much of the windblown dust from the continents.

# THE DESERT ENVIRONMENT

The hot, dry deserts of the world are among the most hostile environments to humans, yet with their strange forms of animal and plant life and their bare rocks and sand dunes, they are fascinating to many people. The desert may be one of the quietest places on Earth. Any sound you hear is the wind. The desert is the environment of Earth's surface where the wind is best able to do its work of erosion and sedimentation.

## Where Deserts Are Found

Rainfall is the major factor determining the location of the world's great deserts. The Sahara and Kalahari deserts of Africa and the Great Australian Desert get some of the lowest amounts of rainfall on Earth, normally less than 25 mm each year, in some places less than 5 mm. These deserts lie in the warmest regions of the globe, within 30° north and 30° south latitude of the equator. The deserts lie under virtually stationary areas of high atmospheric pressure (Figure 14.16), the sun beating down through a cloudless sky week after week and the air maintaining extremely low humidity.

Deserts also form in mid-latitudes, those between 30° and 50° north and 30° and 50° south, typified by the Great Basin and Mohave deserts of the western United States and the deserts of central Asia. Regions of low rainfall exist in these latitudes because moisture-laden winds either are blocked by mountain ranges or must travel great distances from their source of moisture, the ocean. The North American deserts, for example, lie in rain shadows created by the western coastal mountains. As we saw in Chapter 12, wind descending from the mountains warms and dries, leading to low precipitation. The deserts of central Asia are so far inland that the winds reaching them have precipitated all their ocean-derived moisture long before they arrive at the interior of the continent.

Another kind of desert forms in polar regions. Little precipitation is possible in these cold, dry areas because the frigid air can hold only extremely small amounts of moisture. The dry valley region of southern Victoria Land in Antarctica is so dry and cold that its environment resembles that of Mars.

Deserts, in a sense, are reflections of plate tectonics. The mountains that create rain shadows are made by collisions between converging continental and oceanic plates. The great distance of central Asia from the oceans is a consequence of the size of the continent, a huge landmass assembled from smaller plates by continental drift. Large deserts are at low latitudes because continental drift has moved them there from higher latitudes. The interior of Australia, for example, was once in a moist, humid climate when that continent was far to the south of its present position (see Figure 20.25d). Since then Australia has

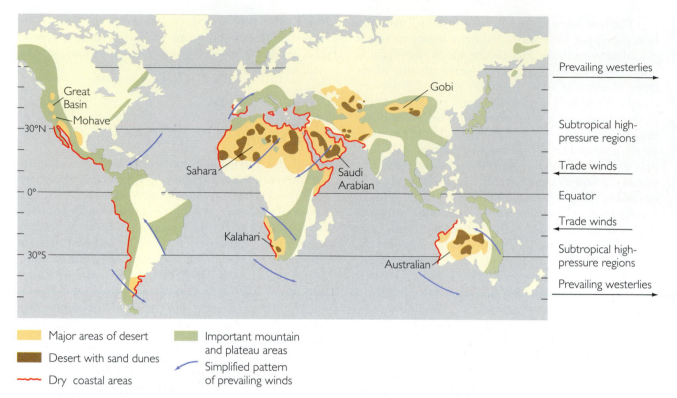

**FIGURE 14.16** Major desert areas of the world (exclusive of polar deserts) in relation to prevailing wind directions and major mountain and plateau areas. Sand dunes are only a small proportion of the total desert area. (After K. W. Glennie, *Desert Sedimentary Environments,* New York, Elsevier, 1970.)

moved northward into an arid subtropical zone, where its interior has become a desert.

Changes in a region's climate, its population, or its population's behavior may transform semiarid lands into deserts by a process called **desertification.** Climatic changes that we do not fully understand may decrease precipitation for time intervals of decades to centuries. Unrestrained growth of human populations and their agriculture and of animals' grazing activities may also result in the expansion of deserts (see Box 14.2).

All told, arid regions amount to one-fifth of Earth's land area, about 27.5 million square kilometers. Semiarid plains regions account for an additional one-seventh. Given the reasons for the existence of large areas of deserts in the modern world—that is, the interaction of mountain building by plate tectonics, transport of continental regions to low latitudes by continental drift, and the climatic belts of the globe—we can be confident that, by the principle of uniformitarianism, extensive deserts have existed throughout geologic time.

## Desert Weathering

As different as deserts seem from more humid regions, the same geological processes operate on both. Weathering and transportation work in the same way, but with a different balance in deserts than in temperate areas. In the desert, the physical aspects of weathering predominate over the chemical. Feldspars and other silicates chemically weather to clay minerals, but slowly, for lack of the water required for the reaction to proceed. What little clay is formed is normally blown away by strong winds before it can accumulate. Slow chemical weathering and rapid wind transport thus combine to prevent the buildup of any significant thickness of soil, even where sparse vegetation binds some of the particles. Thus soils are thin and patchy. Sand, gravel, rock rubble of many sizes, and bare bedrock are characteristic of much of the desert surface.

The common rusty, orange-brown colors of weathered surfaces in the desert come from the ferric iron oxide minerals hematite and limonite. These

## BOX 14.2  DESERTIFICATION IN THE SAHEL

Famine in Ethiopia caused by desertification of sub-Saharan Africa; Mekele Camp, Tigray.
*David Burnett/Woodfin Camp and Associates.*

In 1984 a catastrophic famine struck Ethiopia and other countries bordering the southern Sahara, causing enormous suffering for hundreds of thousands of people. For decades before, the region was racked by a series of droughts, a particularly devastating one in 1973. What were formerly vegetated semiarid lands used by small populations of migrating peoples had become transformed by loss of vegetation and soil into barren desert incapable of supporting even a small population. The process of desertification had ruined the land.

The reasons for desertification in sub-Saharan Africa are complex. This naturally semiarid region had become more desertlike when the earlier droughts killed much natural vegetation. At the same time, the population in the region grew and the people turned to ever more intensive grazing and agriculture, both of which thinned the easily erodible soil, which began to blow away in the fierce winds. Under such conditions of aridity and severe soil erosion, it is doubtful that any conservation methods will be able to restore the original soil before a long time has passed.

Some atmospheric scientists are now predicting that the 20-year drought will end. If it does, additional rainfall will help efforts to restore the land, but the countries are poor and their economic resources are limited. Some scientists estimate that about 35 percent of the Earth's surface, including parts of the western United States and Mexico, central Asia, and Argentina, on which over 850 million people live, are potentially threatened by desertification.

minerals are produced by the slow weathering of iron silicate minerals such as pyroxene. The iron oxides, even when present only in small amounts, stain the surfaces of sands, gravels, and clays.

**Desert varnish** is a distinctive dark-brown, sometimes shiny coating on many rock surfaces in the desert. It is a mixture of clay minerals with smaller amounts of manganese and iron oxides. Desert varnish is hypothesized to form very slowly from the combination of dew, chemical weathering that produces the clay minerals and iron and manganese oxides, and the sticking of windblown dust to exposed rock surfaces. Rocks whose varnish was scratched by Native Americans hundreds of years ago still show the stark contrast between the dark varnish and the light unweathered rock beneath (Figure 14.17).

Though wind is a more significant agent of erosion on the desert than elsewhere, it cannot compete with the erosive power of streams. Even though it rains so seldom that most desert streams flow only intermittently, when streams do flow, they do most of the erosional work in the desert.

Even the driest desert gets occasional rain, sometimes as rare heavy rainstorms. In these cloudbursts, so much water falls in such a short time that even though a fraction soaks into the ground, infiltration

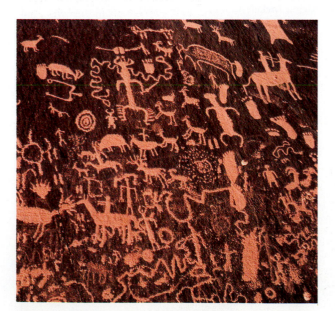

**FIGURE 14.17** Petroglyphs scratched in desert varnish by early Native Americans; Newspaper Rock, Canyonlands, Utah. The scratches are several hundred years old and appear fresh, whereas the varnish accumulated over thousands of years. *Peter Kresan*.

cannot keep pace and the bulk of it runs off into streams. Unhindered by vegetation, the runoff is rapid and may cause flash floods along valley floors that have been dry for years. Thus a large proportion of stream flows in the desert consists of floods.

Because most of the loose debris of weathering is not held by vegetation, the erosive power of flooding streams is great. Desert streams may become so choked with sediment that they sometimes look more like fast-moving mudflows than like rivers. The abrasiveness of this sediment load moving rapidly at flood velocities makes such streams efficient eroders of bedrock valleys.

In sandy, gravelly areas of deserts, infiltration of the occasional rainfall into soil and permeable bedrock temporarily replenishes groundwater in the unsaturated zone. There some of it very slowly evaporates into open pore space between the grains. A smaller amount eventually reaches the groundwater table far below, in some places as much as hundreds of meters below the surface. A desert oasis forms where the groundwater table comes close enough to the surface for the roots of palms and other plants to reach it.

## Desert Sediment and Sedimentation

**ALLUVIAL SEDIMENTS**  As sediment-laden flash floods dry up, they leave distinctive deposits on the floors of desert valleys. A flat fill of coarse debris often covers the entire valley floor, without the ordinary differentiation into channel, levees, and floodplains (Figure 14.18). Large alluvial fans (see Chapter 13) are prominent features at mountain fronts in deserts because desert streams deposit much of their high sediment load on the fans. The rapid infiltration of stream water into the permeable fan material deprives the streams of the water required to carry the sediment load any farther downstream. Debris flows and mudflows make up large parts of the alluvial fans of arid, mountainous regions.

**EOLIAN SEDIMENTS**  By far the most dramatic sedimentary accumulations of the desert are sand dunes, dune fields, and ergs. Though films and TV lead one to think that deserts are mostly sand, actually only one-fifth of the desert areas of the world is covered by sand. The other four-fifths are rocky or covered with desert pavement. Only a little more than one-tenth of the Sahara Desert is covered by sand, and sand dunes are far less common in the arid lands of the southwestern United States.

**FIGURE 14.18** (a) Desert wash flooding during a summer thunderstorm; Saguaro National Monument, Arizona. (b) The same wash a day after flooding. *Peter Kresan.*

**EVAPORITE SEDIMENTS** **Playa lakes** are permanent or temporary lakes that accumulate in arid mountain valleys or basins. As the lake waters evaporate, the dissolved weathering products they hold are concentrated and gradually precipitated. Playa lakes are sources of evaporite minerals: sodium carbonate, borax (sodium borate), and other unusual salts. The water of playa lakes may be alkaline or saline and deadly to drink. The alkalinity or salinity stems from evaporation and high levels of salt contributed by desert streams. The streams may have large amounts of dissolved salts because they redissolve evaporite minerals deposited by evaporation from earlier run-off. If evaporation is complete, the lakes become **playas,** flat beds of clay that are sometimes encrusted with precipitated salts (Figure 14.19).

## Desert Landscape

The landscape of the desert is one of the most varied on Earth. Large, low, flat areas are covered by playas, desert pavements, and dune fields. Uplands are rocky, in many places cut by steep river valleys and gorges. The lack of vegetation and soil makes everything seem sharper and harsher than landscapes in more humid climates (see Chapter 16). The coarse, heterogeneously sized fragments produced by desert weathering form steeper slopes than the soil-covered, vegetated slopes found in more humid regions. In-

**FIGURE 14.19** A desert playa lake; Death Valley, California. *Travis Amos.*

**FIGURE 14.20** A desert landscape; Kofa Butte, Kofa National Wildlife Refuge, Arizona. *Peter Kresan*.

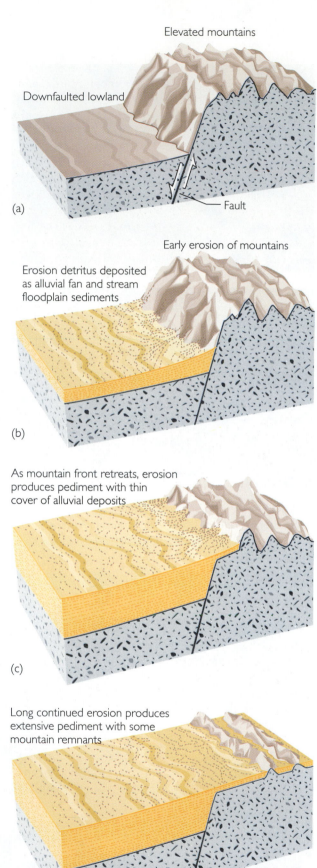

stead of rounded slopes, we see steep cliffs with masses of angular talus slopes at their bases (Figure 14.20).

Valleys in deserts have the same range of profiles as elsewhere, but far more of them have steep valley walls caused by rapid erosion from mass movements and streams. Much of the landscape of deserts, like that of humid regions, is shaped by rivers, but usually the valleys, called **dry washes** in the western United States and **wadis** in the Near East, are dry.

**FIGURE 14.21** Stages in the evolution of a typical pediment, an erosional form produced in arid mountainous settings. (a) First, a downfaulted lowland is formed. (b) Alluvial fan and floodplain sediments are deposited in the lowland. (c) The erosional surface of the pediment, produced by migrating streams, is covered by thin deposits of alluvial sands and gravels. (d) Long-continued erosion makes an extensive pediment with some mountain remnants. The mountain front retreats steadily but keeps the same steep slope angle, in contrast to the gradually rounded slopes that evolve in humid climates.

Even where water is scarce, it is the infrequent rainfall runoff in streams that does most of the basic work of erosion. The wind helps, but it rarely controls. Only in dune fields is the work of the wind the dominant process.

Desert streams are widely spaced because of the relatively infrequent rainfall, but drainage patterns are generally similar to those of other terrains, with one difference. Because of infrequent rainfall and water losses by evaporation and infiltration, many desert streams die out downstream, never reaching across the desert to join larger rivers flowing to the oceans. In mountain basins, lakes may form as drainage pathways are blocked by faults, alluvial fans, or landslides.

A special type of eroded bedrock surface, called a **pediment,** is a characteristic landform of the desert. Pediments are broad, gently sloping platforms of bedrock left behind as an eroding mountain front retreats from its valley. The pediment spreads like an apron around the base of the mountains.

A cross section of a typical pediment and its mountains would reveal a fairly steep mountain slope abruptly leveling into the gentle slope of the pediment, which follows the same general concave-upward longitudinal profile as streams (Figure 14.21). The surface of the pediment is typically covered with thin alluvial sands and gravels. Alluvial fans deposited at the lower edge of the pediment merge with the sedimentary fill of the valley below the pediment.

Although the origin of pediments remains in some doubt, there is much evidence that the pediment is formed by running water that both cuts the erosional platform above and deposits an alluvial fan apron below. At the same time, the mountain slopes at the head of the pediment maintain their steepness as they retreat, instead of becoming the rounded, gentler slopes found in humid regions. We do not know how the specific rock types and erosional processes interact in an arid environment to keep slopes steep while the pediment enlarges.

For many years, film and TV westerns made in southern California and nearby states have shown us the kind of desert landscape formed on horizontal sedimentary rocks. **Mesas** are tablelike uplands capped by erosion-resistant beds bounded by steep-sided erosional cliffs. The capping bed of a mesa tends to maintain the upland-level elevation. Wherever it is cut through, however, erosion cuts down rapidly into the less resistant beds and creates cliffs, as shown in Figure 14.22. As the cliffs retreat, the upland area shrinks, finally leaving only a few isolated mesas above the surrounding lowland.

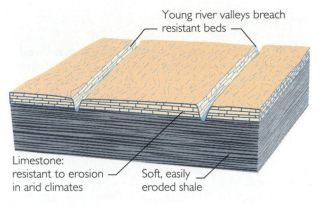

(a)

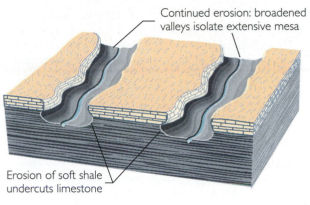

(b)

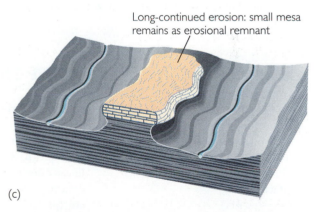

(c)

**FIGURE 14.22** The evolution of a mesa. (a) The process is controlled by an erosion-resistant sedimentary bed overlying weak, easily erodible rocks. (b) The border of the mesa retreats as the lower beds are eroded, undercutting the resistant beds. (c) After long-continued erosion, a small mesa remains elevated above the lowland valley.

# SUMMARY

**How do winds erode and transport sand and finer-grained sediment?** Wind, a horizontal flow of air, can pick up and transport dry sediment particles in the same way as running water does, but air flows are limited in the particle sizes they can carry, rarely larger than coarse-grained sand, and by air's lower viscosity and density. Almost all air flows are turbulent, and high winds may reach a velocity exceeding 100 km per hour, enhancing the wind's ability to carry sediment in suspension. Winds can carry great amounts of sand and dust. Sand grains are moved primarily by saltation, whereas finer-grained silt and clay-sized particles (dust) are carried in suspension. Winds blow sand into cross-bedded ripples and dunes. Windblown materials include volcanic ash, quartz and other mineral fragments such as clay minerals, and organic materials such as pollen and bacteria. Winds erode primarily by deflation and sandblasting, producing desert pavement and ventifacts.

**How do winds deposit sand dunes and dust?** In the course of transporting sand, the wind deposits it in dunes of various shapes and sizes in sandy desert regions, behind beaches, and along sandy floodplains, all places with a ready supply of loose sand and moderate to strong winds. Dunes start as sand drifts in the lee of obstacles and grow to heights of many meters or even hundreds of meters. Dunes migrate downwind as sand grains saltate up the gentler windward slopes and fall over onto the steeper downwind slip faces of the dunes. The various kinds of dunes—transverse, linear, barchan, and blowout—are responses to the speed of

the wind, its constancy or variability of direction, and the abundance of sand. As the velocity of dust-laden winds decreases, the dust settles to form loess, a blanket of dust. Loess layers in many recently glaciated areas were deposited by winds blowing over the floodplains of muddy streams formed by meltwater. Loess can accumulate to great thicknesses downwind of dusty desert regions.

**How do wind and water combine to shape the desert environment and its landscape?** Desert regions are formed in the rain shadows of mountain ranges, in subtropical regions of constant high pressure, and in the interiors of some continents, all places where originally moisture-laden winds become dry and rainfall is rare. Weathering mechanisms are the same in deserts as in more humid regions, but physical breakdown of rocks dominates and chemical weathering is at a minimum because of the lack of water. Most desert soils are thin, and bare rock surfaces are common. Streams may run only intermittently but they are responsible for much of the erosion and sedimentation on the desert, carrying away heavy loads of coarse sediment and depositing them on alluvial fans and floodplains. Naturally dammed rivers in mountainous deserts can form playa lakes, which deposit evaporite minerals as they dry up. Desert landscapes are composed of dunes formed by eolian sedimentation, desert pavements, mesas, and pediments, which are broad, gently sloping platforms eroded from bedrock as mountains retreat while maintaining the steepness of their slopes.

# KEY TERMS AND CONCEPTS

eolian (p. 306)
deflation (p. 310)
desert pavement
  (p. 311)
sandblasting (p. 311)
ventifact (p. 311)
yardang (p. 311)
sand drift (p. 313)

slip face (p. 314)
barchan (p. 316)
transverse dune (p. 316)
blowout (p. 316)
linear dune (p. 316)
draa (p. 316)
erg (p. 316)
loess (p. 316)

desertification (p. 318)
desert varnish (p. 320)
playa lake (p. 321)
playa (p. 321)
dry wash (p. 322)
wadi (p. 322)
pediment (p. 323)
mesa (p. 323)

# EXERCISES

1. What is the range of materials and sizes of particles that the wind can move?

2. What is the difference between the way wind transports dust and the way it transports sand?

3. How is the ability of the wind to blow sedimentary particles linked to climate?

4. What are the main evidences of wind erosion?

5. Where do sand dunes form?

**6.** Name three types of sand dunes and show their relationship to wind direction.

**7.** What typical desert landforms are composed of sediment?

## THOUGHT QUESTIONS

**1.** You have just driven a truck through a sandstorm and discover that the paint has been stripped from the lower parts of the truck but the upper parts are barely scratched. What process is responsible, and why is it restricted to the lower parts of the truck?

**2.** What evidence might you find in an ancient sandstone that would point to its eolian origin?

**3.** Compare the heights to which sand and dust are carried in the atmosphere and explain the differences or similarities.

**4.** Trucks continually have to haul away sand covering a coastal highway. What do you think might be the source of the sand? Could its encroachment be stopped?

**5.** What features of a desert landscape would lead you to believe that it was formed mainly by streams with secondary contributions from eolian processes?

**6.** Which would be a more reliable indication of the direction of the wind that formed a barchan dune, cross-bedding or the orientation of its shape on a map? Why?

**7.** What factors determine whether sand dunes will form on a stream floodplain?

**8.** There are large areas of sand dunes on Mars. What can you infer about conditions on the Martian surface from this fact alone?

## SUGGESTED READINGS

Bagnold, Ralph A. 1941. *The Physics of Blown Sand and Desert Dunes*. London: Methuen.

Brookfield, M. E., and S. Thomas (eds.). 1983. *Eolian Sediments and Processes*. New York: Elsevier.

Cooke, R. U., and A. Warren. 1973. *Geomorphology in Deserts*. Berkeley: University of California Press.

Glennie, K. W. 1970. *Desert Sedimentary Environments*. New York: Elsevier.

Idso, Sherwood B. 1976. Dust storms. *Scientific American* (October):108.

Mabbutt, J. A. 1977. *Desert Landforms*. Cambridge, Mass.: MIT Press.

Sheridan, D. 1981. *Desertification of the United States*. Washington, D.C.: Council on Environmental Quality.

# GLACIERS: THE WORK OF ICE

Ice is an important geological agent shaping the land surface in Earth's frigid regions. Near the poles, snow accumulates and is transformed into continent-sized glaciers, thick sheets of ice that advance slowly outward from the centers of accumulation. Glaciers also form in the cold of high mountains and travel downhill like rivers of ice, sculpting the land surface into deep valleys. As glaciers move into warmer regions and melt, they dump an enormous load of the sediment they have scraped from bedrock, depositing it in a multitude of characteristic shapes. The landforms of the recent past are evidence of the Pleistocene glacial epoch, during which huge areas of North America and Eurasia were covered by glacial ice. What caused these enormous glaciers is among the most vital questions of modern geology.

## 15

Kennicott Glacier,
Wrangell-St. Elias
National Park, Alaska.
These rivers of ice flow
downhill in valleys, like
streams of water. At
left, a tributary glacier
joins the main glacier.
*Tom Bean*.

The vast blue oceans and white clouds of Earth seen from space dramatically emphasize the water covering our planet. But satellite views of polar regions and white-peaked mountain ranges also show us that much of the water is frozen. About 10 percent of Earth's land surface is covered by glacial ice, much of which moves slowly and steadily outward from the centers of the ice caps and downward from the mountain peaks. Over the short term, the ice melts at the edges of glaciers at about the same rate as it advances; thus the total ice area remains the same.

Over a longer span of time, a cooling climate can cause the ice sheets to expand as melting fails to keep pace with the accumulation and movement of glacial ice. As recently (geologically speaking) as 20,000 years ago, snow and ice covered almost three times as much of the land surface as they do now. Yet only a few decades from now, ice on Earth may shrink significantly as global warming melts the ice faster than it can accumulate and move. The impact of these global climate changes on the world may be immense, as melting ice raises sea level and drowns low-lying cities and as climatic zones migrate, changing temperate zones into semiarid ones and vice versa. There is no doubt that a knowledge of Earth's icy regions and how they change is an intensely practical subject.

Glaciers erode steep-walled valleys, scrape bedrock surfaces, and pluck huge blocks from their rocky floors. In the relatively short geological time span of the recent ice ages, glaciers accomplished far more carving of topography than did rivers and the wind. The debris of glacial erosion is enormous—huge tonnages of sediments are transported by the ice to the edge of the glacier, where they are deposited or carried away by meltwater streams. The effects of glacial erosion and sedimentation are widespread over the Earth, affecting the water discharge and sediment loads of major river systems; the quantity of sediment delivered to the oceans; and, through changes in sea level, erosion and sedimentation in coastal areas and on shallow continental shelves.

Many of the landscapes of mountain belts were sculpted by glaciers that have since melted away. The landscapes of large portions of continental lowlands were shaped relatively recently by Pleistocene glaciers that extended far into what are now temperate regions. As these glaciers melted, they left sediments and erosional features that are the evidence of their former extent.

Because snow and prolonged cold are necessary for the formation of glaciers, glacial sediments and erosional forms are indicators of past and present climates. The former large extent of glaciers is evidence of a recent past in which the climate over large regions was much colder and snowier than today's. At some times much farther back in the geologic past, glaciers covered regions that are now tropical jungles. At other times, parts of the globe that are now covered with ice were warm and humid, and there were no polar ice caps. The former extent of glaciers can be used to trace climates of the past and, more important, to monitor future changes in climate.

## ICE AS A MATERIAL

What is ice? To a geologist a block of ice is a rock, a mass of crystalline grains of the mineral ice. Ice is hard, like most rocks, but its composition makes it much less dense. It shares with igneous rocks an origin as a frozen fluid. Like sediments, it is deposited in layers at the surface of the Earth and can accumulate

**FIGURE 15.1** A typical mosaic of crystals of glacier ice. Each small area of uniform color is a single crystal as seen in polarized light. The tiny circular and tubular spots are bubbles of air. *Z. Xie, Lanzhou Institute of Glaciology, Academia Sinica, P.R. China.*

to great thicknesses. Like metamorphic rocks, it is transformed by recrystallization under pressure. Glacial ice forms by the burial and metamorphism of the "sediment" snow. The rock is formed as the loosely packed snowflakes—each one a single crystal of the mineral ice—age and recrystallize into a solid mass (Figure 15.1). The unusual characteristic of ice as a mineral is its extremely low melting temperature, hundreds of degrees lower than the temperatures at which most minerals melt.

## WHAT IS A GLACIER?

A mass of ice, like any other mass of rock or soil at the surface, can slowly move downhill. **Glaciers** are large masses of ice on land that show evidence of being in motion or of once having moved. On the basis of size and shape we divide glaciers into two basic types. Many skiers and mountain climbers are familiar with **valley glaciers,** rivers of ice that flow down bedrock valleys in mountain belts, as seen in the photograph at the beginning of this chapter. Most of the glaciers occupy the complete width of the valley, whose bedrock base may be buried by hundreds of meters of ice. Valley glaciers form in the cold heights of mountain ranges where snow accumulates, usually in preexisting valleys. In warmer, low-latitude climates, valley glaciers may be found only at the heads of valleys on the highest mountain peaks. In colder, high-latitude climates, valley gla-

ciers may descend many kilometers, down the entire length of a valley, in some places extending as broad lobes into lower lands bordering mountain fronts. Where valley glaciers flow down coastal mountain ranges, they may terminate at the ocean's edge, where masses of ice break off and form icebergs.

The other major glacier type is the much larger **continental glacier,** which is an extremely slow moving, thick sheet of ice, such as those covering much of Greenland and the Antarctic continent (Figure 15.2). The glacial ice of Greenland and Antarctica

**FIGURE 15.2** The Prince Charles Mountains project above the surface of the continental glacier covering Antarctica. *D. Parer and E. Parer-Cook/AUSCAPE International.*

is not confined to mountain valleys but covers practically their entire land surface. In Greenland, 2.8 million cubic kilometers of ice covers 80 percent of the island's total area of 4.5 million square kilometers. The upper surface of the ice sheet resembles an extremely wide convex lens. At its highest point, in the middle of the island, the ice is more than 3200 m thick (Figure 15.3). From this central area, the ice surface slopes to the sea on all sides. At the mountain-rimmed coast the ice sheet breaks up into narrow tongues resembling valley glaciers that wind through the mountains to reach the sea. There the ice breaks off to form icebergs.

As large as the Greenland glacier is, it is dwarfed by the Antarctic ice sheet (Figure 15.4). Ice blankets 90 percent of the Antarctic continent, covering an area of about 12.5 million square kilometers and reaching thicknesses of about 3000 m. In Antarctica, as in Greenland, the ice domes in the center and slopes down to the margins. In places, thinner sheets of ice floating on the ocean are attached to the main glacier on land. Of these, the best known is the Ross Ice Shelf, a thick layer of ice about the size of Texas that floats on the Ross Sea.

# GLACIAL BUDGETS: HOW GLACIERS FORM, GROW, AND SHRINK

A glacier starts with abundant winter snowfall that does not melt away in the summer. The snow is gradually converted to ice, and when the ice is thick enough, it begins to flow. Temperatures low enough to keep snow on the ground year-round are found at high latitudes (polar and subpolar regions) and altitudes (mountains). High latitudes are cold because the angle between the Sun's rays and Earth's surface increases toward the poles; the higher the angle, the smaller the amount of solar energy received by a given area. High altitudes are cold because the lowest 10 km of the atmosphere steadily cools with distance from the ground surface. Even in warm climates, glaciers will form if the mountains are high enough. Near the equator, glaciers form only on mountains that are higher than about 4500 m. This minimum altitude steadily decreases toward the poles, where snow and ice stay year-round even at sea level (Figure 15.5).

Snow and glacier formation require moisture as well as cold. Because moisture-laden winds tend to drop most of their snow on the windward side of a high mountain range, the leeward side is likely to be dry and unglaciated. Most of the high Andes Mountains of South America, for example, lie in a belt of prevailing easterly winds. Glaciers are formed on the moist eastern slopes, but the dry western side has little snow and ice. In arid climates glaciers are unlikely to form unless, as in Antarctica, it is so frigid all year that practically no snow melts and all is preserved.

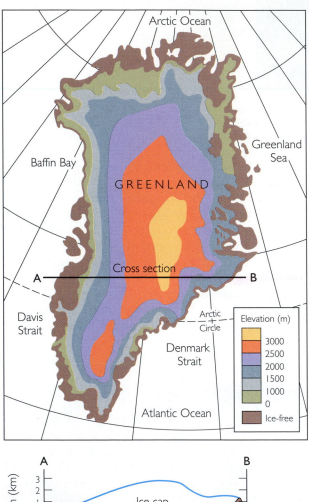

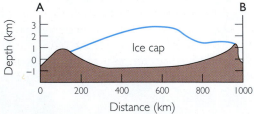

**FIGURE 15.3** The extent of the glacial ice cap and the elevation of the ice surface on Greenland. The generalized cross section of south-central Greenland, A–B, shows the lenslike shape of the ice cap. The ice moves down and out from the thickest section. (Information from R. F. Flint, *Glacial and Quaternary Geology,* New York, Wiley, 1971.)

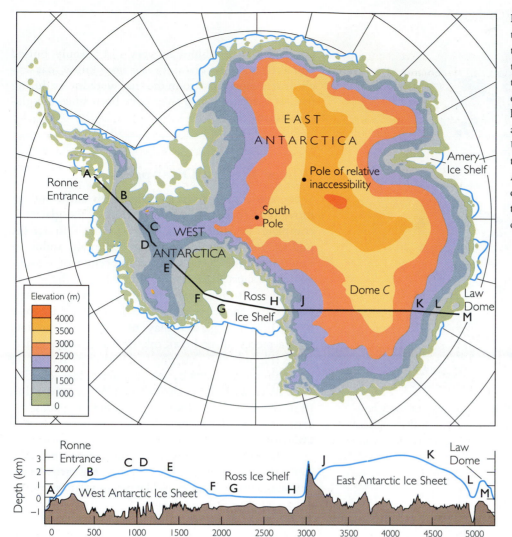

**FIGURE 15.4** This contour map and the cross section of Antarctica show the topography of the continental ice sheet that covers the entire continent and the land beneath it. Ice shelves are shown in white. (After Uwe Radok, "The Antarctic Ice," *Scientific American,* August 1985, p. 100; based on data from the International Antarctic Glaciological Project.)

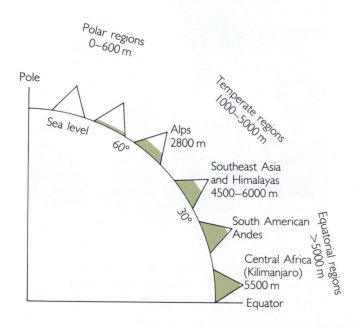

## Snow Becomes Ice

A fresh snowfall is a fluffy mass of loosely packed snowflakes. As the small, delicate crystals age on the ground, they shrink and become equant grains, as in Figure 15.6. During this transformation, the mass of snowflakes is compacted to form a denser granular snow. As new snow falls and buries the older snow, the granular snow further compacts to an even denser form, called **firn** (from a Germanic word applied to

**FIGURE 15.5** The height of the snow line, the altitude above which snow does not melt completely in summer, varies with latitude, from at or near sea level in polar regions to heights of more than 6000 m at the equator.

**FIGURE 15.6** Stages in the transformation of snow crystals, first to granular ice, then to firn, and finally to glacial ice. Accompanying the change of individual crystals is an increase in density by elimination of air. (After H. Bader et al., "Der Schnee und seine Metamorphose," *Beitrage zur Geologie der Schweiz,* 1939.)

analysis of air bubbles in very old, deeply buried Antarctic and Greenland ice, we now know that levels of atmospheric carbon dioxide were lower during the last glaciation than they have been since the glaciers retreated.

## Where the Ice Disappears

When ice accumulates to a thickness sufficient for movement to begin, the formation of the glacier is complete. Ice, like water, flows downhill under the pull of gravity, either down a mountain valley or down from the dome of ice at the center of a continental ice sheet. This trip downhill brings the glacier into lower altitudes, where temperatures are warmer. As a result, the ice starts to melt and the glacier loses material. Or a glacier may move to a shoreline, where pieces of ice break off, or calve, to form icebergs (Figure 15.7). Melting and **iceberg calving** are the main mechanisms by which glaciers lose ice. Glaciers in cold climates also lose ice by **sublimation** (transformation directly from the solid to the gaseous state). The total amount of ice lost annually is called **ablation.**

The difference between accumulation and ablation, the net budget, gives either the growth or shrinkage of a glacier. When accumulation minus

anything related to the previous year). Further burial and aging produce solid glacial ice as the smallest grains recrystallize, cementing all the grains together. One can think of snow as a sediment that is transformed by burial into a metamorphic rock called ice. The whole process may take only a few years, but more likely 10 to 20 years.

A typical glacier grows slightly during the winter, as snow falls on the glacial surface and is converted to ice. The amount of snow added to the glacier annually is termed the **accumulation.**

As it accumulates, glacial ice entraps and preserves valuable relics of Earth's past. In 1992 scientists were excited by the discovery of the preserved body of a prehistoric human in alpine ice on the border between Italy and Austria. Extinct animals such as the woolly mammoth, a great elephantlike prehistoric creature that once roamed icy terrains, have been found frozen and preserved by ancient ice in northern Siberia. At the other extreme, dust particles and bubbles of atmospheric gases are also preserved in glacial ice (see Figure 15.1). From the chemical

**FIGURE 15.7** Iceberg calving; Glacier Bay National Park, Alaska. Huge blocks of ice break off at the edge of the glacier, here projecting over the water. *Tom Bean.*

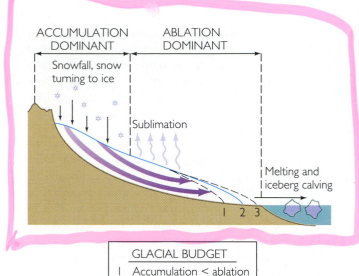

ACCUMULATION DOMINANT | ABLATION DOMINANT

Snowfall, snow turning to ice

Sublimation

Melting and iceberg calving

1  2  3

GLACIAL BUDGET
1  Accumulation < ablation
2  Accumulation = ablation
3  Accumulation > ablation

**FIGURE 15.8** Accumulation of a glacier takes place mainly by snowfall near the colder, upper regions, while ablation takes place mainly in the warmer, lower regions by sublimation or melting and iceberg calving.

by glacial debris, a lake may form at the terminus of the glacier.

The abundance of meltwaters points up the large amounts of fresh water tied up in glacial ice. Glaciers have assumed increasing importance as possible sources of fresh water for water-poor regions, if the transportation from source to users can be made economical. Some geologists have even suggested that icebergs might be towed through the ocean like barges. Though this idea may not be practicable very soon, it might conceivably be put to use in the future.

# HOW GLACIERS MOVE

When ice builds to a thickness sufficient to make gravity overcome the ice's resistance to movement, normally at least several tens of meters, it starts to move and thus becomes a glacier. As glaciers move, the ice deforms and slowly flows downhill. It is important to understand glacial flow, for the motion of glaciers is responsible for the immense amount of

ablation is zero over a long period, the glacier remains of constant size, even as it continues to flow downslope from the area where it formed (Figure 15.8). Such a glacier is accumulating snow and ice in its upper reaches and, in equal measure, ablating in its lower parts. If accumulation exceeds ablation, the glacier grows; if ablation exceeds accumulation, the glacier shrinks. Glacial shrinkage results from warming and melting in the region of the glacier's leading edge. Thus, even though a glacier is advancing downward or outward from its center, the ice edge may be retreating.

Glacier budgets vary from year to year. Some show trends of growth or shrinkage in response to climatic variation over periods of many decades. Yet over the past several thousand years many glaciers have remained constant on the average. Now that many scientists are becoming concerned about the effects of global warming on Earth's climate, geologists have proposed that glacier budgets be carefully monitored. Glacial shrinkage may be a good early warning of climate change.

As a glacier melts, great quantities of meltwater flow out from under the ice and from its edges (Figure 15.9). These meltwaters are the primary sources of cold-water streams that flow in mountain valleys below glaciers. If a glacial valley becomes dammed

**FIGURE 15.9** Meltwater at a glacier terminus; Mount Rainier National Park, Washington. The melting ice edge (at left) supplies the stream of water. *Peter Kresan.*

ice's geologic work. In fact, it was seeing the results of glacial movement—erosion, transportation, and sedimentation—that first clued scientists to the fact that ice does move. Unlike the readily observed rapid flow of a river, glacial motion is so slow that from day to day the ice seems not to move at all, giving rise to the expression "moving at a glacial pace."

The rate of glacial movement increases as the slope steepens or the ice thickens. Even on a flat surface, such as a continental lowland, if the ice builds up to a great thickness, it will flow outward. Just like a viscous fluid on a flat surface—honey on a slice of bread, for example—a continental glacier will flow and spread out as its thickness increases. But how does ice, a solid, flow as if it were a slow-moving, viscous liquid?

## Mechanisms of Glacial Flow

Glaciers flow primarily by two mechanisms. First, ice can slide downslope along the base of a glacier, like a brick sliding down an inclined board. But, unlike a brick, ice deforms and slides internally on a microscopic scale. We call the internal movement plastic flow, for it is like the plastic deformation of deeply buried rock discussed in Chapter 10. Under the great pressure within a glacier, individual crystals of ice slip tiny distances, on the order of a ten-millionth of a millimeter, over short intervals of time (Figure 15.10). The sum total of all the small move-

ments of the enormous number of ice crystals that make up the glacier amounts to a large movement of the whole mass of ice. To visualize this process, think of a random pile of decks of playing cards, each deck held together by a rubber band. The whole pile can be made to shift by many small slips between cards in the individual decks.

The other mechanism of glacial movement is the sliding of a glacier along its base, called **basal slip.** The amount and kind of basal slip depend on the temperature at the boundary between ice and ground in relation to the melting point of ice (Figure 15.11). At the base of a glacier, the ice is under tremendous pressure from the weight of the overlying ice. The melting point of ice decreases as pressure increases. This is the same effect that makes ice skating possible. The weight of the body on the narrow skate blade provides enough pressure to melt just a little ice under the blade, which lubricates the blade so that it can slide easily along the surface. Similarly, the ice at the base of a glacier may melt, even at temperatures lower than the freezing point at the surface. The melting at the ice-ground boundary creates a lubricating layer of water on which the overlying ice can slide.

The temperature of the ice at the base of a glacier depends partly on the surface temperature of the ice and partly on the flow of heat from the Earth below. Where the air temperature at the surface is not too low and the flow of heat from the ground is higher than normal, the temperature at the base of the glacier may be high enough for some melting to occur. In such moderately cold areas, typical of glacier-filled valleys in temperate regions, the ice may be at the melting point not only at the bottom but at higher places in the glacier, particularly near the surface if the air is warmer than freezing. Plastic flow contributes a small amount of internal heating from the friction generated by microscopic slips of crystals. In these "wet" glaciers, water is found in the ice as small drops between crystals and as pools in tunnels in the ice. The water throughout the glacier eases internal slip between layers of ice. In addition, ice may melt a little at the bottom and then refreeze, each time moving downhill a little bit.

In bitterly cold regions, the ice throughout the glacier, including its base, is well below the freezing point. The basal ice is frozen to the ground, so there is no lubricated sliding. What little movement there is rips up any detachable pieces of bedrock or soil. Most of the movement of these cold, dry glaciers takes place above the base by plastic deformation rather than basal slip.

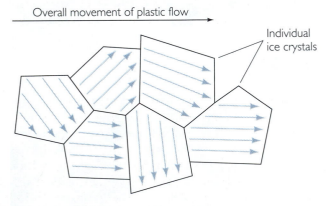

Overall movement of plastic flow

Individual ice crystals

**FIGURE 15.10** The plastic flow of a glacier is accomplished by small slips along the microscopic planes of a great number of ice crystals, which add up to a general flow.

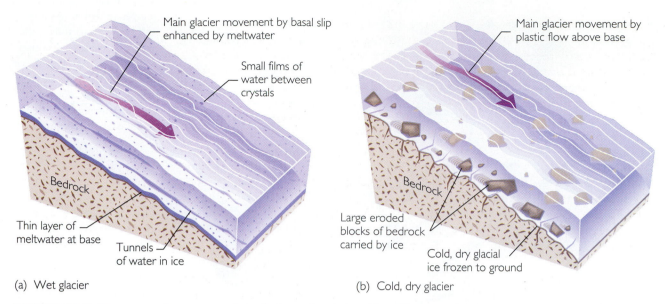

(a) Wet glacier                                         (b) Cold, dry glacier

**FIGURE 15.11** Two types of movement along the base of a glacier. (a) Meltwater at the base may lubricate basal slip. (b) In cold dry regions, ice is frozen to the ground and can move only by plucking blocks of bedrock or soil.

**FIGURE 15.12** Regularly spaced transverse crevasses on the surface of an Antarctic glacier. To some extent the crevasses are mantled with a cover of snow. *I. Allison, Australia Antarctica Division, Glaciology Section.*

The upper parts of glaciers (shallower than about 50 m) have little pressure on them. At these low pressures the ice behaves as a rigid, brittle solid, cracking as it is dragged along by the plastic flow of the ice below. These cracks, called **crevasses,** break up the surface ice into many small blocks in places where glacier deformation is strong (Figure 15.12). Crevasses are more likely to occur where the ice drags against bedrock walls, at curves in the valley, and where the slope steepens sharply. The movement of brittle surface ice at these places is a "flow" resulting from the slipping movements between these irregular blocks, similar in some ways to the microscopic slips of ice crystals but on a much larger scale.

## Flow Patterns and Speeds

The speed with which most valley glaciers move varies with the depth of the ice and with its position in relation to the valley walls. These glaciers flow partly by basal slip and partly by plastic flow within the body of ice. Ice movement is hindered by strong frictional forces at the glacier's solid boundaries at the base and sides of the valley, as shown in Figure 15.13.

The different speeds within alpine valley glaciers were first measured over a century ago by Louis Agassiz, a Swiss zoologist and geologist. As a young professor, not yet 30, he and his students established

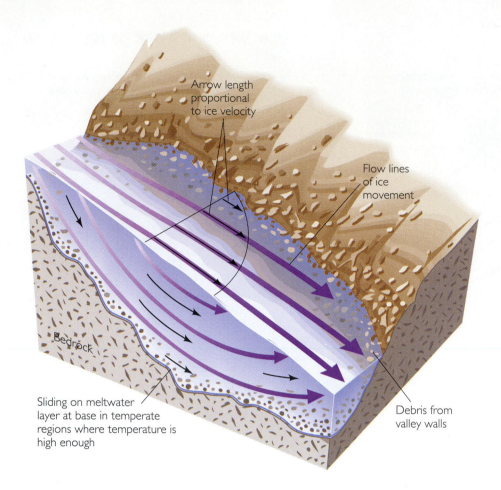

Arrow length
proportional
to ice velocity

Flow lines
of ice
movement

Bedrock

Sliding on meltwater
layer at base in temperate
regions where temperature is
high enough

Debris from
valley walls

**FIGURE 15.13** Ice flow in a typical valley glacier of a temperate region. As a result of frictional forces, the rate of movement decreases toward the base. The glacier moves most rapidly at its center. Friction along the valley walls reduces velocity at the edge of the flow.

a camp on a glacier to monitor its movement. They pounded stakes into the ice and measured their changes in position over a few years. The most rapid movement proved to be along the center line of the glacier, about 75 m in one year. Later it was demonstrated by the deformation of long vertical tubes pounded deep into the ice that the basal ice moved more slowly than the ice in the center. In modern times, geologists have used satellites and airborne radar to map the shapes and overall movements of glaciers. They have found, for example, that the Antarctic continental glacier at the South Pole is moving northward at a rate of 8 to 9 m per year, only a fraction of the highest speed Agassiz measured. The ice of some valley glaciers may move at a uniform speed everywhere, but only where the climate permits motion as a single unit, the glacier sliding entirely by basal slip along a lubricating layer of meltwater next to the ground.

A sudden period of fast movement of a valley glacier, called a **surge,** sometimes occurs after a long period of little movement. Surges may last more than two or three years and the ice may speed along at more than 6 km per year, a thousand times the normal velocity of a glacier. Though the mechanism of surges is not fully understood, it appears that a surge follows a buildup of water pressure in meltwater tunnels at or near the base. This pressurized water greatly enhances basal slip.

Continental glaciers in polar climates, where basal slip may be minor or absent, show the highest rates of movement in the center of the ice. Pressure there is very high, and the only frictional retarding forces are between layers of ice moving at different speeds in laminar flow. Near the surface, where pressure is least, the ice moves more slowly (Figure 15.14).

## GLACIAL LANDSCAPES

Just as you can't see your own footprint in the sand until you move your foot, little can be seen of the effects of an active glacier at its base or sides. Only when the ice melts is its geological work of erosion and sedimentation revealed. Thus it is from the topography of formerly glaciated areas and the distinctive landforms left behind that we can infer the physical processes driven by moving ice.

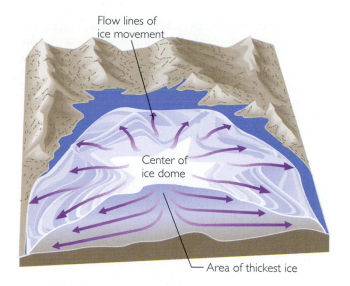

Flow lines of
ice movement

Center of
ice dome

Area of thickest ice

**FIGURE 15.14**  The flow of a continental glacier. The ice moves down and out from the thickest section, as shown by arrows.

## Glacial Erosion and Erosional Landforms

The capacity of glaciers to erode solid rock is amazing. A valley glacier only a few hundred meters wide can tear up and crush millions of tons of bedrock in a single year, enough to keep a fleet of 300 large dump trucks busy every day of the year. This heavy load of sediment is eroded from the floor and the sides of a glacier and carried by the ice to the ice front, where it is deposited as the ice melts (Figure 15.15). It is from

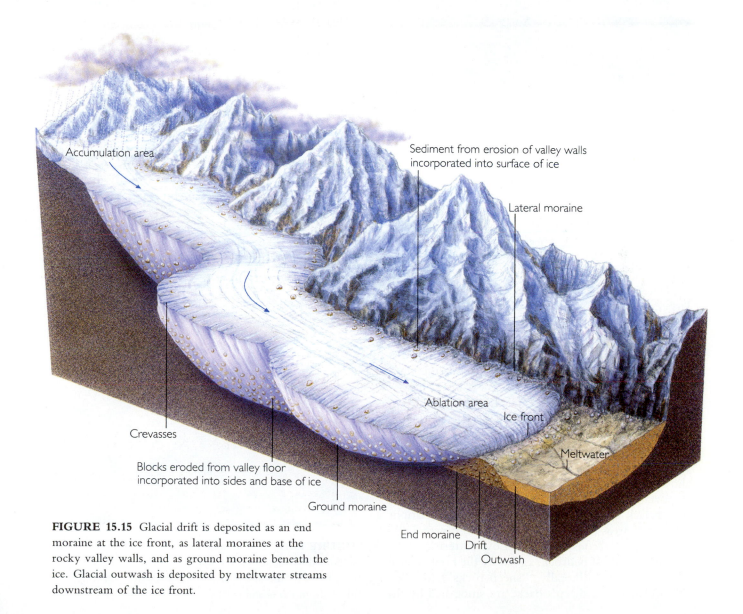

Accumulation area

Sediment from erosion of valley walls
incorporated into surface of ice

Lateral moraine

Crevasses

Blocks eroded from valley floor
incorporated into sides and base of ice

Ground moraine

Ablation area

Ice front

Meltwater

End moraine

Drift

Outwash

**FIGURE 15.15**  Glacial drift is deposited as an end moraine at the ice front, as lateral moraines at the rocky valley walls, and as ground moraine beneath the ice. Glacial outwash is deposited by meltwater streams downstream of the ice front.

**FIGURE 15.16** Glacial polish, striations, and grooves formed on a granite surface; Newfoundland, Canada. *Peter Kresan.*

the size of these deposits that we can estimate how much has been eroded. These estimates tell us that ice is a far more efficient agent of erosion than water or wind. We know that the total amount of sediment deposited in the world's oceans was several times larger during recent glacial times than during nonglacial periods.

At its base and sides, a glacier engulfs jointed, cracked blocks and breaks and grinds them against the rock pavement below. Rocks are fragmented into a great range of sizes, from boulders as big as houses to fine, pulverized material called **rock flour.**

As the glacier drags rocks along its base, the pavement is scratched or grooved (Figure 15.16). **Striations,** as such abrasions are termed, are strong evidence of glacial movement. Their orientation shows us the direction of ice movement, an especially important factor in the study of continental glaciers, which lack obvious valleys. By mapping striations over wide areas formerly covered by continental glaciers, we can reconstruct their flow patterns.

Small hills of bedrock, called by the French term *roches moutonnées* (literally, "sheep rocks") for their resemblance to a sheep's back, are smoothed by the

ice on their upcurrent side and plucked—like the rough surface of a loaf of bread pulled apart—by the advancing ice to a jagged, rough, steep slope on the downcurrent side (Figure 15.17). This, too, gives us an indication of the direction of ice movement.

A valley glacier carves a series of erosional forms as it flows from its origin to its lower edge. At the head of the glacier, the ice tends to carve out an amphitheater-like hollow, called a **cirque,** by a plucking and tearing action (Figure 15.18). With continued erosion, cirques at the heads of adjacent valleys gradually meet at the mountaintops, producing sharp, jagged crests called **aretes** along the divide.

As a valley glacier moves down from its cirque, it excavates a valley or deepens a preexisting river valley, creating a characteristic **U-shaped valley** (Figure 15.19). Glacial valley floors are flat and their walls steep, in contrast to the V-shaped valleys of many mountain rivers.

Another difference between glaciers and rivers lies in the joining of tributaries. Although the ice surface is level at the junction of a tributary glacier with the main valley glacier, the floor of the tributary valley may be much shallower than the main valley. When the ice melts, the tributary valley is left as a **hanging valley,** one whose floor lies high above the main valley floor (Figure 15.20). After the ice has gone and rivers occupy the valleys, the junction is marked by a waterfall as the stream in the hanging valley plunges over the steep cliff separating it from the main valley below.

Unlike rivers, valley glaciers at coastlines may erode their valley floors far deeper than sea level. When the ice retreats, these steep-walled valleys are

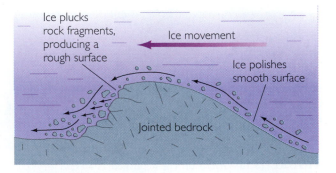

**FIGURE 15.17** A roche moutonnée is a small bedrock hill, smoothed by the ice on the upcurrent side and plucked to a rough face on the downcurrent side as the ice pulls fragments from joints and cracks.

**FIGURE 15.18** Valley glaciers with high snowfields and cirques in the background, tributary glaciers flowing down steep valleys, and their junction with the master glacier in the foreground; Harvard Glacier, Alaska. *B. Washburn.*

**FIGURE 15.19** U-shaped valley and hanging valley; Yosemite National Park, California. *Ric Ergenbright Photography.*

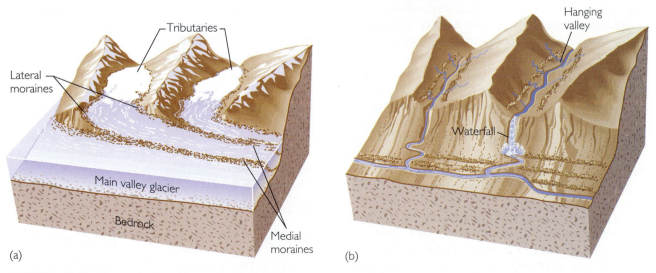

(a)

(b)

**FIGURE 15.20** The evolution of hanging valleys and their waterfalls by tributary valley glaciers. (a) During glaciation, tributary glaciers enter a major glacier at different floor levels. (b) After glaciation ends, the region is left with hanging valleys.

**FIGURE 15.21** A fjord, a drowned glacial valley; McCarthy Fjord, Kenai Fjords National Park, Alaska. *Peter Kresan.*

flooded with seawater (Figure 15.21). These glaciated arms of the sea, called **fjords,** create the spectacular rugged scenery for which the coasts of Alaska and Norway are famous.

## Glacial Sedimentation and Sedimentary Landforms

Eroded rock materials of all kinds and sizes are transported downstream by glaciers, eventually to be deposited where the ice melts. Ice is most effective as a transporter of debris, because the material picked up by ice does not settle out like the load carried by a river. Like water and wind currents, ice has a **competence,** the ability to carry particles of a certain size, and a **capacity,** the total amount of sediment the ice can transport. Ice's competence is extremely high. It can carry huge blocks many meters across that no other transporting agent can budge. The carrying capacity of ice is also tremendous. Some ice is so full of rock material that it is dark and looks like sediment cemented with ice.

When glacial ice melts it drops a poorly sorted, heterogeneous load of boulders, pebbles, sand, and clay. A wide range of particle sizes is the hallmark that differentiates glacial sediment from the much better sorted material deposited by streams and winds. To early geologists who were not aware of its glacial origins, the heterogeneous material was puzzling. They called it **drift,** because it seemed to have drifted in somehow from other areas. The term *drift* is now used for *all* material of glacial origin found anywhere on land or at sea.

As the ice melts and sediment is released, some of it may be picked up by meltwater streams flowing in tunnels within and beneath the ice and in streams at the ice front. The material is then transported and deposited like any other waterborne sediment. It is stratified and well sorted and may be cross-bedded. Drift that has been caught up and modified, sorted, and distributed by meltwater streams is called **outwash.** As we saw in Chapter 14, fine-grained material from outwash floodplains may be picked up by strong winds and deposited as loess. Drift deposited directly by melting ice is called **till** (Figure 15.22). Till is an unstratified and poorly sorted sediment that may be clayey, sandy, or bouldery. Common constituents of till are large boulders, which are termed **erratics.** The name derived from their seemingly random composition, so very different from that of local rocks.

**ICE-LAID DEPOSITS**  An accumulation of rocky, sandy, and clayey material carried by the ice or deposited as till is a **moraine.** There are many kinds of moraines, each named for its position with respect to the glacier that formed it. One of the most prominent in size and appearance is an *end moraine,* formed at the ice front. As the ice steadily flows downhill, it brings more and more sediment to its melting edge, where the unsorted material accumulates as a hilly ridge of till. An end moraine that marks a glacier's farthest advance is called a *terminal moraine* and is a geologist's best guide to the former extent of a valley or continental glacier.

A glacier erodes rock and unconsolidated material from the sides of its valley and gains additional materials from mass movements as the ice undercuts the valley walls above the ice. The eroded dirt and rock, incorporated into the ice as a dark, sediment-laden strip along the sides of the valley, are *lateral moraines* (Figure 15.23). The lateral moraines of joining glaciers merge to form a *medial moraine* in the middle of the larger flow below the junction. Lateral

and medial moraines, like end moraines, are left behind as ridges of till after the glacier melts away.

A layer of glacial drift deposited beneath the ice is called *ground moraine*. Ground moraine ranges from thin and patchy, with exposed bedrock knobs and pavements, to thicknesses great enough to bury bedrock completely. Regardless of the shape or location, moraines of all kinds consist of till.

Prominent landforms of some continental glacier terrains are **drumlins,** large, streamlined hills of till and bedrock that parallel the direction of ice movement (Figure 15.24). Usually found in clusters, drumlins are shaped like long, inverted spoons with the gentle slopes on the downstream sides. They may be 25 to 50 m high and a kilometer long. The precise origin of drumlins is uncertain. They may be formed either by ice erosion of an earlier accumulation of till or by a shaping of accumulated till under the ice.

**FIGURE 15.22** Till deposited on the eastern side of the Sierra Nevada in California during the Pleistocene epoch. Note the differences in particle sizes and the lack of stratification. *Martin Miller.*

**FIGURE 15.23** Lateral moraine; Columbia Icefields, Jasper National Park, Canada. *Gary Meszaros.*

**FIGURE 15.24** Drumlins, such as this one in upstate New York, are composed of till; some have a bedrock base with a streamlined shape formed by ice movement. The steep side commonly faces the direction from which the ice came, but the reverse is also found. *Ward's Natural Science Establishment, Rochester, New York.*

**WATER-LAID DEPOSITS**  Deposits of outwash from glacial meltwaters take a variety of forms. **Kames** are small hills of sand and gravel dumped near or at the edge of the ice (Figure 15.25). Some kames are deltas built into lakes at the ice edge; when the lake drains, these deltas are preserved as flat-topped hills. Kames are often exploited as commercial sand and gravel pits.

Silts and clays are deposited on the bottom of a lake at the edge of the ice in a series of alternating coarse and fine layers called varves (Figure 15.26). A **varve** is a pair of layers formed during one year by seasonal freezing of the lake surface. In the summer, when the lake is free of ice, coarse silt is laid down as abundant meltwater streams flow from the glacier into the lake. In winter, when the surface of the lake is frozen, the water below is stilled and the finest clays settle out.

Some lakes formed by continental glaciers were huge, many thousands of square kilometers in extent. If the till dams that created them were breached and carried away, the lakes drained rapidly, creating huge floods. In eastern Washington, an area called the scablands is covered by broad, dry stream channels, relics of torrential floodwaters draining from glacial Lake Missoula. From giant current ripples, sandbars,

**FIGURE 15.25** Retreat of the ice margin has revealed a kame and an esker formed by a stream that ran under the glacier. The esker is the winding ridge of sand and gravel running from the upper right to the lower left. At the former ice front (*lower left*), the esker ends in a kame as a delta built out into a lake. The grooves at right are in soft glacial sediment and are not bedrock striations. *B. Washburn.*

**FIGURE 15.26** Precambrian varved clays; southern Ontario, Canada. *F. J. Pettijohn.*

## Permafrost

The ground is always frozen in very cold regions, where the summer temperature never gets high enough to melt more than a thin surface layer. Perennially frozen soil, or **permafrost,** today covers as much as 25 percent of Earth's total land area. In addition to the soil itself, permafrost includes aggregates of ice crystals in layers, wedges, and irregular masses. The proportion of ice to soil, as well as the permafrost's thickness, varies from region to region.

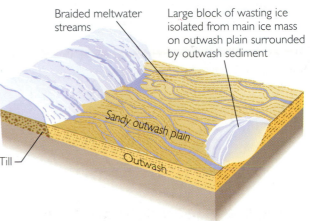

(a)  During ice melting

and the coarseness of the gravels found there, geologists have estimated flow velocities as high as 30 m per second and a discharge of 21 million cubic meters per second.

**Eskers** are long, narrow, winding ridges of sand and gravel found in the middle of ground moraines (see Figure 15.25). They run on for kilometers in a direction roughly parallel to the direction of ice movement. The well-sorted, water-laid character of esker materials and the sinuous, channellike course of the ridge suggest their origin. Eskers were deposited by meltwater streams flowing in tunnels along the bottom of a melting glacier. The tunnels themselves were opened by water seeping through crevasses and cracks in the ice.

Glaciated terrains are dotted with hollows, or undrained depressions, called **kettles.** Many kettles in outwash plains are steep-sided and may be occupied by ponds or lakes. The clue to their origin is the uneven melting-back of modern glaciers, which may leave behind huge isolated blocks of ice. A block of ice a kilometer in diameter might take 30 years or more to melt. During that time the melting block might have become partly buried by outwash sand and gravel carried by meltwater streams coursing around it (Figure 15.27). By the time the block was completely melted, the margin of the glacier would have retreated so far back that little outwash would reach the hole left by the melted block, and the depression would remain unfilled. If the bottom of the kettle were below the groundwater table, a lake would form.

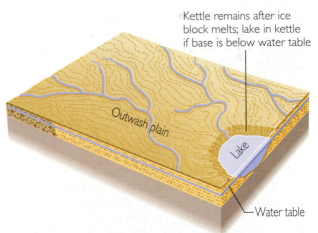

(b)  After complete deglaciation

**FIGURE 15.27** Evolution of an outwash kettle. (a) As a glacier retreats, it may leave behind large blocks of wasting ice that are gradually buried by outwash from the receding ice front. (b) After the front has retreated far enough from the region, outwash sedimentation stops, the ice block melts, and a depression remains, filled with water if it is deep enough to intersect the groundwater table.

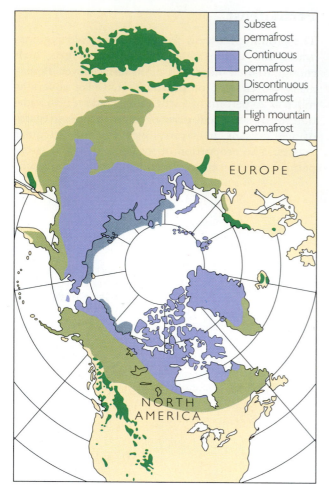

**FIGURE 15.28** The North Pole is at the center of this map of the Northern Hemisphere showing the distribution of permafrost. The large area of high mountain permafrost at the top of the map is on the Tibetan Plateau. (After a map by T. L. Pewe, Arizona State University.)

Subsea permafrost

Continuous permafrost

Discontinuous permafrost

High mountain permafrost

EUROPE

NORTH AMERICA

In Alaska and northern Canada, permafrost may be as much as 300 to 500 m thick. The ground below the permafrost layer, insulated from the bitterly cold temperatures at the surface, remains unfrozen, heated from below by the internal heat of the Earth. Permafrost is a difficult material to handle in engineering projects, such as roads, building foundations, and the Alaska oil pipeline. The difficulty stems from the melting of the surface as it is excavated. The melt-water cannot infiltrate the still-frozen soil below the excavation, so it stays at the surface, waterlogging the soil and leading it to creep, slide, and slump. Engineers decided to build part of the Alaska pipeline aboveground because an analysis showed that in places the pipeline would thaw the permafrost around it and lead to unstable soil conditions.

Permafrost covers about 82 percent of Alaska and 50 percent of Canada, as well as great parts of Siberia (Figure 15.28). Beyond the polar regions it is present in high, mountainous areas, especially on the Tibetan Plateau. Permafrost several hundreds of meters thick occurs in shallow marine areas off Arctic coasts, presenting difficult engineering problems for offshore oil drillers.

# ICE AGES: THE PLEISTOCENE GLACIATION

In the middle of the nineteenth century, the same Louis Agassiz who measured the speed of a Swiss alpine glacier immigrated to the United States and became a professor at Harvard University, continuing his studies in geology and other sciences. For his research, Agassiz visited many places in the northern parts of Europe and North America, from the mountains of Scandinavia and New England to the rolling hills of the American Middle West. In all these diverse regions, Agassiz recognized the signs of glacial erosion and sedimentation. In flat plains country he saw moraines that reminded him of the end moraines of valley glaciers. The heterogeneous material of the drift, including erratic boulders, convinced him of its glacial origin (Figure 15.29).

The areas covered by these glacial deposits were so vast that the ice that produced them must have

**FIGURE 15.29** Irregular hills alternate with lakes in a terrain of glacial till; Coteau des Praires, South Dakota. *John S. Shelton.*

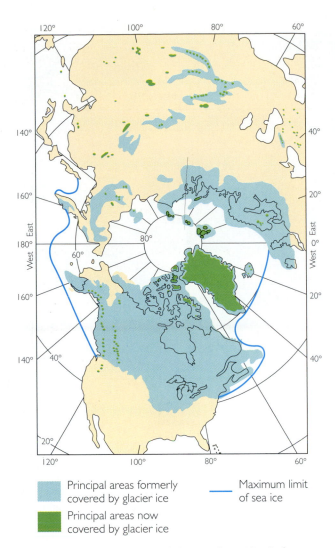

Principal areas formerly
covered by glacier ice

Principal areas now
covered by glacier ice

Maximum limit
of sea ice

**FIGURE 15.30** Glaciation of the Northern Hemisphere. (After R. F. Flint, *Glacial and Quaternary Geology,* New York: Wiley, 1971.)

been a continental glacier larger than Greenland or Antarctica (Figure 15.30). Eventually Agassiz and others convinced geologists and the general public that a great continental glaciation had effectively extended the polar ice caps far into regions that now enjoy temperate climates. For the first time people began to talk about ice ages. It was also apparent that the glaciation occurred in the relatively recent past, for the drift was soft, like freshly deposited sediment. We now know the age accurately from the carbon-14 (radioactive carbon) dating of logs buried in the drift. The drift of the last glaciation was deposited during one of the most recent epochs of geologic time, the Pleistocene, which lasted from 1.6 million to 10,000 years ago.

## Multiple Glacial Ages

It was not long before it became clear that there were multiple glacial ages during the Pleistocene, with warmer intervals between them. As geologists mapped glacial deposits about a century ago, they became aware that there were several layers of drift, the lower ones corresponding to earlier glacial ages. Between the older layers of glacial material were well-developed soils containing fossils of warm-climate plants. These soils were evidence that the glaciers retreated as the climate changed from frigid to warm. By the early part of this century, four distinct glaciations were thought to have affected North America during the Pleistocene epoch.

In the later twentieth century, geologists and oceanographers found fossil evidence of warming and cooling of the oceans that corresponded to interglacial and glacial ages. The fossils, buried in Pleistocene ocean sediments, were of foraminifera, small, single-celled marine organisms that secrete shells of calcium carbonate (calcite, $CaCO_3$). The proportion of ordinary oxygen, oxygen-16, and the heavy oxygen isotope (see Chapter 2), oxygen-18, in the calcite of a foraminifer's shell depends on the temperature of the water it lived in. Different ratios of oxygen-16 to oxygen-18 in the shells preserved in different layers of sediment revealed the temperature changes in the oceans through time in the Pleistocene epoch.

Isotopic analysis of shells also allowed measurement of another glacial effect: the withdrawal of a great deal of water from the ocean by evaporation and its precipitation as snow to form glacial ice affected the isotope ratio in the oceans. The lighter oxygen-16 isotope has a greater tendency to evaporate from the ocean surface than the heavy isotope, leaving more of the heavy isotope behind. Thus, even in parts of the ocean where there was no great change in temperature—around the equator, for example—scientists could trace the growth and shrinkage of continental glaciers. From this analysis of marine sediments we have learned that there were many shorter, more regular cycles of glaciation and deglaciation than geologists had recognized from the glacial tills of the continents alone.

## Changes in Sea Level

We know from the hydrologic cycle (Chapter 12) that the sea is the source of much of the water that falls as snow on the continents. During the Pleistocene epoch, the more water that was tied up as ice in continental glaciers, the less water there was in the

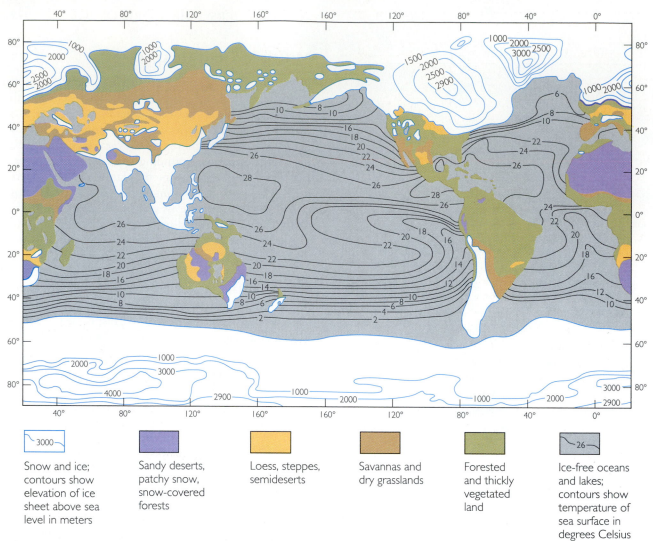

**FIGURE 15.31** Sea surface temperatures (°C), ice extent, and ice elevation on a day in August 18,000 years ago. Continental outlines reflect the lowering of sea level 85 m below present-day levels. (After CLIMAP, "The Surface of the Ice-Age Earth," *Science,* vol. 191, 1976.)

Legend:

- **3000** — Snow and ice; contours show elevation of ice sheet above sea level in meters
- (purple) — Sandy deserts, patchy snow, snow-covered forests
- (yellow) — Loess, steppes, semideserts
- (tan) — Savannas and dry grasslands
- (green) — Forested and thickly vegetated land
- **26** — Ice-free oceans and lakes; contours show temperature of sea surface in degrees Celsius

ocean. As continental glaciers grew, sea level everywhere on the globe lowered. Accumulation of a large amount of ice, say 400,000 km³, corresponds to only a small drop in sea level, 1 m. But during the maximum extent of the most recent glaciation, 18,000 years ago, sea level dropped about 130 m, corresponding to an enormous volume of ice, about 70 million cubic kilometers, or almost three times the amount of ice on the Earth today.

Figure 15.31 shows the temperatures of the oceans and the ice distribution and climatic zones of nonglaciated regions of the continents 18,000 years ago. Masses of ice from 2 to 3.5 km thick were built up over North America, Europe, and Asia. In the Southern Hemisphere, the Antarctic ice expanded and the southern tips of South America and Africa were covered with ice. The continents were slightly larger than they are today because the continental

shelves surrounding them, some more than 100 km wide, were exposed by the large drop in sea level. At this time the climatic zones at low latitudes were not very different from today's, but many temperate regions were heavily forested (see Box 15.1).

Rivers extended their channels across newly emergent continental shelves and began to erode channels in the former seafloor. Early humans roamed these low coastal plains, for early cultures, such as those of prehistoric Egypt, were evolving in the lands beyond the ice sheets.

## Climate Change and Glacial Epochs

Ever since Agassiz's time, geologists have pondered the causes of ice ages. Today the discussion continues, heightened by concern over possible short-term

BOX 15.1 # FUTURE CHANGES IN SEA LEVEL AND THE NEXT GLACIATION

What would happen if the 25 million cubic kilometers of the planet's water now tied up as ice were to melt? The change in sea level would be catastrophic, for the melting of all existing ice would raise the oceans by about 65 m. Major cities of the world—London, New York, Los Angeles, Tokyo, and many others—would be flooded. The bulk of the low-lying areas of the continents, where most of the world's population lives, would be submerged beneath shallow seas.

It is not likely that all of Earth's ice will melt for many millions of years to come. But it is very possible that a warming of climate in the next century will result in a small rise in sea level from melting of glacial ice. Another factor in the rise of sea level is the extremely small expansion of water that accompanies an increase in temperature; this effect, multiplied by the huge amount of ocean water warmed, would cause an additional expansion of the oceans and raise sea level. A global warming of just a few degrees might cause sea level to rise by a few meters. Though this amount may seem negligible, it is not, for at times of storms and tidal surges, this higher sea level would be enough to cause serious flooding of low-lying coastal regions. Millions of people in Bangladesh, for example, live on lands only a few meters above sea level on the deltas of the Ganges and Brahmaputra rivers. These people have already suffered catastrophic floods, and thousands of lives have been lost. A rise of sea level of only 1 m would be devastating.

We can expect the Earth to go back into a period of glaciation over the next 10,000 years, for the current interglacial age has already lasted 10,000 years and the average length of interglacial periods is about 20,000 years. The general scenario is predictable. As polar and temperate regions grow colder, valley glaciers will enlarge and snow in areas of northern North America, Europe, and Asia will stop melting in the summers. Ice will gradually accumulate in large continental glaciers. As the ice accumulates, it will start to flow into the colder parts of formerly temperate regions, engulfing everything in its path. At the same time, sea level will be lowered, leaving the seaports of the world high and dry. A mass migration of people and animals into warmer regions would follow. Forest and plant populations would quickly alter as well. It is hard to envision how the human population of the future will cope with such enormous changes.

warming trends that will melt part of the world's glaciers and long-term cooling trends that might expand them. Theories of ice ages center on climate change and possible factors in the atmosphere and oceans that could provoke glaciation at one time and then reverse course to cause deglaciation. They also have to take into account the fact that the Pleistocene glaciation was not unique in Earth's history. Since the early part of this century we have known from glacial striations and lithified ancient tills, called **tillites,** that glaciations covered parts of the continents numerous times in the geologic past, long before the Pleistocene.

Theories of ice ages have two parts. One is an explanation of the general cooling of polar regions that led to the polar ice caps. There is evidence from glacial sediments in Antarctica and the ocean that the polar ice caps of today began forming about 10 million years ago, in the Miocene epoch, and grew slowly until they culminated in the full-blown glaciers of the Pleistocene. The other part of theories of ice ages focuses on the causes of alternating glacial and interglacial ages, which follow a more or less regular pattern throughout the Pleistocene.

A favored explanation for the general cooling lies in the positions of the continents in relation to one another and to the poles. These positions are determined by continental drift driven by plate tectonics. For most of Earth's history there were no extensive land areas in the polar regions and there were no ice caps. The oceans circulated through the open polar regions, transporting heat from the equatorial regions and helping the atmosphere distribute temperatures fairly evenly over the globe. When large land areas drifted to positions that obstructed efficient transport of heat by the oceans, the differences in

## BOX 15.2 CARBON DIOXIDE AND THE GREENHOUSE EFFECT

The possible relationship between carbon dioxide in the atmosphere and climate warming was proposed by Svante Arrhenius, one of the early winners of the Nobel Prize in chemistry (1903). Arrhenius knew a great deal about geology as well as chemistry, and he was familiar with the effects of glaciation on the terrain of his native Sweden. He was also aware of the growing discussion of the causes of glaciations that enlivened many geologists' meetings at the end of the nineteenth century. He reasoned that the small amount of carbon dioxide in the atmosphere (now about 345 ppm) could affect climate because the carbon dioxide molecules strongly absorb heat rays from the Earth.

Here is how it works. The atmosphere is relatively transparent to the incoming visible rays of the Sun. Much of this radiant energy from the Sun is absorbed by the Earth's surface and then reemitted as invisible infrared heat rays. Just as a hot pavement radiates heat as it is warmed by the Sun, the Earth's surface radiates heat back to the atmosphere. The atmosphere, however, is not transparent to these infrared rays, because carbon dioxide and water molecules strongly absorb the infrared instead of allowing it to escape to space. As a result, the atmosphere is heated and radiates heat back to the surface. This is called the *greenhouse effect,* by analogy to the warming of a greenhouse, whose glass lets in visible light but lets little heat escape.

The more carbon dioxide, the warmer the atmosphere; the less carbon dioxide, the colder. Without any greenhouse effect, Earth's surface temperature would be well below freezing and the oceans would be a solid mass of ice. Geologists now have evidence that, aside from glaciations, climates ranged from warm to cool in the geological past. It is likely that some of these changes were related to changes in the amount of carbon dioxide in the atmosphere.

Two hundred years ago, the Industrial Revolution began, powered largely by the burning of coal. Since then we have been burning carbon-based fossil fuels—coal, oil, and gas—at an increasing rate. Every carbon atom burned ends up as carbon dioxide in the atmosphere. The more carbon we burn, the more carbon dioxide we spew into the atmosphere.

Not all of the increased load of carbon dioxide remains in the atmosphere. About 40 percent of it is absorbed by surface layers of the oceans. Nevertheless, despite the oceans' moderating effect, carbon dioxide levels are expected to reach about 375 ppm by the year 2000 and 600 ppm late in the twenty-first century. This is a tremendous amount in comparison with the 295 ppm present in the middle of the nineteenth century, and it is likely to lead to a significant warming of the climate.

Calculations by large supercomputer programs based on global atmospheric and oceanic dynamics indicate that the increased carbon dioxide levels projected for the middle of the next century may raise the average global temperature by 1.5 to 4.5°C (with allowances made for many uncertainties in the calculations). Such a temperature increase, small though it may seem, could have serious effects on climate and weather patterns, including shifts of arid and temperate zones, shifts in the locations and frequency of droughts and storms, and changes in water supplies. Another consequence would be a small but significant global rise in sea level (70 cm), which would have serious repercussions (see Box 15.1).

Although there are many uncertainties in this picture, most scientists agree that some

temperature between the poles and equator increased. As the poles cooled, the ice caps formed. We can predict that many millions of years from now, if the large polar and subpolar lands of the northern continents and the Antarctic continent drift away from the poles, the ice caps will melt away.

So far, the alternation between glacial and interglacial ages has been explained best in terms of astronomical cycles. The shape of Earth's orbit around the Sun changes periodically, putting us sometimes slightly closer and sometimes slightly farther from the Sun. Also, Earth wobbles slightly on its axis of warming will occur. This climate change will have profound effects on the social, political, and economic life of the next century, however we react to it. The Rio de Janeiro summit conference on the global environment held in June 1992 dramatically highlighted differences among nations. While all the nations of the world agreed that we must move to decrease carbon dioxide emissions by decreasing our dependence on fossil fuels, they did not agree on the rate at which we should decrease emissions. Most nations want to reduce emissions to the 1990 level by the year 2000, but some, such as the United States, emphasize the threat of severe eco-

The lower curve (right-hand scale) shows the increase in atmospheric carbon dioxide since 1860, with a projection to the year 2000. The upper curve (left-hand scale) shows the cumulative input of carbon dioxide. The difference between the two curves represents the amount of carbon dioxide removed by the oceans or by additions to the total carbon in all living organisms. These curves, originally projected in 1970, have proved to predict the actual increases closely. (After B. Bolin, "The Carbon Cycle," *Scientific American,* September 1970, p. 124.)

The greenhouse effect. Just as the glass of a greenhouse transmits light rays but holds in heat, the carbon dioxide of the atmosphere transmits visible radiation from the Sun but absorbs the infrared radiation from the ground surface and reradiates it back to Earth.

nomic dislocations if we move too quickly. Some scientists argue that it is already too late and that we should plan to deal with the consequences, bad as they are. However we attempt to solve the problem, in one way or another the solution will involve every one of us on the globe.

In Chapter 22, we will discuss further the decisions that policymakers face in dealing with the environmental consequences of the greenhouse effect.

rotation. These changes affect the amount of heat that Earth receives from the Sun. Careful calculations of the orbital motions, first worked out in the 1920s and 1930s by Milutin Milankovitch, a Yugoslav geophysicist, showed that variations in the heat received from the Sun due to orbital changes corresponded to glacial and interglacial stages. Recent work on this theory predicts regular changes in climate, with a long-term periodic glaciation every 100,000 years and shorter term ones about every 40,000 and 20,000 years. When the long-term periods are multiples of short-term ones, the changes are greater.

We are now beginning to find out that glacial-interglacial alternations are also related to levels of carbon dioxide in the atmosphere. Analyses of bubbles in glacial ice indicate that carbon dioxide levels were low during the most recent ice age and rose rapidly as the climate warmed and the ice melted. The reasons for this relatively rapid change in carbon dioxide and whether it is a cause or an effect are being hotly debated right now (see Box 15.2). Whatever the outcome of this debate, it seems certain that we will be finding out much more about climate change in relation to glaciation as the nations of the world become increasingly concerned about the impact of future changes in climate.

# SUMMARY

**How do glaciers form and how do they move?** Glaciers form where climates are cold enough that snow, instead of melting away in summer, becomes transformed by recrystallization into ice. As snow accumulates, the ice thickens, either at the tops of mountain valley glaciers or at the domed centers of continental ice sheets, until it becomes so heavy that gravity starts to pull it downhill. During periods of stable climate, the size of a glacier remains constant as it loses ice by melting, sublimation, and iceberg calving in the ablation zone and is replenished by snow in the accumulation zone. During a warming period, a glacier shrinks as ablation exceeds accumulation; conversely, a glacier expands as accumulation outpaces ablation in a cooling climate. Glaciers move by a combination of basal slip and plastic flow, the former being more important in warmer climates, where melting at the glacier's base lubricates the ice's movement over rock. Plastic flow dominates in very cold regions, where the glacier's base is frozen to the ground. The speed of a glacier's movement varies with the level of the ice and its position with respect to the walls of the valley through which it travels.

**How do glaciers erode bedrock, transport and deposit sediment, and shape the landscape?** Glacial erosion is efficient at scraping, plucking, and grinding bedrock into sizes ranging from boulders to finely ground rock flour. Valley glaciers erode cirques and aretes at their heads, U-shaped and hanging valleys in their main courses, and fjords where they end at the ocean, eroding their valleys below sea level. Glacial ice has a high competence and capacity to carry abundant sediments of all sizes, transporting huge quantities to the ice front, where melting releases them. The sediments may be directly deposited from the melting ice as till or picked up by meltwater streams and laid down as outwash. Characteristic depositional forms are moraines, drumlins, kames, eskers, and kettles. Permafrost is perennially frozen soil in very cold regions.

**What are ice ages and what causes them?** Glacial drift of Pleistocene age is widespread over high-latitude regions that now enjoy temperate climates, evidence that continental glaciers once expanded far beyond the polar regions. Studies of the geologic ages of glacial deposits on land and sediments of the seafloor have shown that the Pleistocene glacial epoch consisted of multiple advances (glacial intervals) and retreats (interglacial intervals) of the continental ice sheets. Each of the advances corresponded to a global lowering of sea level that exposed large areas of continental shelf; during interglacial intervals, sea level rose and submerged the shelves. Though the causes of glaciation remain uncertain, it appears that the general cooling of the globe leading to glaciation was the result of continental drift that gradually moved continents to positions where they obstructed the general transport of heat from the equator to the polar regions. A favored explanation for the alternation of glacial and interglacial intervals is the effect of astronomical cycles, by which very small periodic changes in Earth's orbit and axis of rotation alter the amount of sunlight received at the Earth's surface. There is also evidence that decreased levels of carbon dioxide in the atmosphere diminish the greenhouse effect and lead to glaciation. In contrast, the current increase in atmospheric carbon dioxide resulting from the burning of fossil fuels may cause global warming.

# KEY TERMS AND CONCEPTS

glacier (p. 329)
valley glacier (p. 329)
continental glacier (p. 329)
firn (p. 331)
accumulation (p. 332)
iceberg calving (p. 332)

sublimation (p. 332)
ablation (p. 332)
plastic flow (p. 334)
basal slip (p. 334)
crevasse (p. 335)
surge (p. 336)

rock flour (p. 338)
striation (p. 338)
cirque (p. 338)
arete (p. 338)
U-shaped valley (p. 338)
hanging valley (p. 338)

fjord (p. 340)

competence (p. 340)

capacity (p. 340)

drift (p. 340)

outwash (p. 340)

till (p. 340)

erratics (p. 340)

moraine (p. 340)

drumlin (p. 341)

kame (p. 342)

varve (p. 342)

esker (p. 343)

kettle (p. 343)

permafrost (p. 343)

tillite (p. 347)

## EXERCISES

1. How are valley glaciers distinguished from continental glaciers?

2. How is snow transformed into glacial ice?

3. How does glacial growth or shrinkage result from the balance between ablation and accumulation?

4. What are the mechanisms of glacier flow?

5. How do glaciers erode bedrock?

6. Why are striations evidence of a former glaciation?

7. Name three kinds of glacial sediment.

8. Name three landforms made by glaciers.

## THOUGHT QUESTIONS

1. Why is glacial ice stratified in many places?

2. Some parts of a glacier contain much sediment, others very little. What accounts for the difference?

3. Contrast the kinds of till you would expect to find in two glaciated areas, one a terrain of granitic and metamorphic rocks, the other a terrain of soft shales and loosely cemented sands.

4. What geological indication(s) might you find that would tell you the direction of glacial movement in a continentally glaciated region?

5. You are walking over a winding ridge of glacial drift. What evidence would you look for to discover whether you were on an esker or an end moraine?

6. One of the dangers of exploring glaciers is the possibility of falling into a crevasse. What topographic features of a valley glacier or its surroundings would you use to infer that you were on a part of the glacier that was badly crevassed?

7. You are scouting out possible commercial sand and gravel deposits in an area that was once near the edge of a continental glacier. What landforms would you look for?

8. You live in New Orleans, not far from the mouth of the Mississippi River. What might be your first indication that the world was entering a new glacial age?

## SUGGESTED READINGS

Covey, C. 1984. The Earth's orbit and the ice ages. *Scientific American* (February):58.

Denton, George H., and T. J. Hughes. 1981. *The Last Great Ice Sheets.* New York: Wiley.

Flint, Richard F. 1971. *Glacial and Quaternary Geology.* New York: Wiley.

Imbrie, John, and K. P. Imbrie. 1979. *Ice Ages: Solving the Mystery.* Short Hills, N.J.: Enslow Press.

LaChapelle, E. R., and A. S. Post. 1971. *Glacier Ice.* Seattle: University of Washington Press.

Paterson, W. J. B. 1981. *The Physics of Glaciers,* 2d ed. London: Pergamon.

Price, R. J. 1973. *Glacial and Fluvioglacial Landforms.* New York: Macmillan.

# LANDSCAPE EVOLUTION

From high, snowcapped peaks to broad, rolling plains, Earth's landscape comprises a great array of landforms, large and small. The challenge to geologists is to explain how such varied landscapes are formed in different locales. The most dramatic landscapes have been fashioned by tectonics and erosion, but sedimentation plays a role in the shaping of valleys and plains. The carving of landscape is a dynamic process interweaving geologic history, tectonics, climate, weathering, erosion, and sedimentation. Through slow changes, imperceptible on a human time scale, landscapes evolve from the high mountains created by plate-tectonic movements to the low hills and plains formed by erosion as tectonic activity dies away.

## 16

A landscape dominated by tectonics: high mountain peaks seen from a river floodplain in an intermontane basin. Flathead River Valley, Montana. *Tupper Ansel Blake.*

The landscape of the continents is the product of many geological processes, ranging from the erosion of river valleys and the scraping of bedrock by glaciers to the deposition of alluvial fans and glacial moraines. Weathering, erosion, transportation, and deposition combine to sculpt the land surface. Erosion is dominant in high mountains, and sedimentation is the primary agent in low-lying plains.

Erosion and sedimentation are powered by tectonics. Uplifts of Earth's crust raise mountains and expose them to intense erosion, while subsidence of the crust creates low areas that become the sites of deposition. On a smaller scale, the deformation of the crust into folds and faults controls the particular forms of many hills and valleys.

Landscape also depends on geologic history. That history determines what kinds of rocks, with varying resistance to erosion, are exposed at the surface and how the rocks have been affected by structural deformation, igneous activity, and metamorphism. On the other hand, geologic history may encompass millions of years of tectonic stability, allowing weathering and erosion to wear away the land to lower, more level landscapes.

## TOPOGRAPHY, ELEVATION, AND RELIEF

How do we objectively describe the varied surface of the Earth, from the heights of its mountain peaks to the depths of its valleys? The varying heights that give shape to the surface are called its **topography.** We compare heights to sea level, the average height of the oceans around the globe. The altitude, or vertical distance above sea level, is called **elevation.** A topographic map shows the distribution of elevations in an area, which is represented most commonly by **contours,** lines that connect points of equal elevation (Figure 16.1).

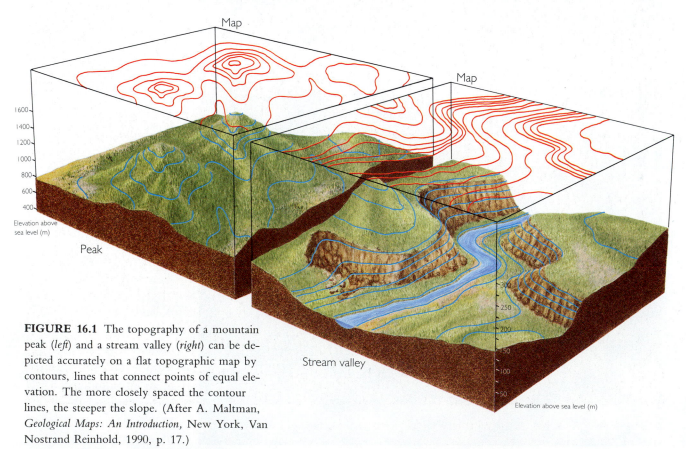

**FIGURE 16.1** The topography of a mountain peak (*left*) and a stream valley (*right*) can be depicted accurately on a flat topographic map by contours, lines that connect points of equal elevation. The more closely spaced the contour lines, the steeper the slope. (After A. Maltman, *Geological Maps: An Introduction,* New York, Van Nostrand Reinhold, 1990, p. 17.)

354

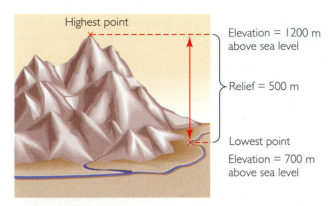

**FIGURE 16.2** Relief is the difference between the highest and lowest elevations in a region.

Another measure of topography, its roughness, is the vertical distance between the highest and the lowest points in a particular area, called **relief** (Figure 16.2). The relief of an area is estimated from contours on a topographic map by subtracting the elevation of the lowest contour, usually at the bottom of a river valley, from the highest, the top of the highest hill or mountain. The higher the relief, the more rugged the topography. Most regions of high elevation also have high relief, and most areas of low elevation have low relief, although there are exceptions. Mountains that rise steeply from the seashore, like some of those on the Pacific coast of North America, may be at relatively moderate elevations yet have high relief. Regions like the Tibetan Plateau of the Himalayan Mountains may lie at high elevations but have relatively low relief.

Relief depends on the area specified. At one extreme is relief on a global scale, from the highest mountain in the world, Mount Everest, at an elevation of 8854 m, to the lowest land area in the world, the shores of the Dead Sea in Israel and Jordan at 392 m below sea level. That global relief of 9246 m dwarfs the modest relief, about 800 m, of areas in the Appalachian Mountains and relief values of a few thousand meters in areas of the Rocky Mountains.

If we fly over the United States, we can see many sorts of topography (Figure 16.3). The moderate elevations and relief of the elongate ridges and valleys of

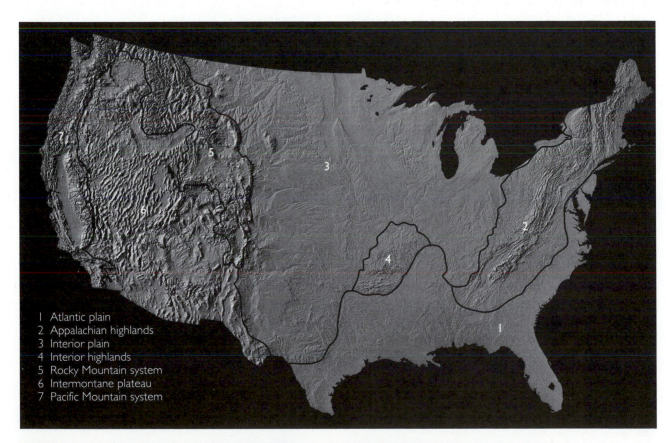

1 Atlantic plain
2 Appalachian highlands
3 Interior plain
4 Interior highlands
5 Rocky Mountain system
6 Intermontane plateau
7 Pacific Mountain system

**FIGURE 16.3** A digital, shaded-relief map of landforms in the contiguous United States. (Richard J. Pike and Gail P. Thelin, USGS, 1989.)

the Appalachian Mountains contrast with the low elevations and relief of the midwestern plains. Even more striking is the contrast between the plains and the Rocky Mountains. As we examine these different types of topography more closely we can characterize them not only by elevation and relief but also by their landforms: the steepness of their slopes, the shapes of the mountains or hills, and the forms of the valleys.

# LANDFORMS: THE COMPONENTS OF LANDSCAPE

Rivers, glaciers, and the wind leave their marks on the surface of the Earth: rugged mountain slopes, broad valleys, floodplains, dunes, and the many other forms created by erosion and sedimentation. Geologists have learned to recognize the shapes of these marks of erosion and sedimentation, which are called **landforms.** The landforms of a region make up its landscape and are guides to the region's geologic structure and history.

## Mountains and Hills

Even the most ordinary landforms are not easy to define precisely. We have used the word *mountain* many times in this book, yet we can define it no more precisely than to say that a mountain is a large mass of rock that projects well above its surroundings. Most mountains are found with others in mountain ranges, where peaks of various heights are easier to distinguish than distinct separate mountains (Figure 16.4). In some regions, isolated volcanoes or erosional remnants of former mountain ranges rise as single peaks above the surrounding lowlands.

We distinguish between mountains and hills only by size and custom. Elevations that would be called mountains in lower terrains would be called hills in high-elevation regions. In general, landforms more than several hundred meters above their surroundings are called mountains.

The steepness of the slopes in mountainous and hilly areas generally correlates with elevation and relief, the steepest slopes generally being found in high mountains with great relief. The slopes of mountains lower in elevation and relief are less steep and rugged. The slopes are gentler in hills and may be barely noticeable in plains. Mountains and hills are reflections of tectonic activity caused directly or indirectly by plate movements. The more recent the activity, the more likely the mountains are to be high. The Himalayas, the highest mountains in the world, are also the youngest.

## Plateaus

A large, broad, flat area of appreciable elevation above the neighboring terrain, at least on one side, is a **plateau.** Smaller plateaus may be called tablelands. In the western United States, a small, flat elevation with steep slopes on all sides is called a **mesa** (Figure 16.5). Most plateaus have elevations of less than 3000 m, but the Altiplano of Bolivia is at an elevation

**FIGURE 16.4** The Alaska Range. All the peaks are sharp aretes in this glacially sculpted terrain. *Peter Kresan.*

FIGURE 16.5 A mesa; Monument Valley, Arizona. The flat tops are held up by erosion-resistant beds. *Raymond Siever.*

of 3600 m and the extraordinarily high Tibetan Plateau has an average elevation of almost 5000 m. Many plateaus owe their flatness to their floors of undeformed sedimentary rocks or layers of lava flows. The tectonics of plateau formation involves general uplift rather than deformation by lateral compression.

## Structurally Controlled Cliffs

The folds and faults produced by rock deformation during mountain building leave their marks on Earth's surface. These topographic expressions of deformation are often a guide to the geologic structures that control them. **Cuestas** are asymmetrical ridges in a tilted and eroded series of beds of alternating weak and strong resistance to erosion. One side of a cuesta has a long, gentle slope determined by the dip of the erosion-resistant bed. The other side is a steep cliff formed at the edge of the resistant bed where it is undercut by erosion of a weaker bed beneath (Figure 16.6). Much more steeply dipping or vertical resistant beds erode to form **hogbacks,** which are narrow, steep ridges similar in shape to cuestas but more prominent (Figure 16.7). Steep cliffs are also produced by nearly vertical faults along which one side has been raised relative to the other.

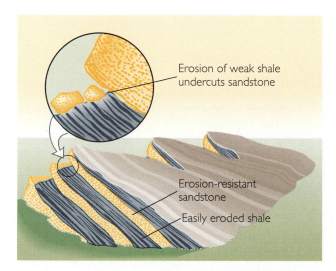

Erosion of weak shale undercuts sandstone

Erosion-resistant sandstone

Easily eroded shale

FIGURE 16.6 Cuestas are formed where gently dipping beds of erosion-resistant rock, such as sandstone, are undercut by erosion of an easily eroded underlying rock, such as shale.

FIGURE 16.7 Hogback ridges formed by tectonically turned up layers of erosion-resistant sedimentary rocks; Rocky Mountains, Colorado. *James M. Soule, Colorado Geological Survey.*

## Structurally Controlled Ridges and Valleys

In young mountains, during the early stages of folding and uplift, the upfolds (anticlines) form ridges and the downfolds (synclines) form valleys (Figure 16.8). But as tectonic activity moderates and erosion bites deeper into the structures, the anticlines may form valleys and the synclines ridges. This happens where the rocks—typically sedimentary rocks such as limestones, sandstones, and shales—exert strong control on weathering and erosion. If the rocks beneath an anticline are mechanically weak, such as shales, the core of the anticline may be eroded to form an anticlinal valley (Figure 16.9). In a region that has been eroded for many millions of years, a pattern of linear anticlines and synclines produces a series of ridges and valleys like those of the Appalachian Mountains in Pennsylvania and adjacent states (Figure 16.10).

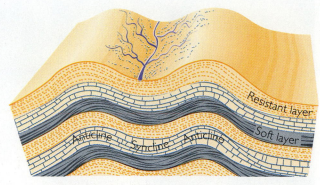

(a) Early stage of folding: ridges over anticlines, streams in synclines

(b) Later stages of erosion: ridges may overlie synclinal axes if capped by resistant beds

**FIGURE 16.9** Two stages in the development of ridges and valleys in folded mountains. (a) In early stages, ridges are formed by anticlines. (b) In later stages, the anticlines may be breached and ridges may be held up by caps of resistant rocks as erosion forms valleys in less resistant rocks.

**FIGURE 16.8** Valley and ridge topography formed on a folded terrain of sedimentary rock. The deformation is so recent (Pliocene) that erosion has not yet significantly modified the original structural forms of anticlines (ridges) and synclines (valleys). Zagros Mountains, Iran. *NASA*.

## River Valleys

Observations of river valleys in various regions led to one of the important early theories of geology: the idea that river valleys were created through erosion by the rivers that flowed in them. Geologists could see that sedimentary rock formations on one side of a valley matched the same kind of formations on the opposite side and that the formations were once deposited as a continuous sheet of sediment. The river had removed enormous quantities of the original formation by breaking up the rock and carrying it away as river sediment.

Many names have been given to river valleys—canyons, gulches, arroyos, and gullies—but all have the same general geometry. A vertical cut through a young mountain river valley with little or no flood-

**FIGURE 16.10** The Appalachian Valley and Ridge province shows the tectonically controlled topography of linear anticlines and synclines produced by millions of years of erosion. *Earth Satellite Corp.*

plain shows a simple **V**-shaped profile (Figure 16.11). A broad, low river valley with a wide floodplain shows a cross section that is more open but still distinct from the **U** shape of a glacial valley. The widths of river valleys vary in regions of different general topography and type of bedrock, ranging from the narrow gorges of mountainous belts and erosion-resistant rock types to the wide, shallow valleys of plains and easily eroded rock types. Between these extremes the width of a valley generally corresponds to the erosion state of the region, being somewhat broader in mountains that have begun to be lowered and rounded from erosion and much broader in low-lying hilly topography.

A **badland** is a deeply gullied topography resulting from the fast erosion of easily erodible shales and clays, such as those of the South Dakota badlands (Figure 16.12). Practically the entire area is a proliferation of gullies and valleys with little flat land between them.

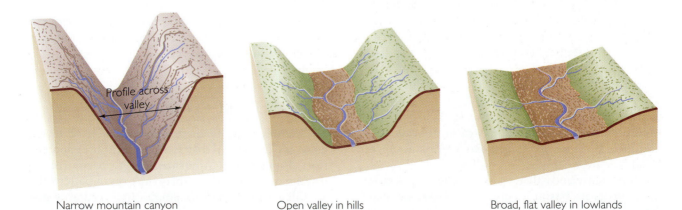

Narrow mountain canyon            Open valley in hills            Broad, flat valley in lowlands

**FIGURE 16.11** River valleys vary from narrow, **V**-shaped profiles in mountains to broader profiles in lowlands.

**FIGURE 16.12** The badlands of South Dakota, produced by gully erosion of easily eroded sedimentary rocks. *William Garnett.*

## Tectonic Valleys

Many valleys formed primarily by tectonic processes are long, narrow, relatively flat-floored, and bounded on one or both sides by faults. Downward movement of the valley caused by crustal subsidence along these faults is responsible for the general valley shape. These tectonic valleys are occupied by rivers and often by lakes. A good example is the Great Valley of California, the route of the San Joaquin and Sacramento rivers. The rivers of a tectonic valley may deposit a great deal of sediment eroded from nearby mountains, filling in the bedrock valley with a broad alluvial plain. This flat floor merges with higher coalesced alluvial fans sloping along the faulted boundaries of the valley. Alternatively, the rivers that run in these valleys may modify the structurally controlled valley form by eroding the valley walls to gentler angles or by eroding the valley deeper than would happen by tectonics alone.

A special type of tectonic valley is the rift valley, formed by the incipient or active spreading apart of lithospheric plates. The great African rift valleys are occupied by large lakes. The River Jordan and the Dead Sea are in another rift valley.

Some lowlands that have subsided by tectonic movements are **basins,** depressions of various shapes from circular to elongate. Basins are often found between mountain ranges. One of the large basins of the United States is the Great Basin, occupying much of western Nevada just east of the Sierra Nevada.

## The Origins of Landforms

As we have seen in previous chapters, there are many other kinds of landforms, both erosional and depositional, that result from the activity of groundwater, winds, and ice. Landforms such as karst topography, dunes, and moraines are discussed in Chapters 12, 14, and 15.

Although landforms are primarily the products of erosion, transportation, or sedimentation, alone or in some combination, they are strongly affected by tectonics and bedrock lithology. Thus river valleys in tectonically elevated high mountains have profiles different from those in tectonically stable low plains. River valleys in easily erodible sediments and rocks are broader and have gentler slopes than those in erosion-resistant rocks.

Climate also has a strong influence on landscape. The landscape of a hot, dry desert, the product of eolian and fluvial processes, is very different from the landscape of a polar region, the product of glacial ice and a frigid climate. The topography of a region, itself the result of tectonics and erosion, strongly influences the rate of weathering and erosion. How all these factors interact tells us how major geologic forces sculpt the surface of the Earth.

# FACTORS THAT CONTROL LANDSCAPE

Broadly speaking, landscape is controlled by the interaction of the internal and external machines of Earth. The internal machine, tectonics, elevates mountains and volcanoes and lowers tectonic valleys and basins. The external machine, powered by the Sun, wears away the mountains and fills in the basins with sediment. Sunlight causes the motions of the atmosphere that make climate, the different temperature regimes of the globe, and the rainwater that runs off the continents as rivers.

## Tectonics and Erosion

In describing the formation and erosion of mountains, we might say that tectonics proposes and erosion disposes. The mechanisms that control the heights of mountains display an overall pattern that we call a *negative-feedback process*. In this kind of process, one action produces an effect (the feedback) that tends to slow the original action and stabilize the process at a lower rate. For example, if you are thirsty you will drink a glass of water quickly at first. But as

**FIGURE 16.13** The negative-feedback loop that relates uplift and erosion to surface elevation. Tectonic uplift causes an increase in erosion rate, which in turn lowers the surface elevation. The elevation is thus a balance between tectonic uplift and erosion rate.

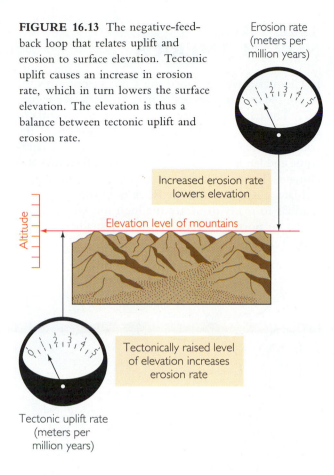

Erosion rate (meters per million years)

Increased erosion rate lowers elevation

Elevation level of mountains

Altitude

Tectonically raised level of elevation increases erosion rate

Tectonic uplift rate (meters per million years)

the drinking reduces your thirst (the feedback), you will drink more and more slowly, until finally your thirst is completely satisfied and you stop drinking. The process of drinking has stabilized at a rate of zero.

In the geologic case, strong tectonic action elevates mountains, provoking intense erosion (Figure 16.13). The higher the mountains grow, the faster erosion wears them down. As long as mountain building continues, elevations stay high or increase. As mountain building slows, perhaps because of a change in the rate of plate-tectonic movements, the mountains rise more slowly or stop rising entirely. As this happens, erosion starts to dominate and elevations begin to decrease. As the lowering of mountains proceeds, the erosion slows too, the whole process eventually tapering off. This explains why old mountains, like the Appalachians, are relatively low compared to the much younger Rockies.

## Topography and Erosion

Topography strongly controls weathering and erosion. High elevation and relief enhance the fragmentation and mechanical breakup of rocks, partly by promoting freezing and thawing. Also, fragmented debris on mountains moves quickly downhill in slides and other mass movements, exposing fresh rock to attack by the weather. Rivers run faster in mountains than in lower topography, and where the climate is cool, mountain glaciers scour bedrock and erode deep valleys. Chemical weathering plays an important part in the erosion of high mountains, but the mechanical breakup of rocks is so rapid that most of the debris appears to be almost unweathered. The products of chemical decay, dissolved material and clay minerals, are carried down from the steep slopes of mountains as soon as they are formed. The intense erosion produces a topography of steep slopes; narrow, deep river valleys; and narrow floodplains and drainage divides.

In lowlands, by contrast, weathering and erosion are slow and the clay mineral products of chemical weathering accumulate as thick soils. Mechanical breakup occurs in lowlands too, but its effects are small compared to those of chemical weathering. Most rivers run on broad floodplains in lowlands and do little mechanical cutting into bedrock. Glaciers, which grind and erode bedrock, are absent except in polar regions. Even on lowland deserts, strong winds merely facet and round rock fragments and outcrops rather than breaking them up. A lowland thus tends to have a gentle topography with rounded slopes, rolling hills, and flat plains.

## Climate, Topography, and Latitude

The effects of climate on weathering (see Chapter 6) include freezing and thawing, expansion and contraction due to heating and cooling, and the chemical dissolving action of water. Rainfall and temperature, the components of climate, affect weathering and erosion through the rain that falls on bedrock and soil, the infiltration of water into the soil, mass wasting, and rivers, all of which help to break up the rock and mineral particles eroded from slopes and carry them downhill. The relationship of climate to topography is well known to mountain climbers. As one climbs higher it gets colder, even in the tropics, and vegetation becomes less abundant. First one passes the timberline, above which no trees grow, and at higher elevations there is hardly any vegetation other than lichens, leafless primitive plants that can survive under these harsh conditions. Lichens aid weathering somewhat, but the general effects of vegetation on weathering are minimal at high elevations. High mountain landscapes are steep and jagged, consisting of bare rock cliffs and talus slopes. Erosion by glaciers contributes to the rugged landscape.

Topography also has other effects on climate. For example, mountains cause rain shadows, which are dry areas on the leeward slopes of mountain ranges (see Chapter 12). Also, the interiors of large continents such as Asia, far from any oceans, tend to have relatively low rainfall.

The climate becomes cooler with increasing latitude, with effects on weathering and erosion much the same as the effects of increasing elevation. Thus an increase in latitude is equivalent to an increase in altitude. At low latitudes near the equator, where tropical rain forests are found, the abundant water, lush vegetation, and warm temperatures promote rapid weathering and soil development. Since these forests are mostly in relatively low-lying areas, mechanical erosion is minimal. The result is a topography of low, rounded hills or, where river sedimentation is rapid, extensive river-bottom plains such as the bottomlands of the Amazon River in the Brazilian jungle.

Where water is scarce at low latitudes, as in deserts such as the Sahara, chemical weathering is slow, while thin soils and the lack of vegetation allow rapid mechanical erosion. Desert landscapes tend to have rough topography, with steep slopes at higher elevations and gentler slopes at lower elevations where the intermittent rivers deposit layers of gravel and sand.

In temperate climates, chemical and mechanical erosion are more balanced. Moderate rainfall and temperatures promote abundant vegetation, particularly forests, at lower elevations, and slopes are moderate. Rugged topography is the rule only at the higher elevations, where cold and dryness prevail.

At high latitudes near the poles, where temperatures are low and moisture may be present mostly as snow, chemical weathering of rocks is restricted while fragmentation and mechanical erosion are extensive. Bare rock and talus slopes with rugged topography are found at high and moderate elevations. Valley bottoms and lower slopes are covered with permanently frozen soil, and vegetation is less abundant than in temperate regions. Large glaciers may erode deep valleys and deposit extensive sheets of erosional debris at their melting edges.

In summary, climate, topography, and latitude—and their interactions—have strong effects on landscape through their enhancement or suppression of chemical and physical weathering and erosion. If one factor can be singled out as dominant it is topography, and that leads us back to tectonics. In all latitudes and climates, the slopes of high mountains are steep and rugged and lowlands tend to have gentler topographies. The effects of folding and faulting on the structure of the terrain are recognizable in any climate or latitude. All these effects on landscape can be seen in North America.

# THE FACE OF NORTH AMERICA

In Chapter 3 we described the kinds of outcrops we might see on a trip across North America (see Figure 3.7). The nature of those outcrops, from sea cliffs to creek bottoms, and the major relief patterns of the continent (Figure 16.14) illustrate clearly that landscape and landforms are the products of tectonics and climate. The major mountain chains are the Appalachians in the east and the Rockies in the west. Less familiar to many of us are the high mountain chains of the far north, in Alaska and the Canadian Northwest Territories.

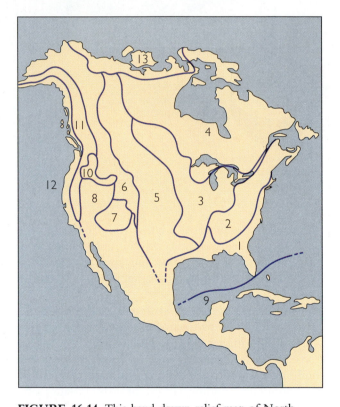

**FIGURE 16.14** This hand-drawn relief map of North America (*right*) clearly shows the regional terrain textures on which physiographic divisions of the continent are based. The map above shows a generalized division of North America into landform provinces: 1. Atlantic Coastal Plain; 2. Applachian Mountains and Plateaus; 3. Central Lowland; 4. Canadian Shield; 5. Great Plains; 6. Rocky Mountains; 7. Colorado Plateau; 8. Basin and Range; 9. Central America; 10. Columbia Plateau; 11. Interior Mountains and Plateaus; 12. Pacific Border; 13. Arctic Lowlands. (Landform relief map-diagram by Erwin Raisz; copyright © by Erwin Raisz, reprinted with permission by Raisz Landform Maps, Melrose, Mass. Landform province map after W. L. Graf, "Geomorphic Systems of North America," Geological Society of America Centennial Special Volume 2, 1987, p. 3.)

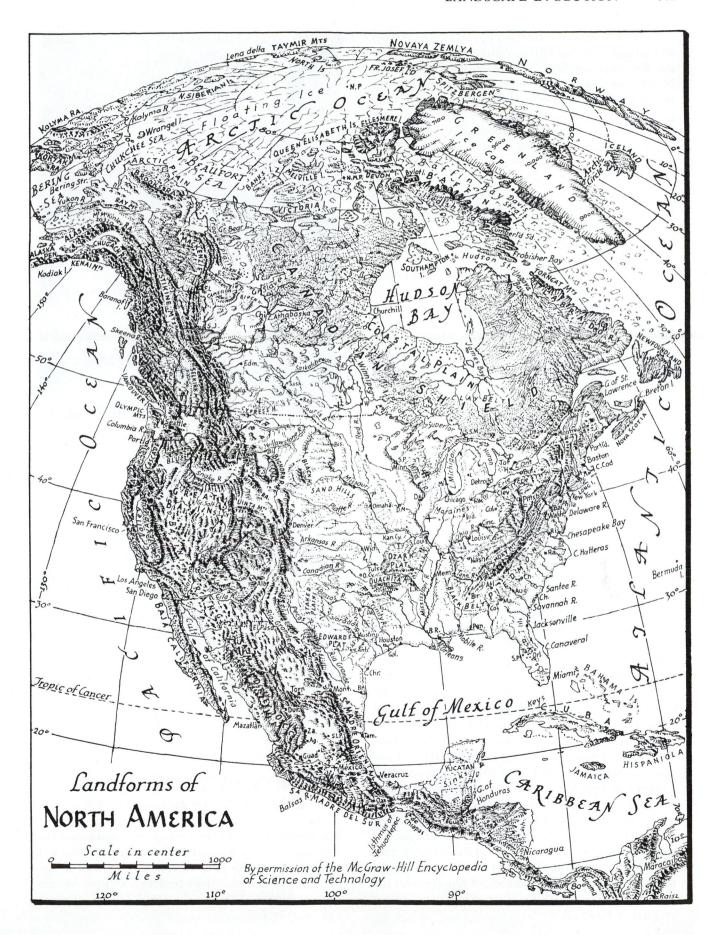

Landforms of
**NORTH AMERICA**

Scale in center

0          1000
Miles

By permission of the McGraw-Hill Encyclopedia
of Science and Technology

These mountain chains are the records of present and former plate boundaries. Along the Pacific coast, the San Andreas fault, the boundary between the North American and Pacific plates, is active today, as is the subduction of the Gorda and Juan de Fuca plates under North America. The other mountain chains offer evidence of movements along plate boundaries that have long since been sutured and immobilized. The high elevations of the western and northern mountains make them steep and rugged even in temperate climates. The Appalachians are much lower, with a gentler topography.

Between the mountain belts are the prairies and plains. In the United States the western plains slope gently eastward from the foothills of the Rockies; the midwestern and southern lowlands slope southward. These slopes are clearly indicated by the paths of the major rivers, flowing eastward from the Rockies like the Missouri and westward from the Appalachians like the Ohio. Both join the Mississippi River, which flows south to the Gulf of Mexico. A low coastal plain lies along much of the central and southern Atlantic coast and the Gulf coast, where crustal subsidence allows continental shelf sedimentation and sea level intermittently rises and falls.

North of the Great Lakes are the low-relief plains and lakes of Ontario and Quebec. To the west are the prairies of Manitoba, Saskatchewan, and Alberta. Most of the plains of the United States and southern Canada are floored by horizontal sedimentary rocks of Paleozoic age. To the north the lowlands of central and northern Canada are floored mostly by deeply eroded and weathered Precambrian metamorphic and igneous rocks (Figure 16.15). In Canada, the rivers of the interior lowlands drain eastward to the Atlantic or northward to the Arctic. All of these lowlands are provinces of the interior of the North American Plate, now far from any plate boundary. But in the Precambrian, 600 million or more years ago, this region was the site of active plate tectonics.

The Basin and Range province, a region of many smaller chains of mountains alternating with elongate basins, lies between the Rocky Mountains and the Sierra Nevada. The topography and landscape of the Basin and Range are the products of recent tectonic activity interacting with a semiarid to arid climate. West of the Basin and Range are the igneous rock terrains of the Sierra Nevada and the Cascades, which are young, high, rugged ranges elevated in the Mesozoic and Cenozoic eras by the convergence of the North American Plate and present and former plates to the west. The western edge of the continent is marked to the south by the Coast Ranges east of the Great Valley of California and to the north by the

**FIGURE 16.15** A low-lying erosional plain developed on a structurally heterogeneous terrain of the Canadian Shield after hundreds of millions of years of tectonic stability. *R. S. Hildebrand, Geological Survey of Canada.*

Coast Ranges of Oregon, Washington, and British Columbia.

These maps of North America emphasize the dominance of tectonics but also suggest the importance of age. The highest and ruggedest mountains are the young western and northern ranges. The lower, older eastern mountains are rounded and gentle. The oldest of all, the expanses of the Precambrian terrain of northern and central Canada, are worn down to a low-lying plain, even though the rocks are complexly deformed. To take age into account in the formation of landscape, we need to know how landscapes evolve over geologic time.

# THE EVOLUTION OF LANDSCAPE

The relationship between geologic age and mountain landscape was emphasized in one of the early theories of landscape development, which held sway in the first part of the twentieth century. William Morris

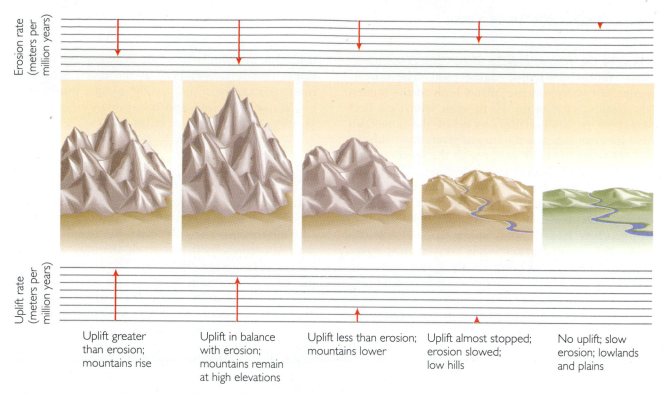

Erosion rate (meters per million years)

Uplift rate (meters per million years)

Uplift greater than erosion; mountains rise

Uplift in balance with erosion; mountains remain at high elevations

Uplift less than erosion; mountains lower

Uplift almost stopped; erosion slowed; low hills

No uplift; slow erosion; lowlands and plains

**FIGURE 16.16** The stages of landscape evolution depend on the balance between uplift and erosion.

Davis, a Harvard geologist of the time, studied mountains and plains all over the world. He proposed a **cycle of erosion** that progresses from the tectonically formed high, rugged mountains of youth to the rounded forms of maturity and the worn-down plains of old age and tectonic stability. Erosion eventually wears down the landscape to a relatively flat surface, leveling all structures and differences in bedrock. Davis saw the flat surfaces of extensive unconformities as evidence of such plains in past geologic times. Here and there an isolated hill might stand as an uneroded remnant of former heights.

Davis's cycle was widely accepted by many geologists, partly because they could find many examples of what seemed to be the different stages of youth, maturity, and old age. There were competing proposals, some of which emphasized somewhat different courses of evolution of slopes and landscape in various climates, but orderly evolution was at the heart of these models, too. Most geologists at that time accepted Davis's assumption that mountains were elevated suddenly over short geologic times and then stayed tectonically fixed as erosion slowly wore them down. We now envision mountain building as an uneven, long-drawn-out, intermittent process driven by continuing plate motions.

If tectonics continues over tens or even hundreds of millions of years, how can we decipher the ways in which landscape evolves? Current views of landscape evolution emphasize the balance between erosion and tectonic uplift (Figure 16.16). If uplift is faster, the mountains will rise; if erosion is faster, the mountains will be lowered. When tectonics dominates, mountains are high and steep, and they remain so as long as the balance is in favor of tectonics. When erosion exceeds uplift, slopes become lower and more rounded. Because few areas of the world remain tectonically quiescent as long as 100 million years, the perfectly flat erosion plain Davis proposed could form only rarely in Earth's history.

## Plate Tectonics

What controls the rate and duration of tectonic activity? Uplift begins when two plates converge. The convergence of two oceanic plates leads to arcs of volcanic islands, whose topography is dominated by active volcanism, with erosion lagging. The convergence of an oceanic and a continental plate leads to a topography like that of the Pacific coast of South America, where the volcanic arc is on land. Extensive volcanism and lateral compression led to the construction of a high mountain chain, the Andes, while sedimentation and subsidence in the forearc region between the Andes and the Peru-Chile Trench offshore formed a narrow coastal plain. The convergence of two continental plates leads to the highest mountains, such as the Himalayas, which are bordered by thick alluvial plains below the mountains and by great deltas where major rivers enter the sea.

The rate of uplift varies with the speed of plate convergence. For reasons geologists do not fully understand, convergence may slow to a halt for periods of time and then resume. At other times plate movements accelerate and then slow. High rates of convergence lead to rapid uplifts that outstrip erosion. Erosion dominates when convergence slows. Erosion may win out permanently if, as happens, the geometry of plate motions changes and new plate boundaries are created far from the mountains. When that happens, the mountains may go through the stages that Davis proposed.

## Climate

Geologists now recognize that climate plays an important modifying role in landscape evolution. A desert landscape evolves differently from a temperate landscape in a similar tectonic situation: the desert's slopes remain steeper even as the topography matures. Glacial landscapes are the unique products of ice, a special erosion and transportation agent that operates only in frigid climates.

## Geologic History

Although there are broad categories of landscape, each region has its own geologic history. That history determined the rock types at the surface and their structures, both of which strongly affect the course of landscape evolution. The latitude of a region is also important. A region may move from low to higher latitudes by continental drift; the different climate in the new location may modify or transform the landscape. There are other, poorly understood reasons for changes in climate. We have recently discovered, for example, that in the relatively recent geologic past some areas of the Sahara Desert were much more humid than they are now. Thus landscape is the mirror of geologic history. To see clearly in that mirror takes a great deal of geological work and understanding.

# SUMMARY

**What are the principal components of landscape?** Landscape is described in terms of topography, the elevation of the surface of the Earth above sea level; relief, the difference between the highest and lowest spots in a region; and the varied landforms produced by rivers, glaciers, mass wasting, and the wind. The most common landforms are mountains and hills, plateaus, and structurally controlled cliffs and ridges, all produced by tectonic activity modified by erosion. Landforms may be erosional or sedimentational: river and glacier valleys and badlands are primarily erosional; river floodplains, dune belts, and glacial moraines are primarily sedimentational.

**What are the major factors that control landscape?** Landscape is determined by tectonics, erosion, climate, and the type of bedrock. Tectonics, driven by plate motions, elevates mountains and lowers tectonic valleys and basins. Erosion carves bedrock into valleys and slopes. Climate affects weathering and erosion and makes possible glacial and desert landscapes. The varying resistance of bedrock types to erosion accounts for differences in slope and valley profiles, steep slopes being found in rocks with greater resistance to erosion.

**How do landscapes evolve?** Landscapes begin their evolution with tectonic uplift, which in turn stimulates erosion. While tectonic activity remains dominant, mountains are high and steep. As tectonic activity slows, erosion becomes more important and the land surface is lowered and slopes are rounded. As erosion becomes dominant, the former mountains are worn down to gentle hills and broad plains. In the unusual event of long-continued tectonic quiescence, the land surface may end as a level plain. Climate and bedrock type strongly modify the evolutionary path in various surface environments, making desert, glacial, and karst landscapes very different.

# KEY TERMS AND CONCEPTS

topography (p. 354)

elevation (p. 354)

contour (p. 354)

relief (p. 355)

landform (p. 356)

plateau (p. 356)

mesa (p. 356)

cuesta (p. 357)

hogback (p. 357)

badland (p. 359)

basin (p. 360)

cycle of erosion (p. 365)

# EXERCISES

1. What is topographic relief and how is it related to altitude?

2. Give three examples of landforms.

3. How does geologic structure control topography?

4. Compare weathering and erosion in topographically high and low areas.

5. How does climate affect topography?

6. In what regions of North America do active plate-tectonic movements currently affect landscape?

7. What kind of landscape evolves from the convergence of a continental plate and an oceanic plate?

# THOUGHT QUESTIONS

1. The heights of two mountain ranges lie at different elevations, A at about 8 km and B at about 2 km. Without knowing anything else about them, could you make an intelligent guess about the relative ages of the mountain-building processes that formed them?

2. If you were to climb 1 km, from a river valley to a mountaintop, in two different regions, one where the mountaintops were about 2 km high and the other where they were about 6 km high, which would probably be the ruggedest climb?

3. A young mountain range of uniform age, rock type, and structure extends from a far northern frigid climate through a temperate zone to a southern tropical rainy climate. How would the landscape of the mountain range differ in each of the three climates?

4. What differences in landscape would you expect between a coastal plain made of young, unlithified sediments and slightly lithified sedimentary rocks and an interior plain whose bedrock is old, very lithified sedimentary rocks?

5. Describe the main landforms in a low-lying humid region where the bedrock is limestone.

6. In what landscapes would you expect to find lakes?

7. What landforms are characteristic of glaciated lowlands?

# SUGGESTED READINGS

Bloom, Arthur L. 1991. *Geomorphology*, 2d ed. Englewood Cliffs, N.J.: Prentice-Hall.

Chorley, Richard J., Stanley A. Schumm, and David E. Sugden. 1984 *Geomorphology*. London: Methuen.

Davis, William M. 1909. *Geographical Essays*. Boston: Ginn.

Hunt, Charles B. 1974. *Natural Regions of the United States and Canada*. San Francisco: W. H. Freeman.

Tuttle, Sherwood D. 1970. *Landforms and Landscapes*. Dubuque, Iowa: W. C. Brown.

# THE OCEANS

The oceans, with their submerged mountains and valleys, underwater volcanoes, and many kinds of rocks and sediments, are central actors in Earth's geology. Waves and tides drive strong currents and shape the edge of the sea into rocky shores and sandy beaches. At continental margins, submarine slumps create currents that carry sediment to the deep seafloor and construct vast plains of muddy deposits. The oceans are a reservoir for the chemical components weathered from continental rocks and for enormous quantities of chemical and biochemical deposits. At mid-ocean ridges, new oceanic lithosphere forms from upwelling basalt, and turbulent springs spew hot chemicals in the ocean's rift valleys. The deepest parts of the oceans are the spectacular trenches of subduction zones, bordered by arcs of volcanic islands.

## 17

Waves dash against an irregular rocky shoreline on the Oregon coast. *Richard Johnston/Tony Stone Worldwide.*

For most of human history, the 71 percent of Earth's surface covered by the oceans was a mystery. Large populations lived at the edge of the sea and knew well the force of waves and the rise and fall of tides. But they could only guess at the nature of the seafloor deeper than the shallowest coastal waters. Then, in 1872, H.M.S. *Challenger,* a small wooden British warship converted and fitted out for the first scientific study of the seas, left England for a four-year voyage over the world's oceans. The 50 thick volumes of reports from that expedition gave the public its first knowledge about the Mid-Atlantic Ridge, one of the longest, highest mountain ranges in the world—all of it underwater. The *Challenger* expedition discovered great areas of submerged hills and flat plains, extraordinarily deep trenches, and submarine volcanoes.

Today, more than a century after that pioneering voyage, hundreds of oceanographic research vessels from many countries ply the seas in search of answers to the questions first raised by the early discoveries. What tectonic forces raised the submarine mountain ranges and depressed the trenches? Why are some areas flat plains and others hilly? Although oceanographers made many important discoveries in the first half of this century, the answers to most of these questions had to await the plate-tectonics revolution of the late 1960s. In fact, it was geological and geophysical observations of the ocean floors, not the continents, that led to the theory of plate tectonics.

We refer to the oceans both as the five major oceans (Atlantic, Pacific, Indian, Arctic, and Antarctic) and as the single connected body of water called the **world ocean.** The term *sea* includes both the oceans and smaller bodies of water set off somewhat from the world ocean. Thus the Mediterranean Sea is narrowly connected with the Atlantic Ocean by the Straits of Gibraltar and with the Indian Ocean by the Suez Canal. Other seas, such as the North Sea and the Atlantic Ocean, are broadly connected. Seawater, the salty water of the oceans and seas, is remarkably constant in its general chemical composition.

We begin our exploration of the oceans with shorelines, where we can observe the constant motion of ocean waters and their effects on the shore.

# THE EDGE OF THE SEA: WAVES AND TIDES

**Coasts,** the broad regions where land meets sea, present striking contrasts of landscape. On the coast of North Carolina, for example, long, straight, sandy beaches stretch for miles along low coastal plains (Figure 17.1). In New England, by contrast, rocky cliffs bound elevated shores, and the few beaches that occur are made of gravel (Figure 17.2). Many of the seaward edges of islands in the tropics, such as those in the Caribbean Sea, are coral reefs, the delight of divers. As we shall see, tectonics, erosion, and sedimentation work together to create this great variety of shapes and materials.

The major geological forces operating at the **shoreline,** the line where the water surface intersects the shore, are waves and tides. Together they erode even the most resistant rocky shores. Waves and tides create currents, which transport sediment produced by erosion of the land and deposit it on beaches and in shallow waters along the shore.

## Wave Motion: The Key to Shoreline Dynamics

Centuries of observation have taught us that waves are changeable. During quiet weather, waves roll regularly into shore with calm troughs between them. In the high winds of a storm, on the other hand, waves are everywhere, moving in a confusion of shapes and sizes. Waves may be low and gentle far from the shore, yet become high and steep as they approach land. High waves can break on the shore

**FIGURE 17.1** A long, straight, sandy beach on South Pea Island, North Carolina. *Peter Kresan*.

with fearful violence, shattering concrete seawalls and tearing apart houses built along the beach. To understand the dynamics of shorelines and to make sensible decisions about shore development, we need to understand how waves work.

Waves are created by the wind blowing over the surface of the water, transferring the energy of motion from air to water. As a gentle breeze of 5 to 20 km per hour starts to blow over a calm sea surface, ripples—little waves less than a centimeter high—take shape (see Table 14.1). As the speed of the wind increases to about 30 km per hour, the ripples grow to full-sized waves. Stronger winds create larger waves and blow off their tops to make white-caps. The height of the waves increases as (1) the wind speed increases, (2) the wind blows for longer times, and (3) the distance over which the wind blows the water increases.

Storms blow up large, irregular waves that radiate outward from the storm area, like the ripples moving outward from a pebble dropped into a still pond. As the waves travel out from the storm center in ever-widening circles, they become more regular, changing to low, broad, rounded waves called **swell,** which can travel hundreds of kilometers. Several storms at different distances from a shoreline, each

**FIGURE 17.2** Small pocket beach (*right foreground*) of pebbles and cobbles; Acadia National Park, Maine. *Ric Ergenbright Photography*.

producing its own pattern of swell, account for the often irregular intervals between waves approaching the shore.

Waves travel as a form while the water stays in the same place. You have probably noticed how a piece of wood or other light material floating on the water moves a little forward as the top of a wave passes and then a little backward as the trough between waves passes. While moving back and forth the wood stays in roughly the same place, and so does the water around it.

Small particles floating on the surface or beneath the waves move in circular vertical orbits, as shown in Figure 17.3. At any given point along the path of a wave, all the water particles are at the same relative positions in their orbits regardless of their depth. The radii of the orbits are large near the water surface but gradually decrease to zero at some depth below. The wave form is made as many water particles move to the top of the orbit; the wave advances as the particles continue around the orbit. The trough is created as the particles reach the bottom of the orbit.

The wave form is defined by its **wavelength,** the distance between crests, and its **wave height,** the vertical distance between the crest and the trough. The wave moves forward with a velocity that is measured by the wavelength divided by the time it takes successive crests to pass, called the **period.** Thus the basic equation of a wave is $V = L/T$, where $V$ is the velocity, $L$ is the wavelength, and $T$ is the period. The periods of most waves range from a few seconds to as long as 15 or 20 seconds, with wavelengths varying from about 6 m to as much as 600 m. Consequently, wave velocities may vary from 3 to 30 m per second. At a depth of about one-half the wave-length, orbital motion stops and wave motion ceases. That is why deep divers and submarines are unaffected by the waves at the surface.

## The Surf Zone

Swell becomes higher as it approaches the shoreline and assumes the familiar sharp-crested wave shape. As the waves come closer to shore they break, forming the foamy, bubbly surface we call **surf;** the region of breaking waves defines the **surf zone.** Breaking waves pound the shore, eroding and carrying away sand, weathering and breaking up solid rock at the shore, and destroying structures built close to the shoreline.

We can explain how waves break by observing how their orbits change where the bottom becomes shallow. The transformation from swell to breakers starts where the bottom shallows to less than one-half the wavelength of the swell. At that point the small orbital motions of the water just above the bottom become restricted because the water can no longer move vertically (Figure 17.4). Right next to the bottom, the water can only move back and forth horizontally. Above that, the water can move vertically just a little, combining with the horizontal motion to give a flat elliptical orbit rather than a circular one. The orbits become more circular the farther they are from the bottom.

The change from circular to elliptical orbits slows the whole wave, because the water particles take longer to travel around ellipses than circles. While the wave slows, its period remains the same because the swell keeps coming in from deeper water

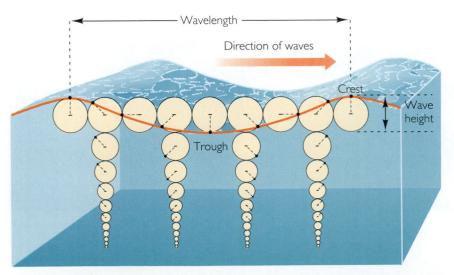

**FIGURE 17.3** The basic parts of a wave and the orbital movement of water particles as a wave advances. Note that orbits decrease in radius with depth.

**FIGURE 17.4** Formation of a breaking wave as swell meets a shallowing bottom. Note that the orbits of water particles become more elliptical as they approach a shallow bottom.

at the same rate. From the wave equation, we know that if the velocity decreases and the period remains constant, the wavelength must also decrease. Thus the waves become more closely spaced, higher, and steeper, with sharper wave crests.

As a wave rolls toward the shore, it becomes steeper until the water can no longer support itself, and the wave breaks with a crash in the surf zone (see Figure 17.4). Gently sloping bottoms cause the waves to break farther out, and steeply sloping bottoms make waves break closer to shore. Where rocky shores are bordered by deep water, the waves break directly on the rocks with a force equivalent to hundreds of tons per square meter, throwing water high into the air. It is not surprising that concrete seawalls built to protect buildings along the shore quickly start to crack and must be repaired constantly.

After breaking at the surf zone, the waves, now reduced in height, continue to move in, breaking again right at the shoreline. They run up onto the sloping front of the beach, forming an uprush of water called **swash.** The water then runs back down again as **backwash.** Swash can carry sand and, if the waves are high enough, large pebbles and cobbles. The backwash carries the particles back down again.

The motion of the water back and forth near the shore is strong enough to carry sand grains and even gravel. Fine sand can be moved by wave action in water up to about 20 m deep. Large waves caused by intense storms can scour the bottom at much greater depths, down to 50 m or more.

## Wave Refraction

Far from shore, the lines of swell are parallel to one another but are usually at some angle to the shoreline. As the waves approach the beach over a shallowing bottom, the rows of waves gradually bend to a direction more parallel to the shore (Figure 17.5). This change in direction is called **wave refraction;** it

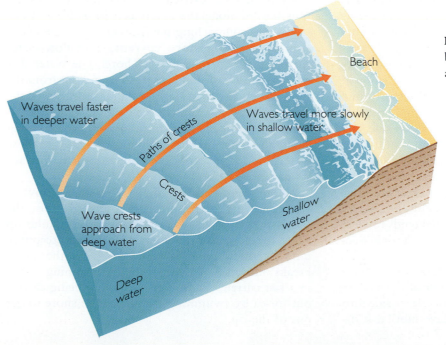

**FIGURE 17.5** Wave refraction is the bending of lines of wave crests as they approach the shore from an angle.

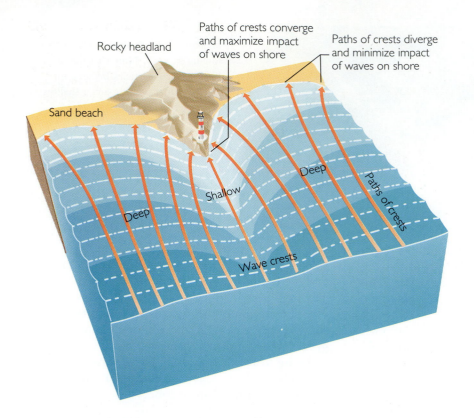

Rocky headland

Paths of crests converge and maximize impact of waves on shore

Paths of crests diverge and minimize impact of waves on shore

Sand beach

Deep

Shallow

Deep

Paths of crests

Wave crests

**FIGURE 17.6** Wave refraction around a headland and bay. Wave energies are concentrated at headlands and dispersed at bays.

is a bending of water waves similar to the bending of light rays in optical refraction that makes a pencil half in and half out of water appear to be bent at the water surface. As a wave approaches the shore at an angle, the part closest to the shore encounters the shallowing bottom first. The orbits of the water particles in that part of the wave become more elliptical, and the front of the wave slows. Then the next part of the wave meets the bottom and also slows. Meanwhile, the parts closest to shore have moved into even shallower water and slowed even more. Thus, in a continuous transition along the wave crest, the line of the wave bends toward the shore as it slows.

Wave refraction results in more intense wave action on projecting headlands and less intense action in indented bays, as Figure 17.6 illustrates. Around headlands, the water becomes shallow more quickly than the surrounding deeper water on either side, and the waves are refracted; that is, they are bent toward the projecting part of the shore from both sides. The waves converge around the point of land and expend proportionately more of their energy breaking there than at other places along the shore. Thus erosion by waves is concentrated at headlands and tends to wear them away more quickly than it does straight sections of shoreline.

The opposite happens as a result of wave refraction in a bay. The waters in the center of the bay are deeper, so the waves are refracted on either side into shallower water. The energy of wave motion is di-

minished at the center of the bay, making bays good harbors for ships.

Although refraction makes waves more parallel to the shore, many waves still approach at some small angle. As the waves break on the shore, the swash moves up the beach slope at this angle. The backwash runs down at a similar angle. The combination of the two motions results in a trajectory that moves the water a short way down the beach (Figure 17.7). Sand grains carried by swash and backwash are thus moved along the beach as a **longshore drift.**

Waves approaching the shoreline at an angle can also cause a **longshore current,** a shallow-water current that is parallel to the shore. The water that moves with swash and backwash in and out from the shore at an angle creates a net transport along the shore in the same direction as the longshore drift. Longshore drift and longshore currents can be strong enough to transport large amounts of sand in shallow waters, and some types can pose a threat to unwary swimmers.

A *rip current,* for example, is a strong flow of water moving perpendicularly outward from the shore. It occurs when a longshore current builds up along the shore and the water piles up imperceptibly until a critical point is reached. There the water breaks out to sea, flowing through oncoming waves in a fast current. Swimmers can avoid being carried out to sea by swimming parallel to the shore to get out of the rip.

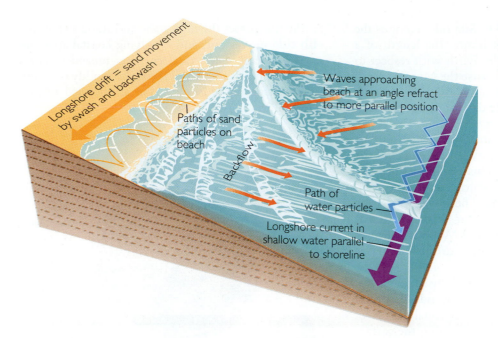

Longshore drift = sand movement by swash and backwash

Waves approaching beach at an angle refract to more parallel position

Paths of sand particles on beach

Backflow

Path of water particles

Longshore current in shallow water parallel to shoreline

# The Tides

The twice daily rise and fall of the sea that we call **tides** have been known to mariners and shoreline dwellers for thousands of years. For much of that time it was also known that there is a relationship between the position and phases of the Moon, the heights of the tides, and the time of day at which the water reaches high tide. It was not until the seventeenth century, however, when Isaac Newton formulated the laws of gravitation, that the tides were understood to result from the gravitational pull of the Moon and the Sun on the water of the oceans.

**THE EFFECT OF THE MOON AND THE SUN ON THE OCEANS**  The Earth and the Moon attract each other strongly with a gravitational force that is slightly greater on the sides of the bodies that face each other. The gravitational attraction between any two bodies decreases as they get farther apart. Thus, the tide-producing force varies on different parts of the Earth, depending on whether they are closer to or farther from the Moon.

The net gravitational attraction between the oceans and the Moon is at a maximum on the side of Earth facing the Moon and at a mimimum on the side facing away from the Moon. As Earth rotates, the high tides pass over it as bulges, one always facing the Moon, the other always directly opposite (Figure 17.8).

The Sun, although much farther away, has so much mass (and thus so much gravity) that it too causes tides. Sun tides are a little less than half the height of Moon tides. Sun tides are not synchronous

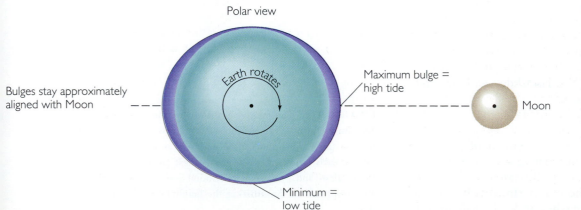

Polar view

Earth rotates

Bulges stay approximately aligned with Moon

Maximum bulge = high tide

Moon

Minimum = low tide

**FIGURE 17.8** The Moon's gravitational attraction causes two bulges of water on the Earth's oceans, one on the side nearest the Moon and the other on the side farthest from the Moon.

with Moon tides (Figure 17.9). Sun tides come as the Earth rotates once every 24 hours, the length of a solar day. The rotation of the Earth with respect to the Moon is a little longer because the Moon is moving around the Earth, giving a lunar day of 24 hours and 50 minutes. In that lunar day there are two high tides, with two low tides between them.

When the Moon, Earth, and Sun line up (see Figure 17.9a), the gravitational pulls of the Sun and the Moon reinforce each other. This produces the highest tides, the **spring tides,** which come every two weeks at full and new Moon. The lowest tides, the **neap tides,** come in between, at first- and third-quarter Moon, when the Sun and Moon are at right angles to each other with respect to the Earth (see Figure 17.9b).

Although the tides occur regularly everywhere, the difference between high and low tides varies in different parts of the ocean. As the tidal bulges of water rise and fall, they also move along the surface of the ocean, encountering obstacles, such as continents and islands, that hinder the flow of water. In the middle of the Pacific Ocean—in Hawaii, for example—the difference between low and high tides is only 0.5 m. On the Pacific coast near Seattle, however, the difference is about 3 m. Extraordinary tides occur in a few places, such as the Bay of Fundy in eastern Canada, where the tidal range can be more than 12 m. Because many people living along the shore need to know when tides will occur, governments publish tide tables showing predicted tide heights and times; these tables are compiled by combining local experience with knowledge of the astronomical motions of Earth and the Moon with respect to the Sun.

Tides may combine with waves to cause extensive erosion of the shore and destruction of shoreline property. When an intense storm passes near the shore during a spring tide, the waves at high tide may overrun the entire beach and batter sea cliffs. These **tidal surges** are not to be confused with the popular but incorrectly termed "tidal waves." Although there are no such waves associated with the tides, there are unusually large ocean waves called *tsunamis* (a Japanese word), which are caused by undersea events such as earthquakes, landslides, and the explosion of oceanic volcanoes.

**TIDAL CURRENTS** The movement of tides near shorelines causes currents that can reach speeds of a few kilometers per hour. As the tide rises, the water flows in toward the shore as a **flood tide,** moving into shallow coastal marshes and up small streams.

As the tide passes the high stage and starts to fall, the **ebb tide** moves out, and low-lying coastal areas are exposed again. Such tidal currents meander across and cut channels into **tidal flats,** the muddy or sandy areas that lie above low tide but are flooded at high tide (Figure 17.10).

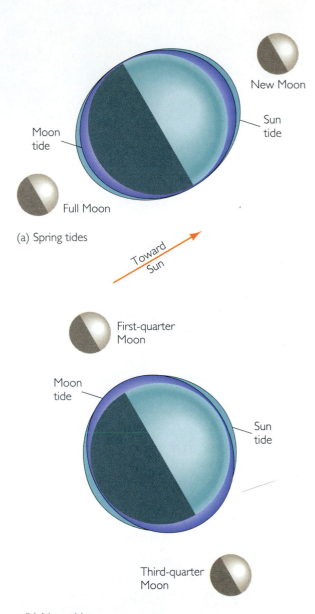

(a) Spring tides

(b) Neap tides

**FIGURE 17.9** The relative positions of the Earth, Moon, and Sun determine the heights of high tide during the lunar month. (a) At new and full Moon, Sun and Moon tides reinforce each other and make the highest (spring) high tides. (b) At first- and third-quarter Moon, Sun and Moon tides are in opposition, minimizing the heights of high (neap) tides.

**FIGURE 17.10** Tidal flats, such as this one at Mont Saint Michel, France, may be narrow strips seaward of the beach or extensive areas covering hundreds of square kilometers. When a very high tide advances on a wide tidal flat, it may move so rapidly that areas are flooded faster than a person can run. The beachcomber is well advised to learn the local tides before wandering. *Patrick Lorne/Explorer.*

# SHORELINES

Combinations of waves, longshore currents, and tidal currents, interacting with the rocks and tectonics of the coast, shape shorelines into a multitude of forms. How these factors operate is illustrated by the most popular of shorelines, beaches.

## Beaches

A beach is a shoreline made up of sand and pebbles. Beaches may change shape from day to day, week to week, season to season, and year to year. Waves and tides sometimes broaden and extend a beach by depositing sand and sometimes narrow it by carrying sand away.

Many beaches are straight stretches of sand that range from a kilometer to over a hundred kilometers long; others are smaller crescents of sand between rocky headlands. Belts of dunes border the landward edge of many beaches; bluffs or cliffs of sediment or rock border others. Beaches may have tide terraces on their seaward sides (Figure 17.11).

Figure 17.12 shows the major parts of a beach, all of which may not be present at all times on any particular beach. Farthest out is the **offshore,** bounded by the surf zone, where the bottom begins to shallow

**FIGURE 17.11** Tide terrace. At low tide, the outer ridge (a sandbar at high tide) is exposed. Also exposed is the shallow depression between the ridge and the upper beach, which is rippled by the tidal flow in many places. *James Valentine.*

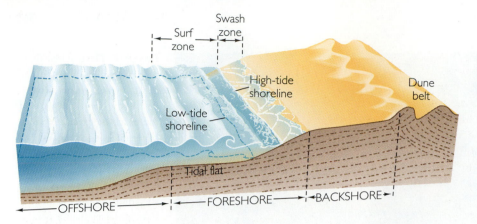

**FIGURE 17.12** A profile of a beach, showing its major parts.

sufficiently for waves to break. The **foreshore** includes the surf zone; the tidal flat; and, right at the shore, the swash zone, a slope dominated by the swash and backwash of the waves. The **backshore** extends from the swash zone up to the highest level of the beach.

A beach is a scene of incessant movement. Each wave moves sand back and forth with swash and backwash. Longshore drift and longshore currents both move sand down the beach. At the end of a beach and to some extent along it, sand is removed and deposited in deep water. In the backshore or along sea cliffs, sand and pebbles are freed by erosion and replenish the beach. The wind that blows over the beach transports sand, sometimes offshore into the water and sometimes onshore onto the land.

The balance between these processes of adding and removing sand results in a beach that may appear to be stable but is actually exchanging its material on all sides. The sand budget of a beach—that is, the inputs to and outputs from any stretch of beach—is illustrated in Figure 17.13. At any point along the beach, sand is gained by the inputs: longshore drift, longshore current, erosion of material in the backshore, and rivers that enter the sea along the shore. From the same stretch of beach, sand is lost by the outputs: longshore drift and current, loss to deep water by wave erosion during storms, and the wind.

If the total input balances the total output, the beach is in equilibrium and keeps the same general form. If input and output are not balanced, the beach either grows or shrinks. Temporary imbalances are

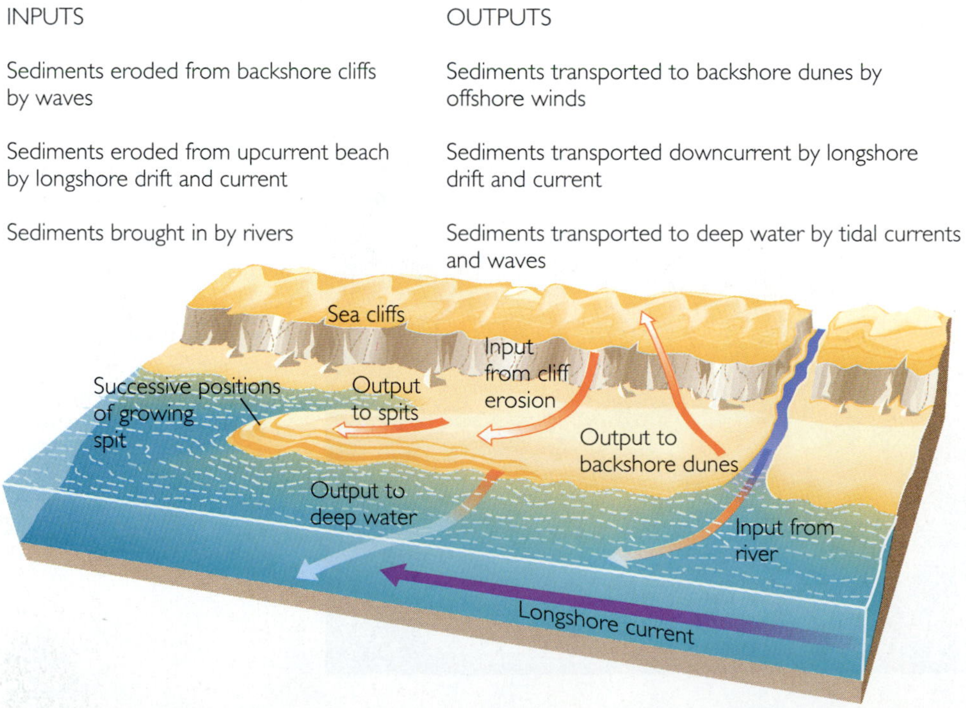

| INPUTS | OUTPUTS |
|---|---|
| Sediments eroded from backshore cliffs by waves | Sediments transported to backshore dunes by offshore winds |
| Sediments eroded from upcurrent beach by longshore drift and current | Sediments transported downcurrent by longshore drift and current |
| Sediments brought in by rivers | Sediments transported to deep water by tidal currents and waves |

**FIGURE 17.13** The beach budget is a balance between inputs and outputs of sand by erosion and sedimentation.

natural over weeks, months, or years. A series of large storms, for example, might move large amounts of sand from the beach to somewhat deeper waters on the far side of the surf zone, narrowing the beach. Then, in a slow return to equilibrium over weeks of mild weather and low waves, the sand might move into shore and rebuild a wide beach.[1]

We can now account for some common beaches. Long, wide, sandy beaches grow where sand inputs are abundant, often where soft sediments make up the coast. Where the backshore is low and the winds blow from onshore, wide dune belts border the beach. If the shoreline is tectonically elevated and the rocks are hard, cliffs line the shore and any small beaches that evolve are composed of material eroded from the cliffs. Where the shore is low-lying, sand is abundant, and tidal currents are strong, extensive tidal flats are laid down (exposed at low tide).

What happens if one of the inputs is cut off—for example, by a concrete wall built at the top of the beach to prevent erosion? Because erosion supplies sand to the beach as one of the inputs, preventing it

---

[1]It is the constant shifting of the sands of a beach that allows it to recover from oil spills washed ashore. Within a year or two the oil will be transported or buried out of sight, although later the tarry residue may be uncovered in spots. The same is true of trash and litter; beaches would clean up rapidly if the littering were to stop.

cuts the sand supply and so shrinks the beach. Attempts to save the beach may actually destroy it (see Box 17.1).

## Erosion and Deposition at Shorelines

The topography of the shoreline, like that of the interior, is the product of tectonic forces elevating or depressing the Earth's crust, erosion wearing it down, and sedimentation filling in the low spots. Thus the factors at work are uplift and subsidence of the coastal region, the nature of its rocks or sediments, changes in sea level, the average and storm wave heights, and the height of the tides.

**EROSIONAL COASTAL FORMS**    Tectonically uplifted rocky coasts, where erosion is active, have a topography of prominent cliffs and headlands that jut into the sea, alternating with narrow inlets and irregular bays with small beaches. Along rocky shorelines, waves undercut cliffs and cause huge blocks to fall into the water, where they are gradually worn away. As the sea cliffs retreat by erosion, isolated remnants called **stacks** are left standing in the sea far from the shore (Figure 17.14). Erosion by waves planes the rocky surface beneath the surf zone and creates a **wave-cut terrace,** sometimes visible at low tide (Figure 17.15). As wave erosion continues over long periods, shorelines may straighten as headlands retreat faster than recesses and bays.

**FIGURE 17.14** Stacks at Crook Point, Oregon. These remnants of shore erosion are left as the shoreline retreats. *Larry Ulrich.*

## BOX 17.1 PRESERVING OUR BEACHES

Orrin Pilkey of Duke University, a geologist and oceanographer, has been in the forefront of scientists concerned about saving our beaches and halting massive development on fragile shorelines. Battered by the waves eroding the shoreline, many houses could be saved by building concrete buttresses, seawalls, and other structures designed to save shoreline property, but these structures would destroy the beach. Pilkey, a well-known researcher on coastal processes, was an advocate for the beaches of the Carolinas, which have come under heavy pressure from commercial developers. Knowing how the beach system works, he believes it is foolish to try to interfere with the natural process by which beaches remain in dynamic equilibrium with the waves and currents.

More and more beaches are being altered from their natural state as people build cottages on the shore, pave beach parking lots, erect seawalls, and construct piers and breakwaters. The consequence of poorly thought out construction is the shrinkage of the beach

A house tilts as sand is eroded from the beach and cliffs (*left background*) by storm waves; Nantucket, Massachusetts. *Steve Rose/Rainbow.*

FIGURE 17.15 Wave-cut terrace; Point Reyes National Seashore. The cliffs have retreated from left to right as erosion planes a terrace at low-tide level. *Travis Amos.*

in one place and its growth in another, usually where no one wants it. The classic example is a narrow pier built out from shore at right angles to it. Over the following months and years, the sand disappears from the beach on one side and greatly enlarges the beach on the other side—much to the surprise of the builders.

The disappearance and enlargement are the predictable results of a longshore current. The waves, current, and drift bring sand toward the pier from the upcurrent direction (usually the dominant wind direction). Stopped at the pier, they dump the sand there. On the downcurrent side of the pier, the current and drift pick up again and erode the beach. On this side, however, there is little replenishment of the sand by the current because it is blocked by the pier, so the beach budget is out of balance and the beach shrinks. If the pier is removed, the beach relaxes to its former state.

The only way to save a beach is to leave it alone. Even if concrete walls and piers can be kept in repair with large expenditures of money, many times at public expense, the beach itself will suffer. Along some beaches, resort hotels truck in sand to replace that lost, but that expensive solution is temporary too. Sooner or later we must learn to let the beaches remain in their natural state.

Construction of piers along a shore to control erosion of a beach may produce erosion downcurrent of the pier and loss of part of the beach (*right of pier*) while sand piles up on the other side (*left of pier*). Longshore current flows from left to right. *Philip Plisson/ Explorer.*

Where relatively soft sediments or sedimentary rocks make up the coastal region, the slopes are gentler and the heights of shoreline bluffs are lower. Waves efficiently erode these softer materials; erosion of bluffs on such shores may be extraordinarily rapid. The high sea cliffs of soft glacial materials along the Cape Cod National Seashore in Massachusetts, for instance, are retreating about a meter per year. Since Henry David Thoreau walked the entire length of the beach below those cliffs in the nineteenth century and wrote of his travels in *Cape Cod,* about 6 km² of coastal land has been eaten by the ocean.

**DEPOSITIONAL COASTAL FORMS** Where tectonic subsidence depresses the crust along a coast,

sediment builds up. Such coasts are characterized by long, wide beaches and wide, low-lying coastal plains of sedimentary strata. Shoreline forms include sandbars, low-lying sandy islands, and extensive tidal flats. Long beaches grow longer as longshore currents carry sand to the downcurrent end of the beach. Here it builds up, first as a submerged bar, then rising above the surface and extending the beach by a narrow addition called a **spit** (Figure 17.16).

Long sandbars offshore may build up and become **barrier islands** forming a barricade between open ocean waves and the main shoreline (Figure 17.17). As the bars are built up above the waves, vegetation takes hold, stabilizing the islands and helping them resist wave erosion during storms. Barrier is-

**FIGURE 17.16** This spit has advanced from the cliffs in the background to the right foreground. *Oregon State Highway Department.*

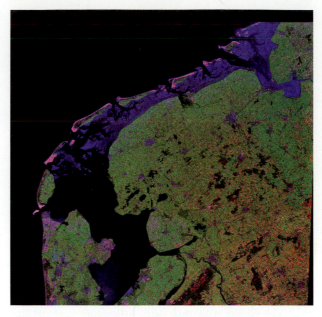

**FIGURE 17.17** Barrier islands separate the North Sea (*left*) from the shallow waters and tidal flats of the main shoreline; the Netherlands. *EOSAT.*

lands are separated from the coast by tidal flats or shallow lagoons. Like beaches on the main shore, barrier islands are in dynamic equilibrium with the forces shaping them. If their equilibrium is disturbed, they may be disrupted or devegetated, leading to increased erosion and even disappearance.

## Changes in Sea Level

Shorelines are sensitive to changes in sea level, which can change the approach of waves, alter tidal heights, and affect the path of longshore currents. Rise and fall of sea level can be local, a result of tectonic subsidence, or global, the result, for example, of continental glacial melting or growth (see Chapter 15).

During periods of lowered sea level, erosion dominates as formerly offshore areas are exposed to agents of erosion. Rivers extend their courses over formerly submerged regions and cut valleys into newly exposed coastal plains. When sea level rises, flooding the lands of the backshore, river valleys are drowned and marine sediments build up along former land areas. Today, long fingers of the sea indent many of the shorelines of the northern and central Atlantic coast. These are drowned river valleys that were flooded as the last glacial age ended about 10,000 years ago and the sea level rose.

Drowned river valleys are one kind of **estuary,** which is a coastal body of water connected to the ocean and also supplied with fresh water from a river. The fresh water comes down the river and mixes with seawater in the estuary long before it reaches the main shoreline. At its upper reaches, in some places many kilometers upstream from the mouth, an estuary is fresh. Downstream, as it gradually mixes with seawater, it becomes saltier. By the time it nears the main shoreline it is entirely seawater.

The shorelines of the world serve as our barometers of impending change as they respond to altered global and local conditions. For example, if warming by the greenhouse effect causes the sea level to rise, we will first see the effects on our beaches. The pollution of our inland waterways sooner or later arrives at our beaches, as sewage from city dumping and oil from ocean tankers wash up on the shore. And as real estate development and construction along shorelines expands, we will see the continuing contraction and even disappearance of some of our finest beaches.

Profound as the geological changes in shorelines may be, they are dwarfed by the geologic processes at work in the vast bulk of the oceans where the water is deep and hides the active seafloor below. In the rest of this chapter we will explore some of the important geologic processes of the deep ocean. We begin with the unusual tools available to measure and map the seafloor.

# SENSING THE FLOOR OF THE OCEAN

The best way to see the seafloor is directly from a deep-diving submersible. Pioneered by the French oceanographer Jacques-Yves Cousteau, these small ships can observe and photograph at great depths. With their mechanical arms they can break off pieces of rock, sample soft sediment, and catch specimens of exotic deep-sea animals. Newer robot submersibles are guided by scientists on the mother ship above (Figure 17.18). But submersibles are expensive to build and operate and cover small areas at best.

For most work, today's oceanographers use instrumentation to sense the seafloor topography indirectly from a ship at the surface. An echo sounder is a shipboard instrument that sends out pulses of sound waves; when the sound waves are reflected back from the ocean bottom, they are picked up by sensitive microphones in the water. By measuring the interval between the time the pulse leaves the ship and the time it returns as a reflection, and using the speed of sound in water, oceanographers can compute the depth. The result is an automatically traced profile of the bottom topography (Figure 17.19 shows an example). Echo sounding is also used to probe the stratigraphy of sedimentary layers beneath the ocean floor (see Box 20.1).

**FIGURE 17.18** The *Benthic Explorer*. This small robot vehicle can explore the seafloor while being directed from shipboard. *T. Kleindinst/Woods Hole Oceanographic Institution.*

Many other instruments are lowered to the bottom to detect the magnetic properties of the seafloor, to scan the shapes of undersea cliffs and mountains by radar, to probe the seafloor with heat detectors, and to take detailed photographs.

Since 1968 the United States–sponsored Deep Sea Drilling Program and its successor, the interna-

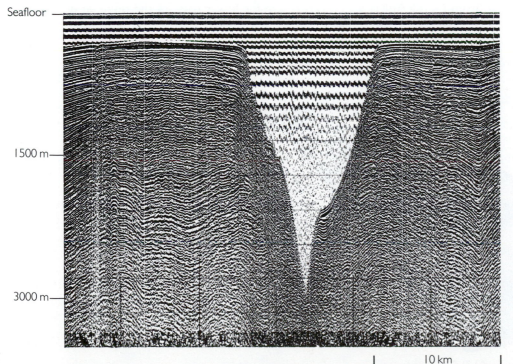

**FIGURE 17.19** An echo-sounding profile of the Congo submarine canyon off the Republic of Zaire on the west coast of Africa. The bottom of the canyon at this point is about 3000 m below the seafloor of the continental shelf; at the top, the canyon is more than 10 km wide. The wavy lines below the seafloor surface are sound reflections from bedding planes in continental shelf sediments, somewhat deformed because of mild tectonic disturbance. *K. O. Emery, Woods Hole Oceanographic Institution.*

Seafloor

1500 m

3000 m

10 km

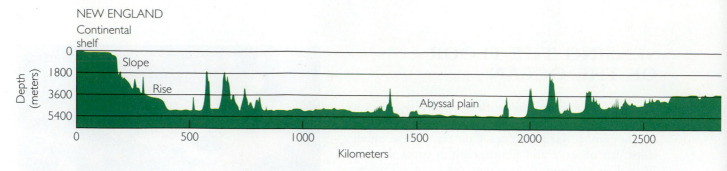

**FIGURE 17.20** Topographic profile of the floor of the Atlantic Ocean from New England (*left*) to Gibraltar (*right*). (After B. C. Heezen, "The Origin of Submarine Canyons," *Scientific American,* August 1956, p. 36.)

tional Ocean Drilling Program, have sunk hundreds of drill holes to depths of many hundreds of meters below the seafloor. These cores have given us an unprecedented three-dimensional picture of the seafloor and provided samples for detailed physical and chemical studies. (In Chapter 20 we will discuss the role of these drilling programs in the study of plate tectonics.)

## PROFILES OF TWO OCEANS

### An Atlantic Profile

Just as we profiled North America in Chapter 3 by taking a hypothetical trip across the continent, we can imagine driving a deep-diving submarine along the floor of the Atlantic Ocean from North America

to Gibraltar. A profile of the Atlantic is shown in Figure 17.20.

Starting from North America, we would descend from the shoreline to depths of 50 to 200 m and travel along the **continental shelf,** a broad, flat, sand- and mud-covered platform that is part of the continent but slightly submerged (Figure 17.21). After traveling about 50 to 100 km across the shelf, down a very gently inclined surface, we would find ourselves at the edge of the shelf, where we would start down a steeper incline, the **continental slope.** This slope, covered mostly with mud, descends at an angle of about 4°, a drop of 70 m over a horizontal distance of 1 km, which would feel like a noticeable grade if we were driving on land.

The continental slope is irregular and marked by gullies and **submarine canyons,** deep valleys eroded into the slope and the shelf behind it (see Figure 17.19). On the lower parts of the slope, at depths of

**FIGURE 17.21** The Atlantic continental shelf, slope, and rise off the eastern coast of North America.

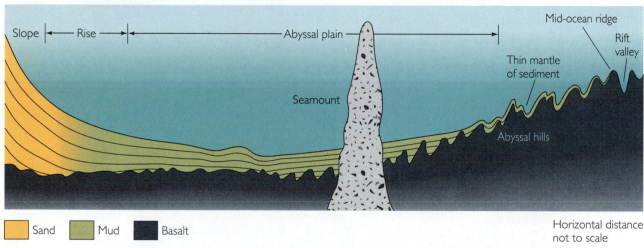

Sand    Mud    Basalt

Horizontal distance
not to scale

**FIGURE 17.22** Profile from continental rise to Mid-Atlantic Ridge.

around 2000 to 3000 m, the incline becomes gentler. Here it merges into a more gradual incline called the **continental rise,** an apron of muddy and sandy sediment extending into the main ocean basin. The rise is broken by an occasional shallow channel or canyon.

The continental rise is hundreds of kilometers wide and grades imperceptibly into a wide, flat **abyssal plain** that covers large areas of the ocean floor at depths of about 4000 to 6000 m (Figure 17.22). These plains are broken by occasional submerged volcanoes, mostly extinct, called **seamounts** (Figure 17.23). A few seamounts extend to the surface as islands. Bermuda is one such volcanic island, capped with coral reef limestones.

As we travel along the abyssal plain, we gradually climb into a province of low abyssal hills whose slopes are covered with fine sediment. Continuing up the hills, the sediment layer becomes thinner and outcrops of basalt appear beneath it. As we rise along this steep, hilly topography to depths of about

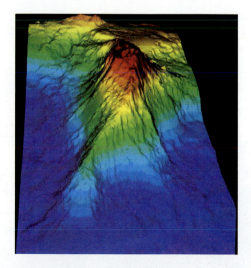

**FIGURE 17.23** Loiki Seamount just south of the Big Island of Hawaii, as imaged by computer-enhanced sidescan radar. Loiki is the newest in the string of hot-spot volcanoes that form the Hawaiian island chain. *Ocean Mapping Development Center, University of Rhode Island.*

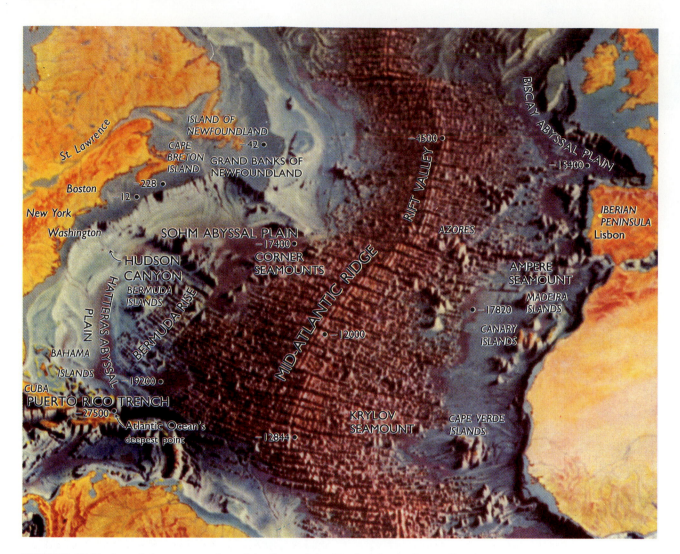

**FIGURE 17.24** An artist's representation of the floor of the North Atlantic Ocean. Depths are shown in feet below sea level. (*World Ocean Floor* based on bathymetric studies by Bruce C. Heezen and Marie Tharp. Painting by Heinrich C. Berann. Copyright © Marie Tharp, 1977.)

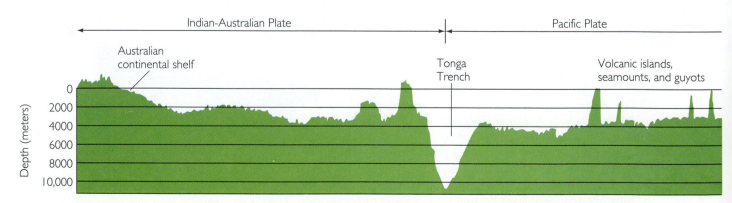

**FIGURE 17.25** Topographic profile of the floor of the Pacific Ocean from Australia to South America. (Data synthesized from various sources.)

3000 m, we are climbing the flanks and then the mountains of the Mid-Atlantic Ridge.

Abruptly, we come to the edge of a deep, narrow valley (about 1 km wide) at the top of the ridge, a narrow cleft marked by active volcanism. This is a rift valley where two plates separate. As we cross the valley and climb the east side, we are moving from the North American Plate to the Eurasian Plate (see the plate map inside the front cover).

Continuing east, we find the same topography as on the west side of the ridge, only in reverse order, for the ocean floor is symmetrical on either side of the ridge. Again we pass over abyssal hills, gradually descend to an abyssal plain, then ascend to the continental rise, slope, and shelf off the coast of Europe.

From many such traverses of the oceans (although not in submarines but with echo sounders), the floor of the Atlantic Ocean has been mapped (Figure 17.24). On this map we can see some of the details of individual ocean-floor provinces and get an impression of the strange submarine landscape that we cannot see from above.

## A Pacific Profile

Just as all continents do not show the same profile, a profile of the Pacific Ocean shows features not seen in the Atlantic profile. If we were to travel westward from South America beginning on the west coast of Peru or Chile, we would, as before, cross a continental shelf. This shelf, however, is only a few tens of kilometers wide and more than 100 m deep. At the edge of the shelf the continental slope is much steeper and extends down to 8000 m as we enter the Peru-

Chile Trench (Figure 17.25), a long, deep, narrow depression in the seafloor that is the surface expression of the subduction of the Nazca Plate, a small plate in the eastern Pacific, under the South American Plate.

Continuing across the trench and up onto the higher hilly region of the Nazca Plate, we soon come to a mid-ocean ridge, the East Pacific Rise. The East Pacific Rise is lower than the Mid-Atlantic Ridge, but it has the characteristic central rift valley and outcrops of basalt. On the west side of the East Pacific Rise, we cross over to the Pacific Plate and drive on westward over its broad central regions. Eventually we come to another trench, the Tonga. This is one of the deepest places in all the oceans, almost 11,000 m deep. Here, in the middle of the ocean, the Pacific Plate subducts beneath the Indian-Australian Plate. On the west side of the trench, an arc of volcanic islands rises from the deep seafloor and erupts basalt and andesite. Leaving the island arc we return to the deep seafloor, now on the Indian-Australian Plate, and soon come to the continental rise, slope, and shelf of Australia, similar to the east coast of North America.

Our traverses of the Atlantic and Pacific have shown some of the differences between oceans, which are related to plate-tectonic movements. The Atlantic, bisected by the Mid-Atlantic Ridge, is primarily a spreading ocean and has only a small subduction zone in the Caribbean Sea. In contrast, the Pacific, crossed by the East Pacific Rise, shows a large number of subduction zones that are currently narrowing this ocean. The entire network of ridges, trenches, and transform faults that bound the world's ocean plates can be seen in the map in Figure 17.26.

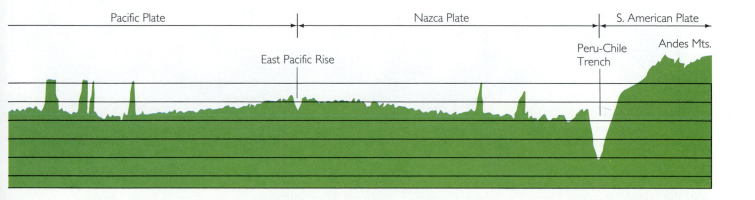

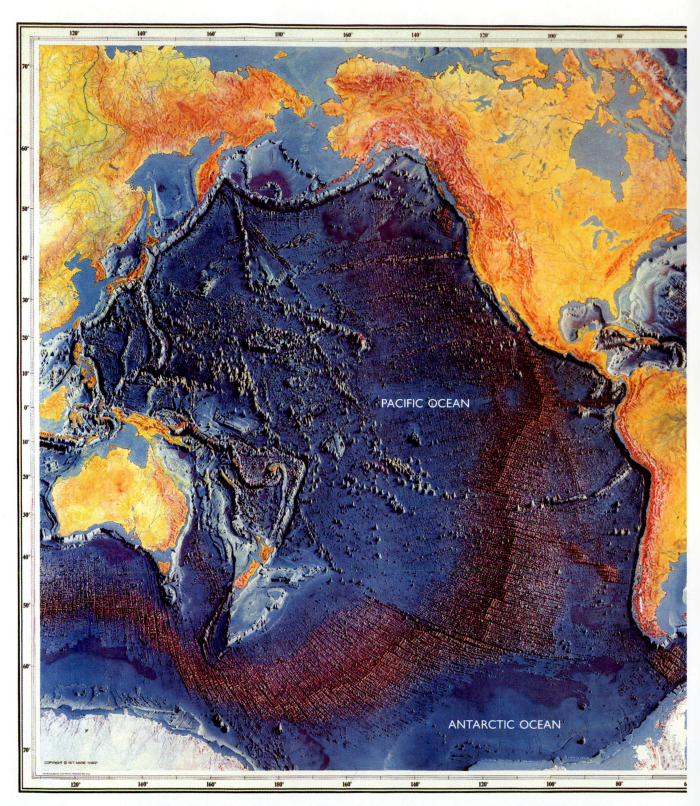

**FIGURE 17.26** This map of the world ocean floor shows the pattern of mid-ocean ridges, trenches, and transform faults that bound lithospheric plates. The Atlantic Ocean is dominated by the Mid-Atlantic Ridge, which bisects the ocean from north to south. In the eastern Pacific, the East Pacific Rise can be clearly seen, as well as the Peru-Chile Trench (black strip along the west coast of South America). The western Pacific is the scene of many trenches, which

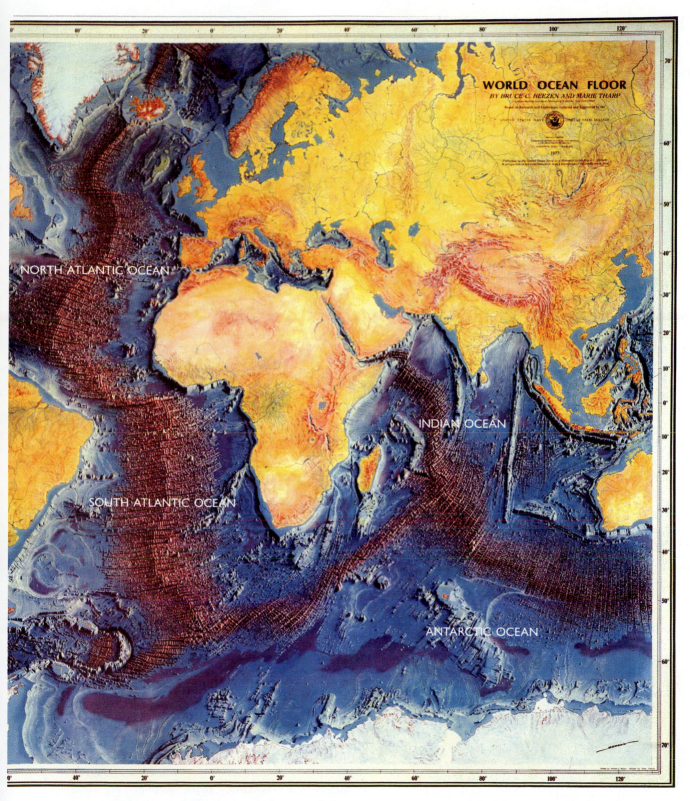

**WORLD OCEAN FLOOR**
*BY BRUCE C. HEEZEN AND MARIE THARP*

NORTH ATLANTIC OCEAN

SOUTH ATLANTIC OCEAN

INDIAN OCEAN

ANTARCTIC OCEAN

mark the subduction zones that separate oceanic lithospheric plates. The Indian Ocean is traversed by two intersecting mid-ocean ridges, the Carlsberg Ridge and the Mid-Indian Ocean Ridge, whose southward extensions circumscribe the continent of Antarctica and the Antarctic Ocean. (*World Ocean Floor* based on bathymetric studies by Bruce C. Heezen and Marie Tharp. Painting by Heinrich C. Berann. Copyright © Marie Tharp, 1977.)

# CONTINENTAL MARGINS

The profiles of the oceans show that the shorelines, shelves, and slopes of the continents, together called **continental margins,** are of two types. One is exemplified by the margin off the east coast of North America. This broad region, associated with a spreading ocean, is a **passive margin,** a continental borderland that is far from a plate boundary (Figure 17.27). Such margins are called passive (implying quiescence) because volcanoes are absent and earthquakes are few and far between.

Continental margins associated with subduction zones and transform faults, in contrast, are **active margins.** The volcanic activity and frequent earthquakes give these narrow and tectonically deformed margins their name. One active margin at a subduction zone is off the west coast of South America (see Figure 17.25). This margin includes the trench offshore, the narrow shelf, and the active volcanic belt of the Andes Mountains. As the subducting Nazca Plate bends below the continent, it heats up, and magmas form and work their way to the surface.

The continental shelves of passive margins consist of essentially flat-lying, shallow-water sediments, both terrigenous and carbonate, several kilometers thick (see Figure 17.27). Although the same kinds of sediment are found on active margin shelves, they are more likely to be structurally deformed and to include ash and other volcanic materials.

## Continental Shelf

The continental shelf is one of the most economically valuable parts of the ocean. George's Bank off New England and the Grand Banks of Newfoundland, for example, have been among the world's most productive fishing grounds for all of this century. In recent years the continental shelf, especially off the Gulf coast of Louisiana and Texas, has housed huge oil-drilling platforms. These are some of the reasons that the international Law of the Sea treaty was signed in 1982 by most of the world's nations (although not by the United States).

Continental shelves are broad and relatively flat at passive continental margins and are narrow and uneven at active margins. As we noted earlier, because continental shelves lie at shallow depths, they are subject to exposure and submergence as a result of changes in sea level. During the Pleistocene glaciation, all of the shelves now at depths of less than 100 m were above sea level, and many of their features were formed then. During that time, shelves at high latitudes were glaciated, producing an irregular topography of shallow valleys, basins, and ridges. The shelves in lower latitudes were left more regular, broken by occasional stream valleys.

## Continental Slope and Rise: Turbidity Currents

The waters of the continental slope and rise are too deep for the seafloor to be affected by waves and tidal currents. As a consequence, muds, silts, and sands that have been carried across the shallow continental shelf come to rest as they are draped over the slope. The slope shows signs of the slumping of sediment and the erosional scars of gullies and submarine canyons. Deposits of sands, silts, and muds on both slope and rise indicate active sediment transport in these deep waters. What kind of current could be responsible for both erosion and sedimentation on the slope and rise at such great depths?

A turbidity current proved to be the answer. A **turbidity current** is a flow of turbid, muddy water down a slope. Because of its suspended load of mud, the turbid water is denser than the overlying clear water and flows beneath it. This kind of flow can both erode and transport sediment. Turbidity currents were first noticed over a century ago where the Rhone River enters Lake Geneva in Switzerland. The

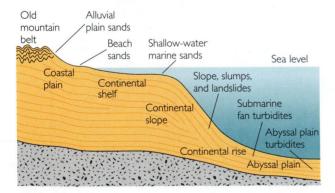

**FIGURE 17.27** A profile of the Atlantic passive continental margin off southern New England. (After K. O. Emery and E. Uchupi, *Atlantic Continental Margin of North America,* American Association of Petroleum Geologists, 1972.)

muddy river water enters the clear water of the lake and flows as a distinct current along the sloping bottom to the level floor of the lake, where it fans out over the bottom.

The role of turbidity currents in ocean processes was first understood by Philip Kuenen, a Dutch geologist and oceanographer. In 1936 he produced and filmed such currents in his laboratory by pouring muddy water into the end of a long, narrow tank with a sloping bottom. He showed that these currents could move at many kilometers per hour and that the speed was proportional to the steepness of the slope and the density of the current. Kuenen reasoned that because of its speed and turbulence, a turbidity current could erode and transport large quantities of sand down the continental slope. He then proposed the idea, revolutionary for that time, that turbidity currents operate widely in the ocean, especially on continental slopes, at depths well below any possible wave or tidal action.

Turbidity currents start when occasional earthquakes trigger slumps of the sediment draped over the edge of the continental shelf and onto the continental slope (Figure 17.28). The sudden slump, or submarine landslide, throws mud into suspension, creating a dense, turbid layer of water near the bottom. This turbid layer starts to flow, accelerating down the slope.

As the turbidity current reaches the foot of the slope and the gentler incline of the continental rise, the current slows and some of the coarser sandy sedi-

ment starts to settle. The layers of sediment from many turbidity flows form a **submarine fan,** a fan-shaped deposit that is something like an alluvial fan on land (Figure 17.29). Many currents continue across the rise, cutting channels in the submarine fans as they go, until they reach the level bottom of the ocean basin, the abyssal plain. There they spread out and come to rest in graded beds of sand, silt, and mud called **turbidites.**

Although the existence of turbidity currents was challenged for several years, dramatic confirmation of the hypothesis came from an unexpected source. Transatlantic telegraph cables that were laid on the ocean floor from North America to Europe were known to break periodically. In 1929, following an earthquake, there had been a particularly large series of cable breaks on the Atlantic continental slope and rise off the Grand Banks of Newfoundland.

In 1952 oceanographers Bruce Heezen and Maurice Ewing of Columbia University, impressed with Kuenen's work, plotted the exact times and positions of the breaks. Could slope slumps and turbidity currents explain these breaks? A rapid breaking of the cables high on the slope was followed by a sequence of breaks going down the slope, farther and farther from the center of the earthquake. The breaks downslope were much later than they would have been if they were caused by earthquake waves. The only reasonable explanation was that the earthquake triggered a slump, which activated a turbidity current fast and powerful enough to snap the cables as it

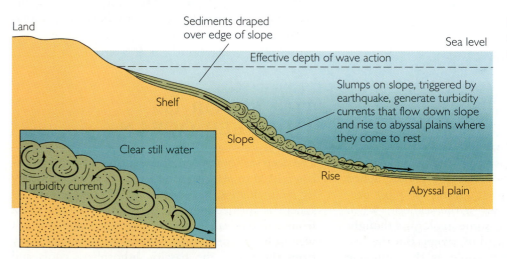

**FIGURE 17.28** How a turbidity current forms in the ocean.

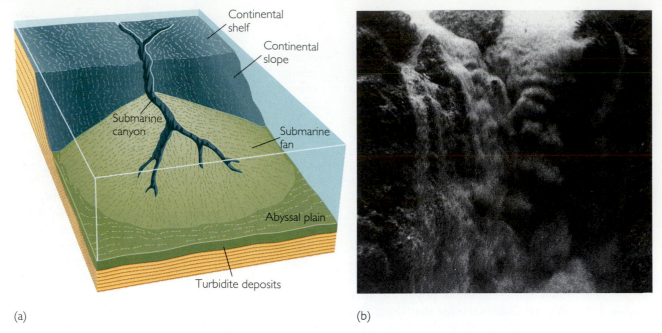

(a)                                                                    (b)

**FIGURE 17.29** (a) How a submarine canyon is formed by turbidity currents. (b) Sandfall at the head of a submarine canyon at the edge of the continental shelf. These falls generate sandy flows, like turbidity currents, that lay down fans of sandy sediment at the foot of the continental slope. *U.S. Navy.*

raced down the slope and rise. Later, graded beds were recovered from the seafloor in the path of the presumed flow. As similar patterns of cable breaks in other places were plotted, the turbidity current theory was confirmed.

Today we recognize turbidity currents as important agents of erosion and sedimentation on continental margins. They are found in trenches along active margins as well as on the continental slopes and rises of passive margins. They form the abyssal plains that cover large areas of the ocean floor.

## Submarine Canyons

Almost immediately after ocean turbidity currents were hypothesized, they were proposed to be the erosive agents that incised submarine canyons into many continental shelves (see Figure 17.29), an idea that remained controversial for many years. Even though submarine canyons have been mapped in detail and their walls and floors amply photographed and sampled, they have been among the most perplexing topographic features of the seafloor. When they were first discovered, some geologists thought they might have been formed by rivers. But this hypothesis soon proved impossible as the complete explanation, for most of the canyon floors are thou-

sands of meters deep. This is far below the approximately 100-m depth to which rivers could erode during the maximum lowering of sea level during the ice ages. Even so, there is no question that the shallower parts of some canyons were river channels during periods of low sea level.

The currently favored explanation for the deeper parts of canyons is turbidity currents, although other types of currents have been proposed. A comparison of modern canyons and their deposits with well-preserved similar deposits of the past, particularly the pattern of turbidites deposited on submarine fans, has reinforced this conclusion.

## THE FLOOR OF THE DEEP OCEAN

The deep seafloor is constructed primarily by volcanism related to plate-tectonic motions and secondarily by sedimentation in the open sea. The oceanic lithosphere is formed where plates grow by spreading from a mid-ocean ridge as huge quantities of basalt well up from the mantle. As the plates spread away from the ridge, the basaltic lithosphere cools and contracts, lowering the seafloor. While this is happening, the basalt surface receives a steady rain of sediment from surface waters and gradually becomes mantled with deep-sea muds and other deposits.

## Mid-Ocean Ridges

Mid-ocean ridges are the sites of the most active volcanic and tectonic activity on the deep seafloor. The central valley is the center of the action. The valley walls are faulted and intruded with basalt sills and dikes (Figure 17.30), and the floor of the valley is covered with flows of basalt and talus blocks from the valley walls, mixed with a little sediment settling from surface waters. Mid-ocean ridges are offset at many places by transform faults that laterally displace the rift valleys (see Figure 17.24).

Hydrothermal springs on the rift valley floor (see Box 17.2) are formed as seawater percolates into cracks and fractures in the basalt on the flanks of the ridge, is heated as it moves down to hotter basalt, and finally exits at the valley floor, where it boils up at temperatures as high as 380°C. Some springs are "black smokers," full of dissolved hydrogen sulfide and metals leached from the basalt by the hot waters. Others are "white smokers" with a different composition and lower temperatures. Hydrothermal springs on the seafloor produce mounds of iron-rich clay minerals, iron and manganese oxides, and large deposits of iron-zinc-copper sulfides.

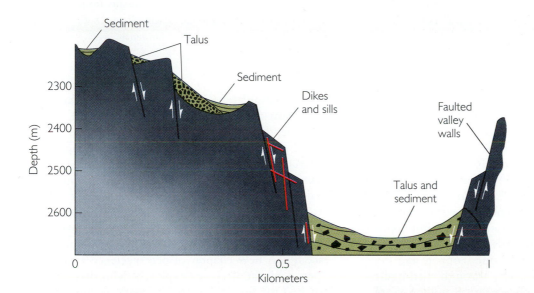

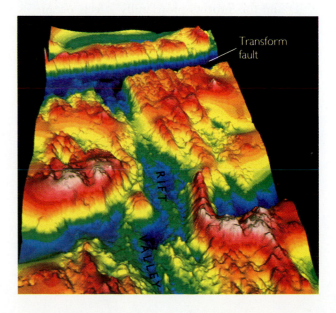

FIGURE 17.30 *Above:* A profile of the central rift valley of the Mid-Atlantic Ridge in the FAMOUS (French-American Mid-Ocean Undersea Study) area southwest of the Azores Islands. The deep valley, where most of the basalt is extruded, is faulted. (After ARCYANA, "Transform Fault and Rift Valley from Bathyscaph and Diving Saucer," *Science,* vol. 190, 1975, p. 108.) *Right:* Computer-generated image (based on data from multibeam sidescan sonar) of the Mid-Atlantic Ridge at about 31° south latitude reveals a 19-km-wide central valley studded with volcanic cones. The valley is bounded by mountains about 2000 m high. A transform fault intersects the valley at the top of the image. (K. C. Macdonald and P. J. Fox, "The Mid-Ocean Ridge," *Scientific American,* vol. 262, 1990, pp. 72–79.)

## BOX 17.2 HOT SPRINGS ON THE SEAFLOOR

Much of the heat coming from Earth's interior is thought to be dissipated when cold seawater percolates into the many fissures associated with seafloor spreading along mid-ocean ridges. The cold water sinks several kilometers, encounters hot basalt, surges upward, and emerges on the seafloor as hot springs rich in dissolved minerals and gases leached from the magma. This hypothesis was first verified in 1977; since then there have been spectacular discoveries of such hydrothermal vents at several places along mid-ocean ridges.

The hot springs seem to occur in two forms. In the Galápagos Islands rift zone, the springs flow gently from cracks at maximum temperatures of about 16°C. On the East Pacific Rise near Baja California, superheated water (380°C) spouts forcefully from mineralized chimneys. The chimneys are built up from the dissolved minerals that precipitate around the hot jet as it mixes with the near-freezing waters on the ocean bottom. The deep-diving submarine *Alvin,* not built for such high temperatures, was nearly destroyed when it first approached a superheated vent.

Hydrothermal vents represent a major ore-forming process, possibly an important new source of minerals on the seafloor, such as sulfide ores rich in zinc, copper, and iron. The entire ocean cycles through such hydrothermal systems once every 8 million years, and this process profoundly affects ocean chemistry by transporting elements from the interior to the ocean. The ecology of these

A plume of hot, mineral-laden water spouts from a hydrothermal vent on the East Pacific Rise. *D. B. Foster, Woods Hole Oceanographic Institution.*

vents is completely different from that of the dark, near-freezing, barren ocean bottom at great depths. Dense colonies of exotic life forms populate the warm water surrounding the vents. Among them are new species of giant worms, clams, and crabs. The vent animals live on unusual bacteria that draw energy from the hydrogen sulfide, carbon dioxide, and oxygen in the vent water rather than from the Sun, as the species at the ocean surface do. A new chapter of ocean science was opened by the deep-sea explorers who discovered the hot springs of the seafloor and verified the dominant role of circulating seawater in cooling the new crust formed at ocean ridges.

## Hills and Plateaus

The floor of the deep oceans away from mid-ocean ridges is a landscape of hills, plateaus, sediment-floored basins, and seamounts. Most of the thousands of volcanoes are submerged, but some rise to the sea surface. Seamounts and volcanic islands may be isolated, in clusters, or in chains. They may be formed along a mid-ocean ridge or where a plate overrides a mantle hot spot. Many seamounts have

flat tops, the result of erosion of an island volcano when it was above sea level. These **guyots,** as they are called, are submerged because the plate they were riding on cooled, contracted, and subsided as it passed away from the hot spot that produced the upwelling basalt from the mantle.

Abyssal hills, plateaus, and low ridges are all accumulations of volcanic rock. Many of these features are formed when the seafloor first opens at a rift valley. Others form as volcanic chains are created over

---

### BOX 17.3  THE OCEANS AS A DEEP WASTE REPOSITORY

People have used the oceans as a garbage dump for millennia, since the first time someone in a dugout canoe threw refuse over the side. Today, enormous quantities of garbage, sewage, and industrial waste find their way into the sea from ships and coastal communities. Only now, spurred by the washing ashore of hypodermic syringes and other medical wastes as well as all sorts of other refuse, are some nations beginning to restrict such dumping.

At the same time, some oceanographers are evaluating how we might safely dump a variety of hazardous materials, from highly toxic sewage sludges to high-level radioactive wastes, for burial in the deep ocean. The goal is to deposit the waste where it will remain undisturbed for the hundreds of thousands of years required for radioactive materials to decay enough to be harmless to humans and animals.

The most likely sites are in the middle of an oceanic plate, far from plate boundaries with their tectonic and volcanic activity. One approach is to design strong drums of corrosion-resistant material that would be embedded many meters below the seafloor in the accumulated sediment. The sediment is expected to seal around the drums and retard or prevent any leakage that might occur after long burial. For this idea to work, we must be able to convert the waste material to a form, such as a dense glass or ceramic, that could be loaded into the drums.

Critics of these proposals point out that the dangers to ocean life, even in the deeps where few of the organisms interact with the bulk of the ocean's populations, are incalculable should the drums prove susceptible to leakage. Although no agreement has been reached, the argument continues because of the hazards of waste disposal on land.

---

hot spots. The traces of transform-fault offsets of the mid-ocean ridges interrupt the abyssal seafloor with long ridges coupled with parallel valleys, both of which are roughly at right angles to the ridge (see Figure 17.24).

Although the deep seafloor is tectonically active at mid-ocean ridges, subduction zones, and hot spots, it is quiescent over much of the vast areas of abyssal hills and abyssal plains. In these areas the only geological activity is the slow rain of sediment from surface waters, a feature that has led to serious consideration of some parts of the deep seafloor as a waste repository (see Box 17.3).

## Coral Reefs and Atolls

For more than 200 years coral reefs have attracted explorers and travel writers. Ever since Charles Darwin sailed the oceans on the *Beagle* from 1831 to 1836, they have been a matter of scientific discussion, too. Darwin was one of the first to analyze the geol-

ogy of coral reefs, and his theory of their origin is still accepted today.

The coral reefs Darwin studied were **atolls,** islands in the open ocean with circular lagoons enclosed within a more or less circular chain of islands (Figure 17.31). Coral reefs also form at continental margins, such as those of the Florida Keys. The outermost part of a reef is a slightly submerged, wave-resistant reef front, a steep slope facing the ocean. The reef front is composed of the interlaced skeletons of actively growing coral and calcareous algae, forming a tough, hard limestone. Behind the reef front is a flat platform extending into a shallow lagoon. An island may lie at the center of the lagoon. Parts of the reef, as well as a central island, are above water and may become forested. A great number of plant and animal species inhabit the reef and the lagoon.

Coral reefs are limited to waters less than about 20 m deep, for seawater does not transmit enough light below that depth for reef-building corals to grow. Darwin explained how coral reefs could be built up from the bottom of the dark, deep ocean.

**FIGURE 17.31** Atolls of the Pacific showing the circular to oval shapes of the reefs outlining the protected lagoons. *NASA.*

The process starts with a volcano building up to the surface from the seafloor (Figure 17.32). As the volcano temporarily or permanently becomes dormant, coral and algae colonize the shore and build **fringing reefs,** coral reefs similar to atolls that grow around the edges of a central volcanic island. Erosion may then lower the volcanic island almost to sea level.

Darwin reasoned that if such a volcanic island were to slowly subside beneath the waves, actively growing coral and algae might keep pace with the subsidence, continuously building up the reef so that the island remained. In this way, the volcanic island would disappear and we would be left with an atoll. More than 100 years after Darwin proposed his theory, deep drilling on several atolls penetrated volcanic rock below the coralline limestone and confirmed the theory. And some decades later, the theory of plate tectonics explained both volcanism and the subsidence that resulted from plate cooling and contraction.

**FIGURE 17.32** Evolution of a coral reef from a subsiding volcanic island, first proposed by Charles Darwin in the nineteenth century.

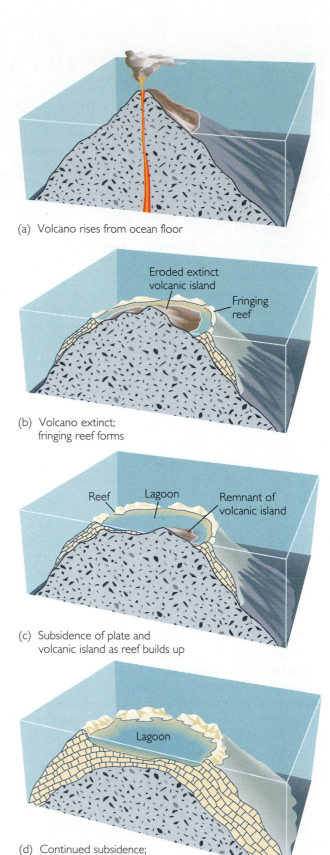

(a) Volcano rises from ocean floor

(b) Volcano extinct; fringing reef forms

(c) Subsidence of plate and volcanic island as reef builds up

(d) Continued subsidence; reef completely covers buried volcanic island

# SEDIMENTATION IN THE OCEAN

Almost everywhere that oceanographers search the seafloor they find a blanket of sediment. The muds and sands round and cover the topography of basalt originally formed at mid-ocean ridges. The ceaseless sedimentation in the world's oceans modifies the structures formed by plate tectonics and creates its own topography at sites of rapid deposition. The sediment is mainly of two kinds: terrigenous muds and sands eroded from the continents, and biochemically precipitated shells of organisms that live in the sea. In parts of the ocean near subduction zones, sediments derived from volcanic ash and lava flows are abundant. In tropical arms of the sea where evaporation is intense, evaporite sediments are deposited.

## Sedimentation on Continental Margins

Terrigenous sedimentation on the continental shelf is produced by the same forces that form beaches: waves and tides. The waves of large storms and hurricanes move sediment over the shallow and moderate depths of the shelf, and tidal currents also flow over the shelf. The waves and currents distribute the sediment brought in by rivers into long ribbons of sand and layers of silt and mud.

Biochemical sedimentation on the shelf results from the buildup of layers of the calcium carbonate shells of clams, oysters, and many other organisms living in shallow waters. Most of these organisms cannot tolerate muddy waters and are found only where terrigenous materials are minor or absent, such as along the extreme southern coast of Florida or off the coast of Yucatan in Mexico. Here coral reefs thrive and organisms build up large thicknesses of carbonate sediment.

## Deep-Sea Sedimentation

Far from the continental margins, fine-grained terrigenous and biochemically precipitated particles suspended in seawater slowly settle from the surface to the bottom. These open-ocean sediments, called **pelagic sediments,** are characterized by great distance from continental margins, fine particle size, and a slow settling mode of deposition. The terrigenous materials are brownish and grayish clays, which ac-

**FIGURE 17.33** Scanning electron micrograph of foraminiferal ooze. *Scripps Institution of Oceanography, University of California, San Diego.*

cumulate on the seafloor at a very slow rate, a few millimeters every 1000 years. A small fraction, about 10 percent, may be blown by the wind to the open ocean. Winds from the Sahara Desert, for example, blow much dust, silt, and fine sand into the eastern Atlantic off the coast of Africa. Windblown volcanic ash may be deposited downwind from subduction-zone volcanoes.

The most abundant biochemically precipitated calcium carbonate particles of pelagic sediments are the shells of foraminifera, tiny single-celled animals that float in the surface waters of the sea (see Figure 7.18). They fall to the bottom after they die and can no longer remain afloat. There they accumulate as **foraminiferal oozes,** sandy and silty sediments composed of foraminiferal shells (Figure 17.33).

Foraminiferal oozes are abundant at depths of less than about 4 km but rare on the deeper parts of the ocean floor. This cannot be because of a lack of shells, for the surface waters are full of them everywhere and the living foraminifera are unaffected by the bottom far below. The explanation for the absence of carbonate oozes below a certain depth, called the **carbonate compensation depth** (usually abbreviated as CCD), is the dissolution of shells in deep seawater (Figure 17.34). The deeper waters of the ocean are colder, are under higher pressure, and contain more dissolved carbon dioxide than shallower waters. These factors make calcium carbonate more

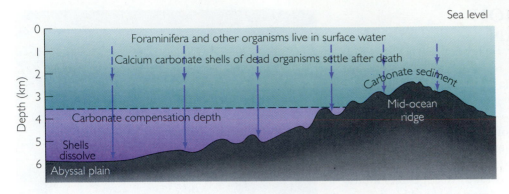

**FIGURE 17.34** The carbonate compensation depth is the level in an ocean below which the calcium carbonate of foraminifera and other shelled organisms that have settled from surface waters dissolves.

soluble in deeper waters than in shallow ones. As the shells of dead foraminifera fall to the bottom below the CCD, they enter an environment undersaturated with respect to calcium carbonate and dissolve.

Another kind of biochemically precipitated sediment, **silica ooze,** is produced by sedimentation of the silica shells of diatoms and radiolaria. Diatoms are green, unicellular algae found in abundance in the surface waters of the oceans. Radiolaria are foraminifera that secrete shells of silica instead of calcium carbonate. After burial on the seafloor, silica oozes are cemented into the siliceous rock, chert.

Some components of pelagic sediments are formed by chemical reactions of seawater with sediment on the seafloor. The most prominent examples are manganese nodules, which are black, lumpy accumulations ranging from a few millimeters to many centimeters across. These nodules cover large areas of the deep ocean floor, as much as 20 to 50 percent of the Pacific. Rich in nickel and other metals, they are a potential commercial resource once we are able to mine them from the seafloor economically.

## DIFFERENCES IN THE GEOLOGY OF OCEANS AND CONTINENTS

Our studies of the ocean floor have given us an appreciation of the differences between continental geology and submarine geology. Whereas tectonics and erosion shape the continents, volcanism and sedimentation predominate in the ocean. Volcanism creates island groups, such as the Hawaiian Islands, in the middle of the oceans; arcs of volcanic islands near deep oceanic trenches; and mid-ocean ridges. Sedimentation shapes much of the rest of the ocean floor.

Soft sediments of mud and calcium carbonate blanket the low hills and plains of the sea bottom and accumulate on oceanic plates as they spread from mid-ocean ridges. As the plates move farther and farther from a ridge, they accumulate more and more sediment. Eventually the plates are swallowed by subduction zones, which destroy the oceanic sediment record by metamorphism and melting.

The oceans have no folded and faulted mountains like those on the continents. Instead, plate-tectonic deformation is restricted to the faulting and volcanism found at mid-ocean ridges and at subduction zones. Weathering and erosion are much less important in the oceans than on the land because there are no efficient fragmentation processes, such as freezing and thawing, or major erosive agents, such as streams. Deep-sea currents can erode and transport sediment but cannot effectively attack basaltic plateaus or hills.

Because tectonic deformation, weathering, and erosion are minimal over much of the seafloor, many more details of the geologic record are preserved in layers of oceanic sediment than in continental sediments. But the oldest parts of the rock record are continuously erased by subduction. The oldest sediments preserved on today's ocean floor are Jurassic, about 150 million years old; they lie at the western edge of the Pacific Plate. In the next million years they too will disappear down a subduction zone. It takes about 150 million years, on average, for the crust created at mid-ocean ridges to spread across an ocean and come to a subduction zone.

This brief survey of the oceans shows that they are geologically complex, with distinctive structures, topographies, and sediments. Our knowledge of the oceans is still in its infancy; much of what we know was only discovered in the past few decades. Much more remains to explore as we invent new ways to sound, map, and sample the ocean floor.

# SUMMARY

**What processes shape shorelines?** At the edge of the sea, waves and tides, interacting with tectonics, control the formation and dynamics of shorelines, from beaches and tidal flats to uplifted rocky coasts. Waves are generated by winds blowing over the sea; as they approach the shore, they are transformed into breakers in the surf zone. Wave refraction results in longshore currents and longshore drift, which transport sand along beaches. Tides, generated by the gravitational attraction of the Moon and Sun on the water of the oceans, are agents of sedimentation on tidal flats.

**What are the major components of continental margins?** Continental margins are made up of shallow continental shelves; continental slopes that descend more or less steeply into the depths of the ocean; and continental rises, gently sloping aprons of sediment deposited at the lower edges of the continental slopes and extending to abyssal plains farther out in the ocean. Waves and tides affect the continental shelves, but continental slopes are shaped by turbidity currents, deep-water currents formed as slumps and slides on the continental slope create turbid suspensions of muddy sediment in bottom waters. Submarine canyons, submarine fans, and abyssal plains are also produced by turbidity currents. Active continental margins are formed where oceanic lithosphere is subducted beneath a continent, and passive continental margins are formed where rifting and seafloor spreading carry continental margins away from plate boundaries.

**How is the deep seafloor formed?** The deep seafloor is constructed by volcanism at mid-ocean ridges and at oceanic hot spots such as Hawaii and by deposition of fine-grained clastic and biochemically precipitated sediments. Mid-ocean ridges are the sites of seafloor spreading and the extrusion of basalt, which produces new oceanic lithosphere. Deep-sea trenches are formed as oceanic lithosphere is pulled downward into a subduction zone. Isolated, submerged seamounts and guyots, volcanic islands, plateaus, and abyssal hills are all accumulations of volcanic rock, most of which are mantled by sediment. Coral reefs form as volcanic islands are fringed with reefs constructed by organisms, which then continue to build the island and keep it at the surface when the volcano becomes extinct and subsides below sea level. Pelagic sediments consist of reddish-brown clays and foraminiferal and silica oozes composed of the biochemically precipitated calcium carbonate and silica shells of microscopic organisms living in surface ocean waters.

# KEY TERMS AND CONCEPTS

world ocean (p. 370)
coast (p. 370)
shoreline (p. 370)
swell (p. 371)
wavelength (p. 372)
wave height (p. 372)
period (p. 372)
surf (p. 372)
surf zone (p. 372)
swash (p. 373)
backwash (p. 373)
wave refraction (p. 373)
longshore drift (p. 374)
longshore current (p. 374)
tide (p. 375)
spring tide (p. 376)
neap tide (p. 376)

tidal surge (p. 376)
flood tide (p. 376)
ebb tide (p. 376)
tidal flat (p. 376)
offshore (p. 377)
foreshore (p. 378)
backshore (p. 378)
stack (p. 379)
wave-cut terrace
    (p. 379)
spit (p. 381)
barrier island (p. 381)
estuary (p. 382)
continental shelf (p. 384)
continental slope (p. 384)
submarine canyon (p. 384)
continental rise (p. 385)

abyssal plain (p. 385)
seamount (p. 385)
continental margin
    (p. 390)
passive margin (p. 390)
active margin (p. 390)
turbidity current (p. 390)
submarine fan (p. 391)
turbidite (p. 391)
guyot (p. 394)
atoll (p. 395)
fringing reef (p. 396)
pelagic sediments (p. 397)
foraminiferal ooze (p. 397)
carbonate compensation depth
    (p. 397)
silica ooze (p. 398)

# EXERCISES

1. How are ocean waves formed?

2. How does wave refraction work to concentrate erosion at headlands?

3. Where do turbidity currents form?

4. Where and how is the deep seafloor created by volcanism?

5. What plate-tectonic process is responsible for deep-sea trenches?

6. Describe the formation of a coral reef in the open ocean.

7. What are pelagic sediments?

8. Along what kinds of continental margins do we find broad continental shelves?

# THOUGHT QUESTIONS

1. Over a 100-year period the southern tip of a long, narrow, north-south beach has become extended about 200 m to the south by natural processes. What shoreline processes would have caused this extension?

2. After a period of calm along a section of the eastern shore of North America, a severe storm with high winds passes over the shore and out to sea. Describe the state of the surf zone before, during, and after the storm passes.

3. Why would you want to know the timing of high tide if you wanted to observe a wave-cut terrace?

4. How might plate tectonics account for the contrast between the broad continental shelf off the eastern coast of North America and the narrow, almost nonexistent shelf off the western coast?

5. A major corporation hires you to find valuable mineral deposits associated with hydrothermal springs on the seafloor. What kinds of places would you explore?

6. There is very little sediment to be found on the floor of the central rift valley of the Mid-Atlantic Ridge. Why might this be so?

7. A plateau rising from the deep ocean floor to within 2000 m of the surface is mantled with foraminiferal ooze, whereas the deep seafloor below the plateau, about 5000 m deep, is covered with reddish-brown clay. How can you account for this difference?

8. Bermuda is a coral reef island in the western Atlantic Ocean. Describe the rocks you might encounter if you were to drill a very deep hole into this island.

9. What kinds of sediment might you expect to find in the Peru-Chile Trench off the coast of South America?

# SUGGESTED READINGS

Anderson, Roger. 1986. *Marine Geology*. New York: Wiley.

Dolan, R., and Lins, H. 1987. Beaches and barrier islands. *Scientific American* (July):146.

Hardisty, J. 1990. *Beaches, Form and Process*. New York: Harper Collins Academic.

Hollister, C. D., A. R. M. Nowell, and P. A. Jumars. 1984. The dynamic abyss. *Scientific American* (March):42.

Kennett, James. 1981. *Marine Geology*. Englewood Cliffs, N.J.: Prentice-Hall.

Macdonald, K. C., and B. P. Luyendyk. 1981. The crest of the East Pacific Rise. *Scientific American* (May):100.

Pilkey, Orrin H., and W. J. Neal. 1988. Coastal geologic hazards. In R. E. Sheridan and J. A. Grow (eds.), *The Geology of North America*, vol. I-2, *The Atlantic Continental Margin: U.S.* Boulder, Colo.: Geological Society of America, pp. 549–556.

# 3

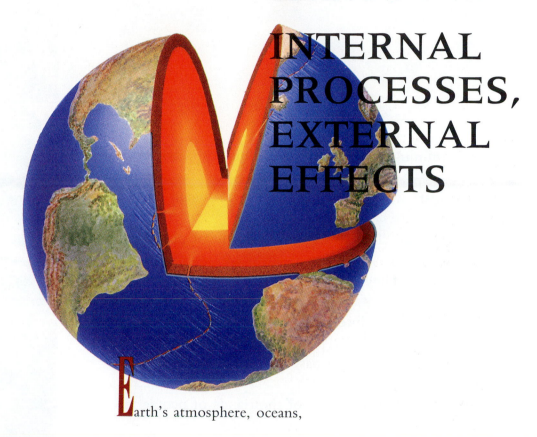

# INTERNAL PROCESSES, EXTERNAL EFFECTS

Earth's atmosphere, oceans, and crust originate in the deep interior. So do the forces that deform the rocky skin of our planet. These forces, driven by the energy supplied by Earth's internal heat, create plates and keep them in motion. We see the effects of these forces in our daily lives as volcanic eruptions and earthquakes, as well as in the grandeur of mountain belts. The geologist studies plate motions, earthquakes, volcanoes, and the deformation of the crust to infer the properties of the deep interior and the nature of the forces at work there.

# EARTHQUAKES

By examining the pattern of earthquakes, seismologists (scientists who study earthquakes) provided one of the essential clues to the development of plate-tectonic theory. They found that most earthquakes occur at plate boundaries, the intensely strained zones where plates collide, split apart, or slide past each other. Because earthquakes can kill hundreds of thousands of people in minutes, many seismologists dedicate their careers to reducing their destructiveness. By estimating the likelihood that an earthquake will occur in specific geographic locations, they can supply information that enables local governments to mandate standards for the location and design of buildings that can withstand earthquakes. A long-term goal is to predict earthquakes; accurate and timely warnings could save millions of lives over a few decades.

## 18

Elevated Interstate Highway 880 in Oakland, California, collapsed following the Loma Prieta earthquake of October 17, 1989. The top deck fell onto the lower deck, killing 41 people trapped in their cars. *James Sugar/ Black Star.*

When Voltaire set out to describe the destructiveness of an earthquake, he drew on the detailed reports that were available on the great earthquake of 1755 in Lisbon, Portugal (Figure 18.1). The Lisbon earthquake, which killed 30,000 people, is remembered not only for its tragic proportions but also for its role in ushering in the Age of Enlightenment, a period in which observation and reasonable explanation of natural events replaced dogma and superstition.

No written description, however, can replace the personal experience of the violent ground movement of a major earthquake. Outdoors, you would be knocked down and the loud noise of buildings being destroyed would be deafening. Bridges would sway and collapse. Indoors, you would be shaken out of bed or thrown wall to wall in a hallway. Furniture would slide all over a room, chandeliers would sway and fall, window glass would break and spray you with shards, dishes and groceries would crash to the floor from shelves. A poorly constructed building would collapse, the floors above falling on you to crush or entrap you. The collapse of buildings is the major cause of earthquake casualties. The video pictures of the death and destruction that accompanied the recent earthquakes in Mexico, Armenia, Iran, California, and the Philippines brought the dimensions of such cataclysms to hundreds of millions of viewers worldwide for the first time (Figure 18.2).

*"The sea rose boiling in the harbour and broke up all the craft harboured there; the city burst into flames, and ashes covered the streets and squares; the houses came crashing down, roofs piling up on foundations, and even the foundations were smashed to pieces. Thirty thousand inhabitants of both sexes and all ages were crushed to death under the ruins.*

*Voltaire,* Candide.

**FIGURE 18.1** A contemporary fanciful print portraying the destruction of Lisbon by the earthquake and tsunami of November 1, 1755. T. C. Lotters geographischem Atlas, nach 1755 (Jan Kozak, Prague). (Suggested by L. C. Pakiser, USGS.)

**FIGURE 18.2** The Hyatt Hotel collapsed in the Digdig earthquake of July 16, 1990, in the Philippines. *Peter Yanev/EQE International.*

# WHAT IS AN EARTHQUAKE?

An **earthquake** is a shaking or vibration of the ground. An earthquake occurs when rocks being deformed suddenly break along a fault. The two blocks of rock on both sides of the fault slip suddenly, setting off the ground vibrations. This slippage occurs most commonly at plate boundaries—regions of the Earth's crust or upper mantle where most of the ongoing deformation takes place.

The earthquake on the San Andreas fault that devastated San Francisco in 1906 received the most detailed study of any earthquake up to that time and provided the information that enabled the investigators of that catastrophe to advance the **elastic rebound theory.** To visualize what happens in an earthquake, according to the theory, imagine the following experiment carried out across a fault between two hypothetical crustal blocks (Figure 18.3). Suppose that surveyors had painted straight lines on the ground, running perpendicular across the fault from one block to the other, as in Figure 18.3a. The two blocks are being pushed in opposite directions by plate motions. Because they are pressed together by the weight of the overlying rock, friction locks them together. They do not move, just as a car does not move when the emergency brake is engaged. Therefore, instead of slipping along the fault, the blocks are deformed or bent near the fault, as shown by the bent lines in Figure 18.3b. As the plate movements continue to push the blocks in opposite directions, the slow movement of deformation continues, perhaps for decades. The strain in the rocks, evidenced by the bending of the survey lines, builds up until the fric-

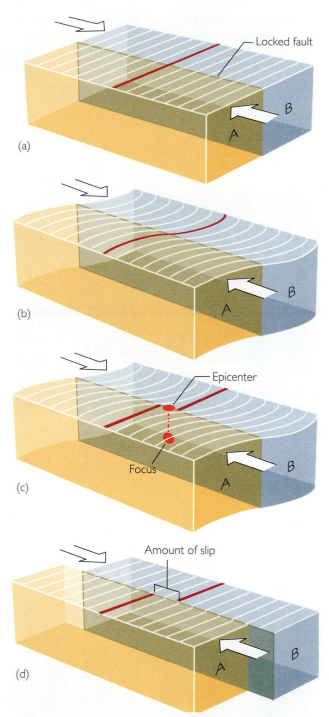

**FIGURE 18.3** The elastic rebound theory of an earthquake. (a) Two crustal blocks, A and B, are slowly forced to slide past each other. (b) Friction along the fault prevents slip, and the crust is deformed. (c) Strain builds up until the "frictional lock" is broken at the focus and a rupture occurs. The focus of an earthquake is the site of an initial slip on the fault. The epicenter is the point on the surface directly above the focus. (d) The rupture spreads and an earthquake slip occurs over a section of the fault.

tional bond that locks the fault can no longer hold at some point on the fault, and it breaks. The rupture extends over a section of the fault, and the blocks slip suddenly along the fault (Figure 18.3c). Figure 18.3d shows that after the earthquake the two blocks have been displaced. The distance of the displacement is called the **slip.** Faulting in a major earthquake can

**FIGURE 18.4** The great San Francisco earthquake of 1906 was caused by slip along the San Andreas fault. The offset fence here, near Bolinas, California, shows a slip of nearly 3 m. *G. K. Gilbert.*

extend for as much as 1000 km, and the slip of the two blocks can be as large as 15 m (Figure 18.4). The point at which the slip initiates is the **focus** of the earthquake (Figure 18.3c). The **epicenter** is the geographic point on the Earth's surface directly above the focus. For example, you might hear in a news report: "Seismologists at the California Institute of Technology report that the epicenter of last night's destructive earthquake in California was located 35 miles southeast of Los Angeles. The depth of the focus was 10 kilometers." Although the focus of the 1988 Armenian earthquake was 10 km below the surface, the fault broke through to the surface and produced the fault scarp shown in Figure 10.29.

When the blocks slip suddenly at the time of the earthquake, intense vibrations called **seismic waves** (from the Greek *seismos,* meaning "shock" or "earthquake") travel outward from the focus much as waves ripple outward from the spot where a stone is dropped in a still pond (Figure 18.5). Near the epicenter the waves can cause the ground to shake violently.

The strain that slowly builds up over decades when two blocks are pushed in opposite directions is analogous to the strain exerted on a rubber band when it is slowly stretched. The sudden release of strain in an earthquake, signaled by slip along a fault and the release of seismic waves, is analogous to the violent movement that occurs when the rubber band breaks. Elastic energy is actually being stored in the

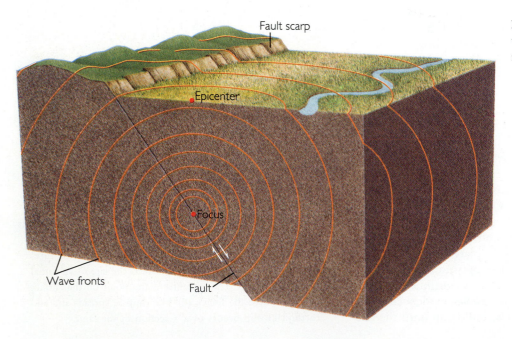

**FIGURE 18.5** Seismic waves radiate from the focus of an earthquake.

stretched rubber band, and it is this energy that is suddenly released in the backlash. In the same way elastic energy accumulates and is stored over many decades in rocks under strain; the energy is released at the moment the fault ruptures and is radiated as seismic waves in the few minutes of an earthquake.

# STUDYING EARTHQUAKES

As in any experimental science, instruments and field observations provide the basic data used to study earthquakes. These data enable the investigator to analyze the seismic waves that originate in earthquakes, locate earthquakes, determine their sizes and numbers, and understand their relationships to faults.

## Seismographs

The modern **seismograph,** which records the seismic waves generated by earthquakes, is the most important tool for studying earthquakes and probing Earth's deep interior. The seismograph is to the Earth scientist what the telescope is to the astronomer—a tool for peering into inaccessible regions. The ideal seismograph would be a device fixed to a stationary frame not attached to the Earth, so that when the ground shook its movements could be measured by the changing distance between the seismograph, which did not move, and the vibrating ground, which did move. Because we have no way to establish a seismograph that is not attached to the Earth, we compromise.[1] A mass is attached to the Earth so loosely that the ground can vibrate without causing much motion of the mass. One way to achieve this loose attachment is to suspend the mass from a spring (Figure 18.6a). When the seismic waves move the ground up and down, the mass tends to remain stationary because of its inertia (an object at rest tends to stay at rest), but the mass and the ground move relative to each other because the spring can compress or stretch. In this way the vertical displacement of the Earth caused by seismic

waves can be recorded by a pen on chart paper. A seismograph that has its mass suspended on hinges like a swinging gate (Figure 18.6b) can record the horizontal motions of the ground. In modern seismographs the most advanced electronic technology is used to amplify this motion before it is recorded, and these instruments can detect ground displacements as small as $10^{-8}$ cm—an astounding feat, considering that such small displacements are of atomic size.

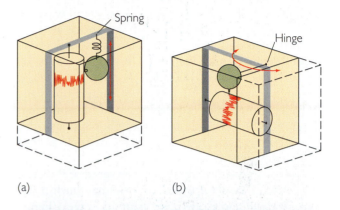

(a)          (b)

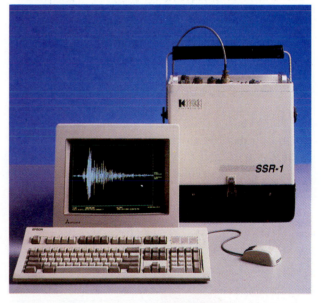

(c)

**FIGURE 18.6** Seismographs record (a) vertical or (b) horizontal motion. Because of its loose coupling to the Earth through the spring (a) or hinge (b) and its inertia, the mass does not keep up with the motion of the ground. The pen traces the differences in motion between the mass and the ground, in this way recording vibrations of seismic waves. Modern seismographs (c) amplify the motion electronically. *Kinemetrics.*

---

[1]Modern space technology may remove this limitation. NASA has placed in orbit Global Positioning Satellites that make it possible for ground-based instruments to determine precise locations on Earth's surface. In 1992, using a network of such instruments, seismologists in southern California were able to determine the actual motion of the ground in the vicinity of the fault during an earthquake.

## Seismic Waves

Install a seismograph anywhere and within a few hours it will record the passage of seismic waves generated by earthquakes somewhere on Earth. The waves will have traveled from the earthquake focus through the Earth and arrived at the seismograph in three distinct groups. The first waves to arrive are called primary waves or **P waves** for that reason. The secondary or **S waves** follow. Both P and S waves travel through the Earth's interior. Finally come the **surface waves,** which travel around the Earth's surface (Figure 18.7).

P waves in rock are analogous to sound waves in air, only P waves travel through solid rock at about 5 km per second, which is faster than sound waves travel through air. Like sound waves, P waves are *compressional waves,* so called because they travel through material as a succession of compressions and expansions; that is, squeezings and unsqueezings (Figure 18.8). P waves can be thought of as push-pull waves: they push or pull particles of rock as they move along the path of travel.

S waves travel at about half the speed of P waves. They are called *shear waves* because they push material at right angles to their path of travel (Figure 18.9). Surface waves are confined to the Earth's surface and outer layers because, like waves on the ocean, they need a free surface to ripple in order to exist (Figure 18.10). Their speed is slightly less than that of S waves.

Seismic waves have been felt and their destructiveness witnessed throughout human history, but not until the close of the nineteenth century were seismologists able to devise instruments to record such waves so that they could be analyzed. Seismic waves enable us to locate earthquakes and determine the nature of faulting, and they provide our most important tool for probing the Earth's deep interior.

## Locating the Epicenter

If we can locate the epicenters of the thousands of earthquakes that occur each year, we will have taken the first step toward understanding them. The principle involved in locating a quake's epicenter is quite similar to the principle used to deduce the distance to a lightning bolt on the basis of the time interval between the flash of light and the sound of thunder. The lightning flash may be likened to the P waves of earthquakes and the thunder to the slower S waves. The time interval between the arrival of P and S waves depends on the distance the waves have traveled, as indicated in Figure 18.11. This relationship is established independently on the basis of data from earthquakes or underground nuclear explosions whose locations are known. To determine the approximate distance to an epicenter, seismologists read from a seismogram the amount of time that elapsed between the arrival of the first P waves and the later arrival of the S waves. They then use a table or a graph like that in Figure 18.11 to get the distance from the seismograph to the epicenter. If they know the distances from three or more stations, they can pinpoint the epicenter. They can also deduce the time of the shock at the epicenter because the arrival time of the P waves at each station is known, and from a graph or table it is possible to determine how long the waves took to reach the station. Actually the process just described is now carried out with repeated iterations in a computer until the data from a large number of seismograph stations agree on where the epicenter is, the time the earthquake began, and the depth of its focus.

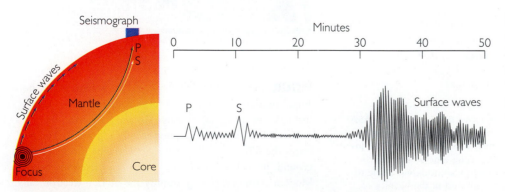

**FIGURE 18.7**
Seismographic recording of P, S, and surface waves from a distant earthquake. The cross section shows the paths followed by the three types of waves.

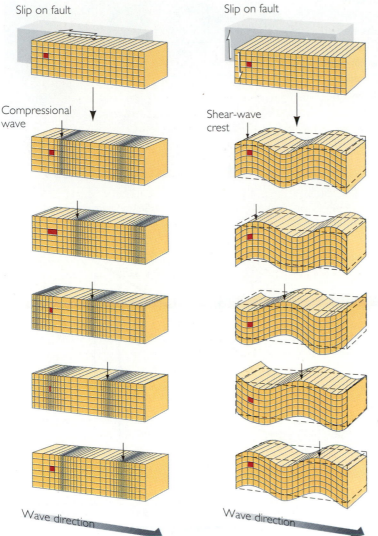

Slip on fault

Compressional wave

Wave direction

Slip on fault

Shear-wave crest

Wave direction

*Far Left:*

**FIGURE 18.8** Stages in the deformation of a block of material with the passage of compressional, or P, waves through it. The undeformed block is shown at the top. In the sequence from top to bottom, a wave of compression, marked by an arrow, moves through the block with the P-wave velocity. It is followed by an expansion, and any small piece of matter, like the marked square, shakes back and forth in response to alternating compressions and expansions as the wave train moves through. A sudden push (or pull) in the direction of wave propagation, indicated by slip on the fault, would set up P waves.

*Left:*

**FIGURE 18.9** Stages in the deformation of a block of material with the passage of shear, or S, waves through it. A wave crest, marked by an arrow, moves through the block with the S-wave velocity. Any small piece of matter, like the marked square, shakes sideways and experiences a shearing deformation (from a square to a parallelogram in the figure) as the shear wave passes through. A sudden shear displacement at right angles to the direction of wave propagation would set up S waves.

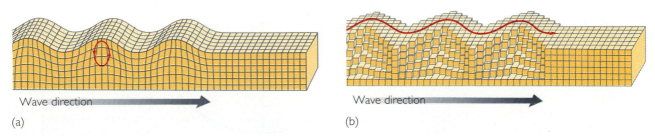

Wave direction

(a)

Wave direction

(b)

**FIGURE 18.10** In one type of surface wave (a), the ground vibrates in a rolling, elliptical motion that dies down with depth beneath the surface. In another type of surface wave (b), the ground shakes sideways, with no vertical motion. (After Bruce A. Bolt, *Nuclear Explosions and Earthquakes: The Parted Veil,* San Francisco, W. H. Freeman, 1976, p. 49.)

(a) The contour numbers in this diagram give the time interval (in minutes) between the arrival of first P and first S waves at successive distances from the epicenter of an earthquake. Because P waves travel about twice as fast as S waves, the further the waves travel, the wider becomes the interval between the arrival of the different waves. Seismographic station A, for example, which is closer to the epicenter, recorded a 3-minute interval; whereas more distant station B recorded an 8-minute interval. The intervals between P and S waves are a critical factor in interpreting seismographic readings.

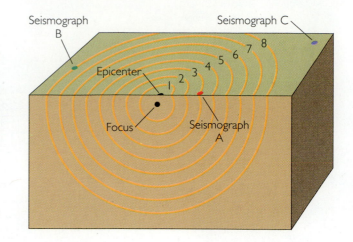

(b) Seismic time-travel curves, like the ones shown in this graph, are basic tools for determining the distance of an earthquake from the seismograph that records it. Geologists at station A, where a 3-minute interval between P and S waves was recorded, can match this interval to the corresponding space between the P and S curves on the graph and determine that their distance from the epicenter was 1500 km. In the same way, seismic recordings at station B, with an 8-minute interval between P and S waves, yields a distance of 5600 km, and at station C, with an 11-minute interval, a distance of 8600 km.

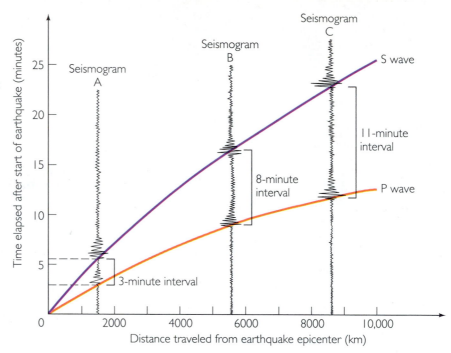

(c) Knowing the distance from the epicenter of three different stations, geologists can pinpoint the location of the epicenter using a map and some simple geometry. They draw three circles, each one centered on one of the three stations and each having a radius equal to the station's distance from the epicenter. The epicenter lies at the single point where the three circles intersect. (The epicenter and depth of focus are now determined by a computer that simulates this graphical method.)

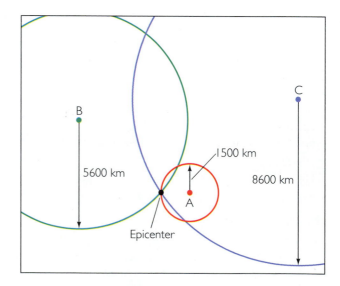

**FIGURE 18.11** From readings at different seismographic stations, geologists can locate the epicenter of an earthquake.

## Measuring the Size of an Earthquake

The ability to locate earthquakes is only one step on the way to understanding them. The seismologist must also determine their sizes, for size is the main factor in destructiveness. The sizes and numbers of earthquakes in a given area indicate the degree of tectonic activity.

In 1935 Charles Richter, a California seismologist, devised a simple procedure that is now used all over the world to measure the size of an earthquake. Richter studied astronomy as a young man and learned that astronomers assign each star a magnitude as a measure of its brightness. He adapted that method to earthquakes, assigning each earthquake a number (now called the **Richter magnitude**) as a measure of its size. The Richter magnitude depends on the amplitude (size) of the ground movement caused by seismic waves. To simplify matters, Richter chose to base his measurement on the largest ground motion recorded by a seismograph (Figure 18.12). Just as the brightnesses of stars vary over a huge range, so do the sizes of earthquakes. For this reason Richter needed to compress his scale. He did so by taking the logarithm of the largest movement of the ground. With this procedure two earthquakes that differ in size of ground motion by a factor of 10 differ in magnitude by 1 Richter unit (see Box 18.1). The ground motion of an earthquake of magnitude 3 is 10 times that of an earthquake of magnitude 2.

A number of adjustments had to be worked into the calculation of magnitude. For instance, just as sound weakens with increasing distance from its source, the seismograph reading must be adjusted by a factor that takes into account the weakening of seismic waves as they spread away from the focus. Richter provided tables to make this adjustment. Thus seismologists all over the world can study their records and in a few minutes come up with nearly the same value for the magnitude of an earthquake no matter how close or far away their instruments are from the focus. Richter also showed how to relate his magnitude scale to the energy released in an earthquake, thus tying his somewhat arbitrary scale to the physical reality of the amount of energy in an explosion or the energy needed to knock a building down.

The Richter magnitude does not necessarily describe the destructiveness of a particular earthquake because a magnitude 8 earthquake 2000 km from the nearest city might cause no damage, while a magnitude 6 quake immediately beneath a city could cause serious damage. To measure an earthquake's intensity by its observed effects on people and structures rather than its actual size, seismologists often use the **Modified Mercalli Scale.** In this scale earthquake intensities range over values from I (very weak) to XII (very intense). (Table 18.1 gives an abbreviated description of the 12 levels of Modified Mercalli intensity.) Intensity I is characterized as not felt by most people, whereas intensity XII represents total damage. Intensities are assigned after field investigations. The intensity assigned to the San Francisco earthquake of 1906 is XI.

## Determining Fault Mechanisms from Earthquake Data

An earthquake occurs. Seismograms from many stations are analyzed. The epicenter and magnitude are determined. Members of a team then go to the epicenter to examine the faulting. They want to see how the orientation of the fault plane and the direction of slip fit into the regional pattern of crustal forces. Was the earthquake the result of normal, thrust, or strike-slip faulting (mechanisms summarized in Figure 18.13)? What can be seen in the field? Nothing, if the focus was so far below the surface that the fault does not show. Even in that case seismologists have learned to deduce the kind of faulting that occurred from information in the seismograms. This ability is especially convenient, for very few earthquake faults break through to the surface so that the slip direction and fault orientation can be observed directly. This is how they do it.

By now so many seismographs have been established around the world that they literally surround

**FIGURE 18.12** The maximum amplitude of the ground shaking indicated on the seismographic record is used to assign a Richter magnitude to an earthquake.

Maximum amplitude

Arrival of seismic waves

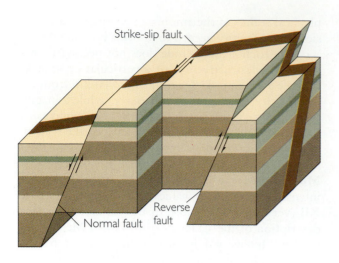

**FIGURE 18.13** The three main types of fault movements that trigger earthquakes. (After Bruce A. Bolt, *Earthquakes,* New York, W. H. Freeman, 1993, p. 76.)

**TABLE 18.1**

## Modified Mercalli Earthquake Intensity Scale

| INTENSITY | DESCRIPTION |
|---|---|
| I | Not felt except by a very few under especially favorable conditions. |
| II | Felt only by a few persons at rest, especially on upper floors of buildings. Delicately suspended objects may swing. |
| III | Felt quite noticeably by persons indoors, especially on upper floors of buildings. Many people do not recognize it as an earthquake. Standing motor cars may rock slightly. Vibration similar to the passing of a truck. Duration estimated. |
| IV | Felt indoors by many, outdoors by few during the day. At night, some awakened. Dishes, windows, doors disturbed; walls make cracking sound. Sensation like heavy truck striking building. Standing motor cars rocked noticeably. |
| V | Felt by nearly everyone; many awakened. Some dishes, windows broken. Unstable objects overturned. Pendulum clocks may stop. |
| VI | Felt by all, many frightened. Some heavy furniture moved; a few instances of fallen plaster. Damage slight. |
| VII | Damage negligible in buildings of good design and construction; slight to moderate in well-built ordinary structures; considerable damage in poorly built or badly designed structures; some chimneys broken. |
| VIII | Damage slight in specially designed structures; considerable damage in ordinary substantial buildings, with partial collapse. Damage great in poorly built structures. Fall of chimneys, factory stacks, columns, monuments, walls. Heavy furniture overturned. |
| IX | Damage considerable in specially designed structures; well-designed frame structures thrown out of plumb. Damage great in substantial buildings, with partial collapse. Buildings shifted off foundations. |
| X | Some well-built wooden structures destroyed; most masonry and frame structures destroyed with foundations. Rails bent. |
| XI | Few, if any (masonry) structures remain standing. Bridges destroyed. Rails bent greatly. |
| XII | Damage total. Lines of sight and level are distorted. Objects thrown into the air. |

# BOX 18.1 EARTHQUAKE MAGNITUDES, GROUND MOTION, AND ENERGY

Earthquake magnitude is a measure of the size of an earthquake. For each unit of magnitude, the amplitude of ground motion (or seismic waves) increases by a factor of 10. Thus a magnitude 6 earthquake produces ground motions that are 10 times greater than those of a magnitude 5 earthquake. The energy released as seismic waves by an earthquake increases even faster, by a factor of 33 for each unit of magnitude. As the table below suggests, there are limits to the sizes of detected earthquakes, but there are no well-established limits to the possible sizes of earthquakes.

| MAGNITUDE | ENERGY RELEASED (MILLIONS OF ERGS) | ENERGY EQUIVALENCE AND EFFECT |
|---|---|---|
| −2 | 600 | 100-watt light bulb left on for a week |
| −1 | 20,000 | Smallest earthquakes detected |
| 0 | 600,000 | Seismic waves from 1 pound of explosives |
| 1 | 20,000,000 | A two-ton truck traveling 75 miles per hour |
| 2 | 600,000,000 | Not felt but recorded |
| 3 | 20,000,000,000 | Smallest earthquake commonly felt |
| 4 | 600,000,000,000 | Seismic waves from 1000 tons of explosives |
| 5 | 20,000,000,000,000 | |
| 6 | 600,000,000,000,000 | Damage varies from slight to great, depending on quality of construction |
| 7 | 20,000,000,000,000,000 | |
| 8 | 600,000,000,000,000,000 | 1906 San Francisco (magnitude = 8.3) |
| 9 | 20,000,000,000,000,000,000 | Largest recorded earthquake (magnitude = 8.9); destruction nearly total |
| 10 | 600,000,000,000,000,000,000 | Approximately all the energy used in the United States in one year |

(Modified from U.S. Geological Survey.)

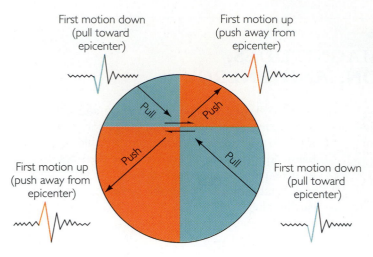

**FIGURE 18.14** The first motion of P waves arriving at seismographic stations is used to determine the orientation of the fault plane and the directions of slip. If the fault and slip were ⇣⇡ or ⇌ the first motions would be the same. Seismologists can choose between the two possibilities because they know the direction of the fault from additional information. For example, other epicenters of earthquakes on the same fault align along the fault and provide a trace of its direction.

the focus of any earthquake. Seismologists have found that in some directions from an earthquake, the very first seismic movement of the ground recorded by a seismograph is a push (an upward motion of the trace on the seismogram) away from the focus. On seismographs located in other directions, the initial ground movement may be a pull (downward motion of the trace) toward the focus. These differences reflect the fact that the slip on a fault looks like a push if you view it from one place but like a pull if you view it from another (Figure 18.14). From the distribution of initial motions, seismologists can deduce the orientation and dip of the fault plane and the direction of slip between the two blocks. Without surface evidence they can deduce from their instruments whether the crustal forces that triggered the earthquake were compressional, tensional, or shear.

## THE BIG PICTURE: EARTHQUAKES AND PLATE TECTONICS

With the tools for locating earthquakes, measuring their magnitudes, and deducing their fault mechanisms, seismologists have discerned patterns in the distribution of earthquakes of various types. They have explained these patterns within the framework of plate-tectonics theory—and thereby provided a tremendous boost to the theory.

A **seismicity** chart (Figure 18.15) shows the locations of the epicenters of almost 30,000 earthquakes that occurred over a six-year period. Earthquakes that originated at depths greater than about 100 km are shown in Figure 18.16. Seismologists have

known for decades that earthquakes tend to occur in "belts." One of the best known belts is the "Ring of Fire" surrounding the Pacific Ocean. In recent years it has become possible to detect the more numerous small earthquakes and to improve methods of locating epicenters. Seismic belts can now be defined so accurately that they can be correlated with geologic features. (Interestingly, the increase in the number of seismic observatories and the use of computers to store and analyze seismic data were stimulated by research done during negotiations for a nuclear test-ban treaty in the 1960s; the purpose was to determine whether small underground nuclear explosions could be detected and distinguished from earthquakes so that a violator of the treaty could not claim that a nuclear test was an earthquake. The seismologists were successful in distinguishing between the two.)

This is what the seismologists have found. The narrow belts of mid-ocean earthquakes coincide with mid-ocean ridge crests, as shown in Figure 18.17. When the topography of mid-ocean ridges is examined in detail, the ridges are often found to be segmented, the segments being offset by transform faults. Earthquake epicenters also line the transform faults between the offset ridge segments. Moreover, the fault mechanisms of the ridge-crest earthquakes, revealed by analysis of the first P-wave motion, are normal, with the faults striking parallel to the trend of the ridges. Normal faulting indicates that tensional (or pull-apart) forces are at work. This is why rift valleys develop in the ridge crests. The seismologists, using the independent evidence of their seismic records, found that mid-ocean ridges define plate boundaries where plates are being pulled apart. They also found that the earthquakes that coincide with transform faults show strike-slip mechanisms—just as one would expect where plates slide past each

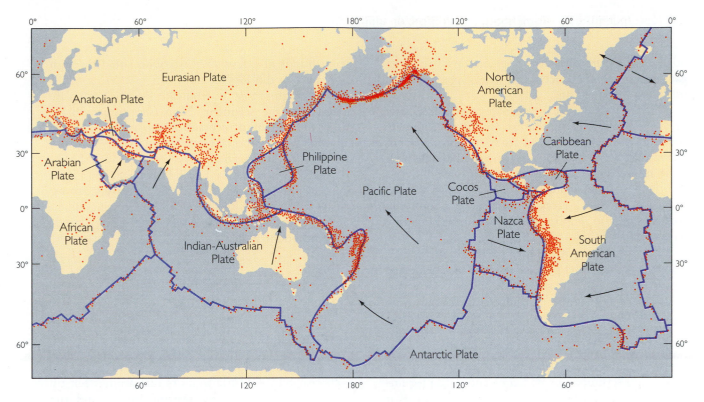

**FIGURE 18.15** Epicenters of some 30,000 earthquakes with focal depths between 0 and 700 km, recorded over a six-year period.

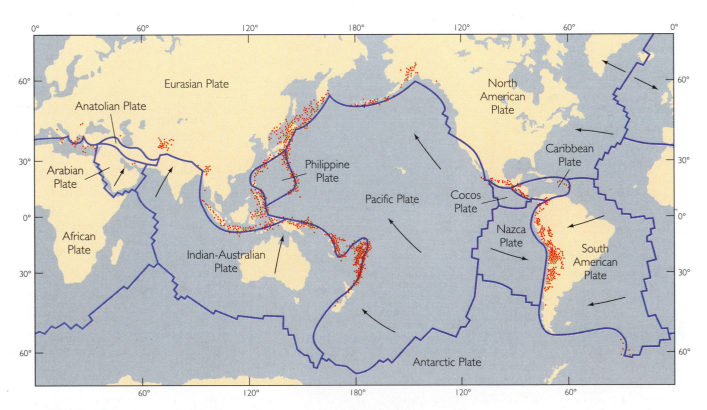

**FIGURE 18.16** Subset of earthquakes from Figure 18.15 with focal depths greater than 100 km. These deep-focus earthquakes typically originate near margins where plates collide and thus serve to identify such plates.

other in opposite directions. Seismology gives elegant support to the concept that plates spread apart at mid-ocean ridge crests.

The earthquakes that originate at depths greater than about 100 km (see Figure 18.16) are found to coincide with continental margins or island chains that are adjacent to ocean trenches and young volcanic mountains. The west coast of South America and the chains of islands that make up Japan and the Philippines are such areas. These features, shown in cross section in Figure 18.18, define a boundary where plates collide. The inclined plane of a subducted plate, where it plunges back into the mantle beneath an overriding plate, is marked by earthquake foci. An analysis of the initial P waves of these earthquakes reveals that the earthquakes in the area are produced by thrust faulting, signifying that compressive forces are at work, as would be the case at a collision boundary.

This global correlation among topography, geology, and seismicity provided essential data for understanding the many complex phenomena that occur at plate boundaries. It may seem straightforward now to draw a line through the seismic belts and so define plate boundaries, but this important advance could not have been made without the knowledge explosion in seismic data and the synthe-

sizing minds of a few scientists who sifted through those data in the 1960s.

Although most earthquakes occur at plate boundaries, the seismicity map (see Figure 18.15) shows that a small percentage originate within plates. Among these earthquakes are some of the most destructive in American history: New Madrid, Missouri (1812); Charleston, South Carolina (1886); Boston, Massachusetts (1755). Apparently, strong crustal forces can still occur and cause faulting, albeit rarely, within the lithospheric plates, far from modern plate boundaries.

# EARTHQUAKE DESTRUCTIVENESS

## Earthquakes: How Big and How Many?

Each year 800,000 little tremors that are not felt by humans are recorded by instruments around the world. About 100 earthquakes occur each year with Richter magnitudes between 6 and 7. A great earthquake, one with a Richter magnitude exceeding 8, occurs somewhere in the world about once every 5 to 10 years. It is a fortunate fact that most earthquakes

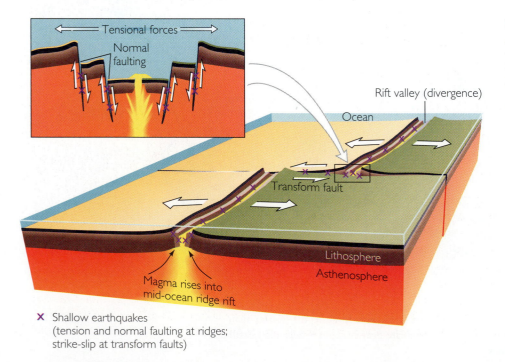

**FIGURE 18.17**
Earthquakes associated with two types of plate boundaries: divergent boundaries at ocean ridges and transform faults.

✗ Shallow earthquakes
(tension and normal faulting at ridges; strike-slip at transform faults)

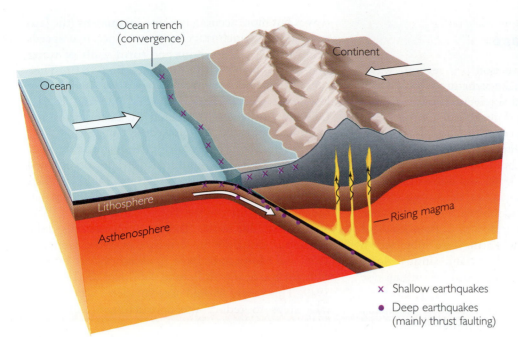

**FIGURE 18.18** Shallow- and deep-focus earthquakes at convergent plate boundaries. Deep-focus earthquakes occur in a subducted plate because of compression forces, and fault mechanisms show corresponding thrust fault patterns.

× Shallow earthquakes

● Deep earthquakes (mainly thrust faulting)

are small. Damage to buildings and other structures near the epicenter begins at magnitude 5 and increases to nearly total destruction when the magnitude is greater than 8. The Loma Prieta earthquake of 1989, which occurred on the San Andreas fault some 80 km south of San Francisco, was one of the costliest disasters in the history of the United States. Damage amounted to more than $6 billion, but fortunately only 62 people were killed. As damaging as this magnitude 6.9 earthquake was, it released less than a hundredth the energy of some truly great earthquakes, such as those in San Francisco (1906, 8.3), Tokyo (1923, 8.2), Chile (1960, 8.6), Alaska (1964, 8.6), and China (1976, 7.8). One California politician announced that the Loma Prieta earthquake had the positive benefit of warning Californians to prepare for the truly great earthquake that is bound to occur there.

Destructive earthquakes are even more frequent in Japan than in California. The recorded history of destructive earthquakes in Japan, going back 2000 years, has left an indelible impression on the Japanese people. Perhaps that is why Japan is the best prepared of any nation in the world to deal with earthquakes. In too many parts of the world authorities take a fatalistic attitude, as though nothing could be done in advance to mitigate the losses of an earthquake. They are wrong.

**FIGURE 18.19** These buildings in Niigata, Japan, toppled intact during an earthquake in 1964. Although the structures themselves were built to withstand earthquakes, the foundations failed because the soil became "liquefied" (see Chapter 11). *G. Hausner, California Institute of Technology; National Science Foundation.*

## How Earthquakes Cause Their Damage

Earthquakes cause destruction in several ways. Ground vibrations can shake structures so hard that they collapse. The ground accelerations near the epicenter of a very great earthquake can approach and even exceed the acceleration of gravity, so an object lying on the surface can literally be thrown into the air. Very few structures built by human hands can withstand such severe shaking, and those that do are severely damaged. People can find the strong motion of an earthquake traumatic in itself, especially when plaster and brick are falling down. Certain kinds of soil behave like a liquid when they are subjected to intense seismic shocks. The ground simply flows away, taking buildings, bridges, and everything else with it (Figure 18.19).

Earthquakes that occur near coasts occasionally generate the awesome waves commonly called tidal waves but more accurately named **tsunamis** (the Japanese term), which travel across the ocean at speeds of up to 800 km per hour and form walls of water that can be higher than 20 m (see Box 18.2). They cause tremendous damage when they sweep over low-lying coastal areas (Figures 18.20 and 18.21). Undersea avalanches triggered by earthquakes and volcanic explosions can also cause tsunamis.

Avalanches triggered by earthquakes take their toll. Fires ignited by ruptured gas lines or downed power lines are especially dangerous. Table 18.2 lists the human losses of some historical earthquakes. Of the 99,000 fatalities in the Tokyo earthquake of 1923, 38,000 were due to fire. Building collapse caused the enormous casualties in Armenia in 1988, China in 1976, and Morocco in 1960. Avalanches were responsible for the deaths in Peru in 1970 and the USSR in 1949. This record should serve to alert authorities to take steps to reduce the destructiveness of earthquakes.

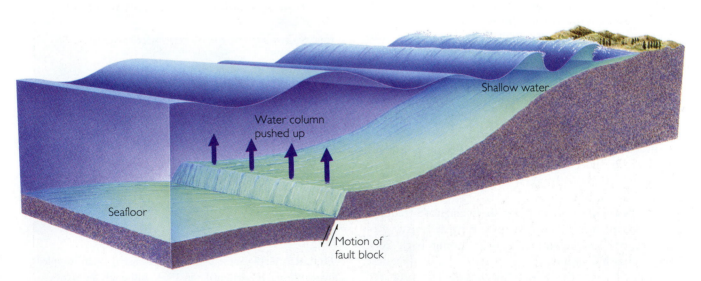

**FIGURE 18.20** Generation of a tsunami by fault movements caused by an earthquake on the seafloor. Movement of the seafloor due to an earthquake produces a surge of water, which oscillates and flows out as a long sea wave or tsunami. Such a wave can increase in height manyfold when it enters shallow coastal waters.

## BOX 18.2 TSUNAMIS

A tsunami is a sea wave that is triggered by an undersea event such as an earthquake or landslide or by the eruption of an oceanic volcano. (A popular name for a tsunami is a "tidal wave"—an unfortunate usage, because tsunamis have nothing to do with tides.) In some complicated way, these events push or displace a large piece of the adjacent ocean, and this disturbance is transformed into a wave that travels across the ocean at speeds of up to 800 km per hour. In mid-ocean, where the water is deep, tsunamis are hardly noticeable, but when they approach shallow coastal waters the waves steepen until they become destructively high, perhaps more than 20 m. As we mentioned in Chapter 5, the tsunami generated by the explosion of Krakatoa in Indonesia in 1883 reached 40 m and drowned 36,000 people on nearby coasts. In 1960 an earthquake off the coast of Chile excited a tsunami that crossed the Pacific and caused property damage and loss of life in Japan. An earthquake off the coast of Alaska in 1964 triggered a tsunami that caused tens of millions of dollars of damage in nearby settlements; it was still destructive when it reached Crescent City, California, where it killed several people.

Disastrous tsunamis are a serious threat to 22 countries surrounding the Pacific Ocean. They have also been known to damage Atlantic coastal countries, though they occur less frequently in the Atlantic than in the Pacific. Because tsunamis can be so destructive, programs have been initiated (and should be expanded) to reduce their damage. The Pacific Tsunami Warning Center rapidly locates subsea earthquakes, estimates their potential for initiating a tsunami, and quickly warns countries that may be in danger. A warning may be broadcast as much as a few hours before the arrival of the tsunami, allowing time for the evacuation of coastal populations. A more difficult situation occurs when a tsunami from a nearby earthquake arrives so quickly that there is no time to warn nearby communities. Some measure of protection can be achieved in such cases by building barrier walls to block the inundation of the sea, but such construction is expensive and is being tried to a significant degree only in Japan.

Some do's and don'ts if you live in an area subject to tsunamis:

- If you feel a strong earthquake, move quickly away from the coastal lowlands to higher ground.
- If you hear that an earthquake has occurred under the ocean or a coastal region, be prepared to move to higher ground.
- Tsunamis sometimes signal their arrival by a precursory rise or fall of coastal water, and this natural warning should be heeded.
- People have lost their lives by going down to the beach to watch for a tsunami. Don't make the same mistake.
- Follow the advice of your local emergency organization.

Tsunami barrier (at left) constructed to protect the town of Taro, Japan. *Frederick Raichlen.*

**FIGURE 18.21** Destruction at Maumere, Flores Island, Indonesia, caused by a tsunami generated during an earthquake on December 12, 1992. Although a warning system exists to alert people on distant coasts to the danger of tsunamis, it cannot yet function rapidly enough to help residents in the epicenter region. *Reuters/Bettman.*

**TABLE 18.2**

## Estimated Numbers of Deaths Caused by Severe Earthquakes, 856–1990

| YEAR | PLACE | DEATHS | YEAR | PLACE | DEATHS |
|------|-------|--------|------|-------|--------|
| 856 | Corinth, Greece | 45,000 | 1939 | Chile | 30,000 |
| 1038 | Shansi, China | 23,000 | 1939 | Erzincan, Turkey | 40,000 |
| 1057 | Chihli, China | 25,000 | 1948 | Fukui, Japan | 5000 |
| 1170 | Sicily | 15,000 | 1949 | Ecuador | 6000 |
| 1268 | Silicia, Asia Minor | 60,000 | 1949 | Khait, USSR | 12,000 |
| 1290 | Chihli, China | 100,000 | 1950 | Assam, India | 1500 |
| 1293 | Kamakura, Japan | 30,000 | 1954 | Northern Algeria | 1500 |
| 1456 | Naples, Italy | 60,000 | 1956 | Kabul, Afghanistan | 2000 |
| 1531 | Lisbon, Portugal | 30,000 | 1957 | Northern Iran | 2500 |
| 1556 | Shen-shu, China | 830,000 | 1960 | Southern Chile | 5700 |
| 1667 | Shemaka, Caucasia | 80,000 | 1960 | Agadir, Morocco | 12,000 |
| 1693 | Catania, Italy | 60,000 | 1962 | Northwestern Iran | 12,000 |
| 1693 | Naples, Italy | 93,000 | 1963 | Skopje, Yugoslavia | 1000 |
| 1731 | Peking, China | 100,000 | 1968 | Dasht-e Bayaz, Iran | 11,600 |
| 1737 | Calcutta, India | 300,000 | 1970 | Peru | 20,000 |
| 1755 | Northern Persia | 40,000 | 1972 | Managua, Nicaragua | 10,000 |
| 1755 | Lisbon, Portugal | 30,000–60,000 | 1976 | Guatemala | 23,000 |
| 1783 | Calabria, Italy | 50,000 | 1976 | Tangshan, China | 250,000 |
| 1797 | Quito, Ecuador | 41,000 | 1976 | Philippines | 3100 |
| 1822 | Aleppo, Asia Minor | 22,000 | 1976 | New Guinea | 9000 |
| 1828 | Echigo (Honshu), Japan | 30,000 | 1976 | Iran | 5000 |
| 1847 | Zenkoji, Japan | 34,000 | 1977 | Romania | 1500 |
| 1868 | Peru and Ecuador | 25,000 | 1978 | Iran | 15,000 |
| 1875 | Venezuela and Colombia | 16,000 | 1980 | Algeria | 3500 |
| 1896 | Sanriku, Japan | 27,000 | 1980 | Italy | 4000 |
| 1897 | Assam, India | 1500 | 1981 | Iran | 3000 |
| 1898 | Japan | 22,000 | 1982 | West Arabian Peninsula | 2800 |
| 1906 | Valparaiso, Chile | 1500 | 1983 | Turkey | 1400 |
| 1906 | San Francisco | 500 | 1985 | Mexico | 30,000 |
| 1907 | Kingston, Jamaica | 1400 | 1986 | El Salvador | 1200 |
| 1908 | Messina, Italy | 160,000 | 1987 | Ecuador | 1000 |
| 1915 | Avezzano, Italy | 30,000 | 1988 | Nepal | ~1000 |
| 1920 | Kansu, China | 180,000 | 1988 | China | ~1000 |
| 1923 | Tokyo, Japan | 99,000 | 1988 | Armenia | 25,000 |
| 1930 | Apennine Mountains, Italy | 1,500 | 1990 | Iran | 40,000 |
| 1932 | Kansu, China | 70,000 | 1990 | Philippines | 1700 |
| 1935 | Quetta, Baluchistan | 60,000 | | | |

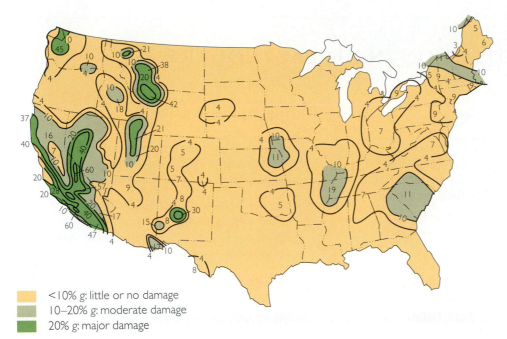

**FIGURE 18.22** Expected level of earthquake-shaking hazards. The levels of ground shaking for different regions are shown by contour lines that express the maximum amount of shaking likely to occur at least once in a 50-year period as a percentage of the acceleration of gravity (*g*). Damage begins to occur at about 10 percent *g*. An acceleration of 0.1 percent *g* or more is perceptible to people. (Modified from USGS chart.)

<10% *g*: little or no damage
10–20% *g*: moderate damage
20% *g*: major damage

## Mitigating the Destructiveness of Earthquakes

Can anything be done to reduce earthquake hazards? A good first step is to prepare a map showing the likelihood of an earthquake based on the number that have occurred in the past. A seismic-risk map of the United States is shown in Figure 18.22. You may be surprised to find that you live in a zone where there is risk of earthquake damage.

A seismic-risk map provides a basis for organizing local earthquake protection programs consonant with the degree of danger. In high-risk areas, building codes should require engineers to design structures that can resist earthquake damage. Figure 18.23 shows something that is unwise—a residential area built in an active fault zone, the San Andreas, the most dangerous in the United States. Construction on unstable soils or in avalanche-prone areas should also be regulated. As a homeowner or tenant you should inspect your home for hazards. The house should be bolted firmly to the foundation. Appliances connected to gas lines should be bolted down, and the lines should be flexible to avoid breaking and becoming a source of fire. Shelves should be attached to walls and heavy objects placed at low levels. Beds should not be near windows. If your home was constructed before modern building codes were introduced, added bracing and structural support may be required. See Box 18.3 for steps you can take to protect yourself and your family in an earthquake.

**FIGURE 18.23** Housing tracts constructed within the San Andreas fault zone, San Francisco Peninsula. The white line indicates the approximate fault trace, along which the ground ruptured and slipped about 2 m during the earthquake of 1906. *R. E. Wallace, USGS.*

## BOX 18.3 PROTECTION IN AN EARTHQUAKE

### Before an Earthquake

**AT HOME**

Have a battery-powered radio, flashlight, and first-aid kit in your home. Make sure everyone knows where they are stored. Keep batteries on hand.

Learn first aid.

Know the location of your electric fuse box and the gas and water shut-off valves (keep a wrench nearby). Make sure all responsible members of your family learn how to turn them off.

Don't keep heavy objects on high shelves.

Securely fasten heavy appliances to the floor, and anchor heavy furniture, such as cupboards and bookcases, to the wall.

Devise a plan for reuniting your family after an earthquake in the event that anyone is separated.

**AT SCHOOL**

Urge your school board and teachers to discuss earthquake safety in the classroom and secure heavy objects from falling. Have class drills.

**AT WORK**

Find out if your office or plant has an emergency plan. Do you have emergency responsibilities? Are there special actions for you to take to make sure that your workplace is safe?

### During an Earthquake

Stay calm. If you are indoors, stay indoors; if outdoors, stay outdoors. Many injuries occur as people enter or leave buildings.

If you are indoors, stand against a wall near the center of the building, or get under a sturdy table. Stay away from windows and outside doors.

If you are outdoors, stay in the open. Keep away from overhead electric wires or anything that might fall (such as chimneys, parapets and cornices on buildings).

Don't use candles, matches, or other open flames.

If you are in a moving car, stop away from overpasses and bridges and remain inside until the shaking is over.

**AT WORK**

Get under a desk or sturdy furniture. Stay away from windows.

In a high-rise building, protect yourself under sturdy furniture or stand against a support column.

Evacuate if told to do so. Use stairs rather than elevators.

**AT SCHOOL**

Get under desks, facing away from windows.

If on the playground, stay away from the building.

If on a moving school bus, stay in your seat until the driver stops.

## After an Earthquake

Check yourself and people nearby for injuries. Provide first aid if needed.

Check water, gas, and electric lines. If damaged, shut off valves.

Check for leaking gas by gas odor (*never* use a match). If it is detected, open all windows and doors, shut off gas meter, leave immediately, and report to authorities.

Turn on the radio for emergency instructions. Do not use the telephone—it will be needed for high-priority messages.

Do not flush toilets until sewer lines are checked.

Stay out of damaged buildings.

Wear boots and gloves to protect against shattered glass and debris.

Approach chimneys with caution.

### AT SCHOOL OR WORK

Follow the emergency plan or instructions given by someone in charge.

Stay away from beaches and waterfront areas where tsunamis could strike, even long after the shaking has stopped.

Do not go into damaged areas unless authorized. Martial law against looters has been declared after a number of earthquakes.

Expect aftershocks: they may cause additional damage.

From Bruce A. Bolt, *Earthquakes,* newly revised and expanded ed., New York: W. H. Freeman, 1993, pp. 220–222.

A few decades ago only astrologers, mystics, and religious zealots were concerned with earthquake prediction. Today seismologists in many countries are actively working on this problem and can claim a few successes. In February 1975 an earthquake was predicted five hours before it occurred near Haicheng, in northeast China. The Chinese seismologists used what they considered to be premonitory (precursor) events to make their predictions, such as the occurrence of swarms of tiny earthquakes and a rapid deformation of the ground several hours before the main shock. They also took seriously an ancient piece of peasant wisdom: that animals, as if sensing coming danger, behave strangely just before an earthquake. Several million people, prepared in advance by a public education campaign, evacuated their homes and factories in the hours before the shock. Although many towns and villages were totally destroyed, only a few hundred lives were lost. Western scientists who have since visited the region estimate that tens of thousands of lives were saved.

The Chinese have successfully predicted other earthquakes, but unfortunately they were able to provide only a long-term warning (within five years) of the great Tangshan earthquake of August 1976—not enough accuracy to save the 250,000 people estimated to have lost their lives. Smaller earthquakes have been predicted in the United States and the former Soviet Union, but most go unpredicted.

Scientists in the United States, Japan, China, Russia, and other countries are engaged in an intensive search for premonitory indicators to help them predict the time and place of a forthcoming destructive earthquake. One possible indicator being examined is an unusual increase in the frequency of smaller earthquakes in a region before a main shock. Figure 18.24 shows how such an indicator could be used to predict when a great earthquake might occur. Another possible indicator is a rapid tilting of the ground or other form of surface deformation. An unusual aseismic slip (smooth, slow sliding along a fault rather than the sudden slip of an earthquake) on a fault might signal a coming earthquake. Some seismologists propose that an episode of stretching of the crust across a fault—that is, tensional strain—might serve to pull apart two blocks in contact along a fault. This stretching would tend to reduce the friction between the two facing blocks, "unlocking" the fault. Perhaps changes in the physical properties of rock in the vicinity of a fault, such as its ability to conduct an electric current or its P-wave velocity, might show up hours or days before an earthquake. Changes in

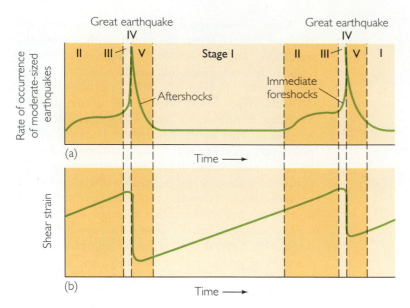

**FIGURE 18.24** Model of cycle of repeating great earthquakes along a plate boundary such as the San Andreas fault. (a) Great earthquakes are followed by several years of declining after-shocks (stage V); most of the 50- to 500-year intervals between great earthquakes are characterized by low levels of seismic activity (stage I). Stage II, an increased level of earthquakes, occurs some decades before the main shock. Stage III may be measured in years, days, or hours before the main shock. Some seismologists believe that northern California is entering stage II and that southern California may be entering stage III. (b) Movement of plates results in buildup of shear strain along the fault, which is released suddenly at the moment of a great earthquake (stage IV). (After C. B. Raleigh, K. Sieh, L. R. Sykes, and D. L. Anderson, "Forecasting Southern California Earthquakes," *Science,* vol. 217, 1982, pp. 1097–1104.)

the level of water in wells might precede an earthquake because the porosity of rock may increase or decrease as a result of premonitory changes in strain. Actually all of these phenomena have been observed in various combinations, but not with the consistency and reliability that a useful prediction method requires.

Ask a seismologist to predict the time of the next great earthquake and the response is likely to be "The longer the time since the last big one, the sooner the next great shock." This simple statement is the basis of the **seismic gap method,** which has successfully forecast the locations and magnitudes of more than six major earthquakes within a few years of their occurrence.

The seismic gap method has been most successful in predicting earthquakes on plate boundaries. The basic idea is that earthquakes result from the accumulation of strain caused by the steady motion of

the plates along faults. On reaching some critical level of strain, the brittle lithosphere breaks. The cycle of slow accumulation of strain and its sudden release in an earthquake recurs again and again. The average interval varies from place to place. According to the seismic gap method, the most likely place for an earthquake to occur is at a locked portion of a fault where an earthquake is "due"—that is, where the time since the last earthquake has reached or exceeds the average interval between earthquakes in this location. The average recurrence interval can be estimated in several ways. We can estimate the timing of great earthquakes several thousand years in the past by finding soil layers that are offset by fault displacements and then dating these layers. In another method, the interval is given by the number of years of steady plate motion it take to accumulate, or "store," the fault displacements that occurred in earlier earthquakes. For example, it would take 150 years to accumulate a fault displacement of 6 m if the plate motion were 4 cm per year. Although these methods give similar results, the uncertainty of the prediction is unfortunately large—as much as 50 percent of the average recurrence time. The time of uncertainty can amount to many decades. However, the concurrent use of the seismic gap method and one or more of the premonitory indicators described earlier could sharpen the prediction.

The part of the San Andreas fault that runs through southern California provides a good example of how these methods might be applied. The recurrence time between great earthquakes in this region, measured by the two methods, is 100 to 150 years. The last great earthquake occurred in 1857; therefore an earthquake can be expected at any time— tomorrow or some decades from now. However, southern California has had a larger number of moderate-sized earthquakes since 1978 than in the preceding 25 years. Southern California may have entered stage III, the last stage of seismic activity before a great earthquake (see Figure 18.24). Geologists have also observed surface deformation of the kind that stretches the distance between points on opposite sides of the San Andreas, as if the fault were beginning to unlock. For these reasons seismologists now estimate that there is a 60 percent chance that a great earthquake will occur in California, probably in southern California, in the next 30 years. This prospect is particularly worrisome because of the high density of population in the region. A 1980 government report estimates that if an earthquake on the order of the 1857 earthquake (whose magnitude exceeded 8) were to recur, 10,000 to 15,000 people would die, 50,000 would be hospitalized, and $17 billion in property would be destroyed. These estimates might be too low or too high by a factor of 2 or 3. If the prediction could be improved so that a warning could be issued hours or days before the shock, the casualties could be reduced significantly.[2]

Because of these concerns, a major earthquake prediction and hazard mitigation program was initiated by the U.S. government in recent years. Hundreds of instruments are being placed adjacent to the San Andreas fault in search of premonitory indicators. An official board of scientists has been named with the authority to alert public officials if they believe that the data foretell an earthquake. Similar activities are getting under way in other countries with high seismic risk. Let us hope that these initiatives will reduce the destructiveness of earthquakes, one of nature's most feared hazards.

---

[2]Since radio waves can outrace seismic waves, it is technically feasible to radio a warning that an earthquake has occurred tens of seconds before the destructive seismic vibrations arrive. Some scientists have recently proposed that installation of an automated system could save lives and property. Gas lines could be shut off, nuclear reactors deactivated, people alerted to seek safety under tables, and so forth.

# SUMMARY

**What is an earthquake?** An earthquake is a shaking of the ground caused by seismic waves that emanate from a fault that breaks suddenly. The strain built up over years of slow deformation of rocks is released in a few minutes as seismic waves at the time that the fault breaks.

**Where do most earthquakes occur?** Most earthquakes originate in the vicinity of plate boundaries, although a smaller number occur within plates.

**What governs the type of faulting that occurs in an earthquake?** In most earthquakes, the fault mechanism is governed by the kind of plate boundary: normal faulting under tensile stress occurs at boundaries of divergence, thrust faulting under compressive stress at boundaries of convergence, and strike slip along transform faults.

**What is the Richter magnitude and how is it measured?** The Richter magnitude measures the size of an earthquake. It is determined from the amplitude of the ground motions, as measured when seismic waves are recorded on seismographs.

**What are the three types of seismic waves?** The three types of seismic waves are P (primary) waves, S (secondary) waves, and surface waves.

**What causes the destructiveness of earthquakes?** Ground vibrations can damage or destroy buildings and other structures and can also trigger avalanches. Earthquakes on the seafloor can excite tsunamis, which sometimes cause widespread destruction when they reach shallow coastal waters.

**What can be done to mitigate the damage of earthquakes?** Construction in earthquake zones can be regulated so that buildings and other structures will not be located on soils that are unstable and will be strong enough to withstand destructive vibrations. People living in earthquake-prone areas should be in-

formed about what to do when an earthquake occurs, and public authorities should plan ahead and be prepared with emergency supplies, rescue teams, evacuation procedures, fire-fighting plans, and other steps to minimize the consequences of a severe earthquake.

**Can scientists predict earthquakes?** Not yet with the degree of accuracy that would be needed to alert a population hours or days before an earthquake occurred. However, they can characterize the degree of risk in a region. They are searching for premonitory phenomena that may be used to pinpoint more accurately the time and place of an earthquake.

# KEY TERMS AND CONCEPTS

earthquake (p. 405)

elastic rebound theory (p. 405)

slip (p. 406)

focus (p. 406)

epicenter (p. 406)

seismic wave (p. 406)

seismograph (p. 407)

P wave (p. 408)

S wave (p. 408)

surface wave (p. 408)

Richter magnitude (p. 411)

Modified Mercalli Scale (p. 411)

seismicity (p. 414)

tsunami (p. 418)

seismic gap method (p. 424)

# EXERCISES

1. What is an earthquake? How is its magnitude measured? How many earthquakes cause serious damage each year?

2. How does the distribution of earthquake foci correlate with the three types of plate boundaries?

3. What kinds of earthquake faults occur at the three types of plate boundaries?

4. Destructive earthquakes occasionally occur within plates, far from plate boundaries. Why?

5. Seismograph stations report the following S − P time differences for an earthquake: Dallas, S − P

= 3 minutes; Los Angeles, S − P = 2 minutes; San Francisco, S − P = 2 minutes. Use a map of the United States and travel-time curves (see Figure 18.11) to obtain a rough epicenter.

6. At a place along a boundary fault between the Nazca Plate and the South American Plate, the relative plate motions are 11.1 cm per year. The last great earthquake, in 1880, showed a fault slip of 12 m. When should local residents begin to worry about another great earthquake?

# THOUGHT QUESTIONS

1. Taking into account the possibility of false alarms, reduction of casualties, mass hysteria, economic depression, and other possible consequences of earthquake prediction, do you think the objective of predicting earthquakes is a worthwhile goal?

2. Would you rather live on a planet without earthquakes (implying no plate tectonics)?

3. Which is the more useful scale, Richter or Mercalli?

# SUGGESTED READINGS

Bolt, Bruce A. 1993. *Earthquakes,* newly revised and expanded ed. New York: W. H. Freeman.

*Earthquakes and Volcanoes.* A bimonthly periodical. Washington, D.C.: U.S. Government Printing Office.

National Research Council. 1993. Hazards, land use, and environmental change. Chap. 5 in *Solid-Earth Sciences and Society*. Washington, D.C.: National Academy Press.

Pakiser, L. C. 1991. *Earthquakes*. Washington, D.C.: U.S. Government Printing Office.

Raleigh, C. B., K. Sieh, L. R. Sykes, and D. L. Anderson. 1982. Forecasting southern California earthquakes. *Science* 217:1097–1104.

Richter, C. F. 1958. *Elementary Seismology*. San Francisco: W. H. Freeman.

# EXPLORING EARTH'S INTERIOR

Because volcanism and deformation bring rocks to the surface of the Earth from depths as great as 50 to 100 km, geologists can get an idea of some of the properties of the Earth at those depths by sampling these rocks. The only way to study Earth to its full depth of 6400 km, however, is indirectly. Geologists have found certain phenomena at Earth's surface that are direct reflections of the properties and behavior of matter deep inside the Earth. One of the major ways that geologists study the interior is by measuring the speed of seismic waves, which varies with the kinds of materials they pass through. Geologists also learn about the interior by measuring the flow of heat from great depths to the surface and by measuring the properties of Earth's magnetic field.

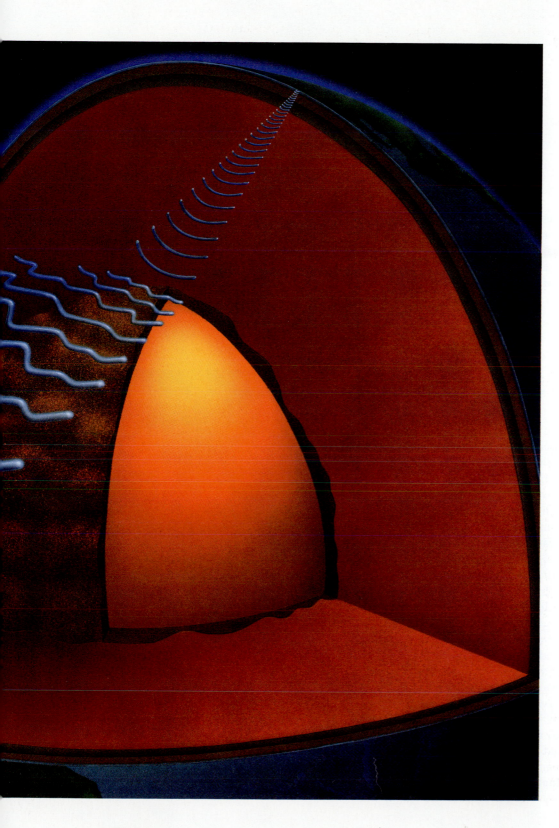

19

Artist's rendition of
seismic waves bouncing
off Earth's molten core,
revealing an irregular
boundary between the
core and the bottom
of the mantle.
*Tomo Narashima.*

In Chapter 18 we saw one result of seismic waves: the destruction they cause when they shake the ground. In this chapter we will see another facet of seismic activity: how geologists use the behavior of seismic waves to infer the kinds of materials that make up Earth's interior.

ond material as in the first. Figure 19.1 shows a laser light beam whose path bends as it goes from air into water, much as a P or S wave bends as it travels from one material to another. Seismologists study how seismic waves are refracted and reflected, and this behavior reveals certain features of Earth's interior.

## EXPLORING THE INTERIOR WITH SEISMIC WAVES

Geologists know that waves of different types have a common feature: the velocity at which they travel depends on the material they pass through. Thus light travels more quickly through air than through glass, and seismic P and S waves travel more rapidly through basalt than through granite. Remember that P (primary) waves are compressional waves and S (secondary) waves are shear waves, and that compressional waves move faster than shear waves when they pass through the Earth. Geologists calculate the velocity of a P or S wave by measuring the time of travel and dividing the distance traveled by this elapsed time. Any differences detected for different paths can be used to infer the properties of materials the waves have encountered along these paths.

The principle sounds simple enough, but as usual there are some complications. When waves encounter the boundary between two materials, some of the waves bounce off—that is, they are reflected—and others are transmitted into the second material, just as light is partly reflected and partly transmitted when it strikes a windowpane. When waves cross the boundary between two materials, they bend or refract because their velocity is not the same in the sec-

**FIGURE 19.1** A beam of light is refracted (bent) or reflected when it crosses the boundary between air and water. Seismic waves behave similarly at boundaries within the Earth. They reflect from the boundary and are refracted as they are transmitted across it. *Susan Schwartzenberg/The Exploratorium.*

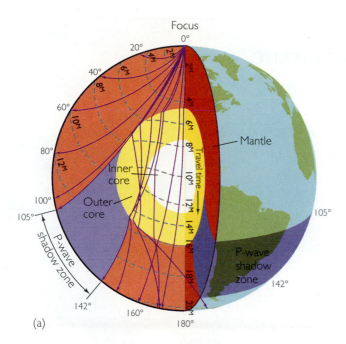

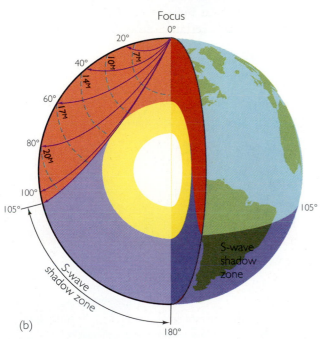

**FIGURE 19.2** (a) The pattern of P-wave paths through Earth's interior. Dashed lines show the progress of wave fronts through the interior at 2-minute intervals. The shadow zone is a region not reached by P waves (for this hypothetical earthquake at the North Pole) because they are deflected by Earth's core. The shadow zone for P waves extends from 105° angular distance from the focus to 142°. It takes 20 minutes for a P wave generated from the earthquake focus to travel through the core and emerge at the opposite position on the surface at an angular distance of 180°. (b) S waves cannot penetrate the liquid core and therefore never emerge beyond 105° from the focus.

# Paths of Seismic Waves in the Earth

If Earth were made of a single material with constant properties from the surface to the center, P and S waves would travel along a straight line through the interior from the focus of an earthquake to a distant seismograph. But in reality the Earth is layered and made up of many materials that conduct P and S waves at different velocities. The waves bend as they go from layer to layer, and as a result their paths through the interior are curved.

**WAVES TRAVELING THROUGH THE EARTH**
Figure 19.2 shows the curved paths of P waves as they travel from an earthquake focus to distant points where they may be recorded on a seismograph. Geologists have determined these wave paths and their corresponding travel times by studying the seismographic records of earthquakes all over the world.

Follow the path of a wave that just misses the core and emerges at the surface at an angular distance of 105°. Next follow the wave that just strikes the core. Note that it is bent downward when it enters the core and bent again when it leaves. Because of this bending at the mantle-core boundary, the wave emerges at the surface at 142° angular distance from the focus. Thus no P wave reaches the surface between 105° and 142°—as though the core had cast a shadow across this zone. This is in fact called the **shadow zone.**

Discovery of the shadow zone led geologists to surmise that the Earth has a core made of a different material than the overlying mantle. They could even conclude that the core is liquid, because the waves are bent downward rather than upward when they enter the core, much as the light beam bends downward after it enters the water in Figure 19.1. This means that the waves travel more slowly in the core than in the mantle. P waves are known to move much more slowly in liquids than in solids, so it was reasonable to guess that the existence of the shadow zone implies a molten core. This guess was verified by one other piece of evidence: the behavior of S waves. Whenever S waves strike the core, they fail to emerge from the other side and are not seen at all beyond the 105° angular distance from the focus. Only liquids are known not to transmit shear waves; therefore geologists concluded that a molten core must be blocking S waves from reaching these distances.

**WAVES REFLECTED IN THE EARTH**    Since a reflected wave implies a boundary between two materials, a good way to find boundaries in Earth's interior is to find reflections. Let us see what happens to

## BOX 19.1 FINDING OIL WITH SEISMIC WAVES

Exploration for oil is an important application of seismology. In offshore prospecting, a ship tows a sound source and underwater phones. P waves (sound waves) are generated by a pneumatic device that works like a balloon burst. The sound waves bounce off rock layers below the seafloor and are picked up by the phones. In this way subsurface sedimentary structures that trap oil, such as faults, folds, and domes, are "mapped" by the reflected waves. This technique is used extensively to explore the submerged continental shelves and shallow seas for oil and gas deposits. Oceanographers use this method to study the sedimentary layers on the continental slope and on the floor of the deep sea.

P and S waves when they bounce off the boundary between two layers. In Figure 19.3, for example, follow the wave PcP as it bounces like a radar beam from the Earth's core back to the surface. Because the velocity of a P wave is known, the round-trip travel time for PcP is used to determine the depth to the core, just as the time it takes an echo to return can be

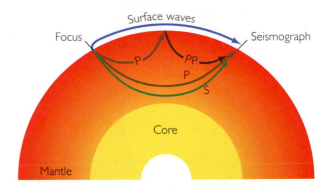

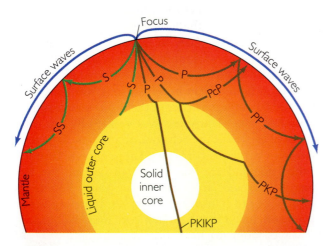

**FIGURE 19.3** P and S waves radiate from an earthquake focus in many directions. A PcP wave bounces off the core. PP or SS waves are reflected from Earth's surface. A PKP wave is transmitted through the liquid outer core. A PKIKP wave traverses the solid inner core.

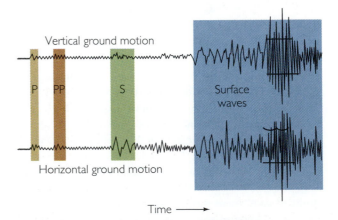

**FIGURE 19.4** Seismographic recording of P, S, and surface waves from a distant earthquake.

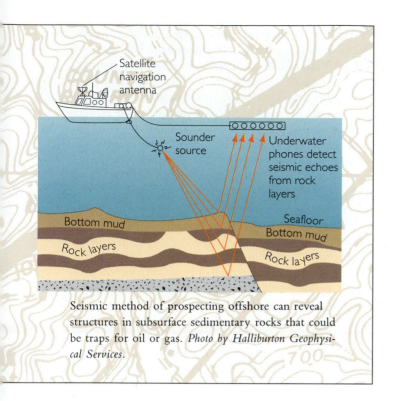

Seismic method of prospecting offshore can reveal structures in subsurface sedimentary rocks that could be traps for oil or gas. *Photo by Halliburton Geophysical Services.*

in the interior of the Earth. The key step then is to convert the travel-time information to a graph or table that shows how the velocity of seismic waves changes with depth in the Earth. Solving this problem is something like guessing which of several possible routes a driver took on a trip from Los Angeles to San Francisco and how fast he traveled along the route if you know that the trip took 6.1 hours and know the speed limits along the way.

The solution that seismologists have come up with is shown in Figure 19.5. The illustration shows both the changes in the velocities of P and S waves with depth and the geological interpretation of that information: the structure of Earth's interior. The major layers that make up Earth's interior are a very thin outer crust, a mantle of rock extending to a

used to measure the distance to the wall on the other side of a valley. P waves can also bounce off the Earth's surface back into the Earth, as shown by the PP waves in Figure 19.3. Similarly, S waves that reflect back into the Earth are called SS. The P waves that penetrate the outer core—called PKP—or the inner core—called PKIKP—are useful for exploring those regions. Figure 19.4 shows seismograms in which the P, PP, S, and surface waves are indicated. They are differentiated from one another by their amplitude and frequency. One practical application of reflected seismic waves is their use in exploring for oil (see Box 19.1).

## Composition and Structure of the Interior

Thousands of sensitive seismographs and highly accurate clocks enable seismologists around the world to measure precisely the travel times of P, S, and surface waves. Nuclear explosions set off underground at nuclear test sites also excite seismic waves and add valuable data to those derived from earthquakes. From these measurements seismologists can plot travel-time curves of the kind shown in Figure 18.11 for the various kinds of seismic waves. The travel times depend on the velocities of compressional and shear waves in the materials the waves pass through

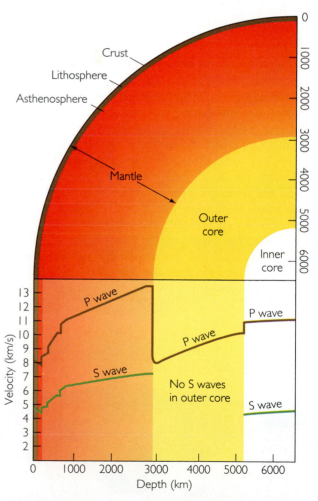

**FIGURE 19.5** Changes in P- and S-wave velocities with depth in the Earth reveal the sequence of layers that make up Earth's interior.

depth of 2900 km, a liquid outer core extending another 2200 km, and a solid inner core extending to the center of the Earth at a depth of 6370 km.

**THE CRUST** The crust, the outermost layer, has been explored extensively by means of seismic waves. It varies in thickness (Figure 19.6)—thin under oceans (about 5 km), thicker under continents (about 35 km), and thickest under high mountains (ranging up to 65 km). P waves move through crustal rocks at about 6 to 7 km per second. By sampling various types of materials from the crust and mantle and by making laboratory measurements on these materials, we can compile a library of seismic speeds through all sorts of materials that compose the Earth. The velocities of P waves in igneous rocks, for example, are as follows:

- Felsic (granitic): 6 km per second
- Mafic (gabbro): 7 km per second
- Ultramafic (peridotite): 8 km per second

We know from the correlations of wave velocity and rock type that the continental crust consists mostly of granitic rocks, with gabbro appearing near the bottom, and that no granite occurs on the floor of the deep ocean. The crust there consists entirely of basalt and gabbro. Below the crust the velocity of P waves increases abruptly to 8 km per second. This is an indication of a sharp boundary between crustal rocks and underlying rocks, as shown in Figure 19.6. The velocity of 8 km per second indicates that the deeper rocks are probably the denser ultramafic rock peridotite. The boundary between the crust and the mantle is called the **Mohorovičić discontinuity** (**Moho** for short), after the Yugoslav seismologist who discovered it in 1909. The indications that the

crust is less dense than the underlying mantle are consistent with the theory that the crust is made up of lighter materials that floated up from the mantle.

The idea that continents are less dense and float on a denser mantle, as a life jacket or an iceberg floats in the ocean, is the **principle of isostasy.** Life jackets float because the lightweight flotation material inside is less dense than seawater; icebergs float because ice is less dense than seawater. The flotation, or buoyancy, occurs because the volume of the life jacket or of the ice below the sea surface weighs less than the volume of water it displaces. Bigger icebergs stand higher above the sea surface but also have a deeper root below the surface to provide greater buoyancy. Continents float because the large volume of less dense continental crust that projects into the denser mantle provides the buoyancy, as shown in Figure 19.6. Note that the crust is thicker under a mountain because a deeper root is needed to float the additional weight of the mountain.

The idea that continents float suggests that there is a liquid in which they float. Yet from seismological data we know that the mantle just beneath the crust is solid rock. How can continents float on solid rock? As we saw earlier, rocks, which we know to be solid and strong over the short term (seconds or years), are weak over the long term (thousands to millions of years) and flow very slowly like a viscous fluid when forces are applied. The principle of isostasy predicts that over long periods the mantle has little strength and behaves like a viscous liquid when it is forced to support the weight of continents and mountains. Isostasy also implies that as a large mountain range forms, it slowly sinks as the crust bends downward. When enough of a root bulges into the mantle, the mountain floats. If the mass of a mountain range is depleted by erosion, the weight on the crust is lightened, less root is needed for flotation, and the moun-

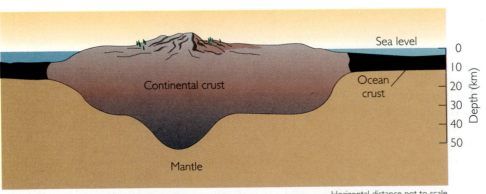

**FIGURE 19.6** Seismic waves reveal the boundary between the crust and underlying mantle and variations in the thickness of the crust.

Horizontal distance not to scale

## BOX 19.2  THE UPLIFT OF SCANDINAVIA: NATURE'S EXPERIMENT WITH ISOSTASY

If you depress a cork floating in water with your finger and then release it, the cork pops up almost instantly. A cork floating in molasses would rise more slowly; the drag of the viscous fluid would slow the process. If we could perform a similar experiment on the Earth, we could learn much about how isostasy works, in particular about the viscosity of the mantle and how it affects rates of uplift and subsidence. How convenient it would be if we could push the crust down somewhere, remove the force, and then sit back and watch it rise!

Nature has been good enough to perform this experiment for us in less time than it takes to build and erode a mountain. The weight is a continental glacier—an ice sheet 2 to 3 km thick. Such ice sheets can appear with the onset of an ice age in the geologically short period of a few thousand years. The crust is depressed by the ice load, and a downward bulge develops on its underside to provide buoyant support. At the onset of a warming trend, the glacier melts rapidly. With the removal of the weight, uplift of the depressed crust begins. We can discover the rate of uplift by dating ancient beaches that are now well above sea level. Such raised beaches can tell us how long ago a particular stretch of land was at sea level.

Such depression and uplift have occurred in Norway, Sweden, Finland, and Canada, as well as elsewhere in glaciated regions. The ice cap retreated from these regions some 10,000 years ago, and the land has been rising ever since.

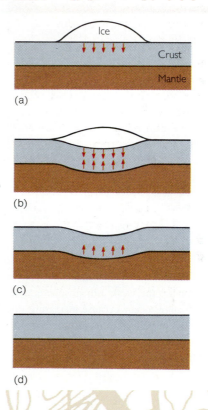

Isostasy and postglacial uplift. (a) A continental glacier grows, forming a weight on the crust. (b) The crust sags, and a root develops to support the ice load isostatically. (c) The glacier disappears, but the root remains because of the viscosity of the mantle. (d) Buoyancy of the root leads to slow uplift. As the root disappears, the surface assumes its original level. Arrows depict the direction of forces exerted by the ice load and the root. Scandinavia is now between stages (c) and (d). (The illustration is not to scale: the crust is about 40 km thick; a 3-km-thick glacier would produce a root about 1 km thick.)

tain range and root disappear. This is occurring in Scandinavia at the present time, not because a mountain is eroding but because the load of a continental glacier was removed at the end of the last ice age (see Box 19.2).

**THE MANTLE**  In the 1970s geologists the world over concentrated research efforts on an international project to study the upper 1000 km of the Earth. They made many discoveries, especially about zones at various depths in the mantle revealed by changes in

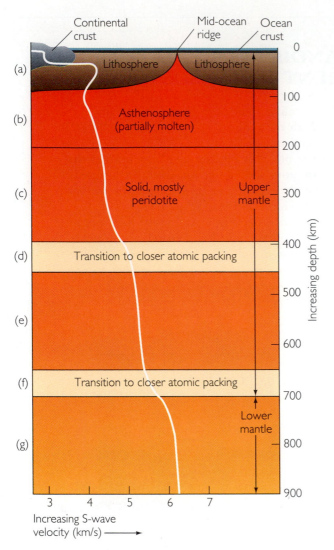

**FIGURE 19.7** The structure of the upper mantle, the outermost 900 km of the Earth, is illustrated by a plot of S-wave velocity against depth. Changes in velocity mark the strong lithosphere, the weak asthenosphere, and two zones in which changes occur because increasing pressure forces a rearrangement of the atoms into denser or more compact crystalline structures. (After D. P. McKenzie, "The Earth's Mantle," *Scientific American,* September 1983, p. 66.)

the velocity of S waves (Figure 19.7). The outermost zone, the lithosphere (a), is a slab about 70 km thick in which the continents are embedded. S waves pass easily through the lithosphere without being absorbed—a sign of solidity. As we mentioned earlier, the tectonic plates of the Earth are large fragments of the lithosphere.

In the zone below the lithosphere, the velocity of S waves decreases and the waves are partially absorbed. Laboratory experiments show that both of these phenomena are characteristic of S waves passing through a solid that contains a small amount of liquid. This is the asthenosphere (zone b in Figure 19.7), or zone of weakness. It rises close to the surface at mid-ocean ridges where plates separate and is found at depths below 70 km elsewhere. Geologists figure that the asthenosphere contains a small quantity of melt, perhaps a few percentage points. This idea fits nicely with the evidence (cited in Chapter 5) that the probable source of much basaltic magma is this region of the mantle. The idea that a small fraction of the asthenosphere is liquid is also consistent with other speculations about plate tectonics, especially the picture of solid lithospheric plates moving more easily because the underlying asthenosphere is partially molten and weak. The asthenosphere ends at a depth of about 200 km, where the velocity of S waves has increased to a value that fits that of solid peridotite.

From about 200 to 400 km (zone c in Figure 19.7), the velocity of S waves increases gradually with depth, but not enough to make us think that any material other than peridotite is present. About 400 km below the surface (zone d in Figure 19.7), however, the velocity of S waves increases rapidly. The increase is too great to be explained by a simple change in chemical composition. What could explain it is a repacking of atoms more closely, which could be caused by the high pressure at this depth. Geologists turned to the laboratory to test this theory. Peridotite is made up mostly of the mineral olivine. When scientists squeezed olivine in the laboratory, they found that when the temperatures and pressures reached values corresponding to depths of about 400 km, the atoms took up a more compact arrangement and assumed a crystal structure consistent with the seismological observations.

In the region from 450 to 650 km (zone e in Figure 19.7), properties change little as depth increases. Near 670 km (zone f in Figure 19.7), however, the velocity of S waves increases again, this time so greatly that atoms must be packed even more closely.

Two entirely different approaches, the analysis of seismic waves and the study of materials at high pressures, led to the same conclusion about the collapse of olivine to more compact materials at these depths.

The lower mantle, extending from 700 km to the core at a depth of 2900 km, is a region that changes little in composition and crystal structure with depth, as revealed by the fact that in this region the velocity of S waves increases gradually.

**THE CORE** Although Earth's core is 2900 km from the surface, it is still within reach of seismic waves. Earlier in this chapter we saw how geologists infer that the core is fluid. But it is not fluid to the very center of the Earth. P waves that penetrate to depths of 5100 km suddenly speed up. The interpretation of this increase in velocity is that Earth's innermost core is solid (see Figure 19.5).

Geologists still know nothing about the composition of the core from direct observation. But information derived from astronomical data, laboratory experiments, and seismological data have led them to form some ideas. First, to be consistent with the theory that the core is made up of material that sank during the initial formation and differentiation of the Earth, geologists searched for substances that were dense. Second, because the core contains one-third of the mass of the Earth, geologists considered substances that were abundant in the universe. Astronomers study abundances of elements in the cosmos, and their data point to iron as a plentiful heavy element. Laboratory measurements of the speed of P waves in liquid and in solid iron at pressures and temperatures corresponding to those of the inner core approximately fit the observed speed of P waves in the inner core. From these observations, geologists concluded that Earth's core is composed mostly of iron, molten in the outer core and solid in the inner core, as determined by the temperatures and pressures at which iron changes from a liquid to a solid. This conclusion is buttressed by discoveries of meteorites that are made almost entirely of iron and that presumably came from the breakup of a planetary body that also had an iron core.

Seismological observations and laboratory measurements of the properties of materials combine to paint a picture of Earth's interior: a zoned planet whose major components are a metallic iron core and a rocky mantle. The mantle includes two transition zones in which atoms are forced into closer packing, a partially molten asthenosphere, and most of the lithosphere. A thin, lightweight crust—the end product of the differentiation process—caps the mantle.

The future holds even greater promise for exploration of Earth's interior as new tools are developed. The newest tool is an adaptation of computerized tomography (CAT scanners), used in medicine to reconstruct images of organs as a computer calculates small differences in X rays that sweep the organ in many directions. Geologists are using seismic waves that sweep the mantle to construct images of pieces of subducted slabs, the rising plumes of hot spots, and other discrete structures down to Earth's center. In this work, hot matter is correlated with slower P- and S-wave velocities and cool matter with faster velocities (Figure 19.8).

## EARTH'S INTERNAL HEAT

The evidence of Earth's internal heat is everywhere: volcanoes, hot springs, and the elevated temperatures in mines and boreholes. Even global plate motions, earthquake activity, and the uplift of mountains are driven by this internal heat.

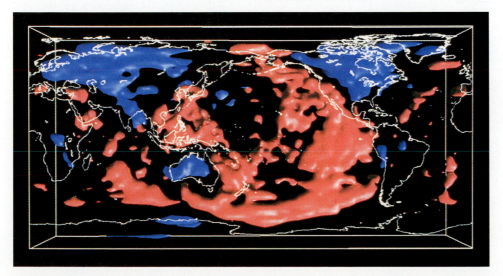

**FIGURE 19.8** Seismic tomography of the Earth's upper mantle reveals hot matter (red) around the rim of the Pacific Ocean basin and cool matter (blue) under the older continental interiors. *T. Tanimoto and Y. Zhang.*

Earth's interior is hot for several reasons, as we mentioned in Chapter 1. Briefly, the violent origin of Earth by the infalling of chunks of matter made the interior hot, and disintegration of the radioactive elements uranium, thorium, and potassium also produces a significant amount of heat.

## Heat Flow from Earth's Interior

As soon as the Earth heated up, it began to cool, and it is cooling to this day as heat flows from the hot interior to the cool surface. The Earth cools in two main ways: slowly by conduction and more rapidly by convection.

**CONDUCTION**   Think of heat as energy in transit from a hot place to a cool place. Heat energy exists in a material as the vibration of atoms; the higher the temperature, the more intense the vibrations. The **conduction** of heat occurs when thermally agitated atoms and molecules jostle one another, thus mechanically transferring the vibrational motion from a hot region to a cool one (Figure 19.9).

Materials vary in their ability to conduct heat. Metal is a better conductor than plastic (think of how rapidly the metal handle of a frying pan heats up in comparison with one made of plastic). Rock and soil are very poor heat conductors; that is why underground pipes are less susceptible to freezing than those aboveground and why underground vaults have a nearly constant temperature despite large seasonal temperature changes at the surface. Because of the poor conductivity of rock, a lava flow 100 m thick would take about 300 years to cool from 1000°C to surface temperatures. Heat that entered one side of a plate of rock 400 km thick would take about 5 billion years to reach the other side, longer than the Earth has existed. In other words, if the 4.6-billion-year-old Earth cooled by conduction only, heat from depths greater than about 400 km would not yet have reached the surface. The mantle, which was molten in Earth's early history (as we saw in Chapter 1), would still be liquid. We know from seismic waves that this is not the case. Therefore we must look for some means of removing heat from the interior more efficient than conduction to account for the cooling of the Earth and solidification of the mantle over the past 4 billion years. Convection is that mechanism.

**CONVECTION**   **Convection,** a rather common phenomenon, occurs when a heated fluid, either liquid or gas, expands and rises because it becomes less dense than the surrounding material. Convection moves heat more efficiently than conduction because the heated material itself moves, carrying its heat with it. Colder material flows in to take the place of the hot rising fluid, is itself heated, and then rises to continue the cycle. This is the process by which water is heated in a kettle. Because liquids conduct heat poorly, a kettle of water would take a long time to heat to the boiling point if convection did not distribute the heat rapidly. Convection moves heat when a chimney draws, when warm tobacco smoke rises, or when clouds form on a hot day.

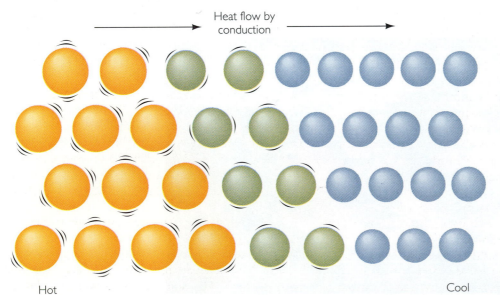

Heat flow by conduction

Hot

Cool

**FIGURE 19.9** Heat flow by conduction through a solid. Heat applied at the left induces thermal agitation of the atoms. Heat is conducted as the vibrations gradually spread to the right.

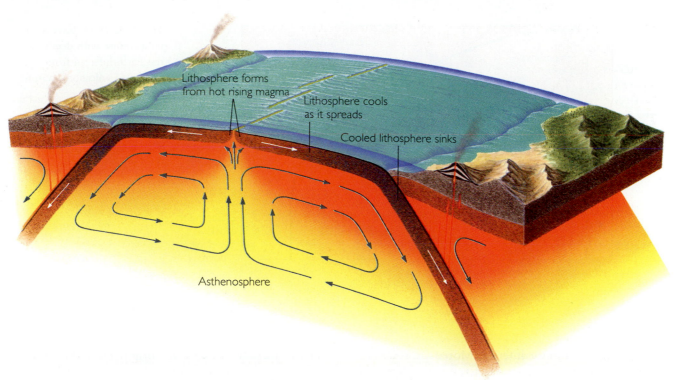

Lithosphere forms
from hot rising magma

Lithosphere cools
as it spreads

Cooled lithosphere sinks

Asthenosphere

**FIGURE 19.10** Some geologists believe that plate-tectonic movements can be explained by convection in the upper mantle. Hot matter rises and spreads laterally under the ocean ridges; it cools as it spreads and solidifies to form the cold, rigid lithosphere. The descending matter in the subduction zone is the cooled lithosphere. Other geologists believe that convection occurs in the entire mantle.

Although in general solids cool only by conduction, convection can occur in solids that "flow" over longer periods. The silicone compound known as Silly Putty demonstrates how a solid can flow. Silly Putty can be bounced like a ball or broken by a sudden blow, but overnight a ball of it flows into a pancake shape under its own weight. In the case of the Earth, the mantle behaves as a rigid solid over the short term, from seconds to years. But when forces are applied over millions of years, at conditions of high pressure and temperature, the mantle behaves as an extremely viscous substance and "creeps" or flows. Thus convection in the mantle is indeed possible and prompts geologists to debate some key questions: Is convection an important process by which heat is transferred within the Earth? Is convection occurring now? Has it occurred at any time in the past?

It turns out that seafloor spreading and plate tectonics may in fact be driven by convection. The rising hot matter under mid-ocean ridges builds new lithosphere, which cools as it spreads away, eventually to sink back into the mantle where it is resorbed (Figure 19.10). This is convection; heat is carried from the interior to the surface by the motion of matter. Some geologists believe that only the upper few hundred kilometers of the mantle are subject to convection, as in Figure 19.10. Others think that the whole mantle is involved. Still others believe that the rising narrow plumes beneath hot spots (see Chapter 5) provide the driving force for convection. Regardless of the specifics, geologists now believe that the movement of heat from the interior to the surface as the seafloor spreads is an important mechanism by which Earth has cooled over geologic time.

The British geologist Arthur Holmes was among the first to propose convection as the driving mechanism of continental drift. When he advanced this theory in the 1930s, Holmes was 30 years ahead of his time; corroboration had to wait for the extensive exploration of the seafloor that began after World War II, which led in 1963 to the concept of seafloor spreading.

Although the heat energy transferred to the surface from the interior is enough to create and move plates and to raise mountains and make earthquakes, it is puny in comparison with the energy received from the Sun. The Sun delivers 5000 times the en-

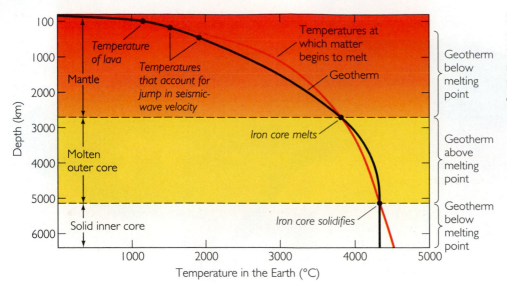

**FIGURE 19.11** Increase of temperature with depth in the Earth inferred from studies of volcanoes, seismic waves, and laboratory experiments.

ergy received from the interior. Moreover, solar heat has its geologic consequences, for it is the dominant controller of climate; it drives the atmosphere and hydrosphere, causing rain and winds—the chief agents of erosion. In a real sense, Earth's internal heat engine builds mountains, and its external heat engine, the Sun, destroys them.

## Temperatures in the Earth

Just how hot Earth gets in its interior is a matter of considerable importance to geologists. Temperature and pressure determine whether matter is solid or molten, the degree to which solid matter can creep, and how atoms are packed together in crystals. The higher the temperature at depth, the more rapidly convecting matter will move.

At present all geologists can do is draw certain conclusions from the limited information on temperature they have. They know that the average increase of temperature with depth, as measured in boreholes and mines, is about 2 or 3°C per 100 m. But how can they estimate temperatures in the Earth at depths greater than those they can reach with a thermometer—that is, below about 8 km? They can't simply assume that the rising temperatures they observe as depth increases near the surface continue at the same rate all the way to the center of the Earth. In that case, temperatures near the center would be so high (tens of thousands of degrees) that most of the interior would be molten, which seismology tells us is not the case.

One possible **geotherm,** or temperature-depth curve, showing temperatures in the interior arrived at by some geologists is illustrated in Figure 19.11.

The temperature of lava that originates in the mantle and emerges from volcanoes, laboratory data on the temperatures at which rocks and iron begin to melt, and information from seismology were combined to infer the geotherm from the surface to the very center of Earth, where the temperature rises to between 4000 and 5000°C.

# THE INTERIOR REVEALED BY EARTH'S MAGNETIC FIELD

Earth has a magnetic field, and geologists have learned how to use its characteristics as tools to examine Earth's interior, particularly the core and the lithosphere.

## The Earth as One Big Magnet

In 1600 William Gilbert, physician to Queen Elizabeth I, first explained how a magnetic compass works. He offered the proposition that "the whole Earth is a big magnet" whose field acts on the small magnet of the compass needle to align it in the north-south direction.

Earth's magnetic field behaves as if a small but powerful permanent bar magnet were located near the center of the Earth and inclined about 11° from the geographic axis (Earth's rotational axis), as in Figure 19.12. Magnetism can be visualized as lines of force of a magnetic field that indicate the presence of a magnetic force at each point in space. A compass needle that is free to swing under the influence of this magnetic force rotates into a position parallel to the local line of force, approximately in the north-south direction.

Unfortunately, although a good description of the field can be given if we assume a permanent magnet at the center of the Earth, this model has a fatal defect. Laboratory experiments show that heat destroys magnetism, and materials lose their permanent magnetism when temperatures exceed about 500°C. Material below depths of about 20 or 30 km in the Earth, then, cannot be magnetized because the temperatures are too high.

Another way to create a magnetic field is with electric currents. Dynamos in power plants make electricity by means of an electrical conductor in the form of a coil of copper wire rotated through a magnetic field. The rotation is driven by steam or falling water. Where inside the Earth is there a dynamo with the capacity to generate enough current to explain the magnetic field we observe at the surface?

Scientists theorize that the place to look might be the Earth's fluid iron core. Because fluid iron can move readily and iron is a good conductor, the core might be the moving conductor required of a dy-

namo. Scientists speculate that the fluid iron is stirred into convective motion by heat generated from radioactivity in the core. By a process not completely understood, this motion is thought to produce both the electric currents and the magnetic field needed to sustain a dynamo in the core. A magnetic field emanates to the surface from the dynamo in the core, as if there were a bar magnet at the center of the Earth. This idea offers the best explanation so far of Earth's magnetic field. The existence of the magnetic field itself is in turn further evidence that Earth's core is liquid iron.

## Paleomagnetism

In the early 1960s an Australian graduate student found a fireplace in an ancient campsite where the aborigines had cooked their meals. The stones were magnetized. He carefully removed several stones that had been baked by the fires, first noting their physical orientation. Then he measured the direction of the stones' magnetization and found that it was exactly

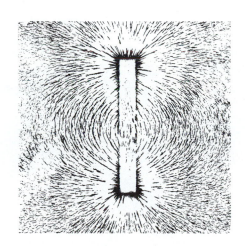

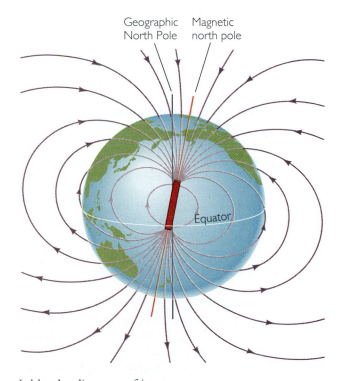

**FIGURE 19.12** *Left:* The magnetic field of a bar magnet is revealed by the alignment of iron filings on paper. (From *PSSC Physics,* 3d ed., Lexington, Mass., D. C. Heath, 1971.) *Right:* Earth's magnetic field is much like the field that would be produced if a giant bar magnet were placed at the Earth's center and slightly inclined (11°) from the axis of rotation. Lines of magnetic force produced by such a bar magnet are shown. A compass needle points to the north magnetic pole because it orients in the direction of the local line of force.

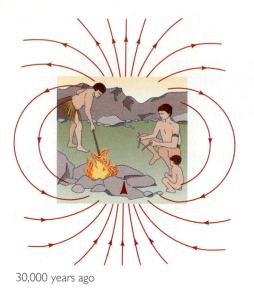
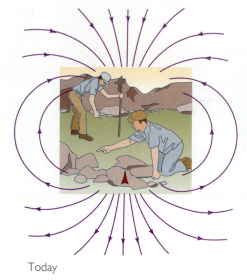

30,000 years ago                                    Today

**FIGURE 19.13** Earth's magnetic field 30,000 years ago was the reverse of to-day's. One way we know this is from the discovery of reversely magnetized rocks found in the fireplace of an ancient campsite. The rocks, cooling after the last fire, became magnetized in the direction of the ancient magnetic field, leaving a permanent record of it, just as a fossil leaves a record of ancient life.

the reverse of Earth's present magnetic field. He proposed to his disbelieving professor that as recently as 30,000 years ago, when the campsite was occupied, the magnetic field was the reverse of the present one—that is, a compass needle would have pointed south rather than north!

Scientists have discovered how to determine the direction of Earth's magnetic field in the past, not just thousands but millions of years before there were instruments to record it. Recall that high temperatures destroy magnetism. An important property of many very hot, magnetizable materials is that as they cool below about 500°C they become magnetized in the direction of the surrounding magnetic field. The reason is that groups of atoms of the material align themselves in the direction of the magnetic field when the material is hot. Once the material cools, these atoms are locked in place and therefore are always magnetized in the same direction. This is called **thermoremanent magnetization,** because the magnetization is "remembered" by the rock long after the magnetizing field has disappeared. Thus the Australian student was able to determine the direction of the field when the stones cooled after the last fire and took on the magnetization of Earth's magnetic field of that time (Figure 19.13). As another example, imagine an ancient volcano in eruption, say 100 million years ago. When the lava solidified and cooled, it became magnetized, leaving us with a permanent record of the geomagnetic field in mid-Cretaceous time, just as a fossil leaves a record of ancient life.

Some sedimentary rocks can also take on a remanent magnetization. Recall that marine sedimentary rocks are formed when particles of sediment that have settled through the ocean to the seafloor become

lithified. Magnetic grains among the particles—chips of the mineral magnetite, for example—would become aligned in the direction of Earth's field as they fell through the water, and this orientation would be incorporated into the rock when the particles became lithified. The **depositional remanent magnetization** of a sedimentary rock would then be due to the

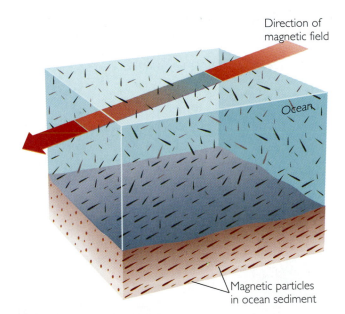

Direction of magnetic field

Ocean

Magnetic particles in ocean sediment

**FIGURE 19.14** Newly formed sedimentary deposits can become magnetized in the same direction as the contemporaneous Earth's magnetic field. Magnetic mineral grains transported to the ocean with other erosion products become aligned with the Earth's magnetic field while settling through the water. This orientation is preserved in the lithified rock, which thus "remembers" the field that existed at the time of deposition.

parallel alignment of all these tiny magnets, as if they were compasses pointing in the direction of the field prevailing at the time of deposition (Figure 19.14).

Ancient magnetism, called **paleomagnetism** or **fossil magnetism,** has become an important tool for understanding the history of the Earth. Scientists collect old rocks on every continent and determine their magnetism and ages in order to reconstruct the history of the magnetic field. The oldest magnetized rocks found so far indicate that 3.5 billion years ago Earth had a magnetic field not unlike the present one. The presence of magnetism in rocks that old implies that a fluid core probably existed for at least three-fourths of Earth's 4.6-billion-year history.

## Magnetic Stratigraphy

The Australian student's discovery of a reversely magnetized rock is consistent with worldwide observations of rocks demonstrating that Earth's magnetic field reverses periodically. The compass needle on which navigators and hikers depend is not so stable as it seems. Erratically, but roughly every half-million years, Earth's magnetic field changes polarity, taking perhaps a few thousand years to reverse its direction. Reversals are clearly indicated in the fossil magnetic record of layered lava flows, as shown in Figure 19.15. Each layer of rocks from the top down represents a progressively earlier period of geologic time, whose age can be determined by radiometric dating methods. The direction of remanent magnetism can be obtained for each layer, and in this way the time sequence of flip-flops of the field—that is, the **magnetic stratigraphy**—can be deduced. The detailed history of reversals over more than 5 million years has been worked out in this way (Figure 19.16). This

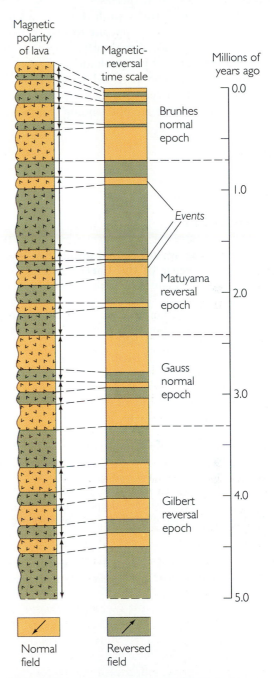

FIGURE 19.16 Magnetic polarities of lava flows are used to construct the time scale of magnetic reversals over the past 5 million years. Within epochs of magnetic polarity, there are short-term flip-flops of the field called *events*. In no one place is the entire sequence found; the sequence is worked out by patching together the ages and polarities from lava beds all over the world.

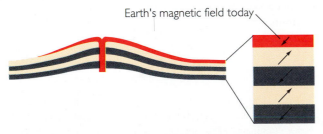

FIGURE 19.15 Lava beds become magnetized in the direction of the Earth's magnetic field existing at the time the beds solidified and cooled. In this way, they preserve the record of reversals of Earth's magnetic field. The modern flow at the top shows the direction of the field today. Underlying beds record the directions of ancient fields.

information is of use to archeologists and anthropologists as well as geologists. For example, the magnetic stratigraphy of continental sediments has been used to date sediments containing the remains of predecessors of our own species.

About half of all rocks studied are found to be magnetized in a direction opposite to that of Earth's present magnetic field. Apparently, then, the field has flipped frequently over geologic time, and normal (same as now) and reversed fields are equally likely. Normal and reverse magnetic epochs (each of which is named after an outstanding specialist in magnetism) seem to last on the order of a half-million years. Superimposed on the major epochs are transient, short-lived reversals of the field, known as magnetic events, which may last anywhere from several thousand to 200,000 years. The Australian graduate student apparently found a new reversal event within the present normal magnetic epoch.

The cause of reversals remains for future scientists to explain. We will see in Chapter 20, however, that their occurrence has made possible an important discovery that enables geologists to determine rates of seafloor spreading.

# SUMMARY

**What do seismic waves reveal about the layering in the interior of the Earth?** Seismic waves reveal that beneath Earth's felsic crust lies a denser ultramafic mantle. The crust and outer mantle to a depth of about 70 km make up the slablike lithosphere, which is broken into large, mobile plates. Beneath the lithosphere lie a partially molten asthenosphere, a primary source of basaltic magma; two transition zones where atoms are forced into closer packing by high pressures; a thick lower mantle; a fluid outer core, made mostly of iron; and an inner core made mostly of solid iron.

**What is the principle of isostasy and what evidence supports it?** The principle of isostasy proposes that continents float on the denser mantle, supported by a buoyant root that projects into the mantle. Mountains require even deeper roots to support their weight. Seismic waves reveal the existence of these roots.

**Where does the energy that drives geological processes come from?** Earth heated up in the process of becoming a planet, and its temperature was increased further by the heat released by the disintegration of radioactive elements. The cooling process takes place primarily by convection in the mantle. The convective movements are responsible for the motions of plates. Most geological activities, such as mountain making, volcanism, and earthquakes, occur at the boundaries where plates collide, separate, or slide by each other. The Sun's energy is responsible for climate, wind, and rain—all controlling factors in the erosion of rocks.

**What is paleomagnetism and what is its importance?** Geologists have discovered that rocks can become magnetized in the direction of Earth's magnetic field at the time they were formed. This remanent magnetization of rocks can be preserved for millions of years. Paleomagnetism tells us that Earth's magnetic field has reversed (flipped back and forth) over geologic time. The chronology of reversals has been worked out so that the direction of remanent magnetization of a rock formation is often an indicator of stratigraphic age.

# KEY TERMS AND CONCEPTS

shadow zone (p. 431)
Mohorovičić discontinuity (Moho) (p. 434)
principle of isostasy (p. 434)

conduction (p. 438)
convection (p. 438)
geotherm (p. 440)
thermoremanent magnetization (p. 442)

depositional remanent magnetization (p. 442)
paleomagnetism (fossil magnetism) (p. 443)
magnetic stratigraphy (p. 443)

# EXERCISES

1. How does the speed of P waves differ in granite, gabbro, and peridotite?

2. What evidence suggests that the asthenosphere is probably partially molten?

3. What evidence indicates that Earth's outer core is molten and composed mostly of iron?

4. What is the depth to the core and how do we know it?

5. What is the difference between heat conduction and convection?

6. How is convection in the mantle related to plate tectonics?

7. How can a solid rock creep?

8. How can a mountain float on the mantle when both are composed of rock?

9. How do rocks become magnetized when they form?

10. Was the direction of Earth's magnetic field what we think of as normal or reversed 4.7 million years ago?

# THOUGHT QUESTIONS

1. Mars and the Moon show no evidence of tectonic plates or of their motions. What does that observation imply about the state and temperature of the interiors of these planetary bodies?

2. How does the existence of Earth's magnetic field, iron meteorites, and the abundance of iron in the cosmos support the concept that Earth's core is mostly iron and the outer core is liquid?

3. How would you use seismic waves to find a chamber of molten magma in the crust?

# SUGGESTED READINGS

Bolt, B. A. 1982. *Inside the Earth*. San Francisco: W. H. Freeman.

Bolt, B. A. 1993. *Earthquakes and Geological Discovery*. New York: Scientific American Library.

Bott, M. H. P. 1982. *The Interior of the Earth: Its Structure, Constitution, and Evolution*, 2d ed. London: Edward Arnold.

Ernst, W. G. 1990. *The Dynamic Planet*. New York: Columbia University Press.

Fowler, C. M. R. 1990. *The Solid Earth*. Cambridge, England: Cambridge University Press.

McKenzie, D. P. 1983. The Earth's mantle. *Scientific American* (September):66.

Olson, P., P. G. Silver, and R. W. Carlson. 1990. The large-scale structure of convection of the Earth's mantle. *Nature* 344:209–215.

Sclater, J. G., C. Jaupart, and D. Galson. 1980. The heat flow through oceanic and continental crust and the heat loss of the Earth. *Reviews of Geophysics and Space Physics* 18:269–311.

# PLATE TECTONICS: THE UNIFYING THEORY

Geologists believe that Earth's lithosphere is broken into about a dozen plates, which slide by, collide with, or separate from each other as they move over Earth's interior. Plates are created where they separate and are recycled where they collide, in a continuous process of creation and destruction. Continents, embedded in the lithosphere, drift along with the moving plates. The theory of plate tectonics describes the movement of plates and the forces acting between them. The concept explains the distribution of many large-scale geologic features— mountain chains, structures on the seafloor, volcanoes, and earthquakes—which result from movements at plate boundaries. Plate tectonics provides the conceptual framework for this book, and indeed for the entire field of geology.

## 20

The Red Sea (*lower right*) divides to form the Gulf of Suez on the left and the Gulf of 'Aqaba on the right. The Arabian Peninsula on the right, splitting away from Africa on the left, has opened these great rifts, which are now flooded by the sea. The Nile River (*left center*) flows north into the Mediterranean Sea (*top*). *Earth Satellite Corp.*

Just a few years after plate-tectonics theory was proposed, on the occasion of an international scientific meeting in Moscow, an interesting exchange took place between two Western participants: a younger man who had achieved prominence because of his work on plate tectonics and a well-known older scientist. The setting was a party in the apartment of a Soviet geologist, and the conversation was well lubricated by vodka. The din of cocktail-party chatter stopped suddenly when the younger man called out to his older colleague, "Dr. ———, everyone tells me how brilliant you were in your younger days. If that's the case, why didn't you discover seafloor spreading and plate tectonics twenty years ago?" The explosive response of the older man needn't be recorded, but the question, properly generalized, is indeed thought-provoking. Why did this concept, which unifies so much of geological thought, arrive so late on the scene?

## FROM CONTROVERSIAL HYPOTHESIS TO RESPECTABLE THEORY

Some of the basic ideas of plate tectonics have a long history of support by a few visionary scientists and rejection by most of the established scientific leaders. The concept of **continental drift,** for example— large-scale movements of continents over the globe— had been around for a long time. The jigsaw-puzzle fit of the coasts on both sides of the Atlantic, as if the Americas, Europe, and Africa were at one time assembled together, had not escaped the notice of early natural philosophers. The English philosopher-scientist Sir Francis Bacon remarked on the parallelism of the facing shores of the Atlantic in 1620. At the close of the nineteenth century the Austrian geologist Ed-

uard Suess put some of the pieces of the puzzle together and postulated the former existence of a single giant continent: Gondwanaland, made up of the combined present-day southern continents. In 1915 Alfred Wegener, a German meteorologist, cited as further evidence of the breakup and drift of continents the remarkable similarity of rocks, geologic structures, and fossils on opposite sides of the Atlantic. In the years that followed, Wegener continued to build the case for continental drift; he postulated that a supercontinent called **Pangaea** (Greek for "all lands") once comprised all of the present continents, as in Figure 20.1. Pangaea began to break up in the Mesozoic, some 200 million years ago, with ocean filling the widening gaps between the continents as we know them today.

After about a decade of vociferous debate, the theory of continental drift was ignored by all but a few geologists in Europe, South Africa, and Australia. Although they could point not only to geographic matching but also to geological similarities in rock ages and trends in geologic structures when the continents are reassembled, the proponents could not come up with a plausible driving force that would split Pangaea and move the continents apart. Drift advocates buttressed their speculations with special pleading, selecting evidence patently favorable to their views even though it was far from incontrovertible. But they also offered significant arguments— accepted now as good evidence of drift—based on fossil and climatological data (Figure 20.2). The evolution of vertebrates and land plants showed similarities in development on different continents up to the supposed breakup time; thereafter these organisms followed divergent evolutionary paths, supposedly because of the isolation and changing environments of the separating fragments. The distribution of Permian glacial deposits in South America, Africa, India, and Australia was difficult to explain in terms of separate glaciers, some close to the equator. Drift

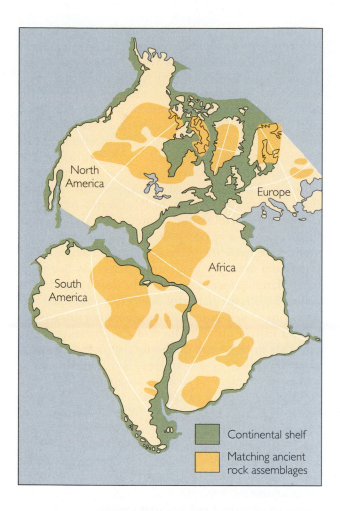

**FIGURE 20.1** The jigsaw-puzzle fit of continents bordering the Atlantic Ocean is a feature noted by scientists since the seventeenth century. Alfred Wegener called this supercontinent Pangaea and cited as additional evidence the similarity of geologic features on opposite sides of the Atlantic. The matchup of ancient crystalline rocks is shown by orange in adjacent regions of South America and Africa and of North America and Europe. (Geographic fit from data of E. C. Bullard; geological data from P. M. Hurley.)

advocates noted that if the southern continents are reassembled into Gondwanaland in the South Polar region, a single continental glacier could account for all the glacial deposits. However, an influential English scientist, Sir Harold Jeffreys, argued in 1929: "It has always happened that after several distinguished paleontologists have presented evidence favourable to continental drift, some other equally distinguished ones have proceeded to point out other facts that are made more difficult to explain." Geology and paleontology were not enough. Additional independent, diverse, corroborative evidence from other parts of geological science would be needed to persuade the scientific establishment to abandon prevailing ideas and elevate an unorthodox speculation to the level of a generally accepted theory.

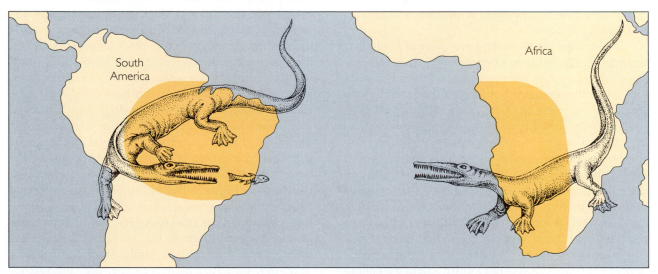

**FIGURE 20.2** Fossils of the late Paleozoic reptile *Mesosaurus* are found in South America and Africa and nowhere else in the world. If *Mesosaurus* could swim across the South Atlantic Ocean, it could have crossed other oceans and it should have spread more widely. That it did not suggests that South America and Africa must have been joined at that time. (After A. Hallam, "Continental Drift and the Fossil Record," *Scientific American,* November 1972, pp. 57–66.)

In 1928 Arthur Holmes invoked the mechanism of thermal convection in the mantle as the driving force of continental drift. Holmes proposed that subcrustal convection currents "dragged the two halves of the original continent apart, with consequent mountain building in the front where the currents are descending, and the ocean floor development on the site of the gap, where the currents are ascending." Holmes came close to expressing the modern notions of plates, divergence, and subduction when he speculated that a subcrustal basaltic layer serves as a conveyor belt that carries a continent along to the place where the belt turns downward into the mantle, leaving the continent resting on top. Nevertheless, Holmes himself recognized that "purely speculative ideas of this kind, specially invented to match the requirements, can have no scientific value until they acquire support from independent evidence."

Convincing evidence began to emerge as a result of extensive exploration of the seafloor after World War II. The mapping of the Mid-Atlantic Ridge and the discovery of the deep, cracklike valley, or rift, running down its center sparked much speculation. In the early 1960s Harry Hess of Princeton University and Robert Dietz of the University of California suggested that the seafloor separates along the rifts in mid-ocean ridges and that new seafloor forms by upwelling of hot mantle materials in these cracks, followed by lateral spreading. Thus was born the theory of seafloor spreading. Within a few years abundant confirmation was available from many independent lines of evidence, which we will investigate later.

It remained for the next generation of geologists to broaden the concept of continental drift and seafloor spreading into the more general theory of plate tectonics. Beginning about 1967, they extended the ideas of Hess and of the Canadian geologist J. T. Wilson about the mobility of the lithosphere by identifying the separate lithospheric plates and discussing their relative motions and the phenomena that occur at their boundaries. By the end of the 1960s the evidence became so persuasive that most Earth scientists embraced these concepts. Textbooks were revised, and specialists began to think of the implications that the new discoveries held for their own fields. From time to time in the history of science a fundamental concept appears that serves to unify a field by pulling together diverse theories and explaining a large body of observations. Such a concept in physics is the theory of relativity; in chemistry, the nature of the chemical bond; in biology, DNA; in astronomy, the Big Bang; and in geology, plate tectonics.

Let us return to the question of why these new concepts became generally accepted so late in the history of geology. Scientists work in different styles. Some scientists—those with particularly inquiring, uninhibited, and synthesizing minds—perceive great truths before others. Although their perceptions may turn out to be false, these individuals are often the first to see the great generalizations of science. Most scientists, however, proceed more cautiously and wait out the slow process of gathering supporting evidence. The concepts of continental drift and seafloor spreading were slow to be accepted simply because the audacious ideas came so far ahead of the firm evidence. The oceans had to be explored, a new worldwide network of seismographs had to be installed and used, the magnetic stratigraphy had to be painstakingly worked out, and the deep sea had to be drilled before the majority could be convinced. Scientists in a well-known European laboratory compiled a list (in good humor) of the names of Earth scientists in the order of the date they accepted seafloor spreading as a confirmed phenomenon. The names of scientists of distinction appear at both the top and the bottom of the list.

# OVERVIEW

As graduate students we were taught that the ocean basins were permanent and that continental drift was a wild hypothesis not to be taken seriously. The notion of a stable Earth with never-changing geographic features was a main tenet of geology before the emergence of plate tectonics. We now know that on the geologic time scale the seafloor is far from permanent. The present ocean basins are being created by seafloor spreading and destroyed and recycled by subduction on a time scale of about 200 million years. This is only about 4 percent of the age of the Earth, a measure of how geologically young the oceans are. Geologists have drilled into the floor of the deep oceans of the world in efforts to find remnants of seafloor older than 200 million years—without success. In 1990 the oldest rocks were found after a 20-year search in the western Pacific. They were of mid-Jurassic age, only some 175 million years old. Older rocks don't exist; they are lost by subduction. Small fragments of older seafloor are found embedded in continents—relics of old oceans that have disappeared.

Continents are more permanent than the seafloor. They are too buoyant to be subducted. They may be fragmented, moved, aggregated, and deformed by the movement of plates, but the pieces survive. The old core of North America, for example, was assembled by plate collisions about 2 billion years ago from pieces of even older continents, some as old as 4 billion years. Continents can be eroded and fragmented, but they can also grow over time by the gradual accumulation of materials along their margins. New continental strips can be added on here and there from time to time as plates separate and collide, fragment, move about, and reassemble.

With the emergence of these revolutionary ideas, geologists are rethinking Earth history. Most of the evidence we now have for plate tectonics comes from the seafloor. Seafloor spreading is reasonably well understood and is a relatively simple mechanism in comparison with the many phenomena that shape the enormously complex evolution of continents. Just how plate tectonics explains continental geology is now receiving much attention. Continental rock assemblages, volcanism, metamorphism, the evolution of mountain chains—all are being reexamined in the framework of plate tectonics. Most geologists now believe that the geology of continents has been dominated by plate tectonics throughout most of geologic history.

## THE MOSAIC OF PLATES

According to the theory of plate tectonics, the lithosphere is broken into a dozen or so rigid plates. Their outlines are shown on the maps of plates in Figure 20.3 and inside the front cover. The plates slide over a partially molten, weak asthenosphere, and the continents, embedded in the moving plates, are carried along passively. Continental drift is basically a consequence of plate movements. Plate tectonics works because Earth's rigid lithosphere enables plates with horizontal dimensions of thousands of kilometers to move as distinct, rigid units with very little buckling or breaking, except at boundaries. In earlier chapters we have described three types of plate boundaries according to the relative motions of adjacent plates: (1) boundaries of divergence or spreading, typically ocean ridges and rifted continents; (2) boundaries of

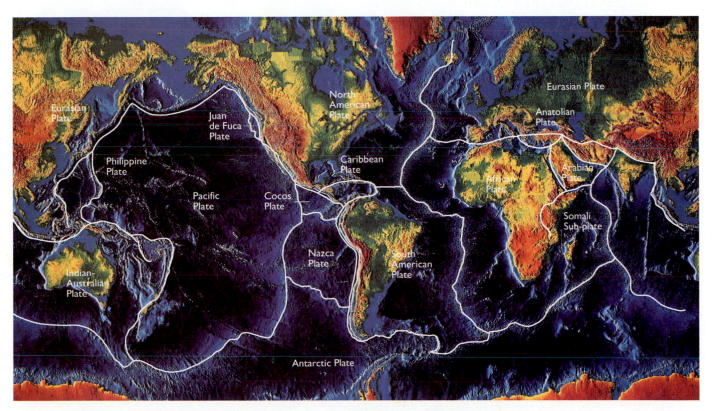

**FIGURE 20.3** Mercator-projection, color shaded-relief map of Earth's crustal plate boundaries. (Digital image by Dr. Peter W. Sloss, National Geophysical Data Center, Boulder, Colo.)

convergence or collision, typically deep-sea trenches, mountain ranges, and magmatic belts; and (3) transform boundaries.

## Divergent Boundaries

Divergent boundaries occur where plates move apart. When plates separate, partially molten mantle material rises and fills the gap between them. This material becomes new lithosphere added to the trailing edges of the diverging plates.

On the seafloor the boundary between separating plates is marked by a mid-ocean ridge that exhibits active basaltic volcanism, shallow-focus earthquakes, and normal faulting caused by tensional or stretching forces created by the pulling apart of two plates. Figure 20.4a shows what happens there. (A detailed portrait of the Mid-Atlantic Ridge may be seen in Figure 17.24.) The process by which plates separate and ocean crust is created is called **seafloor spreading.** The Mid-Atlantic Ridge and the East Pacific Rise are spreading centers that have created millions of square kilometers of seafloor. As we mentioned in Chapter 1, Iceland is an exposed segment of the Mid-Atlantic Ridge. It provides an opportunity to view the process of plate separation and seafloor spreading directly (see Figure 1.16).

Early stages of plate separation can be found on continents. Such sites are characterized by long

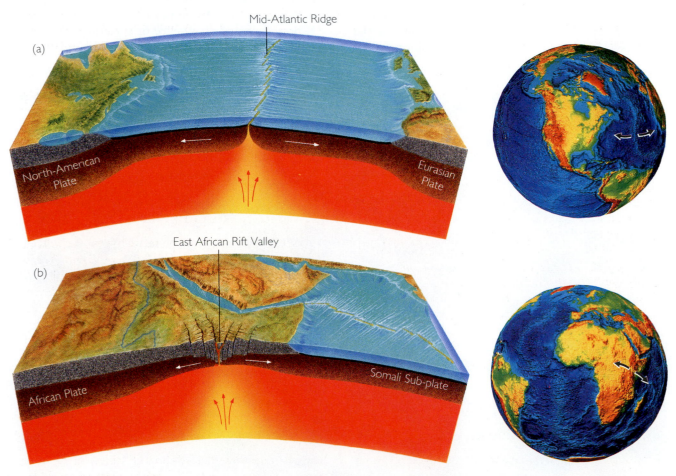

**FIGURE 20.4** (a) Rifting and seafloor spreading along the Mid-Atlantic Ridge create a mid-ocean volcanic mountain chain and a coincident earthquake belt. (b) Initiation of rifting and plate separation within a continent. Characteristic features are rift valleys, with multiple normal faults; volcanism; and earthquakes. (Color shaded-relief globes courtesy of National Geophysical Data Center, Boulder, Colo.)

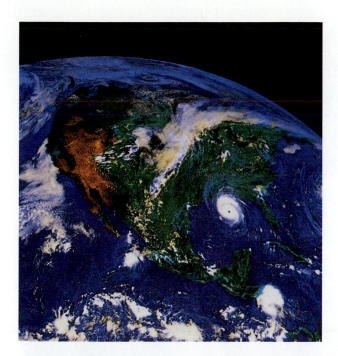

**FIGURE 20.5** Color-composite satellite view of Earth showing Gulf of Baja California, an opening ocean due to plate motions. Note Hurricane Andrew in the Gulf of Mexico. *Hasler/Palaniappan/Pierce/Manyin, NASA/Goddard Space Flight Center, Laboratory for Atmospheres.*

downfaulted valleys, volcanic activity, and earthquakes (Figure 20.4b). The Great Rift Valley of East Africa, labeled on the plate maps, is thought to represent an early stage of plate separation within a continent. On occasion continental rifting may slow down or stop before the continent splits apart and a new ocean basin opens up. The Great Rift Valley and the Rhine Valley, both of which are still mildly active, may be such places. The Red Sea and the Gulf of California are rifts that are farther along in spreading; because of their greater width and depth, they have been flooded by the ocean. The Arabian Peninsula is splitting away from Africa at the Red Sea (see the photograph at the beginning of this chapter), and Baja California is separating from the Mexican mainland (Figure 20.5).

## Convergent Boundaries

Plates collide along convergent boundaries. The overridden plate is subducted, or thrust downward into the mantle below, where eventually it is recycled. Collision and subduction produce deep-sea trenches, adjacent mountain ranges of folded and faulted rocks, and magmatic belts, as illustrated in Figure 20.6a, b. The magmatic belt can be a mountain range on land or a chain of volcanic islands, called an **island arc,** on the seafloor. Once subducted lithosphere is heated and partially melted, it joins hot mantle materials as a source of magma to feed and build the overlying chain of volcanoes. The collision of two plates generates very large forces in the region, and in a general way these forces must result in the faulting that triggers the shallow- and deep-focus earthquakes that occur in subduction zones, as we explained in Chapter 18. By the time the lithospheric plate has migrated from a divergent to a convergent boundary, it has cooled to become denser than the mantle below. Some geologists believe that the weight of the sinking part of the plate helps to pull the entire plate down and thus serves as an important part of the driving mechanism of plate tectonics.

A convergent boundary that exhibits all of these features is the one between the Nazca Plate and the South American Plate (see plate maps). The Peru-Chile deep-sea trench offshore of these countries, the Andes Mountains with their many volcanoes, and some of the world's largest shallow- and deep-focus earthquakes occur here.

When a plate collision involves two continents, both tend to remain afloat (see Figure 20.6c). The collision of India and Asia is a good example. In this case the Eurasian Plate is overriding the Indian Plate, creating a double thickness of crust and forming the highest mountain range in the world, the Himalayas.

## Transform Boundaries

At transform boundaries plates slide past each other, neither creating nor destroying lithosphere. Transform faults occur where the continuity of a divergent boundary is broken and offset, as Figure 20.7 shows. The San Andreas fault in California, where the Pacific Plate slides by the North American Plate, is a prime example of a transform boundary on land. Shallow-focus earthquakes with horizontal slips occur on transform boundaries.

## Modern Plate Boundaries

Each plate is bounded by some combination of these three kinds of boundaries. As the plate maps show, the Nazca Plate in the Pacific is bounded on three sides by zones of divergence, along which new lithosphere forms, and on one side by the Peru-Chile subduction zone, where lithosphere is consumed. Most

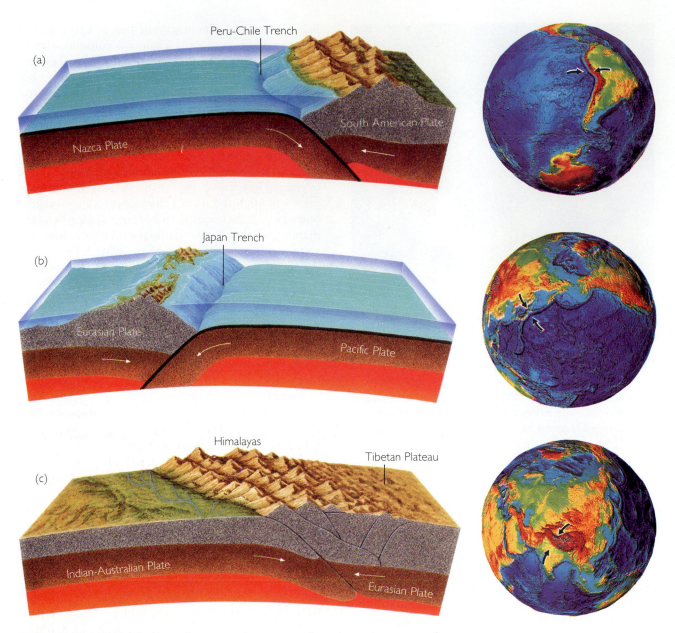

**FIGURE 20.6** (a) Subduction of an ocean plate at a continental margin creates a deep-sea trench, a volcanic belt at the margin of the continent, and shallow- and deep-focus earthquakes. (b) Subduction of an ocean plate beneath another ocean plate forms a volcanic island arc. (c) A continent-continent plate collision creates multiple thrusting and folding, double thickening of the continental crust, and high mountains. (Color shaded-relief globes courtesy of National Geophysical Data Center, Boulder, Colo.)

of the North American Plate is bounded by the divergent Mid-Atlantic Ridge on the east, the San Andreas fault and other transform boundaries on the west, and zones of subduction and transform that run from Oregon to the Aleutians on the northwest. Figure 20.8 depicts some relationships among modern plates, oceans, and continents.

# RATES OF PLATE MOTION

How fast do plates move? Do some plates move faster than others, and if so, why? Is the velocity of plate movements today the same as it was in the geologic past? Geologists have tried to answer these questions in recent years to gain a better understand-

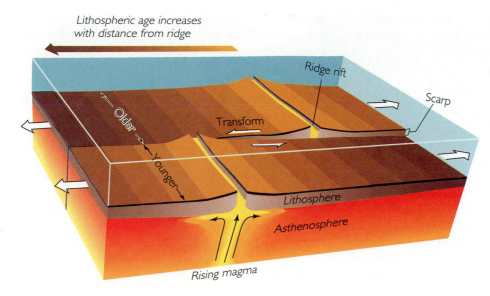

Lithospheric age increases with distance from ridge

**FIGURE 20.7** On the sea-floor, plates move in opposite directions across a transform boundary between offset sections and in the same direction across other segments of the transform. A scarp can occur because of the age difference across the transform boundary.

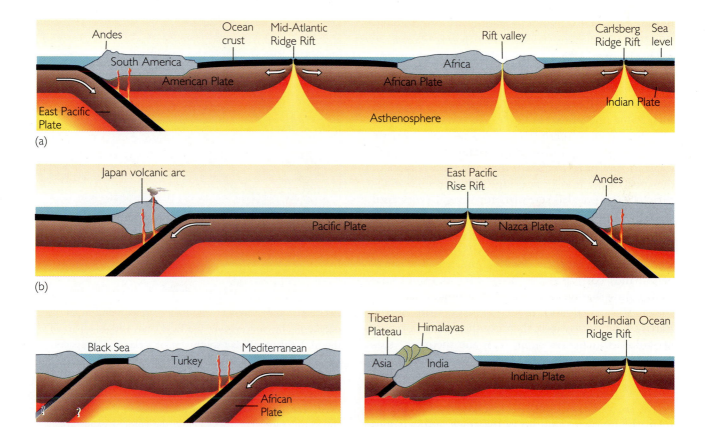

**FIGURE 20.8** Several plates and the boundaries between them as they exist today. Seafloor spreading in the Atlantic (a), Pacific (b), and Indian (d) oceans is depicted along with plate collisions of the types ocean-ocean (Japan, b), ocean-continent (South America, b, and Turkey, c), and continent-continent (India, d).

ing of plate tectonics. In doing so they have developed some ingenious methods to study the motions of plates.

## The Seafloor as a Magnetic Tape Recorder

During World War II extremely sensitive instruments were developed to detect submarines by the magnetic fields emanating from their steel hulls. Geologists made some slight modifications to these instruments and towed them behind research ships so that they could measure the local magnetic field created by magnetized rocks beneath the sea. In Chapter 19 we saw that rocks can become magnetized by the Earth's magnetic field in whatever direction the field is oriented at the time the rocks formed. The current direction of the magnetic field is referred to as "normal." The opposite orientation is referred to as "reverse." In the geologic past, Earth's magnetic field switched back and forth erratically between normal and reverse. If the ship was above rocks magnetized in the normal direction, geologists found a positive local magnetic field, or a *positive anomaly;* reversely magnetized rocks below the seafloor created a *negative anomaly.*

Steaming back and forth across the ocean, seagoing scientists discovered magnetic anomaly patterns, such as the one shown in Figure 20.9, with a regularity that surprised them. In many areas, long narrow bands of positive and negative magnetic anomalies showed an almost perfect symmetry with respect to the crest of the mid-ocean ridge. This peculiar magnetic pattern puzzled scientists for several years until two Englishmen, F. J. Vine and D. H. Mathews, and, independently, two Canadians, L. Morley and A. Larochelle, made a startling proposal in 1963. They reasoned that the positive and negative magnetic bands correspond to bands of rock on the seafloor below that were magnetized during ancient episodes of normal and reversed magnetism of the Earth's field. If this were the case, the magnetic bands provided evidence in support of the theory of seafloor spreading, which had already been proposed (Figure 20.10). They argued that the ocean progressively widens as new seafloor is created along a crack on the crest of a mid-ocean ridge. Magma flowing up from the interior solidifies in the crack and becomes magnetized in the direction of Earth's field at the time. As the seafloor splits and moves away from the ridge, approximately half of the newly magnetized material moves to one side and half to the other, forming two symmetrical magnetized bands. Newer

material fills the crack, continuing the process. In this way the seafloor acts like a tape recorder that encodes, by magnetic imprinting, the history of the opening of the oceans in terms of the history of reversals of Earth's magnetic field.

Remember that the ages of reversals have been worked out from magnetized lavas on land (see Figure 19.16). Using this known sequence of reversals over time, geologists could assign ages to the bands of magnetized rocks on the seafloor. Since they now knew the age of a band of magnetized rocks on the seafloor and knew the distance from a mid-ocean ridge crest where the magnetized rocks were created, they could calculate how fast the ocean opened up—that is, the velocity of plate movements. For exam-

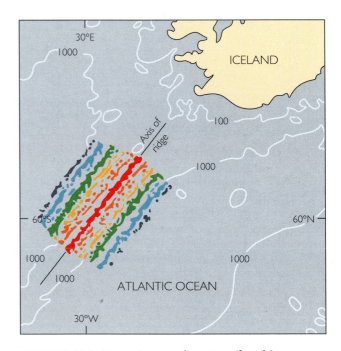

**FIGURE 20.9** Magnetic anomaly pattern found in an oceanographic survey over the Reykjanes Ridge, a part of the Mid-Atlantic Ridge southwest of Iceland. The spaces between the colored bands show where the survey ship found negative magnetic anomalies and correspond to rock formations on the seafloor below that are reversely magnetized. The ship found positive anomalies in the bands shown in color. The rocks below the colored bands are magnetized in the normal direction; that is, similar to the present-day direction. The pattern, with its almost perfect symmetry with respect to the ridge axis, puzzled geologists. When the pattern was explained (see Figure 20.10), it provided strong support for the concept of seafloor spreading.

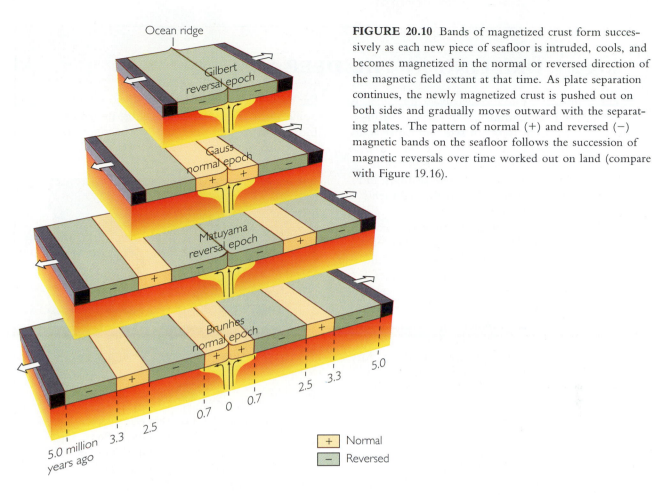

**FIGURE 20.10** Bands of magnetized crust form successively as each new piece of seafloor is intruded, cools, and becomes magnetized in the normal or reversed direction of the magnetic field extant at that time. As plate separation continues, the newly magnetized crust is pushed out on both sides and gradually moves outward with the separating plates. The pattern of normal (+) and reversed (−) magnetic bands on the seafloor follows the succession of magnetic reversals over time worked out on land (compare with Figure 19.16).

☐ + Normal
☐ − Reversed

ple, rocks on the crest of the ridge would be modern, hence normally magnetized, because they were extruded during the current normal magnetic epoch. Conversely, magnetized rocks corresponding to a magnetic epoch of about 1 million years ago have been displaced some distance from the ridge—say, about 20 km on each side of the ridge crest if the plates are spreading apart at a rate of 2 cm per year, or 40 km if the spreading rate is 4 cm per year. Using this method, geologists found that the highest rate of seafloor spreading is 10 to 12 cm per year at the East Pacific Rise and the lowest is 2 cm per year at the Mid-Atlantic Ridge.

The normal-reversal time scale can be followed through many oscillations of Earth's magnetic field. The corresponding magnetic bands on the seafloor, which can be thought of as age bands, extend from the ridge crests across the ocean basins over a time span exceeding 100 million years.

The power and convenience of using seafloor magnetism to work out the history of ocean basins cannot be overemphasized. Simply by steaming back

and forth over the ocean, measuring the magnetic field of the magnetized rocks of the seafloor, and correlating the pattern of reversals with the time sequence worked out by the methods just described, geologists determined the ages of various regions of the seafloor without even examining rock samples. In effect, they learned how to "replay the tape." The simplicity and elegance of seafloor magnetism made it an attractive tool. But it was an indirect method in that rocks were not recovered from the seafloor and their ages were not directly determined in the laboratory. Corroborative evidence would still be needed to convince the remaining skeptics. Deep-sea drilling supplied it.

## Deep-Sea Drilling

In 1968 a program of drilling into the seafloor was launched (see Box 20.1). This joint project of major oceanographic institutions and the National Science Foundation aimed to drill through, retrieve, and

## BOX 20.1  DRILLING IN THE DEEP SEA

The deep-sea drilling vessel *Joides Resolution* is 143 m long, and amidships it carries a drilling derrick 61 m high. In size and capacity to drill in the deepest ocean it is the only ship of its kind in the world. It can lower drill pipe thousands of meters to the seafloor and drill thousands of meters into the sediments and underlying basaltic crust.

Before the ship could accomplish such a feat, a technological breakthrough was required. A way had to be found to hold the ship stationary during the drilling process, regardless of currents and winds; otherwise, the drill pipe would break off. The problem was solved by development of a positioning device that uses sound waves transmitted by acoustic beacons planted on the seafloor. Any change in the ship's position is sensed by a computer that monitors changes in the time of arrival of the sound pulses from each beacon. The same computer controls the speed and steering to keep the vessel stationary.

Deep-sea drilling was the answer to those who said, when lunar exploration started, "Better to explore the ocean's bottom than the back side of the Moon." We ended up doing both. The deep-sea drilling program is more than 25 years old and has become international in scope.

Scientists on board a drilling ship take samples from cores of sediment recovered from the seafloor. These samples can be analyzed to reveal the history of ocean basins and ancient climatic conditions. *Texas A&M University*.

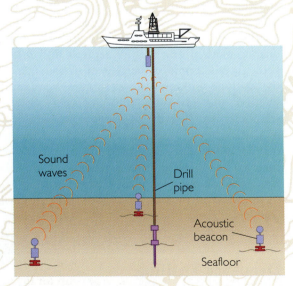

Acoustic beacons on the seafloor are used to keep a deep-sea drilling vessel stationary.

The deep-sea drilling vessel *Joides Resolution*. *Texas A&M University*.

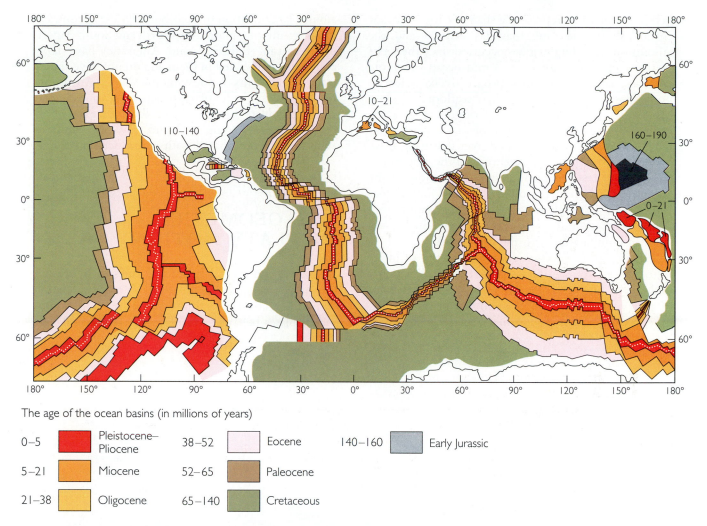

The age of the ocean basins (in millions of years)

| | | | | |
|---|---|---|---|---|
| 0–5 | Pleistocene–Pliocene | 38–52 | Eocene | 140–160 — Early Jurassic |
| 5–21 | Miocene | 52–65 | Paleocene | |
| 21–38 | Oligocene | 65–140 | Cretaceous | |

**FIGURE 20.11** The worldwide pattern of seafloor spreading is revealed by the isochrons (contours separating the bands of different colors and textures) that give the age of the seafloor in millions of years since its creation at ridges. Mid-ocean ridges, along which new seafloor is extruded, coincide with the youngest seafloor (red). The Atlantic Ocean is symmetrical about the Mid-Atlantic Ridge. Asymmetry of the pattern in the Pacific is caused partly by subduction in the Aleutian Trench south of Alaska, in the Peru-Chile Trench along the west coast of South America, and in many trenches in the western Pacific. (After map prepared by J. Sclater and L. Meinke.)

study seafloor sediments from many places in the world's oceans. Geologists thus had an opportunity to work out the history of the ocean basin directly.

One of the most important facts geologists sought was the age of each sample. Since sedimentation begins as soon as ocean crust forms, the age of the oldest sediments in the core, those immediately on top of the basaltic crust, tells the geologist how old the ocean floor is at that spot. The age is obtained from the fossil skeletons of tiny animals that live in the ocean and sink to the bottom when they die. It

was found that the oldest sediments in the cores become older with increasing distance from mid-ocean ridges and that the age of the seafloor at any one place agrees almost perfectly with the age determined from magnetic reversal data. This agreement validated magnetic dating of the seafloor and clinched the concept of seafloor spreading. What a coup for the scientists who discovered this tool!

Figure 20.11 shows the ages of the seafloor of the world's oceans. The contours that represent ages of the seafloor are called **isochrons.** The colored bands between isochrons represent a span of time. The isochrons show the time that has elapsed and the amount of spreading that has occurred since the rocks were injected as magma into a mid-ocean rift. Note how the seafloor becomes progressively older on both sides of the ridge rifts, where seafloor spreading originates. The distance from a ridge axis to a 65-million-year isochron, for example, indicates the extent of new ocean floor created over that time span. The more widely spaced isochrons of the eastern Pacific signify faster spreading rates than those in the Atlantic. No sediments older than the Jurassic period, about 200 million years ago, have been found. This observation attests to the "youth" of the seafloor in comparison with the continents.

Figure 20.12 summarizes the velocities of plates with respect to one another—their relative velocities—and shows the directions of plate motions. Geologists have noted that the fast-moving plates (the Pacific, Nazca, Cocos, and Indian) are being subducted along a large fraction of their boundaries. In contrast, the slow-moving plates (the North and South American, African, Eurasian, and Antarctic) have large continents embedded in them and do not have significant attachments of downgoing slabs. An attractive hypothesis that explains these observations associates rapid plate motions with the pull exerted by large-scale downgoing slabs and slow plate motions with the drag associated with embedded continents.

## THE GEOMETRY OF PLATE MOTION

Geologists use several geometric rules to figure out the direction of the movement of one plate in relation to another. These rules follow from the fact that individual plates behave as rigid bodies. By "rigid" we

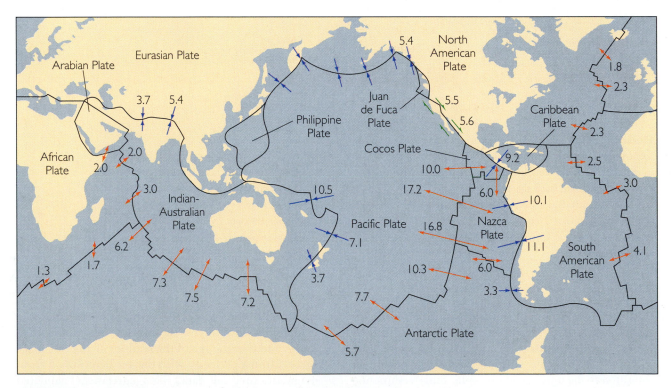

**FIGURE 20.12** Relative velocities (in centimeters per year) and directions of plate separation and convergence. Opposed arrowheads indicate convergence. Diverging arrowheads indicate plate separation at ocean ridges. Parallel arrowheads, as along the San Andreas fault in California, indicate transform faults, where plates slide past each other. Spreading is fastest between the Pacific and Nazca plates and slowest between the North American and Eurasian plates. (Data from C. Demets, R. G. Gordon, D. F. Argus, and S. Stein, Model Nuvel-1, 1990.)

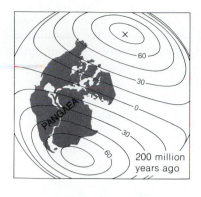

× = Ancient geographic pole

**FIGURE 20.13** Plate movements have led to the northward drift of the continents and the opening of the Atlantic Ocean over the past 200 million years. The central Atlantic, the Caribbean, and the Gulf of Mexico began to form about 200 million years ago in Triassic time, when Pangaea began to break up and Africa and South America drifted away from North America. The South Atlantic opened about 150 million years ago with the separation of South America from Africa. As the continents drifted apart, they also migrated in a northerly direction to their present positions. Note that the equator passed through the southern parts of the United States and Europe in Triassic time. (After J. D. Phillips and D. Forsyth, "Plate Tectonics, Paleomagnetism, and the Opening of the Atlantic," *Bulletin of the Geological Society of America,* vol. 83, no. 6, 1972, pp. 1579–1600.)

mean simply that the distances between three points on the same plate—say, New York, Miami, and Bermuda on the North American Plate—do not change, no matter how the plate moves. But the distance between New York and Lisbon, say, increases because the two cities are on different plates that are being separated along a narrow zone of spreading on the Mid-Atlantic Ridge. Here are some geometric principles that govern the sliding of plates on our planet.

1. No overlap, buckling, or separation occurs along typical transform boundaries; the two plates merely slide past each other without creating or destroying plate material. Look for a transform boundary if you want to deduce the direction of plate motions, because the orientation of the fault is the direction in which one plate slides with respect to the other, as Figure 20.7 shows.

2. Magnetic anomaly bands and age bands (isochrons) on the seafloor are roughly parallel and are symmetrical with respect to the ridge axis along which they were created. Figures 20.10 and 20.11 illustrate this observation. Since each isochron was at the plate boundary of separation at an earlier time, isochrons that are of the same age but on opposite sides of an ocean ridge can be brought together to show the positions of the plates and the configuration of the continents embedded in them as they were in that earlier time. By this means we can reconstruct, for example, the opening of the Atlantic Ocean, as shown in Figure 20.13.[1]

By applying such geometric principles, geologists are able to deduce spreading rates from spreading directions and magnetic anomalies and can work out the history of the motions of all the lithospheric plates. Some results have already been pictured in Figures 20.11 and 20.12. However, geologists are searching for other ways to measure the motions of plates. If the hot spots discussed in Chapter 5 turn out to be fixed in the mantle below plates, then the string of extinct volcanoes trailing from the hot spot records the movement of individual plates as they glide over the mantle (see Figure 5.29).

---

[1]Archeologists believe that the builders of the Great Pyramid of Egypt designed it to aim due north. Today it is aimed slightly east of north. Did the ancient Egyptian astronomers make a mistake in orienting the pyramid 40 centuries ago? Probably not. Over this period Africa drifted enough to rotate the pyramid out of alignment with true north.

An exciting new method for measuring plate motions involves bouncing pulses of light from ground-based lasers off the orbiting *Laser Geodynamics Satellite (LAGEOS).* Because the motion of the satellite is known precisely, scientists can determine the positions of the ground sites with respect to one another by timing the round trip of the laser pulse. The measurements are repeated thousands of times over a period of a few years to detect changes in the distances between the ground sites. NASA scientists have recently announced preliminary results for continental motions that agree overall in magnitude and direction with those found by the geological methods described earlier. In a sense the satellite serves as an outside observer, independently validating the theories and methods of Earth-bound geologists as they reconstruct plate motions from the geologic record. Taken together, the short-term *LAGEOS* observations and the long-term geologic data indicate that the plate motions taking place today are roughly the same as those that have been occurring over the past few million years. Plate motions are now being measured on a yearly basis in many places over the globe.

We used to depend solely on ships to explore the topography and structure of the seafloor. But ships would require decades to explore every corner of every ocean. Fortunately, another new geological tool made possible by advances in technology is now available to speed up the process. It is a method of imaging the seafloor from an orbiting satellite, as if the seawater were drained away (see Box 20.2). All of the major structures associated with seafloor spreading are revealed, such as mid-ocean ridges, deep-sea trenches, and transform boundaries, including many not previously found by ship surveys.

# ROCK ASSEMBLAGES AND PLATE TECTONICS

The only record we have of past geologic events is the incomplete one found in the rocks that have survived erosion or subduction. Since only seafloor younger than 200 million years has survived subduction, we must focus on the continents to find the old rocks that provide the evidence for most of Earth's history. Some of the methods of reading the rock record have been described in earlier chapters. They include interpreting unconformities, faults, and other structures; finding evidence for uplift and erosion; deducing the environment in which sediments were deposited; and reconstructing the original condition of rocks that have been deformed or metamorphosed. In Chapter 4 we investigated the relationship between plate boundaries and various types of igneous rocks. Here we explore in more detail the types and combinations of rocks that are characteristic of the various kinds of plate boundaries. Geologists who study continents use these characteristic **rock assemblages** to identify ancient episodes of plate separations and collisions.

## Rocks at Divergent Boundaries

Before the advent of plate tectonics, geologists were puzzled by unusual assemblages of rocks found on land consisting of deep-sea sediments, submarine basaltic lavas, and mafic igneous intrusions (Figure 20.14). Such assemblages are known as **ophiolite suites.** Using data gathered from deep-diving sub-

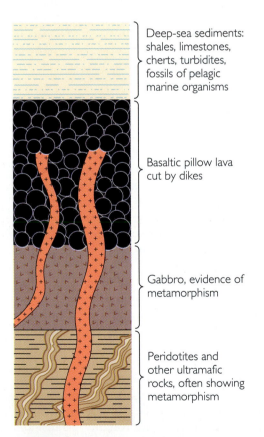

Deep-sea sediments: shales, limestones, cherts, turbidites, fossils of pelagic marine organisms

Basaltic pillow lava cut by dikes

Gabbro, evidence of metamorphism

Peridotites and other ultramafic rocks, often showing metamorphism

**FIGURE 20.14** Idealized section of an ophiolite suite. The combination of deep-sea sediments, submarine pillow lavas, sheeted basaltic dikes, and mafic igneous intrusions indicates a deep-sea origin. Ophiolites are fragments of ocean lithosphere emplaced on a continent as a result of plate collisions.

## BOX 20.2 CHARTING THE SEAFLOOR BY SATELLITE

The rich geology of the seafloor—its ridges, trenches, seamounts, and transform boundaries—became apparent only after decades of ship soundings. Our knowledge of regions where few ships travel remains fragmentary. Recently, however, scientists have developed a tool that enables a satellite to "see through" the ocean and chart the topography of the seafloor, gathering data in mere months.

The new method makes use of an altimeter mounted on a satellite. The altimeter sends pulses of radar beams that are reflected back from the ocean below, giving measurements of the distance between the satellite and the sea surface with a precision of a few centimeters. The height of the sea surface depends not only on waves and ocean currents but also on changes in gravity caused by the topography and composition of the underlying seafloor. The gravitational attraction of a seamount, for example, can cause water to "pile up" above it, producing a bulge in the sea surface as much as 5 m above average sea level. Similarly, the diminished gravity over a deep-sea trench would show as a depression of the sea surface of as much as 60 m. In this way, features of the ocean floor can be inferred from satellite data and displayed as if the seas were drained away.

In the satellite photograph below, shallow regions, deep regions, and intermediate depths can be distinguished. Also clearly visible are the raised stripe between Europe and North America that marks the Mid-Atlantic Ridge and its associated transform boundaries, the trail of a hot spot in the Pacific marked by the Emperor-Hawaiian seamount chain, and the major deep-sea trenches at subduction boundaries. New features not revealed by ship surveys have already been found, and future surveys may reveal even deeper structures, such as convection currents in the mantle. The methodology for making geotectonic images such as this one was developed by W. F. Haxby of Columbia University, who supplied the photograph.

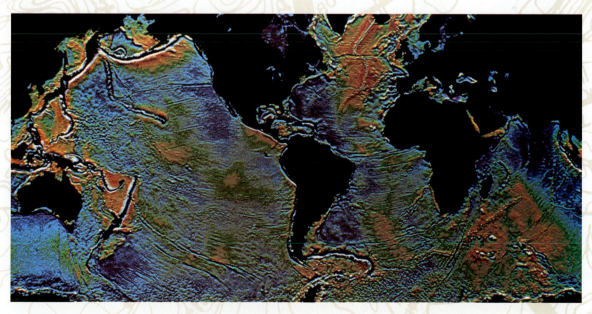

Satellite chart of the seafloor. *W. F. Haxby, Columbia University.*

marines, dredging, deep-sea drilling, and seismic exploration, geologists can now explain these exotic rocks as fragments of oceanic crust that were originally formed on the seafloor, were transported by seafloor spreading, and were then raised above sea level in an episode of plate collision.

The creation of oceanic crust along the axis of a mid-ocean ridge is intimately related to the separation of plates there. Figure 20.15 is a highly simplified picture of the way ocean crust forms. As the plates separate, hot mantle rises and begins to melt. It is a mush by the time it reaches shallow depths. Basaltic melt fills a shallow magma chamber. Magma from the chamber repeatedly intrudes the rift between the spreading plates and solidifies as vertical sheets of dikes. Dikes intrude dikes to form a structure that has been likened to a pack of cards standing on edge. Basalt spilling out on the seafloor freezes as pillow lavas (see Figure 5.4)—the characteristic form of undersea volcanism—and forms a cover over the sheeted dikes. The roof of the magma chamber, cooled by circulating seawater, cools the adjacent magma, which crystallizes and sticks to the roof as the coarse-grained basaltic rock gabbro. It forms a layer below the sheeted dikes. Within the magma chamber, denser minerals settle out to form peridotites. The Moho is the boundary between the gabbro and the peridotite. A thin blanket of deep-sea sediments covers the ocean crust. As the seafloor spreads, the zones of lavas, dikes, gabbros, and peridotites are transported away from the mid-ocean ridge, where

this characteristic sequence of rocks that make up the oceanic crust and uppermost mantle is assembled—almost like a production line. The scale of this "factory" where ocean crust is created is some 10 km wide and 10 km deep, and it extends along the thousands of kilometers of mid-ocean ridge. The magma chambers along the length of the ridge are periodically replenished by fresh injections of basaltic magma to keep the process going.

The mid-ocean ridge is also a factory for the formation of massive ore bodies rich in iron, copper, and other minerals. The ores are formed when seawater sinks through porous volcanic rocks, becomes heated, and leaches these elements from the underlying hot rocks. When the heated seawater enriched with dissolved minerals rises and reenters the cold ocean, the ore-forming minerals precipitate (see Figure 23.14).

Ophiolites found on land have helped geologists to reconstruct the deeper features of the process by which ocean crust is formed. We can literally walk across rocks that used to lie along the Moho of the ocean crust on some of the more complete ophiolite sequences preserved on land (Figure 20.16).

When plate separation is initiated within a continent, the ancient site of rifting can often be found by another characteristic assemblage of rocks and structures, including rifts and downdropped blocks of continental crust, volcanic intrusions, and thick sedimentary basins along continental margins. The process begins when the continental crust and underly-

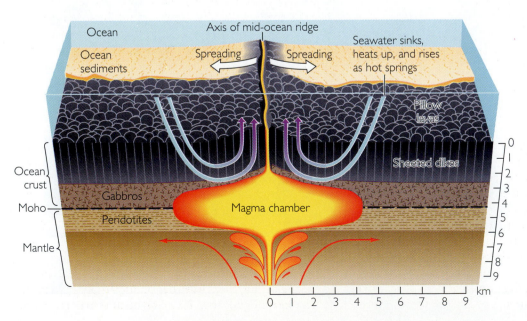

**FIGURE 20.15**

Oceanic crust forms at an ocean spreading center. (After G. I. Bass, "Ophiolites," *Scientific American,* August 1982, p. 122.)

**FIGURE 20.16**
Precambrian ophiolite suite, northern Quebec, Canada. The basalt pillows and other rocks typical of plate separation in this 2-billion-year-old slice of seafloor are compelling evidence that plate tectonics was occurring in Precambrian time. *M. St. Onge/Geological Survey of Canada.*

ing lithosphere are stretched and thinned by the forces of separation (Figure 20.17). A long, narrow rift develops, with great downdropped crustal blocks. Hot ductile mantle rises and fills the space created by the thinned crust, initiating the volcanic eruption of basaltic rocks in the rift zone.

If the divergence continues until the two segments of continent separate, the widening rift is flooded by the sea and a new ocean basin forms and grows. The receding continental margins subside gradually as the underlying lithosphere cools and contracts, forming offshore basins that can receive sediments eroded from the adjacent land. These are **continental shelf deposits**—sedimentary rock assemblages that are laid down in an orderly sequence under the quiet conditions of a slowly subsiding basin along a receding continental margin. The sedimentary basins off the Atlantic coasts of North and South America, Europe, and Africa are products of this process. These basins began to form when the supercontinent Pangaea split about 200 million years ago and the American plates separated from the European and African plates. Figure 20.17 shows the wedge-shaped deposit of sediments underlying the Atlantic continental shelf and margin of the United States, which were formed in this way. Because the trailing edge of the continent slowly subsides, the offshore basins continue to receive sediments for a long time. The load of the growing mass of sediment further depresses the crust, so that the basins can re-

ceive still more material from land. The result of these two effects is that the deposits can accumulate in an orderly fashion to thicknesses of 10 km or more.

## Rocks at Convergent Boundaries

*Orogeny* means "mountain making," particularly by the folding and thrusting of rock layers, often with accompanying magmatic activity. Plate tectonics has provided new insights into this fundamental process of geology and the rock assemblages associated with it. As we saw earlier (see Figure 20.6), there are three types of convergent boundaries: ocean-ocean, ocean-continent, and continent-continent.

**OCEAN-OCEAN**  When an ocean plate collides and overrides another ocean plate, several complex processes are set in motion (Figure 20.18). The ocean sediments of the downgoing plate are mostly scraped off and accrete along the margin of the overriding plate. The cool subducted lithosphere descends into the hot mantle below. At depths of 50 to 100 km it encounters temperatures in the range of 1200–1500°C. Some of the subducted plate melts, and water and other volatiles are released. This process in turn causes peridotite, the major constituent in the wedge of mantle above the subducted plate, to melt. The heated and now buoyant mantle material rises,

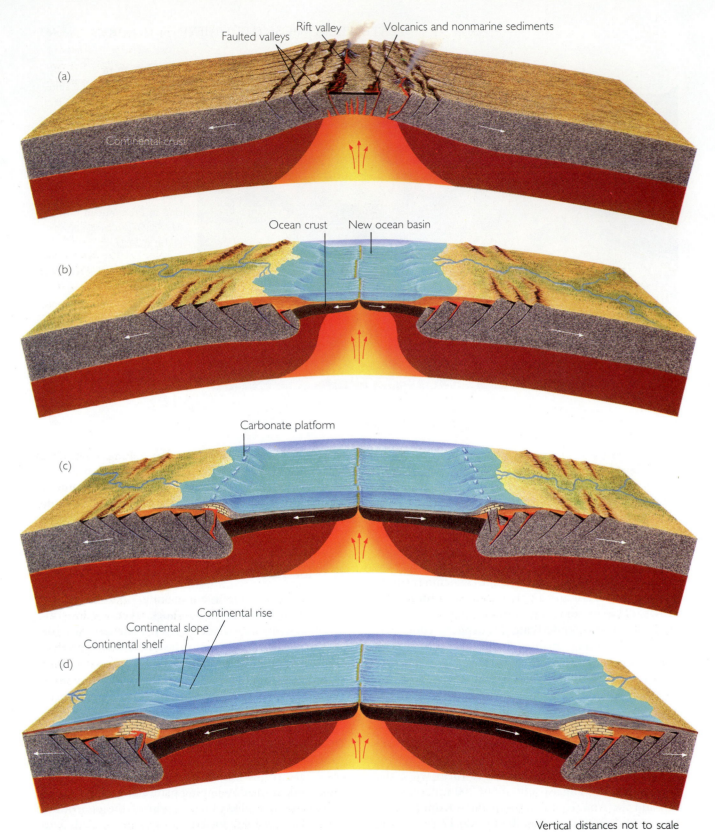

Vertical distances not to scale

**FIGURE 20.17** The development of sedimentary basins on a rifted continental margin. A rift develops in Pangaea as hot mantle materials upwell and the ancient continent stretches and thins. Volcanics and Triassic nonmarine sediments are deposited in the faulted valleys (a). Seafloor spreading begins (b). The lithosphere cools and contracts under the receding continental margins, which subside below sea level. Evaporites, deltaic deposits, and carbonates (c) are deposited and then covered by Jurassic and Cretaceous sediments derived from continental erosion (d). The Atlantic margins of Europe, Africa, and North and South America have histories similar to this.

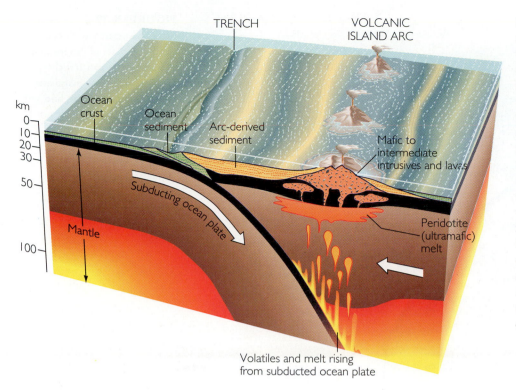

km

0 —
10 —
20 —
30 —

50 —

100 —

TRENCH

VOLCANIC
ISLAND ARC

Ocean
crust

Ocean
sediment

Arc-derived
sediment

Subducting ocean plate

Mantle

Mafic to
intermediate
intrusives and lavas

Peridotite
(ultramafic)
melt

Volatiles and melt rising
from subducted ocean plate

**FIGURE 20.18** Rock assemblages associated with the collision of two ocean plates and subduction. Water, other volatiles, and melt rise from the heated subducted plate and cause melting of peridotite in the overlying wedge of mantle. Ocean crust is intruded by magma to form a volcanic arc, which erupts mafic and intermediate lavas. (After National Research Council, "Margins," *Proceedings of Workshop of National Research Council,* Washington, D.C., National Academy of Sciences, 1989.)

and further melting occurs as the pressure eases. The result is ultramafic magma, which accumulates at the base of the crust of the overriding plate; some of it intrudes into the crust. Crystalline fractionation occurs, and some crustal rocks are assimilated into the magma, as we mentioned in Chapter 4. In this way the ultramafic magma evolves into mafic and more silicic magmas and lavas such as basalt, andesite, and dacite. The intrusions and volcanic eruptions build an arcuate chain of volcanic islands on the seafloor. The West Indies, the Aleutians, the Philippines, and the Marianas are islands of this sort.

**OCEAN-CONTINENT** The rock assemblages that form when a plate carrying a continent on its leading edge collides with, overrides, and subducts an oceanic plate are shown in Figure 20.19. A mountain belt develops on the continental margin, built up by multiple intrusions of igneous rock and eruptions of lava. Multiple thrusts may occur, with sections of rock thrust on top of one another, contributing to the mountain-building episode.

As when one ocean plate collides with another, the primary source of magma is the melting of peridotite in the wedge of mantle above the subducted plate. The characteristic igneous rocks produced are basaltic and andesitic lavas, along with a few dacites and rhyolites. Granitic batholiths intrude within the continental crust. The assemblage of igneous rocks is more silicic than those found on island arcs, perhaps because the magma derived from the mantle is con-

taminated by (assimilates) the melting continental crust.

Metamorphic rocks are found in these magmatic belts, typically the result of recrystallization under high temperatures and low pressures. These conditions occur because the hot fluids rise close to the surface, delivering much heat to this low-pressure environment.

Thick marine sediments, many of them turbidites, eroded from the continent rapidly fill the adjacent depressions in the seafloor. In descending, the cold oceanic slab stuffs the region below the inner wall of the trench (the wall closer to land) with these sediments and with deep-sea sediments and ophiolite shreds scraped off the descending plate. Regions of this sort between the magmatic arc on the continent and the offshore trench are enormously complex and variable. The deposits are all highly folded, intricately sliced, and metamorphosed. They are difficult to map in detail but are recognizable by their distinctive mix of materials and structural features. Such a chaotic mess has been called a **mélange**. The metamorphism is the kind characteristic of high pressure and low temperature because the material may be carried relatively rapidly to depths as great as 30 km, where recrystallization occurs in the environment of the still-cold subducting slab. Somehow, as part of the subduction process, the material rises back to the surface.

Find a paired belt of mélange and magmatism and you have a relic left behind by an ancient episode

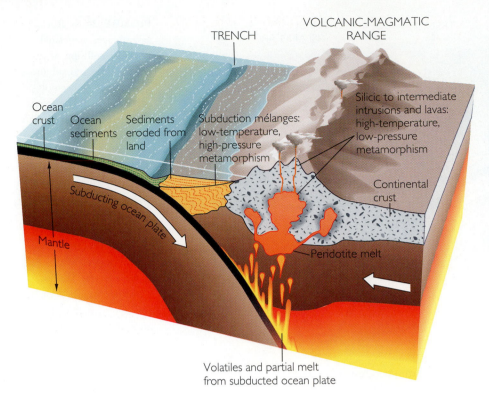

**FIGURE 20.19** Rock assemblages associated with ocean-continent plate collisions and subduction: ocean trenches, mélange deposits, magmatic belts, metamorphism, and volcanism. The drawing is not to scale; the thickness of the lithosphere is about 70 km, the depth of the ocean trench is about 10 km, and the distance from trench to magmatic belt is 300 to 400 km.

of subduction: the mélange formed offshore near where the subducted plate turned downward; the magmatic belt formed landward, from the melt that rose from greater depths along the downthrust plate. The essential elements of these rock assemblages of collision have been found in many places in the geologic record. One can see mélange in the Franciscan formation of the California Coast Ranges and magmatism in the parallel belt of the Sierra Nevada to the east (Figure 20.20). This paired belt marks the Mesozoic boundary of a collision between the North American Plate and the Farallon Plate, which has disappeared. The location of mélange on the west and magmatism on the east even shows that the now-absent Farallon Plate was the subducted one, overridden by the North American plate on the east. Other paired belts can be found along the continental margins framing the Pacific basin—in Japan, for instance. The central Alps were produced by the convergence of a Mediterranean plate with the European continent.

The Andes Mountains (from which the name of the volcanic rock andesite is derived), near the west coast of South America, are products of a collision between ocean and continental plates. Here the Nazca Plate collides with and is subducted under the South American plate.

**CONTINENT-CONTINENT** Since plates may have continents embedded in them, a continent can collide with another continent, as shown in Figure 20.21.

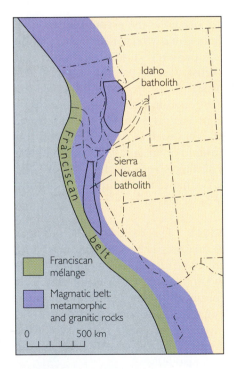

**FIGURE 20.20** The geology of the western United States at the beginning of Tertiary time, some 60 million years ago. The paired mélange and magmatic belts indicate a collision of the North American Plate to the east and the Farallon Plate to the west; the Farallon Plate was subducted. The Franciscan mélange formation and the batholiths of the Sierra Nevada to the east as they exist today are indicated. (After W. Hamilton and W. B. Myers, "Cenozoic Tectonics," *Reviews of Geophysics,* vol. 4, 1966, p. 541.)

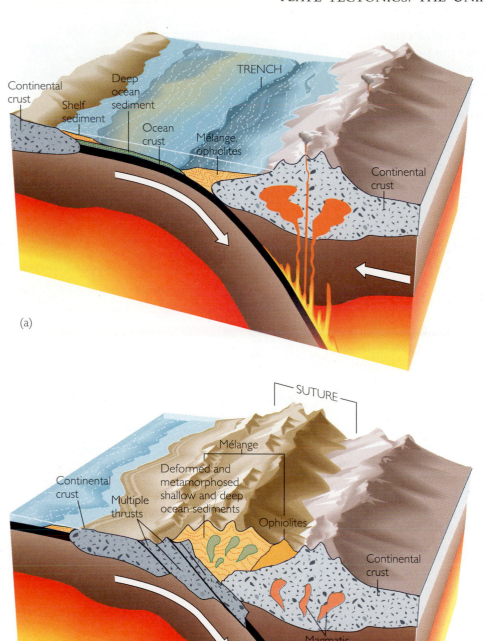

(a)

(b)

**FIGURE 20.21** (a) A plate carrying a continent is subducted under a plate with a continent at its leading edge. (b) The two continents collide. The continent at left breaks into several thrust sheets, thickening the continental crust and raising a high mountain range. Other rocks associated with plate convergence are caught up in the collision zone: magmatic intrusions, deformed and metamorphosed shelf and deep-water sediments, and ophiolite fragments.

Both continents may stay afloat. A wide zone of intense deformation develops at the boundary where the continents grind together. It is marked by a mountain range, in which sections of deep- and shallow-water sediments are found highly folded and broken by multiple thrust faults. The buildup of thrust sheets, one atop another, results in a much-thickened continental crust in the collision zone. Often there is a magmatic belt within the mountain range. The remnant of such a zone left behind in the geologic record is called a **suture.** Ophiolites are frequently found near the suture; they are relics of an ancient ocean that disappeared in the convergence of two plates.

A prime example of a collision of continents is the Himalayas, which began to form some 50 million years ago when a plate carrying India collided with the Eurasian Plate. The collision continues: India is moving into Asia at a rate of 5 cm per year and the uplift is still going on, together with faulting and many great earthquakes.

Geologists now believe that much of the geology of continents can be explained by episodes of continental rifting and plate separation and by plate collisions of the types just described.

## MICROPLATE TERRANES AND PLATE TECTONICS

Geologists have argued for decades about blocks as large as hundreds of kilometers across within orogenic belts of continents, with assemblages of rocks that are alien to their surroundings. These **microplate terranes**[2]—sometimes called displaced, exotic, or suspect terranes—contrast sharply with adjacent provinces in the assemblages of rocks, the nature of folding and faulting, and the history of magmatism of metamorphism. Fossils indicate that these blocks originated in environments and at times other than those of the surrounding terrain. These microplate terranes are now believed to be fragments of other continents, seamounts, island volcanic arcs, or slices of ocean crust that were swept up and plastered onto a continent when plates collided (Figure 20.22). The Appalachian orogenic belt, from Newfoundland to the southeastern United States, contains microplate terranes—slices of ancient Europe, Africa, oceanic islands, and crust welded onto North America during ancient collisions. Most of Florida is probably a piece of Africa left behind when North America and Africa parted about 200 million years ago. Florida's oldest rocks and fossils are more like those in Africa than like those found in the rest of the United States. As many as 100 terranes are regarded as microplate in the Cordilleran orogenic belt of western North America (Figure 20.23). Microplate terranes have also been found in Japan, Southeast Asia, China, and Siberia, but their original locations have yet to be worked out. Over the past 3.8 billion years (the age of the oldest known continental rocks) the continents have grown at an average rate of about 2 km³ per year.

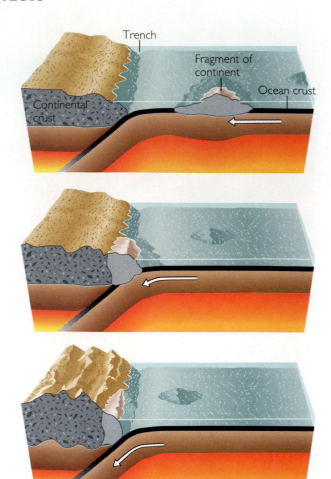

**FIGURE 20.22** Origin of a microplate terrane. An oceanic island arc or a fragment of continent is carried into a plate collision zone. Instead of being subducted, it is welded onto the overriding plate. Because the fragment may have originated thousands of kilometers away, it differs in its rock assemblages from the surrounding geological terrain.

## THE GRAND RECONSTRUCTION

One of the great triumphs of modern geology is the reconstruction of events that led to the assembly of the supercontinent Pangaea and its later fragmentation into the continents we know today. Pangaea was the only continent existing at the close of the Paleozoic, some 250 million years ago. It stretched from pole to pole (see Figure 20.13) and was made up of smaller continents that collided during the Paleozoic—not the same continents we know today but continents that existed earlier in the Paleozoic. The ocean-floor record for this period has been destroyed by subduction, so we must rely on the older evidence

---

[2]The familiar term *terrain* is synonymous with region, area, or territory. *Terrane* is used to signify a region in which rocks that were formed elsewhere have been accreted after being carried great distances by plate movements.

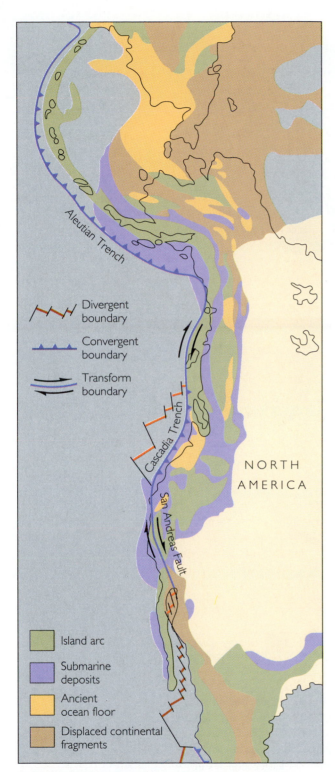

Divergent
boundary

Convergent
boundary

Transform
boundary

Island arc

Submarine
deposits

Ancient
ocean floor

Displaced continental
fragments

Aleutian Trench

Cascadia Trench

San Andreas Fault

NORTH
AMERICA

**FIGURE 20.23** Microplate terranes added to western North America in the past 200 million years. They are made up of island arcs, ancient seafloor, continental fragments, and submarine deposits, as indicated by different colors in the map. (After D. R. Hutchison, "Continental Margins," *Oceanus,* vol. 35, Winter 1992–1993, pp. 34–44; modified from work of D. G. Howell, G. W. Moore, and T. J. Wiley.)

preserved on continents today to identify and chart the movements of these paleocontinents. Old mountain belts such as the Appalachians of North America and the Urals, which separate Europe from Asia, help us locate ancient collisions of the paleocontinents. In many places alien rock assemblages reveal ancient episodes of rifting and subduction. Rock types and fossils also indicate the distribution of ancient seas, glaciers, lowlands, mountains, and climatic conditions. A knowledge of ancient climates enables geologists to locate the latitudes at which the continental fragments formed, which in turn helps us assemble the jigsaw fragments of ancient continents.

One of the first efforts to depict the pre-Pangaean configuration of continents by these methods is shown in Figure 20.24. It is truly impressive that modern science can recover the geography of this strange world of hundreds of millions of years ago and portray continents now gone, such as Gondwana, Laurussia, and the others shown in the illustration. Geologists are continuing to sort out more details of this complex jigsaw puzzle, whose individual pieces change shape over geologic time.

More is known about the most recent **breakup of Pangaea** because much of the evidence is still available on the seafloor, such as isochrons that provide the time and the directions of spreading. Figure 20.25 (see pp. 474 and 475) reconstructs this record of the breakup of Pangaea as we now understand it. Figure 20.25a shows the world as it looked at the close of the Paleozoic. Pangaea was an irregularly shaped landmass surrounded by a universal ocean called Panthalassa ("all seas"), the ancestral Pacific. The Tethys Sea, between Africa and Eurasia, was the ancestor of part of the Mediterranean. Permian glacial deposits have been found in widely separated areas, such as South America, Africa, India, and Australia. This distribution is explained by postulating a single continental glacier flowing over the South Polar regions of Gondwanaland in Permian time, before the breakup of the continents.

The breakup of Pangaea was signaled by the opening of rifts from which basalt poured. Rock assemblages that are relics of this great event can be found today in Triassic dikes and sills from Nova Scotia to Virginia and in the great Palisades sill along the Hudson River. The radioactivity in these rocks tells us that the breakup and the beginning of drift occurred about 200 million years ago.

The geography of the world in early Jurassic time, after 20 million years of drift, is sketched in Figure 20.25b. The Atlantic has partially opened, the Tethys has contracted, and the northern continents (Laurasia) have all but split away from the southern continents (Gondwana). New ocean floor has also

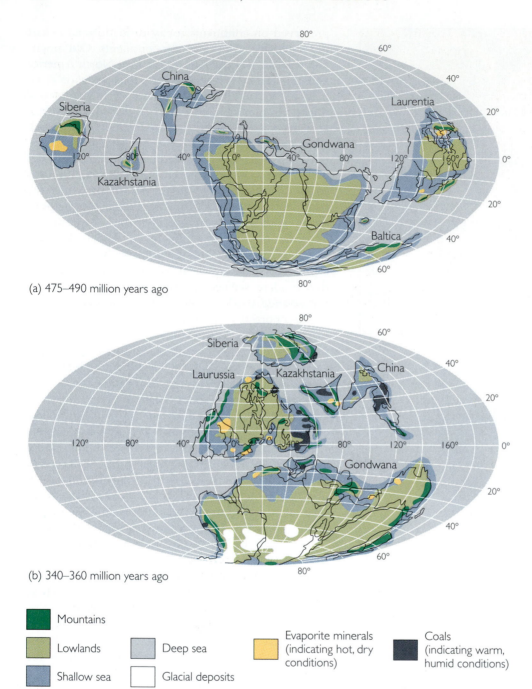

(a) 475–490 million years ago

(b) 340–360 million years ago

■ Mountains

■ Lowlands    ■ Deep sea    ■ Evaporite minerals (indicating hot, dry conditions)    ■ Coals (indicating warm, humid conditions)

■ Shallow sea    □ Glacial deposits

**FIGURE 20.24** (a) Paleocontinents in the middle Ordovician, about 475 to 490 million years ago. At that time the continents consisted of Gondwana (made up of South America, southern Europe, Africa, the Near East, India, Australia, New Zealand, and Antarctica), Laurentia (North America and Greenland), Baltica (most of northern Europe and European Russia), Kazakhstania (central Asia), China (China and Malaysia), and Siberia. (b) Paleocontinents in the early Carboniferous, about 340 to 360 million years ago. Gondwana has moved across the South Pole, entering the opposite hemisphere; Baltica has collided with Laurentia to form a larger continent, Laurussia. The continents are assembling for the collisions that formed the supercontinent Pangaea at the end of the Paleozoic. (After R. K. Bambach, C. R. Scotese, and A. M. Ziegler, "Before Pangaea: The Geographies of the Paleozoic World," *American Scientist,* vol. 68, January 1980, pp. 26–38.)

separated Antarctica-Australia from Africa-South America. India is off on a trip to the north.

By the end of the Jurassic period, 140 million years ago, drift had been under way for 60 million years. The big event at that time was the splitting of South America from Africa, which signaled the birth of the South Atlantic (Figure 20.25c). The North Atlantic and Indian oceans were enlarged, but the Tethys Sea continued to close. India continued its northward journey.

The close of the Cretaceous period, 65 million years ago, saw a widened South Atlantic, the splitting of Madagascar from Africa, and the close of the Tethys to form an inland sea, the Mediterranean (Figure 20.25d). After 135 million years of drift, the modern configuration of continents becomes discernible.

The modern world, produced over the past 65 million years, is shown in Figure 20.25e. India has collided with Asia, ending its trip across the ocean, although it is still pushing northward. Australia has separated from Antarctica. Nearly half of the present-day ocean floor was created in this period.

Figure 20.8 summarizes the relationships of the modern American, African, Eurasian, and Indian plates. Most of the modern Pacific Ocean basin consists of the Pacific Plate on the west side of the East Pacific Rise spreading zone, as can be seen in Figures 20.8b and 20.11. But where is the corresponding plate that should be on the east side of the rise? The implication is that an area equal to most of the present Pacific Ocean has disappeared by subduction under the Americas in the past 130 million years. As much as 7000 km of Pacific seafloor may have been thrust under North America!

Hardly any branch of geology remains untouched by this grand reconstruction of the continents. Economic geologists are using the fit of the continents to find mineral and oil deposits by correlating the formations in which they occur on one continent with their predrift continuations on another continent. Paleontologists are rethinking some aspects of evolution in the light of continental drift, as we mentioned earlier. Geologists are extending their sights from regional mapping to the world picture, for the concept of plate tectonics provides a way to interpret such geologic processes as sedimentation and orogeny in global terms. One of the longest mountain belts in Earth's history, for example, was formed when the ancient continents collided to form Pangaea. The belt was torn apart with the opening of the modern Atlantic Ocean. The modern remnants are now on different continents: the Caledonian mountain belt that runs along the northwest margin of Europe and the Appalachian belt of North America. Similarly, the trend of the Andes may be followed into Antarctica and Australia, as the reconstructed Pangaea in Figure 20.26 shows.

Oceanographers are reconstructing currents as they might have existed in the ancestral oceans to understand better the modern circulation and to account for the variations in deep-sea sediments that are affected by such currents. Scientists are "forecasting" backward in time to describe temperature, winds, the extent of continental glaciers, and the level of the sea as they were in predrift times. They hope to learn from the past so they can predict the future better—a matter of great urgency because of the possibility of greenhouse warming triggered by human activity. What better testimony to the triumph of this once outrageous hypothesis than its ability to revitalize and shed light on so many diverse topics!

## THE DRIVING MECHANISM OF PLATE TECTONICS

Up to this point everything we have discussed might be called descriptive plate tectonics. But a description is not an explanation. We will not fully understand plate tectonics until we can explain why plates move. The International Geodynamics Project, a global program of coordinated experiments, enlisted the efforts of thousands of scientists in many countries to discover the underlying cause of plate motions.

In Chapter 19 we described the mantle as a hot solid capable of flowing like a liquid at a speed of a few centimeters per year, about the rate at which your fingernails grow. The plates of the lithosphere somehow respond in their motions to the flow in the underlying mantle. As is generally the case when we have an abundance of data in search of a theory, many hypotheses have been advanced. Some scientists would have plates pushed by the weight of the ridges at the zones of spreading or pulled by the weight of the sinking slab at subduction zones. Others hold that the plates are dragged along by currents in the underlying asthenosphere. Hot spots, the jets of hot matter rising from the deep interior, may spread laterally when they reach the lithosphere, dragging the plates. Figure 20.27 shows some of these ideas. We agree with those who view the process not piecemeal but as a highly complex convective flow involving rising, hot, partially molten materials and sinking, cool, solid materials under a variety of conditions ranging from melting to solidification and remelting. A significant part of the man-

**FIGURE 20.25** The breakup of Pangaea. (After R. S. Dietz and J. C. Holden, "The Breakup of Pangaea," *Scientific American,* October 1970, p. 30.)

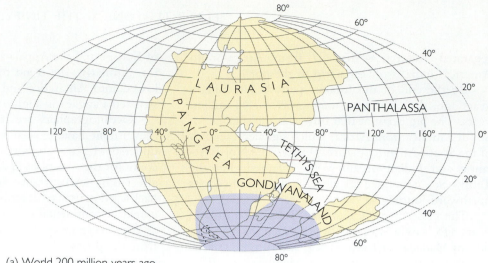

(a) World 200 million years ago

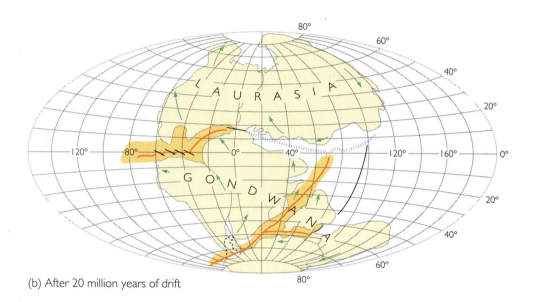

(b) After 20 million years of drift

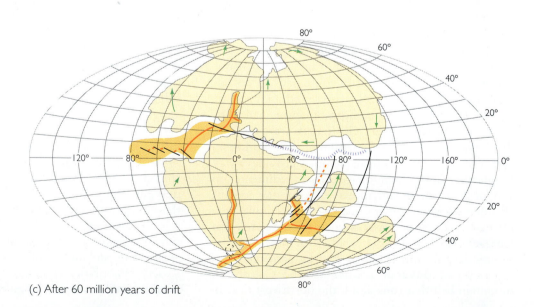

(c) After 60 million years of drift

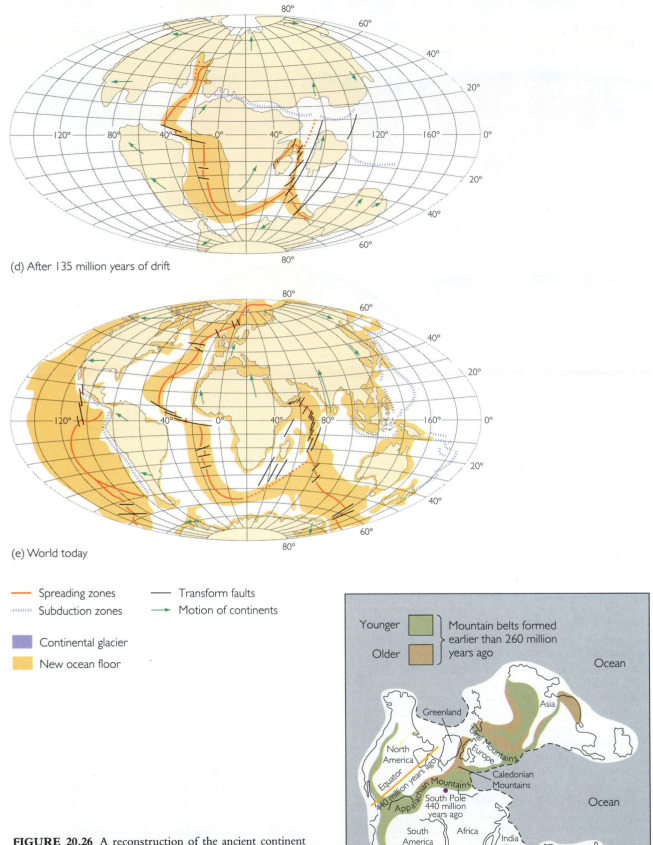

(d) After 135 million years of drift

(e) World today

— Spreading zones     — Transform faults

······· Subduction zones     → Motion of continents

◼ Continental glacier

◼ New ocean floor

Younger ◼ / Older ◼ } Mountain belts formed earlier than 260 million years ago

**FIGURE 20.26** A reconstruction of the ancient continent of Pangaea. The Urals and other old mountains contain ophiolite zones, marking these sutures as the sites of vanished oceans. (After J. F. Dewey, "Plate Tectonics," *Scientific American*, May 1972, p. 56.)

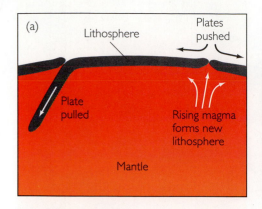

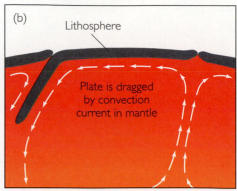

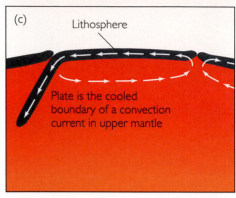

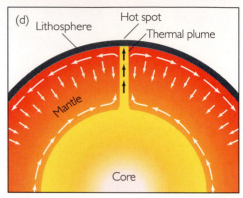

**FIGURE 20.27** Possible driving mechanism of plate tectonics. (a) The plates are pushed by the weight of the ridges at the centers of spreading or pulled by a cool, heavy, downgoing slab, or both. (b) Plates are dragged by convection currents in the mantle. (c) A plate is the cooled upper boundary region of a convection current in the hot, plastic upper mantle. (d) Jetlike thermal plumes rise from great depth, cause hot spots, and spread laterally, dragging the plates. Downward return flow occurs throughout the rest of the mantle.

tle must be involved, for slabs are known to penetrate to depths of some 700 km before being completely resorbed. Among the problems left to the next generation of Earth scientists is the incorpora-tion of such important details as the shapes of plates, the history of their movements, and the formation and growth of continents into an explanation involving convective currents in the interior.

# SUMMARY

**What is the theory of plate tectonics?** According to the theory of plate tectonics, the lithosphere is broken into about a dozen rigid moving plates. Three types of plate boundaries are defined by the relative motion between plates: boundaries of divergence, boundaries of convergence, and transform boundaries.

**What are some of the geologic characteristics of plate boundaries?** In addition to earthquake belts, many large-scale geologic features are associated with plate boundaries, such as narrow mountain belts and chains of volcanoes. Boundaries of convergence are recognized by deep-sea trenches, inclined earthquake belts, mountains and volcanoes, and paired belts of mélange and magmatism. The Andes Mountains and the trenches of the west coast of South America are modern examples. Divergent boundaries typically show as seismic, volcanic mid-ocean ridges (such as the Mid-Atlantic Ridge). A characteristic deposit of this environment is the ophiolite suite. Transform boundaries, along which plates slide past each other, can be recognized by their topography, seismicity, and offsets in magnetic anomaly bands. Ancient conver-

gences may show as old mountain belts, such as the Appalachians.

**How can the age of the seafloor be determined?** The age of the seafloor can be measured by means of magnetic anomaly bands and the stratigraphy of magnetic reversals worked out on land. The procedure has been verified and extended by deep-sea drilling. Isochrons can now be drawn for most of the Atlantic and for large sections of the Pacific, enabling geologists to reconstruct the history of the opening and closing of these oceans. On the basis of this method and of geological and paleomagnetic data, the fragmentation of Pangaea over the last 200 million years can be sketched.

**What is the mechanism that causes plates to form and move?** Although plate motions can now be described in some detail, the driving mechanism is still a puzzle. An attractive hypothesis proposes that the upper mantle is in a state of convection, with hot material rising under divergence zones and cool material sinking in subduction zones.

# KEY TERMS AND CONCEPTS

continental drift (p. 448)

Pangaea (p. 448)

seafloor spreading (p. 452)

island arc (p. 453)

isochron (p. 459)

rock assemblage (p. 462)

ophiolite suite (p. 462)

continental shelf deposits (p. 465)

mélange (p. 467)

suture (p. 469)

microplate terrane (p. 470)

breakup of Pangaea (p. 471)

# EXERCISES

1. Give a geographic example of each of the three types of plate boundaries.

2. What evidence suggests that Pangaea ever existed?

3. How can the rate of motion between plates be calculated?

4. What is the rate of separation between the North American and Eurasian plates? Between the Nazca and Pacific plates?

5. What kinds of rocks would you expect to find near a boundary of divergence? Near a boundary of convergence?

6. What happens when two continents collide in a boundary of convergence?

7. What are the circumstances that led to the formation of the Himalayas, the Andes, and the Urals?

8. What are the driving forces of plate tectonics?

# THOUGHT QUESTIONS

1. What would Earth be like if plate tectonics did not exist?

2. If plate tectonics explains so much of geology, why was it not until the 1960s that most geologists accepted the concept?

3. Would you characterize plate tectonics as a hypothesis, a theory, or a fact? Why?

4. Can you conceive an experiment that would demonstrate what mechanism provides the driving force of plate tectonics?

# SUGGESTED READINGS

Anderson, R. N. 1986. *Marine Geology*. New York: Wiley.

Bambach, R. K., C. R. Scotese, and A. M. Ziegler. 1980. Before Pangaea: geographies of the Paleozoic world. *American Scientist* 68 (no. 1):26–38.

Beloussov, V. V. 1979. Why I do not accept plate tectonics. *EOS* 60:207–210. (See also comments on this paper by A. M. S. Senger and K. Burke on the same pages.)

Courtillot, V., and G. E. Vink. 1983. How continents break up. *Scientific American* (July):42.

Cox, A. (ed.). 1973. *Plate Tectonics and Geomagnetic Reversals*. San Francisco: W. H. Freeman.

Hallam, A. 1973. *A Revolution in the Earth Sciences: From Continental Drift to Plate Tectonics*. New York: Oxford University Press (Clarendon Press).

Kearey, P., and F. J. Vine. 1990. *Global Tectonics*. Oxford, England: Blackwell Scientific Publications.

Le Pichon, X., J. Francheteau, and I. Bonnin. 1973. *Plate Tectonics*. London, New York: Elsevier.

Menard, H. W. 1986. *Islands*. New York: Scientific American Library.

Phillips, J. D., and D. Forsyth. 1972. Plate tectonics, paleomagnetism, and the opening of the Atlantic. *Bulletin of the Geological Society of America* 83 (no. 6):1579–1600.

Sclater, J. G., and C. Tapscott. 1979. The history of the Atlantic. *Scientific American* (June):156–174.

Siever, R. (ed.). 1983. The dynamic Earth. *Scientific American* (September):46.

Uyeda, S. 1978. *The New View of the Earth*. San Francisco: W. H. Freeman.

Wessel, G. R. 1986. *The Geology of Plate Margins*. Geological Society of America, Map and Chart Series MC-59.

# DEFORMATION OF THE CONTINENTAL CRUST

The rock layers of the continental crust record some four billion years of vertical movements, folding, faulting, igneous intrusion, and metamorphism. The geologic fabric of continents exhibits a pattern: eroded remnants of very old deformed rocks in the interior, more recent deformation in the mountain systems closer to margins. Mountain building occurs where plates collide, as sediments of the continental margin are thrown into a series of folds and faults. A subducted plate melts and magma rises into the deformed belt. Plate movements bring in foreign fragments and weld them to the deformed belt. Upward and downward movements within the continent create interior basins and domes and lift old, worn-down mountains high again. Offshore, downward movements create basins on the continental shelves.

## 21

Folds exposed in
2-billion-year-old
limestone and shale
beds; Wopmay orogenic
belt, northern Canada.
*Robert S. Hildebrand,
Geological Survey of
Canada.*

If a movie camera snapped one frame of the Earth's surface every thousand years, almost 50 million years of geologic time could be condensed to a 35-minute film. Geologic movements that stretch over thousands of years, far too slow to be observed in a human lifetime, could be studied and understood. The film would show continents drifting apart to open one ocean, colliding to close another. We could see regions of the Earth being uplifted, tilted, faulted, and folded to create plateaus, mountains, rift valleys, and other geologic structures. In other places subsidence would lower portions of a continent beneath the sea, forming basins in which sediments would accumulate. Actually, such movies have been made with the use of animation. Each frame of these films is a sketch based on an interpretation of the geologic record at a particular time. Simulations of plate movements are also available for your personal computer.

Most of Earth's surface consists of oceans whose underlying crust is younger than 200 million years. The seafloor crust has a relatively simple structure that is easy to interpret because it is being continuously created at mid-ocean ridges and destroyed at subduction zones. For this reason, however, the existing seafloor provides a record of only about 4 percent of Earth's 4.6-billion-year history. Only the continents have rocks as old as 4 billion years, and it is there that we must turn to look back over most of geologic time. The long span of geologic evolution recorded in the continental crust is complex, but we are beginning to understand it better as geologists use concepts from plate tectonics to interpret old geologic features, such as eroded mountain belts and ancient rock assemblages.[1]

# SOME REGIONAL TECTONIC STRUCTURES

The rocks that make up the continental crust can be grouped in two distinct categories: the veneer of sedimentary rocks deposited in an orderly process and not yet significantly deformed; and the deformed regions of sedimentary, igneous, and metamorphic rocks that have been subjected to intense crustal forces during various geologic periods. Most of the continental crust, either exposed rocks or basement buried beneath the layered sedimentary (or sometimes volcanic) cover, fall into the second category; that is, rocks that have been deformed and altered by crustal forces, such as the rocks shown in the photograph at the beginning of this chapter. Thus **orogeny**—the mountain-building processes of folding, faulting, magmatism, and metamorphism—and the evolution of continents are intimately related.

The map in Figure 21.1 shows the pattern of deformed continental rocks, colored according to the geologic period in which the deformation occurred. Note that the geologic fabric of the continents is not random. Rocks that reveal the most ancient episodes of deformation tend to be found in the interior of continents, which are now generally stable and eroded flat. External to these old terrains are the more recently active mountain belts, where most of the present-day mountain systems are found. They occur as long, narrow topographic features at the margins of continents, such as the cordillera that runs down the western edges of North and South America; the Appalachian belt, which trends southwest to northeast on the eastern margin of North America; and the Alpine-Himalayan chain, which runs across southern Europe and Asia. The key to deciphering the process by which the ancient terrains were deformed lies in these younger belts, because much of the record of deformation is still preserved uneroded in the rocks.

We will see that the theory of plate tectonics provides a framework for understanding how continents

---

[1] We draw on the work of B. C. Burchfiel of MIT, one of the geologists who is reinterpreting the geology of the continental crust in the framework of plate tectonics.

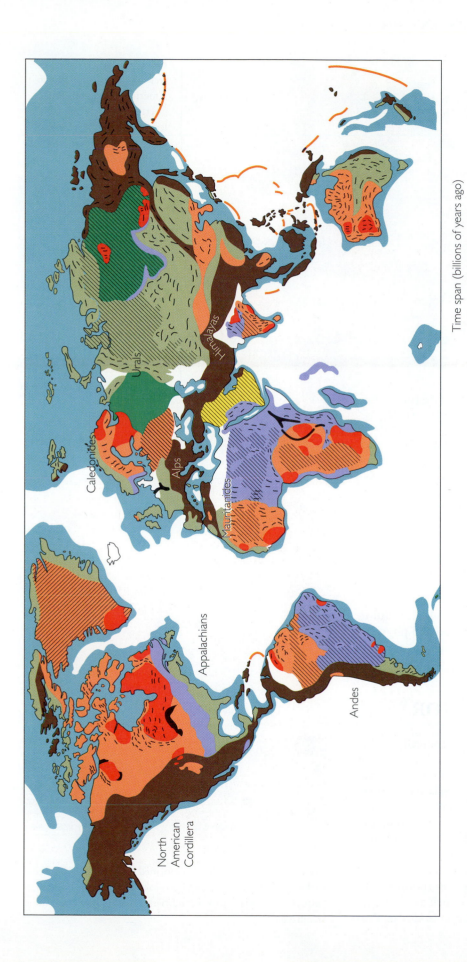
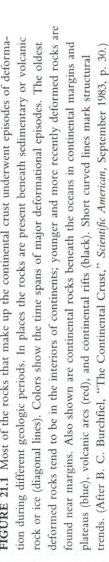

**FIGURE 21.1** Most of the rocks that make up the continental crust underwent episodes of deformation during different geologic periods. In places the rocks are present beneath sedimentary or volcanic rock or ice (diagonal lines). Colors show the time spans of major deformational episodes. The oldest deformed rocks tend to be in the interiors of continents; younger and more recently deformed rocks are found near margins. Also shown are continental rocks beneath the oceans in continental margins and plateaus (blue), volcanic arcs (red), and continental rifts (black). Short curved lines mark structural trends. (After B. C. Burchfiel, "The Continental Crust," *Scientific American*, September 1983, p. 30.)

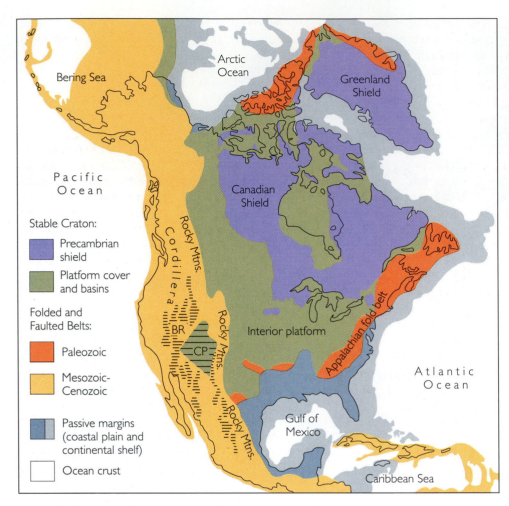

**FIGURE 21.2** Major tectonic features of North America: Canadian Shield; interior platform; Cordilleran orogenic belt, including Basin and Range (BR and short dashes); Colorado Plateau (CP); Appalachian fold belt; coastal plain and continental shelves of passive margins. (After A. W. Bally, C. R. Scotese, and M. I. Ross, *Geology of North America,* vol. A, Boulder, Colo., Geological Society of America.)

evolve. In the following sections it will be useful to refer to Figure 16.14 to see the location and physiography (surface landforms) of the regional tectonic structures of the United States.

# THE STABLE INTERIOR

**Cratons** are the extensive, flat, tectonically stable interiors of the continents, composed of ancient rocks that underwent intense deformational episodes in Precambrian time and have been relatively quiescent since then. Typically the cratons include large areas, called **shields,** that consist of very old, exposed crystalline basement rocks. The Canadian Shield (blue in Figure 21.2) is typical. It is dominated by granitic and metamorphic rocks, such as gneisses, together with highly deformed metamorphosed sedimentary and volcanic rocks. These rock assemblages indicate intense mountain-building episodes in Pre-

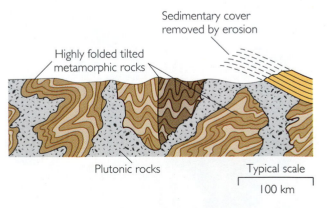

Labels: Sedimentary cover removed by erosion; Highly folded tilted metamorphic rocks; Plutonic rocks; Typical scale 100 km

**FIGURE 21.3** The highly deformed and metamorphosed rocks in the Canadian Shield indicate that intense orogenic episodes took place in Precambrian times, before conditions stabilized.

cambrian time followed by a long period of stability, indicated by a lack of evidence of more recent deformation. This primitive region, idealized in Figure 21.3, contains one of the oldest records of geologic history, much of it still known only in outline. It is famous for major deposits of iron, gold, copper, and nickel. Other shields are found in Scandinavia and Finland, Siberia, central Africa, Brazil, and Australia.

South of the Canadian Shield is a sediment-covered, almost level region, or **platform** (the "interior platform" in Figure 21.2). It forms the central stable region and Great Plains. This platform is in a sense a subsurface continuation of the Canadian Shield, for it contains similar Precambrian basement rocks, but here they are covered by a veneer of Paleozoic sedimentary rocks typically less than about 2 km thick. The North American platform sediments were laid down on the deformed and eroded Precambrian basement under a variety of conditions. The rock assemblages indicate sedimentation in extensive shallow inland seas (marine sandstones, limestones, shales, deltaic deposits, evaporites) and deposition on alluvial plains or in lakes or swamps (nonmarine sediments, coal deposits). Many of the continent's deposits of uranium, coal, oil, and gas are contained in the platform's sedimentary cover. Within the platform broad **sedimentary basins** are defined by roughly oval depressions where the sediments are thicker than on the surrounding platform. The Michigan Basin, a circular area of about 500,000 km² that covers much of the lower peninsula of Michigan, subsided throughout much of Paleozoic time and received sediments more than 3 km thick in its central, deepest part (Figure 21.4). The sediments, laid down under quiet conditions, have remained unmetamorphosed and only slightly deformed to this day. The sedimentary formations in this basin have been likened to a pile of saucers. As we mentioned in Chapter 20,

some geologists believe that basins subsided in an episode of stretching and thinning of the continental lithosphere.

## OROGENIC BELTS

Fortunately, the details of orogeny are still preserved in the younger orogenic belts. We need that information if we are to understand the incomplete record found in the eroded remnants of Precambrian mountains that now make up the basement rocks of shields and platforms. Fringing the great stable interior of the continent are younger orogenic belts, regions that were deformed by folding and faulting and were subjected to plutonism and metamorphism at various times in the Paleozoic, Mesozoic, and Cenozoic eras (see Figures 21.1 and 21.2).

Most geologists now believe that orogenies— that is, periods of mountain building—involve plate collisions. Most of the motion of collision is absorbed by the subduction of one of the plates. However, a basic tenet of plate tectonics—the rigidity of plates—must be modified when plate collisions bring together buoyant fragments of crust—in other words, continents. These fragments are embedded atop the plates on which they ride. Both colliding fragments tend to float and resist being subducted with the plates that carry them. But the collisional forces are so great that the continental crust loses its rigidity, deforms, and breaks in a variety of ways. By a combination of intense folding and faulting, the crust absorbs much of the motion of the collision within a zone of intense deformation extending hundreds of kilometers into the continent. The faulting breaks the crust into multiple thrust sheets up to 20 km thick, stacked one above the other by subhorizontal (low-angle) thrusting (Figure 21.5). The indi-

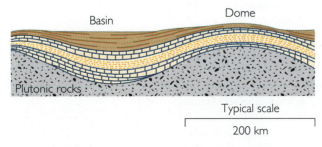

**FIGURE 21.4** Idealized section of the Michigan Basin provides evidence of vertical movements in the relatively undeformed, stable interior platform of the United States.

**FIGURE 21.5** When plates bearing continents collide, the continental crust can break into multiple thrust fault sheets stacked one above the other.

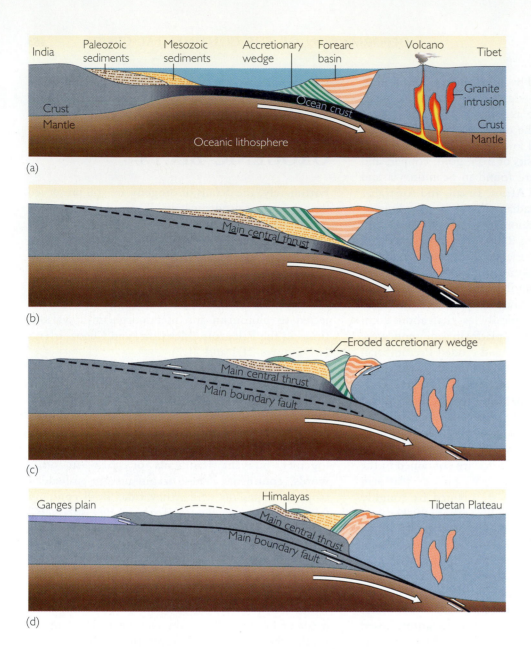

(a)

(b)

(c)

(d)

**FIGURE 21.6** The Himalayan orogeny, simplified and vertically exaggerated. (a) Subduction of the Indian Plate under the Eurasian Plate began some 60 million years ago. The Indian Plate moved northward, carrying the Indian subcontinent with its leading edge of Paleozoic and Mesozoic continental shelf sediments. Magma rising from the subducted plate formed granitic intrusions and volcanoes, thickening the crust. An accretionary wedge was formed from the pile of sediments and oceanic crust scraped off the descending plate. A forearc basin formed behind the wedge, trapping sediments eroded from Tibet. (b) The Indian subcontinent collided with Tibet sometime between 40 and 60 million years ago. India was too buoyant to be subducted into the mantle, and the Indian crust broke in a thrust fault—the main central thrust. (c) As the collision continued, with the motion taken up along the thrust fault, a slice of Indian crust and shelf sediments was stacked onto the oncoming subcontinent. The accretionary wedge and the forearc basin sediments were thrust northward onto Tibet. (d) About 10 to 20 million years ago a new thrust fault developed—the main boundary fault. A second slice of crust was stacked onto India, lifting up the first slice. The two overthrust slices make up the bulk of the Himalayas, including the Paleozoic sediments that are found in the peaks. (After P. Molnar, "The Structure of Mountain Ranges," *Scientific American,* July 1986, p. 70.)

**FIGURE 21.7** The Himalayan Mountains in east-central Tibet. The dark-colored rocks forming the crest of the ridge are early Cenozoic sedimentary rocks thrust over younger (light-colored) Cenozoic sedimentary rocks. Overthrusting of several kilometers reflects a shortening of the crust in Tibet, after India collided with Eurasia. *Peter Molnar.*

vidual thrust sheets are often themselves deformed and metamorphosed. Wedges of continental shelf sediments can be detached from the basement on which they were deposited and thrust inland. Figure 21.6, depicting the orogeny that formed the Himalayas, shows most of these features. Figure 21.7 is a photograph of thrust faulting in the Tibetan Himalayas. The Himalayan orogeny is still going on, with India moving into Asia at a rate of about 5 cm per year (see Box 21.1).

We saw in Chapter 20 that a significant fraction of an orogenic belt can be made up of a succession of displaced or microplate terranes, fragments of crust that may have traveled thousands of kilometers from distant parts of the world before they accreted to the orogenic belt. These microplate terranes can be fragments of oceanic crust (ophiolites), island arcs, or pieces of continental crust carried by the subducting plate but driven up over the edge of the continent rather than down into the mantle (see Figure 20.23).

Orogeny thickens the continental crust and extends it in several ways. One is by microplate accretion. Another is by the addition of batholiths and volcanic rocks derived from melting of the mantle. In these ways continents can grow by a succession of orogenies over geologic time. In Figures 21.1 and 21.2 we can see this growth in the pattern of younger orogenies fringing older ones.

## Case History: The Appalachians

The old, eroded Appalachian Mountains are a classical fold-and-thrust belt that extends along eastern North America from Newfoundland to Alabama. The rock assemblages and structures found there today are the primary data geologists have used to reconstruct the tectonic history of the belt. The Appalachians consist of a sequence of physiographic subregions, as Figure 21.8a shows. The underlying rock structures are shown in cross section in Figure 21.8b.

1. *Valley and Ridge province.* Thick Paleozoic sedimentary rocks laid down on an ancient continental shelf were folded and thrust to the northwest by compressional forces from the southeast. The rocks show that deformation occurred in three orogenic episodes, one at the end of the Ordovician, one at the end of the Devonian, and one in Permian-Carboniferous time.
2. *Blue Ridge province.* These eroded mountains are composed largely of Precambrian and Cambrian crystalline rock, showing much metamorphism. The Blue Ridge rocks were not intruded and metamorphosed in place, but were thrust as sheets over the sedimentary rocks of the Valley and Ridge province (see Figure 21.8b).
3. *Piedmont.* This region contains Precambrian and Paleozoic metamorphosed sedimentary and volcanic rocks intruded by granite that have now been eroded to low relief. Volcanism began in the late Precambrian and continued into the Cambrian. The Piedmont was thrust over Blue Ridge rocks along a major thrust fault, overriding them to the northwest. At least two episodes of deformation are evident, one at the end of the Devonian, the other in the early Carboniferous.
4. *Coastal Plain.* Relatively undisturbed sediments of Jurassic age and younger are underlain by rocks similar to those of the Piedmont. The continental shelf is the offshore extension of the Coastal Plain.

Figure 21.9 shows a possible plate-tectonic reconstruction of southern Appalachian history consistent with present-day geologic features. In late Precambrian time (a) a supercontinent split apart, leaving what was to be North America on one side of the rift, Africa on the other, and the ancestral Atlantic Ocean in between, growing as the plates separated. A continental fragment that would become the Blue Ridge and part of the Piedmont was left between the

## BOX 21.1  THE COLLISION BETWEEN INDIA AND EURASIA

Some 40 to 60 million years ago the Indian subcontinent collided with Eurasia. The collision slowed India's advance, but it continued to drive northward. So far it has penetrated about 2000 km into Eurasia. Since continental crust does not subduct readily, geologists must explain what happened to a piece of crust as wide as India and 2000 km long. Using earthquake data and satellite photographs, Peter Molnar and Paul Tapponier prepared this tectonic map and offered a startling proposal to account for the vast area of Eurasia displaced by the collision.

The Himalayas, the world's highest mountains, were formed from overthrust slices of the old north portion of India, stacked one atop the other. This process took up some of the compression. Horizontal compression in Tibet found some relief in vertical expansion; this process contributed to the uplift of the high Tibetan Plateau. Compression by thrust faulting is the pattern of deformation in the Tien Shan Mountains. But these and other zones of compression could account for perhaps only half of India's penetration into Eurasia. Molnar and Tapponier account for the other half by suggesting that China and Mongolia were pushed eastward, out of India's way, like toothpaste squeezed from a tube. The movement took place along the enormously long Altyn Tagh fault and other strike-slip faults shown on the map. The mountains, plateaus, faults, and great earthquakes of Asia, thousands of kilometers from the Indian-Eurasian suture, are thus influenced by the continuing collision of the two continents.

Tectonic features associated with the collision between India and Eurasia: large-scale faulting, uplift, and earthquakes. (After P. Molnar and P. Tapponier, "The Collision between India and Eurasia," *Scientific American,* April 1977, p. 30.)

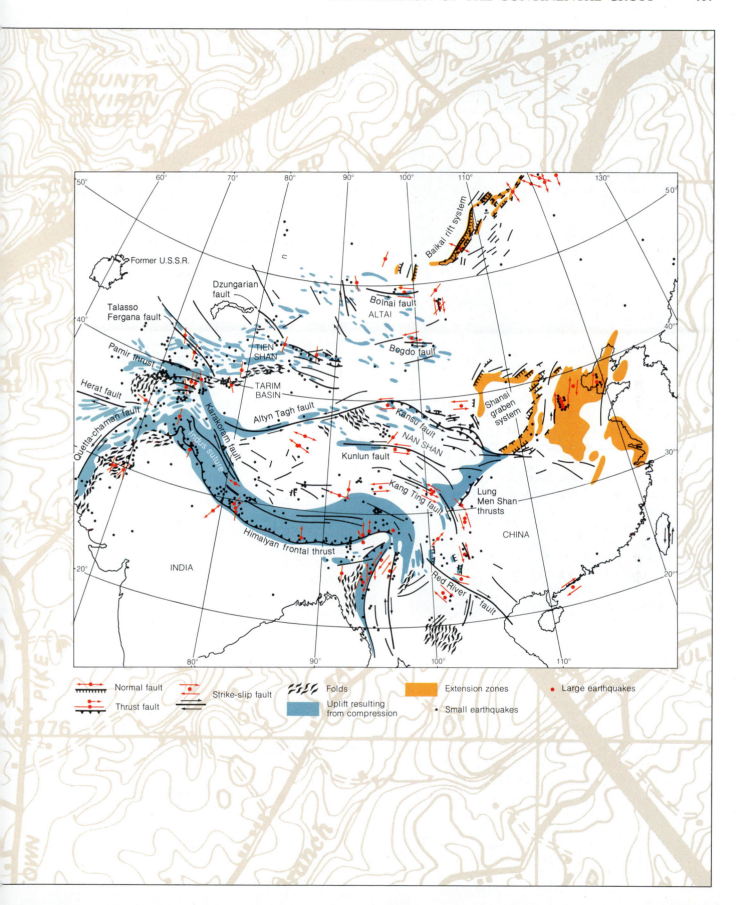

Normal fault

Thrust fault

Strike-slip fault

Folds

Uplift resulting
from compression

Extension zones

Large earthquakes

Small earthquakes

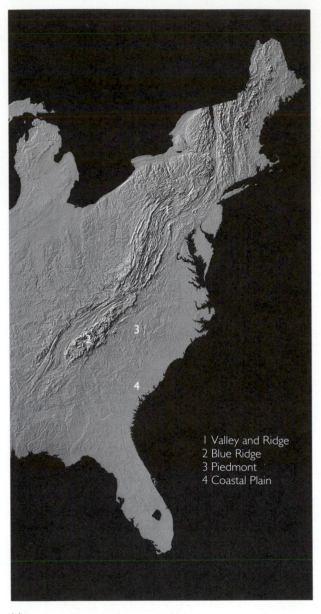

(a)

diverging North America and Africa at this time. A marginal sea separated the fragment from North America. Continental shelf sediments were deposited along the margins of the receding plates.

In early Cambrian time (b) plate convergence began to close the ancestral Atlantic. (Africa is out of view to the right of Figure 21.9b, on the other side of the Atlantic.) The ancient Atlantic seafloor was subducted under North America and the continental fragment, and an island arc developed adjacent to the subduction zone. The fragment was somewhere between the island arc and North America.

Convergence continued from the middle Cambrian (c) through the Ordovician to early Silurian (d), when the continental fragment collided with North America. The fragment was thrust as sheets of crys-

FIGURE 21.8 (a) The Appalachian Mountain region of the eastern United States. The major physiographic subregions are the Valley and Ridge province, consisting of long mountain ridges and valleys and erosional features in folded sedimentary rocks; the Blue Ridge province, consisting of high mountains of crystalline rock; the Piedmont, rough to gentle hilly terrain on crystalline rock; and the Coastal Plain, a terrain of low hills shading to flat plains on sediments. (Digital shaded-relief map by Richard J. Pike and Gail P. Thelin, USGS, 1989, with added physiographic regions.) (b) Multiple plate collisions in the Paleozoic thrust the Blue Ridge and Piedmont sheets over the sediments of an ancient North American shelf and deformed the sedimentary rock of the Valley and Ridge province. The modern shelf developed after the Triassic-Jurassic splitting of North America from Africa. (The vertical scale is 10 times the horizontal.) (After F. A. Cook et al., "Thin-Skinned Tectonics in the Crystalline Southern Appalachians," *Geology,* vol. 7, pp. 563–567.)

1 Valley and Ridge
2 Blue Ridge
3 Piedmont
4 Coastal Plain

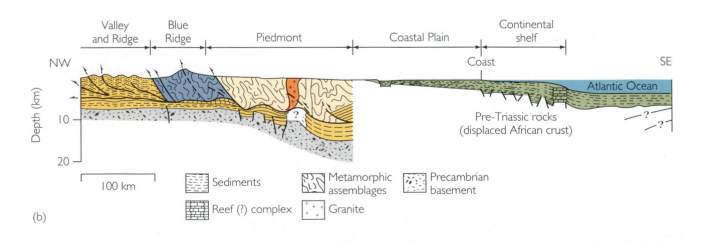

(b)

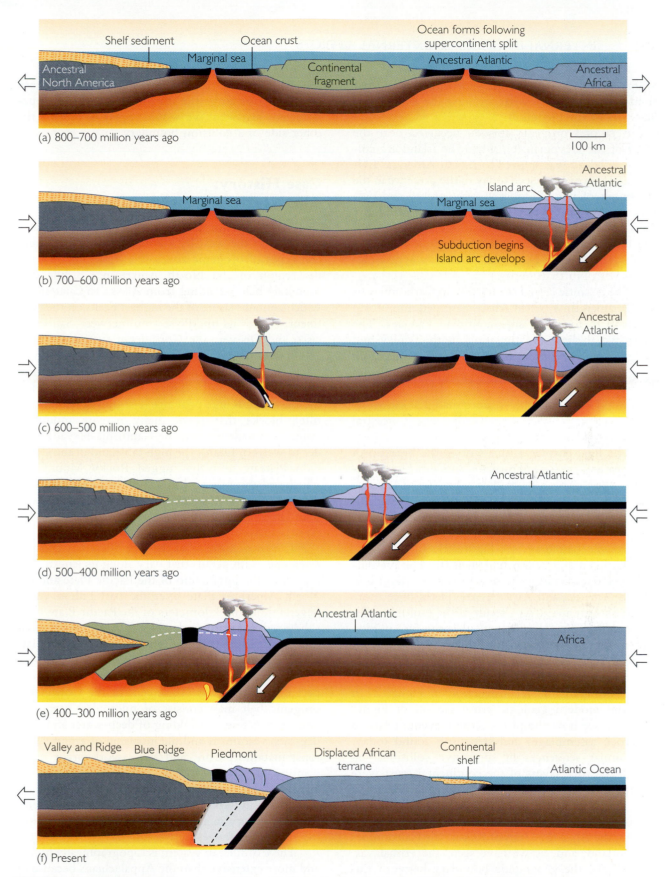

(a) 800–700 million years ago

100 km

(b) 700–600 million years ago

(c) 600–500 million years ago

(d) 500–400 million years ago

(e) 400–300 million years ago

(f) Present

**FIGURE 21.9** The Appalachian orogenic belt can be interpreted as the result of plate divergences and convergences from late Precambrian to the present. In the southern Appalachians, successive collisions resulted in subduction and overthrusting of the continental margin of North America. (After F. A. Cook et al., "Thin-Skinned Tectonics in the Crystalline Southern Appalachians," *Geology,* vol. 7, pp. 563–567.)

talline rock over the younger North American shelf sediments to become what are now the Blue Ridge and part of the Piedmont.

A second orogenic episode occurred near the close of the Devonian (e), when the island arc collided with and was sutured to North America. Evidence of this event is seen today in the presence in the Piedmont of metamorphosed sedimentary and volcanic rocks of the type characteristic of an island arc. This collision pushed the Blue Ridge and Piedmont sheets farther west. The Atlantic continued to close. (Africa reappears in Figure 21.9e, on its way to a collision with North America.)

The Atlantic closed completely in Carboniferous and Permian time, when Africa slammed into North America to form the supercontinent Pangaea. The extensive folding and thrusting in the Valley and Ridge province occurred during this culminating orogeny. The accumulated thrusting of all three deformational episodes may have pushed the Blue Ridge and Piedmont sheets more than 250 km westward over the continental shelf sediments of ancestral North America.

The final chapter in this story—the breakup of Pangaea and the opening of the modern North Atlantic—began some 200 million years ago in Triassic-Jurassic times, when rifting occurred east of the Piedmont. Africa split away from North America, leaving behind a piece of itself that today is the pre-Triassic basement (see Figure 21.8b) under the Coastal Plain and continental shelf (f). The present-day shelf deposits began to develop on the margins of North America and Africa as the plates once more began to drift apart. As a result of the multiple collisions and rifting over a span of some 500 million years, North America grew by the accretion of a continental fragment, an island arc, a piece of Africa, other foreign fragments, and a modern continental shelf. Compare the reconstruction in Figure 21.9f and the modern geologic cross section of Figure 21.8b to see how the plate-tectonic events relate to the rock assemblages and structures found today.

Similar sequences of successive collisions—in which the rock layers of the continental crust were compressed and deformed by folding and breaking into a stack of thrust sheets—were also important in the orogenies that formed the North American Cordillera; the Alps; the Urals of Russia; the Caledonides of Scotland; and, as we have seen, the Himalayas. The Urals, the geographic boundary between Europe and Asia, represent an ancient suture, or boundary line, between colliding continental plates in the middle of a continent that has not rifted since it

formed. The Mauritanide mountain belt of western Africa appears to be the mirror image of the Appalachians—just what we would expect to find on the other side of a collision boundary.

## Case History: The North American Cordillera

The stable interior platform of North America is bounded on the west by a younger complex of several types of orogenic zones (Figure 21.10). This is the region of the North American Cordillera, a mountain belt extending from Alaska to Guatemala, which contains some of the highest peaks on the continent. Across its middle section, between San Francisco and Denver, the Cordilleran system is about 1600 km wide and includes several contrasting physiographic provinces: the Coast Ranges along the Pacific Ocean; the lofty Sierra Nevada and Cascades; the Basin and Range province (a region of faulted and tilted blocks that form many narrow mountain ranges and valleys from the California-Nevada border to western Utah); the high tableland of the Colorado Plateau; and the rugged Rocky Mountains, which end abruptly at the edge of the Great Plains on the stable interior. Some of the types of mountains found in orogenic belts such as the Cordillera and the Appalachians are depicted in Figure 21.11.

The entire history of the Cordillera is a complicated one, with details that vary along its length. It is a story of the interaction of the Pacific Plate and the North American Plate over the past billion years. Several thousand kilometers of the Pacific Plate were subducted eastward under North America over this time span. All of the phenomena that go with plate collisions occurred: the accretion of foreign continental and oceanic fragments (some geologists now suspect that as much as 70 percent of the Cordilleran orogenic belt may consist of microplate terranes); intense thrusting and folding of deep-water and shelf deposits; volcanism and the intrusion of granitic plutons; metamorphism; and the reworking or overprinting of features of one episode of orogeny by later orogenic events. Together these processes have led to the juxtaposition of deformed rock assemblages that vary in age and origin and the overprinting that characterize the Cordillera.

The Cordilleran system is topographically higher and more extensive than the Appalachians because its main orogeny was more recent, having occurred in the last half of Mesozoic and early Tertiary time. The form and height of the Cordillera we see today are

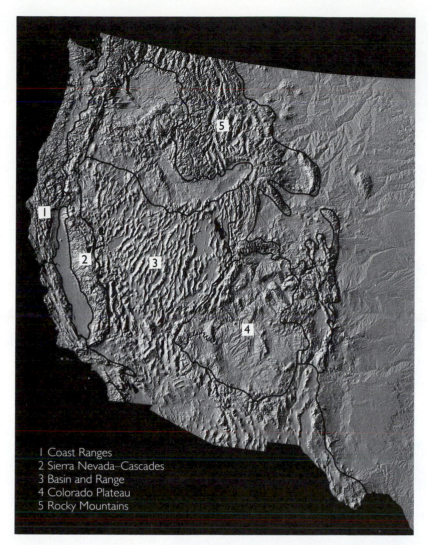

**FIGURE 21.10** The Cordilleran orogenic belt and adjacent areas of the United States. (Digital shaded-relief map by Richard J. Pike and Gail P. Thelin, USGS, 1989, with added physiographic regions.)

1 Coast Ranges
2 Sierra Nevada–Cascades
3 Basin and Range
4 Colorado Plateau
5 Rocky Mountains

manifestations of even more recent events in Tertiary and Quaternary time, over the last 15 or 20 million years, which resulted in **rejuvenation** of the mountains; that is, the mountains were brought back to a more youthful stage. At that time, the central and southern Rockies attained much of their present height as a result of a broad regional upwarp, or uplift (see Figure 21.11b). The cause of upwarps is a matter of current debate among geologists. The region was raised 1500 to 2000 m as Precambrian basement rocks and their veneer of later-deformed sediments were pushed above the level of their surroundings. Stream erosion was accelerated, the mountain topography sharpened, and the canyons deepened. Other **upwarped mountains** are the Adirondacks of New York, the Labrador Highlands, and the mountains of Scandinavia and Finland. Reju-

venated deformed belts are also seen in the modern topography of the Alps, the Urals, and the Appalachians.

The late structural imprinting responsible for the present-day features of the Basin and Range province and the tilted uplift of the Sierra Nevada of California (Figure 21.12) also occurred in Tertiary and Quaternary time. They took the form of a spectacular episode of faulting in a region extending southeast from southern Oregon to Mexico and encompassing Nevada, western Utah, and parts of eastern California, Arizona, New Mexico, and western Texas. Thousands of nearly vertical faults sliced the crust into innumerable upheaved and downdropped blocks, forming hundreds of nearly parallel **fault-block mountain ranges** (see Figure 21.11c) separated by alluvium-filled rift valleys bounded by normal faults.

**FIGURE 21.11** Mountain structures vary in form and origin. (a) Mountains formed by volcanic action, such as the Cascades, extending from northern California through western Oregon and Washington into British Columbia. (b) Up-warped mountains, with reverse faults, such as the Front Range of the Rocky Mountains in Colorado. (c) Mountains formed from fault blocks, bounded by normal faults, such as the Teton Range in Wyoming and the mountains of the Basin and Range province. (d) Mountains resulting from folded layers of rock such as the Appalachian fold belt. (After U.S. Geological Survey.)

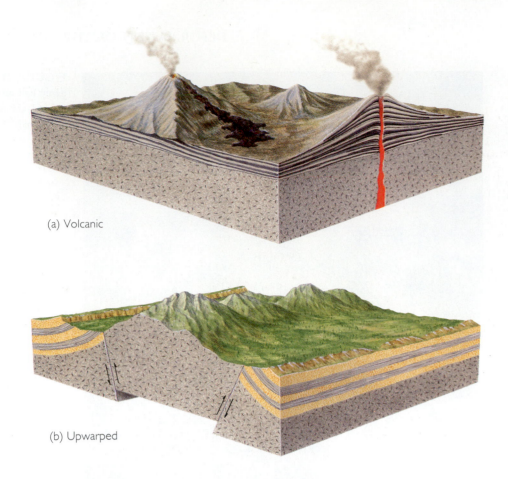

(a) Volcanic

(b) Upwarped

Sierra Nevada

Sierra Nevada fault scarp

Argus Mountains

Panamint Valley

Shallow fault valley

**FIGURE 21.12** The Sierra Nevada (skyline) viewed from the Panamint Mountains, California. The Sierra Nevada fault scarp (steep cliffs and slopes formed by faulting) is in the distance. In the middle ground are the Argus Mountains, formed by a late Quaternary upfaulting of early Quaternary basalts. A small fault-block valley, far side up, cuts the alluvium of Panamint Valley (below the center of the picture.) *W. B. Hamilton, U.S. Geological Survey.*

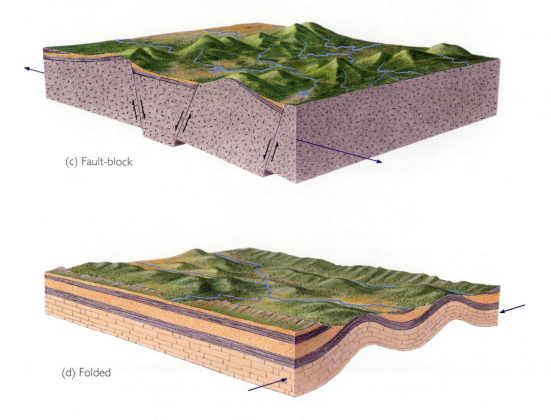

(c) Fault-block

(d) Folded

Some geologists think that the explanation is to be found in a crustal stretching or extension caused by flow in the mantle below. Other fault-block mountains are the Wasatch Range of Utah, the Teton Range of Wyoming, and the tilted edges of the rift valleys of east Africa and the Dead Sea–Jordan Valley rift of Israel.

The Colorado Plateau seems to be an island of the central stable region, cut off from the interior by the Rocky Mountain orogenic belt. Since the late Precambrian it has been a stable area, experiencing no thick basin deposits and no major orogeny. Its rock formations, exposed in the Grand Canyon (see Box 9.1), reflect mainly up-and-down movements.

## COASTAL PLAIN AND CONTINENTAL SHELF

The Atlantic coastal plain and the continental shelf, its offshore extension (see Figure 21.2), began to develop in Triassic time with the rifting that preceded the opening of the modern Atlantic Ocean. The rift valleys formed basins that trapped a thick series of nonmarine deposits. As these deposits accumulated, they were intruded by contemporaneous basaltic sills and dikes. The Connecticut River valley and the Bay of Fundy are such rift valleys (Figure 21.13).

In early Cretaceous time the deeply eroded and beveled coastal plain and continental shelf began to

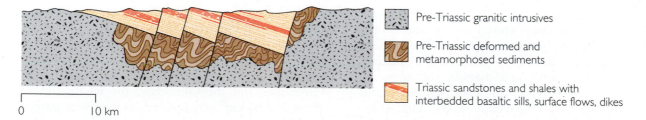

Pre-Triassic granitic intrusives

Pre-Triassic deformed and metamorphosed sediments

Triassic sandstones and shales with interbedded basaltic sills, surface flows, dikes

0        10 km

**FIGURE 21.13** Triassic rift valleys of Connecticut. Tilted fault blocks formed rift valleys. Nonmarine sediments were trapped in basins formed by rift valleys. Basaltic flows intruded and covered these deposits.

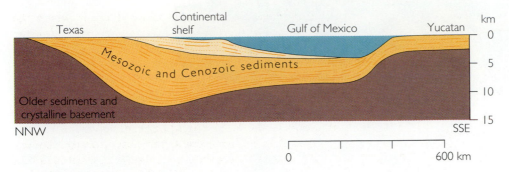

**FIGURE 21.14** Inferred thicknesses of Mesozoic and Cenozoic sediments in the Gulf of Mexico from Texas to Yucatan. Maximum thickness near the Texas coast may exceed 10 km. (After P. B. King, *The Evolution of North America,* Princeton, Princeton University Press, 1977.)

subside and to receive sediments from the continent. Cretaceous and Tertiary sediments up to 5 km thick filled the slowly subsiding trough, and even more material was dumped into the deeper water of the continental rise (see Figure 20.17). This living off-shore sedimentary basin (it's still receiving sediments) may be the site of some future orogeny. If the present stage of opening of the Atlantic is reversed some millions of years from now, the sediments in this basin will be folded and faulted in the same process that produced the Appalachians.

The coastal plain and shelf of the Gulf of Mexico are continuous extensions of the Atlantic ones, interrupted only briefly by the Florida peninsula. The Mississippi, Rio Grande, and other rivers that drain the interior of the North American continent have delivered sediments to fill a trough some 10 to 15 km deep running parallel to the coast (Figure 21.14). The Gulf coastal plain and shelf are a rich reservoir of petroleum and natural gas. The Atlantic shelf is now being actively explored for these and other resources.

# REGIONAL VERTICAL MOVEMENT

So far our discussion of crustal movements has emphasized orogeny, which originates in plate collisions and involves compressive deformation (folding, thrust faulting), intrusion of magma, volcanism, and metamorphism. All over the world, however, sedimentary rock sequences record another kind of history: gradual downward and upward movements of the crust without significant deformation, called **epeirogeny.** Though many of the vertical movements are connected with orogeny, epeirogenic

movements commonly affect large regions without extensive folding or faulting. Fossil trees and other plants embedded in coal deposits now mined deep in the Earth tell of earlier times when they grew on the surface. Great thicknesses of sediments deposited on the seafloor and then buried hundreds or thousands of meters beneath it have been raised to hundreds and thousands of meters above sea level, where they are now found. These deposits give evidence of continued slow subsidence during sedimentation. In many cases, later elevation to their present position above the sea can be attributed to simple uplift, with no significant disturbance of the deposits.

A typical product of this kind of epeirogenic downward movement, which is usually slow and intermittent, is a basin in the stable interior, such as the Michigan Basin. Upward movement with little or only moderate faulting or folding can produce broad uplands and plateaus. The Colorado Plateau and the Black Hills of South Dakota were both produced by general upward movements. The Black Hills structure is an oval dome—an area of uplift that slopes more or less uniformly in all directions from the highest point—rising more than 2 km above the surrounding Great Plains. The region in which the Black Hills are located was once much lower; it was a terrain of Precambrian igneous and metamorphic rocks that subsided beneath the sea and was covered by a blanket of Paleozoic and Mesozoic sediment more than 2 km thick. Sometime between the late Cretaceous and the Oligocene, the whole area of about 50,000 km$^2$ was uplifted, as if it were pushed up from below by a piston, without being extensively crumpled or broken. Subsequent erosion has stripped away the sediments that overlay the central part of the uplift, exposing the Precambrian igneous

**FIGURE 21.15** The Black Hills of South Dakota, a dome structure in the Great Plains. Erosion of the central part exposes the core of Precambrian igneous and metamorphic rocks and the succession of Paleozoic and Mesozoic sediments that slope away from the center. (After C. B. Hunt, *Geology of Soils,* San Francisco, W. H. Freeman, 1972.)

and metamorphic rocks below. Here is where the famous Homestake gold mine was discovered in the Precambrian rocks near Lead, South Dakota (Figure 21.15).

Geologists have no ready explanation for most of these slow and broad epeirogenic movements, but they have some hypotheses that may account for some of them. The uplift of Scandinavia and Finland (see Box 19.2) represents the slow upward recovery of the crust after the removal of the glacial load that had depressed it (Figures 21.16 and 21.17a). The formation of deep basins on both sides of mid-ocean ridges is believed to be caused by the cooling and contraction of the newly formed ocean plate (Figure 21.17b). Heating of the lithosphere from below can result in thinning and upwarping (Figure 21.17c, top panel). Movements in the mantle can stretch and thin the lithosphere above, without breaking the plate. This may explain the subsidence of basins within continents (Figure 21.17c, middle panel). If stretching continues and a rift occurs, two continents result with an ocean growing between them. The sediment-filled basins on continental margins (such as those on the east coasts of the Americas and the west coasts of Europe and Africa) reflect the subsidence of the edge of the continent after rifting. This subsidence is caused by contraction as the margin cools and erodes during its withdrawal from the rift (Figure 21.17c, bottom panel), as we explained in Chapter

**FIGURE 21.16** Raised beaches, evidence of upward recovery of the crust after removal of glacial load; Mansel Island, Canada. *Dilabio/Shilts, Geological Survey of Canada.*

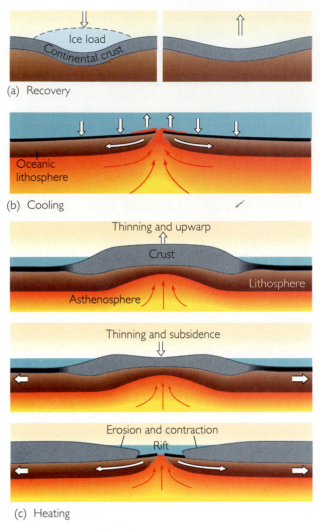

(a) Recovery

(b) Cooling

(c) Heating

(d) Intrusion

**FIGURE 21.17** Some proposed mechanisms for vertical movements (not to scale). (a) A glacial ice load buckles the crust; slow uplift follows removal of the ice. (b) Formation of lithosphere at mid-ocean ridges; arrows show uplift at the crest of the ridge and subsidence of the seafloor as the spreading ocean plate cools and contracts. (c) Thinning and upwarping of continental lithosphere as a result of heating. Stretching can thin the continental crust as it subsides to form a basin within the continent. If the plate rifts, two continents result, with an ocean growing between them. Receding edges of the continents erode at the top and cool and contract within, forming subsiding continental margins. (d) The crust thickens with the addition of magma.

20. Intrusion of magma can thicken the continental crust and result in uplift (Figure 21.17d). In general, however, warping within continents or anywhere far from plate margins is not completely understood.

Up-and-down movement of large regions is not restricted to the geologic past; it is occurring at a measurable rate in our own time. The city of Venice, for example, is slowly sinking into the Adriatic Sea at a rate of 4 mm per year. The process consists mostly of coastal downwarping, though subsidence caused by the withdrawal of water and natural gas from the underlying sediments is, unfortunately, hastening the death of this beautiful city (Figure 21.18).

Topographic surveys across the United States have revealed the pattern of vertical movements shown in Figure 21.19. Large regions are sinking or rising at rates of 1 to 15 mm per year. Not all of these movements will continue. If the pattern persists for the geologically short time of only a million years, however, much of New England, the Gulf coast, and part of California will sink to the seafloor, and plateaus a few kilometers high will grow from the lowlands in the Midwest. These seemingly slow rates of displacement, unnoticed by the inhabitants, are all that it takes to produce plateaus and basins.

**FIGURE 21.18** Venice is slowly subsiding into the Adriatic Sea. The water now reaches the base of the columns of this old building. The raised sidewalk, of recent construction, will be awash in several decades if subsidence continues. *R. Frassetto, National Research Council of Italy.*

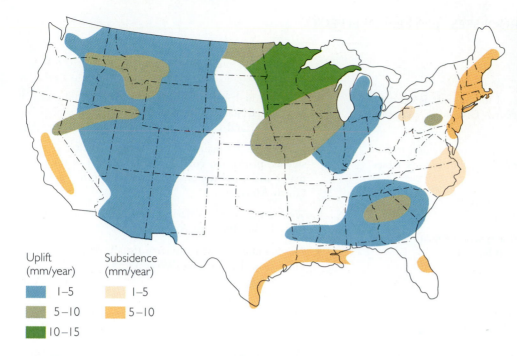

**FIGURE 21.19** Areas of present-day uplift and subsidence in the United States. (After S. P. Hand, National Oceanic and Atmospheric Administration.)

Uplift (mm/year)
- 1–5
- 5–10
- 10–15

Subsidence (mm/year)
- 1–5
- 5–10

# SUMMARY

**What are the major belts of deformation of North America?** Most of the continental crust can be divided into belts that have been deformed during various geologic periods. North America contains a large central stable region that has been relatively undisturbed (except for gentle vertical movements and erosion) since episodes of intense deformation in the Precambrian. Surrounding the stable interior are younger orogenic belts—the mountainous Cordillera and Appalachians, which were deformed in plate collisions at various times in the Paleozoic, Mesozoic, and Cenozoic eras. Continents can grow by a succession of orogenies over geologic time in several ways: by sedimentation and buildup of continental shelves, by the addition of batholiths and volcanic rocks derived by melting in the mantle in subduction zones, and by microplate accretion.

**What events typify an orogeny caused by plate convergence?** A typical orogenic episode is preceded by a stage in which plates separate and an ocean widens between them. Continental margins subside and accumulate sediments. Plate convergence initiates deformation in a belt that extends hundreds of kilometers from the collision site. The marginal sediments are disrupted by folding and faulting. Thrust sheets 10 to 20 km thick slide over one another over distances of tens to hundreds of kilometers. Foreign fragments (microplates), brought in with the subducting plate, accrete to the continent. Intrusion of batholiths and metamorphism typically accompany orogeny. The mountains raised in the deformed belt erode after the orogeny ends. A renewed stage of uplift or block faulting that again raises the regions accounts for many of the mountainous features we see today.

**What are epeirogeny and orogeny?** Forces within the crust can deform large regions of the continents. Some regional movements are simple up-and-down displacements without severe deformation of the rock formations (epeirogeny); examples are the Colorado Plateau and the postglacial uplift of Scandinavia and central Canada. In other cases, horizontal forces connected mainly with plate collisions can produce extensive and complex folding and faulting (orogeny), as in the Cordillera and Appalachians of North America, the Alps of Europe, and the Himalayas of Asia. The Rockies and the Alps are mountains that were eroded to low relief only to be rejuvenated by recent broad regional uplift.

# KEY TERMS AND CONCEPTS

orogeny (p. 480)
craton (p. 482)
shield (p. 482)

platform (p. 483)
sedimentary basin (p. 483)
rejuvenation (p. 491)

upwarped mountain (p. 491)
fault-block mountain range (p. 491)
epeirogeny (p. 494)

# EXERCISES

1. Evidence of vertical crustal movements is often found in the geologic record. Give some examples of such evidence.

2. Describe the geologic evolution of a region that leads to the formation of mesas, buttes, and table-lands bounded by high cliffs.

3. From 1925 to 1977 the San Joaquin Valley of California subsided 9 m. Guess why.

4. Summarize the stages of a typical orogenic episode in a series of sketches with legends.

5. Draw a rough topographic profile of the contiguous United States from San Francisco to Washington, D.C., and label the major geologic features.

6. Summarize the main features (orogenic belts, shields, platforms, coastal plains, shelves) of the major structural regions of a continent other than North America.

7. Why is the North American Cordillera topographically higher than the Appalachians?

8. Are the interiors of continents usually younger or older than the margins? Why?

# THOUGHT QUESTIONS

1. How would you recognize a microplate? How could you tell if it originated far away or nearby?

2. How would you identify a region where active orogeny is taking place today? Give an example.

3. Would you prefer to live on a planet with orogenies or without them? Why?

# SUGGESTED READINGS

Bally, A. W., and A. R. Palmer (eds.). 1989. *The Geology of North America: An Overview*. Boulder, Colo.: Geological Society of America.

Burchfiel, B. C. 1983. The continental crust. *Scientific American* (September):130.

Cook, F. A., D. S. Albaugh, L. D. Brown, S. Kaufmann, J. E. Oliver, and R. D. Hatcher, Jr. 1979. Thin-skinned tectonics in the crystalline southern Appalachians. *Geology* 7:563–567.

Jones, D. L., A. Cox, P. Coney, and M. Beck. 1982. The growth of western North America. *Scientific American* (November):70.

Kearey, P., and F. J. Vine. 1990. *Global Tectonics*. London: Blackwell Scientific Publications.

Twiss, R. J., and E. M. Moores. 1992. *Structural Geology*. New York: W. H. Freeman.

# BOUNTIFUL EARTH

**E**arth's internal processes have given us our life-sustaining environment: a breathable atmosphere, oceans, rich soils, a moderate climate. Humans living during the Stone Age survived in this environment at subsistence levels. Humankind progressed to a better quality of life when we learned how to draw and use Earth's minerals. These resources create wealth and comfort by providing the materials and energy to grow and process food, build structures, transport things, and manufacture goods of all kinds. Profligate use of Earth's finite resources without regard for the fragility of the Earth system can lead to their depletion and the hazardous accumulation of wastes; it can also trigger climatic change, with serious consequences.

# ENERGY RESOURCES FROM THE EARTH

Humans have steadily increased the amounts of coal, oil, natural gas, and uranium we have taken from the Earth to power our complex societies. How do these fuels form? Where are they found? Who does and will control them? How long will the supplies of these vital nonrenewable resources last, and what will we do when they are exhausted? Such questions have aroused concern about the environmental impact of the residues of the fuels we use so recklessly, about the need to conserve our resources, and about the development of substitutes for them. These concerns reflect a new and deeper understanding that we cannot continue to draw wealth from the Earth indefinitely without concern for our habitat and the generations to come.

## 22

Windmills generate electricity from the wind. This is one form of renewable solar energy. *Pacific Gas and Electric.*

We rely on oil, coal, natural gas, and uranium, resources derived from the Earth, for the energy to run our factories, construct our buildings, heat our homes, generate our electricity, produce our food, and fuel our transportation. In our increasingly systematic search of the globe for new sources of the fuels we depend on, we use our geological knowledge of how known natural deposits form to determine where we may find more of them. At the same time we are becoming more sensitive to the finiteness of Earth's resources and the delicacy of its environment. We are paying more attention to extracting and using natural resources more efficiently, without damaging the environment. We are beginning to think about how to achieve a sustainable society that both satisfies its needs and preserves the prospects of future generations.

## RESOURCES AND RESERVES

Two major questions arise in all discussions of materials we draw from the Earth: How much is left? How long will it last? The amount left consists of more (we hope) than what we know as reserves. **Reserves** are deposits that have already been discovered and that can be mined economically and legally at the present time. When we speak of **resources,** we are referring to the entire amount of a given material that eventually may become available for use. Resources include reserves, plus discovered deposits that are too poor in quality or quantity to be worth mining now or that are difficult to mine and await new technology or higher prices before they can be mined profitably, plus undiscovered deposits that geologists think they may be able to find eventually (Figure 22.1).

Reserves are considered a dependable assessment of supply as long as economic and technological conditions remain as they are now. As conditions change, some resources become reserves and vice versa. The assessment of resources is much less certain than the assessment of reserves. Any figure cited as the resources of a particular material represents only an educated guess of how much will be available in future decades.

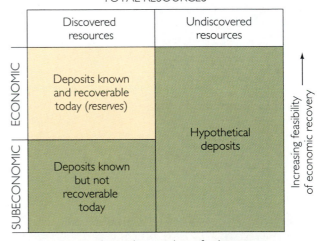

FIGURE 22.1 Categories that constitute total resources. Discovered resources consist of reserves, which are known deposits that are economically minable today, and deposits that are known but that are currently subeconomic. Undiscovered resources are hypothetical deposits that we may find. We can reasonably depend on reserves, but the areas that include known subeconomic deposits and estimates of undiscovered deposits represent only an educated guess of what might become available in the future.

502

**FIGURE 22.2** Early oil wells; Signal Hill, California. *Chevron Corp.*

Most useful geological materials are *nonrenewable* in the sense that geologic processes produce them more slowly than civilization uses them up. Coal and oil, for example, will be depleted faster than nature can replenish them. This fact puts a premium on the development of *renewable* resources such as solar energy, which is essentially infinite in supply, and fuels such as alcohol, derived from crops that can be replanted after they are harvested.

# ENERGY

Energy is fundamental to everything. A crisis in the supply of energy can bring a modern society to a halt. Wars have been fought over access to supplies of fuel resources; economic recession and destructive currency inflation have resulted from gyrations in the price of oil.

As the world industrialized, the demand for energy increased and the types of energy used changed. The Industrial Revolution of the eighteenth and nineteenth centuries was powered by the energy from coal—in Britain from the coalfields of England and Wales, in continental Europe from the coal basins of western Germany and bordering countries, and in North America from the Appalachian coalfields of Pennsylvania and West Virginia. As industrialization expanded, so did the hunger for coal. Geological exploration for this fuel spread over much of the world, as coal use climbed at an ever-accelerating pace.

Half a century after the first oil well in America was drilled in 1859, oil and gas were beginning to displace coal as the fuels of choice (Figure 22.2). Not only did they burn more cleanly, producing no ash, but they could be transported by pipeline as well as by rail and ship.

The last quarter of the twentieth century saw the introduction of nuclear energy, with expectations that it would provide a low-cost, environmentally benign, large new source of energy. These expectations were not realized, however, because safety concerns, the inability to handle nuclear wastes, and escalating costs slowed the growth of nuclear energy.

The advanced nations depend primarily on oil, coal, natural gas, and nuclear energy, the precise mix depending on how well they are endowed with fossil fuels. In poorer countries wood is an important source of fuel. The history of energy use in the United States is summarized in the graph in Figure 22.3. Today oil, coal, and natural gas supply about 90

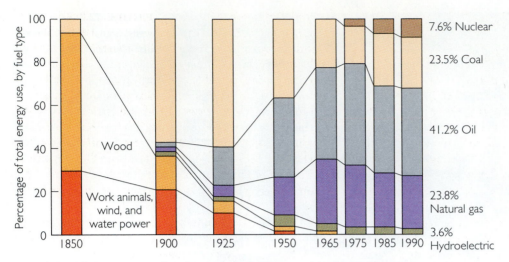

**FIGURE 22.3** Percentages of various types of energy used in the United States from 1850 to 1990. (Data from U.S. Energy Information Agency, 1991.)

percent of the energy used in the United States. Consumption of this supply is distributed among residential and commercial uses (35.8 percent), industrial use (37.0 percent), and transportation (27.1 percent). Approximately half of the energy produced is lost in distribution and inefficient use.

A century and a half ago, most of the energy used in the United States came from the burning of wood. A wood fire, in chemical terms, releases energy by the combustion of organic matter consisting of compounds of carbon with hydrogen attached to the carbon atoms. Organic matter is produced by plants or animals. The organic matter in this case is a tree produced from carbon dioxide and water, with energy for the transformation supplied by sunlight. This is the familiar process of photosynthesis, discussed in Chapter 9 (see Box 9.3). Thus we can look upon a piece of wood or any piece of plant matter as a photosynthetic product that can be returned by burning or decay to the carbon dioxide and water from which it was made. If we burn wood that was buried and transformed into the combustible rock known as coal 300 million years ago, we are using energy stored by photosynthesis from late Paleozoic sunlight. We are recovering "fossilized" energy. Crude oil and natural gas were also created by a process of burial and chemical transformation of dead organic matter to a combustible liquid and gas, respectively. We refer to all such resources derived from natural organic materials, from coal to oil and natural gas, as **fossil fuels** (Figure 22.4).

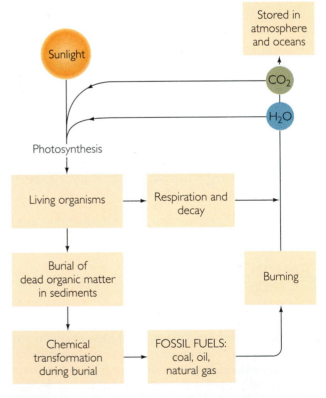

**FIGURE 22.4** Photosynthesis produces organic matter from carbon dioxide ($CO_2$) and water ($H_2O$). If dead organic matter is buried and transformed into coal, oil, or natural gas, it becomes a fossilized product of photosynthesis—a fossil fuel. The burning of fossil fuels releases the carbon dioxide and water from which they were made.

# OIL AND NATURAL GAS

## How Oil and Gas Form

Crude oil (petroleum) and natural gas in minable form develop under special conditions in the environmental and geologic history of a region. Both are the organic debris of former life—plants, bacteria, algae and other microorganisms—that has been buried, transformed, and preserved in marine sediments. Oil and gas begin to form when more organic matter is produced than is destroyed by scavengers and decay. This condition is met in environments where the production of organic matter is high—as in the coastal waters of the sea, where large numbers of organisms thrive—and where the supply of oxygen in bottom sediments is inadequate to decompose all organic matter. Many offshore sedimentary basins on continental shelves satisfy both of these conditions; in such environments, and to a lesser degree in some river deltas and inland seas, organic matter is buried and protected from decomposition. During millions of years of burial, chemical reactions trig-

gered by the elevated temperatures at depth slowly transform some of the organic material into liquid and gaseous compounds of hydrogen and carbon (hydrocarbons). The hydrocarbons are the combustible materials of oil and natural gas. Compaction of muddy organic sediment into **source beds** forces the hydrocarbon-containing fluids and gases into adjacent beds of permeable rock (such as sandstones or porous limestones), which we call **oil reservoirs.** The low density of oil and gas causes them to rise to the highest place they can reach, where they float atop the water that almost always occupies the pores of permeable formations.

Geological conditions that favor the large-scale accumulation of oil and natural gas are combinations of structure and rock types that create an impermeable barrier to upward migration—an **oil trap.** One type of oil trap is formed by an anticline in which a permeable bed of sandstone is overlain by an impermeable shale (Figure 22.5a). The oil and gas accumulate at the crest of the anticline, the gas highest, the oil next, both floating on the groundwater that saturates the sandstone. An oil trap caused by a structural

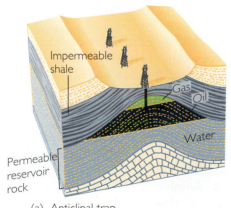

(a) Anticlinal trap

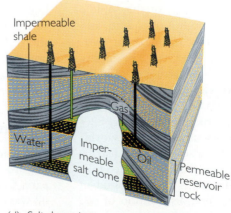

(b) Fault trap

**FIGURE 22.5** Oil traps: (a) anticlinal trap; (b) fault trap; (c) stratigraphic trap; (d) oil trapped by a salt dome. Natural gas (green) and oil (black) are trapped by an impermeable layer above the permeable oil-producing formation. Oil floats above the water line.

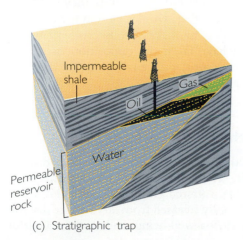

(c) Stratigraphic trap

(d) Salt dome trap

deformation like the one that created the anticline is called a *structural trap*. Similarly, displacement at a fault may place a dipping permeable limestone bed opposite an impermeable shale, creating a structural trap for oil (Figure 22.5b). An oil trap created by the original pattern of sedimentation, as when a dipping permeable sandstone bed thins out against an impermeable shale (Figure 22.5c), is called a *stratigraphic trap*. Oil can also be trapped against an impermeable mass of salt, as in a salt dome (Figure 22.5d).

In their search for oil, geologists have mapped thousands of structural and stratigraphic traps all over the world. Only a fraction of them have proved to contain any oil or gas, because the traps alone are not enough. A trap will contain oil only if source beds were present, if the necessary chemical reactions took place, and if the oil could migrate into the trap and stay there without being disturbed by subsequent severe heating or deformation. Although oil and gas are not rare, most of the easy-to-find deposits have already been located, and new fields are becoming more difficult to find.

## The World Distribution of Oil and Natural Gas

If you were to visit the exploration and research offices of any large oil company, you would find maps and reports of all the regions containing geologic sections of sedimentary rock in which the company has operated, for where there is a thick section of sedimentary rocks, there may be oil. Thirty-one of the 50 states of the United States produce oil for the market, and small, noncommercial occurrences are known in most of the others. Many of Canada's provinces produce oil.

Two of the richest and most important oil-producing regions in the world are the Middle East and the area around the Gulf of Mexico and the Caribbean (Figure 22.6). The oil fields of the Middle East, including Iran, Kuwait, Saudi Arabia, Iraq, and the Baku region in Azerbaijan, contain about two-thirds of the world's known reserves. The highly productive Gulf coast–Caribbean area includes the Louisiana-Texas region, Mexico, Colombia, Venezuela, and Trinidad. Saudi Arabia holds the largest reserves. The United States ranks eighth. Figure 22.7 summarizes oil reserves in various parts of the world.

Why are oil and natural gas so unevenly distributed about the globe? A political scientist once jokingly asked, "Why is it that oil is always found in underdeveloped countries?" The distribution of oil, of course, is determined not by political boundaries

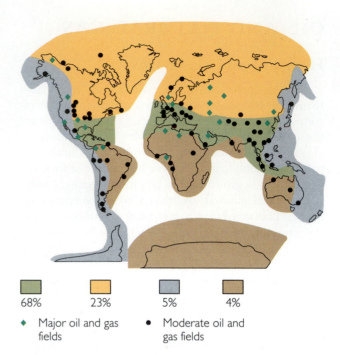

| | | | |
|---|---|---|---|
| 68% | 23% | 5% | 4% |

◆ Major oil and gas fields     ● Moderate oil and gas fields

**FIGURE 22.6** Four petroleum provinces of the world and the oil and gas content of each as a percentage of world discovered oil and gas. Locations of major and moderate oil and gas fields are also shown. (After C. D. Masters, D. H. Root, and E. D. Attanasi, USGS.)

but by the geologic history of a region. For example, continental shelves are good places to explore for oil. They easily meet the conditions required for rich oil deposits: the sediments are old enough to accommodate the slow process of oil formation and young enough to be preserved in an environment that is little eroded, metamorphosed, or deformed. Yet most shelf deposits are still undeveloped, even those of the U.S. continental shelf. One reason is concern about the environment. Another is the difficulty and expense of drilling in deep water. Many oil companies are waiting for oil prices to rise before they make the huge investments that would be required. Some companies willing to make the investment in offshore drilling have faced the opposition of a local population concerned with environmental damage, as in California.

## Oil and the Environment

Pollution is the major problem of offshore drilling (Figure 22.8). The environment around Santa Barbara, California, suffered great damage in 1969 when oil was accidentally released from an offshore drilling platform. In 1979 a well being drilled in the Gulf of

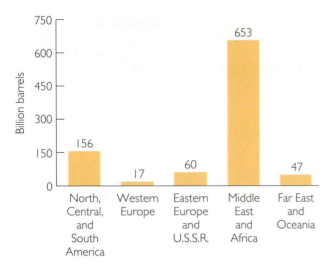

**FIGURE 22.7**  Estimated world reserves of crude oil, 1990. (Data from World Oil.)

of equipment and safety procedures can greatly reduce the chances of a serious accident. There are large resources of oil and gas under the seabed, they argue, and eventually they will have to be drilled to satisfy the world's growing energy needs. A skeptical public, however, is not convinced that offshore drilling can be done without serious threat to the oceans and nearby coasts.

Deliberate oil pollution as an act of war was carried out on a large scale for the first time in the Gulf War of 1991, resulting in great damage to marine life in the Persian Gulf. This destructive action by Iraq is too recent to permit a full assessment of its environmental, economic, and military consequences.

## Oil: The Exhaustible Supply

Inconceivable as it may seem to the landowner who sees a gusher spurting oil from a derrick on his property, that oil well will eventually run dry. What the world wants to know is: How soon will *all* the wells run dry? Forty years ago, many oil producers were optimistic; supplies seemed so immense, and so many scientific and engineering innovations held promise of tapping new sources, that no one saw a need to worry about the future. A more sober analysis of the total quantity of oil remaining on Earth presents a different picture, one of the steady dwindling of a fixed supply. World demand has accelerated rapidly. Twice as much oil was removed from the ground in the past 20 years as in the previous 100 years. The process of oil formation takes millions of

Mexico off the Yucatan coast "blew out," spilling as much as 100,000 barrels of oil a day for many weeks before it could be capped. In 1988 an explosion destroyed a drilling platform in the North Sea, killing many oil workers and marine animals. The grounding of the tanker *Exxon Valdez* off the coast of Alaska in 1989, with the release of 240,000 barrels of crude oil in pristine coastal waters, was covered widely by TV and newspapers and heightened public awareness of the severe ecological damage that can result from an oil spill. Despite such incidents and the difficulty of guaranteeing the safety of a well, proponents of offshore oil development believe that careful design

**FIGURE 22.8**  The effect on wildlife of an oil spill from an offshore well; Northumberland, UK. *David Woodfall/NHPA.*

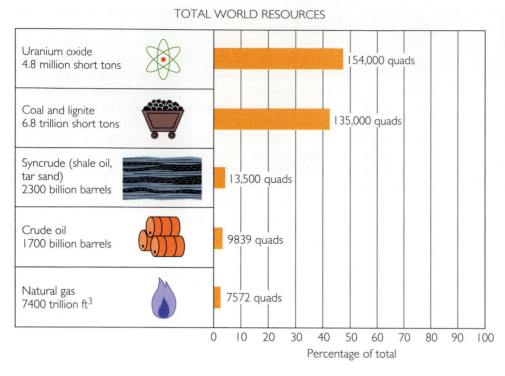

TOTAL WORLD RESOURCES

| | | |
|---|---|---|
| Uranium oxide 4.8 million short tons | | 154,000 quads |
| Coal and lignite 6.8 trillion short tons | | 135,000 quads |
| Syncrude (shale oil, tar sand) 2300 billion barrels | | 13,500 quads |
| Crude oil 1700 billion barrels | | 9839 quads |
| Natural gas 7400 trillion ft³ | | 7572 quads |

Percentage of total

**FIGURE 22.9** Remaining nonrenewable world energy resources amount to about 320,000 quads. Amounts are given in conventional units of weight (short tons), volume (barrels), and energy content (quads). A short ton is 2000 pounds or 907.20 kilograms; a barrel of oil is 42 gallons. Coal and lignite resources, for example, amount to 6.8 trillion short tons, equivalent to 135,000 quads, or 42 percent of total energy resources.

years and humankind is using it up in centuries. Natural processes cannot replenish the oil supply as quickly as we are using it.

Figure 22.9 shows that the world's remaining energy resources of all types are estimated to be about 320,000 quads.[1] Of this total, nearly 10,000 quads (1700 billion barrels) are oil. How long this supply of oil will last depends on the rate at which we use oil and on our success in converting resources into reserves. At current rates of world use (20 billion barrels per year), most of the remaining oil will be depleted in about 85 years. To round off an uncertain number, say 100 years. The remaining oil supply will last longer if consumption declines, as it did because of conservation and the industrial slowdown that came with the recession of the early 1980s, or it may increase if use picks up.

Here are some facts to ponder. The United States consumes more oil than it produces. It produces only

about 17 percent of the world's oil and consumes about 30 percent. Of the total energy consumed in the United States each year, oil supplies about 40 percent. Most of the oil (about 60 percent) is used for transportation. The total amount of oil used in the United States is about 6 billion barrels per year, about half of it imported. These annual imports cost anywhere from $30 billion to $50 billion, depending on the current price of a barrel of oil, and account for a large part of the U.S. trade deficit—the excess of imports over exports. At current use rates, remaining U.S. oil reserves (26 billion barrels) amount to roughly a 10-year supply. Resource estimates vary considerably. One estimate is about 100 billion barrels, which would carry the country for another 30 years or so—not much time before the United States would become dependent on imports for most of its oil, at a cost of perhaps several hundred billion dollars annually.

New technology might be developed to find oil deposits that have been missed by current methods of exploration and to enhance the recovery of oil from existing domestic oil fields. About half of the oil in a field is now left in the ground because of the difficulty of pumping it out. Water can be pumped to the periphery of a field to force more of this oil out of a central well. To loosen up very viscous oils and allow them to flow more easily, steam or carbon dioxide can be injected into the oil reservoir under pressure.

---

[1]Fuel resources are measured in units appropriate to the material; for example, barrels of oil, tons of coal. To make it easier to compare the energy available from different fuel resources, however, a common unit called the quad is used. A **quad** is a measure of the energy that can be extracted from a given amount of fuel. The quad is based on a standard measure of energy called the British thermal unit (Btu). One Btu is the amount of energy needed to raise the temperature of 1 pound of water by 1°F. One quad equals $10^{15}$ Btu. The United States used about 80 quads of energy in 1989.

A large amount of oil may become economically recoverable in the United States by these enhancement techniques, possibly increasing domestic resources significantly. The amount recovered could be equivalent to a few decades of imports.

Many in the United States view the dependence on foreign sources as potentially destabilizing to the economy. Many oil-exporting nations have not been reliable suppliers and have cut off exports because of political disagreements and wars. Other industrial nations that depend on oil imports, such as Japan, France, and Germany, have similar concerns.

Oil-importing nations have options to reduce their vulnerability. They can change their pattern of oil use to reduce their need for imports. Automobile and airplane engines can be designed to use fuel much more efficiently. If the efficiency of the U.S. automobile fleet were to improve by 5 miles per gallon of gasoline, current oil imports could be reduced by almost half. Automobiles can run on alcohol produced from biomass (grain, sugarcane), on natural gas, and on electric batteries. These fuels also would be less polluting than gasoline. Brazil fuels almost all of its automobiles with alcohol made from sugarcane and so does not need to import oil for transportation. New technology is needed to improve these alternative fuels, however, particularly to improve on the engines that use them, to lower their costs, and to arrange for efficient, large-scale production and distribution.

With about 100 years' worth of oil remaining beneath the Earth at current rates of use before nature's oil legacy is gone, we cannot yet proclaim the end of the age of oil. A century may be barely enough time to plan and put into place an orderly transfer to new systems of transportation and other energy uses in the home and in industry. We may have to make greater use of mass transportation rather than automobiles, design more fuel-efficient automobiles and airplanes, exploit alternative sources of fuel, and place greater emphasis on energy conservation in industry, commerce, and the home. Without such planning for alternatives and conservation, many nations may face social and economic disruption when transportation systems grind to a halt, production lines stop, and homes grow cold in winter because oil has become so scarce that its price has risen out of reach.

### Natural Gas

The resources of natural gas are comparable to those of crude oil (see Figure 22.9) and may exceed them in the decades ahead. Estimates of natural gas resources have been rising in recent years. Exploration for this clean fuel has increased and geologic traps in new geological settings have been identified, such as very deep formations, overthrust belts, coal beds, tight (that is, somewhat impermeable) sandstones, and shales. The world's resources of this fuel are less depleted than oil resources because it is a relative newcomer on the energy scene; it has been used on a large scale only in the United States and the former Soviet Union.

The burning of natural gas releases less carbon dioxide per unit of energy than the combustion of coal or oil. It is less polluting (little ash or acid rain precursors are released), and it is easily transportable. Natural gas from fields in Siberia, for example, are piped to factories and homes in Germany. For these reasons, natural gas is a premium fuel. Natural gas accounts for about 25 percent of all fossil fuels consumed in the United States each year, most of it for industry and commerce (52 percent), followed by residential use (28 percent) and generation of electric power (18 percent). More than half of American homes and a great majority of commercial and industrial buildings are connected to a network of underground pipelines that draw gas from fields in the United States, Canada, and Mexico. American resources should last about 35 years at current rates of use, probably longer with the tapping of accumulations in the new settings we mentioned earlier.

## COAL

The abundant plant fossils found in coal beds indicate that coal is formed from large accumulations of plant materials, of the sort that would occur in wetlands. As the luxuriant plant growth of a wetland dies, it falls to the waterlogged soil. Rapid burial by falling leaves and immersion in water protect the dead twigs, branches, and leaves from complete decay because the bacteria that decompose vegetative matter are cut off from the oxygen they need. The vegetation accumulates and gradually turns into peat, a porous brown mass of organic matter in which twigs, roots, and other plant parts can still be recognized (Figure 22.10). The accumulation of peat in an oxygen-poor environment can be seen in modern swamps and peat bogs. Peat burns readily when it is dried because it is 50 percent carbon.

Over time, with continued burial, the peat is compressed and heated. Chemical transformations increase the peat's already high carbon content, and the peat becomes *lignite,* a very soft brownish-black

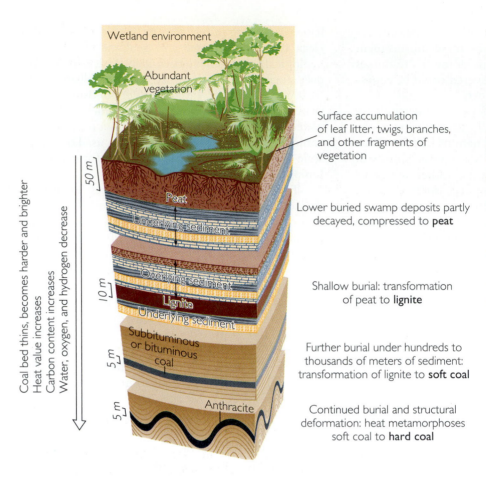

Coal bed thins, becomes harder and brighter
Heat value increases
Carbon content increases
Water, oxygen, and hydrogen decrease

Wetland environment

Abundant vegetation

Surface accumulation of leaf litter, twigs, branches, and other fragments of vegetation

Peat

Underlying sediment

Overlying sediment

Lignite

Underlying sediment

Subbituminous or bituminous coal

Anthracite

Lower buried swamp deposits partly decayed, compressed to **peat**

Shallow burial: transformation of peat to **lignite**

Further burial under hundreds to thousands of meters of sediment: transformation of lignite to **soft coal**

Continued burial and structural deformation: heat metamorphoses soft coal to **hard coal**

**FIGURE 22.10** The process by which coal beds form begins with the deposition of vegetation. Protected from complete decay and oxidation in a wetland environment, the deposit is buried and compressed into peat. Subjected to further burial, peat undergoes mild metamorphism, which transforms it successively into lignite, subbituminous and bituminous (soft) coal, and anthracite (hard coal), as the deposit becomes more deeply buried, temperature rises, and structural deformation progresses.

coallike material, about 70 percent carbon. The higher temperatures and structural deformation that accompany greater depths of burial may metamorphose the lignite into *subbituminous* and *bituminous coal,* or soft coal, and ultimately to *anthracite,* or hard coal. The greater the metamorphism, the harder and brighter the coal and the higher its carbon content, which increases its heat value. Anthracite is over 90 percent carbon.

According to some estimates (see Figure 22.9), about 6.8 trillion short tons of coal remain in the world. The leading producers are the United States (Figure 22.11), the former Soviet Union, and China, which together hold about 85 percent of the world's coal resources (former Soviet Union, 50 percent; China, 20 percent; United States, 15 percent). Domestic coal resources in the United States would last for 300 to 400 years at current rates of use—about a billion tons a year. Coal has supplied an increasing portion of the energy needs of the United States since 1975, when the price of oil began to rise, and currently accounts for about 24 percent of the energy consumed (see Figure 22.3).

If oil supplies become scarce, coal could be converted to liquid or gaseous fuels comparable to those obtained from crude oil today. The cost of synthetic oil, called **syncrude,** derived from coal is higher than that of crude oil at today's prices, but with depletion of world oil reserves and political instability in the Middle East, the gap between the two could decrease in the years ahead. A few commercial-scale synthetic oil and gas plants were constructed in the United States in the 1970s, when oil-exporting nations in the Middle East cut back on oil production for political reasons. With today's low cost of imported oil, these plants are not economical and most have been shut down. If prices increase in the years ahead, however, as most observers expect, several million barrels of synthetic oil per day may well be produced from coal in the United States in the next century.

There are serious problems with the recovery and use of coal—problems that make it less desirable than oil or gas, whether the coal is burned or converted to synthetic liquid fuel. Much coal contains appreciable amounts of sulfur, which vaporizes during combustion and liberates noxious sulfur oxides to the atmosphere. Acid rain, formed when these gases combine with rainwater, is becoming a serious problem in the northeastern United States, Canada, Scandinavia, and eastern Europe (see Box 6.1). Coal ash is

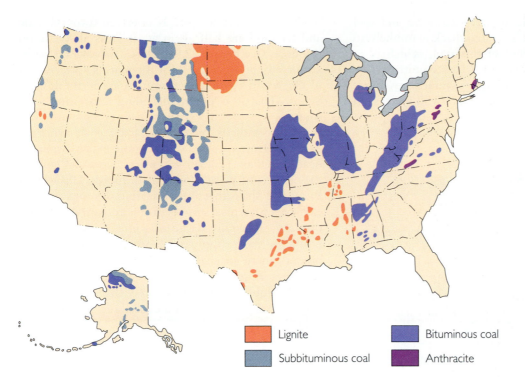

**FIGURE 22.11**
Coalfields of the United States. (Data from U.S. Bureau of Mines.)

Lignite

Subbituminous coal

Bituminous coal

Anthracite

the inorganic residue that remains after coal is burned. It contains metal impurities from the coal, some of which are toxic. Ash can amount to several tons of matter for every 100 tons of coal burned and poses a significant disposal problem. It can escape from smokestacks, posing a health risk to people downwind. Strip mining, or the removal of soil and surface sediments to expose coal beds, can ravage the countryside if the land is not restored (Figure 22.12).

Underground mining accidents take the lives of miners each year, and many more suffer from black lung, a debilitating inflammation of the lungs caused by the inhalation of coal particles. These human costs are as important to consider as the possible dangers of nuclear reactors.

The technologies exist for the "clean" combustion or liquefaction of coal, for the restoration of land ravaged by strip mining, and for the reduction of

**FIGURE 22.12** Coal strip mine near Shamokin, Pennsylvania. *Left:* Strip mining of anthracite. *Right:* After reclamation. *Peter Kresan.*

danger to miners. But they are expensive and will add to the cost of coal. This drawback is unlikely to prevent the increased use of this fuel, which is so much more abundant than oil. Many countries have no other fuel resources, and other countries will not be able to afford to import oil, which will increase in cost as the supply diminishes.

## OIL SHALE AND TAR SANDS

Syncrude can also be extracted from oil shale and tar sands. **Oil shales** are fine-grained sedimentary rocks containing a large proportion of solid organic matter that can be distilled by heating to yield oil. This organic matter has the same general origin as the other common organic components of fossil fuels, all of which ultimately derive from plant and animal matter. Some oil shales can produce up to 150 gallons of oil per ton, but most yield from 25 to 50 gallons per ton (Figure 22.13).

**Tar sand** is a sandy deposit of organic matter impregnated with a tarry substance made up almost entirely of hydrocarbons. Oil of a type similar to liquid petroleum can be recovered from tar sand by heating, and geological evidence suggests that its tarry components are a transformed product of a formerly liquid petroleum. Some deposits of tar sands are oil pools that have dried up and become tarry by loss of the more volatile hydrocarbons. One of the largest deposits, the Athabaska tar sands of Cretaceous age, is found in Alberta, Canada. Other major deposits are known to exist in Venezuela.

World syncrude resources derived from oil shale and tar sand are half again as great as all crude oil resources (see Figure 22.9). The Athabaska tar sands are already producing oil for the market. Nearly half of the world's syncrude resources are in the United States. This is a potentially significant future resource.

Like synthetic oil derived from coal, syncrude extracted from oil shale will be more expensive in the near term than crude oil, but prices could converge in the next century as oil supplies diminish. Disposal of the waste shale that remains after the oil has been extracted presents a serious problem. And the extraction process requires large amounts of water, a commodity in short supply in Wyoming, Utah, and Colorado, where oil shales are plentiful. The difficulty and expense of extraction and associated environmental problems make it unlikely that syncrude will displace much crude oil in the fossil fuel market in the next few decades.

## THE FUTURE OF FOSSIL FUELS

If crude oil and gas continue to be used as the major resources for satisfying the world's voracious appetite for energy, the great bulk of the world's supply will be exhausted within a century. Oil shale and tar sands will add some decades, but coal will eventually become the predominant fossil fuel in many countries. It may be reassuring to know that at modest annual energy growth rates, say 3 percent per year, coal and other fossil fuels can meet the world's en-

**FIGURE 22.13** Syncrude production in the principal oil-shale group of the Green River formation of Colorado, Utah, and Wyoming. Almost half of the world's supply of oil shale is found in this formation. At this facility oil shale mined from the cliffs is heated to produce syncrude oil, which is subsequently refined into common petroleum products. The plant can produce 10,000 barrels of syncrude each day. *Unocal Corp.*

ergy needs for about 100 years or longer. This security, however, may be false. Carbon dioxide released during the combustion of fossil fuels may trigger climatic changes that could force a shift away from these traditional fuels before they are depleted, to avoid a climatic crisis (see Box 22.1).

These estimates do not take into account the possibility that we may learn to meet much of our growing energy need in nontraditional ways: through increased efficiency in the use of fossil fuels and by the development and use of alternative energy sources such as nuclear energy, solar energy, geothermal energy, and energy derived from biomass. To the extent that these sources can be used, the pressure on our fossil fuel resources can be relaxed and their life extended.

# NUCLEAR ENERGY FUELED BY URANIUM

Although uranium 235 ($^{235}$U) was first used in an atomic bomb in 1944, the nuclear physicists who first observed the vast energy released when its nucleus splits spontaneously (called *fission*) foresaw the possibility of peaceful applications of this new energy source. After World War II, their predictions were fulfilled as countries all over the world built nuclear reactors to produce **nuclear energy:** the fission of $^{235}$U releases heat to make steam, which then drives turbines to create electricity. The fission of 1 gram of $^{235}$U liberates energy equivalent to that in about 3 short tons of coal or 13.7 barrels of crude oil. In the United States some 113 nuclear reactors now provide about 20 percent of the electric energy we use. France derives 75 percent of its electric energy from nuclear power. Today more than 400 nuclear reactors are producing electricity in 25 countries. If its full potential is realized, nuclear energy can meet the world's needs for electric energy for hundreds of years.

One aspect of nuclear energy is most definitely in the province of geology: the question of reserves of uranium. Uranium is present in very small amounts in Earth's crust, constituting only 0.00016 percent of the average crustal rock. The isotope that fissions and releases energy, $^{235}$U, constitutes only 1 of every 141 atoms of uranium mined. In terms of energy content, however, uranium is potentially our largest minable energy resource (see Figure 22.9). It is typically found as small quantities of the uranium oxide mineral uraninite (also called pitchblende) in veins in granites and other felsic igneous rocks. Uranium can also be found in sedimentary rocks. Under near-surface groundwater conditions, uranium in igneous rocks may become oxidized and dissolved, be transported in groundwater, and later be reprecipitated as uraninite in sedimentary rocks.

The United States is known as a major source of uranium partly as a result of government support of exploration in the 1950s and 1960s to ensure a uranium supply during the nuclear arms race. The richest ores in the United States are found in the Triassic and Jurassic sedimentary rocks of the Colorado Plateau, in western Colorado and adjacent parts of Utah, Arizona, Wyoming, and New Mexico. Large resources are also found in Canada, Australia, South Africa, and Brazil, in that order. Little is known publicly about the uranium resources of the former Soviet Union and China. At one time there was uncertainty about the adequacy of uranium resources. Currently there is a glut of uranium and its price has plummeted. Not only has the demand for uranium slowed along with the lack of growth of nuclear power, but substantial new deposits have been discovered in Australia and elsewhere. The uranium and plutonium from demilitarized Russian nuclear weapons may be brought to the civilian market in competition with mined uranium.

The debate over the adequacy of uranium resources will undoubtedly be rekindled if enthusiasm for nuclear power revives, but such a revival may not occur because of the serious problems that have emerged with the use of nuclear energy. At one time nuclear energy was promoted as the cheap, clean alternative to fossil fuels. The costs of building and maintaining reactors have since proved prohibitive. A more important obstacle to revival has been two accidents that created a public mind-set against nuclear energy. The first was an accident at the Three Mile Island reactor in Pennsylvania in 1979. A reactor was destroyed; radioactive debris was released but confined within the containment building. Although no one was harmed, most experts agree that it was a close call. Much more serious was the destruction of a nuclear reactor in the town of Chernobyl in the Ukraine in 1986. The reactor went out of control because of poor design and human error and was destroyed. Radioactive debris spilled into the atmosphere and was carried by winds over Scandinavia and western Europe. Contamination of buildings and soil has made hundreds of square miles of land surrounding Chernobyl uninhabitable. Food supplies in many countries were contaminated by the fallout and had to be destroyed. Excess deaths from cancer caused by exposure to the fallout may be in the thousands over the next 40 years.

The uranium "used up" in nuclear reactors leaves behind dangerous radioactive wastes that must be

## BOX 22.1 CLIMATIC CATASTROPHE

As we discussed in Chapter 15 (see Box 15.2), humankind has been dumping carbon dioxide ($CO_2$) into the atmosphere at an accelerating rate by our burning of coal, oil, and gas and our destruction of the forests.[1] At the present rate at which we burn fossil fuels and destroy forests we may expect the amount of $CO_2$ in the atmosphere to be double the preindustrial level in the second half of the next century. Other greenhouse gases introduced to the atmosphere by human activity are methane, nitrous oxides, and the chlorofluorocarbons (CFCs) used as refrigerants, cleaning agents, and propellants. The combined effect of all these emissions may be a global warming of 1 to 5°C toward the middle of the next century.

A potentially serious consequence of global warming may be regional changes in wind and rainfall patterns and soil moisture that could convert some of our most productive agricultural regions into semiarid lands. (The drought and unusually hot weather of 1988 and 1991, though not necessarily related to the greenhouse effect, served to raise public consciousness of the seriousness of climatic change in the decades ahead.) Because the changes would be more rapid than any that occur naturally, it would be difficult for many plant and animal species to adjust or migrate. Those that could not cope would become extinct.

At the same time, the oceans would warm and expand, raising sea level as much as 60 cm, a serious problem in low-lying countries such as Bangladesh. If global warming started to melt the continental ice sheets, the problem would be much more serious (see Box 15.1). At the extreme, a global disaster would result if the continental ice sheets in Antarctica and Greenland melted entirely; sea level would then rise some 65 m. Most experts, however, believe this will not happen for several hundred years, if at all.

Most atmospheric scientists expect the climate to change. No one can say how much or to what effect, but we must reckon with the possibility that a climatic crisis could prevent us from fully using the remaining resources of coal and other fossil fuels.

This uncertainty presents a problem for policymakers: How much money should be spent now to reduce the emissions of greenhouse gases and prepare to shift to nonfossil fuels? Policymakers must also face a fact of international politics: the rich, industrially advanced nations (which have been the greatest emitters of greenhouse gases) will be able to adjust to climatic change more easily than the poor countries. These developing nations want financial resources and technology transferred to them to help them cope with climatic change. The issue is currently being debated in the United Nations.

The consequences of climatic change could be so serious that it is not too soon for our leaders to start planning for reduced reliance on fossil fuels. The United States could reduce emissions of greenhouse gases by as much as 40 percent from 1990 levels at no or relatively low cost by improving the efficiency of energy use in buildings (for example, by replacing a few incandescent lights in each home with fluorescent lights, which use less energy), increasing the fuel efficiency of automobiles a few miles per gallon, making greater use of natural gas (which emits less $CO_2$ when burned than coal), and replacing CFCs with more benign substitutes. The extent of global warming may be uncertain, but so is the possibility that a fire will destroy our home. These modest steps would be a low-cost insurance policy.

---

[1]Growing trees absorb $CO_2$. When the wood is burned or decays, it releases $CO_2$.

disposed of (see Box 22.2). A system of safe long-term waste disposal is not yet available, and reactor wastes are being held in temporary storage at reactor sites. In a few years the limits of space available for temporary storage in the United States will be reached. Although many scientists believe that geological containment—that is, burial of nuclear wastes in deep, stable, impermeable rock formations—is a workable solution, there is as yet no generally approved plan for storage of the most dangerous wastes for the hundreds of thousands of years required before they cease to be radioactive. France and Sweden have built underground nuclear waste depositories, but the United States is still in the stage of research, development, and testing. It is also embroiled in litigation as some states battle the federal government to keep waste repositories from being built within their borders. No one wants to leave to the next generations the combined legacy of depleted energy resources and a possibly unmanageable environmental hazard.

In the United States these unresolved problems have essentially halted the installation of new plants, and new installations have been slowed in other countries as well. Concern about the safety of nuclear reactors, their high cost, and the safe disposal of radioactive waste must be allayed before we can light up the world with nuclear energy.

## SOLAR ENERGY

Since all of our fossil energy sources ultimately come from the Sun anyway, why not convert its rays to energy? In principle, the Sun can provide us with all the energy we need, in all the forms we use. Light from the Sun can be converted to heat and electricity. It can even be used to obtain hydrogen, which can be used as a gaseous fuel, from water. Solar energy is risk-free and nondepletable—the Sun will continue to shine for at least the next several billion years. Unfortunately, the technology currently available for large-scale conversion of solar energy to useful forms is inefficient and expensive, although the situation is improving.

In the near term, the only form of solar energy likely to be available at costs nearly competitive with those of other sources is heat for homes, water, and industrial and agricultural processes. Many homes and factories are beginning to use solar energy for these purposes, motivated in part by government tax credits and other incentives. **Solar energy** can be used to generate electricity in several ways. One way is by photovoltaic conversion (solar cells of the kind that power space satellites collect and convert light to electricity directly) (Figure 22.14). Another is by solar thermal conversion; for example, sunlight can be concentrated by means of mirrors and lenses on a tank of water to convert the water to steam, which drives an electric generator. These systems are being used for installations where costs are no impediment, such as demonstration projects, or in remote areas where alternatives are not available. But large-scale solar-generated electricity is not yet an alternative for conventional sources in general use. It must await further technological advances to become competitive. Commercial-scale solar electric power plants can present significant environmental problems. A plant with 100-megawatt electric capacity (about 10 percent of the capacity of a nuclear power plant) located in a desert region would require at least a square mile of land and might alter the local climate by significantly changing the balance of solar radiation in the area.

Solar energy is also stored in **biomass**—material of biological origin, such as wood, grain, sugar, and municipal wastes. Energy can be extracted by direct combustion or by processes that convert biomass to gaseous or liquid fuels, such as methane and alcohol. Converting the biomass in garbage also puts waste material to good use and eases the growing problem of finding sites for its disposal. The technology for some of these processes is well developed, but their price is not yet competitive without some form of

**FIGURE 22.14** Solar cells convert sunlight, a renewable resource, to electric energy at this utility. *Siemens Solar Industries.*

## BOX 22.2  RADIOACTIVE WASTE DISPOSAL

There are 415 nuclear power plants in the world generating about 17 percent of the total electricity used, with 113 of these in the United States. Each 1000-megawatt nuclear electric power plant produces about 30 metric tons of highly radioactive spent fuel each year. Regardless of one's views on the future of nuclear energy, the nation faces the unresolved problem of permanent and safe storage of wastes already generated. Power plant wastes are now stored temporarily on the surface at each reactor site, and most sites will reach maximum capacity for such storage in a few years. We face the additional problem of finding long-term storage for the highly radioactive wastes generated by the military nuclear reactors that made the fissionable materials used in bombs. These are stored temporarily at sites in South Carolina, Idaho, and Washington.

There is general agreement among geologists and other scientists that deep geological containment is the best long-term solution. Some thought is being given to storage on the bottom of the deep ocean (see Box 17.3), but this option is still speculative. Geological containment involves construction of a facility deep underground in stable geological formations (that is, ones with no tectonic or volcanic activity, no faults or fractures penetrating to the surface, and very well understood hydrological conditions that demonstrate negligible flow of groundwater through the repository). After treatment to reduce the volume of the radioactive materials and transform them to a nonsoluble, relative inert substance such as glass, they would be encapsulated and placed in engineered tunnels or caverns. Such a system of multiple barriers would decrease the chances of leakage into the environment. The geology and groundwater conditions would be selected to prevent materials from moving rapidly through rock formations, even if they escaped from their containers and came into contact with the rock.

Federal regulations limit the release of radioactive materials to the environment during the first 10,000 years after containment: fewer than 1000 deaths must result from inhalation or ingestion of any materials that reach the environment for each 100,000 tons of stored material. This is equivalent to an average of less than one death every 10 years.

Congress has designated Miocene volcanic ash-flow tuffs under Yucca Mountain in Nevada as the repository for civilian reactor wastes. Construction has not yet started, and the state of Nevada is fighting this decision. The Waste Isolation Pilot Plant (WIPP) in Carlsbad, New Mexico, has been constructed in Permian salt beds 655 m below the surface as a test repository for military wastes. Regulatory agencies have not granted approval to store wastes there. Facilities able to receive wastes from temporary storage sites will not be ready as planned.

Environmentalists and the state of Nevada support the federal standards. Others feel that they require scientists to predict the behavior of rock for thousands of years into the future—something that realistically cannot be done. They call for more flexible standards like those in place in other countries such as Switzerland and Canada. For example, without relaxing overall safety requirements, waste repositories could be constructed and used, with the expectation of learning from realistic conditions. If an unanticipated problem occurs, as almost always happens and cannot be guaranteed against, the wastes could be safely retrieved if necessary or the construction design modified.

There are risks associated with both points of view. On the one hand, compliance with the existing standards will cause long delays and may never be achieved, posing the risks of relying on temporary storage at many sites on the surface. On the other hand, approving a flexible process that requires trust may not satisfy those who live near a repository. The debate continues.

government subsidy. The growing of crops that can be converted to energy also presents ecological problems: the need for water, fertilizer, and pesticides. An ethical issue also arises if food crops are displaced by "energy crops" when so many people in the world are going hungry.

In a sense, **hydroelectric energy** is a form of solar energy because it depends on rainfall, and the energy that drives weather comes from the Sun. Hydroelectric energy is derived from water that falls by the force of gravity and is made to drive electric turbines. Waterfalls or artificial reservoirs behind dams provide the water. Hydroelectric energy is clean and relatively riskless and cheap. Significant expansion of the present capacity would be resisted in the United States, however, because it would involve the drowning of farmlands and wilderness areas under reservoirs behind dams. Hydroelectric energy delivers about 3 quads annually, or about 4 percent of the annual energy consumption in the United States. **Wind power,** or the use of a windmill to drive an electric generator, is also a form of solar energy (see the photograph at the beginning of this chapter); its use is slowly growing in some places as designs improve and costs are brought down. Solar energy (exclusive of hydroelectric power) supplies only a few tenths of 1 percent of consumption.

How much solar energy can we depend on in the years ahead? Solar energy enthusiasts believe that some 20 quads per year (exclusive of hydroelectric power) might be supplied in the United States by the year 2010. This is equivalent to about half of the oil we now use. Others think it more realistic to figure on less than 10 quads per year. All agree that under either scenario, important social benefits would be realized: conservation of other energy resources, diversification of energy supply so that we are not overly dependent on a single source, and reduction of fuel imports. With adequate research and development, solar energy can probably become economically competitive and a major source of energy in the next century.

## GEOTHERMAL ENERGY

In Chapter 19 we saw that Earth's internal heat, fueled by radioactivity, provides the energy for plate tectonics and continental drift, mountain building and earthquakes. The same internal heat can be harnessed to drive electric generators and heat homes. **Geothermal energy** is produced when underground heat is transferred by water that is heated as it passes through a subsurface region of hot rocks (a

**FIGURE 22.15** The Geysers, the world's largest supply of natural steam. The geothermal energy is converted into electricity for San Francisco, 120 km to the south. *Pacific Gas and Electric.*

**heat reservoir**) that may be hundreds or thousands of feet deep. The water is brought to the surface as hot water or steam through boreholes drilled for the purpose. The water is usually naturally occurring groundwater that seeps down along fractures or, less typically, water artificially introduced by being pumped down from the surface.

By far the most abundant form of geothermal energy occurs at the relatively low temperatures of 80 to 180°C. Water-circulating heat reservoirs at this temperature range are able to extract enough heat to warm residential, commercial, and industrial spaces. More than 20,000 apartments in France are now heated by warm underground water drawn from a heat reservoir in a geologic structure near Paris called the Paris Basin. Iceland sits on the Mid-Atlantic Ridge, a volcanic structure that was discussed in Chapters 5 and 20. Reykjavík, the capital of Iceland, is entirely heated by geothermal energy derived from volcanic heat.

Geothermal reservoirs with temperatures above 180°C are useful for generating electricity. They occur primarily in regions of recent volcanism as hot, dry rock, natural hot water, or natural steam (Figure 22.15). The latter two sources are limited to those few areas where surface water seeps down through underground faults and fractures to reach deep rocks heated by recent magmatic activity. Naturally occur-

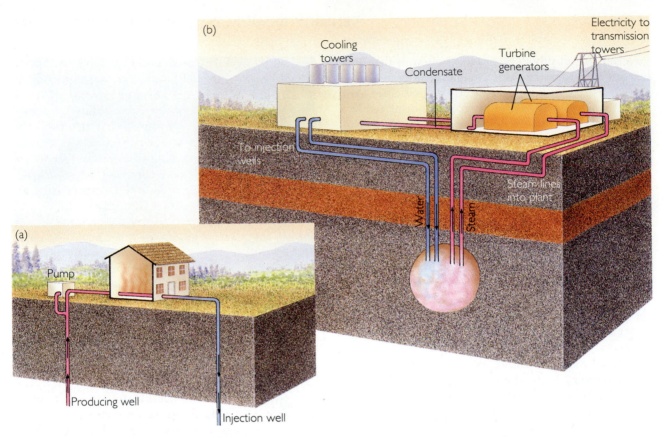

**FIGURE 22.16** Applications of geothermal power. (a) Low-temperature hot water is used to heat a building and is then reinjected into the reservoir. (b) High-temperature steam drives turbines to generate electricity. Water condensed from steam is reinjected into the reservoir.

ring water heated above the boiling point and naturally occurring steam are highly prized resources for which geologists are searching (Figure 22.16). The world's largest supply of natural steam occurs at The Geysers, 120 km north of San Francisco. Over 600 megawatts of electricity (about half the needs of San Francisco) is currently being generated there. This field is now in its third decade of production and is beginning to show signs of decline, perhaps because of overdevelopment. In 1904 Italian engineers first put the natural steam vents at Larderello, about 125 km northwest of Rome, to work producing electricity. These fields are still producing 190 megawatts of electric power annually.

Extracting heat from very hot, dry rocks presents a more difficult problem: the rocks must be fractured at depth to permit the circulation of water, and the water must be artificially provided. The rocks are fractured by water pumped down at very high pressures. Experiments are under way to develop technologies for exploiting this resource.

Like most of the other energy sources we have looked at, geothermal energy presents some environmental problems. Regional subsidence can occur if hot groundwater is withdrawn without being replaced. In addition, geothermally heated waters can contain salts and toxic materials dissolved from the hot rock. These waters present a disposal problem if they are not reinjected.

The contribution of geothermal energy to the world's energy future is difficult to estimate. Only $\frac{1}{40}$ of a quad per year of geothermal energy is currently produced in the United States, and perhaps twice that amount in the entire world. The resource is in a sense nonrenewable, because in most places the heat is being drawn out of a reservoir more rapidly than it can be replenished by slow geological processes, and because heat flows very slowly through solid rock. In many places, however—California, Hawaii, Japan, the rift valleys of Africa, Mexico—it is potentially so large a resource that its future will depend on the economics of production. At present we know how to use only naturally occurring hot water or steam deposits. Our guess is that in the near term geothermal energy can make important local contributions where the proximity of the resource to the user and the economics are favorable, as they are in California, New Zealand, and Iceland. Geothermal energy probably will not make large-scale contributions to the world energy budget until well into the next century, if ever.

# CONSERVATION

In a real sense, using energy more efficiently is like discovering a new source of fuel. It has been calculated that since the rise in oil prices in 1973 turned us all into conservationists, the world has saved more energy than it has gained from all new sources discovered in the same time. Some experts believe that conservation alone could cut the energy used by the industrialized nations in half. The savings in the United States could amount to some $200 billion a year, much of it on imported oil. Savings of this magnitude would reduce the cost of U.S. products and make them more competitive in the world market, reduce the dependence on imported oil, and lower the trade deficit significantly. The kinds of practices that could lead to these savings require the application of mostly familiar technologies: changes from incandescent to fluorescent lighting; better home insulation; more efficient refrigerators, air conditioners, furnaces, and other appliances; more efficient motors, pumps, and other industrial devices; and better-performing automobile engines. Political leadership and public education will be required to induce us to undertake these worthwhile changes today. In the years ahead dwindling reserves and higher energy prices will force us to do so.

# ENERGY POLICY

In view of the diversity and abundance of energy resources, you may well wonder why the world faces an energy crisis. The critical problem for the next few decades is fluid fuels, because world oil production will peak and begin to decline in this period even though world demand could increase, especially if conservation policies are not implemented. "Shortages"—the disruption of supply on political grounds by large oil-producing nations—and the resultant escalation of prices can shock the world economic and political order severely. Even in the absence of such manipulations, global climatic change may restrict our ability to use fossil fuels, precipitating an energy crisis of even greater severity.

All nations aspire to grow economically and improve the quality of their people's lives. The challenge is to provide the energy to fuel this growth in the face of the depletion of oil resources and the possible need to cap or reduce the emission of carbon dioxide by reducing the use of fossil fuels. A partial solution, and the cheapest one, is to reduce the waste of energy by using it more efficiently. In the next few decades, however, we may also have to shift to a different mix of energy supply—drawn from natural gas, synthetic fuels for transportation, nuclear energy, renewable resources such as solar and biomass energy, and a gradually decreasing dependence on oil and coal. It is to be hoped that nuclear technology will improve in safety and regain public confidence and that advances in the technology for renewable energy sources will lower their costs. Some experts believe that even today the United States can meet 30 percent of its energy demand with renewable resources at competitive prices.

In a sense we are in a race against time—to develop the several options of renewable or indefinitely sustainable energy resources before remaining oil resources are depleted or before we are forced to shift away from fossil fuels because of their harmful effects on climate. Whether the transition to an era of energy security is smooth or rough depends on our determination, technological skill, and ability to solve the attendant complex social and political problems. In our view, the technological advances required are achievable if we start to develop them now. Whether the sociopolitical problems will be resolved in time is less certain.

# SUMMARY

**What is the origin of oil and natural gas?** Oil and gas form from organic matter deposited in marine sediments, typically in the coastal waters of the sea. The organic materials are buried as the sedimentary layers grow in thickness. Under heat and high pressure, the organic matter is transformed into liquid and gaseous hydrocarbons, compounds of carbon and hydrogen. Oil and gas accumulate in geologic traps that confine the fluids within impermeable barriers.

**Are any environmental concerns connected with the production of oil?** Pollution in the production and transportation of oil is a major problem. Proponents of oil production believe that careful design of equipment and safety procedures can greatly reduce the chances of oil spills and satisfy the energy needs of our civilization. A significant body of public opinion is unconvinced, however, and exploration for new reserves is therefore restricted in many coastal regions and pristine wilderness areas, such as northern Alaska.

**Why is there concern about the world's oil supply?** Oil is a finite resource: it will be depleted faster than nature can replenish it. Therefore, as the supply is withdrawn from the oil reservoirs of the world, its availability diminishes, its price rises, and other types of energy sources will have to be found. At current rates of use, the remaining oil available for transportation, heating, and generation of electricity will be depleted in about 100 years.

**Why is natural gas a premium fuel?** Natural gas burns cleanly, releases less carbon dioxide per unit of energy than other fossil fuels, and is comparable in supply to oil resources.

**What is the origin of coal and how big a resource is it?** Coal is formed by the compaction and chemical alteration of wetland vegetation. It is present as huge resources in sedimentary rocks. We have used only about 2.5 percent of the world's minable coal.

**What is the trade-off between risk and benefit in the use of coal?** Coal mining and pollution caused by coal burning are risky to human life and to the environment. Coal combustion is a major source of carbon dioxide and the acid emissions that are the precursors to acid rain. Because of its abundance and low cost, however, the use of coal is likely to increase in the next decades for generation of electric power and for conversion to liquid and gaseous fuels.

**Are there any other fossil fuels that can be drawn upon as oil reserves disappear?** Reserves of oil shale and tar sands are great and may be extensively exploited in the next few decades as oil costs rise and technology improves. These fossil fuels will last well into the next century, when alternative energy sources should be available. Combustion of these fuels, however, like that of coal and oil, has environmental consequences of some concern.

**Is nuclear energy a solution to the world's energy problem?** Nuclear power from the fission of uranium 235 can be a major energy source but only if its costs do not keep escalating and the public can be assured of its safety. It has the advantage that it does not release carbon dioxide and the disadvantage that safe repositories must be found to store radioactive wastes for hundreds of thousands of years. Known high-grade reserves of uranium 235 ore can support the projected use of conventional nuclear power plants for a few decades, longer if advanced reactors are installed early in the next century. The use of nuclear energy could extend our resources of fossil fuels.

**What are the prospects for alternative energy sources?** Alternative energy sources include hydroelectric power, solar energy, biomass energy, and geothermal power, none of which has any immediate prospect of being an adequate response to world energy needs. With advances in technology and reduction in cost, however, solar and biomass energy could become major energy sources in the next century.

**What should be the goal of energy policy?** The goal of energy policy should be to guide the nations of the world through the transition from oil to the more plentiful fossil fuels and to nuclear energy if its safety can be improved, thus to buy time to develop nonpolluting and unlimited energy sources to replace fossil fuels in the next century. An important component of this policy must be the more efficient use of energy to conserve resources and reduce environmental damage. The resources and technology exist to achieve this goal peacefully and without economic dislocation if the sociopolitical problems can be worked out.

# KEY TERMS AND CONCEPTS

reserves (p. 502)
resources (p. 502)
fossil fuels (p. 504)
source bed (p. 505)
oil reservoir (p. 505)
oil trap (p. 505)

quad (p. 508)
syncrude (p. 510)
oil shale (p. 512)
tar sand (p. 512)
nuclear energy (p. 513)
solar energy (p. 515)

biomass (p. 515)
hydroelectric energy
    (p. 517)
wind power (p. 517)
geothermal energy (p. 517)
heat reservoir (p. 517)

# EXERCISES

1. What sedimentary environments favor the formation of sediments containing organic matter that might later be transformed into petroleum? Give some modern examples.

2. Do you think the continental slope and rise might be good places to drill for oil if we can invent the necessary technology? Why or why not?

3. Which of the following factors are most important in estimating the future supply of oil and gas: (a) rate of oil accumulation, (b) rate of natural seepage of oil, (c) rate of pumping of oil from known reserves, (d) rate of discovery of new reserves, (e) total amount of oil now present in the Earth?

4. Taking benefits and risks into consideration, rank according to relative importance all of the forms of fossil fuels and explain how their rankings might differ at the end of the next century.

5. Name four regions of the world that are major sources of oil.

6. Which three countries have the largest coal reserves?

7. Contrast the risks and benefits of nuclear fission and coal combustion as energy sources.

8. How would you use knowledge of the distribution of plate boundaries to make a map showing the areas of Earth most likely to be sources of geothermal power?

9. In what important respects do uranium reserves differ from fossil fuel reserves?

10. Give three examples of energy derived from biomass.

11. What do you think will be the major sources of energy in the year 2050? In the year 3000?

# THOUGHT QUESTIONS

1. A tax of $1 per gallon on gasoline would generate $100 billion of income for the federal government and reduce the budget deficit substantially. It would force drivers to switch to more fuel-efficient automobiles, thereby reducing imports of oil. Even with this increase, the price of gasoline would still be lower in the United States than in many European countries. Despite all these advantages, such a tax is unlikely to be enacted. What are the arguments against it?

2. Should we stockpile oil against future shortages caused by political disruptions? What are the costs and benefits of doing so?

3. Taking into account the costs of oil imports to the economy, the risks of relying on foreign exporters, and the environmental consequences, do you think we should remove restrictions on the production of oil from the continental shelves and wildlife preserves?

4. Explain the statement that increased conservation is the cheapest new energy source.

5. Would you rather live near a nuclear reactor or a plant that generates electricity by burning coal?

# SUGGESTED READINGS

Abelson, P. H. 1987. Energy futures. *American Scientist* 75:584–593.

Department of Energy. 1991. *National Energy Strategy*. Washington, D.C.: U.S. Government Printing Office.

Energy Information Administration. 1993. *Annual Energy Review*. Washington, D.C.: U.S. Department of Energy.

*Energy for Planet Earth*. 1990. Special issue of *Scientific American* (September).

Gibbons, J. H., and P. D. Blair. 1991. U.S. energy transition: On getting from here to there. *Physics Today* (July):21–30.

MacKenzie, J. J. 1989. *Breathing Easier: Taking Action on Climate Change, Air Pollution, and Energy Insecurity*. Washington, D.C.: World Resources Institute.

National Research Council. 1991. *Policy Implications of Greenhouse Warming*. Washington, D.C.: National Academy Press.

National Research Council. 1992. *Radioactive Waste Repository Licensing*. Washington, D.C.: National Academy Press.

National Research Council. 1993. *Solid-Earth Sciences and Society*. Washington, D.C.: National Academy Press.

Resource Reserve Definitions. 1980. Washington, D.C.: U.S. Geological Survey, Circular 831.

# MINERAL RESOURCES FROM THE EARTH

We humans have learned to use the materials of the Earth to vastly improve the quality of our lives. In fact, our evolving ability to use Earth's materials is a measure of the progress of our civilization, from the Stone Age to the Bronze Age to the Iron Age. Today we mine mineral deposits deep in the Earth. We strip away the surface deposits on land and evaporate seawater to obtain dissolved metals. As we explore the oceans, we are finding minerals deposited by hot springs on the seafloor and nodules that are rich in manganese and other metals. As with fuel materials, our growing understanding of the finiteness of resources and the need to protect the environment leads us to think of conservation, recycling, and substitution as we proceed with geological exploration for new mineral resources.

## 23

Limestone blocks cut from limestone formation in an Indiana quarry. Limestone is an important source of building stone. Code numbers are used to match colors and textures. *Jeffrey A. Wolin.*

Like the biologically derived fossil fuel materials we discussed in Chapter 22, minerals are vital to the functioning of a modern nation. Just about everything we use comes from the ground, including all of our metals and the thousands of products made from the materials and chemicals refined from natural deposits. From time to time we have mentioned the practical uses of various minerals and rocks. Metallic mineral deposits are the source of the metals we use. Nonmetallic mineral deposits provide salt, clay, gravel, building stone, the limestone from which cement is made, the sand from which glass and transistors are made, and many other important products. In this chapter we will survey a broad range of useful Earth materials. We will discuss them in an economic context, because the moment commercial utility enters the picture, matters of availability, estimates of reserves and resources, and price become important.

# MINERALS AS ECONOMIC RESOURCES

Although mining contributes only a small part of the gross domestic product (GDP) of the United States, in a sense most of that GDP depends on the materials extracted from the Earth because modern factories and farms could not function without them. If we consider the uses of stone for buildings, phosphates for fertilizers, cement for construction, clays for ceramics, sand for silicon transistors and fiber-optic cables, and metals for just about everything, the large annual U.S. per capita consumption of minerals becomes understandable (Figure 23.1).

The chemical elements of the crust—the portion of the Earth that is readily accessible to us for mining and drilling—are widely distributed in many kinds of minerals, and those minerals are found in a great variety of rocks. Throughout this book, in discussing

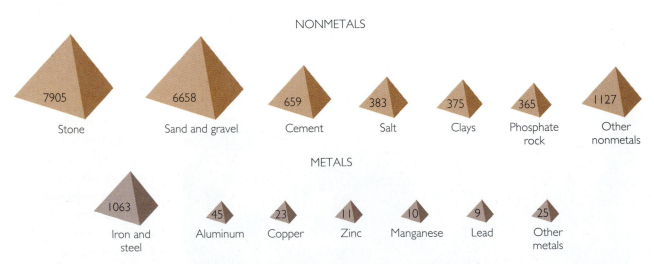

**FIGURE 23.1** Average annual per capita consumption of nonfuel minerals in the United States (in pounds). During the 1980s, each person in the United States used more than 18,000 pounds of minerals each year. (Data from U.S. Bureau of Mines, 1992.)

**FIGURE 23.2** Iron ores (clockwise from left): magnetite, siderite, iron pyrite, hematite. *Chip Clark.*

erosion, dissolution, transportation by water, and the melting and crystallization of igneous rocks, we have seen how nature sometimes homogenizes materials and sometimes segregates them. In most places, a particular element will be found in amounts close to its average concentration in the crust—that is, it is homogenized with the other elements. An ordinary granitic rock, for example, may contain a few percentage points of iron, close to the average concentration of iron in the Earth's crust. Elements that occur in higher concentrations—those that have undergone some geologic process that has segregated much larger quantities of the element than normal—are found in a smaller number of geological settings. These are the settings that interest us, because the higher the concentration of a resource in a given deposit, the cheaper it is to recover it.

Rich deposits of minerals from which valuable metals may be recovered profitably are called **ores;** the minerals containing these metals are **ore minerals.** Ore minerals include sulfides (the main group), oxides, and silicates. Ore minerals in each of these groups are compounds of metallic elements together with sulfur, oxygen, and silicon oxide, respectively. The copper ore covelite, for example, is a copper sulfide ($CuS$). The iron ore hematite is an iron oxide ($Fe_2O_3$) (Figure 23.2). The nickel ore garnierite is a nickel silicate, $Ni_3Si_2O_5(OH)_4$. In addition, some metals, such as gold, are found in their native state—that is, uncombined with other elements (Figure 23.3).

The **concentration factor**—that is, the ratio of an element's abundance in a mineral deposit to its average abundance in the crust—varies from element to element (Table 23.1). For example, iron, one of the common elements of the crust, has an average

**FIGURE 23.3** Gold occurring in the free state (native gold) on a quartz crystal. *Chip Clark.*

**TABLE 23.1**

## Economical Concentration Factors of Some Commercially Important Elements

| ELEMENT | CRUSTAL ABUNDANCE (PERCENT BY WEIGHT) | CONCENTRATION FACTOR[1] |
|---------|----------------------------------------|-------------------------|
| Aluminum | 8.00 | 3–4 |
| Iron | 5.8 | 5–10 |
| Copper | 0.0058 | 80–100 |
| Nickel | 0.0072 | 150 |
| Zinc | 0.0082 | 300 |
| Uranium | 0.00016 | 1,200 |
| Lead | 0.00010 | 2,000 |
| Gold | 0.0000002 | 4,000 |
| Mercury | 0.000002 | 100,000 |

[1] Concentration factor = abundance in deposit divided by crustal abundance.
SOURCES: Data from B. J. Skinner, *Earth Resources,* Prentice-Hall, 1969; D. A. Brobst and W. P. Pratt, *Mineral Resources of the U.S.,* USGS Prof. Paper 820, 1973.

abundance in crustal rocks of 5.8 percent. An *economical* iron ore—one that is profitable to mine under current costs of extraction, costs of transportation, and selling prices—would contain at least 50 percent iron, about 10 times the average crustal abundance. In other words, an iron ore becomes economical when its concentration factor is about 10. A less abundant metal, such as copper, which has a crustal abundance of 0.0058 percent, is concentrated by factors of at least 80 to 100 in its economical ores. Ore deposits of the rarer elements, such as mercury and gold, require concentration factors in the thousands to hundreds of thousands to be economical.

Because the elements are so widely distributed in many common rocks, whether a particular deposit should be considered a resource or a reserve (see Chapter 22) depends on the costs of recovery and the selling price. Theoretically, given enough money and energy, we could take almost any rock and extract both abundant and rare elements from it. So from this point of view, it is clear that we will never run out of any vital minerals. Our practical concern is the exhaustion of reserves, the identified mineral deposits that are profitable to mine and purify. We can reasonably expect to add to current reserves by new dis-

coveries, but at an uncertain rate. Once the highest grade deposits are mined out, we will be forced to rely on deposits of lower grades, which will be more expensive to recover.

It is not too soon to start planning for the future. In the United States, for example, only a few minerals are available in quantities adequate to last for more than a few hundred years. The U.S. dependence on mineral imports is already great and is growing, sometimes because imports are cheaper, at other times because the country lacks reserves of the mineral in question. Some people think that the United States should increase access to its federal lands for exploitation of mineral resources, but others think that such lands should be preserved for public recreational use (see Box 23.1). Figure 23.4 shows that the United States satisfies more than 60 percent of its needs for 20 of the 42 listed minerals through imports. There is nothing intrinsically bad about importing materials as long as the supply is secure, the price is reasonable, and overall exports and imports are in balance. However, some people who are concerned with U.S. economic and defense security, and mining industry representatives worried about their sales, are troubled by the heavy dependence on im-

Major sources (1988–1991)

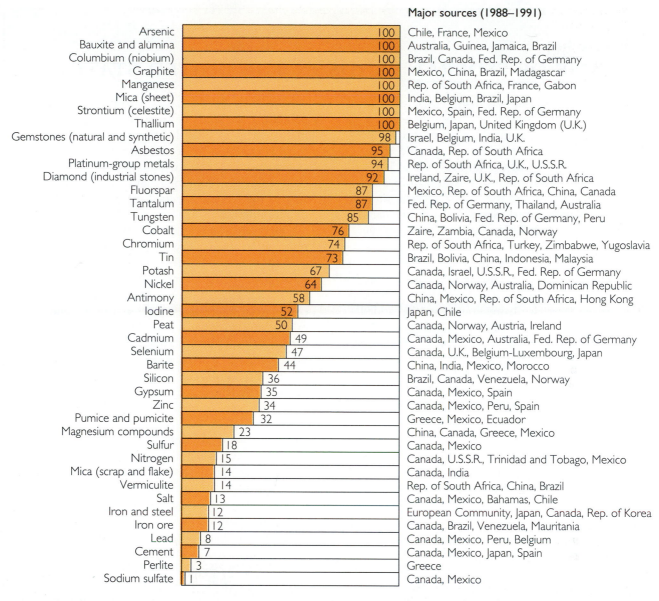

| Mineral | Percentage | Major sources (1988–1991) |
|---|---|---|
| Arsenic | 100 | Chile, France, Mexico |
| Bauxite and alumina | 100 | Australia, Guinea, Jamaica, Brazil |
| Columbium (niobium) | 100 | Brazil, Canada, Fed. Rep. of Germany |
| Graphite | 100 | Mexico, China, Brazil, Madagascar |
| Manganese | 100 | Rep. of South Africa, France, Gabon |
| Mica (sheet) | 100 | India, Belgium, Brazil, Japan |
| Strontium (celestite) | 100 | Mexico, Spain, Fed. Rep. of Germany |
| Thallium | 100 | Belgium, Japan, United Kingdom (U.K.) |
| Gemstones (natural and synthetic) | 98 | Israel, Belgium, India, U.K. |
| Asbestos | 95 | Canada, Rep. of South Africa |
| Platinum-group metals | 94 | Rep. of South Africa, U.K., U.S.S.R. |
| Diamond (industrial stones) | 92 | Ireland, Zaire, U.K., Rep. of South Africa |
| Fluorspar | 87 | Mexico, Rep. of South Africa, China, Canada |
| Tantalum | 87 | Fed. Rep. of Germany, Thailand, Australia |
| Tungsten | 85 | China, Bolivia, Fed. Rep. of Germany, Peru |
| Cobalt | 76 | Zaire, Zambia, Canada, Norway |
| Chromium | 74 | Rep. of South Africa, Turkey, Zimbabwe, Yugoslavia |
| Tin | 73 | Brazil, Bolivia, China, Indonesia, Malaysia |
| Potash | 67 | Canada, Israel, U.S.S.R., Fed. Rep. of Germany |
| Nickel | 64 | Canada, Norway, Australia, Dominican Republic |
| Antimony | 58 | China, Mexico, Rep. of South Africa, Hong Kong |
| Iodine | 52 | Japan, Chile |
| Peat | 50 | Canada, Norway, Austria, Ireland |
| Cadmium | 49 | Canada, Mexico, Australia, Fed. Rep. of Germany |
| Selenium | 47 | Canada, U.K., Belgium-Luxembourg, Japan |
| Barite | 44 | China, India, Mexico, Morocco |
| Silicon | 36 | Brazil, Canada, Venezuela, Norway |
| Gypsum | 35 | Canada, Mexico, Spain |
| Zinc | 34 | Canada, Mexico, Peru, Spain |
| Pumice and pumicite | 32 | Greece, Mexico, Ecuador |
| Magnesium compounds | 23 | China, Canada, Greece, Mexico |
| Sulfur | 18 | Canada, Mexico |
| Nitrogen | 15 | Canada, U.S.S.R., Trinidad and Tobago, Mexico |
| Mica (scrap and flake) | 14 | Canada, India |
| Vermiculite | 14 | Rep. of South Africa, China, Brazil |
| Salt | 13 | Canada, Mexico, Bahamas, Chile |
| Iron and steel | 12 | European Community, Japan, Canada, Rep. of Korea |
| Iron ore | 12 | Canada, Brazil, Venezuela, Mauritania |
| Lead | 8 | Canada, Mexico, Peru, Belgium |
| Cement | 7 | Canada, Mexico, Japan, Spain |
| Perlite | 3 | Greece |
| Sodium sulfate | 1 | Canada, Mexico |

**FIGURE 23.4** Net U.S. reliance on imports of selected nonfuel minerals as a percentage of consumption in 1992. (U.S. Bureau of Mines, 1993.)

ports of several strategically critical metals. These are metals—such as cobalt, manganese, chromium, titanium, and the platinum group—without which an entire industry, such as the aircraft or chemicals industry, could collapse if they suddenly became unavailable. For this reason, since 1939 the U.S. government has established stockpiles of minerals for use in an emergency, such as an economic boycott or a wartime cutoff of supply. Many economists point out that the United States has always been dependent on imports of many minerals, that alternative mar-

kets are always available, and that interruptions of supply have been rare.

At some point it may become less expensive to extract a certain element from manufactured goods that have been discarded than from deposits of mineral ores. This is the practice popularly known as **recycling.** Gold and platinum, among other valuable metals, have been recycled for years. Growing fractions of iron, lead, copper, and aluminum are being recycled. Deciding whether or not to recycle can be a complex process. Sometimes the only reasonable

## BOX 23.1
# USE OF FEDERAL LANDS IN THE UNITED STATES

The land owned and managed by the federal government (today amounting to about a quarter of the country) represents an important part of the political, social, and economic history of the United States. The stewards are primarily the Bureau of Land Management and the Forest Service, though the National Park Service and the Department of Defense also manage some of the land.

Over the years, these lands were bought at bargain prices, won in wars, or acquired by negotiations with other nations. The Louisiana Purchase from France in 1903 added huge land areas from the mouth of the Mississippi to what is now the state of Montana at a cost of $15 million. Florida was bought from Spain in 1819 for $7 million. The treaty of 1848 following the war with Mexico added most of the Southwest to the public domain, including parts of Arizona and New Mexico and all of California, Nevada, and Utah. Great Britain gave us the lands that are now the states of Washington, Oregon, and Idaho. The last major acquisition was the purchase of Alaska from Russia for $7 million in 1867.

In addition to these lands, the continental shelf beyond a few miles offshore is owned and managed by the federal government. It is valuable for its oil and gas resources, fisheries, and mineral resources. Over our history some 60 percent of lands in the public domain were transferred to states, to homesteaders, and to other organizations to aid education (the land grant colleges), build the privately owned railroads, reward veterans, and for other purposes.

The federal lands have many uses, some in conflict with others. The commercial value is very large: rents and royalties from private users amounted to $6.2 billion in 1992. About 80 percent of this is from oil and gas producers (mostly offshore), followed by timber (less than 20 percent) and mineral (about 1 percent) interests. Cattle ranchers (grazing) and recreational users contribute less than 1 percent each. Recreational users constitute by far the largest group of individuals using the land. Federal ownership and management of these holdings is one of the most contentious political issues of the day, as evidenced by congressional debates, election campaigns, lobbying, and suits brought against the federal government by states and private interests.

Here are some of the policy issues that must be resolved. How much, if any, federal land should be sold off, to whom, and at what price? If the government retains ownership, what kind of access should be given commercial interests, recreational users, and others, and on what terms? How much of the land should be set aside for wilderness preservation? To what degree should state and local governments participate in management of the lands?

A case can be made for federal retention of the lands: The federal government can look after national interests better than state and local governments. Public agencies can take a longer term view of conservation and the needs of future generations than private interests driven by short-term profits. The federal government is best suited to organize and monitor the multiple diverse uses of the land.

There are also arguments for selling off much of the land: Private owners would manage the lands more efficiently than federal bureaucrats who lack management experience and are subject to political pressure. The greatest benefits to society come from the accumulated effect of individual entrepreneurs who work to achieve the greatest personal gain. There is not much evidence that government administration of the land for any purpose is more exemplary than that of the private sector. Jobs are more important than some unimportant subspecies.

The time has come to try to achieve a national consensus on the uses of federal lands.

SOURCE: Much of this discussion was drawn from Marion Clawson, *The Federal Lands Revisited*, Washington, D.C., Resources for the Future, 1983.

way to decide is to calculate costs and compare the costs of mining, smelting, and transporting natural ores, disposing of wastes, and controlling pollution with the costs of processing recycled metals.

# THE GEOLOGY OF MINERAL DEPOSITS

Mineral deposits are created by various kinds of geologic processes, most of which have already been discussed in earlier chapters. In general, a mineral deposit forms when three conditions are satisfied: (1) a source of the minerals exists in a place where it is accessible to a natural transport mechanism; (2) a natural transport mechanism is available to move the minerals away from the source; and (3) a site exists where the transport mechanism can deposit the minerals. It is not luck that places ore deposits close to Earth's surface, where humans can reach them. Rocks near the surface contain cracks and open fractures (pressure closes them at greater depths), allowing easier transportation of ore-bearing fluids. Also, rocks near the surface are cooler, so that ore minerals precipitate from the hot fluids that carry them.

## Hydrothermal Deposits

Many of the richest ore deposits known were deposited from hot, aqueous solutions called *hydrothermal solutions*. These hot waters are the transport mechanism. They can emanate directly from the magma of an igneous intrusion (the source), carrying away in solution ore constituents derived from the soluble components of the magma. Hydrothermal solutions can also form when circulating groundwater comes into contact with a hot intrusion, reacts with it, and carries off ore constituents released by the reaction. Fractured rock is a good site for the deposition of ore constituents. The hot fluids can flow easily through the fractures and joints, cooling rapidly in the process. Quick cooling causes fast precipitation of the ore constituents. When the fractures and joints are filled with precipitated minerals, they are called **hydrothermal vein deposits,** or simply **veins.** Some ores are found in the veins themselves; others are found in the rock adjacent to the veins (country rock) that has been altered by heating and infiltration by the vein-forming solutions. As the solutions react with surrounding rocks, they may precipitate ore minerals together with quartz, calcite, or other common vein-filling minerals (Figures 23.5 and 23.6). Hydrothermal vein deposits are among the most important sources of metal ores. Typically metals occur as sulfides, such as iron sulfide (pyrite), lead sulfide (galena), zinc sulfide (sphalerite), mercury sulfide (cinnabar), and copper sulfide (covelite and chalcocite) (Figures 23.7 and 23.8). When hydrothermal solutions reach the surface, they become hot springs and geysers, many of which precipitate mineral ores—including lead, zinc, and mercury—as they cool.

Mineral deposits that are scattered through volumes of rock much larger than veins are called **disseminated deposits.** In igneous and sedimentary

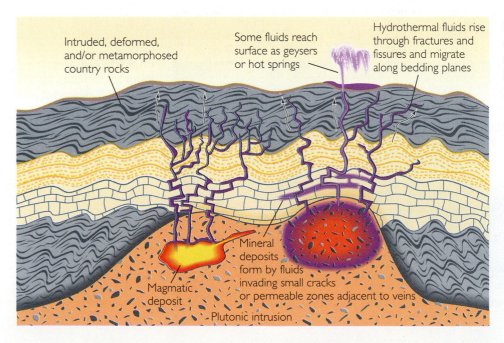

Intruded, deformed, and/or metamorphosed country rocks

Some fluids reach surface as geysers or hot springs

Hydrothermal fluids rise through fractures and fissures and migrate along bedding planes

Mineral deposits form by fluids invading small cracks or permeable zones adjacent to veins

Magmatic deposit

Plutonic intrusion

**FIGURE 23.5** Many ore deposits are found in hydrothermal veins formed by hot solutions rising from magmatic intrusions.

**FIGURE 23.6** Quartz vein deposit (about 1 cm thick) containing gold and silver ores; Oatman, Arizona. *Peter Kresan.*

**FIGURE 23.7** Metal sulfide ores (from left to right): galena (lead sulfide), cinnabar (mercury sulfide), iron pyrite, sphalerite (zinc sulfide). *Chip Clark.*

**FIGURE 23.8** Copper ores (from left to right): chalcopyrite, malachite, chalcocite. *Chip Clark.*

rocks alike, minerals are disseminated along abundant cracks and fractures. Among the economically important disseminated deposits are the porphyry-copper deposits of the southwestern United States and Chile. These deposits became mineralized when ore-forming minerals were introduced into a great number of tiny fractures in porphyritic felsic intrusives (granitic rocks containing large feldspar or quartz crystals in a finer-grained matrix) and country rocks surrounding the higher parts of the plutons. Some unknown process associated with the intrusion or its aftermath broke the rocks into millions of pieces. Hydrothermal solutions penetrated and recemented the rocks by precipitating ore minerals throughout the extensive network of tiny fractures. This widespread dispersal produced a low-grade but very large resource of many millions of tons of ore, which can be mined economically by large-scale methods (Figure 23.9). The most common copper mineral in porphyry is chalcopyrite, copper sulfide.

Extensive disseminated hydrothermal deposits may also occur in sedimentary rocks, such as in the lead-zinc province of the upper Mississippi River valley, which extends from southwestern Wisconsin to Kansas and Oklahoma. Because the ores in this province are not associated with a known magmatic intrusion that could have been a source of hydrothermal fluids, the origin of these ores is unknown. Some geologists speculate that the ores were deposited by groundwater. Groundwater may have penetrated hot crustal rocks at great depths and extracted soluble ore constituents, then moved upward into the overlying sedimentary rocks, where it precipitated its mineral load as fillings in cavities. In some cases ore fluids appear to have dissolved some carbonates when they infiltrated limestone formations and then replaced them with equal volumes of new crystals of sulfide. The major minerals of the hydrothermal deposits in this province are lead sulfide (galena) and zinc sulfide (sphalerite).

**FIGURE 23.9** Open-pit copper mine south of Tucson, Arizona. *Bob Lynn, Cyprus Minerals.*

## Igneous Ore Deposits

The most important **igneous ore deposits**—deposits of ores in igneous rocks—are found as segregations of ore minerals near the bottom of intrusions. The deposits are formed when minerals crystallize from molten magma, settle, and accumulate on the floor of a magma chamber (see Chapter 4). Most of the chromium and platinum ores in the world, such as the deposits in South Africa and Montana, are found as layered accumulations of minerals that formed this way (Figure 23.10). One of the richest ore bodies ever found is at Sudbury, Ontario. This large mafic intrusive formation contains great quantities of layered nickel, copper, and iron sulfides near its base. These sulfide deposits are believed to have formed from crystallization of a dense, sulfide-rich liquid that separated from the rest of the cooling magma and sank to the bottom of the chamber before it congealed.

One of the most valuable minerals, diamond, occurs chiefly in ultramafic igneous rocks called kimberlites. These rocks were forcefully intruded to the surface from deep in the crust and upper mantle in the form of long, narrow pipes. We know these diamond-bearing kimberlites originate at great depths because diamonds and other minerals found in them can be formed only under the conditions of extremely high pressure that exist in the upper mantle. How kimberlites erupt to the surface is uncertain.

## Sedimentary Ore Deposits

**Sedimentary ore deposits** include some of the world's most valuable sources of materials. Many economically important minerals segregate by chemical and physical means as an ordinary result of sedimentary processes (see Chapter 7). Limestones, for instance, chemically precipitated mainly by marine organisms, are used for cement, agricultural lime, and building stone. Pure quartz sands, left behind when mixed-mineral sands are abraded and winnowed by waves and currents so that all materials other than quartz are removed, are the raw materials for glassmaking and for the fiber-optic cables that are replacing copper wires in communication lines. Coarse sand and gravel, suitable for construction purposes, have been abundantly distributed in many

**FIGURE 23.10** Chromite (chrome ore, dark layer) occurring in a layered igneous intrusive; Bushveld Complex, South Africa. *Spence Titley.*

**FIGURE 23.11** Precambrian banded iron beds. Rust-colored layers are limonite, interbedded with hematite and chert. Hammerslee, Australia. *Spence Titley.*

areas of the northern United States and southern Canada by the Pleistocene glaciers; these materials are also widely distributed in channels and former channels of many rivers. Clays of high purity produced by prolonged weathering are used for ceramics in both home and industry. Evaporite deposits of gypsum, separated from seawater by precipitation, are used for plaster, and sodium and potassium salts from evaporites have varied uses, from table salt to fertilizer. Phosphate rocks—marine shales and limestones enriched in phosphate by chemical reaction with deep seawaters—are raw materials for the world's fertilizer industry.

Sedimentary ore deposits are also important sources of copper, iron, and other metals. These deposits are chemical precipitates formed in sedimentary environments to which large quantities of metals were transported in solution. Some of the important sedimentary copper ores, such as those of the Permian Kupferschiefer (German for "copper slate") beds of Germany, may have precipitated from hot brines of hydrothermal origin, rich in metal sulfides, that interacted with sediments on the ocean bottom.

The major iron ores have been found in Precambrian sedimentary rocks. Earth's atmosphere was poor in oxygen early in its history (see Chapter 1), and it is now thought that the low availability of oxygen allowed an abundance of iron in its soluble, lower oxidation or ferrous form ($Fe^{2+}$) to be leached

in great quantities from the land surface. The ferrous iron was transported in solution by groundwater to broad, shallow marine environments where it could be oxidized to its insoluble ferric form ($Fe^{3+}$) and precipitated. (Chapter 6 discusses oxidation states of iron.) In many of these basins the iron was deposited in thin layers alternating with layers of siliceous sediments (cherts). Such iron ores are called *banded iron formations* (Figure 23.11). The Lake Superior iron deposits, for many years the source of iron for the U.S. steel industry, are of this type.

Possibly the most publicized (and romanticized) type of mineral prospecting is panning for gold: the gold seeker dredges and sifts a flat pan of river sediment in the hope of turning up the glint of a nugget. Many rich deposits of gold, diamonds, and other heavy minerals such as magnetite and chromite are found in **placers,** ore deposits that have been concentrated by the mechanical sorting action of river currents (Figure 23.12). Because heavy minerals settle out of a current more quickly than lighter minerals such as quartz and feldspar, the heavy minerals tend to accumulate on river bottoms and sandbars, where the current is strong enough to keep the lighter minerals suspended and in transport but too weak to move the heavier material. In a similar manner ocean waves preferentially deposit heavy minerals on the beach or on shallow offshore bars. The gold panner accomplishes the same thing: the shaking of a water-

**FIGURE 23.12** Gold mining in the Sierra Nevada, California, 1852. *California State Archives.*

filled pan allows the lighter minerals to be washed away, leaving the heavier gold in the bottom of the pan. Because of the recent high prices of gold, old-fashioned gold panning is undergoing a revival.

Some placers can be traced upstream to the location of the original mineral deposit, usually of igneous origin, from which the minerals were eroded. Erosion of the Mother Lode, an extensive gold-bearing vein system lying along the western flanks of the Sierra Nevada batholith, produced the placers that were discovered in 1848 and led to the California gold rush. The placers were discovered first, then their source. This was also the sequence of events that led to the discovery of the Kimberley diamond mines of South Africa two decades later.

Although there is probably an abundance of ore bodies on the deep seafloor, most known ore bodies are found on the continental crust. They either originated on the continent or occur as remnants of mineralized pieces of ocean crust thrust onto the continent in plate collisions. Figure 23.13 shows the locations of some of the major metallic ore deposits on the same map of continental deformation over geologic time that appears in Figure 21.1. Note that iron ores tend to be found in older parts of the crust and that ore deposits tend to be associated with orogenic belts.

# ORE DEPOSITS AND PLATE TECTONICS

With the advent of plate-tectonics theory, the various types of igneous activity could be explained in terms of the interactions of plates at boundaries where they separate or collide. Since igneous processes bring chemical elements and their mineral compounds from the interior to the surface, the theory of plate tectonics provides a foundation for understanding the origin of ore deposits. Such an understanding not only helps to explain existing ore deposits but also can lead to the discovery of new ones.

In 1979 geologists exploring the seafloor at a plate-separation center (the East Pacific Rise) made one of the most important geological discoveries in many decades. They found hot springs, laden with dissolved minerals, venting on the seafloor (see Box 17.2). These hot springs have their origin in seawater that circulates through fractures near the rift where the plates separate. The seawater is heated to temperatures of several hundred degrees Celsius when it comes in contact with magma or hot rocks deep in the crust. The heated seawater leaches minerals from the hot rocks and rises to the seafloor. When the hot waters, now loaded with dissolved minerals, reach

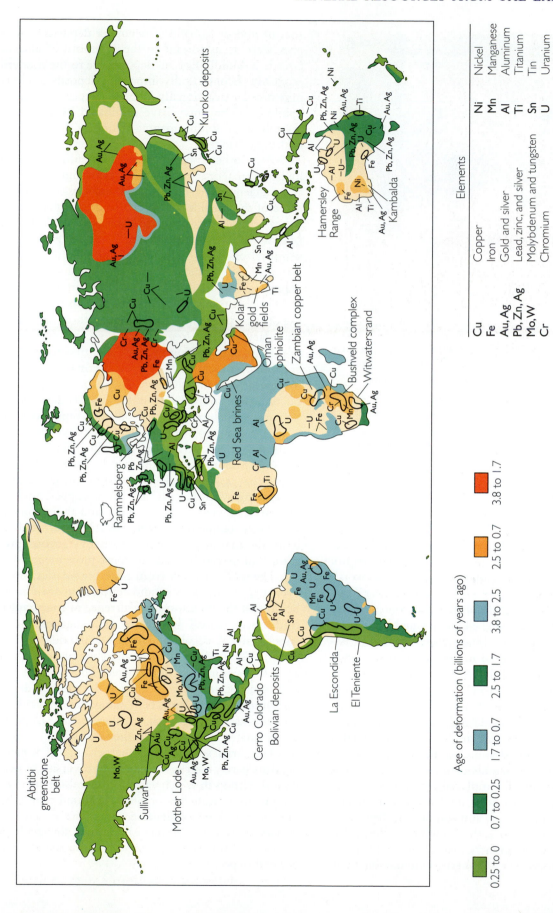

**FIGURE 23.13** Locations of major metallic ore bodies on continents. Iron is concentrated primarily in Precambrian time. Note also the association of ore deposits with orogenic belts. Some famous ore bodies are identified by name. (After G. Brimhall, "The Genesis of Ores," *Scientific American*, May 1991, p. 84; based on a map by B. Clark Burchfiel.)

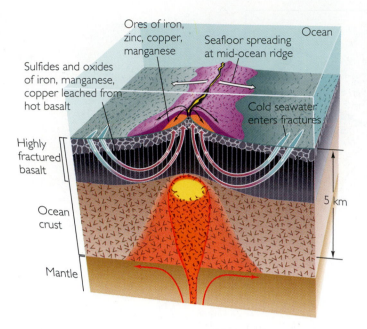

Ores of iron, zinc, copper, manganese

Seafloor spreading at mid-ocean ridge

Ocean

Sulfides and oxides of iron, manganese, copper leached from hot basalt

Cold seawater enters fractures

Highly fractured basalt

Ocean crust

5 km

Mantle

**FIGURE 23.14** Cold seawater percolates through fractured volcanic rocks at mid-ocean ridges and is heated when it approaches the magma chamber below. The hot fluid leaches metals from the basaltic rock and rises to the sea-floor. When the hot fluid with its dissolved metals vents into the cold ocean-bottom waters, the metals it is carrying in solution precipitate as rich sulfides of iron, zinc, copper, and other ores.

the cooler upper crust and near-freezing ocean-bottom waters, the minerals precipitate (Figure 23.14). This is the origin of the "black smokers" shown in the photo in Box 17.2. In this manner enormous quantities of sulfide ores rich in zinc, copper, iron, and other metals are being deposited along mid-ocean spreading centers. High-grade deposits of native gold have been found in one submarine deposit on the Mid-Atlantic Ridge.

Once it was understood that current spreading centers are rich sources of mineral deposits, geologists began to look on land for the geological remains of ancient seafloor, which might also hold valuable resources. They found some of them in plate-collision zones where fragments of ancient oceanic lithosphere are occasionally emplaced on land. These deposits, known as ophiolites, are discussed in Chapter 20. The rich copper, lead, and zinc sulfide deposits in the ophiolites of Cyprus, the Philippines, the Apennines in Italy, and elsewhere probably owe their origin to the process of hydrothermal circulation along

ancient mid-ocean rifts. The copper deposits of Cyprus were as important to the economy of ancient Greece as Middle East oil deposits are to the modern economy. Economically important deposits of chromite ores are occasionally found in deeper portions of ophiolites. They may have originated by fractional crystallization within magma chambers that underlie mid-ocean ridges.

Many other types of deposits of sulfide ores, of hydrothermal or igneous origin, are found at modern and ancient plate-collision boundaries, including those of the cordillera of North and South America, the eastern Mediterranean to Pakistan, the Philippine Islands, and Japan (see Figure 23.13). Figure 23.15 summarizes some of the associations between plate tectonics and mineral deposits. Deposits found in magmatic arcs are thought to result from the igneous activity that typically occurs in collision zones. One hypothesis proposes that some of these collision-boundary deposits represent the second stage in a two-stage ore-forming process. The first stage is the creation of mineral ores by hydrothermal activity at a mid-ocean spreading center. The second stage, separated in time and space from the first, is the subduction and partial melting at a collision zone of oceanic sediments and crust containing these previously concentrated minerals. As the plate descends into increasingly hot regions of the mantle, the metals "boil off"—that is, they melt and rise into the overriding plate along with magma. The iron, copper, molybdenum, lead, zinc, tin, and gold found along convergent plate boundaries could have been created by hydrothermal activity and reborn by igneous processes, all driven by plate-tectonic movements.

The seafloor away from plate boundaries, however, may be the first candidate for deep-sea mining because of the widespread occurrence of **manganese nodules,** spherical aggregates of manganese, iron, copper, nickel, cobalt, and other metal oxides (Figure 23.16). The nodules vary in size, but most of them are a few centimeters in diameter. They are potentially valuable not only because of the gradual depletion of high-grade deposits of manganese on land but also because they are rich in other metals. Deposits are estimated to be in the trillions of tons.

This brief summary of the geology of mineral deposits barely touches on the great diversity of geological settings in which various minerals of value are found. Some minerals or ores are found mainly or only in one kind of deposit; others are found in a variety of settings. Table 23.2 shows the geologic occurrence and uses of some of the principal kinds of mineral deposits.

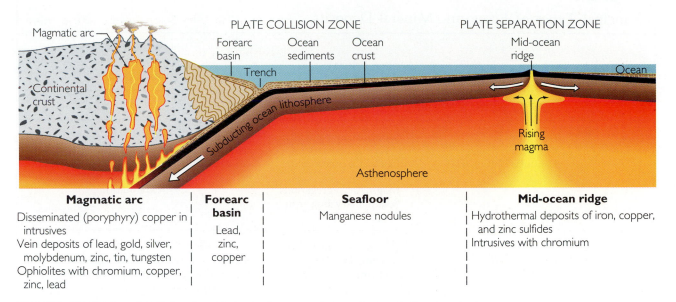

| **Magmatic arc** | **Forearc basin** | **Seafloor** | **Mid-ocean ridge** |
|---|---|---|---|
| Disseminated (poryphyry) copper in intrusives | Lead, zinc, copper | Manganese nodules | Hydrothermal deposits of iron, copper, and zinc sulfides |
| Vein deposits of lead, gold, silver, molybdenum, zinc, tin, tungsten | | | Intrusives with chromium |
| Ophiolites with chromium, copper, zinc, lead | | | |

**FIGURE 23.15** The role of plate boundaries in the accumulation of mineral deposits. Ocean sediment and crust are enriched in metals by hydrothermal ore deposition along a mid-ocean ridge. Rising magma in the subduction zone is the source of ores that form the metal-bearing provinces of a magmatic belt such as the cordillera of North and South America. The melting of subducted sediment and crust may contribute ore constituents. Mineral-bearing oceanic fragments (ophiolites) accrete to the continent in the collision zone.

**FIGURE 23.16** Manganese nodules are small concretions found on the deep seafloor that contain as much as 20 percent manganese and smaller amounts of iron, copper, and nickel. (This nodule is about 7.5 cm in diameter.) They have been dredged from the ocean floor to demonstrate the feasibility of mining the seabed. *Chip Clark.*

**TABLE 23.2**

# Principal Types of Economic Mineral Deposits

| MINERAL DEPOSIT | TYPICAL MINERALS | GEOLOGICAL OCCURRENCE | USES | MAJOR DEPOSITS / REMARKS |
|---|---|---|---|---|
| **METALS PRESENT IN MAJOR AMOUNTS IN EARTH'S CRUST** | | | | |
| Iron | Hematite, $Fe_2O_3$<br>Magnetite, $Fe_3O_4$<br>Limonite, $FeO(OH)$ | Sedimentary banded iron formation<br>Contact metamorphic rocks<br>Magmatic segregations<br>Sedimentary bog iron ore | Manufactured materials, construction, etc. | *Mesabi, Minn.; Cornwall, Pa.; Kiruna, Sweden*<br>Resources immense; economics determines exploitation |
| Aluminum | Gibbsite, $Al(OH_3)$<br>Diaspore, $AlO(OH)$ | Bauxite: residual soils formed by deep chemical weathering | Lightweight manufactured materials | *Jamaica*<br>Resources great, but expensive to smelt |
| Magnesium | Dolomite, $CaMg(CO_3)_2$<br>Magnesite, $MgCO_3$ | Dissolved in seawater<br>Hydrothermal veins, limestones | Lightweight alloy metal, insulators, chemical raw material | Most extracted from seawater; unlimited supply |
| Titanium | Ilmenite, $FeTiO_3$<br>Rutile, $TiO_2$ | Magmatic segregations<br>Placers | High-temperature alloys; paint pigment | *Allard Lake, Quebec; Kerala, India*<br>Reserves large in relation to demand |
| Chromium | Chromite, $(Mg,Fe)_2CrO_4$ | Magmatic segregations of mafic and ultramafic rocks | Steel alloys | *Bushveldt, South Africa*<br>Extensive reserves in a number of large deposits |
| Manganese | Pyrolusite, $MnO_2$ | Chemical sedimentary deposits, residual weathering deposits, seafloor nodules | Essential to steelmaking | *Ukraine*<br>World's land resources moderate, but seafloor deposits immense |
| **METALS PRESENT IN MINOR AMOUNTS IN EARTH'S CRUST** | | | | |
| Copper | Covelite, $CuS$<br>Chalcocite, $Cu_2S$<br>Digenite, $Cu_9S_5$<br>Chalcopyrite, $CuFeS_2$<br>Bornite, $Cu_5FeS_4$ | Porphyry copper deposits<br>Hydrothermal veins<br>Contact metamorphic rocks<br>Sedimentary deposits in shales (Kupferschiefer type) | Electrical wire and other products | *Bingham Canyon, Utah; Kuperschiefer, Germany; Poland* |
| Lead | Galena, $PbS$ | Hydrothermal (replacement) deposits<br>Contact metamorphic rocks<br>Sedimentary deposits (Kupferschiefer type) | Storage batteries, gasoline additive (tetraethyl lead) | *Mississippi Valley; Broken Hill, Australia*<br>Large resources; many lower grade deposits |

**TABLE 23.2** (*Continued*)

## Principal Types of Economic Mineral Deposits

| MINERAL DEPOSIT | TYPICAL MINERALS | GEOLOGICAL OCCURRENCE | USES | MAJOR DEPOSITS / REMARKS |
|---|---|---|---|---|
| | METALS PRESENT IN MINOR AMOUNTS IN EARTH'S CRUST | | | |
| Zinc | Sphalerite, $ZnS$ | Same as lead | Alloy metal | Same as lead |
| Nickel | Pentlandite, $(Ni, Fe)_9S_8$ Garnierite, $Ni_3Si_2O_5(OH)_4$ | Magmatic segregations Residual weathering deposits | Alloy metal | *Sudbury, Ontario* High-grade ores limited; large resources of low-grade ores; also in seafloor manganese nodules |
| Silver | Argentite, $Ag_2S$ In solid solution in copper, lead, and zinc sulfides | Hydrothermal veins with lead, zinc, and copper | Photographic chemicals; electrical equipment | Most produced as by-product of copper, lead, and zinc recovery |
| Mercury | Cinnabar, $HgS$ | Hydrothermal veins | Electrical equipment, pharmaceuticals | *Almadén, Spain* Few high-grade deposits with limited reserves |
| Platinum | Native metal | Magmatic segregations (mafic rocks) Placers | Chemical and electrical industry; alloying metal | *Bushveldt, South Africa* Large reserves in relation to demand |
| Gold | Native metal | Hydrothermal veins Placers | Coinage; dentistry; jewelry | *Witwatersrand, South Africa* Reserves concentrated in a few larger deposits |
| | NONMETALS | | | |
| Salt | Halite, $NaCl$ | Evaporite deposits Salt domes | Food; chemicals | Resources unlimited; economics determines exploitation |
| Phosphate rock | Apatite, $Ca_5(PO_4)_3OH$ | Marine phosphatic sedimentary rock Residual concentrations of nodules | Fertilizer | *Florida* High-grade deposits limited but extensive resources of low-grade deposits |
| Sulfur | Native sulfur Sulfide ore minerals | Caprock of salt domes (main source) Hydrothermal and sedimentary sulfides | Fertilizer manufacture; chemical industry | *Texas; Louisiana; Sicily* Native sulfur reserves limited but immense resources of sulfides |
| Potassium | Sylvite, $KCl$ Carnallite, $KCl\cdot MgCl_2\cdot 6H_2O$ | Evaporite deposits | Fertilizer | *Carlsbad, New Mexico* Great resources of rich deposits |
| Diamond | Diamond, $C$ | Kimberlite pipes Placers | Industrial abrasives | *Kimberley, South Africa* Synthetic diamond now commercially available |

**TABLE 23.2** (*Continued*)

## Principal Types of Economic Mineral Deposits

| MINERAL DEPOSIT | TYPICAL MINERALS | GEOLOGICAL OCCURRENCE | USES | MAJOR DEPOSITS / REMARKS |
|---|---|---|---|---|
| | | **NONMETALS** | | |
| Gypsum | Gypsum, $CaSO_4 \cdot 2H_2O$ Anhydrite, $CaSO_4$ | Evaporite deposits | Plaster | Immense resources widely distributed |
| Limestone | Calcite, $CaCO_3$ Dolomite, $CaMg(CO_3)_2$ | Sedimentary carbonate rocks | Building stone, agricultural lime; cement | Widely distributed; transportation a major cost |
| Clay | Kaolinite $Al_2Si_2O_5(OH)_4$ Smectite[1] Illite[1] | Residual weathering deposits; sedimentary clays and shales | Ceramics: china, electrical; structural tile | Many large pure deposits; immense reserves of all grades |
| Asbestos | Chrysotile, $Mg_3Si_2O_5(OH)_4$ | Ultramafic rocks altered and hydrated in near-surface crustal zones | Nonflammable fibers and products | *Southeastern Quebec* Limited high-grade reserves but great low-grade reserves |

[1] Formula highly variable; a hydrous aluminum silicate with other cations, such as $Na^+$, $K^+$, $Ca^{2+}$, $Mg^{2+}$.

# FINDING NEW MINERAL DEPOSITS

As the Earth's human population grows at an ever-increasing rate and people all over the world seek higher standards of living (more food, raw materials, energy, and manufactured goods), the demand for mineral resources shoots upward. The total dollar value of all mineral resources, including fuels, produced in the United States has grown from less than $5 billion in 1952 to about $100 billion in recent years. This 20-fold growth in less than four decades (which includes some inflationary increases because of the lowering of the value of the dollar in that time period) took place in a highly industrialized society that had already built a huge technological capability and whose population grew by only about 60 percent in the same period. The rate of increase of production and consumption in countries that are rapidly developing their industries has been even faster, and all countries aspire to speed their growth even further. The recovery of many nations after World War II and the growth in their economies led to an enormous increase in consumption and production in the rest of the world, reducing the U.S. percentage share. Figure 23.17 shows the relative decline in production and consumption of nonfuel minerals by the United States in comparison with the rest of the world.

Unequal sharing of world mineral resources has far-reaching economic and political repercussions. For instance, North America, with less than 10 percent of the world's population, consumes almost 75 percent of the world's production of aluminum, whereas Asia and Africa, with about two-thirds of the world's population, together use only a little over 5 percent of the production. The same extreme imbalance is associated with other materials as well. International relations are now deeply affected by struggles over the control of resources. One widespread cause of conflict is the demand by some mineral-rich developing nations for a greater degree of

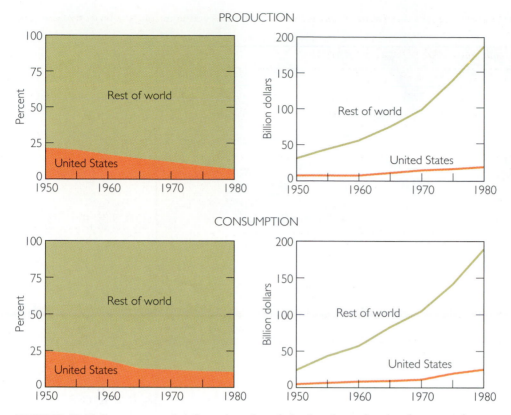

**FIGURE 23.17** Percentage and dollar value of nonfuel minerals produced and consumed over a 30-year period in the United States and the rest of the world. (Data from U.S. Bureau of Mines.)

control over resources mined on their land by corporations based in North America or Europe. In recent years several countries in Africa and South America have nationalized foreign-owned mining companies.

Plate-tectonics theory may serve the economic geologist in the exploration for new high-grade ores and mineral deposits, but how well remains to be seen. How extensive undersea exploration will become is also an open question. The answer depends in part on the development of efficient marine technology for deep-sea exploration and mining and the resolution of legal issues regarding ownership of deep-sea deposits. It is generally agreed that a nation has exclusive rights to mineral deposits in the offshore area within 200 nautical miles of its coast—the so-called **exclusive economic zone** (Figure 23.18). Still in question is who owns the mineral deposits on the seafloor beyond this zone. Are the deposits owned by all nations in common? By the discoverer and developer? By both? Because the economies of many nations that export minerals are threatened by

competition from new sources on the seafloor, these nations would like to limit such development and obtain a share of the profits.

These issues of international law have been debated for more than 20 years. In 1982 the United Nations adopted an agreement known as the Law of the Sea Treaty, providing a legal-regulatory system for the development of deep-sea resources, by a vote of 130 for, 4 against. The United States was one of the four nations opposed, because the treaty placed limits on seabed mineral production, and the United States feared that private companies that developed the technology and invested in production would be inadequately compensated.

A hard fact of life is that an equal per capita sharing of the world's available resources would not be enough to bring everyone to a "satisfactory" level of consumption, certainly not to a level anywhere near that of an affluent country in Europe or North America. The problem is compounded by the uncontrolled growth of the world's population (see Box 23.2).

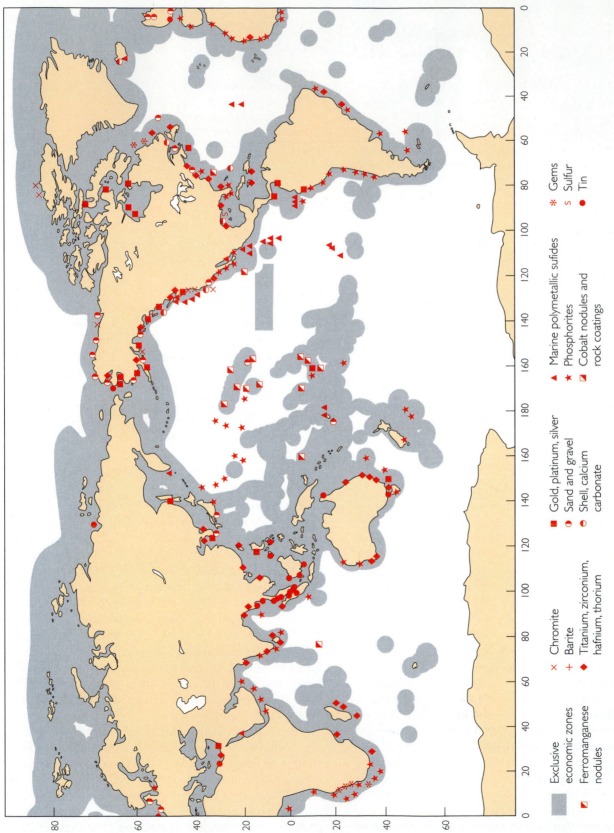

**FIGURE 23.18** Locations of some major nonfuel seabed ore deposits. Note the concentrations at plate boundaries and continental shelves. The gray areas outline exclusive economic zones claimed by individual nations—the seas within 200 nautical miles of their territory. (After J. M. Broadus, "Sea Bed Materials," *Science*, vol. 235, 1987, pp. 853–859.)

Legend:

- Exclusive economic zones
- Ferromanganese nodules
- × Chromite
- + Barite
- ◆ Titanium, zirconium, hafnium, thorium
- ■ Gold, platinum, silver
- ○ Sand and gravel
- ◐ Shell, calcium carbonate
- ◀ Marine polymetallic sulfides
- ★ Phosphorites
- ◪ Cobalt nodules and rock coatings
- ✳ Gems
- s Sulfur
- ● Tin

BOX 23.2
# CAN WE HAVE A SUSTAINABLE WORLD WITH UNLIMITED POPULATION GROWTH?

The world population of 5.5 billion in 1993 is expected to double by 2050. About 95 percent of this growth will occur in less developed countries (LDCs) where some 77 percent of Earth's inhabitants live. The developed countries, with only 23 percent of the world's population, produce 85 percent of the world's economic output and withdraw the majority of the minerals and fossil fuels from Earth's finite resources. These resources fuel industry and provide the basis for food, shelter, transportation, recreation, and all the other aspects of a high standard of living. The use of these resources is also responsible for a significant part of the pollution that has occurred.

The world faces several dilemmas. How do we reduce the greenhouse gases, wastes, and other pollutants released by the developed nations without halting their economic growth? (Even a developed nation like the United States needs economic growth to provide the funds to solve some of its social problems.) How can we meet the legitimate aspirations of the LDCs for economic growth without adding to the load of pollutants borne by our planet? If the LDCs were to achieve the high standard of living of the developed nations and consume Earth's resources in the same manner, the load on the biosphere could not be sustained. Profound changes in the global environment would occur, and human health would be threatened. The world would suffer an irreversible loss of biodiversity with the large-scale destruction of plant and animal species. There would be political unrest and large-scale migrations of people. The population explosion in the LDCs compounds the problem.

Some believe that the only alternative is for the developed nations to revert to the simpler lifestyles of earlier times, with a slower economic pace and much lower consumption of resources. In practice, this could mean smaller homes without air conditioning; fewer automobiles, televisions, and telephones; no packaged food—in short, fewer conveniences of all kinds. This philosophy has a small following, however, and is not likely to be adopted by leaders sensitive to electoral politics.

Is it possible to have *sustainable* economic growth in both the developed nations and the LDCs while still preserving the environment and biodiversity and recognizing the finiteness of Earth's resources? The stabilization of population growth and advances in science and technology might offer a solution. Here is how it could work.

High rates of population growth are characteristic of societies in poverty, where women have low status and lack education and access to contraceptives and family planning services and there is high infant mortality. A cooperative world effort could be undertaken by political leaders of both the developed nations and the LDCs to address these problems democratically and with sensitivity to human rights.

The single gravest threat to the environment is the use of energy. Science and technology can enable us to use fuels more efficiently in transportation, homes, and factories. Alternative energy sources—such as safe nuclear energy, solar and biomass energy, and other renewable sources—could be developed. Runoff of water contaminated by fertilizers and pesticides used in agriculture, a serious source of chemical pollution, could be eliminated. New crop species could be developed that need fewer artificial fertilizers and chemical pesticides. With new technology, industrial processes could be modified to produce little or no effluents. Recycling and substitution would reduce the demand for materials.

Only by addressing population growth *and* using Earth's resources wisely can the world avoid a calamity as serious as any humankind has ever faced.

The solution is probably a combination of conservation, recycling, and more efficient use of materials; increased discovery and exploitation of mineral and energy resources; and development of substitutes for scarce minerals, such as fiber-optic cables made of glass instead of copper wire. In a sense, we are all dependent on design and manufacturing engineers, economic geologists, and scientists who develop new materials from plentiful resources. Their skills will be of no avail, however, if population growth continues undiminished.

The world clearly faces major readjustments in the decades ahead, and it is certainly not too soon to start working out equitable, humane, and lasting approaches to building a sustainable world. How well we use our planet will depend on how well we understand how it works and the extent to which the people of the world cooperate intelligently in the development and use of its resources—and do so in a manner that protects the environment. We have only one Earth. To continue to live on it, we must learn to appreciate it better.

# SUMMARY

### What distinguishes an economical mineral deposit from one that is not economical?
Mineral deposits of economic value are those in which an element occurs in much greater abundance than in the average crustal rock—great enough to make the deposit economically worthwhile to mine when the costs of mining, processing, and transportation and the market price are taken into account.

### What is hydrothermal mineral deposition?
Hydrothermal deposits, which are some of the most important ore deposits, are formed by hot water that emanates from igneous intrusions or by heated circulating groundwater or seawater. The heated water leaches soluble minerals in its path and transports them to cooler rocks, where they are precipitated in fractures, faults, and voids. These ores may occur in veins or in such disseminated deposits as the copper-porphyry type.

### What are some of the processes that lead to the formation of sedimentary ore deposits?
Ordinary chemical and mechanical sedimentary processes segregate such economically important minerals as limestone, sand and gravel, and evaporite salt deposits. Sedimentary ores of copper and iron have formed as precipitates in special sedimentary environments, the iron ores chiefly in Precambrian times. Placers are ore deposits, rich in gold or other heavy minerals, that were laid down by currents.

### How do igneous ore deposits form?
Igneous ore deposits typically occur when minerals crystallize from molten magma, settle, and accumulate on the floor of a magma chamber. They are often found as layered accumulations of minerals. The rich ore body at Sudbury, Ontario, for example, is a mafic intrusive formation that contains great quantities of layered nickel, copper, and iron sulfides near its base.

### What insights into the formation of ores are provided by plate-tectonics theory?
Whether it is mineral precipitation from hydrothermal fluids or crystal settling in a magmatic body, igneous activity is the ultimate source of many minerals. Igneous activity is a feature of plate spreading and subduction, so modern and ancient plate boundaries are places where many concentrations of minerals should be found.

### What are some of the mineral policy issues faced by modern society?
The discovery of new mineral deposits is vital to support an increasingly industrialized world civilization. Prospects for finding new resources are good. The sea represents a largely untapped resource. Although technological advance can postpone the day of reckoning, stocks will become increasingly scarce. Conservation, recycling, and substitution of alternative materials will therefore become increasingly important in the years ahead. However, these advances will be unable to satisfy the demands of a world population that increases without limits.

# KEY TERMS AND CONCEPTS

ores (p. 525)
ore minerals (p. 525)
concentration factor (p. 525)
recycling (p. 527)

hydrothermal vein deposits
   (veins) (p. 529)
disseminated deposits (p. 529)
igneous ore deposits (p. 532)

sedimentary ore deposits (p. 532)
placers (p. 533)
manganese nodules (p. 536)
exclusive economic zone (p. 541)

# EXERCISES

1. What are the characteristics of an economical ore deposit?

2. Describe the creation of an ore body by hydrothermal activity.

3. Give some examples of useful hydrothermally deposited minerals.

4. How do useful minerals become concentrated in an igneous ore deposit?

5. List several examples of minerals in igneous ore bodies.

6. What is the role of physical and chemical processes in the formation of sedimentary ore deposits?

7. Identify some sites where sedimentary ore deposits may be found and what the deposits consist of.

8. Contrast the process of ore formation in the sediment and crust of the deep sea with that in a convergence zone.

9. What useful minerals might be found in the exclusive economic zone of the United States? Of Canada? Of England?

10. Why are conservation and recycling of useful materials drawn from the Earth and the creation of substitutes important to the future of humankind?

# THOUGHT QUESTIONS

1. Why should two advanced nations, such as the United States and Canada, take opposite positions with regard to the Law of the Sea Treaty?

2. Should the United States spend large sums to stockpile strategically important minerals (those essential for economic or defense needs) for use in an emergency in the event that foreign supply is cut off?

3. In 1992 many nations with interests in Antarctica signed a treaty agreeing not to mine for minerals on that continent. Why did they do so? Do you agree with their decision?

4. Will plate-tectonics theory contribute to the search for ore bodies? How?

# SUGGESTED READINGS

Bailly, P. A. 1984. Geologists and GNP: Future prospects. *Bulletin of the Geological Society of America* 95:257–264.

Broadus, J. M. 1987. Sea bed materials. *Science* 235:853–859.

Carr, D. D., and N. Herz (eds.). 1988. *Concise Encyclopedia of Mineral Resources.* Cambridge, Mass.: MIT Press.

Clark, J. P., and F. R. Field III. 1985. How critical are critical materials? *Technology Review* (August/September).

Dorr, Ann. 1987. *Minerals: Foundations of Society.* Alexandria, Va.: American Geological Institute.

Guilbert, J. M., and C. F. Park, Jr. 1986. *The Geology of Ore Deposits.* New York: W. H. Freeman.

Heaton, George, Robert Repetto, and Rodney Sobin. 1991. *Transforming Technology: An Agenda for Environmentally Sustainable Growth in the 21st Century.* Washington, D.C.: World Resources Institute.

Keary, P., and F. J. Vine. 1990. Plate tectonics and economic geology. Chap. 11 in *Global Tectonics.* London: Blackwell Scientific Publications.

National Research Council. 1990. *Competitiveness of the U.S. Minerals and Metals Industry.* Washington, D.C.: National Academy Press.

Sawkins, F. J. 1984. *Metal Deposits in Relation to Plate Tectonics.* New York: Springer-Verlag.

*Science Summit on World Population.* Report of conference of national academies of sciences of 80 countries, New Delhi, October 24–27, 1993.

U.S. Bureau of Mines. 1983. *The Domestic Supply of Critical Minerals.* Washington, D.C.: U.S. Government Printing Office.

U.S. Bureau of Mines. 1993. *Minerals Yearbook.* Washington, D.C.: U.S. Government Printing Office.

# APPENDIX 1

# CONVERSION FACTORS

## METRIC-ENGLISH

### LENGTH

| | |
|---|---|
| 1 centimeter | 0.3937 inch |
| 1 inch | 2.5400 centimeters |
| 1 meter | 3.2808 feet |
| 1 foot | 0.3048 meter |
| 1 meter | 1.0936 yards |
| 1 yard | 0.9144 meter |
| 1 kilometer | 0.6214 mile (statute) |
| 1 kilometer | 3281 feet |
| 1 mile (statute) | 1.6093 kilometers |
| 1 mile (nautical) | 1.8531 kilometers |
| 1 fathom | 6 feet |
| 1 fathom | 1.8288 meters |
| 1 angstrom | $10^{-8}$ centimeter |
| 1 micrometer | 0.0001 centimeter |

### VELOCITY

| | |
|---|---|
| 1 kilometer/hour | 27.78 centimeters/second |
| 1 mile/hour | 17.60 inches/second |

### AREA

| | |
|---|---|
| 1 square centimeter | 0.1550 square inch |
| 1 square inch | 6.452 square centimeters |
| 1 square meter | 10.764 square feet |
| 1 square meter | 1.1960 square yards |
| 1 square foot | 0.0929 square meter |
| 1 square kilometer | 0.3861 square mile |
| 1 square mile | 2.590 square kilometers |
| 1 acre (U.S.) | 4840 square yards |

## VOLUME

| | |
|---|---|
| 1 cubic centimeter | 0.0610 cubic inch |
| 1 cubic inch | 16.3872 cubic centimeters |
| 1 cubic meter | 35.314 cubic feet |
| 1 cubic foot | 0.02832 cubic meter |
| 1 cubic meter | 1.3079 cubic yards |
| 1 cubic yard | 0.7646 cubic meter |
| 1 liter | 1000 cubic centimeters |
| 1 liter | 1.0567 quarts (U.S. liquid) |
| 1 gallon (U.S. liquid) | 3.7853 liters |

## MASS

| | |
|---|---|
| 1 gram | 0.03527 ounce |
| 1 ounce | 28.3495 grams |
| 1 kilogram | 2.20462 pounds |
| 1 pound | 0.45359 kilogram |

## PRESSURE

| | |
|---|---|
| 1 kilogram/square centimeter | 0.96784 atmosphere |
| 1 kilogram/square centimeter | 0.98067 bar |
| 1 kilogram/square centimeter | 14.2233 pounds/square inch |
| 1 bar | 0.98692 atmosphere |
| 1 bar | $10^5$ pascals |
| 1 kilometer of granite | ~265 kilograms/square centimeter |

## ENERGY

| | |
|---|---|
| 1 erg | $2.39006 \times 10^{-8}$ calorie (gram) |
| 1 erg | $9.48451 \times 10^{-11}$ Btu |
| 1 erg | $10^{-7}$ joule |
| 1 quad | $10^{15}$ Btu |

## POWER

| | |
|---|---|
| 1 watt | $10^7$ ergs/second |
| 1 watt | 0.001341 horsepower (U.S.) |
| 1 watt | 0.05688 Btu/minute |

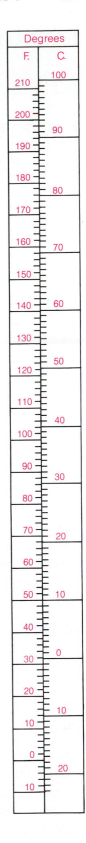

# ENERGY UNITS

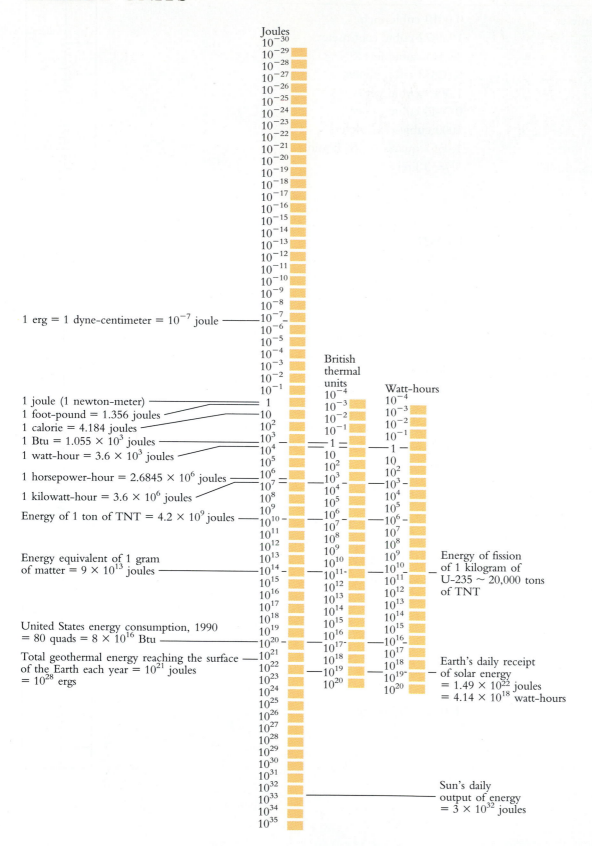

Joules

$10^{-30}$
$10^{-29}$
$10^{-28}$
$10^{-27}$
$10^{-26}$
$10^{-25}$
$10^{-24}$
$10^{-23}$
$10^{-22}$
$10^{-21}$
$10^{-20}$
$10^{-19}$
$10^{-18}$
$10^{-17}$
$10^{-16}$
$10^{-15}$
$10^{-14}$
$10^{-13}$
$10^{-12}$
$10^{-11}$
$10^{-10}$
$10^{-9}$
$10^{-8}$
1 erg = 1 dyne-centimeter = $10^{-7}$ joule —— $10^{-7}$
$10^{-6}$
$10^{-5}$
$10^{-4}$
$10^{-3}$
$10^{-2}$
$10^{-1}$

British thermal units

Watt-hours

1 joule (1 newton-meter) —— 1 | $10^{-4}$ | $10^{-4}$
1 foot-pound = 1.356 joules —— 10 | $10^{-3}$ | $10^{-3}$
1 calorie = 4.184 joules —— $10^{2}$ | $10^{-2}$ | $10^{-2}$
1 Btu = 1.055 × $10^{3}$ joules —— $10^{3}$ | $10^{-1}$ | $10^{-1}$
1 watt-hour = 3.6 × $10^{3}$ joules —— $10^{4}$ | 1 | 1
$10^{5}$ | 10 | 10
1 horsepower-hour = 2.6845 × $10^{6}$ joules —— $10^{6}$ | $10^{2}$ | $10^{2}$
$10^{7}$ | $10^{3}$ | $10^{3}$
1 kilowatt-hour = 3.6 × $10^{6}$ joules —— $10^{8}$ | $10^{4}$ | $10^{4}$
Energy of 1 ton of TNT = 4.2 × $10^{9}$ joules —— $10^{9}$ | $10^{5}$ | $10^{5}$
$10^{10}$ | $10^{6}$ | $10^{6}$
$10^{11}$ | $10^{7}$ | $10^{7}$
$10^{12}$ | $10^{8}$ | $10^{8}$

Energy equivalent of 1 gram
of matter = 9 × $10^{13}$ joules —— $10^{13}$ | $10^{9}$ | $10^{9}$
$10^{14}$ | $10^{10}$ | $10^{10}$   Energy of fission
$10^{15}$ | $10^{11}$ | $10^{11}$   of 1 kilogram of
$10^{16}$ | $10^{12}$ | $10^{12}$   U-235 ~ 20,000 tons
$10^{17}$ | $10^{13}$ | $10^{13}$   of TNT
$10^{18}$ | $10^{14}$ | $10^{14}$
$10^{19}$ | $10^{15}$ | $10^{15}$

United States energy consumption, 1990
= 80 quads = 8 × $10^{16}$ Btu —— $10^{20}$ | $10^{16}$ | $10^{16}$
$10^{17}$ | $10^{17}$

Total geothermal energy reaching the surface —— $10^{21}$
of the Earth each year = $10^{21}$ joules
= $10^{28}$ ergs
$10^{22}$ | $10^{18}$ | $10^{18}$   Earth's daily receipt
$10^{23}$ | $10^{19}$ | $10^{19}$   of solar energy
$10^{24}$ | $10^{20}$ | $10^{20}$   = 1.49 × $10^{22}$ joules
= 4.14 × $10^{18}$ watt-hours
$10^{25}$
$10^{26}$
$10^{27}$
$10^{28}$
$10^{29}$
$10^{30}$
$10^{31}$   Sun's daily
$10^{32}$   output of energy
$10^{33}$   = 3 × $10^{32}$ joules
$10^{34}$
$10^{35}$

# APPENDIX 2

# NUMERICAL DATA PERTAINING TO EARTH

| | |
|---|---|
| Equatorial radius | 6378 kilometers |
| Polar radius | 6357 kilometers |
| Radius of sphere with Earth's volume | 6371 kilometers |
| Volume | $1.083 \times 10^{27}$ cubic centimeters |
| Surface area | $5.1 \times 10^{18}$ square centimeters |
| Percent surface area of oceans | 71 |
| Percent surface area of land | 29 |
| Average elevation of land | 623 meters |
| Average depth of oceans | 3.8 kilometers |
| Mass | $5.976 \times 10^{27}$ grams |
| Density | 5.517 grams/cubic centimeter |
| Gravity at equator | 978.032 centimeters/second/second |
| Mass of atmosphere | $5.1 \times 10^{21}$ grams |
| Mass of ice | $25–30 \times 10^{21}$ grams |
| Mass of oceans | $1.4 \times 10^{24}$ grams |
| Mass of crust | $2.5 \times 10^{25}$ grams |
| Mass of mantle | $4.05 \times 10^{27}$ grams |
| Mass of core | $1.90 \times 10^{27}$ grams |
| Mean distance to Sun | $1.496 \times 10^{8}$ kilometers |
| Ratio: Mass of Sun/mass of Earth | $3.329 \times 10^{5}$ |
| Ratio: Mass of Earth/mass of Moon | 81.303 |

# APPENDIX 3

# PROPERTIES OF THE MOST COMMON MINERALS OF EARTH'S CRUST

| Category | Structure | Mineral or group name | Varieties and chemical composition | Form, diagnostic characteristics | Cleavage, fracture | Color | Hardness |
|---|---|---|---|---|---|---|---|
| LIGHT-COLORED MINERALS, VERY ABUNDANT IN EARTH'S CRUST IN ALL MAJOR ROCK TYPES | FRAMEWORK SILICATES | FELDSPAR | *POTASSIUM FELDSPARS* $KAlSi_3O_8$ *Sanidine Orthoclase Microcline* | Cleavable coarsely crystalline or finely granular masses; isolated crystals or grains in rocks, most commonly not showing crystal faces | Two at right angles, one perfect and one good; pearly luster on perfect cleavage | White to gray, frequently pink or yellowish; some green | 6 |
| | | | *PLAGIOCLASE FELDSPARS* $NaAlSi_3O_8$ *Albite* $CaAl_2Si_2O_8$ *Anorthite* | | Two at nearly right angles; one perfect, one good; fine parallel striations on perfect cleavage | White to gray, less commonly greenish or yellowish | |
| | | QUARTZ | $SiO_2$ | Single crystals or masses of 6-sided prismatic crystals; also formless crystals and grains or finely granular or massive | Very poor or nondetectable; conchoidal fracture | Colorless, usually transparent; also slightly colored smoky gray, pink, yellow | 7 |
| | SHEET SILICATES | MICA | *MUSCOVITE* $KAl_3Si_3O_{10}(OH)_2$ | Thin, disc-shaped crystals, some with hexagonal outlines; dispersed or aggregates | One perfect; splittable into very thin, flexible, transparent sheets | Colorless; slight gray or green to brown in thick pieces | 2–2½ |
| DARK-COLORED MINERALS, ABUNDANT IN MANY KINDS OF IGNEOUS AND METAMORPHIC ROCKS | | | *BIOTITE* $K(Mg,Fe)_3AlSi_3O_{10}(OH)_2$ | Irregular, foliated masses; scaly aggregates | One perfect; splittable into thin, flexible sheets | Black to dark brown; translucent to opaque | 2½–3 |
| | | | *CHLORITE* $(Mg,Fe)_5(Al,Fe)_2Si_3O_{10}(OH)_8$ | Foliated masses or aggregates of small scales | One perfect; thin sheets flexible but not elastic | Various shades of green | 2–2½ |
| | DOUBLE CHAINS | AMPHIBOLE | *TREMOLITE–ACTINOLITE* $Ca_2(Mg,Fe)_5Si_8O_{22}(OH)_2$ | Long, prismatic crystals, usually 6-sided; commonly in fibrous masses or irregular aggregates | Two perfect cleavage directions at 56° and 124° angles | Pale to deep green Pure tremolite white | 5–6 |
| | | | *HORNBLENDE* Complex Ca, Na, Mg, Fe, Al silicate | | | | |

| Mineral or group name | | Varieties and chemical composition | Form, diagnostic characteristics | Cleavage, fracture | Color | Hardness |
|---|---|---|---|---|---|---|
| PYROXENE | SINGLE CHAINS | *ENSTATITE–HYPERSTHENE* $(Mg,Fe)_2Si_2O_6$ | Prismatic crystals, either 4- or 8-sided; granular masses and scattered grains | Two good cleavage directions at about 90° | Green and brown to grayish or greenish white | 5–6 |
| | | *DIOPSIDE* $(Ca,Mg)_2Si_2O_6$ | | | Light to dark green | |
| | | *AUGITE* Complex Ca, Na, Mg, Fe, Al silicate | | | Very dark green to black | |
| OLIVINE | ISOLATED TETRAHEDRA | $(Mg,Fe)_2SiO_4$ | Granular masses and disseminated small grains | Conchoidal fracture | Olive to grayish green and brown | 6½–7 |
| GARNET | | Ca, Mg, Fe, Al silicate | Isometric crystals, well-formed or rounded; high specific gravity, 3.5–4.3 | Conchoidal and irregular fracture | Red and brown, less commonly pale colors | 6½–7 |
| CALCITE | CARBONATES | $CaCO_3$ | Coarsely to finely crystalline in beds, veins, and other aggregates; cleavage faces may show in coarser masses; calcite effervesces rapidly, dolomite slowly, only in powders | Three perfect cleavages, at oblique angles; splits to rhombohedral cleavage pieces | Colorless, transparent to translucent; variously colored by impurities | 3 |
| DOLOMITE | | $CaMg(CO_3)_2$ | | | | 3½–4 |
| CLAY MINERALS | HYDROUS ALUMINO-SILICATES | *KAOLINITE* $Al_2Si_2O_5(OH)_4$ | Earthy masses in soils; bedded; in association with other clays, iron oxides, or carbonates; plastic when wet; montmorillonite swells when wet | Earthy, irregular | White to light gray and buff; also gray to dark gray, greenish gray, and brownish depending on impurities and associated minerals | 1½–2½ |
| | | *ILLITE* Similar to muscovite + Mg, Fe | | | | |
| | | *SMECTITE* Complex Ca, Na, Mg, Fe, Al silicate + $H_2O$ | | | | |
| GYPSUM | SULFATES | $CaSO_4 \cdot 2H_2O$ | Granular, earthy, or finely crystalline masses; tabular crystals | One perfect, splitting to fairly thin slabs or sheets; two other good cleavages | Colorless to white; transparent to translucent | 2 |
| ANHYDRITE | | $CaSO_4$ | Massive or crystalline aggregates in beds and veins | One perfect, one nearly perfect, one good; at right angles | Colorless, some tinged with blue | 3–3½ |
| HALITE | | $NaCl$ | Granular masses in beds; some cubic crystals; salty taste | Three perfect cleavages at right angles | Colorless, transparent to translucent | 2½ |
| OPAL–CHALCEDONY | | $SiO_2$ [Opal is an amorphous variety; chalcedony is a formless microcrystalline quartz.] | Beds in siliceous sediments and chert; in veins or banded aggregates | Conchoidal fracture | Colorless or white when pure, but tinged with various colors by impurities in bands, especially in agates | 5–6½ |

LIGHT-COLORED MINERALS, TYPICALLY AS ABUNDANT CONSTITUENTS OF SEDIMENTS AND SEDIMENTARY ROCKS

| Group | Name | Formula | Crystal habit / Description | Cleavage / Fracture | Color | Hardness |
|---|---|---|---|---|---|---|
| DARK-COLORED MINERALS, COMMON IN MANY ROCK TYPES | MAGNETITE | $Fe_3O_4$ | Magnetic; disseminated grains, granular masses; occasional octahedral isometric crystals; high specific gravity, 5.2 | Conchoidal or irregular fracture | Black, metallic luster | 6 |
| | HEMATITE | $Fe_2O_3$ | Earthy to dense masses, some with rounded forms, some granular or foliated; high specific gravity, 4.9–5.3 | None; uneven, sometimes splintery fracture | Reddish brown to black | 5–6 |
| | "LIMONITE" | GOETHITE [the major mineral of the mixture called "limonite," a field term] $HFeO_2$ | Earthy masses, massive bodies or encrustations, irregular layers; high specific gravity, 3.3–4.3 | One excellent in the rare crystals; usually an early fracture | Yellowish brown to dark brown and black | 5–5½ |
| LIGHT-COLORED MINERALS, MAINLY IN IGNEOUS AND METAMORPHIC ROCKS AS COMMON OR MINOR CONSTITUENTS | KYANITE | $Al_2SiO_5$ | Long, bladed or tabular crystals or aggregates | One perfect and one poor, parallel to length of crystals | White to light-colored or pale blue | 5 parallel to crystal length 7 across crystals |
| | SILLIMANITE | $Al_2SiO_5$ | Long, slender crystals or fibrous, felted masses | One perfect parallel to length, not usually seen | Colorless, gray to white | 6–7 |
| | ANDALUSITE | $Al_2SiO_5$ | Coarse, nearly square prismatic crystals, some with symmetrically arranged impurities | One distinct; irregular fracture | Red, reddish brown, olive-green | 7½ |
| | FELDSPATHOIDS | NEPHELINE $(Na,K)AlSiO_4$ | Compact masses or as embedded grains, rarely as small prismatic crystals | One distinct; irregular fracture | Colorless, white, light gray; gray-greenish in masses, with greasy luster | 5½–6 |
| | | LEUCITE $KAlSi_2O_6$ | Trapezohedral crystals embedded in volcanic rocks | One very imperfect | White to gray | 5½–6 |
| | SERPENTINE | $Mg_6Si_4O_{10}(OH)_8$ | Fibrous (asbestos) or platy masses | Splintery fracture | Green; some yellowish brownish, or gray; waxy or greasy luster in massive habit; silky luster in fibrous habit | 4–6 |
| | TALC | $Mg_3Si_4O_{10}(OH)_2$ masses or aggregates | Foliated or compact masses or aggregates | One perfect, making thin flakes or scales; soapy feel | White to pale green; pearly or greasy luster | 1 |
| | CORUNDUM | $Al_2O_3$ | Some rounded, barrel-shaped crystals; most often as disseminated grains or granular masses (emery) | Irregular fracture | Usually brown, pink, or blue; emery black Gemstone varieties: ruby, sapphire | 9 |

| Category | Mineral or group name | Class | Varieties and chemical composition | Form, diagnostic characteristics | Cleavage, fracture | Color | Hardness |
|---|---|---|---|---|---|---|---|
| DARK-COLORED MINERALS, COMMON IN METAMORPHIC ROCKS | EPIDOTE | SILICATES | $Ca_2(Al,Fe)Al_2Si_3O_{12}(OH)$ | Aggregates of long prismatic crystals, granular or compact masses, embedded grains | One good, one poor at greater than right angles; conchoidal and irregular fracture | Green, yellow-green, gray, some varieties dark brown to black | 6–7 |
| | STAUROLITE | SILICATES | $Fe_2Al_9Si_4O_{22}(O,OH)_2$ | Short prismatic crystals, some cross-shaped, usually coarser than matrix of rock | One poor | Brown, reddish, or dark brown to black | 7–7½ |
| METALLIC LUSTER, COMMON IN MANY ROCK TYPES, ABUNDANT IN VEINS | PYRITE | SULFIDES | $FeS_2$ | Granular masses or well-formed cubic crystals in veins and beds or disseminated; high specific gravity, 4.9–5.2 | Uneven fracture | Pale brass-yellow | 6–6½ |
| | GALENA | | $PbS$ | Granular masses in veins and disseminated; some cubic crystals; very high specific gravity, 7.3–7.6 | Three perfect cleavages at mutual right angles, giving cubic cleavage fragments | Silver-gray | 2½ |
| | SPHALERITE | | $ZnS$ | Granular masses or compact crystalline aggregates; high specific gravity, 3.9–4.1 | Six perfect cleavages at 60° to one another | White to green, brown, and black; resinous to submetallic luster | 3½–4 |
| | CHALCOPYRITE | | $CuFeS_2$ | Granular or compact masses; disseminated crystals; high specific gravity, 4.1–4.3 | Uneven fracture | Brassy to golden-yellow | 3½–4 |
| | CHALCOCITE | | $Cu_2S$ | Fine-grained masses; high specific gravity, 5.5–5.8 | Conchoidal fracture | Lead-gray to black; may tarnish green or blue | 2½–3 |
| MINERALS FOUND IN MINOR AMOUNTS IN A VARIETY OF ROCK TYPES AND IN VEINS OR PLACERS | RUTILE | TITANIUM OXIDES | $TiO_2$ | Slender to prismatic crystals; granular masses; high specific gravity, 4.25 | One distinct, one less distinct; conchoidal fracture | Reddish brown, some yellowish, violet, or black | 6–6½ |
| | ILMENITE | TITANIUM OXIDES | $FeTiO_3$ | Compact masses, embedded grains, detrital grains in sand; high specific gravity, 4.79 | Conchoidal fracture | Iron-black; metallic to submetallic luster | 5–6 |
| | ZEOLITES | SILICATES | Complex hydrous silicates; many varieties of minerals, including analcime, natrolite, phillipsite, heulandite, and chabazite | Well-formed radiating crystals in cavities in volcanics, veins, and hot springs; also as fine-grained and earthy bedded deposits | One perfect for most | Colorless, white, some pinkish | 4–5 |

# APPENDIX 4

# TOPOGRAPHIC AND GEOLOGIC MAPS

A map is a quantitative representation of the spatial distribution of some attribute or property of the Earth. It is a kind of graph in which the axes are lines of latitude and longitude and the positions of points on the surface (or beneath it) are plotted in relation to those axes or some other established reference. Geologists often need to show the configuration and nature of the geological materials at or near the surface in a meaningful way, so that they can construct a three-dimensional mental picture of the geology from this two-dimensional graph. Once the nature of maps becomes familiar, the map reader can become practiced at deducing much of the geologic structure and history of an area.

The use of topographic maps (see Chapter 16) and geologic maps (see Chapter 10) has spread widely throughout our culture. To the more traditional users of such maps—geologists and surveyors—have been added city planners, industrial zoning commissions, and many members of the public seeking recreational areas for hiking, camping, fishing, and other activities. In 1993 the U.S. Geological Survey distributed nearly 7 million copies of its 70,000 published topographic maps and 18,000 copies of geologic and hydrologic maps from its open file reports. Maps are a necessity for all kinds of geological and mineral resource studies, as well as for studies of groundwater, flood control, soil management, and such environmental concerns as land-use planning, which involves the locations of highways, industrial areas, oil and gas pipelines, and recreational areas.

## Topographic Maps

Because the size of the area covered, and thus the amount of detail that can be shown, is always important, we use the concept of *scale*—that is, the relationship of a distance (or area) on the map to the true distance on the Earth. This is simply done by stating a ratio, such as 1:24,000, which indicates that a distance of one unit on the map represents a distance of 24,000 such units on the Earth. It does not matter what the units are: a map of scale 1:24,000 is the same whether we use metric or English systems. The scale can be thought of in any convenient units desired: 1 in. = 2000 ft, or 1 m = 24 km, or 10 cm = 2.4 km. For convenience, maps have a graphic scale, usually at the bottom margin, in which a distance such as 1 km or 1 mile, usually with subdivisions, is shown as it would appear on the map. A common scale for detailed topographic and geologic maps is 1:24,000, used by the U.S. Geological Survey for most modern maps. The scale used for regional maps covering much larger areas is 1:250,000. Scales of 1:1,000,000 are used for aeronautical charts. In 1976 the U.S. Geological Survey introduced the first of a new series of 1:100,000 all-metric topographic maps on which graphic scales show both kilometers and miles.

On most maps, natural and constructed features of the surface are represented by conventional symbols. Those used by the U.S. Geological Survey are typical: rivers, lakes, and oceans are shown in blue; topography is shown in brown; constructed features are shown in black, with main highways and urban areas in red; green shaded areas show wooded land. Some special symbols may be shown on the explanation, or **legend,** of the map, which is usually displayed along the bottom margin. The most complex representations are the topographic elevations of the surface, usually shown on North American maps by contours (see Chapter 16). Special maps are sometimes prepared to show environmental variables, such as the distribution of slopes of various steepness.

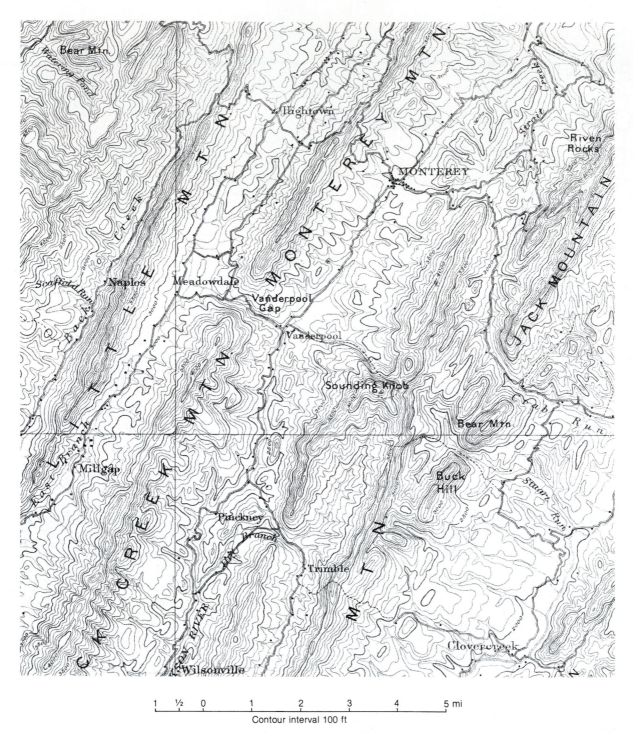

1  ½  0     1     2     3     4     5 mi

Contour interval 100 ft

**FIGURE 1** Topographic map *(above)* and geologic map with cross sections *(facing page)* of folded sedimentary rocks in the Valley and Ridge province of the Appalachian Mountains. Contours show the pronounced trends of valleys and ridges that reflect the parallel folds. The ridges have developed along the formations that are resistant to erosion, some at the crests of anticlines, such as Jack Mountain north of Crab Run, and others along the flanks of folds. The valleys are in the easily eroded formations, some in synclines, such as Jackson River, some on anticlines, such

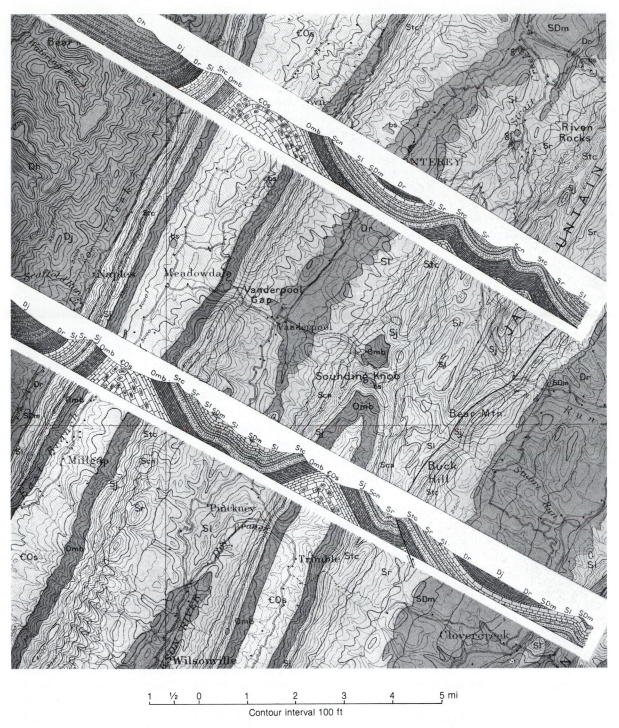

1  ½  0        1        2        3        4      5 mi

Contour interval 100 ft

as East Branch, and some in the flanks of folds, such as Back Creek. On the geologic map the pattern of anticlines and synclines can be read from the positions of formations of different age, such as at East Branch, where the oldest rocks, Cambrian and Ordovician formations (COs), are at

the surface bordered on both sides by younger formations of Ordovician and Silurian age (Omb, Stc, Sj, and others). The cross sections make these relationships clearer and add some detail. For the same map indicating the rock formations in color, see Figure 2. (From U.S. Geological Survey.)

# Geologic Maps

Geologic maps are a representation of the distribution of rocks and other geologic materials of different lithologies and ages over the Earth's surface or below it. The geologist perceives the Earth not only in its surface expression of topography and patterns of land and water but in terms of its pattern of subsurface structures, stratigraphic sequences, igneous intrusions, unconformities, and other geometric relationships of rocks. Just as an anatomist can visualize the muscles and bones beneath the skin, so can a geologist visualize details of the Earth's subsurface.

Detailed geologic maps are normally constructed on a topographic map base (Figure 1). This makes it easy to locate geologic structures with respect to surface features of the Earth. It is also important because topography is so often related to the nature of the underlying rocks and their structures. Because it contains so much more information than a topographic map alone, a geologic map is the most valuable for many of the purposes for which maps are used.

Geologic maps are ordinarily made by a geologist who roams over the area and notes the kinds of rocks, sediments, and soils and their structural and stratigraphic relationships. In modern times this is supplemented or even entirely supplanted by remote sensing by aerial and satellite photography or geophysical instruments. Remote, inaccessible areas, such as those in some polar or desert regions, may be mapped almost entirely by this method, with the geologist ground-checking in scattered places. The mapping of the Moon is an extreme example of this approach. Mars is being mapped with no ground check at all except for the area immediately surrounding the *Viking* landing site. The best and most accurate maps, however, are those made by traditional means: geologists cover the ground on foot to see most if not all of the outcrops; they are helped enormously by the automobile, sometimes by a helicopter, and in some places even by a horse or donkey.

The mapping proceeds by the following steps:

1. *Principal observations.* Description of outcrop location, lithology, age, fossil content, and structural attitude as measured by dip and strike, direction of fault movement, fold axes, and so forth (see Chapter 10). Plotting observations on work map.
2. *First integration.* Conceptualizing the spatial relationship of one outcrop to another by stratigraphic correlation of rocks of the same age, facies, degree of metamorphism and deformation. Grouping of mappable rock units into formations. Drawing of lines on the primitive geologic map of inferred connections where formations are hidden. Compilation of the complete or composite stratigraphic sequences and ages of deformational or igneous intrusive events.
3. *Synthesizing the map.* Visualizing the larger pattern of geologic relationships and constructing the map, together with geologic cross sections made both to help geologists in their thinking and to illustrate more detail and inference from the map.

Analyses of rock composition, absolute age, and seismic, gravity, and magnetic data may be incorporated into the map. The geologist further draws on the geologic literature or personal experience of the geology of nearby and similar kinds of regions. The final result is the finished geologic map, a codified mass of information displayed in a form in which anyone familiar with geology can quickly read the nature of the Earth's crust in the area and a good deal of its geologic history.

There are many kinds of geologic maps. The most common shows the bedrock geology and gives a picture of what the land would look like if all soil were stripped away. Surficial geologic maps, on the other hand, emphasize the nature of soils, unconsolidated river sediment, sand dunes, and whatever other materials, including outcrops, appear at the surface. A special kind of surficial geologic map is used for environmental hazards. One kind shows areas of high-angle or unsupported slopes that are likely to slump or slide (see Chapter 11). Whatever the geologic purpose, there is a map that can be made to show the relevant data. There is no question that the map is both the best device for geological research into the origin of the distribution of important geologic characteristics over the Earth and the best way to illustrate the patterns discovered from such research.

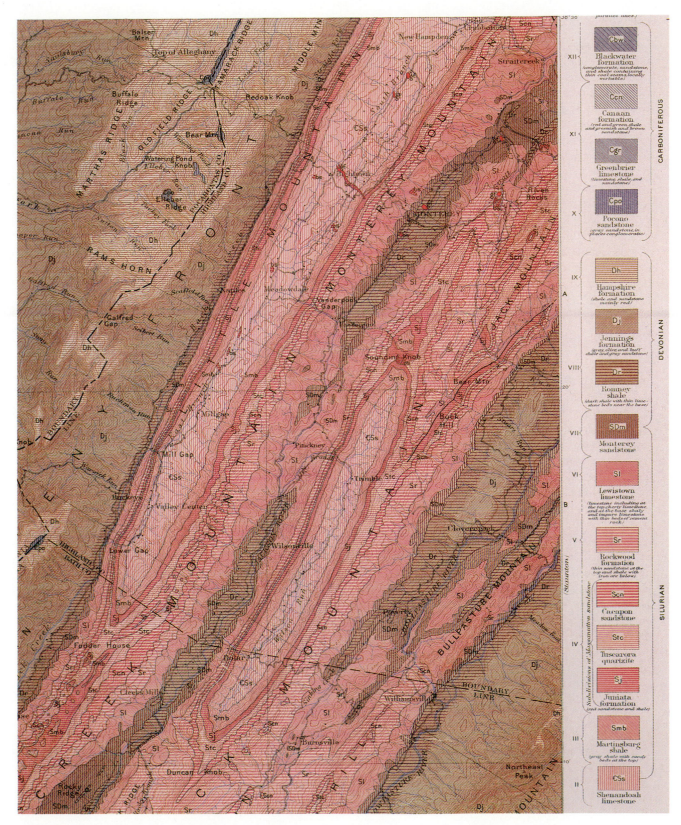

**FIGURE 2** A part of the geologic map of the Monterey quadrangle, including the same area shown in Figure 1.

(From N. H. Darton, *Monterey: Virginia–West Virginia,* U.S. Geological Survey Folio 61, 1899.)

# GLOSSARY

Words in *italic* are defined elsewhere in the glossary.
Specific minerals are defined in Appendix 3.

**AA:** A blocky and fragmented form of *lava* occurring in flows with fissured and angular surfaces.

**ABLATION:** The annual amount of snow lost from a *glacier* by the processes of melting, *sublimation*, wind *erosion*, and iceberg calving.

**ABSOLUTE AGE:** The age in years of a particular geologic event or feature, generally obtained with *radiometric dating* techniques. (Compare *Relative age*.)

**ABYSSAL PLAIN:** A flat, sediment-covered province of the seafloor that slopes at less than 1:1000.

**ACCUMULATION (GLACIAL):** The amount of snow added to a *glacier* annually. Most accumulation occurs in the upper reaches of the glacier.

**ACTIVE MARGIN:** A *continental margin* characterized by *earthquakes*, igneous activity, and/or uplifted mountains resulting from convergent or transform *plate* motion.

**A-HORIZON:** The uppermost layer of a *soil*, containing organic material and leached *minerals*.

**ALLUVIAL FAN:** A low, cone-shaped deposit of *terrigenous sediment* formed where a *stream* undergoes an abrupt reduction in slope.

**AMPHIBOLITE:** A *metamorphic rock* containing mostly amphibole and plagioclase feldspar.

**ANDESITE:** A volcanic rock type intermediate in composition between *rhyolite* and *basalt*.

**ANGLE OF REPOSE:** The steepest slope angle at which a particular *sediment* will lie without cascading down.

**ANGULAR UNCONFORMITY:** An *unconformity* in which the *bedding* planes of the rocks above and below are not parallel.

**ANION:** Any negatively charge *ion;* the opposite of *cation*.

**ANTECEDENT STREAM:** A *stream* that existed before the present *topography* was created, thereby maintaining its original course despite changes in the structure of the underlying rocks and in topography.

**ANTICLINE:** A large upfold of strata, usually from 100 m to 300 km in width, whose *limbs* are lower than its center. (Compare *Syncline*.)

**AQUICLUDE:** A stratum with low *permeability* that acts as a barrier to the flow of *groundwater*. Also called "confining layer."

**AQUIFER:** A permeable *formation* that stores and transmits *groundwater* in sufficient quantity to supply wells.

**ARETE:** The sharp, jagged crest along the divide between glacial *cirques*, resulting from the headward *erosion* of the walls of adjoining cirques.

**ARGILLITE:** A low-grade *metamorphic rock* made from a shaly *sedimentary rock*, characterized by irregular fracture and lack of *foliation*.

**ARTESIAN FLOW:** Flow in a *confined aquifer*, in which the *groundwater* is at a greater pressure than in an *unconfined aquifer* at similar depths, thereby causing water in a well that penetrates a confined aquifer (an artesian well) to rise above the level of the *aquiclude*.

**ASEISMIC RIDGE:** A linear submarine ridge characterized by the absence of seismic activity, thereby distinguishing it from a *mid-ocean ridge*, which is seismically active.

**ASH:** See *Volcanic ash*.

**ASH-FLOW DEPOSIT:** See *Volcanic ash-flow deposit*.

**ASTHENOSPHERE:** The weak layer below the *lithosphere* that is marked by low *seismic wave* velocities and high seismic wave attenuation. Movement in the asthenosphere occurs by plastic deformation.

**ASYMMETRICAL FOLD:** A *fold* of strata in which the *dips* of the two *limbs* are unequal.

**ATOLL:** A continuous or broken circle of coral *reefs* and low coral islands surrounding a central lagoon.

**ATOMIC NUMBER:** The number of *protons* in the nucleus of an atom.

**ATOMIC WEIGHT:** The sum of the masses of the *protons* and *neutrons* in the atomic nucleus of an element.

**AXIAL PLANE:** In *folds*, the plane that most nearly separates two symmetrical *limbs*. In a simple *anticline*, it is vertical; in complex folding, it is perpendicular to the direction of compression.

**AXIS (FOLD):** Within each stratum involved in a *fold*, the axis connecting all the points in the center of the fold, from which both *limbs* bend.

**BACKSHORE:** The upper, generally dry, zone of the shore, extending landward from the upper limit of wave wash at high *tide* to the upper limit of shore-zone processes.

**BACKWASH:** The return flow of water down a beach after a wave has broken.

**BADLAND:** *Topography* characterized by intricate patterns of stream *erosion* developed on surfaces with little or no vegetative cover overlying *unconsolidated* or poorly cemented *clays*, silts, or sands.

**BARCHAN:** A crescent-shaped *eolian* sand *dune* that moves across a flat surface with its convex face upwind and its concave *slip face* downwind.

**BARRIER ISLAND:** A long, narrow island parallel to the shore, composed of sand and built by wave action.

**BASAL SLIP:** The sliding of a *glacier* along its base.

**BASALT:** A fine-grained, dark, *mafic igneous rock* composed largely of plagioclase feldspar and pyroxene.

**BASE LEVEL:** The level below which a *stream* cannot erode: usually sea level, sometimes locally the level of a lake or resistant *formation*.

**BASEMENT:** The oldest rocks recognized in a given area, a complex of *metamorphic* and *igneous rocks* that underlies all the sedimentary *formations*. Usually Precambrian or Paleozoic in age.

**BASIN (SEDIMENTARY):** See *Sedimentary basin*.

**BASIN (TECTONICS):** A circular, synclinelike depression of strata.

**BATHOLITH:** A great irregular mass of coarse-grained *igneous rock* with an exposed surface of more than 100 km$^2$, which has either intruded the *country rock* or been derived from it through metamorphism.

**BAUXITE:** A rock composed primarily of hydrous aluminum oxides and formed by intense *chemical weathering* in tropical areas with good drainage; a major ore of aluminum.

**BEDDING:** A characteristic of *sedimentary rocks* in which parallel planar surfaces separating different grain sizes or compositions indicate successive depositional surfaces formed at the time of sedimentation.

**BEDDING SEQUENCE:** A pattern of interbedding of different *sedimentary rock* types or sedimentary rocks with different *sedimentary structures* that is characteristic of a certain *sedimentary environment*.

**BED LOAD:** The *sediment* that a *stream* moves along the bottom of its *channel* by rolling and bouncing (*saltation*).

**BEDROCK:** The solid rock underlying *unconsolidated* surface materials, such as *soil*.

**B-HORIZON:** The intermediate layer in a *soil*, below the *A-horizon* and above the *C-horizon*, consisting of *clays* and oxide materials.

**BIOCHEMICAL SEDIMENT, ROCK:** A *sediment* or rock containing the mineral remains of organisms, such as shells, or *minerals* precipitated as a result of biological processes, such as in *iron formations*.

**BIOMASS:** Organic carbon-containing material of biological origin, including living and dead animals and plants.

**BIOTURBATION:** The reworking of existing *sediments* by organisms.

**BLOWOUT:** (1) A parabola-shaped *eolian* sand *dune*, typically one blown back from a beach, that has its convex *slip face* oriented downwind. (2) A shallow circular or elliptical depression in sand or dry soil formed by wind erosion. (See also *Deflation*.)

**BLUESCHIST:** A *metamorphic rock* formed under conditions of high pressure (in excess of 5000 bars) and relatively low temperature, often containing the blue *minerals* glaucophane (an amphibole) and kyanite.

**BOTTOMSET BED:** A flat-lying bed of fine *sediment* deposited seaward of a *delta* and then buried by continued delta growth.

**BOWEN'S REACTION SERIES:** A simple schematic description of the order in which different *minerals* crystallize during the cooling and progressive *crystallization* of a *magma*.

**BRAIDED STREAM:** A *stream* so choked with *sediment* that it divides and recombines numerous times, forming many small and meandering *channels*.

**BRECCIA:** See *Sedimentary breccia; Volcanic breccia*.

**BRITTLE MATERIAL:** A material that breaks abruptly when its elastic limit is reached; the opposite of a *ductile material*.

**BURIAL METAMORPHISM:** A low-grade metamorphism in which buried *sedimentary rocks* are metamorphosed by the heat and pressure exerted by overlying *sediments* and sedimentary rocks; *bedding* and other *sedimentary structures* are preserved.

**CALDERA:** A large basin-shaped volcanic depression, typically originating through volcanic explosion and/or collapse. Potentially catastrophic eruptions of a "resurgent caldera" can occur when fresh *magma* reenters the collapsed volcanic *magma chamber*.

**CAPACITY (STREAM):** The amount of *sediment* and detritus a *stream* can transport past any point in a given amount of time. (Compare *Competence*.)

**CARBONATE COMPENSATION DEPTH:** The depth in the oceans below which the solution rate of calcium carbonate ($CaCO_3$) becomes so great that no carbonate organisms or *sediments* are preserved on the seafloor.

**CARBONATE SEDIMENT, ROCK:** A *sediment* or *sedimentary rock* formed from the accumulation of carbonate *minerals* precipitated organically or inorganically. Rocks are chiefly *limestone* and *dolostone*.

**CATACLASTIC METAMORPHISM:** High-pressure, low-temperature metamorphism occurring primarily by the crushing and shearing of rock during tectonic movements and resulting in the formation of powdered rock.

**CATION:** Any positively charged *ion*; the opposite of *anion*.

**CEMENTATION:** A *lithification* process in which *minerals* are precipitated in the pore space of *sediments*, often binding the grains.

**CENTRAL VENT:** The largest vent of a volcano, situated at the center of its cone.

**CHANNEL:** The trough through which water flows in a stream *valley;* sometimes reserved for the deepest part of the streambed, in which the main current flows.

**CHEMICAL SEDIMENT, ROCK:** A *sediment* or *sedimentary rock* that is formed at or near its place of deposition by chemical precipitation, usually from seawater.

**CHEMICAL WEATHERING:** The total set of all chemical reactions that can act on rock exposed to water and the atmosphere and so dissolve the *minerals* or change them to more stable forms.

**CHERT:** A sedimentary form of poorly or extremely finely crystalline silica, commonly quartz; usually a *chemical* or *biochemical sediment*.

**C-HORIZON:** The lowest layer of a *soil,* consisting of fragments of rock and their chemically weathered products.

**CINDER CONE:** A steep, conical hill built up about a volcanic vent and composed of coarse *pyroclastic rock* fragments expelled from the vent by escaping gases.

**CIRQUE:** The head of a glacial valley, usually with the form of one half of an inverted cone. The upper edges have the steepest slopes, approaching the vertical, and the base may be flat or hollowed out. The base is commonly occupied by a small lake or pond after deglaciation.

**CLASTIC SEDIMENT, ROCK:** A *sediment* or *sedimentary rock* formed from particles (clasts) derived from the *erosion* of preexisting rocks and mechanically transported.

**CLAY:** Any of a number of hydrous aluminosilicate *minerals* with sheetlike *crystal* structure, formed by the *weathering* and *hydration* of other silicates. Also, any mineral fragments smaller than 0.0039 mm.

**CLEAVAGE (MINERAL):** The tendency of a *crystal* to break along certain preferred planes in the crystal lattice; also, the geometric pattern of such a breakage.

**CLEAVAGE (ROCK):** The tendency of a rock to break along certain planes induced during deformation or metamorphism, usually in the direction of preferred orientation of the *minerals* in the rock.

**COAL:** The metamorphic product of stratified plant remains. It contains more than 50 percent carbon compounds and burns readily.

**COAST:** The strip of land adjacent to an ocean or sea and extending from low *tide* landward to the point of major change in landscape features.

**COMPACTION:** The decrease in volume and *porosity* of a *sediment* caused by burial beneath other sediments.

**COMPETENCE (STREAM):** A measure of the largest particle a *stream* is able to transport, not the total volume. (Compare *Capacity.*)

**COMPOSITE VOLCANO:** A volcanic cone containing layers of both *lava* flows and *pyroclastic rocks.* Synonym of *stratovolcano.*

**COMPRESSIVE FORCES:** Forces that squeeze together or shorten a body. Compressive forces dominate at *convergent plate boundaries.*

**CONCENTRATION FACTOR:** The ratio of the abundance of an element in a *mineral* deposit to its average abundance in the *crust.*

**CONCORDANT INTRUSIVE:** An *intrusive igneous rock* having contacts with the *country rock* that are parallel to *bedding* or *foliation* planes. (Compare *Discordant intrusive.*)

**CONDUCTION:** See *Heat conduction.*

**CONFINED AQUIFER:** An *aquifer* overlain by relatively impermeable strata (*aquicludes*), thereby causing the water to be contained under pressure. (Compare *Unconfined aquifer.*)

**CONGLOMERATE:** A *sedimentary rock,* a significant fraction of which is composed of rounded pebbles, cobbles, and boulders. The lithified equivalent of *gravel.*

**CONSOLIDATED MATERIAL:** *Sediment* that is lithified; that is, compacted and bound together by *mineral* cements.

**CONTACT METAMORPHISM:** Changes in the *mineralogy* and *texture* of rock resulting from the heat and pressure of an igneous intrusion in the near vicinity.

**CONTINENTAL DRIFT:** The horizontal displacement or rotation of continents relative to one another.

**CONTINENTAL GLACIER:** A continuous, thick *glacier* covering more than 50,000 km² and moving independently of minor topographic features. (Compare *Valley glacier.*)

**CONTINENTAL MARGIN:** The portion of the ocean floor extending from the *shoreline* to the landward edge of the *abyssal plain* and including the *continental shelf, slope,* and *rise.*

**CONTINENTAL RISE:** A broad and gently sloping ramp that rises from an *abyssal plain* to the *continental slope* at a rate of less than 1:40.

**CONTINENTAL SHELF:** The gently sloping submerged edge of a continent, extending commonly to a depth of about 200 m or the edge of the *continental slope.*

**CONTINENTAL SHELF DEPOSITS:** *Sediments* laid down in a tectonically quiet *syncline* at a *passive continental margin.*

**CONTINENTAL SLOPE:** The region of steep slopes between the *continental shelf* and *continental rise.*

**CONTINUOUS REACTION SERIES:** A *reaction series* in which the same *mineral* crystallizes throughout the range of temperatures in question, but in which there is gradual change in the chemical composition of the mineral with changing temperature.

**CONTOUR:** A curve on a topographic map that connects points of equal *elevation.*

**CONVECTION:** A mechanism of heat transfer in a flowing material in which hot material from the bottom rises because of its lesser *density* while cool surface material sinks. (Compare *Heat conduction.*)

**CONVERGENT PLATE BOUNDARY:** A boundary at which the *plates* collide and area is lost either by shortening and crustal thickening or by *subduction* of one plate beneath the other. The site of volcanism, *earthquakes, trenches,* and mountain building. (See also *Subduction zone.*)

**CORE:** The central part of the Earth below a depth of 2900 km. It is thought to be composed of iron and nickel and to be molten on the outside with a central solid inner core.

**COUNTRY ROCK:** The rock into which an *igneous rock* intrudes or a *mineral* deposit is emplaced.

**COVALENT BOND:** A bond between atoms in which outer *electrons* are shared.

**CRATON:** A portion of a continent that has not been subjected to major deformation for a prolonged time, typically since Precambrian or early Paleozoic time.

**CREEP:** Slow, downhill *mass movement* of *soil* and *regolith* under gravitational force.

**CREVASSE:** Any large vertical crack in the surface of a *glacier* or snowfield.

**CROSS-BEDDING:** Inclined beds of depositional origin in a *sedimentary rock.* Formed by currents of wind or water in the direction in which the bed slopes downward.

**CRUST:** The outermost layer of the *lithosphere* consisting of relatively light, low-melting-temperature materials. The continental crust consists largely of *granite* and *granodiorite;* the oceanic crust is mostly *basalt.*

**CRYSTAL:** A form of matter in which the atoms, *ions,* or molecules are arranged regularly in all directions to form a regular, repeating network.

**CRYSTAL HABIT:** The general shape of a *crystal;* for example, cubic, prismatic, or fibrous.

**CRYSTALLIZATION:** The formation of crystalline solids

from a gas or liquid, such as in the formation of crystalline *minerals* in *magma*.

**CUESTA:** An asymmetrical ridge with one steep and one gentle face formed where gently dipping beds of erosion-resistant rocks are undercut by *erosion* of a weaker bed underneath.

**CYCLE OF EROSION:** A proposed sequence of changes in a landscape that progresses from high, rugged, tectonically formed mountains to low, rounded hills and finally to worn-down, tectonically stable plains.

**DACITE:** Volcanic equivalent of *granodiorite*.

**DEBRIS AVALANCHE:** A fast downhill *mass movement* of *soil* and rock.

**DEBRIS FLOW:** A fluid *mass movement* of rock fragments supported by a muddy matrix. Debris flows differ from *earthflows* in that they generally contain coarser material and move faster than earthflows.

**DEBRIS SLIDE:** A *mass movement* of rock material and *soil* largely as one or more units along planes of weakness at the base of or within the rock material.

**DEFLATION:** The removal of *clay* and dust from dry *soil* by strong winds.

**DELTA:** A body of *sediment* deposited in an ocean or lake at the mouth of a *stream*.

**DENDRITIC DRAINAGE:** A *stream* system that branches irregularly, resembling a branching tree.

**DENSITY:** The mass per unit volume of a substance, commonly expressed in grams per cubic centimeter.

**DEPOSITIONAL REMANENT MAGNETIZATION:** A weak magnetization created in *sedimentary rocks* by the rotation of magnetic *crystals* into line with the ambient field during settling.

**DESERTIFICATION:** The process by which arid lands become transformed by loss of vegetation and *soil* into barren desert.

**DESERT PAVEMENT:** A residual deposit produced by continued *deflation*, which removes the fine grains of a *soil* and leaves a surface covered with close-packed cobbles.

**DESERT VARNISH:** A dark coating commonly found on the surface of rock in the desert. It consists of *clays*, iron oxides, and magnesium oxides produced during *weathering*.

**DIAGENESIS:** The physical and chemical changes undergone by a *sediment* during *lithification* and *compaction*, excluding *erosion* and *metamorphism*.

**DIATREME:** A volcanic vent filled with *volcanic breccia* by the explosive escape of gases.

**DIKE:** A roughly planar body of *intrusive igneous rock* that has *discordant intrusive* contact with the surrounding rock.

**DIORITE:** A *plutonic* rock with composition intermediate between *granite* and *gabbro*; the intrusive equivalent of *andesite*.

**DIP:** The maximum angle by which a stratum or other planar feature deviates from the horizontal. The angle is measured in a plane perpendicular to the *strike*.

**DISCHARGE (GROUNDWATER):** The exit to surface water bodies of *groundwater*; the opposite of *recharge*.

**DISCHARGE (STREAM):** The rate of water movement through a *stream*, measured in units of volume per unit time.

**DISCONTINUOUS REACTION SERIES:** A *reaction series* in which the end members have different *crystal* structures (are distinct *mineral phases*). (See also *Continuous reaction series*).

**DISCORDANT INTRUSIVE:** An *intrusive igneous rock* that has contacts with the *country rock* cutting across *bedding* or *foliation* planes. (Compare *Concordant intrusive*.)

**DISSEMINATED DEPOSIT:** An *ore deposit* in which the metal is distributed in small amounts throughout the rock, not concentrated in *veins*.

**DISTRIBUTARY:** A smaller branch of a large *stream* that receives water from the main *channel*; the opposite of a *tributary*.

**DIVERGENT PLATE BOUNDARY:** A boundary at which the *plates* move apart and new *lithosphere* is created; the site of *mid-ocean ridges*, shallow-focus *earthquakes*, and volcanism.

**DIVIDE:** A ridge of high ground separating two *drainage basins* emptied by different *streams*.

**DOLOSTONE:** A *sedimentary rock* composed primarily of dolomite, a carbonate *mineral* with the general formula $CaMg(CO_3)_2$.

**DOME:** In structural geology, a round or elliptical upwarp of strata resembling a short *anticline*. See also *Volcanic dome*.

**DRAAS:** Extremely large (1 km or more long and over 100 m high) composite of sand *dunes* found in deserts.

**DRAINAGE BASIN:** A region of land surrounded by *divides* and crossed by *streams* that eventually converge to one *river* or lake.

**DRAINAGE NETWORK:** A map of the pattern of *tributaries*, large and small, of a *stream* system.

**DRIFT (GLACIAL):** A collective term for all the rock, sand, and *clay* that is deposited by a *glacier* either as *till* or as *outwash*.

**DRUMLIN:** A smooth, streamlined hill composed of *till*.

**DRY WASH:** An intermittent streambed in a desert canyon that carries water only briefly after a rain.

**DUCTILE MATERIAL:** A material that can undergo considerable change in shape without breaking after its elastic limit is reached; the opposite of a *brittle material*.

**DUNE:** An elongated mound of sand formed by wind or water.

**EARTHFLOW:** A fluid *mass movement* of mainly fine-grained material, along with some broken rock, at slow or moderate speeds.

**EARTHQUAKE:** The violent oscillatory motion of the ground caused by the passage of *seismic waves* radiating from a *fault* along which sudden movement has taken place.

**EBB TIDE:** The part of the *tide* cycle during which the water level is falling.

**ECLOGITE:** An extremely high pressure *metamorphic rock* containing the *minerals* garnet and pyroxene.

**EFFLUENT STREAM:** A *stream* or portion of a stream that receives some water from groundwater *discharge* because the stream's *elevation* is below the *groundwater table*. (Compare *Influent stream*.)

**ELASTIC REBOUND THEORY:** A theory of *fault* movement and *earthquake* generation holding that faults remain locked while strain energy accumulates in the rock and then suddenly slip and release this energy.

**ELECTRON:** A negatively charged atomic particle with a mass of $9.1 \times 10^{-28}$ gram and a negative charge of $1.6 \times 10^{-19}$ coulomb. The position of an electron about an atomic nucleus is not fixed but is described by a probability statement.

**ELEVATION:** The vertical height of one point on the Earth above a given plane, usually sea level.

**ENVIRONMENT OF DEPOSITION:** See *Sedimentary environment*.

**EOLIAN:** Pertaining to or deposited by wind.

**EON:** The largest division of geologic time, embracing several *eras;* for example, the Phanerozoic eon, from 600 million years ago to the present.

**EPEIROGENY:** Large-scale, primarily vertical, movement of the *crust*. It is characteristically so gradual that rocks are little folded and faulted.

**EPICENTER:** The point on the Earth's surface directly above the *focus* or hypocenter of an *earthquake*.

**EPOCH:** One subdivision of a geologic *period,* often chosen to correspond to a *stratigraphic sequence*. Also used for a division of time corresponding to a *paleomagnetic* interval.

**ERA:** A division of geologic time including several *periods,* but smaller than an *eon*. Commonly recognized eras are Precambrian, Paleozoic, Mesozoic, and Cenozoic.

**ERG:** Extensive region, or "sea," of wind-transported sand found in major deserts.

**EROSION:** The set of all processes by which *soil* and rock are loosened and moved downhill or downwind.

**ERRATIC:** Rock fragment (especially boulder-sized) carried by a *glacier* away from the *outcrop* from which it was derived, often into an area underlain by a rock type different from that of the rock fragment.

**ESKER:** A glacial deposit in the form of a continuous, winding ridge, formed from the deposits of a *stream* flowing beneath the ice.

**ESTUARY:** A body of water along a coastline, open to the ocean but diluted by fresh water.

**EVAPORITE:** See *Marine evaporite*.

**EXCLUSIVE ECONOMIC ZONE:** The zone from a country's *coast* to 200 miles offshore in which the country has exclusive rights to *mineral* deposits.

**EXFOLIATION:** A *physical weathering* process in which sheets of rock are fractured and detached from an *outcrop*.

**EXTRUSIVE IGNEOUS ROCK:** An *igneous rock* formed from *lava* or from other volcanic material spewed out onto the surface of the Earth.

**FACIES:** See *Metamorphic facies*.

**FAULT:** A planar or gently curved fracture in the Earth's *crust* across which there has been relative displacement.

**FAULT-BLOCK MOUNTAIN:** A mountain or range formed as an upthrown block when it was elevated (or as the surrounding region sank) between two normal *faults*.

**FAULTING:** See *Fault*.

**FELSIC:** An adjective used to describe a light-colored *igneous rock* that is poor in iron and magnesium and contains abundant feldspars and quartz.

**FERRIC IRON:** Iron with a +3 charge ($Fe^{3+}$).

**FERROUS IRON:** Iron with a +2 charge ($Fe^{2+}$).

**FIRN:** Old, dense, compacted snow.

**FISSURE ERUPTION:** A volcanic eruption emanating from an elongate fissure rather than a *central vent*.

**FJORD:** A former glacial valley with steep walls and a U-shaped profile, now occupied by the sea.

**FLINT:** See *Chert*.

**FLOOD BASALT:** A plateau *basalt* extending many kilometers in flat, layered flows originating in *fissure eruptions*.

**FLOODPLAIN:** A level plain of stratified, *unconsolidated* sediment on either side of a *stream;* submerged during floods and built up by silt and sand carried out of the main *channel*.

**FLOOD TIDE:** The part of the *tide* cycle during which the water is rising or leveling off at high water.

**FOCUS (EARTHQUAKE):** The point along a *fault* at which the rupture occurs. Also called the "hypocenter."

**FOLD:** A bent or warped stratum or sequence of strata that was originally horizontal, or nearly so, and was subsequently deformed.

**FOLD BELT:** Synonym of *Orogenic belt*.

**FOLDING:** See *Fold*.

**FOLIATION:** A set of flat or wavy planes in a *metamorphic rock,* produced by structural deformation.

**FORAMINIFER:** An order of oceanic protozoa most of which have shells composed of calcite.

**FORAMINIFERAL OOZE:** A calcareous *pelagic sediment* composed of the shells of dead *foraminifera*.

**FORESET BED:** One of the inclined beds found in *cross-bedding;* also an inclined bed deposited on the outer front of a *delta*.

**FORESHORE:** The marine zone between the upper limit of wave wash at high *tide* and the low-tide mark.

**FORMATION:** The basic unit for the naming of rocks in *stratigraphy:* a set of rocks that are or once were horizontally continuous, that share some distinctive feature of lithology, and that are large enough to be mapped.

**FOSSIL FUEL:** A general term for combustible geologic deposits of carbon in reduced (organic) form and of biologic origin, including *coal,* oil, natural gas, *oil shales,* and *tar sands*.

**FOSSIL MAGNETISM:** See *Paleomagnetism*.

**FRACTIONAL CRYSTALLIZATION:** The separation of a cooling *magma* into components by the successive formation and removal of *crystals* at progressively lower temperatures.

**FRACTURE (MINERALOGY):** The irregular breaking of a *crystal* along a surface not parallel to a crystal face; serves to group and classify *minerals*.

**FRINGING REEF:** A coral *reef* that is directly attached to a landmass not composed of coral.

**GABBRO:** A black, coarse-grained, *intrusive igneous rock,* composed of calcic feldspars and pyroxene. The intrusive equivalent of *basalt*.

**GEOLOGIC TIME:** The time from the formation of the Earth to the present, divided into periods of time during which known geological events have taken place.

**GEOLOGIC TIME SCALE:** The division of geologic history into *eras, periods,* and *epochs* accomplished through *stratigraphy* and *paleontology*.

**GEOTHERM:** A curve on a temperature-pressure or temperature-depth graph that describes how temperature in the Earth changes with depth. Different tectonic provinces are characterized by more or less rapid increases of temperature with depth.

**GEOTHERMAL ENERGY:** Energy generated by using the heat energy of the *crust,* especially in volcanic regions.

**GLACIAL STRIATIONS:** Scratches left on *bedrock* and

boulders by overriding ice and showing the direction of glacial motion.

**GLACIER:** A mass of ice and surficial snow that persists throughout the year and flows downhill under its own weight. The size range is from 100 m to 10,000 km. (See also *Continental glacier; Valley glacier*.)

**GLACIER SURGE:** A period of unusually rapid movement of a glacier, sometimes lasting more than a year.

**GLASS:** A rock formed when *magma* is cooled too rapidly to allow *crystal* growth.

**GLASSY:** Adjective indicating that a material does not have an orderly, repeating, three-dimensional array of atoms.

**GNEISS:** A coarse-grained *regional metamorphic* rock that shows banding and parallel alignment of *minerals*.

**GRADED BEDDING:** A bed in which the coarsest particles are concentrated at the bottom and grade gradually upward into fine silt, the whole bed having been deposited by a waning current.

**GRADED STREAM:** A stream whose smooth *longitudinal profile* is unbroken by resistant ledges, lakes, or waterfalls and that exactly maintains the slope, velocity, and *discharge* required to carry its *sediment* load in equilibrium without *erosion* or sedimentation.

**GRANITE:** A coarse-grained *intrusive igneous rock* composed of quartz, orthoclase feldspar, sodium-rich plagioclase feldspar, and micas. Also sometimes a metamorphic product.

**GRANITIZATION:** The formation of metamorphic *granite* from other rocks by recrystallization with or without complete melting.

**GRANODIORITE:** A *plutonic* rock similar to *granite* in composition, except that plagioclase feldspar is present in greater abundance than orthoclase feldspar.

**GRANULITE:** A *regional metamorphic* rock with coarse interlocking grains, generally formed under conditions of relatively high pressure and temperature.

**GRAVEL:** The coarsest *clastic sediment,* consisting mostly of particles larger than 2 mm and including cobbles and boulders.

**GREENSCHIST:** A metamorphic *schist* containing chlorite and epidote (which are green) and formed by low-temperature, low-pressure metamorphism.

**GREENSTONE:** A field term applied to any altered or metamorphosed *mafic igneous rock* (for example, *basalt, gabbro,* or diabase).

**GROUNDWATER:** The mass of water in the ground (below the *unsaturated zone*) occupying the total pore space in the rock and moving slowly downhill where *permeability* allows.

**GROUNDWATER TABLE:** The upper surface of the *saturated zone* of *groundwater*. Also called the "water table."

**GUYOT:** A flat-topped submarine mountain or *seamount*.

**HALF-LIFE:** The time required for half of a sample of a given radioactive *isotope* to decay.

**HANGING VALLEY:** A former glacial tributary valley that enters a larger glacial valley above its base, high up on the valley wall.

**HEAT CONDUCTION:** The transfer of the vibrational energy of atoms and molecules, which constitutes heat energy, by the mechanism of atomic or molecular impact. (Compare *Convection*.)

**HEAT RESERVOIR:** A subsurface region containing enough heat to be used for *geothermal energy*.

**HOGBACK:** A *formation* similar to a *cuesta* in that it is a ridge formed by slower *erosion* of hard strata, but having two steep, equally inclined slopes.

**HORNFELS:** A high-temperature, low-pressure *metamorphic rock* of uniform grain size showing no *foliation*. Usually formed by *contact metamorphism*.

**HOT SPOT:** The surface expression of a mantle *plume*.

**HUMUS:** The decayed part of the organic matter in a *soil*.

**HYDRATION:** The absorption of water by a *mineral*, usually in *weathering*.

**HYDROLOGIC CYCLE:** The cyclical movement of water from the ocean to the atmosphere, through rain to the surface, through *runoff* and *groundwater* to *streams,* and back to the sea.

**HYDROLOGY:** The science of that part of the *hydrologic cycle* between rain and return to the sea; the study of water on and within the land.

**HYDROTHERMAL ACTIVITY:** Any process involving high-temperature *groundwater,* especially the alteration and emplacement of *minerals* and the formation of hot springs and geysers.

**HYDROTHERMAL VEIN DEPOSIT:** A cluster of *minerals* precipitated by *hydrothermal activity* in a rock cavity.

**ICEBERG CALVING:** The breaking off of blocks of ice from a *glacier,* forming icebergs.

**IGNEOUS ROCK:** A rock formed by the solidification of a *magma*.

**INFILTRATION:** The movement of *groundwater* or hydrothermal water into rock or *soil* through pores and *joints*.

**INFLUENT STREAM:** A *stream* or portion of a stream that *recharges groundwater* through the stream bottom because its *elevation* is above the *groundwater table*. (Compare *Effluent stream*.)

**INTRUSIVE IGNEOUS ROCK:** *Igneous rock* that forced its way in a molten state into the *country rock*. Also called an "intrusion."

**ION:** An atom or group of atoms that has gained or lost *electrons* and so has a net electric charge.

**IONIC BOND:** A bond formed between atoms by electrostatic attraction between oppositely charged *ions*.

**IRON FORMATION:** A *sedimentary rock* containing much iron, usually more than 15 percent, as sulfide, oxide, hydroxide, or carbonate; a low-grade ore of iron.

**ISLAND ARC:** A linear or arcuate chain of volcanic islands formed at a *convergent plate boundary*. The island arc is formed in the overriding plate from rising melt derived from the subducted plate and from the *asthenosphere* above that plate.

**ISOCHRON:** A line connecting points of equal age.

**ISOSTASY, PRINCIPLE OF:** The mechanism whereby areas of the *crust* rise or subside until the mass of their *topography* is buoyantly supported or compensated by the thickness of crust below, which "floats" on the denser *mantle*. The theory contends that continents and mountains are supported by low-density crustal "roots."

**ISOTOPE:** One of several forms of one element, all having the same number of *protons* in the nucleus, but differing in their number of *neutrons* and thus in their *atomic weight*.

**JOINT:** A large and relatively planar fracture in a rock across which there is no relative displacement of the two sides.

**KAME:** A ridgelike or hilly local glacial deposit of coarse *clastic sediment* formed as a *delta* at the *glacier* front by meltwater *streams*.

**KARST TOPOGRAPHY:** An irregular *topography* characterized by *sinkholes*, caverns, and lack of surface *streams*; formed in humid regions because an underlying carbonate *formation* has been riddled with underground drainage channels that capture the surface streams.

**KETTLE:** A small hollow or depression formed in glacial deposits when *outwash* was deposited around a residual block of ice that later melted.

**LAHAR:** A *mudflow* of *unconsolidated volcanic ash*, dust, *breccia*, and boulders mixed with rain or the water of a lake displaced by a *lava* flow.

**LAMINAR FLOW:** A flow in which streamlines are straight or gently curved and parallel. (Compare *Turbulent flow*).

**LANDFORM:** A characteristic landscape feature on the Earth's surface that attained its shape through the processes of *erosion* and sedimentation; for example, hill, valley.

**LATERITE:** A distinctive *soil* formed in very humid regions, characterized by high alumina and iron oxide content, and produced by rapid *chemical weathering* of feldspar *minerals*.

**LAVA:** *Magma* that has reached the surface.

**LEVEE:** A ridge along a *stream* bank, formed by deposits left when floodwater slowed on leaving the *channel;* also an artificial barrier to floods built in the same form.

**LIMB (FOLD):** The relatively planar part of a *fold* or of two adjacent folds (for example, the steeply dipping part of a stratum between an *anticline* and a *syncline*).

**LIMESTONE:** A *sedimentary rock* composed mainly of calcium carbonate ($CaCO_3$), usually as the *mineral* calcite.

**LINEAR DUNE:** A long, narrow *eolian* sand *dune* that is aligned parallel to the direction of the prevailing wind.

**LITHIFICATION:** The processes that convert a *sediment* into a *sedimentary rock.*

**LITHOSPHERE:** The outer, rigid shell of the Earth, situated above the *asthenosphere* and containing the *crust,* the uppermost part of the *mantle,* the continents, and the *plates*.

**LOESS:** An unstratified, wind-deposited, dusty *sediment* rich in *clay* minerals.

**LONGITUDINAL PROFILE:** A cross section of a *stream* from its mouth to its head, showing *elevation* versus distance to the mouth.

**LONGSHORE CURRENT:** A current that flows parallel to the *shoreline;* the summed longshore components of water motion of waves that break obliquely with respect to the shore.

**LONGSHORE DRIFT:** The movement of *sediment* along a beach by *swash* and *backwash* of waves that approach the shore obliquely.

**LUSTER:** The general quality of the shine of a *mineral* surface, described by such subjective terms as dull, glassy, or metallic.

**MAFIC:** Adjective describing dark-colored *minerals* rich in iron and magnesium (for example, pyroxene, amphibole or olivine); also, describing rocks rich in mafic minerals.

**MAGMA:** Molten rock material that forms *igneous rocks* upon cooling. Magma that reaches the surface is *lava*.

**MAGMA CHAMBER:** A magma-filled cavity within the *lithosphere*.

**MAGMATIC DIFFERENTIATION:** The process by which a uniform parent *magma* may lead to rocks of a variety of compositions. Magmatic differentiation occurs because different *mineral* phases crystallize from a melt at different temperatures.

**MAGNETIC STRATIGRAPHY:** The study and correlation of polarity epochs and events in the history of the Earth's magnetic field as contained in magnetic rocks.

**MAGNITUDE:** A measure of *earthquake* size, determined by taking the common logarithm (base 10) of the largest ground motion observed during the arrival of a *P wave* or *seismic surface wave* and applying a standard correction for distance to the *epicenter*.

**MANTLE:** The main bulk of the solid Earth, between the *crust* and the *core*, ranging from depths of about 40 km to 2900 km. It is composed of dense, *mafic* silicates and divided into concentric layers by phase changes that are caused by the increase in pressure with depth.

**MARBLE:** The metamorphosed equivalent of *limestone* or other *carbonate rock.*

**MARINE EVAPORITE SEDIMENT, ROCK:** A *sediment* or *sedimentary rock* consisting of *minerals* precipitated by evaporating seawater. Includes salt and gypsum.

**MASS MOVEMENT:** A downhill movement of *soil* or fractured rock under the force of gravity.

**MASS WASTING:** See *Mass movement*.

**MEANDER:** A broad, semicircular curve in a *stream* that develops as the stream erodes the outer bank of a bend and deposits *sediment* (as *point bars*) against the inner bank.

**MECHANICAL WEATHERING:** See *Physical weathering*.

**MÉLANGE:** A formation found at *convergent plate boundaries* consisting of a heterogeneous mixture of rock materials. Fragments of diverse composition, size, and *texture* have been mixed and consolidated by tremendous deformational pressure.

**MERCALLI SCALE:** A measure of *earthquake* intensities as gauged by the observed effects on people and structures rather than by actual size. It ranges from I (not felt by people) to XII (destruction nearly total).

**MESA:** A flat-topped, steep-sided upland topped by a resistant *formation*.

**METALLIC BOND:** A type of *covalent bond* in which freely mobile *electrons* are shared and dispersed among *ions* of metallic elements, which have the tendency to lose electrons and pack together as *cations*.

**METAMORPHIC FACIES:** Characteristic assemblages of *minerals* in *metamorphic rocks* that are indicative of the range of pressures and temperatures experienced during metamorphism.

**METAMORPHIC ROCK:** A rock whose original *mineralogy, texture,* or composition has been changed by the effects of pressure, temperature, or the gain or loss of chemical components.

**METASOMATISM:** A change in the bulk chemical composition of a rock by fluid transport of some chemical components into or out of the rock.

**METEORIC WATER:** Rainwater, snow, hail, and sleet.

**MICROPLATE TERRANE:** A block within an *orogenic belt* containing rock assemblages that contrast sharply with those in the surrounding areas, thereby leading to the interpretation that the blocks are small continents, *seamounts,* or *island arcs* that were accreted onto the larger continent at a *convergent plate boundary.*

**MID-OCEAN RIDGE:** A major elevated linear feature of the seafloor consisting of many small, slightly offset segments, with a total length of 200 to 20,000 km. A mid-ocean ridge occurs at a *divergent plate boundary,* a site where two *plates* are being pulled apart and new oceanic *lithosphere* is being created.

**MIGMATITE:** A rock with both igneous and metamorphic characteristics that shows large *crystal* and laminar flow structures. Probably formed metamorphically in the presence of water and without complete melting.

**MINERAL:** A naturally occurring, inorganic, crystalline solid substance with a specific chemical composition.

**MINERALOGY:** The study of *mineral* composition, structure, appearance, stability, occurrence, and associations. The mineralogy of a rock is the mineral assemblage contained within that rock.

**MODIFIED MERCALLI SCALE:** A revision of the *Mercalli Scale* adapted for North American conditions.

**MOHOROVIČIĆ DISCONTINUITY, MOHO:** The boundary between *crust* and *mantle,* at a depth of 5 to 45 km, marked by a rapid increase in *seismic wave* velocity to more than 8 km per second.

**MOHS SCALE OF HARDNESS:** An empirical, ascending scale of *mineral* hardness. (See Table 2.4.)

**MORAINE:** A glacial deposit of *till* left at the margins of an ice sheet. Subdivided into ground moraine, lateral moraine, medial moraine, and terminal moraine.

**MUDFLOW:** A mass *movement* of material mostly finer than sand, along with some rock debris, lubricated with large amounts of water. The water tends to make mudflows move faster than *earthflows* or *debris flows.*

**MUDSTONE:** The lithified equivalent of mud; a fine-grained *sedimentary rock* similar to *shale* but less finely laminated.

**MYLONITE:** A very fine grained *metamorphic rock* commonly found in major thrust *faults* and produced by shearing and rolling during fault movement.

**NATURAL LEVEE:** See *Levee.*

**NEAP TIDE:** A *tide* cycle of unusually small amplitude that occurs twice monthly when the lunar and solar tides are opposed—that is, when the gravitational pull of the Sun is at right angles to that of the Moon. (Compare *Spring tide.*)

**NEUTRON:** An electrically neutral elementary particle in the atomic nucleus, having the mass of one *proton.*

**OBSIDIAN:** Dark volcanic *glass* of *felsic* composition.

**OFFSHORE:** The marine zone extending from the breaker zone to the edge of the *continental shelf.*

**OIL SHALE:** A dark-colored *shale* that contains organic material and that can be crushed and heated to liberate mostly gaseous hydrocarbons.

**OIL RESERVOIR:** A bed of permeable and porous rock that contains commercially producible oil.

**OIL TRAP:** See *trap.*

**OPHIOLITE SUITE:** An assemblage of *mafic* and *ultramafic igneous rocks* with deep-sea *sediments;* believed to be associated with *divergent plate boundaries* and the seafloor environment.

**ORE DEPOSIT:** A *sedimentary, igneous,* or *metamorphic rock* containing *minerals,* commonly metallic oxides or silicates, that can be commercially mined.

**ORGANIC SEDIMENT, ROCK:** A *sediment* or *sedimentary rock* consisting entirely or in part of organic carbon-rich deposits formed by the decay of once-living material after burial. Includes *coal* and organic carbon-rich *shales.*

**ORIGINAL HORIZONTALITY, PRINCIPLE OF:** The proposition that all sedimentary *bedding* is horizontal at the time of deposition.

**OROGENIC BELT:** A linear region that has been subjected to folding and other deformation in a mountain-building episode.

**OROGENY:** The tectonic process in which large areas are folded, thrust-faulted, metamorphosed, and subjected to *plutonism.* The cycle ends with uplift and the formation of mountains.

**OUTCROP:** A segment of *bedrock* exposed to the atmosphere.

**OUTWASH:** A *sediment* deposited by meltwater *streams* emanating from a *glacier.*

**OVERTURNED FOLD:** A *fold* in which a *limb* has tilted past vertical so that the older strata are uppermost.

**OXBOW LAKE:** A long, broad, crescent-shaped lake formed when a *stream* abandons a *meander* and takes a new course.

**OXIDATION:** A chemical reaction in which *electrons* are lost from an atom and its charge becomes more positive.

**PAHOEHOE:** A basaltic *lava* flow with a *glassy,* smooth, and ropy surface.

**PALEOMAGNETIC STRATIGRAPHY:** A branch of *stratigraphy* in which the *remanent magnetization* recorded in a rock is used to place the rock on the "magnetic" time scale constructed from known temporal variations in the Earth's magnetic field.

**PALEOMAGNETISM:** The *remanent magnetization* recorded in ancient rocks; allows the reconstruction of Earth's ancient magnetic field and the positions of the continents.

**PALEONTOLOGY:** The science of fossils of ancient life forms and their evolution.

**PANGAEA:** Supercontinent that coalesced in the latest Paleozoic and comprised all present continents. The breakup of Pangaea began in Mesozoic time, as inferred from *paleomagnetic* and other data.

**PARTIAL MELT:** The product produced during *partial melting.*

**PARTIAL MELTING:** A process in which heating causes a mass of rock to become partially molten. Partial melting occurs because the *minerals* that compose a rock melt at different temperatures.

**PASSIVE MARGIN:** A *continental margin* characterized by thick, flat-lying, shallow-water *sediments* with only limited tectonism related to divergent *plate* motion.

**PEAT:** A marsh or swamp deposit of water-soaked plant remains containing more than 50 percent carbon.

**PEDALFER:** A common *soil* type in humid regions, characterized by an abundance of iron oxides and *clay* minerals deposited in the *B-horizon* by leaching.

**PEDIMENT:** A planar, sloping rock surface forming a ramp up to the front of a mountain range in an arid region. It may be covered locally by thin alluvial deposits.

**PEDOCAL:** A common *soil* type of arid regions, characterized by accumulation of calcium carbonate in the *A-horizon*.

**PELAGIC SEDIMENT:** A deep-sea *sediment* composed of fine-grained detritus that slowly settles from surface waters. Common constituents are *clay*, *foraminiferal ooze*, and *silica ooze*.

**PERCHED WATER TABLE:** The upper surface of an isolated body of *groundwater* that is perched above and separated from the main body of groundwater by an *aquiclude*.

**PERIDOTITE:** A coarse-grained *mafic igneous rock* composed of olivine with small amounts of pyroxene and amphibole.

**PERIOD (GEOLOGIC):** The most commonly used unit of geologic time, representing one subdivision of an *era*.

**PERIOD (WAVE):** The time interval between the arrival of successive crests in a homogeneous wave train.

**PERMAFROST:** A permanently frozen aggregate of ice and *soil* occurring in very cold regions.

**PERMEABILITY:** The ability of a *formation* to transmit *groundwater* or other fluids through pores and cracks.

**PHENOCRYST:** A large *crystal* surrounded by a finer matrix in an *igneous rock*. Igneous rocks that contain abundant phenocrysts are called *porphyrys*.

**PHOSPHORITE:** A *sedimentary rock* composed largely of calcium phosphate, usually as a variety of the mineral apatite and largely in the form of concretions and nodules. Primary ore of phosphate *minerals* and elemental phosphorus.

**PHREATIC EXPLOSION:** A volcanic eruption of steam, mud, and debris caused by the expansion of steam formed when *magma* comes in contact with *groundwater*.

**PHYLLITE:** A *metamorphic rock* that is intermediate in grade between slate and mica *schist*. Small *crystals* of micas give a silky sheen to the *cleavage* surfaces.

**PHYSICAL WEATHERING:** The set of all physical processes by which an *outcrop* is broken up into smaller particles.

**PILLOW LAVA:** A type of *lava* formed underwater, in which many small, pillow-shaped tongues break through the chilled surface and quickly solidify, leading to a rock *formation* resembling a pile of sandbags.

**PLACER:** A *clastic* sedimentary deposit of a valuable *mineral* or native metal in unusually high concentration, usually segregated because of its greater *density*.

**PLANETARY EVOLUTION:** The process by which a differentiated planet is formed.

**PLASTIC FLOW:** Deformation of the shape or volume of a substance without fracturing.

**PLATE:** One of the dozen or more segments of the *lithosphere* that ride as distinct units over the *asthenosphere*.

**PLATEAU:** An extensive upland region at high *elevation* with respect to its surroundings.

**PLATE TECTONICS:** The theory and study of *plate* formation, movement, interactions, and destruction; the attempt to explain *seismicity*, volcanism, mountain-building, and *paleomagnetic* evidence in terms of plate motions.

**PLATFORM:** A sediment-covered, tectonically stable, almost level region of a continent.

**PLAYA, PLAYA LAKE:** The flat floor of a closed basin in an arid region. It may be occupied by an intermittent lake.

**PLUME:** Hypothetical rising jet of hot, partially molten material, perhaps emanating from the *mantle*, believed by some to be responsible for intraplate volcanism.

**PLUNGING FOLD:** A *fold* whose *axis* is not horizontal but *dips*. Thus, progressively younger strata are found at the center of the fold as one travels along the direction of plunge, and the geologic map pattern is one of nested V-shaped *outcrops* of *formations*.

**PLUTON:** A large igneous intrusion, formed at depth in the *crust*.

**PLUTONIC:** Relating to igneous activity at depth in the *crust*.

**PLUTONISM:** Igneous activity at depth in the *crust*.

**POINT BAR:** A deposit of *sediment* on the inner bank of a *meander* that forms because the *stream* velocity is lower against the inner bank.

**POLYMORPH:** One of two or more alternative possible structures for a single chemical compound; for example, the minerals calcite and aragonite are polymorphs of calcium carbonate ($CaCO_3$).

**POROSITY:** The percentage of the total volume of a rock that is pore space (not occupied by *mineral* grains).

**PORPHYROBLAST:** A large *crystal* in a finer-grained matrix in a *metamorphic rock*; analogous to a *phenocryst* in an *igneous rock*.

**PORPHYRY:** An *igneous rock* containing abundant *phenocrysts*.

**POTABLE WATER:** Water that is agreeable to the taste and not dangerous to the health.

**POTHOLE:** A hemispherical hole in the *bedrock* of a streambed, formed by abrasion of small pebbles and cobbles in a strong current.

**PRINCIPLE OF ISOSTASY:** See *Isostasy, principle of*.

**PRINCIPLE OF ORIGINAL HORIZONTALITY:** See *Original horizontality, principle of*.

**PRINCIPLE OF SUPERPOSITION:** See *Superposition, principle of*.

**PRINCIPLE OF UNIFORMITARIANISM:** See *Uniformitarianism, principle of*.

**PROTON:** An elementary particle in the atomic nucleus with a positive charge of $1.602 \times 10^{-19}$ coulomb and a mass of 1836 *electrons*.

**PROVEN RESERVES:** See *Reserves*.

**PUMICE:** A form of volcanic *glass*, usually of *felsic* composition, so filled with holes from the escape of gas during quenching that it resembles a sponge and has very low *density*.

**P WAVE:** The primary or fastest wave traveling away from a seismic event through the solid rock and consisting of a train of compressions and dilations of the material.

**PYROCLASTIC FLOW:** A mixture of *volcanic ash*, fragments of volcanic rock, and gases that moves rapidly downhill away from the eruptive center during a volcanic eruption. Synonym of *volcanic ash flow*, but used in a more general sense.

**PYROCLASTIC ROCK:** A rock formed by the accumulation of fragments of volcanic rock scattered by volcanic explosions.

**PYROXENE GRANULITE:** A coarse-grained *regional metamorphic* rock containing pyroxene; formed at high temperatures and pressures deep in the *crust*.

**QUAD:** A unit of energy equal to $10^{15}$ British thermal units (Btu).

**QUARTZITE:** A very hard, nonfoliated, white metamorphic rock formed from a *sandstone* rich in quartz sand grains and quartz cement.

**RADIAL DRAINAGE:** A system of *streams* running in a radial pattern away from the center of a circular *elevation*, such as a volcano or *dome*.

**RADIOACTIVITY:** The emission of energetic particles and/or radiation during radioactive decay.

**RADIOMETRIC DATING:** The method of obtaining ages of geological materials by measuring the relative abundances of radioactive parent and daughter *isotopes* in them.

**RAIN SHADOW:** An area of low rainfall on the leeward slope of a mountain range.

**REACTION SERIES:** A series of chemical reactions occurring in a cooling *magma* by which a *mineral* formed at high temperature becomes unstable in the melt and reacts to form another mineral.

**RECHARGE:** In *hydrology*, the replenishment of *groundwater* by infiltration of *meteoric water* through the *soil*.

**RECTANGULAR DRAINAGE:** A system of *streams* in which each straight segment of each stream takes one of two characteristic perpendicular directions, usually following sets of *joints*.

**RECURRENCE INTERVAL:** The average time interval between occurrences of a geological event, such as floods and *earthquakes*, of a given or greater *magnitude*.

**REEF:** A mound or ridge-shaped organic structure that is built by calcareous organisms, is wave resistant, and stands in relief above the surrounding seafloor.

**REGIONAL METAMORPHISM:** Metamorphism occurring over a wide area and caused by deep burial or strong tectonic forces of the Earth.

**REGOLITH:** The layer of loose, heterogeneous material lying on top of *bedrock*; includes *soil*, unweathered fragments of parent rock, and rock fragments weathered from the bedrock.

**REJUVENATION (OF MOUNTAINS):** Renewed uplift in a mountain chain on the site of earlier uplifts, returning the area to a more youthful stage of the *cycle of erosion*.

**RELATIVE AGE:** The age of a geologic event or feature relative to other geologic events or features and expressed in terms of the *geologic time scale*. (Compare *Absolute age*.)

**RELATIVE HUMIDITY:** The amount of water vapor in the air, expressed as a percentage of the total amount of water vapor that the air could hold at that temperature if saturated.

**RELIEF:** The maximum regional difference in *elevation*.

**REMANENT MAGNETIZATION:** See *Depositional remanent magnetization*; *Thermoremanent magnetization*.

**RESERVES:** Deposits of *minerals*, *coal*, or oil and gas that have been shown to be extractable profitably with existing technology. "Proven reserves" are those for which good estimates of the quantity and quality have been made. (See also *Resources*.)

**RESERVOIR:** A source or place of residence for elements in a chemical cycle or *hydrologic cycle*.

**RESOURCES:** Discovered and undiscovered deposits of *minerals*, *coal*, or oil and gas that are or may become available for use in the future; includes *reserves*, plus discovered deposits not now commercially or technologically extractable, plus undiscovered deposits that may be inferred to exist. (See also *Reserves*.)

**RETROGRADE METAMORPHISM:** Metamorphism in which a rock that has been metamorphosed to a fairly high grade is later remetamorphosed at lower temperature and pressure to a lower grade.

**RHYOLITE:** The fine-grained volcanic or extrusive equivalent of *granite*, light-brown to gray and compact.

**RICHTER MAGNITUDE:** See *Magnitude*.

**RIFT VALLEY:** A *fault* trough formed at a *divergent plate boundary* or other area of tension.

**RIPPLE:** A very small *dune* of sand or silt whose long dimension is formed at right angles to the current.

**RIVER:** A general term for a relatively large *stream*, or the main branches of a stream system.

**ROCK AVALANCHE:** The rapid downhill *mass movement* of broken rock material, during which further breakage of the material may occur.

**ROCK CYCLE:** The geologic cycle, with emphasis on the rocks produced: *sedimentary rocks* are metamorphosed to *metamorphic rocks* or melted to create *igneous rocks*, and all rocks may be uplifted and eroded to make *sediments*, which lithify to sedimentary rocks.

**ROCKFALL:** The relatively free falling of a newly detached segment of *bedrock* from a cliff or other steep slope.

**ROCK FLOUR:** A glacial *sediment* of extremely fine (silt- and clay-size) ground rock formed by abrasion of rocks at the base of the *glacier*.

**ROCKSLIDE:** The *mass movement* of large blocks of detached *bedrock* sliding more or less as a unit.

**RUNOFF:** The amount of rainwater directly leaving an area in surface drainage, as opposed to the amount that seeps out as *groundwater*.

**SALTATION:** The movement of sand or fine *sediment* by short jumps above the ground or streambed under the influence of a current too weak to keep it permanently suspended.

**SANDBLASTING:** A *physical weathering* process in which rock is eroded by the impact of sand grains carried by the wind, frequently leading to *ventifact* formation of pebbles and cobbles.

**SANDSTONE:** A *clastic rock* composed of grains from 0.0625 to 2 mm in diameter, usually quartz, feldspar, and rock fragments, bound together by a cement of quartz, carbonate, or other *minerals*, or by a matrix of *clay* minerals.

**SATURATED ZONE:** The zone of *soil* and rock in which pores are completely filled with *groundwater*.

**SCHIST:** A *metamorphic rock* characterized by strong *foliation* or *schistosity*.

**SCHISTOSITY:** The parallel arrangement of sheety or prismatic *minerals* like micas and amphiboles resulting from metamorphism.

**SEAFLOOR SPREADING:** The mechanism by which new seafloor is created at ridges at *divergent plate boundaries* and adjacent *plates* are moved apart to make room. This process may continue at a few centimeters per year through many geologic *periods*.

**SEAMOUNT:** An isolated tall mountain on the seafloor that may extend more than 1 km from base to peak.

**SEDIMENT:** Any of a number of materials deposited at Earth's surface by physical agents (such as wind, water, and ice), chemical agents (precipitation from oceans, lakes, and rivers), or biological agents (organisms, living and dead).

**SEDIMENTARY BASIN:** A region of considerable extent (at least 10,000 km²) that is the site of accumulation of a large thickness of *sediments*.

**SEDIMENTARY BRECCIA:** A *clastic rock* composed mainly of large angular fragments.

**SEDIMENTARY ENVIRONMENT:** A geographically limited area where *sediments* are preserved; characterized by its *landforms*, relative energy of currents, and chemical equilibria.

**SEDIMENTARY ROCK:** A rock formed by the accumulation and *cementation* of *mineral* grains by wind, water, or ice transportation to the site of deposition or by chemical precipitation at the site.

**SEDIMENTARY STRUCTURE:** Any structure of a sedimentary or weakly metamorphosed rock that was formed at the time of deposition; includes *bedding, cross-bedding, graded bedding, ripples,* scour marks, mudcracks.

**SEISMIC GAP METHOD:** A predictive model for *earthquake* occurrences along active *fault* zones based on the study of segments that have experienced little or no movement and are thought to be under high *stress*.

**SEISMICITY:** The worldwide or local distribution of *earthquakes* in space and time; a general term for the number of earthquakes in a unit of time.

**SEISMIC SURFACE WAVE:** A *seismic wave* that follows the Earth's surface only, with a speed less than that of *S waves*.

**SEISMIC WAVE:** A general term for the elastic waves produced by *earthquakes* or explosions. See *P wave, S wave,* and *Seismic surface wave*.

**SEISMOGRAPH:** An instrument for magnifying and recording the motions of the Earth's surface that are caused by *seismic waves*.

**SETTLING VELOCITY:** The rate at which a sedimentary particle of a given size falls through water or air.

**SHADOW ZONE:** A zone 105° to 142° from the *epicenter* of an *earthquake* in which there is no penetration of *seismic waves* through the Earth because of *wave refraction* or because the waves are not transmitted upon entering the liquid *core*.

**SHALE:** A very fine grained *clastic rock* composed of silt and *clay* that tends to part along *bedding* planes.

**SHEARING FORCES:** Forces that deform a body so that parts of the body on opposite sides of a plane slide past each other; that is, forces acting tangentially to the plane. Shearing forces dominate at *transform fault plate boundaries*.

**SHIELD:** A large region of stable, ancient *basement* rocks within a continent.

**SHIELD VOLCANO:** A large, broad volcanic cone with very gentle slopes built up by nonviscous basaltic *lavas*.

**SHORELINE:** The straight or sinuous, smooth or irregular interface between land and sea.

**SILICA OOZE:** A *pelagic sediment* consisting of the remains of tiny organisms that have shells made of amorphous silica.

**SILICEOUS ROCK:** A rock containing abundant free silica of either organic or inorganic origin, formed by biochemical, chemical, or physical deposition of silica.

**SILL:** A horizontal tabular intrusion with *concordant intrusive* contacts.

**SILTSTONE:** A *clastic rock* that contains mostly silt-sized material, from 0.0039 to 0.062 mm.

**SINKHOLE:** A small, steep depression caused in *karst topography* by the dissolution and collapse of subterranean caverns in carbonate formations.

**SLIP (FAULT):** The motion of one face of a *fault* relative to the other.

**SLIP FACE:** The steep downwind face of a *dune* on which sand is deposited in *cross-beds* at the *angle of repose*.

**SLUMP:** A slow *mass movement* of *unconsolidated materials* that slide as a unit.

**SOIL:** The surface accumulation of sand, *clay*, and humus that composes the *regolith*, but excluding the larger fragments of unweathered rock.

**SOLIFLUCTION:** The *creep* of soil saturated with water and/or ice, caused by alternate freezing and thawing; most common in polar regions.

**SORTING:** A measure of the homogeneity of the sizes of particles in a *sediment* or *sedimentary rock*.

**SOURCE BED:** *Organic sediment* or rock that liberates oil or gas when heated during burial. Usually source beds are organic-rich "black" *shales* or *limestones*.

**SPECIFIC GRAVITY:** The ratio of the *density* of given substance to the density of water.

**SPHEROIDAL WEATHERING:** The formation of spherical residual inner cores by the *weathering* of boulders.

**SPIT:** A long range of sand deposited by *longshore currents* and *longshore drift* where the *coast* takes an abrupt inward turn. It is attached to land at the upstream end.

**SPRING TIDE:** A *tide* cycle of unusually large amplitude that occurs twice monthly when the lunar and solar tides are in phase. (Compare *Neap tide*.)

**STACKS:** Isolated rocky prominences or pinnacles left standing above a marine platform as erosional remnants.

**STALACTITE:** An icicle- or toothlike deposit of calcite or aragonite hanging from the roof of a cave. It is deposited by evaporation and precipitation from solutions seeping through *limestone*.

**STALAGMITE:** An inverted icicle-shaped deposit that builds up on a cave floor beneath a *stalactite* and is formed by the same process as a stalactite.

**STOCK (VOLCANIC):** An intrusion with the characteristics of a *batholith* but less than 100 km² in area.

**STRATIFICATION:** The characteristic layering or *bedding* of *sedimentary rocks*.

**STRATIGRAPHIC SEQUENCE:** A set of deposited beds that reflects the changing conditions and *sedimentary environments* that define the geologic history of a region.

**STRATIGRAPHY:** The science of description, correlation, and classification of strata in *sedimentary rocks,* including the interpretation of the *sedimentary environments* of those strata.

**STRATOVOLCANO:** A volcanic cone consisting of both *lava* and *pyroclastic rocks*. Synonym of *Composite volcano*.

**STREAK:** The fine deposit of mineral dust left on an abrasive surface when a *mineral* is scraped across it; especially the characteristic color of the dust.

**STREAM:** A general term for any body of water that moves under the force of gravity in a relatively narrow *channel*. (Compare *River*.)

**STREAM PIRACY:** The *erosion* of a *divide* between two *streams* by the more *competent* stream, leading to the capture of all or part of the drainage of the slower stream by the faster.

**STRESS:** A quantity describing the forces acting on each part of a body in units of force per unit area.

**STRIATION:** See *Glacial striation*.

**STRIKE:** The angle between true north and the horizontal line contained in any planar feature (inclined bed, *dike*, fault plane, and so forth); also the geographic direction of this horizontal line.

**SUBDUCTION:** The sinking of an oceanic *plate* beneath an overriding plate.

**SUBDUCTION ZONE:** The zone between a sinking oceanic *plate* and an overriding plate, descending away from a *trench* and characterized by high *seismicity*. (See also *Convergent plate boundary*.)

**SUBLIMATION:** A phase change between the solid and gaseous states without passing through the liquid state.

**SUBMARINE CANYON:** An underwater canyon in the *continental shelf*.

**SUBMARINE FAN:** A *terrigenous*, cone- or fan-shaped deposit located at the foot of a *continental slope*, usually seaward of large *rivers* and *submarine canyons*.

**SUBSIDENCE:** A gentle *epeirogenic* movement where a broad area of the *crust* sinks without appreciable deformation.

**SUPERPOSED STREAM:** A *stream* that flows through resistant *formations* because its course was established at a higher level on uniform rocks before downcutting began.

**SUPERPOSITION, PRINCIPLE OF:** The principle that, except in extremely deformed strata, a bed that overlies another bed is always the younger.

**SURF:** The foamy, bubbly surface of water waves as they break close to shore.

**SURFACE TENSION:** The attractive force between molecules at a surface.

**SURFACE WAVE:** See *Seismic surface wave*.

**SURF ZONE:** An offshore belt along which the waves collapse into breakers as they approach the shore.

**SURGE:** See *Glacier surge; Tidal surge*.

**SUSPENDED LOAD:** The fine *sediment* kept suspended in a *stream* because the *settling velocity* of the sediment is lower than the upward velocity of eddies.

**SUTURE:** A zone of intensely deformed rocks that marks the boundary where two continents collided.

**SWASH:** The landward rush of water from a breaking wave up the slope of the beach.

**S WAVE:** The secondary *seismic wave*, which travels more slowly than the *P wave* and consists of elastic vibrations transverse to the direction of travel. S waves cannot penetrate a liquid.

**SWELL:** An oceanic water wave with a *wavelength* on the order of 30 m or more and a *wave height* of approximately 2 m or less that may travel great distances from its source.

**SYNCLINE:** A large downfold, whose *limbs* are higher than its center. (Compare *Anticline*.)

**SYNCRUDE:** Synthetic oil produced from *coal*.

**TALUS:** A deposit of large angular fragments of physically weathered *bedrock*, usually at the base of a cliff or steep slope.

**TAR SAND:** A sandy deposit of organic matter impregnated with a tarry substance made up mostly of hydrocarbons, from which petroleum can be extracted.

**TENSIONAL FORCES:** Forces that stretch a body and pull it apart. Tensional forces dominate at *divergent plate boundaries*.

**TERRACE (STREAM VALLEY):** A flat, steplike surface above the *floodplain* in a stream *valley*, marking a former floodplain that existed at the higher level before regional uplift or an increase in *discharge* caused the *stream* to erode into the former floodplain.

**TERRANE:** See *Microplate terrane*.

**TERRIGENOUS SEDIMENT:** *Sediment* eroded from the land surface.

**TEXTURE (ROCK):** The rock characteristics of grain or *crystal* size, size variability, rounding or angularity, and preferred orientation.

**THERMOREMANENT MAGNETIZATION:** A permanent magnetization acquired by *minerals* in *igneous rocks* during *crystallization*.

**TIDAL FLAT:** A broad, flat region of muddy or sandy *sediment*, covered and uncovered in each *tide* cycle.

**TIDAL SURGE:** Waves that overrun a beach and batter sea cliffs when an intense storm passes near the shore during a *spring tide*.

**TIDE:** The rise and fall of the water level of the ocean that occurs twice a day and is caused by the gravitational attraction of the Moon and, to a lesser degree, the Sun, with greater force on the parts of the Earth facing and opposite the Moon (and Sun).

**TILL:** An unstratified and poorly sorted *sediment* containing all sizes of fragments from *clay* to boulders, deposited by glacial action.

**TILLITE:** The lithified equivalent of *till*.

**TOPOGRAPHY:** The shape of the Earth's surface, above and below sea level; the set of *landforms* in a region; the distribution of *elevations*.

**TOPSET BED:** A horizontal sedimentary bed formed at the top of a *delta* and overlying the *foreset beds*.

**TRACE ELEMENT:** An element that appears in a *mineral* in a concentration of less than 1 percent (often less than 0.001 percent).

**TRANSFORM FAULT PLATE BOUNDARY:** A boundary at which *plates* slide horizontally past each other.

**TRANSPIRATION:** The release of water vapor by plants into the atmosphere.

**TRANSVERSE DUNE:** A *dune* that has its axis perpendicular (transverse) to the prevailing winds or to a current. The upwind or upcurrent side has a gentle slope, and the downwind or downcurrent side lies at the *angle of repose*.

**TRAP (OIL):** A tectonic or *sedimentary structure* that impedes the upward movement of oil and gas and allows it to collect beneath the barrier.

**TRELLIS DRAINAGE:** A system of *streams* in which *tributaries* tend to lie in parallel valleys formed in steeply dipping beds in folded belts.

**TRENCH:** A long, narrow, deep trough in the seafloor; marks the line along which a *plate* bends down into a *subduction zone*.

**TRIBUTARY:** A *stream* that discharges water into a larger stream.

**TSUNAMI:** A large destructive wave caused by seafloor movements in an *earthquake*.

**TUFF:** See *Volcanic tuff*.

**TURBIDITE:** The sedimentary deposit of a *turbidity current*, typically showing *graded bedding*.

**TURBIDITY CURRENT:** A mass of mixed water and *sediment* that flows downhill along the bottom of an ocean or lake because it is denser than the surrounding water. It may reach high speeds and erode rapidly.

**TURBULENT FLOW:** A high-velocity flow in which streamlines are neither parallel nor straight but curled into small tight eddies. (Compare *Laminar flow*.)

**ULTRAMAFIC ROCK:** An *igneous rock* consisting mainly of *mafic* minerals and containing less than 10 percent feldspar. Includes *peridotite, amphibolite, dunite,* and pyroxenite.

**UNCONFINED AQUIFER:** An *aquifer* that is not overlain by an *aquiclude,* thereby causing the level of water in a well that penetrates the aquifer to be at the level of the surrounding *groundwater table.* (Compare *Confined aquifer*.)

**UNCONFORMITY:** A surface that separates two strata. It represents an interval of time in which deposition stopped, *erosion* removed some *sediments* and rock, and deposition resumed. (See also *Angular unconformity*.)

**UNCONSOLIDATED MATERIAL:** Unlithified *sediment* that has no *mineral* cement or matrix binding its grains.

**UNIFORMITARIANISM, PRINCIPLE OF:** The concept that the processes that have shaped Earth through geologic time are the same as those observable today.

**UNSATURATED ZONE:** The region in the ground between the surface and the *groundwater table* in which pores are not filled with water.

**UPWARPED MOUNTAINS:** Mountains elevated by uplift of broad regions without *faulting*.

**U-SHAPED VALLEY:** A deep valley with steep upper walls that grade into a flat floor; typical shape of a valley eroded by a *glacier*.

**VALLEY (STREAM):** The entire area between the top of the slopes on either side of a *stream*.

**VALLEY GLACIER:** A *glacier* that is smaller than a *continental glacier* or an icecap and that flows mainly along well-defined valleys in mountainous regions.

**VARVE:** A thin pair of sedimentary layers grading upward from coarse to fine and light to dark, found in a glacial lake and representing one year's deposition.

**VEIN:** A deposit of foreign *minerals* within a rock fracture or *joint*.

**VENTIFACT:** A rock that exhibits the effects of *sandblasting* or "snowblasting" on its surfaces, which become flat with sharp edges in between.

**VISCOSITY:** A measure of a liquid's resistance to flow.

**VOLCANIC ASH:** A volcanic *sediment* of rock fragments, usually *glass*, less than 4 mm in diameter that is formed when escaping gases force out a fine spray of *magma*.

**VOLCANIC ASH-FLOW DEPOSIT:** A layer of *volcanic ash* and debris deposited during a *pyroclastic flow*.

**VOLCANIC BRECCIA:** A *pyroclastic rock* in which all fragments are more than 2 mm in diameter.

**VOLCANIC DOME:** A rounded accumulation around a volcanic vent of congealed *lava* too viscous to flow away quickly; hence usually *rhyolite* lava.

**VOLCANIC TUFF:** A consolidated rock composed of *pyroclastic rock* fragments and fine *volcanic ash*. If particles are melted slightly together from their own heat, it is a "welded tuff."

**WADI:** A steep-sided valley containing an intermittent *stream* in an arid region.

**WATER TABLE:** See *Groundwater table*.

**WAVE-CUT TERRACE:** A level surface formed by wave *erosion* of coastal *bedrock* to the bottom of the turbulent breaker zone. May appear above sea level if uplifted or if sea level drops.

**WAVE HEIGHT:** The vertical distance from the trough to the crest of a wave.

**WAVELENGTH:** The distance between two successive peaks, or between troughs, of a wave.

**WAVE REFRACTION:** The bending of water waves as they encounter different depths and bottom conditions, or of other waves as they pass from one medium to another of different properties.

**WEATHERING:** The set of all processes that decay and break up *bedrock,* by a combination of physical fracturing and chemical decomposition.

**WORLD OCEAN:** The combination of all the individual oceans (Atlantic, Pacific, and so on) considered as a single interconnected body of water.

**YARDANG:** A streamlined, sharp-crested ridge aligned with the direction of the prevailing wind in arid regions. Yardangs appear to have been carved by wind *erosion* and abrasion by silt and dust carried by the wind.

**ZEOLITE:** A class of silicate *minerals* containing water in cavities within the *crystal* structure. Formed by alteration at low temperature and pressure of other silicates, often volcanic *glass*.

**ZONED CRYSTAL:** A single *crystal* of one *mineral* that has a different chemical composition in its inner and outer parts; formed in minerals that can have variation in abundance of some elements and caused by the changing concentration of elements in a cooling *magma*.

# INDEX

Page numbers in **boldface** refer to a definition; page numbers followed by an asterisk indicate an illustration or table. The appendixes and glossary are not covered by this index.

Aa, 93–94, **94,** 94★
Ablation, **332,** 332★, 332–333, 333★
Abyssal hills, **385,** 385★
Abyssal plain, **385,** 385★
Accumulation, 332, 333★
Acid, **125**
Acid rain, 126–127, 127★
Active margins, **390,** 390★
Adirondack Mountains, New York, 491
  and acid rain, 126
Adriatic Sea, 496
African Plate
  and Appalachian orogeny, 486, 488★, 488–490, 489★
  rate of motion, 460, 460★
Agassiz, Louis, glacial study by, 335–336, 344–345
Age of Enlightenment, 404
Ages (of geologic phenomena)
  absolute, **189,** 197, 200–204
  absolute, and geologic time scale, 203–204
  absolute, from paleomagnetism, 203
  of Earth, early estimates, 197
  relative, **189**
  relative, and field relationships, 194–195, 195★
  relative, and fossils, 190–192
  and stratigraphic record, 189–190
Agriculture
  and soils, 136–138
  and water use, 256
Alaska oil pipeline and permafrost, 344
Alaska Range, 356★
Alberta, Canada, tar sands in, 512
Alcohol (as fuel), 503, 509
Aleutian Islands, Alaska, 87
Algae, and carbonate sedimentation, 163
Alluvial fans, **294,** 294★
  in deserts, 320
Alps, (European), 468, 490, 491
Altiplano, Bolivia, 356–357
Aluminum, as resource, 524, 524★
Aluminum ore, bauxite, **128**
Aluminum oxides, and soil classification, 137–138, 138★
Amazon, soils, 138–139
Amazon River, 361
  delta, 298
  flows in, 258, 258★
Ammonites, 191★

Amphiboles
  chemical stability and weathering, 131★
  chemical weathering of, 128–130
  and contact metamorphism, 182–183
  and discontinuous reaction series, 77–78
  and foliation, 175
  in gneisses, 177
  in metamorphic rocks, 180
  and metamorphism, 171
Amphibolites, 179★, **180,** 180★, 181★
Anak Krakatoa, Indonesia, 95★, 105★
Anchorage, Alaska, landslide from earthquake of 1964, 236, 236★
Andalusite, and contact metamorphism, 182–183
Andes Mountains, 16, 453, 468, 473
  debris avalanches in, 242
  glaciation in, 330, 331★
  mudflows in, 232
  and volcanism, 111
Andesite, **71**
  distribution, and plate tectonics, 462–470
  and island arcs, 110
Angle of repose, **233,** 234★
Angular unconformity, **192,** 192★, 194★
Anhydrite, 40
Anions, **27**
  and mineral classification, 46
  size of, 34, 34★
Antarctica
  chemical weathering in, 130
  continental glacier in, 329★, 329–330, 331★
Antarctic Plate, rate of motion, 460, 460★
Anticlines, **217,** 217★, 217–219
  as oil traps, 505★, 505–506
  plunging, **218,** 219★★
Appalachian Mountains, 473, 491. *See also* Valley and Ridge belt, Appalachians.
  microplate terranes in, 470
  orogeny of, 486, 488★, 488–490, 489★
  and plate-tectonic reconstruction, 471
Aquicludes, **262,** 262–263, 263★
Aquifers
  confined, **262,** 262–263, 263★
  unconfined, **262,** 262–263, 263★

Aragonite. *See also* Carbonate minerals, rocks.
  in sedimentary rocks, 162
Archean eon, **196**
Arches National Park, Utah, 222★
Argillites, **178**
Argon, radiometric dating with, 200, 201★
Arid regions. *See* Deserts.
Arkose, **160,** 160★
Armenia fault scarp, 227★
Arrhenius, Svante, 348
Artesian flows, **262,** 262–263, 263★
Artesian wells, **263,** 263★
Asbestos, **44,** 44–45, 45★
  and health, 44–45
Aseismic ridges, **113**
Ash. *See* Volcanic ash.
Asia, and Himalayan orogeny, 3★, 484★, 485, 485★, 486, 487★
Asthenosphere, **14**
  studying with seismic waves, 433★, 436★, 436–437
  temperatures in, 92
Athabaska tar sands, 512
Atlantic Ocean
  coastal plain, 493–494, 494★
  continental shelf deposits, 465, 466★
  creation of, 461★
  passive margin profile, 390, 390★
  profile of, 384★, 384–387, 385★, 386★
Atlantis (lost continent), 114
Atmosphere
  carbon dioxide, abundance of, 125–126
  circulation of, 307, 307★
  circulation, and deserts, 317–318, 318★
  early composition, 12–13
  evolution of oxygen abundance, 196
  origin of, 11, 12–13
  and volcanic gases, 107–108
Atolls, **395,** 395–396, 396★
Atomic number, **25**
  and periodic table, 28–29
Atomic structure, 25–31
Atomic weight, **26**
Atoms, **25,** 25–26, 26★

Bacon, Sir Francis, and continental drift, 448
Bacteria, effect on weathering, 121

Badlands, **359**
Badlands, South Dakota, 360★
Bahama Islands, carbonate
　　sedimentation in, 162–164
Baja California, 453, 453★
Banded iron formations, **533**, 533★
Banff Sulfur Springs, 272
Bangladesh
　　floods in, 290
　　and possible sea level rise, 347
Barchans, 315★, **316**
Barrier islands, **381**, 382★
Basal slip, of glaciers, **334**, 335
Basalt, 70★, **71**
　　chemical analysis of, 57★
　　columnar, 65★
　　at convergent boundaries, 111
　　distribution, and plate tectonics,
　　　462–470
　　at divergent boundaries, 462, 462★,
　　　464★, 464–465, 465★
　　flood basalts, **106**, 106★
　　Iceland, 110
　　and island arcs, 110
　　lava flows, 93–94
　　and magmatic differentiation, 80–82
　　metamorphism of, 180, 180★
　　ocean ridges, 110
Basement rocks, **216**
Basin and Range province
　　and Cordilleran orogeny, 490–493,
　　　491★, 492★
　　interpretating deformation in,
　　　225–226, 226★
Basins (sedimentary), **483**, 483★. *See
　　also* Epeirogeny.
　　and continental shelf deposits, 465,
　　　466★
　　as landforms, **360**
Batholiths, **83**
　　and continental growth, 485
　　distribution, and plate tectonics, 467
Bauxite, **128**
Beaches, **377**, 377★, 377–382, 378★,
　　382★
　　backshore, **378**, 378★
　　barrier islands, **381**, 382★
　　and eolian dunes, 312★, 312–317
　　foreshore, **378**, 378★
　　offshore, **377**, 377–378, 378★
　　sand budget, 378★, 378–379
　　spits, **381**, 382★
　　surf zone, **372**, 372–373, 373★, 378,
　　　378★
Becquerel, Henri, 197
Bed load, **280**, 280★
Bedding, **55**, **156**, 156–157
　　and foliation, 174–176
Bedding sequences, **158**, 158★
Bedforms, in streams, **281**, 281–282,
　　282★
Bermuda, 385
　　and windblown carbonate sand, 310
Bicarbonate, chemical weathering of
　　carbonate, 130

Big Bang, 4
Big Bend National Park, Texas, 84, 85★
Biochemical sediments, rocks. See
　　Sediments, sedimentary rocks:
　　biochemical.
Biomass, **515**, 517
Bioturbation, **157**, 157–158, 158★
Black Hills, South Dakota, 220
　　and epeirogeny, 494, 495★
Black smokers, **393**
Blowout, 315★, **316**
Blue Ridge province, Appalachians,
　　486, 488★, 488–490, 489★
Blueschists, **180**, 180★, 181★
　　and plate tectonics, 183
Boston, Massachusetts, earthquake of
　　1755, 416
Bowen, N. L., 78–82
Bowen's reaction series. **81**, 81★
　　chemical stability and weathering,
　　131★, 131–132
Brahmaputra River, floods of, 290
Braided streams, **173**, 173★
Breccias
　　sedimentary, **161**
　　volcanic, **96**, 96★
Brittle materials, **216**
Burchfiel, B. C., 480
Burgess Shale, Canadian Rockies, 187★
Burial. *See also* Epeirogeny.
　　and metamorphism, 60
　　and rock cycle, 59★, 60

Calcite, 38, 39★. *See also* Carbonate
　　minerals, rocks.
　　acid test, 46★, 47
　　chemical stability and weathering,
　　130, 131★
　　and cleavage, 41–42, 42★
　　from hydrothermal waters, 272
　　in sedimentary rocks, 162–164
　　structure, 39★
Calcium, ocean chemistry and
　　sedimentation, 149
Calderas, 102★, **103**. *See also* Volcanoes.
　　Crater Lake, Oregon, 103, 114★
　　Kilauea, Hawaii, 99, 99★, 103, 103★,
　　　114★
　　Long Valley, California, 103, 114★
　　Valles, New Mexico, 103
　　Yellowstone, 103, 114★
Caledonian Mountains, 473, 490
California, earthquake prediction, 424★,
　　424–425
California, Gulf of, 453, 453★
Canada
　　permafrost in, 344
　　soils, 137–138
Canadian Rockies, 19★
Canadian Shield, 364★, **482**, 482★,
　　482–483
Cape Cod, Massachusetts, saltwater
　　intrusion, 265
Cape Hatteras, North Carolina, 380★,
　　380–381

Carbohydrates, and photosynthesis,
　　204–205
Carbon, organic
　　burial, 204–205
　　and photosynthesis, 204–205
Carbon, radiometric dating with, 200,
　　201★
Carbon dioxide
　　in atmosphere, and glacial ice, 332
　　and dissolution of limestone, 269
　　and global warming, 514
　　and greenhouse effect, 348–349, 349★
　　and organic matter combustion, 205
　　and petroleum combustion, 504,
　　　504★, 509
　　and photosynthesis, 204–205
　　and weathering reactions, 121,
　　　125–130
Carbonate compensation depth, **397**,
　　398★
Carbonate minerals, rocks, 38, 39★, 156★
　　chemical weathering of, 130
　　classification, **154**, 155★
　　diagenesis, 153–154
　　in evaporites, 152
　　precipitation in seawater, 163
　　sand dunes of, 310
　　and sediment deposition, 145–146
　　sediments, sedimentary rocks,
　　　162–164
　　and soil classification, 137–138, 138★
Carbonate sedimentary environments,
　　**151**, 151–152
Carbonic acid, **125**, 125–127
Carboniferous period, plate-tectonic
　　reconstruction, 472★
Caribbean, carbonate sedimentation in,
　　162, 164
Carlsbad Caverns, New Mexico, 269
Cascade Mountains, rain shadow in,
　　**255**, 255★
Cascade Range, Pacific Northwest, 17,
　　87, 111, 114★, 115
　　and Cordilleran orogeny, 490–493,
　　　491★, 492★
Cations, **27**
　　size, 34, 34★
　　substitution, 34–35
Caves, formation of, 267★, 267, 269
Cement, as resource, 524, 524★,
　　532–533
Cementation, **153**
Cenozoic era, **196**
Cerro Negro, Nicaragua, 101★
Chalcocite, hydrothermal vein deposits
　　of, 529, 530★
Chalcopyrite, porphyry deposits, 531
*Challenger,* H.M.S., 370
Charleston, South Carolina, earthquake
　　of 1886, 416
Chemical bonding, 31
　　chemical stability and weathering, 132
　　electron sharing, **28**, 28★, 31
　　and hardness, 41
Chemical reactions, **27**, 27–31

Chemical sediments, rocks. *See* Sediments, sedimentary rocks: chemical.
Chemical stability
  and bonding, 132
  and Bowen's reaction series, 131★
  relative weathering rates, **130,** 130–132, 131★
  of various minerals, 131★
Chemical weathering, **120,** 122–132. *See also* Weathering.
  carbonate minerals and rocks, 130
  and chemical bonding, 132
  in deserts, 318, 320
  dissolution rates, **131**
  hydration, **125**
  oxidation, **129**
  physical weathering, relation to, 120–122, 123★, 132
  rates, and chemical stability, 130–132, 131★
  rates, and soil/outcrop composition, 131
  relative rates of, various minerals, 131★
  and sandstone mineralogy, 160
  soil, effect of, 126–128
  and soils, 135–139
  solubility, **131**
  in streambeds, 282–283
  surface area, effect of, 124, 125★
  temperature, effect of, 128
  transportation, during, 147
Chernobyl, Ukraine, nuclear accident, 513
Chert, **165,** 398
China, loess deposits, 317
Chlorite
  in foliated rocks, 175, 176–177
  in greenstones, 178
  as metamorphic index mineral, 179, 179★
Cinnabar, hydrothermal vein deposits of, 529, 530★
Clastic sedimentary environments, **150,** 150★, 150–151
Clastic sediments, sedimentary rocks. *See* Sediments, sedimentary rocks: clastic.
Clay minerals
  chemical stability and weathering, 131★
  clay type and parent silicate, 128
  and fine-grained sediments, 161
  kaolinite, **124,** 124–126
  montmorillonite, 128
  and soil classification, 137–138, 138★
  as weathering products, 123–129
Clay sediments, **154,** 155★, 158–159, **161**
  as resource, 524, 524★, 532–533
Clean Air Act of 1991, and acid rain, 127
Cleavage (mineral), 41–43, **41,** 42★, 43★
Cleavage (rock), 174–175, **175**

Climate
  and carbonate sedimentation, 152
  change, 514
  change, and carbon dioxide, 348–349
  change, and glaciation, 346–350
  change, and volcanism, 108
  and glaciation, 328
  and hydrology, 254–255, 257★
  as landscape control, 360–362
  and landscape evolution, 366
  and plate-tectonic reconstruction, 471
  soil, effects on, 135–139
  topography, effect on, 361
  weathering, effect on, **121**
Coal, **149,** 166, 509–512
  acid rain production by burning, 126–127
  anthracite, **510**
  bituminous, **510**
  combustion of, 504, 504★
  diagenesis, 154
  distribution, U.S., 511★
  environmental issues, 510–512
  formation of, 509–510, 510★
  and Industrial Revolution, 503, 504★
  lignite, **509,** 509–510
  and sedimentary rock classification, 156
  subbituminous, **510**
  and syncrude, **510**
Coast Ranges. *See* Pacific Coast Ranges.
Coastal Plain, Appalachians, 486, 488★, 488–490, 489★
Coastal plains, formation of, 493–494, 494★
Coasts, **370.** *See also* Shorelines.
Cocos Plate, rate of motion, 460, 460★
Colorado Plateau
  and Cordilleran orogeny, 490–493, 491★, 492★
  and epeirogeny, 494
Colorado River
  drainage basin map, 295★
  Marble Canyon, 143★
Columbia Plateau, Pacific Northwest, 93, 94★, 106, 106★, 113
  eruptive style, 97
Compaction, **153**
Compasses. *See* Magnetism (Earth's).
Compressive forces, **215,** 215★, 215–216
Concentration factor, **525,** 525–526, 526★
Conduction (heat), from Earth's interior, **438,** 438★
Conglomerates, **154,** 155★, 158–159, **161**
Connecticut River valley, 493, 493★
Consolidated materials, **232,** 232–233
  mass wasting in, 235
Contact aureoles, **181,** 181–183
Contact metamorphism, **56,** 56★
  and granite intrusion, 67–68
Continental crust deformation, 478–497. *See also* Orogeny.
  deformed rocks, distribution of, 480, 481★

mountain building. *See* Orogeny.
regional uplift. *See* Epeirogeny.
stable interior epeirogeny. *See* Epeirogeny.
"stable" interiors, 482–483
Continental Divide, 295
Continental drift, **448,** 448–450, 449★
  fossil evidence, 448–449
  glacial evidence, 448–449
Continental glaciers, **329,** 329★
Continental margins, **390,** 390–392
  active margins, **390,** 390★
  economic assets, 390
  passive margins, **390,** 391★
  sedimentation on, 390–392
  submarine fans, **391,** 392★
  turbidites, **391,** 392★
  turbidity currents, **390,** 390–392, 391★
Continental rise, **385,** 385★
Continental shelf, **384,** 384★
  deposits, **465**
  formation of, 493–494, 494★
  and petroleum, 506
Continental slope, **384,** 384★
Continents
  cratons, **482,** 482★, 482–483
  deformation of. *See* Continental crust deformation.
  divergence in continental crust, 464–465, 466★
  formation of, 12
  growth of, 470, 470★, 471★
  platforms, 482★, **483**
  sedimentary basins, **483,** 483★
  shields, **482,** 482★, 482–483
Contours, **354,** 354★
Convection
  heat, from Earth's interior, **438,** 438–440, 439★
  mantle, **14,** 15★
  mantle, and plate tectonics, 473, 476, 476★
Convergent boundaries, **14,** 14–17, 15★, 16★, 453, 454★, 455★
  continent-continent, rocks at, 468–470, 469★
  earthquakes along, 415★, 416, 417★
  metamorphism at, 183
  and mineral resources, 536, 537★
  ocean-continent, rocks at, 467–468, 468★
  ocean-ocean, rocks at, 465, 467, 467★
  and orogenies, 483–493
  rocks at, 465, 466–470
  tectonic forces at, 215
  volcanism along, 109★, 110–111
Copper
  porphyry deposits, 531
  as resource, 524, 524★
Coral reefs, 162–164, 163★, 395–396, 396★
Cordillera, North American
  microplate terranes in, 470, 471★
  orogeny, 490–493, 491★, 492★

Core, **10,** 10★
  seismic waves and composition of,
    431
  studying with seismic waves, 433★,
    437
Coseguina, Nicaragua, 111
Cotopaxi, Ecuador, 111
Country rock, **83**
Cousteau, Jacques-Yves, 383
Covalent bond, **31**
Covelite, hydrothermal vein deposits of,
  529
Crater Lake, Oregon, 103, 114★
Cratons, **484,** 482★, 482–483
Creep, 238★, **240★, 241,** 241★
Cretaceous period, plate-tectonic events
  of, 473, 475★
Cross-bedding, **156,** 156★, 156–157, 281
Cross-cutting relationships, 193–194,
  194★
  and absolute ages, 204
Cross sections, 214, 214★
Crust, **10,** 10★
  continental, deformation of. *See*
    Continental crust deformation.
  ocean, formation of, 464, 464★
  and plate tectonics, 13–14
  studying with seismic waves, 433★,
    433–435, 434★
Crystallization, **32,** 32–34
  fractional, 78–80, **79,** 79★, 532, 536
  of magma, 76–82
Crystals, **32,** 32–34
Cuestas, **357,** 357★
Curie, Marie, 197
Cycle of erosion, 364–365, 365★

da Vinci, Leonardo, and fossils, 191
Dacite, **71**
  distribution, and plate tectonics,
    462–470
Dalton, John, 25
Dams
  and flood control, 290–291
  longitudinal profile, effect on, 293,
    293★
Darcy, Henry, 265–266
Darcy's law, 265–266, 266★
Darwin, Charles, 191–192
  and coral reefs, 396–397
Davenport, Iowa, flood of 1993, 288★
Davis, William Morris, and cycle of
  erosion, 364–365
Dead Sea, 355, 360
Dead Sea–Jordan Valley, Israel, 493
Death Valley, California
  alluvial fans in, 294★
  dune field in, 313★
  playa lake in, 321★
Debris avalanche, 241★, **242**
Debris flow, 240★, **241,** 241–242
Debris slide, 238★, **241, 243**
Deep-Sea Drilling Program, 57,
  383–384, 457–460, 458★
Deflation, **310,** 310–311, 311★

Deforestation, and mass wasting, 235,
  235★
Deformation, **212,** 213★
  of continental crust. *See* Continental
    crust deformation.
  at different depths in crust, 216
  faulting. *See* Faulting.
  folding. *See* Folding.
  forces, major types of, **215,** 215★,
    215–216
  interpreting deformed sequences,
    225–226, 226★
  and oil traps, 505★, 505–506
Deltas, 298–301, **299**
  Amazon River, 298
  bottomset beds, **299,** 299★
  distributaries, **300**
  foreset beds, **299,** 299★
  growth of, 300, 300★
  Mississippi River, 298, 299★, 300★
  sedimentation in, 298–300
  topset beds, **299,** 299★
  waves and tides, effect of, 300–301
Desertification, **318,** 319, 319★
  in the Sahel, 319, 319★
Deserts. *See also* Wind.
  deposition in, 312–317
  dry washes, **322**
  dunes. *See* Dunes.
  erosion in, 320
  and hydrology, 258
  landscape, 321–323, 322★, 323★
  mesas, **323,** 323★
  oases, 320
  pavement, **311,** 311★, 320
  pediments, 322★, **323**
  Sahara, 308, 320
  in Saudi Arabia, 305★, 308, 316
  sedimentary environments in, 320–321
  varnish, **320,** 320★
  ventifacts, **311,** 312★
  wadis, **322**
  weathering in, 132
  weathering and erosion in, 318, 320,
    320★
  and wind, 306–323
  world distribution of, 317–318
  yardangs, **311,** 312★
Detroit, Michigan, salt mines in, 164
Development of shorelines, 380–381
Diagenesis, **152,** 152–154, 153★
  and burial metamorphism, 173
  carbonate, 153–154
  cementation, **153**
  chert formation, 165
  clastic, 153
  coal, 154
  compaction, **153**
  dolostone formation, 164
  lithification, **153**
  organic matter, 154
  petroleum, 154
  phosphorite formation, 165–166
  porosity, **153**
  and sandstone composition, 160–161

Diamonds
  covalent bonding in, 31, 31★
  and igneous ore deposits, 532
  and polymorphs, 35, 35★
Diatremes, **103,** 103–104, 104★
  Kimberley Mines, South Africa, 105
  Ship Rock, New Mexico, 103, 104★
Dietz, Robert, and seafloor spreading,
  450
Differentiation. *See* Planetary
  differentiation.
Digdig, Philippines, earthquake of 1990,
  405★
Dikes, 83–85, **84**
Diorite, **71**
Dip, 212–215, 213★, 214★
  dipping beds, 214★
  and slope stability, 236
Displaced terranes. *See* Microplate
  terranes.
Disseminated deposits (of ores), **529,**
  531
Dissolution (of minerals and rocks),
  rates of, **131**
Divergent boundaries, **14,** 14–17, 15★,
  16★, 452–453, 452★, 455★
  earthquakes along, 414, 415★, 416★
  metamorphism at, 183
  and mineral resources, 534–537
  rocks at, 464, 462★, 464★, 464–465,
    465★
  tectonic forces at, 215
  volcanism along, 109★, 110
Dolomite, 38. *See also* Carbonate
  minerals, rocks.
  chemical weathering of, 130
  in evaporites, 164–165
  in sedimentary rocks, 162–165
Dolostone, **162,** 162★, 162–164
Domes, **218,** 219★, 218–220
Draas, 315★, **316**
Drainage (soil), mass wasting
  prevention, 237
Drainage (stream)
  basins, **295,** 295★, 295–296
  dendritic, **296,** 296★, 296–297
  networks, 294–298, **296**
  patterns and geologic history,
    297–298, 298★
  radial, **297,** 297★
  rectangular, 296★, **297**
  trellis, **297,** 297★
Drill holes, 57, 384
Droughts, 309
  and hydrology, 255, 256
Dry washes, **322**
Ductile materials, **216**
Dunes
  barchans, 315★, **316**
  blowout, 315★, **316**
  in deserts, 320
  draas, 315★, **316**
  eolian sand, 312–317, 312★, 313★,
    314★, 315★
  ergs, 315★, **316**

formation and movement, 313★, 313–314, 314★
linear, 315★, **316**
recognizing in sedimentary rocks, 314
sand drift, **313,** 313★
and sedimentary structures, 157, 157★
slip face, **314,** 314★
and stream velocity, 281–282, 282★
in streams, **281,** 281–282, 282★
transverse, 315★, **316**
wind, 308

Dust
falls, 316–317
loess, **316,** 316★, 316–317
mineral composition, 310
storms, 308–310, 309★
and volcanic ash, 317
Dust Bowl, 309
Dynamo theory, 441

Earth
cooling of, 438–440
early history, 6–13, 10★
Earthflow, 240★, **241,** 241–242
Earthquakes, **405,** 405–425
Anchorage, Alaska, 1964, 236
Boston, Massachusetts, 1755, 416
Charleston, South Carolina, 1886, 416
along convergent boundaries, 415★, 416, 417★
damage, causes of, 418–420
deaths caused by major, 420★
deep-focus, 414, 415★
destruction by, 404, 416–425
destruction mitigation, 421–425
Digdig, Philippines, 1990, 405★
along divergent boundaries, 414, 415★, 416★
elastic rebound theory, **405,** 405★, 405–406
energy released by, 413★
epicenter, 405★, **406,** 406★
epicenter, locating, 408, 410★
fault mechanisms, determining, 411, 414, 414★
and fault types, 411, 412★
focus, 405★, **406,** 406★
frequency and size, 416–417
global distribution, 415★
Haicheng, China, 1975, 423
intraplate, 415★, 416
in Japan, 417, 417★
Lisbon, Portugal, 1755, 404
Loma Prieta, California, 1989, 403★, 417
Modified Mercalli Scale, **411,** 412★
New Madrid, Missouri, 1812, 416
Niigata, Japan, 1964, 417★
and plate tectonics, 414–416, 415★, 416★, 417★
prediction of, 423–425, 424★
Richter magnitude, **411,** 411★, 413★
San Francisco, 1906, 405, 406★, 411
seismic gap method, 424–425
seismic waves. See Seismic waves.

seismicity, **414,** 415★
seismic-risk maps, 421, 421★
seismograph, **407,** 407★
size, measuring, 411, 411★, 412★, 413★
slip, 405★, **406**
studying, 407–414
Tangshan, China, 1976, 423
along transform fault boundaries, 414, 415★, 416★
triggers of mass wasting, 236
and tsunamis, **104, 376, 418,** 418★, 419, 420★
and volcanoes, 99, 112–115

Earth's interior, 430–444
composition and structure, 433–437
conductive heat flow, **438,** 438★
convective heat flow, **438,** 438–440, 439★
exploring with seismic waves, 430–437
geotherms, **440,** 440★
heat in, 437–440, 439★, 440★
heat flow mechanisms, 438–440, 439★
isostasy, principle of, **434,** 434–435
Mohorovičić discontinuity, **434**
study by magnetism, 440–444, 441★, 442★, 443★
temperatures in, 440, 440★
East Africa Rift Valley, 113, 225, 360, 453, 493
East Pacific Rise, 386★, 387★, 387, 452
hydrothermal springs, 394
Eclogites, **180,** 180★, 181★
and metamorphic facies, 181★
Economic geology, 6
and plate tectonics, 473
Ecuador, landslides in, 244
Einstein, Albert, 7
El Chichón, Mexico, 108
Elastic rebound theory, **404,** 405★, 405–406
Electrons, **25,** 26★
gain or loss, **27,** 27–31
sharing, **28,** 28★, 31
shells, orbitals, **25,** 26★, 28–29, 30★
Elements
abundance in Earth, 10, 11★
atomic structure, **25,** 25–26
chemical properties, 28–31
crustal abundance, 36, 36★
metallic, 31, 38–39
periodic table of, 28–31
Elevation, **354**
Elizabeth I, Queen, 440
Emperor seamounts, 109★, 113
Energy resources, 500–519
alternative, 509
biomass, **515,** 517
coal. See Coal.
conservation of, 509, 519
energy policy, 519
fossil fuels, **504,** 504★, 512–513
geothermal energy. See Geothermal energy.

and global warming, 514
hydroelectric, 517
nuclear energy. See Nuclear energy.
petroleum. See Petroleum.
and population growth, 543
quads, **508**
solar energy, 503, **515,** 515★, 517
syncrude, 517
and use of federal lands, 528
wind power, 517
Environmental issues, 5
acid rain, 126–127, 127★, 510
alternative energy, 509
asbestos, 44–45
beach erosion, 380–381, 381★
carbon dioxide atmospheric increase, 514
carbon dioxide and climate change, 348–349
Clean Air Act of 1991, 127
coal combustion, 510–512
desertification, **318,** 319, 319★
droughts, 309
earthquake protection, 419
energy conservation, 509, 519
energy policy, 519
exploration for mineral resources, 540–544
federal lands, use of, 528
floodplains, development of cities on, 288
floods, and flood control, 290–291
global warming (greenhouse effect), 205, 348–349, 349★, 514
groundwater, depletion of, 268
groundwater, excessive pumping of, 264★, 264–265, 265★
groundwater use, 266–267
hazardous wastes, storage on seafloor, 395
landslides, prevention of, 237
landslides, prevention of losses from, 244–245
mass wasting. See Mass wasting; Mass movements.
mass wasting from deforestation, 235, 235★
nuclear energy use, 513, 515, 516
oil spills, 379
oil well fires, 6★
petroleum use, 506–507
population growth, 543
radioactive waste, 513, 515, 516
radon, 202–203
recycling of mineral resources, **527,** 529
sea level, changes in, 347
shoreline construction, 380–381, 381★
soil erosion, 136–137, 137★
strip mining, 511, 511★
tsunamis, 419
volcano hazards, reduction of, 115
water allocation policies, 256
water, contamination of, 271★, 271–272
water quality, 270–272

Eolian, **306.** *See also* Wind; Deserts.
sands, composition of, 310
sediments, in deserts, 320
Eons, **196,** 196–197
Epeirogeny, **494,** 494–497, 495★, 496★, 497★
Epidote, in metamorphic rocks, 180
Epochs, **196,** 196–197
Eras, **196,** 196–197
Ergs, 315★, **316**
Erosion, 120
cycle of, 364–365, 365★
in deserts, 320
and geological maps, 214–215, 214★
glacial, 328, 337–340
by groundwater, 267★, 267, 269
headward, **282**
jointing, resulting in, 221
and landforms, **356,** 356–360, 362–364, 363★
as landscape control, 360–362
mass wasting by, 232–249
of mesas, **323,** 323★
of pediments, 322★, **323**
rates of, 206–207
by running water, 282–283, 283★
sandblasting, **311,** 312★
at shorelines, 379★, 379–381, 380★
of soil, 136–137
and soils, 135
topographic control of, 360
by wind, 310–311, 311★, 312★
Erratics, **340**
Eskers, 342★, **343**
Estuary, **382**
Etna, Mount, Sicily, 91★, 100
Eurasia, and Himalayan orogeny, 484★, 485, 485★, 486, 487★
Eurasian Plate, 453, 469–470, 473
rates of motion, 460, 460★
Evaporation, crystallization by, 33–34
Evaporite minerals, rocks, **156, 164,** 164–165, 165★
in deserts, 321
in playas and playa lakes, **321,** 321★
Evaporite sedimentary environments, 152, 321
Everest, Mount, Himalayas, 211★, 355
Everglades, Florida, 251★
Evolution, 192
and plate tectonics, 473
Ewing, Maurice, 391
Exclusive economic zones, **541,** 542★
Exfoliation, **133,** 134★
Exotic terranes. *See* Microplate terranes.
Extrusive igneous rocks, **53,** 53★, 53–54, 71
and contact metamorphism, 172
cooling rate, 66–68
*Exxon Valdez,* oil spill of 1989, 507

Farallon Plate, 468, 468★
Fault blocks, 225, 225★
Fault-block mountains, **491,** 493★
Faulting, **212,** 213★. *See also* Continental crust deformation; Orogeny.
dip, **223,** 223★

dip-slip, **223,** 223★
and earthquakes, 405–425
fault plane, **223,** 223★
left lateral, **223,** 223★
mechanisms, determining, 411, 414, 414★
normal, 223★, **224**
normal, and earthquakes, 411, 412★
oblique slip, 223★, **224**
overthrust, 224★, **225**
reverse, 223★, **224**
right lateral, 223★, **224**
and stress, 216
strike, **223,** 223★
strike-slip, **223,** 223★
strike-slip, and earthquakes, 411, 412★
thrust, 224★, **225,** 483★, 483–486, 485★
thrust, and convergent boundaries, 467–469
thrust, and earthquakes, 411, 412★
transform, 223. *See also* Transform fault boundaries.
Faults
and cross-cutting relationships, 193–194, 194★
as oil traps, 505★, 505–506
Faunal succession, **192**
Federal lands, use of, 528
Feldspar
chemical stability and weathering, 131, 131★
chemical weathering of, 123–126
dissolution in laboratory, 124–125
in foliated rocks, 176–177
in metamorphic rocks, 180
and weathering intensity, 145, 145★
Feldspar (plagioclase)
and Bowen's reaction series, 81
and continuous reaction series, 76★, **77**
and fractional crystallization, 80
Felsic lavas, 111
Felsic magmas, fissure erruptions of, 106
Felsic minerals, rocks, **70,** 70–71
distribution, and plate tectonics, 462–470
and igneous rock compositions, 70–72, 71★
seismic wave velocities in, 434
Ferric iron, **129**
Ferrous iron, **129**
Field relationships of strata
interpreting deformation, 212–215
and relative dating, 194–195, 195★
Finland
isostatic uplift after glaciation, 495
mountains, 491
Firn, **331,** 331–332, 332★
Fjords, **340,** 340★
Flathead River valley, Montana, 353★
Flint. *See* Chert.
Flood basalts, **106,** 106★
Columbia Plateau, Pacific Northwest, 106, 106★
Floods, 289–291

and cities on floodplains, 288, **288**
control of, 258★, 258–259, 290–291
and floodplains, 286–287
frequency curves, 291★
historic, 290–291
landslides, caused by, 244–248
of Mississippi, 1973, 288
of Mississippi, 1993, 288, 288★
probability of, 290
recurrence interval, 290
Florida
microplate terranes in, 470
sinkholes in, 269★
Florida Keys, 395, 397
Fluid flow, 278–279, 279★
laminar, **278,** 278–279, 279★
turbulent, **278,** 278–279, 279★
viscosity, **278,** 278–279
wind flow processes, 306–307
Folding, **212,** 213★, 217–225. *See also* Continental crust deformation; Orogeny.
anticlines. *See* Anticlines.
asymmetrical folds, **218,** 218★
axis of folds, **218,** 218★
basins. *See* Basins.
and cross-cutting relationships, 193–194, 194★
domes. *See* Domes.
fold belts, 220
folds, **221,** 222★, 223–225
limbs of folds, **218,** 218★
and oil traps, 505★, 505–506
overturned folds, **218,** 218★
and plate tectonics, 467–469
plunging folds, **218,** 219★
scales of, 217
and stress, 216
synclines. *See* Synclines.
Foliation, **56,** 174★, 174–175, **175,** 175★
Foraminifera, **162,** 162★
paleotemperatures from, 345
Foraminiferal oozes, **397,** 397★, 397–398
Forces (tectonic). *See* Tectonic forces.
Formation, **192**
Fossil fuels, **504,** 504★. *See also* Petroleum.
future use of, 512–513
and global warming, 514
Fossil magnetism. *See* Paleomagnetism.
Fossils, 189–193
ammonites, 191★
and continental drift, 448–449
faunal succession, **192**
geologic time scale construction, 190–192
petrified forest, 191★
and plate tectonics, 459
and relative ages, 189
stratigraphic correlation by, 190–192
trilobites, 187★, 198
Fractional crystallization, 78–80, **79,** 79, 532, 536
Fracturing, 220–225. *See also* Faulting.
faults, **221,** 222★, 223–225
and stress, 216
Franciscan formation, California, 468

Fringing reef, **396**
Fujiyama, Japan, 97, **100**, 101★, 110
Fundy, Bay of, 493

Gabbro, 70★, **71**
    at divergent boundaries, 462, 462★,
        464★, 464–465, 465★
Galápagos Islands, hydrothermal
    springs, 394
Galena, hydrothermal vein deposits of,
    529, 530★, 531
Ganges River, India, 287, 290
Garnet
    and contact metamorphism, 182
    as metamorphic index mineral, 179,
        179★
    in metamorphic rocks, 180
    porphyroblasts, 178, 178★
Gas, 149, 509
Genesis (Bible), and geologic time scale,
    197
Geologic processes, estimation of rates,
    204–207
Geologic time, **18**, 190
Geologic time line, 207★, inside back
    cover★
Geologic time scale, 188–189, 194–197,
    196★
    and absolute ages, 203–204
    units, **196**
Geological containment of radioactive
    waste, 516
Geological maps, 214★, 214–215
Geomorphology, 352–366
Georges Bank, 390
Georgia, chemical weathering and soils,
    129–130
Geothermal energy, **517**, 517★,
        517–519, 518★
    from hydrothermal waters, 273
    and volcanism, 115
Geotherms, **440**, 440★
Geyser Hot Springs, Nevada, 108★
Geysers, 108, 108★, 273
    and ore deposits, 529
Geysers, The, California, 517★
    and geothermal energy, 518
Gilbert, William, 440
Glaciation and continental drift,
    448–449
    and loess, 316★, 316–317
Glacier National Park, potholes in
    McDonald River, 283★
Glaciers, **329**, 326–350
    ablation, **332**, 332★, 332–333, 333★
    accumulation, **332**, 333★
    Antarctica, speed of, 336
    aretes, **338**, 339★
    basal slip, **334**, 335★
    and braided streams, **286**, 286★
    calving of, **332**, 332★
    capacity of, **340**
    cirques, **338**, 339★
    competence of, **340**
    continental, **329**, 329★
    crevasses, **335**, 335★
    drift, **340**

drumlins, **341**, 342★
"dry", 334, **335**
and epeirogeny, 495, 495★, 496★
erosion, 328
erosion, and landforms, 337–340
erratics, **340**
eskers, 342★, **343**
extent of, during Pleistocene, 346★
firn, **331**, 331–332, 332★
fjords, **340**, 340★
flow directions, 338
flow mechanisms, 333–336
flow patterns and speeds, 335–336,
    336★, 337★
formation and growth of, 330–332
glacial budgets, 330–333, 333★
hanging valleys, **338**, 339★
Harvard Glacier, Alaska, 339★
ice, air bubbles in, 329★, 332
ice, formation of, 331–332, 332★
ice, properties of, 328–329, 329★
ice age theories, 347–350
ice ages, and carbon dioxide, 348–349
ice ages, and orbital cycles, 348–349
ice ages, and plate tectonics, 347–348
and isostasy, 435, 435★
Jasper National Park, Canada, 341★
kames, **342**, 342★
Kennicott, Alaska, 327★
kettles, **343**, 343★
lake deposits, **342**, 343★
Lake Missoula, 342
landscapes, 336–344
meltwater, 333, 333★
moraines, **340**, 340★, 340–341
outwash, **340**
permafrost, **343**, 343–344, 344★
Permian, 471
plastic flow, **334**, 334★, 335★
Pleistocene ice ages, 344–347, 345★
roches moutonées, **338**, 338★
rock flour, 338
sea temperatures during ice ages, 346,
    346★
sediment sorting, 340
sedimentation, and landforms,
    340–343
and soil formation, 136
striations, **338**, 338★
surges, **336**
temperature and flow, 334, 335★
till, **340**, 341★
tillites, 347
U-shaped valleys, **338**, 339★
valley, 327★, **329**
varves, **342**, 343★
"wet", 334, **335**
world distribution, 330, 331★
Glassy materials, **25**
    formation of, 33
Global Positioning Satellites, 407
Global warming, 205, 347, 348–349, 514
Gneiss, **176**, 176–177, 177★
    in Canadian Shield, 482–483
Gobi Desert, 317
Gold, ore deposits, 533–534
Gondwanaland, 448, 471, 472★

Grain size
    and current velocity, 280–281, 281★
    and physical weathering, 132
Grand Banks, Newfoundland, 390
Grand Canyon, 493
    Great Unconformity, 194★
    and sedimentary rocks, 144
    sedimentary sequence, interpretation,
        198–199, 199★
    springs in, 259★
Granite, 67★, 68★, **70**, 70★, 70–71
    in Canadian Shield, 482–483
    distribution, and plate tectonics,
        462–470
    granitic gneisses, 177
    and magmatic differentiation, 80–82
    origin of, 67–68
    weathering of, 123, **123**
Granitization, **83**
Granodiorite, **71**
Granulite, **178**
    and metamorphic facies, 181★
Gravel, **154**, 155★, 158–159, **161**
    as resource, 524, 524★, 532–533
Gravity, and tides, 375–376
Graywacke, **160**, 160★
Great Basin, Nevada, 106, 360
Great Pyramid of Egypt, 461
Great Rift Valley, East Africa, 113, 225,
    360, 453, 493
Great Salt Lake, Utah, chemical
    sedimentation in, 162, 165
Great Valley, California, 360
Green River, Wyoming and Utah, 206★
Greenhouse effect (global warming),
    205, 348–349, 349★, 514
Greenland, continental glacier in, 329★,
    329–330, 330★
Greenschists, 179★, **180**, 180★, 181★
    and metamorphic facies, 181★
Greenstones, **178**
    and plate tectonics, 183
Gros Ventre, Wyoming, landslide of
    1925, 246, 246★, 247★
Gross domestic product, 524
Groundwater, **253**, 259–269. See also
        Hydrology; Water.
    aquicludes, **262**, 262–263, 263★
    aquifers, **259**
    aquifers, confined, **262**, 262–263,
        263★
    aquifers, unconfined, **262**, 262–263,
        263★
    artesian flows, **262**, 262–263, 263★
    artesian wells, **263**, 263★
    artificial recharge, 265, 266
    and cave formation, 267★, 267, 269
    contamination, 271★, 271–272
    Darcy's law, 265–266, 266★
    deep crustal, 272★, 272–273
    depletion of, 268
    discharge, **262**, 262★
    dissolved substances in, 270★,
        270–272
    and effluent streams, **262**, 262★
    flow, 259–260
    and hydrologic cycle, 253★, 253–254

Groundwater (continued)
  hydrothermal, 272–273, 273★
  in igneous rocks, 272★, 272–273
  and influent streams, **262,** 262★
  karst topography, **269**
  in metamorphic rocks, 272★, 272–273
  mining, 268
  Ogallala aquifer, 268, 268★
  ore deposition from, 531
  perched water table, **263,** 263★,
      263–264
  and porosity, **260,** 260★, 261★
  quality, 270–272
  recharge, **262,** 262★, 263
  recharge and discharge balance, 264★,
      264–265, 265★
  resources, 266–267, 268
  saltwater intrusion, 264–265, 265★
  saturated zone, **261,** 261★
  and sinkholes, **269,** 269★
  speed of flows, 265–266, 266★
  springs, **259,** 259★
  table, **261,** 261★
  table elevation and Darcy's law,
      265–266, 266★
  unsaturated zone, **261,** 261★
  use of, 266★, 266–267
  use of, excessive, 256, 264★, 264–265,
      265★
Gulf coast, continental margin, 390
Gulf of Mexico
  coastal plain, 494, 494★
  and petroleum, 506
Gulf War, 1991, and oil spills, 507
Guthrie, Woody, "Dust Bowl Ballads",
    309
Guyots, **394**
Gypsum, 39★, 40, **164.** See also
    Evaporite minerals, rocks.
  in evaporites, 152
  sand dunes of, 310

Haicheng, China, earthquake of 1975,
    423
Half-life, **201,** 201–203
Halite. See also Evaporite minerals,
    rocks.
  chemical stability and weathering,
      131★
  in evaporites, 152, 164–165
Harvard Glacier, Alaska, 339★
Hawaiian Islands, and intraplate
    volcanism, 111
Haxby, W. F., 463
Hazardous wastes, storage on seafloor,
    395
Heat, in Earth
  control on surface processes, 229
  external, 12
  internal, 12, 437–440, 439★, 440★
  and plate tectonics, 17
  reservoir, **517**
  sources, 9–10
  Sun's, and hydrologic cycle, 253
Heezen, Bruce, 391
Hematite, **129**
  chemical stability and weathering, 131★

and desert weathering, 318, 320
  as an ore mineral, 525★
Hess, Harry, and seafloor spreading,
    450
Hills, **356**
Himalayas, 3★, 356, 453, 469–470
  deformation, expression of, 226–227
  orogeny, 484★, 485, 485★, 486, 487★
  Tibetan Plateau, 355
Hogbacks, **357,** 357★
Holmes, Arthur, and continental drift,
    439, 450
Homestake Gold Mine, 495
Hornfels, **177**
Hot spots. See Plumes (mantle).
Hot springs, 108, 272–273. See also
    Hydrothermal springs.
  mid-ocean ridge, and ore deposits,
      534–537
  and ore deposits, 529
Hot Springs, Arkansas, 272, 273
Huang Ho, China, flood of 1931, 290
Human population growth, 543
Humidity, relative, **254**
  and hydrology, 254–255
Humus, **134**
Hutton, James
  and geologic time scale, 197
  and origin of granite, 67–68
  and rock cycle, 59
  and uniformitarianism, 13
Hydration, **125**
Hydroelectric energy, **517**
Hydrologic cycle, 253★, 253–254, 256
  infiltration, 253★, **254**
  precipitation, 253★, **254**
  runoff, 253★, **254**
  sublimation, 253★, **254**
  transpiration, 253★, **254**
Hydrology, **252.** See also Water;
    Groundwater.
  and climate, 254–256, 257★
  runoff, 253★, **254,** 257★, 257–259,
      259★
  runoff, storage of, 258★, 258–259
  runoff and reservoirs, 256, 258★, 258–259
  runoff and wetlands, 258★, 258–259
Hydrothermal processes, and
    metamorphism, 171–172
Hydrothermal solutions, ore deposits
    from, 529★, 529–531, 530★
Hydrothermal springs, along mid-ocean
    ridges, **393,** 393–394. See also
    Hot springs.
Hydrothermal veins, 85★, 85–87, **87**
Hydrothermal water, 272–273, 273★

Iceland, 105–106, 110, 110★
  control of volcanism in, 114
  geothermal energy, 115
Igneous rocks, 52★, **53,** 53–54, 272★,
    272–273
  chemical and mineral composition,
      69★, 69–72
  classification, 69★, 69–72
  composition, and plate tectonics, 74,
      74★

crystal size, 53, 66–67
  distribution, and plate tectonics,
      462–470
  extrusive. See Extrusive igneous
      rocks.
  and feldspar composition, 71
  felsic. See Felsic minerals, rocks.
  fractional crystallization, 78–80, **79,**
      79★
  intrusive. See Intrusive igneous rocks.
  mafic. See Mafic minerals, rocks.
  magmatic differentiation, 76–82
  major elemental composition, 70–72,
      71★
  melting and crystallization, 73–82
  mineral composition, 54, 54★
  ore deposits in, **532,** 532★
  partial melting, **75**
  radiometric dating of, 200
  reaction series, **77,** 77★, 77–78
  and rock cycle, 59★, 59–60
  temperatures of crystallization, 70–72
  texture, 66–67
  texture, and classification, 70–72
  ultramafic. See Ultramafic minerals,
      rocks.
  volcanic, 71–72
Iguaco River, Brazil, waterfall on, 283★
India
  and Himalayan orogeny, 3★, 484★,
      485, 485★, 486, 487★
  soils, 138
Indian Plate, 453, 469–470, 473
  and Himalayan orogeny, 3★, 484★,
      485, 486, 487★
  rate of motion, 460, 460★
Indiana, karst topography in, 269
Industrial Revolution, 503
Infiltration, 253★, **254**
Interior of Earth. See Earth's interior.
Internal friction, **235**
International Geodynamics Project, 473
Intraplate earthquakes, 415★, 416
Intrusions (igneous)
  and contact metamorphism, 172–173,
      173★, 181–183
  and cross-cutting relationships,
      193–194, 194★
  and domes, 220
  forms, 82–87
  mechanisms of emplacement, 83
Intrusive igneous rocks, 53, 53★, 70–71
  concordant intrusives, **84**
  cooling rate, 66–68
  discordant intrusives, **83**
  and plate tectonics, 87
Ionic bonds, **31**
Ions, **27,** 27–31
  complex, **27**
  size and packing, 34
  size and substitution, 34–35
  stability, 28
Irazú, Costa Rica, 111
Iron
  ore, as weathering product, 129★,
      129–130
  as resource, 524, 524★

sedimentary ore deposits, 533, 533★
Iron formations, **156,** 166, 212★, 533,
    533★
Iron oxide minerals
    and desert weathering, 318, 320
    and soil classification, 137–138, 138★
Island arcs, 110, **453,** 454★, 455★, 467,
    467★
Isochrons, **459,** 459★, 459–460
Isograds, **179,** 179★
Isostasy, principle of, **434**
    and epeirogeny, 495–496, 496★
    and glacial rebound, 435, 435★
Isotopes, **26,** 26★
Itkillik River, Alaska, 286★

Jackson Hole, Wyoming, Gros Ventre
    landslide of 1925, 246, 246★, 247★
Japan
    earthquakes in, 417, 417★
    and igneous rocks, 86, 86★
    and island arcs, 110
    landslide loss prevention, 244–245,
        245★
Jasper National Park, Canada, lateral
    moraines, 341★
Jeffreys, Sir Harold, and continental
    drift, 449
Johnston, David A., 112–113
Johnstown, Pennsylvania, floods in,
    290, 291★
*Joides Resolution,* 458, 458★
Jointing
    joints, **220,** 220–223, 222★
    and weathering, **133**
Jordan River, 360
Jurassic period, **196**
    plate-tectonic events of, 473, 473★

Kaolinite, **124,** 124–126
    and fine-grained sediments, 161
Kao-ling, China, 124
Karst topography, **269,** 269★
Kenai Fjords National Park, Alaska,
    340★
Kennicott Glacier, Alaska, 327★
Kentucky, karst topography in, 269
Kilauea, Hawaii, 99, 99★, 103, 103★, 114★
Kimberlites, diamond deposits in, 532
Kimberley Mines, South Africa, 105,
    534
Kissimmee River, Florida, 266
Kofa Butte, Arizona, 322★
Krafft, Maurice and Katia, 97, 107★
Krakatoa, Indonesia, 103, 104–105
Kuenen, Philip, 391
Kupferschiefer, Germany, sedimentary
    copper ores, 533
Kyanite, as metamorphic index mineral,
    179, 179★

Labrador Highlands, 491
LAGEOS. *See* Laser Geodynamics
    Satellite.

Lahars, **106,** 106–107
    and Mount St. Helens, 112–113
Lake Superior, iron deposits, 533, 533★
Lakes, playa, 321, 321★
Landforms, **356,** 356–360, 362–364,
    363★
    badlands, **359,** 360★
    basins, **360**
    and climate, 366
    controls on, 360–362
    cuestas, **357,** 357★
    cycle of erosion, 364–365, 365★
    deformation, expression of, 226–227
    and geologic history, 366
    hills, **356**
    hogbacks, **357,** 357★
    mesas, **356,** 356–357, 357★
    mountains, **356**
    of North America, 362–364, 363★
    origin of, 360
    and plate tectonics, 365–366
    plateaus, **356,** 356–357
    river valleys, 358–359, 359★, 360★
    structural valleys and ridges, 358,
        358★, 359★
    structurally controlled cliffs, 357, 357★
    tectonic valleys, 360
Landscape, of North America, 362–364,
    363★
Landscape evolution, 354–366
    and climate, 366
    controls on, 354, 360–362
    cycle of erosion, 364–365, 365★
    and geologic history, 366
    and plate tectonics, 365–366
Landslides. *See* Mass movements.
Larderello, Italy, and geothermal
    energy, 517
Larochelle, A., 456
Laser Geodynamics Satellite, measuring
    plate motion with, 462
Lassen Peak, California, 114★, 115
Laurussia, 471, 472★
Lava
    aa, 93–94, **94,** 94★
    basaltic, and eruptive style, 97–98
    composition, and eruptive style,
        97–108
    composition, and landforms, 93–97
    felsic, and eruptive style, 97–98
    gas bubbles in, 94–95
    pahoehoe, **93,** 93–94, 94★
    pillow, **94,** 94–95, 95★
    pillow, and ophiolites, **462,** 462★,
        464–465, 465★
    plateaus, 93
    temperatures of, 82
    vesicles in, **95,** 95★
Lava flows, 91★, 97–98
    along ocean ridges, 110
    temperature influence, 93–94
Law of the Sea Treaty, 541
Lead
    radiometric dating with, 200, 201★
    as resource, 524, 524★
Limestone, **162,** 162★, 162–164, 164★
    chemical weathering of, 130, 130★

contact metamorphism in, 182, 182★
weathering rate, 121
Liquefaction, **236**
Lisbon, Portugal, earthquake of 1755,
    404
Lithic sandstones, **160,** 160★
Lithification, **55,** 152–154, **153,** 153★
Lithosphere, **14**
    formation of ocean crust, 464, 464★
    and plate tectonics, 13★, 14
    studying with seismic waves, 433★,
        436★, 436–437
Loess, **316,** 316★, 316–317
Loiki Seamount, 385★
Loma Prieta, California, earthquake of
    1989, 403★, 417
London, England, windstorm of 1990,
    306
Long Island, New York
    artificial groundwater recharge, 266
    saltwater intrusion, 265
Long Valley, California, volcanism in,
    92
Long Valley Caldera, California, 103,
    114★
Longshore currents, **374,** 375★
Longshore drift, **374,** 375★
Louisiana, delta sedimentation in, 300,
    300★
Lyell, Charles, and geologic time scale,
    197

Mafic minerals, rocks, **70,** 70–71
    and convergent boundaries, 465, 467,
        467★
    crystallization of, 77★, 77–78
    in gneisses, 177
    and igneous rock compositions,
        70–72, 71★
    metamorphism of, 178
    seismic wave velocities in, 434
    volcanic, metamorphism of, 180,
        180★
Magma, **10.** *See also* Melting of rocks;
        Continental crust deformation;
        Orogeny.
    composition, and plate tectonics,
        110–113, 462–470
    continuous reaction series, **77**
    and convergent boundaries, 465, 467,
        467★
    crystallization of, 76–82
    discontinuous reaction series, **77,** 77★,
        77–78
    fractional crystallization, 78–80, **79,**
        79★, 532, 536
    hydrothermal waters, 273, 273★
    and igneous rock formation, 53
    intrusions. *See* Intrusions (igneous).
    magma chambers, **73**
    magma chambers, formation of, 75
    "magma ocean", 10
    magmatic differentiation, 76–82
    ocean crust, formation of, 464, 464★
    origin of, 72–76
    and Palisades intrusion, 79–90, 80★
    at plate boundaries, 16

Magma (*continued*)
　reaction series and crystal structure,
　　78
　from varied source rocks, 82
Magnetism (Earth's), 440–444, 441*,
　　442*, 443*
　depositional remanent magnetization,
　　**442,** 442*, 442–443
　dynamo theory of, 441
　magnetic stratigraphy, **443,** 443*,
　　443–444
　paleomagnetism, 441–444, 442*,
　　443*, **443**
　paleomagnetism and plate tectonics,
　　456*, 456–457, 457*
　thermoremanent magnetization, **442,**
　　442*
Magnetite, as an ore mineral, 525*
Maiandros River, Turkey, 284
Mammoth Cave, Kentucky, 267, 269
Mammoth Hot Springs, Yellowstone,
　　273*
Manganese
　nodules, **398,** 536, 537*
　as resource, 524, 524*
Mansel Island, Canada, 495*
Mantle, **10,** 10*, 14
　convection, and plate tectonics, 473,
　　476, 476*
　and plate tectonics, 13*, 13–14
　studying with seismic waves, 433*,
　　435–436, 436*, 437*
Mantle plumes. *See* Plumes (mantle).
Marble, **177,** 177*, 177–178
Marine evaporites. *See* Evaporite
　　minerals, rocks.
Mars, 11
　color of, 43
Marshes. *See* Wetlands.
Mass movements, 231*, **232,** 236. *See
　　also* Mass wasting.
　classification of, 238–243, 238*
　creep, 238*, 240*, **241,** 241*
　debris avalanche, 241*, **242**
　debris flow, 240*, **241,** 241–242
　debris slide, 238*, 241*, **243**
　earthflow, 240*, **241,** 241–242
　Gros Ventre, Wyoming, landslide of
　　1925, 246, 246*, 247*
　human-influenced, 247
　landslides, 5*
　mudflow, 240*, **242,** 242*
　mudslide, 244–246, 245*
　physical weathering and erosion, 134
　prevention, 237
　prevention of losses from, 244–245
　of rock, 238*, 238–240, 239*
　rock avalanche, **239,** 239*, 239–240
　rockfall, 238*, **239,** 239–240
　rockslide, **239,** 239*, 239–240
　slump, 241*, **243**
　solifluction, **243,** 243*
　Spanish Fork Canyon, Utah,
　　mudslide of 1983, 244–246, 245*
　Tadzhikistan, mudflow of 1989, 242*

of unconsolidated materials, 240*,
　　240–243
Vaiont, Italy, debris slide of 1963,
　　248*
Mass spectrometers, 200
Mass wasting, **232,** 232–249. *See also*
　　Mass movements.
　in consolidated materials, 235
　factors influencing, 232–233, 233*
　and internal friction, **235**
　from liquefaction, **236**
　prevention, 237
　on rock slopes, 235–236
　triggers of, 236
　in unconsolidated materials, 233–234
Mathew, D. H., 456
Mauna Loa, Hawaii, 97–99, 102*
Mauritanide Mountains, 490
McDonald River, Montana, potholes in,
　　283*
McPhee, John, 232
Mélanges, **467,** 467–468, 468*
Melting of rocks
　composition change, 75, 76–82
　composition dependence, 73–74
　geographic occurrence, 73–74
　magmatic differentiation, 76–82
　melt processes, 74–76
　and mid-ocean ridges, 73–74, 74*
　partial melt, **75**
　sedimentary rocks, 75–76
　and subduction zones, 73–74, 74*,
　　75–76
　and tectonics, 73–74
　temperature/pressure dependence,
　　73–74
　varied source rocks, 82
　water, effects of, 75–76
Mercury, 11
Mesas, **323,** 323*, **356,** 356–357, 357*
Mesozoic era, **196**
Metallic bond, **31**
Metamorphic facies, **180,** 180–181, 181*
Metamorphic rocks, 52*, **53,** 55–56,
　　**170,** 170–185, 272*, 272–273. *See
　　also* Metamorphism.
　amphibolites, 179*, **180,** 180*, 181*
　amphibolites, and metamorphic facies,
　　181*
　argillites, **178**
　banding, 176*, 176–177
　blueschists, **180,** 180*, 181*
　blueschists, and plate tectonics, 183,
　　183*
　in Canadian Shield, 482–483
　classification, 174*, 174–178
　cleavage, and physical weathering,
　　133
　cleavage (rock), 174–175, **175**
　contact aureoles, **181,** 181–183
　deformational textures, 178
　distribution, and plate tectonics,
　　462–470
　eclogites, **180,** 180*, 181*
　eclogites, and metamorphic facies, 181*

foliated rocks, classification, 176–177
foliation, **56,** 174*, 174–175, **175,**
　　175*
gneiss, **176,** 167–177, 177*
grade, 179–181
grade, and foliation, 176*, 176–177
granulites, **178**
granulites, and metamorphic facies,
　　181*
greenschists, 179*, **180,** 180*, 181*
greenschists, and metamorphic facies,
　　181*
greenstones, **178**
greenstones, and plate tectonics, 183,
　　183*
high-grade, **170,** 171*
hornfels, **177**
low-grade, **170,** 171*
marble, **177,** 177*, 177–178
migmatites, **181**
mineral composition, 54*, 56
mylonites, **178**
nonfoliated rocks, classification,
　　177–178
phyllite, **176**
porphyroblasts, **178**
preferred orientation of minerals, **175,**
　　176*
pyroxene granulites, 179*, **180,** 180*,
　　181*
quartzites, **177,** 177*
radiometric dating of, 200
and rock cycle, 59*, 59–60
schists, **176,** 177*
schistosity, **176,** 176*
slates, **176,** 177*
slates, and foliation, 175
slaty cleavage, 176, 176*
textures, 174–178
Metamorphism. *See also* Metamorphic
　　rocks; Continental crust
　　deformation; Orogeny.
　burial, **173,** 173*
　cataclastic, **173,** 173*
　chemical changes, 171–172
　contact, **56,** 56*, **172,** 172–173, 173*
　contact, and granite intrusions, 67–68
　contact metamorphic zones, 181–183
　depths of, 170, 171*
　and fluids, 172
　grade of, 179–181
　hydrothermal, **173,** 173*
　and hydrothermal fluids, 171–172
　and igneous intrusions, 172–173,
　　173*
　limestones, contact, 182, 182*
　of mafic volcanics, 180, 180*
　and magma intrusion, 171–172
　metasomatism, 171–172, **172**
　mineral isograds, **179,** 179*
　parent rock composition, influence of,
　　180
　physical and chemical controls,
　　171–172
　and plate tectonics, 173*, 183, 183*

pressures of, 170–171, 171★, 179–180, 180★
pressure's relation to temperature, 179–180, 180★
regional, **55**, 55–56, 56★, **172**, 173★
regional, and metamorphic grade, 179–181
retrograde, **170**
sandstones, contact, 182★, 182–183
of shales, 179, 179★
shales, contact, 181–183, 182★
structural deformation, 178
temperature and deformation, 171
temperature range, 55
temperatures of, 170–171, 171★, 179–180, 180★
types of, 172–173, 173★
Metasomatism, 171–172, **172**
Meteor Crater, Arizona, 19★
Meteorites
    chemical weathering of, 130
    and metallic iron, 129
Methane, 28, 28★
Micas
    biotite, as index mineral, 179
    biotite, and reaction series, 77–78
    chemical stability and weathering, 131★
    and cleavage, 41–42, 42★
    and contact metamorphism, 182–183
    in foliated rocks, 176–177
    and foliation, 175
    and metamorphism, 171
Michigan Basin, 220, 483, 483★
    and epeirogeny, 494
Microplate terranes
    and continental growth, 485
    and plate tectonics, **470**, 470★, 471★
Middle East, and petroleum, 506
Mid-Atlantic Ridge, 15, 17★, 387, 452
    gold deposits on, 536
    and volcanism, 110, 110★
Mid-ocean ridges, **393**, 393★, 452–453, 452★, 455★
    earthquakes along, 414, 415★, 416★
    hydrothermal metamorphism at, 173
    hydrothermal springs along, **393**, 393–394
    metamorphism at, 183
    Mid-Atlantic Ridge, 15, 17★, 387, 452
    and mineral resources, 534–537
    volcanism along, 109★, 110
Migmatites, **181**
Milankovitch, Milutin, 349
Mineral resources, 524–544
    concentration factor, **525**, 525–526, 526★
    crustal abundance of elements, 525–526
    disseminated deposits, **529**, 531
    exclusive economic zones, **541**, 542★
    geology of, 529–534
    geology and use of major minerals, 538★, 539★, 540★
    global distribution of, metallic, 535★
    hydrothermal, at mid-ocean ridges, 464

of hydrothermal springs, 393–394
hydrothermal vein deposits, **529**, 529★, 529–531, 530★
igneous ore deposits, **532**, 532★
Law of the Sea Treaty, 541
manganese nodules, **398**, 536, 537★
new sources, 540–544
per capita consumption of, 524, 524★
and placers, **533**, 533–534
and plate tectonics, 534–537
and population growth, 543
porphyry deposits, 531
recycling of, **527**, 529
resource use, and politics, 540–544
sedimentary ore deposits, 531, **532**, 532–534, 533★
U.S. reliance on imports, 527★
and use of federal lands, 528
and volcanism, 115
Mineralogy, **24**. See also Minerals.
Mineralogy (of a rock), 52
Minerals, **24**, 24–47, 25★
    atomic structure, 32–35
    carbonates. See Carbonate minerals, rocks.
    chemical properties, 46–47
    chemical stability and weathering, 130–132, 131★
    classification, 46, 46★
    cleavage, 41–43, **41**, 42★, 43★
    color, **43**
    crystal faces, **33**, 33★
    crystal habit, **45**
    crystalline structure, 32–34
    crystallization, **32**, 32–34
    crystals, **32**
    density, **44**, 44–45
    dissolution rates, **131**
    fracture, **43**
    as geobarometers, 179–180
    as geothermometers, 179–180
    grains, **33**
    hardness, **41**
    index (metamorphic), **179**
    luster, **43**, 43★
    Mohs scale of hardness, **41**, 41★
    ore, **525**. See also Mineral resources.
    oxides, 38–39, **39**
    physical properties, 40★, 40–45
    polymorphs, **35**
    rock-forming, 35–40
    silicates, 35, 36–38, 37★, **38**★
    solubility, **131**
    specific gravity, 44–45, **45**
    streak, **44**
    sulfates, 39★, 40
    sulfides, 39★, 39–40
    and trace elements, **46**
Mining. See Mineral resources.
Mississippi River, 288, 288★
    competence and capacity, 280
    delta, 298, **299**, 300★
    discharge of, 289
    floods of, 288, 288★
    flows in, 258, 258★

Mississippi Valley ore deposits, 531
Mohorovičić discontinuity, **434**
    below ocean crust, 464, 464★
Mohs scale of hardness, **41**, 41★
Molnar, Peter, 486
Molecular, 128
Mono-Inyo Craters, California, 114★, 115
Monsoons, and hydrology, 254
Montmorillonite, 128
Monument Valley, Arizona, mesas, 357★
Moon, 11
    and tides, 375–376
    weathering on, 132
Morley, L., 456
Mother Lode (gold, California), 534
Mountains, **356**
    building of. See Orogeny.
    roots of, and isostasy, 434–435
Mudflow, 240★, **242**, 242★
Muds, **154**, 155★, 158–159, **161**
Mudstones, **154**, 155★, 158–159, **161**
Mylonites, **178**

NASA, 407
National Science Foundation, 457
Native elements, **46**
Natural disasters, 5
Natural gas, **149**, 509
Nazca Plate, 453, 468
    rate of motion, 460, 460★
Nebulae, **7**, 7–8. See also Solar system, origin of.
Negative feedback processes, **360**
Neutron, **25**, 26★
Nevada del Ruiz, Colombia, 107
New Madrid, Missouri, earthquake of 1812, 416
New Mexico, groundwater resources, 268
Newton, Isaac, and tides, 375
Niagara Falls, 283
Niigata, Japan, earthquake of 1964, 417★
Nile River, 278, 287
    delta, 299
Nitrogen, radiometric dating with, 200, 201★
North America
    age of craton, 451
    landscape of, 362–364, 363★
North American Plate, 453, 468, 468★, 473
    and Appalachian orogeny, 486, 488★, 488–490, 489★
    and Cordilleran orogeny, 490–493, 491★, 492★
    rate of motion, 460, 460★
North Sea, oil platform explosion in 1988, 507
Nuclear energy, **513**, 515
    history of use, 503, 504★
    radioactive waste from, 513, 515, 516
    uranium resources, 508★, 513
Nuclear reactions, fusion, 7

Nuclear test ban treaty, and seismology, 414
Nucleus, **25**, 26★

Obsidian, **72**, 72
Ocean Drilling Program, 57, 383–384, 457–460, 458★
Ocean crust, cooling and subsidence, 495–496, 496★
Oceans, **368**, 368–398
  abyssal hills, **385**, 385★
  abyssal plain, **385**, 385★
  age of floor, and plate tectonics, 450–451, 456–460, 459★
  Atlantic profile, 384★, 384–387, 385★, 386★
  atolls, **395**, 395–396, 396★
  carbonate compensation depth, **397**, 398★
  chemical composition, and sedimentation, 149
  chemistry, and hot springs, 149
  continental margins. See Continental margins.
  continental rise, **385**, 385★
  continental shelf, **384**, 384★
  continental slope, **384**, 384★
  coral reefs, 395–396, 396★
  coring sediments in, 384
  deep ocean floor, 392–396
  as deep waste repository, 395
  drilling in, and plate tectonics, 457–460, 458★
  echo sounder profiles, **383**, 384–385
  floor, volcanic features, 394–395
  geology compared to continents', 398
  guyots, **394**
  hills on floor, 394
  hot springs on floor, 384
  imaging floor with satellites, 462, 463, 463★
  longshore currents, **374**, 375★
  longshore drift, **374**, 375★
  mapping of floor, 383–384
  Mid-Atlantic Ridge, 387
  mid-ocean ridges. See Mid-ocean ridges.
  mineral resources of, 536, 537★, 541, 542★
  ocean crust, formation of, 464, 464★
  origin of, 11, 12–13
  Pacific profile, 386★, 387★, 387
  pelagic sediments, **397**, 398★, 397–398, 398★
  profiles of, 384–387, 384★, 385★, 386★, 387★
  rift valleys, **387**
  salinity, 149
  seamounts, **385**, 385★, 394
  sedimentation, 397–398, 398★
  shorelines. See Shorelines.
  submarine canyons, **384**, 384★, 392, 392★
  submarine fans, **391**, 392★
  submersibles, **383**, 383★

surf, **372**, 372–373, 373★
tides. See Tides.
trenches, **387**
turbidites, **391**, 392★
turbidity currents, 391
and volcanic gases, 107–108
waves. See Waves.
world ocean, **370**
world ocean map, 388★, 389★
Ogallala aquifer, 268, 268★
Oil. See Petroleum: oil.
Oil reservoirs, **505**, 505–506
Oil shales, **512**, 512★
Oil traps, **505**, 505★, 505–506
Old Faithful Geyser, Wyoming, 103
Olivine
  and Bowen's reaction series, 81
  and cation substitution, 34–35
  chemical stability and weathering, 131★
  chemical weathering of, 128–130
  and contact metamorphism, 182
  and discontinuous reaction series, 77–78
  and fractional crystallization, 80
  and mantle composition, 436
Opal, **165**
  from hydrothermal waters, 272
Ophiolite suites, **462**, 462★, 464–465, 465★
  and mineral resources, 536
Orbitals (electron), **25**, 26★
Ordovician period, plate-tectonic reconstruction, 472★
Ore minerals, **525**. See also Mineral resources.
  from hydrothermal waters, 273
  of oxide minerals, 39
Organic matter
  burial, 204–205
  combustion of, 504, 504★
  diagenesis, 154
  and photosynthesis, 204–205
  productivity, and petroleum, 505
  and soil, 134–138
Organic sediments, rocks, 149, **156**
Original horizontality, principle of, **189**, 189–190, 190★
  and deformation, 214–215
Orogeny, **59**, **465**, 467–470, 484–493
  Appalachian, 486, 488★, 488–490, 489★
  fault-block mountains, **491**, 493★
  Himalayan, 3★, 484★, 485, 485★, 486, 487★
  North American Cordilleran, 490–493, 491★, 492★
  rejuvenation, **490**
  unwarped mountains, **491**, 492★
Outcrops, **57**, 58★, 57–58
  and chemical weathering rates, 131
  and deformation, 212–215
  geographic distribution, 57–58
  subsurface reconstruction from 221★
Overthrusts, 224★, **225**, 483★, 483–486, 485★
  and convergent boundaries, 467–469
Oxbow lakes, 285–286, 285★, **286**

Oxidation, **129**
  of iron silicates, 129
Oxides, 39, **39**
Oxygen
  early atmosphere, and iron ores, 532
  evolution in atmosphere, 196, 205
  isotopes, and paleotemperatures, 345
  and photosynthesis, 204–205
  and weathering reactions, 121, 129–130

P waves, **408**, 408★, 409★
Pacific Coast Ranges
  and Cordilleran orogeny, 490–493, 491★, 492★
  expression of deformation, 226–227
Pacific Ocean
  active margin profile, 390, 391★
  coral reefs in, 162
  profile of, 386★, 387★, 387
Pacific Plate, 453, 473
  and Cordilleran orogeny, 490–493, 491★, 492★
  rate of motion, 460, 460★
Pahoehoe, **93**, 93–94, 94★
Paleoclimates, 198
Paleomagnetic stratigraphy, 203
Paleomagnetism, 441–444, 442★, 443★, **443**
  and plate tectonics, 456★, 456–457, 457★
Paleontology, 190–192, **191**
  and continental drift, 448–449
  and plate tectonics, 459, 473
Paleozoic era, **196**
  plate-tectonic events of, 471–473, 472★
Palisades intrusion, New York, 79–80, 80★, 471
Pangaea, 17, 18★, **448**, 448–449, 449★, 465, 466★
  breakup of, **471**, 471–473, 472★, 474★, 475★
  reconstruction of, 470–473, 472★
Panthalassa, 471
Parícutin, Mexico, 100★, 111, 115
Paris Basin, and geothermal energy, 517
Partial melt, **75**
Passive margins, **390**, 391★
Peat, **149**, **509**, 509–510
Pediments, 322★, **323**
Pelagic sediments, **397**, 398★, 397–398
  foraminiferal oozes, **397**, 397★, 397–398
  silica oozes, 398
Pelée, Mont, Martinique, 96, 97★, 110
Peridotites, **71**
  and mantle composition, 436
Period (geological), **196**, 196–197
Period (wave), **372**
Periodic table (of elements), 28–31, 29★
Permafrost, **343**, 343–344, 344★
Permeability, **260**, 261★
  and Darcy's law, 265–266, 266★
  of various sediments and rocks, 261★

Persian Gulf, and oil spills, 507
Peru-Chile Trench, 387, 388★, 453
Petrified forest, 191★
Petroleum, 166
    in coastal plain, 494
    combustion of, 504, 504★
    and deformation of strata, 220
    diagenesis, 154
    enhanced recovery, 508–509
    and environment, 506–507
    formation of, 505–506, 505★
    future use of, 512–513
    gas, **149,** 509
    and global warming, 514
    history of use, 503, 504★
    natural gas, **149,** 509
    oil, **149**
    oil, exhaustible supply of, 507–509
    oil reservoirs, **505,** 505–506
    oil shales, **512,** 512★
    oil traps, **505,** 505★, 505–506
    sandstone reservoirs, 161
    and sedimentary rock classification,
        156
    and sedimentary rocks, 145
    seismic exploration for, 432★,
        432–433, 433★
    source beds, **505,** 505–506
    syncrude, **510**
    tar sands, **512**
    U.S. importation of, 508
    world distribution of, 506, 506★,
        507★
Phanerozoic eon, **196**
Phenocrysts, **72**
Philippines, and island arcs, 110
Phosphates, as resource, 524, 524★,
    532–533
Phosphorites, **156,** 165–166
Photosynthesis, 504, 504★
    and organic carbon, 204–205
Phyllite, **176**
Physical weathering, **120,** 132–134. *See
    also* Weathering.
    chemical weathering, relation to,
        120–122, 123★, 132
    by crystallization of minerals,
        133
    in deserts, 318, 320
    and erosion, 134
    exfoliation, **133,** 134★
    by ice, 133
    and joints, 133
    and rock structures, 133
    spheroidal weathering, **133,** 134★
    in streambeds, 282–283
    by tectonic forces, 133
    by temperature change, 133
    by tree roots, 132–133, 133★
Piedmont, Appalachians, 486, 488★,
    488–490, 489★
Pilkey, Orrin, 380
Pillow lava, **94,** 94–95, 95★
    and ophiolites, **462,** 462★, 464–465,
        465★

Pinatubo, Mount, Philippines, 96, 108
    ash transport, 310
    eruption in 1991, 71
Placers, **533,** 533–534
Planetary differentiation, 10★, 10–11
    and origin of atmosphere and oceans,
        12–13
    and origin of continents, 12–13
Planetary evolution, **9,** 9–13
    accretion, 9★
    gravitational compression, 9★, 9–10
    heating mechanism, 9★, 9–10
    radioactivity, 9★, 9–10
Planets
    early history, 9–12
    inner, or terrestrial, **8,** 11
    origin of, 6–9
    outer, **8,** 11–12
Plastic flow, of glaciers, **334,** 334★, 335★
Plate interiors, volcanism in, 109★,
    110–113
Plate map, 451★, inside front cover★
Plate tectonics, **13,** 13–19, 446–476
    African Plate, 460, 460★, 473
    and age of ocean floor, 456–460, 459★
    and Andes Mountains, 453
    Antarctic Plate, 460, 460★, 473
    and Appalachian orogeny, 486, 488★,
        488–490, 489★
    Atlantic Ocean, creation of, 461★
    and Baja California, 453, 453★
    and California, Gulf of, 453, 453★
    Cocos Plate, 460, 460★
    collisions and orogeny, 483–493
    continental drift, **448,** 448–450, 449★
    and continental margins, 390–392
    continent-continent collisions,
        468–470, 469★
    convective heat flow drive, 439, 439★
    convergent boundaries, **14,** 14–17,
        15★, 16★, 453, 454★, 455★
    convergent boundaries, rocks at, 465,
        467–470
    and deep-sea drilling, 457–460, 458★
    deep-sea trench, **15,** 16★
    direction and velocity of plates, 113
    divergence in continental crust,
        464–465, 466★
    divergence on continents, 452–453
    divergent boundaries, **14,** 14–17, 15★,
        16★, 452–453, 452★, 455★
    divergent boundaries, rocks at, 462,
        462★, 464, 464–465, 465★
    driving mechanisms of, 14, 473, 476,
        476★
    and earthquakes, 414–416, 415★,
        416★, 417★
    East Pacific Rise, 452
    Eurasian Plate, 453, 460, 460★,
        469–470, 473, 484★, 485, 486,
        487★
    Farallon Plate, 468, 468★
    fossil evidence, 448–449
    and Great Rift Valley, East Africa,
        453

and Himalayan orogeny, 3★, 453,
    469–470, 484★, 485, 485★, 486,
    487★
history of concept, 18, 448–450
and igneous rock compositions,
    73–74, 74★
imaging seafloor with satellites, 462,
    463, 463★
Indian Plate, 453, 460, 460★,
    469–470, 473, 484★, 485, 486,
    487★
island arcs, **453,** 454★, 455★, 467,
    467★
isochrons on ocean floor, 458★
LAGEOS, and plate motion, 462
and landscape evolution, 365–366
and magma composition, 110–113
mantle plumes and plate motion, 461
and mélanges, **467,** 467–468, 468★
and metamorphism, 55–56, 173★,
    183, 183★
and microplate terranes, **470,** 470★,
    471★
Mid-Atlantic Ridge, 452
mid-ocean ridges, 452–453, 452★,
    455★
and mineral resources, 534–537
Nazca Plate, 453, 460, 460★, 468
North American Cordilleran orogeny,
    490–493, 491★, 492★
North American Plate, 453, 460,
    460★, 468, 468★, 473, 486, 488★,
    488–490, 489★
and ocean floor, 393–393
and ocean study, 370
ocean-continent collisions, 467–468,
    468★
ocean-ocean collisions, 465, 467, 467★
and ophiolite suites, **462,** 462★,
    464–465, 465★
and orogenies, 483–493
Pacific Plate, 453, 460, 460★, 490
paleomagnetism, and rates of motion,
    456★, 456–457, 457★
Pangaea, breakup of, **471,** 471–473,
    472★, 474★, 475★
Pangaea, reconstruction of, 470–473,
    472★
and Peru-Chile Trench, 453
plate map, 451★, inside front cover★
plate motion geometry, 460–462,
    461★
and plutonism, 87
rates of motion, 18, 454–460, 460,
    460★
and Red Sea, 453
and Rhine Valley, 453
rift valleys, 14, 452–453
and rock assemblages, 462–470
and rock cycle, 59–60, 60★, 61★
rock types, distribution of various,
    462–470
and San Andreas fault, 453
and sandstone mineralogy, 160
and seafloor spreading, **15,** 450, 452★

Plate tectonics (*continued*)
  and sedimentary environments, 150
  South American Plate, 453, 460,
      460★, 468, 473
  subduction, **15,** 15–17, 16★
  subduction zones, **16,** 16–17, 453,
      454★, 455★
  sutures, **469,** 469
  and tectonic forces, 215
  transform fault boundaries, **14,**
      14–17, 15★, 16★, 453, 455★
  and volcanoes, 108–113, 109★, 111★
Plateaus, **356,** 356–357
Platforms, **482★, 483**
Platte River, Colorado, Nebraska,
    longitudinal profile of, 292★
Playas, playa lakes, **321,** 321★
Plumes (mantle), **74,** 111★, **113**
  driving force for convection, 439
  and intraplate volcanism, 111–113
  and plate motion direction, 461
  and plate tectonics, 439, 473, 476,
      476★
Plutonic episode, **59**
Plutonism, and plate tectonics, 87
Plutons, **83,** 83★
Polymorphs, **35**
Population growth, 543
Porosity, **153, 260,** 260★, 261★
  changes with depth in crust, 272,
      272★
  and groundwater flow, 260
  of various sediments and rocks, 261★
Porphyroblasts, 178
Porphyry, **72,** 73
Porphyry deposits (ores), 531
Positive feedback processes, **122**
Potassium, radiometric dating with,
    200, 201★
Potholes, **282,** 283★
Powderhouse Cave, West Virginia,
    267★
Precambrian, and cratons, 482–483
Precipitation, 253★, **254**
  amounts in United States, 257★
  relation to runoff, 257★, 257–258
Pressure
  confining, **171,** 216
  confining, and jointing, 221
  and deformation, 216, 216
  directed, **171,** 216
Proterozoic eon, **196**
Proton, **25,** 26★
Pumice, **72,** 72★
Pyrite, 39★, 40
  as an ore mineral, 525★, 529
Pyroclastic debris, and Mount St.
    Helens, 112–113
Pyroclastic deposits, 95★, 95–97
  ash. *See* Volcanic ash.
Pyroclastic eruptions, 100
Pyroclastic flow, **96,** 96★
Pyroclastic rocks, **71,** 71–72, **72**
Pyroxene
  and Bowen's reaction series, 81

chemical stability and weathering,
    131★
  chemical weathering of, 128–130
  and contact metamorphism, 182
  and discontinuous reaction series,
      77–78
  and fractional crystallization, 80
  in hornfels, 177
  in metamorphic rocks, 180
Pyroxene granulites, 179★, **180,** 180★,
    181★

Quad, **508**
Quartz
  arenite, **160,** 160★
  chemical stability and weathering,
      131★
  in chert, **165**
  in foliated rocks, 176–177
  and polymorphs, 35
  and soil classification, 137–138, 138★
  and weathering intensity, 145, 145★
Quartzites, **177,** 177★

Radioactive decay, absolute dating by.
    *See* Radiometric dating.
Radioactive waste disposal, 516
Radioactivity, **197,** 197–203
  dating by. *See* Radiometric dating.
  discovery of, 197
  as heat source, 9★, 9–10
Radiometric dating, 189, **197,** 197–203
  major elements used, 201★
Radium, radioactivity of, 197
Radon, 200, 202–203
  geographic distribution of hazard,
      203★
Rain shadows, **255,** 255★
Rainfall. *See also* Precipitation.
  distribution of, 255
  process of, 254–255
Rainier, Mount, Washington, 114★, 115
  and debris avalanches, 243
Rainstorms, triggers of mass wasting,
    236
Rancho Mirage, California, landslide of
    1979, 231★
Rawlins, Wyoming, dome in, 220★
Reaction series, of igneous rocks, **77,**
    77★, 77–78
Recycling of mineral resources, **527,** 529
Red Sea, 453
Red Sea Rift, 225, 225★, 447★
Redwood National Park, California, soil
    erosion in, 137★
Reefs
  atolls, **395,** 395–396, 396★
  coral, 162–164, 163★, 395–396, 396★
  fringing, **396**
Regional metamorphism, **55,** 55–56,
    56★
Regolith, **134**
Rejuvenation (mountain belt), **490**
Relative ages. *See* Ages (of geologic
    phenomena): relative.

Relief, **355,** 355★
Reserves, **502,** 502★. *See also* Mineral
    resources.
Resources, **502,** 502★
  energy. *See* Energy resources.
  mineral. *See* Mineral resources.
  nonrenewable, **503**
  renewable, **503**
Respiration, 204–205
Retaining walls, and slope
    oversteepening, 237
Reykjavík, Iceland, 272, 517
Rhine River valley, 225, 453
Rhone River, turbidity currents, 391
Rhyolite, 70★, **71**
  distribution, and plate tectonics,
      462–470
  lavas, 93
Richter, Charles, 411
Richter magnitude, **411,** 411★, 413★
Rift valleys, **225,** 225★, 360, 452–453,
    493, 493★
  at mid-ocean ridges, 387
Rifting of plates, 113
Ring of Fire, 108, 109★, 414
Ripples
  formation and movement of,
      313–314, 314★
  and sedimentary structures, **157,**
      157★, 158★
  slip face, **314,** 314★
  and stream velocity, 281–282, 282★
  in streams, **281,** 281–282, 282★
  wind, 308, 308★
Rivers, **278,** 278–301. *See also* Streams.
  bedding sequence, 158, 158★
  deltas. *See* Deltas.
  flows in major, 258★
  sediment and dissolved load, 300–301
  valleys. *See* Streams: valleys.
Rock assemblages, **462**
  at different plate boundaries, 462–470
Rock avalanche, **239,** 239★, 239–240
Rock cycle, **59,** 59★, 60★, 59–60
  and plate tectonics, 59–60, 60★, 61★
  and sedimentary rocks, 144
  weathering and erosion in, 120
Rockfall, 238★, **239,** 239–240
Rocks
  chemical analysis, 56–57, 57★
  chemical composition, 56–57
  crustal abundance of major types, 55,
      55★
Rockslide, **239,** 239★, 239–240
Rocky Mountains
  and Cordilleran orogeny, 490–493,
      491★, 492★
  expression of deformation, 226–227
  hogbacks in, 357★
Ross Ice Shelf, Antarctica, 330
Rubidium
  radioactivity of, 200, 200★
  radiometric dating with, 200–203
Runoff. *See* Hydrology: runoff.
Rutherford, Ernest, 197

S waves, **408**, 408★, 409★
Saguaro National Monument, Arizona, 321★
Sahara Desert, 320
  and droughts, 255
  sand transport, 308
St. Helens, Mount, Washington, 87, 98, 98★, 100, 103, 111, 112–113
  and debris avalanches, 243, 243★
  weathering rate of volcanic rocks, 122
Salt. *See* Halite.
  as resource, 524, 524★, 532–533
Salt domes, as oil traps, 505★, 505–506
Saltation, **280**, 280★
  in wind, 307, 307★
Salton Sea, California, 83
San Andreas fault, 17, 17★, 222★, 453
  earthquakes along, 405, 406★, 411, 417
  housing near, 421★
  landslides along, 244
  rates of movement along, 205–206, 206★
San Francisco, earthquake of 1906, 405, 406★, 411
San Gabriel Mountains, California, debris flow of 1978, 232
Sand, 158–161, **154**, 155★
  angle of repose, **233**, 234★
  composition of windblown, 310
  dunes. *See* Dunes.
  grain frosting, **310**, 310★
  mass wasting in, 233–244
  as resource, 524, 524★, 532–533
  transport at beaches, 378★, 378–379, 381–382
  wind transport, 307–310
Sand drift, **313**, 313★
Sand structures, cross-bedding formation, 314, 314★
Sandblasting, **311**, 312★
Sandstones, **154**, 155★, 158–161
  contact metamorphism in, 182★, 182–183
  diagenesis of, 153, 153★
  major types, **160**, 160★
  metamorphism of, 177
  as oil reservoirs, 505–506
  slopes on, 235
Santa Barbara, California, oil spill of 1969, 462, 463, 463★
Satellites
  imaging seafloor, 462, 463, 463★
  measuring plate motion with, 462
Saudi Arabia
  dunes in, 305★, 314
  ergs, 316
  sand transport, 308
Scandinavia
  isostatic rebound after glaciation, 435, 435★
  isostatic uplift after glaciation, 495
  mountains, 491
Scarps, 227
Schist, **176**, 177★

Schistosity, **176**, 176★
Scientific method, **18**, 18–19
  hypothesis, **18**
  theory, **18**
Sea, **370**
Sea level
  changes, from glaciation, 345–346, 346★, 347
  changes, and shorelines, 382
  glaciation, effect of, 328
  and global warming, 514
  possible future changes in, 347
  sea surface elevation variations, 463
  and stream base level, 292–294
Seafloor spreading, **15**, **452**
  convective heat flow drive, 439, 439★
  origin of theory, 450
  rates of, 204–205
  and volcanism, 110, 110★
Seamounts, **383**
  Emperor, 109★, 113
  Loiki, 385★
Seawater, **370**
Sedimentary basins. *See* Basins (sedimentary).
Sedimentary breccias, **161**
Sedimentary environments, **149**, 149–152, 150★, 151★
  alluvial fans, **294**, 294★, 323
  alluvial fans, in deserts, 320
  beaches. *See* Beaches.
  carbonate, **151**, 151–152
  chemical and biochemical, **151**, 151–152, 152★
  clastic, **150**, 150★, 150–151
  continental margins, 390–392
  coral reefs and atolls, 162–164, 163★, 395–396, 396★
  on cratons, 483
  deep sea, 150, 152
  deltas. *See* Deltas.
  deserts. *See* Deserts.
  eolian, in deserts, 320
  evaporite, in deserts, 321
  glacial, 340–343
  marine evaporite, 152
  oceans, 397–398, 398★
  and organic matter preservation, 505
  pediments, 322★, **323**
  and plate tectonics, 150
  reefs, 152
  shorelines. *See* Shorelines.
  siliceous, 152
Sedimentary rocks. *See* Sediments, sedimentary rocks.
Sedimentary sequences, interpretation of Grand Canyon, 198–199, **199**
Sedimentary structures, 156–158, **157**
  bedding, **156**, 156–158
  bedforms in streams, **281**, 281–282, 282★
  and bioturbation, **157**, 157–158, 158★
  cross-bedding, **156**, 156★, 156–157, 281
  dunes, 157, 157★

dunes, wind, 308
  dunes in streams, **281**, 281–282, 282★
  graded bedding, **157**
  ripples, **157**, 157★, 158★
  ripples, wind, 308, 308★
  ripples in streams, **281**, 281–282, 282★
Sediments, sedimentary rocks, 142–166, **149**, **533**, 533–534
  in alluvial fans, 294
  bedding and physical weathering, 133
  bedding sequences, **158**, 158★
  bedding (stratification), **156**, 156–158
  biochemical, **55**, **145**, 161–166, **162**
  biochemical, in oceans, **397**, 397–398, 398★
  black shales, **161**
  breccias, **161**
  burial and diagenesis, 59–60, 153–154
  calcareous shales, **161**
  carbonate, **156**
  carbonate, classification, **154**, 155★
  cementation, **153**
  chemical, **55**, **145**, 161–166, **162**
  chemical, in lakes, 162
  chemical and biochemical sedimentation, 149
  chert, **165**, 398, 533, 533★
  classification, 154★, 154–156
  clast composition and weathering, 145★, 145
  clastic, **54**, **55**, **145**, 158–161, 159★
  clastic, classification, **154**, 155★
  clastic, in oceans, 397–398, 398★
  clastic sedimentation, 148–149
  clays, **154**, 155★, 158–159, **161**
  coal. *See* Coal.
  compaction, **153**
  composition of windblown material, 310
  conglomerates, **154**, 155★, 158–159, **161**
  of continental margins, 390–392
  continental sedimentation, 148
  continental shelf deposits, **465**
  coral reefs, 162–164, 163★, 395–396, 396★
  cross-bedding, **156**, 156★, 156–157, 281
  deep ocean, 162
  deformation of. *See* Continental crust deformation; Orogeny.
  diagenesis. *See* Diagenesis.
  distribution, and plate tectonics, 462–470
  dolostone, **162**, 162★, 162–164
  environments of sedimentation. *See* Sedimentary environments.
  evaporites. *See* Evaporite minerals and rocks.
  flint. *See* Chert.
  foraminiferal oozes, **397**, 397★, 397–398
  gas, **149**
  glacial, 340–343

Sediments, sedimentary rocks (*continued*)
  glacial transportation and sorting, 146–147
  graded bedding, **157**
  grain shape and size, 145
  grain size reduction, 147–148
  grain size and settling velocity, 146
  gravels, **154,** 155★, 158–159, **161**
  iron formations, **156,** 166, 212★, 533, 533★
  and landforms, **356,** 356–360, 362–364, 363★
  limestone, **162,** 162★, 162–164, 164★
  lithification, 152–154, **153,** 153★
  loads of and transport by streams, 279–282
  manganese nodules, **398**
  muds, **154,** 155★, 158–159, **161**
  mudstones, **154,** 155★, 158–159, **161**
  ocean chemistry, and sedimentation, 149
  in oceans, 148, 397–398, 398★
  oil, 149
  oil shales, **512,** 512★
  ore deposits in, 531, **532,** 532–534, 533★
  organic, 149, **156**
  paleomagnetism in, 442–444
  peat, 149
  pelagic, **397,** 398★, 397–398
  petroleum. *See* Petroleum.
  petroleum, and classification, 156
  phosphorites, **156,** 165–166
  and plate tectonics, 145
  porosity, **153**
  practical value, 145
  radiometric dating of, 200
  reefs, 162–164, 163★
  relative abundances of major types, 158, 159★
  and rock cycle, 144, 144★
  rounding, 147–148, 148★
  sands, **154,** 155★, 158–161
  sandstones, **154,** 155★, 158–161. *See also* Sandstones.
  sandstones, major types, **160,** 160★
  sedimentary structures. *See* Sedimentary structures.
  sedimentation, 148
  settling velocity, **146**
  shales, **154,** 155★, 158–159, **161**
  silica oozes, **298**
  silts, **154,** 155★, 158–159, **161**
  siltstones, **154,** 155★, 158–159, **161**
  sorting, **146,** 147★
  tar sands, **512**
  terrigenous, **151,** 397
  transportation, 146–148
  transportation's effect on clasts, 147–148, 148★
  turbidites, **391,** 392★
  weathering, production by, 139
  weathering during transportation, 147
  wind transport, 307–310
Seismic gap method, 424–425

Seismic shadow zone, **431,** 431★
Seismic tomography, 437
Seismic waves, **406,** 406★, 406–408
  Earth's composition and structure, 433–437
  exploring Earth's interior with, 430–437
  P waves, **408,** 408★, 409★
  paths of, in Earth, 431★, 431–433, 432★
  reflection, 430, 431–433, 432★, 433★
  refraction, 430, 430★, 431
  S waves, **408,** 408★, 409★
  surface waves, **408,** 408★, 409★
  velocity changes with depth, 433–434, 433★
Seismicity, **414,** 415
Seismograph, **407,** 407★
Semiarid environments, weathering in, 128
Serpentine, and contact metamorphism, 182
Settling velocity, **280,** 281
Shadow zone (seismic), **431,** 431★
Shales, **154,** 155★, 158–159, **161**
  black (organic-rich), **161**
  calcareous, **161**
  contact metamorphism in, 181–183, 182★
  metamorphism of, 170, 174★, 176, 179, 179★
  slopes on, 235
Shasta, Mount, California, 114★, 115
Shearing forces, **215,** 215★, 215–216
Shields, **482,** 482★, 482–483
Ship Rock, New Mexico, 103, 104★
Shorelines, **370,** 377–382
  barrier islands, **381,** 382★
  beaches. *See* Beaches.
  deposition along, 381–382, 382★
  erosion along, 379★, 379–381, 380★
  estuaries, **382**
  rocky, 379★, 379–381, 380★
  and sea level changes, 382
  spits, **381,** 382★
  stacks, **379,** 379★
  tide terraces, 377★
  and tides, 375–376
  wave-cut terraces, **379,** 380★
  and waves, 370–374
Siberia, permafrost in, 344
Siderite, as an ore mineral, 525★
Sierra Nevada, **468,** 468★
  and Cordilleran orogeny, 490–493, 491★, 492★
  and gold, 534
  and lahar deposits, 107
  till in, 341★
Silica
  chert, **165**
  dissolved, as weathering product, 124–129
  flint, **165**
  and igneous rock classification, 69–72
Silica oozes, **398**

Silica tetrahedra, and discontinuous reaction series, 78
Silicate ion, **36**
Silicate minerals, 35, 36–38, 37★, 38★
  chemical weathering of, 123–130
Silicates
  iron-rich, and weathering, 129–130
  weathering rates, 130
Siliceous sedimentary environments, 152
Sillimanite, as metamorphic index mineral, 179, 179★
Sills, 83–84, **84**
Silts, **154,** 155★, 158–159, **161**
Siltstones, **154,** 155★, 158–159, **161**
Sinkholes, **269,** 269★
Sklowdowska-Curie, Marie, 197
Skykomish River, Washington, flood frequency curve, 291★
Slate, **176,** 177★
  and foliation, 175
Slip face, **314,** 314★
Slopes
  and angles of repose, **233,** 234★
  and construction, 237
  rock slopes, 235–236
  stability, 233, 235–236. *See also* Mass wasting.
Slump, 241★, **243**
Smith, William, 192
Snowflake crystals, 33★
Sodium chloride. *See also* Halite; Evaporite minerals, rocks.
  crystal structure, 32, 32★
  ionic bonding in, 31
  ocean chemistry and sedimentation, 149
Soil, **120, 134,** 134–139
  carbonic acid in, 126
  chemical composition, 137–138, 138★
  and chemical weathering rates, 131
  classification, 137–138, 138★
  climate, effect of, 135–139
  color, 134
  environment, and economy, 135, 136–137
  erosion, 136–137, 137★
  horizons, **135,** 135★
  humus, **134**
  laterites, **138,** 138★
  liquefaction during earthquakes, 418
  mineralogy, 137–138, 138★
  parent rock type, effect of, 136
  pedalfers, **137,** 138★
  pedocals, **137,** 138★
  profiles, 135, 135★
  regolith, **134**
  residual, **135**
  time, effect of, 135–139
  transported, **135**
  vegetation, effect of, 135–139
  volcanic, 115
  weathering, effect on, 121–122, 126–128
Solar energy, 503, **515,** 515★, 517
Solar heat, 439–440

Solar system, origin of, **6,** 6–9
  condensation of planets, 7–8
  nebular hypothesis, 7, 7★
Solifluction, **243,** 243★
Solubility (mineral), **131**
Solutions, 33–34
Somerset, U.K., 214★
Source beds, **505,** 505–506
South American Plate, 453, 468
  rate of motion, 460, 460★
Spanish Fork Canyon, Utah, mudslide
  of 1983, 244–246, 245★
Sphalerite, hydrothermal vein deposits
  of, 529, 530★, 531
Specific gravity, 44–45, **45**
Spinel, 39
Spits, **381,** 382★
Springs, **259,** 259★
Stacks, **379,** 379★
Stalactites, **269,** 269★
Stalagmites, **269,** 269★
Staurolite
  as metamorphic index mineral, 179,
    179★
  porphyroblasts, 178
Steel, as resource, 524, 524★
Steinbeck, John, *The Grapes of Wrath,*
  309
Steno, Nicolaus, and fossils, 191
Stocks, **83**
Stone, building, as resource, 524, 524★,
  532–533
Storms, and desert erosion, 320
Stratigraphic sequences, **190**
  composite, **192,** 192★
  correlation of, 190–192
  gaps in, **190**
Stratigraphic units, formations, **192**
Stratigraphy, **190**
  geologic time scale construction, 189
  magnetic, **443,** 443★, 443–444
  magnetic, and seafloor spreading,
    456★, 456–457, 457★
  original horizontality, principle of,
    **189,** 189–190, 190★
  principles of, 189–190
  seismic, 432–433
  superposition, principle of, **189,**
    189–190, 190★
Streams, **278,** 278–301, **281,** 281–282,
  282★. *See also* Rivers.
  and alluvial fans. **294,** 294★
  antecedent, **297,** 297–298, 298★
  badland topography, **359,** 360★
  base level, **292,** 292★, 292–294, 293★
  bed load, **280,** 280★
  bedforms, **281,** 281–282, 282★
  braided, **286,** 286★
  capacity, **280**
  channel patterns, 284–286
  channel stability, 286
  channels, **284,** 284★
  competence, **280**
  deltas. *See* Deltas.
  dendritic drainage, **296,** 296★, 297

discharge, **287,** 289, 289★
distance, changes with, 287, 292–294
distributaries, **300**
divides, **294,** 294–296, 295★
drainage basins, **295,** 295★, 295–296
drainage networks, 294–298, **296**
drainage patterns, 297–298, 298★
dry washes, **322**
dunes, **281,** 281–282, 282★
effluent, **262,** 262★
erosion of rock by, 282–283, 283★
floodplains, **284,** 284★, 286–287
floodplains, cities on, 288
floods. *See* Floods.
flow, general aspects of, 278–279,
  279★
grade, 292–294
graded, **293,** 293–294
headward erosion of, **282**
influent, **262,** 262★
laminar flow, **278,** 278–279, 279★
longitudinal profiles, **292,** 292★,
  292–294, 293★
meander migration, 285★, 285–286
meanders, **284,** 284–286, 284★, 285★
natural levees, **287,** 287★
oxbow lakes, 285–286, 285★, **286**
piracy, **296**
point bars, **285,** 285★
potholes, **282,** 283★
radial drainage, **297,** 297★
rapids, and erosion, 283
rectangular drainage, 296★, **297**
ripples, **281,** 281–282, 282★
saltation, **280,** 280★
sediment loads and transport,
  279–282
settling velocity, **280,** 281
streamlines, 278–279
superposed, **298,** 298★
suspended load, **279,** 279–280, 280★
terraces, **294,** 295★
time, changes with, 287, 289–291
trellis drainage, **297,** 297★
tributaries, **300**
turbulent flow, **278,** 278–279, 279★
valley shapes, 359, 359★
valleys, **283,** 283–284, 284★
valleys in deserts, 322–323
valleys as landforms, 358–359, 359★,
  360★
velocity and discharge, 289, 289★
velocity and grain size, 280–281, 281★
viscosity of fluids, **278,** 278–279
wadis, **322**
waterfalls and erosion, 283, 283★
Stress, **171**
  and deformation, 216, 216★
  and metamorphism, 171
  preferred orientation of minerals, **175,**
    176★
Strike, 212–215, 213★, 214★
Strontium
  radioactivity of, 200, 200★
  radiometric dating with, 200–203

Subduction, **15,** 15–17
Subduction zones, **16,** 16–17, 453,
  454★, 455★
  earthquakes along, 415★, 416, 417★
  and magma generation, 87
  metamorphism at, 183
  and mineral resources, 536, 537★
  and orogenies, 483–493
Sublimation, 253★, **254**
  and glacial ablation, 332, 333★
Submarine canyons, **384,** 384★, 392,
  392★
Submarine fans, **391,** 392★
Subsidence, **60.** *See also* Epeirogeny.
Sudbury, Ontario, and igneous ore
  deposits, 532
Suess, Eduard, and continental drift,
  448
Sulfates, 39★, 40
Sulfides, 39★, 39–40
Sulfur, in coal, and acid rain, 510
Sulfuric acid, and acid rain, 126–127
Sun, and tides, 375–376
Sun's heat, compared to interior heat,
  439–440
Superposition, principle of, **189,**
  189–190, 190★
  and deformation, 215
Surface tension, **233,** 234★
  effect on angle of repose, 233
Surface waves (seismic), **408,** 408★,
  409★
Suspect terranes. *See* Microplate
  terranes.
Susquehanna River, Pennsylvania, 219★
  floods of, 290, 291★
Sutures, **469,** 469★
Swamps. *See* Wetlands.
Synclines, **217,** 217★, 217–219
  plunging, **218,** 219★
Syncrude, **510**

Tadzhikistan, mudflow of 1989, 242★
Talus, **239,** 322
Tambora, Mount, Indonesia, 108
Tangshan, China, earthquake of 1976,
  423
Tanner Creek, Oregon, 277★
Tapponier, Paul, 486
Tar sands, **512**
Tectonic forces, 212
  compressive forces, **215,** 215★,
    215–216
  and earthquakes, 405–425
  and faulting, 223–225
  and folding, 217–220
  and jointing, 220–223
  and plate tectonics, 215
  shearing forces, **215,** 215★, 215–216
  tensional forces, **215,** 215★, 215–216
Tectonics, as landscape control, 360–362
Temperatures (in Earth), 440, 440★
Tensional forces, **215,** 215★, 215–216
Terrigenous sediments, **151,** 397
Tethys Sea, 471

Teton Range, Wyoming, 492★, 493
Tetrahedra, **31,** 31★
   silicate, **36,** 36★, 36–37
Texas, groundwater resources, 268
Texture (rock), **52**
Thera (Santorini), Aegean Sea, 114
Thoreau, Henry David, 6
   *Cape Cod,* 381
Three Mile Island, Pennsylvania,
      nuclear accident, 513
Tibetan Plateau, and Himalayan
      orogeny, 484★, 485, 485★, 486,
      487★
Tidal waves. *See* Tsunamis.
Tide terraces, 377★
Tides, **375,** 375–376
   deltas, effect on, 300–301
   ebb tide, **376**
   flood tide, **376**
   Moon and Sun, effects on, 375★,
      375–376, 376★
   neap tides, **376,** 376★
   spring tides, **376,** 376★
   tidal currents, **376**
   tidal flats, **376,** 377★
   tidal surges, **376**
Time scale, *See* Geologic time scale.
Tonga Trench, 387
Topography, **354,** 354★
   climate, effect on, 361
   contours, **354,** 354★
   elevation, **354**
   as landscape control, 360–362
   relief, **355,** 355★
   topographic maps, 354, 354★
   of United States, 355★
Trace elements, **46**
Transatlantic telegraph cables, 391
Transform fault boundaries, **14,** 14–17,
      15★, 16★, 453, 455★
   earthquakes along, 414, 415★, 416★
   and faulting, 223
   tectonic forces at, 215
Transpiration, 253★, **254**
Travertine, 108, **272,** 272–273, 273★
Trenches, **387**
Triassic period, plate-tectonic events of,
      471–473, 474★
Triassic basalts, 471
Trilobites, 187★, 198
Tsunamis, **104,** 376, **418,** 418★, 419,
      420★
Tuff, volcanic, **72, 96,** 106★
Turbidites, **391,** 392★
   distribution, and plate tectonics, 467
Turbidity currents, **390,** 390–392, 391★

Ultramafic minerals, rocks, **70,** 70–71
   and convergent boundaries, 465, 467,
      467★
   seismic wave velocities in, 434
Unconformity, **192,** 192★, 194★
   angular, **192,** 192★, 194★

Unconsolidated materials, **232,** 232–233
   mass wasting in, 233–234
Uniformitarianism, principle of, **13,** 197
   and deserts, 318
United States
   soils, 137–138
   topographic map, 355★
Universe
   Big Bang, 4
   origin of, 4
Unzen, Mount, Japan, 96★, 97
Uplift, *See also* Orogeny.
   and longitudinal profiles, 294
   rates of, 206
   and rock cycle, 59, 59★
Uplift, regional. *See* Upwarped
      mountains; Epeirogeny.
Upwarped mountains, **491,** 492★
Ural Mountains, 490, 491
   and plate-tectonic reconstruction, 471
Uraninite, 513
Uranium
   for nuclear energy. *See* Nuclear
      energy.
   and radioactivity, 197
   radiometric dating with, 200–203,
      201★
Urbanization, effect on groundwater
      recharge, 267


Vaiont, Italy, debris slide of 1963, 247,
      248★
*Valdez, Exxon,* oil spill of 1989, 507
Valles Caldera, New Mexico, 103
Valley and Ridge belt, Appalachians,
      219★, 220, 221★, 358, 358★, 359★
Valley and ridge topography, 358, 358★,
      359★
Valley glaciers, 327★, **329**
Valleys. *See* Streams: valleys.
Varves, **342,** 343★
Vegetation
   and desert landscape, 321
   and mass wasting, 235, 235★
Veins, **85,** 85★, 85–87, 529★, 529–531,
      530★
Venice, Italy, subsidence of, 496, 496★
Ventifacts, **311,** 312★
Venus, 11
Vesicles, **72**
Vesuvius, Mount, 100, 114
Vicksburg, Mississippi, 287
Vine, F. J., 456
Viscosity
   of air, 306–307
   of fluids, **278,** 278–279
Volcanic arcs
   metamorphism at, 183
   and subduction zones, 87
Volcanic ash, **71,** 71–72, 72★, **95**
   and dust, 317
   and Mount St. Helens, 112–113
   weathering of, 128

Volcanic ash-flow deposits, **106,** 106★,
      107★
Volcanic breccias, **96,** 96★
Volcanic gases, 107–108
Volcanic glass, and igneous rock
      composition, 71–72
Volcanic island arcs
   Japan, 110
   Martinique, 110
   Philippines, 110
Volcanic rocks
   classification, 71–72
   metamorphism of, 178, 180, 180★
   paleomagnetism in, 442–444
Volcanic tuff, **72, 96,** 106★
Volcanism
   and atmospheric gases, 12, 12★
   and continental growth, 485
   and continent-continent collisions,
      468–470, 469★
   and debris avalanches, 242–243, 243★
   and igneous rock type, 71–72
   and ocean-continent collisions,
      467–468, 468★
   and ocean-ocean collisions, 465, 467,
      467★
   and plate tectonics, 453, 454★, 455★
Volcanoes
   Anak Krakatoa, Indonesia, 95★, 105★
   Andes Mountains, 111
   and aseismic ridges, **113**
   beneficial aspects, 115
   calderas, 102★, **103**
   Cascade Range, Pacific Northwest,
      111, 114★, 115
   central eruptions, 97–105
   central vent, **97**
   Cerro Negro, Nicaragua, 101★
   cinder cones, **100,** 100★, 101★
   Columbia Plateau, Pacific Northwest,
      93, 94★, 97
   composite, **100,** 101★, 104
   at convergent boundaries, 110–111
   coral reefs and atolls, 395–396, 396★
   Coseguina, Nicaragua, 111
   Cotopaxi, Ecuador, 111
   Crater Lake, Oregon, 103, 114★
   craters, 100
   deposits, 93–97
   diatremes, **103,** 103–104, 104★
   and earthquakes, 99, 112–115
   El Chichón, Mexico, 108
   Emperor seamounts, 109★, 113
   eruptive styles, 97–108
   Etna, Mount, Sicily, 91★, 100
   fissure eruptions, **105,** 105–106, 106★
   flood basalts, **106,** 106★
   Fujiyama, Japan, 97, **100,** 101★, 110
   geothermal energy, 115
   global pattern of, 108–113
   Great Basin, Nevada, 106
   Iceland, 105–106, 110, 110★, 114
   and igneous rock, 66–67
   instrumentation of, 99

intraplate, 110–113
Irazú, Costa Rica, 111
Kilauea, Hawaii, 99, 99★, 103, 103★, 114★
Krakatoa, Indonesia, 103, 104–105
lahars, **106,** 106–107
Lassen Peak, California, 114★, 115
lava. *See* Lava.
Long Valley, California, 92
Long Valley Caldera, California, 103, 114★
Mauna Loa, Hawaii, 97–99, 102★
and mineral resources, 115
Mono-Inyo Craters, California, 114★, 115
Nevada del Ruiz, Colombia, 107
at ocean ridges, 110
and Old Faithful Geyser, Wyoming, 103
Parícutin, Mexico, 100★, 111, 115
Pelée, Mont, Martinique, 96, 97★, 110
phreatic (steam) explosions, **103,** 103★, 104
Pinatubo, Mount, Philippines, 71, 96, 108, 310
and plate tectonics, 108–113, 109★, 111★
prediction of eruptions, 99, 112–115
pyroclastic deposits, 95★, 96–97
pyroclastic eruptions, 100
Rainier, Mount, Washington, 114★, 115
resurgent calderas, 102★, **103**
Ring of Fire, 108, 109★
St. Helens, Mount, Washington, 87, 98, 98★, 100, 103, 111, 112–113
Shasta, Mount, California, 114★, 115
shield, 97–98, **98,** 98★
soils, 115
stratovolcano, **100,** 101★, 104
Tambora, Mount, Indonesia, 108
Unzen, Mount, Japan, 96★, 97
Valles Caldera, New Mexico, 103
Vesuvius, Mount, 100, 114
volcanic ash-flow deposits, **106,** 106★, 107★
volcanic domes, **97,** 98★
Yellowstone, Wyoming, 92, 106
Yellowstone Caldera, 103, 114★
Voltaire, 404

Wadis, **322**
Wasatch Range, Utah
fault-block mountains, 493
mudslides in, 244–246
Wastes
disposal of radioactive, 516
hazardous storage on seafloor, 395
Water. *See also* Hydrology;
Groundwater.
and agriculture, 256
allocation policies, 256
amount on Earth, 253
available for human use, 254

contamination, 271★, 271–272
deep crustal, 272★, 272–273
dissolved substances in, 270★, 270–272
distribution of, 252★, 252–253
droughts, 255, 256
flows in major rivers, 258, 258★
hardness, **270**
hydrologic cycle, 253★, 253–254, 256
hydrothermal, 272–273, 273★
and industry, 256
infiltration, 253★, 254★
in metamorphic rocks, 172
meteoric, **273**
in minerals of rocks, 57
potable, **270,** 270–271
precipitation, 253★, **254**
quality, 270–272
rainfall, 254–255
rainfall, distribution of, 255
reservoirs of, 252★, 252–253, 253★
runoff, 253★, **254.** *See also*
Hydrology: runoff.
shortages, 256
sublimation, 253★, **254**
surface. *See* Hydrology: runoff.
taste, 270
transpiration, 253★, **254**
use of, 274
use, in U.S., 256
and weathering reactions, 124–132
Water table, **261,** 261★
elevation of, and Darcy's law, 265–266, 266★
Waterfalls, and glacial erosion, 338, 339★
Wave-cut terraces, **379,** 380★
Waves, 370–374
backwash, **373,** 373★
deltas, effect on, 300–301
generation of, 371–372
longshore currents, **374,** 375★
longshore drift, **374,** 375★
motion of, 372, 372★, 373★
period, **372**
refraction of, **373,** 373★, 373–374, 374★
surf, **372,** 372–373, 373★
surf zone, **372,** 372–373, 373★, 378, 378★
swash, **373,** 373★
swell, **371,** 371–372
wave height, **372,** 372★
wavelength, **372,** 372★
Weathering, **54,** 54★, 120, 120–139. *See
also* Chemical weathering;
Physical weathering.
climate, effect of, 121
general controls on, 120–122, 122★
mass balance, 139
rates of 120–122, 122★
and rock cycle, 59★, 59–60
rock type, effect of, 121, 121★
sediment production, 139
and sedimentary rock abundances, 158

soil, effect of, 121–122, 126–128
and soils, **134,** 134–139
spheroidal, **133,** 134★
time, effect of, 122
topographic control on, 360
Wegener, Alfred, and continental drift, 448
Wellfleet Bay, Cape Cod, 189★
Wells, artesian, **263,** 263★
Wetlands
and coal, 509–510
Everglades, Florida, 251★
storage of runoff, 258★, 258–259
White Sands National Monument, gypsum sand dunes, 310
White smokers, **393**
Wilson, J. T., and plate-tectonics theory, 450
Wind. *See also* Deserts.
capacity, 308, 308★
composition of windblown material, 310
deflation, **310,** 310–311, 311★
as depositional agent, 312–317
dust storms, 308–310, 309★
erosion by, 310–311, 311★, 312★
erosion and desert pavement, **311,** 311★
flow processes, 306–307
rates of sand transport, 308, 308★
saltation, **307,** 307★
sandblasting, **311,** 312★
speeds, 306★
trades, **307,** 307★
as transport agent, 307–310
ventifacts, **311,** 312★
westerlies, **306,** 307★
yardangs, **311,** 312★
Wind power, 517
Wollastonite, and contact metamorphism, 182
Wopmay orogenic belt, Canada, 479★
World ocean, **370**

Yardangs, **311,** 312★
Yellowstone, Wyoming, 106
Mammoth Hot Springs, 273★
volcanism in, 92
Yellowstone Caldera, 103, 114★
Yellowstone National Park, forest fires and mass wasting, 235★
Yosemite National Park, exfoliation, Half Dome, 134★
hanging and U-shaped valleys, 339★
Yucatan, Mexico
karst topography in, 269
oil spill of 1979, 507
Yugoslavia, karst topography in, 269★

Zagros Mountains, Iran, structural valleys and ridges, 358★
Zeolites, as index minerals, **180**
Zinc, as resource, 524, 524★
Zoned crystals, **78,** 78★, 78–79

# THE GEOLOGIC TIME SCALE

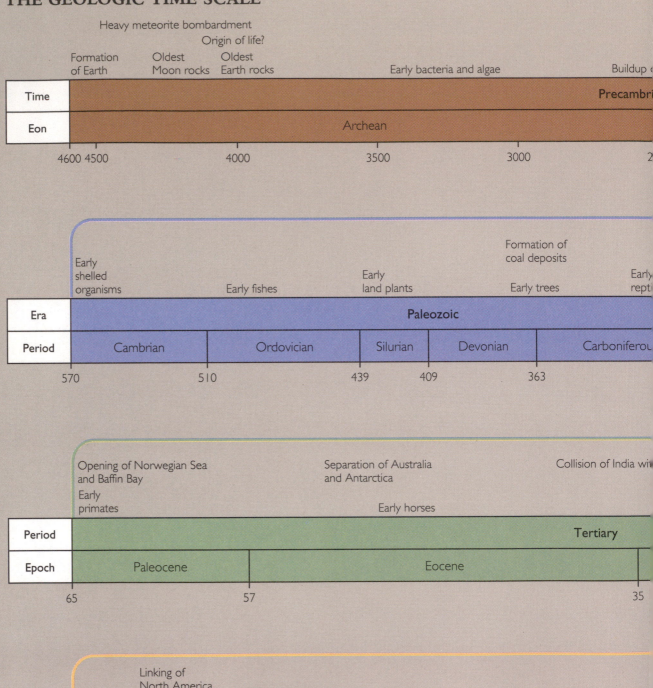

Heavy meteorite bombardment

Origin of life?

| Formation of Earth | Oldest Moon rocks | Oldest Earth rocks | Early bacteria and algae | Buildup o |

| Time | Precambri |
|---|---|
| Eon | Archean |

4600 4500     4000     3500     3000     2

Formation of coal deposits

| Early shelled organisms | Early fishes | Early land plants | Early trees | Early rept |

| Era | Paleozoic |
|---|---|
| Period | Cambrian | Ordovician | Silurian | Devonian | Carboniferou |

570     510     439   409     363

Opening of Norwegian Sea and Baffin Bay

Separation of Australia and Antarctica

Collision of India wi

Early primates

Early horses

| Period | Tertiary |
|---|---|
| Epoch | Paleocene | Eocene |

65     57     35

Linking of North America and South America

Worldwide glaciations

Oldest stone tools

*Homo erectus*

| Epoch | Pleistocene |
|---|---|

1.6   1.5   1.4   1.3   1.2   1.1   1   0.9

Time before present (millions of

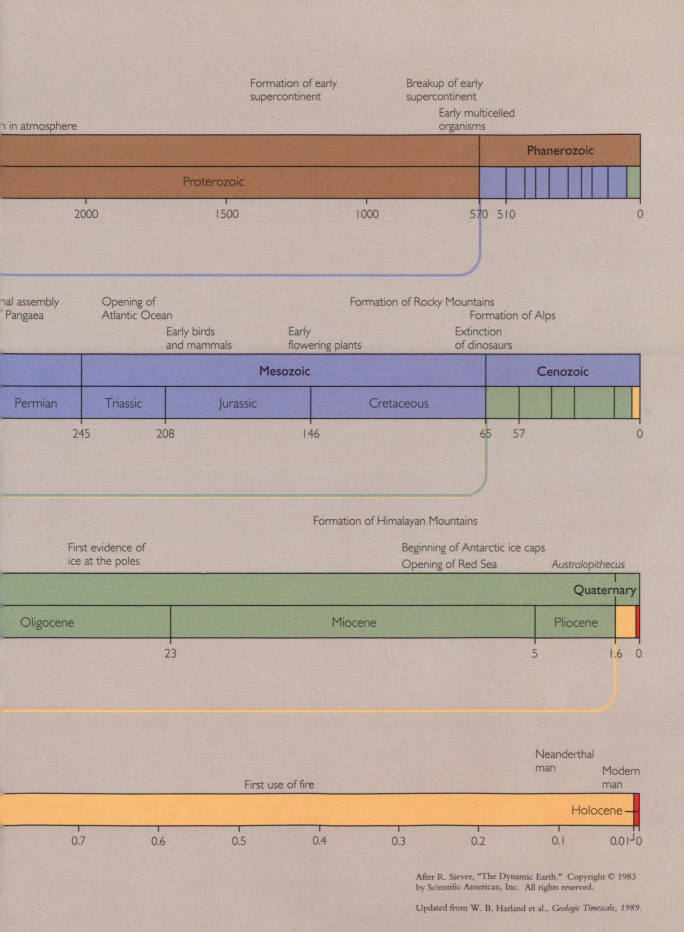

Formation of early
supercontinent

Breakup of early
supercontinent

Early multicelled
organisms

n in atmosphere

**Phanerozoic**

**Proterozoic**

2000     1500     1000     570   510     0

nal assembly
Pangaea

Opening of
Atlantic Ocean

Formation of Rocky Mountains

Formation of Alps

Early birds
and mammals

Early
flowering plants

Extinction
of dinosaurs

**Mesozoic**     **Cenozoic**

Permian | Triassic | Jurassic | Cretaceous

245     208     146     65   57     0

Formation of Himalayan Mountains

First evidence of
ice at the poles

Beginning of Antarctic ice caps

Opening of Red Sea     *Australopithecus*

**Quaternary**

Oligocene     Miocene     Pliocene

23     5     1.6   0

Neanderthal
man     Modern
man

First use of fire

Holocene

0.7     0.6     0.5     0.4     0.3     0.2     0.1     0.01  0